FERTILIZERS AND ENVIRONMENT

Developments in Plant and Soil Sciences

VOLUME 66

The titles published in this series are listed at the end of this volume.

Fertilizers and Environment

Proceedings of the International Symposium "Fertilizers and Environment", held in Salamanca, Spain, 26–29, September, 1994

Edited by

C. RODRIGUEZ-BARRUECO

Partly reprinted from *Fertilizer Research*, Volume 43, Nos. 1–3 (1995/1996).

CSIC

Universidad Salamanca

Ministerio de Industria y Energía

Miner

MINISTERIO DE AGRICULTURA PESCA Y ALIMENTACION

KLUWER ACADEMIC PUBLISHERS
DORDRECHT / BOSTON / LONDON

Library of Congress Cataloging-in-Publication Data

Fertilizers and environment / edited by C. Rodriguez-Barrueco.
p. cm. -- (Developments in plant and soil sciences ; v. 66)
Includes index.

1. Fertilizers--Congresses. 2. Fertilizers--Environmental aspects--Congresses. I. Rodriguez-Barrueco, C. II. Series.
S631.3.F478 1995
631.8--dc20 95-35896

ISBN-13: 978-94-010-7210-6 e-ISBN-13: 978-94-009-1586-2
DOI: 10.1007/978-94-009-1586-2

Published by Kluwer Academic Publishers,
P.O. Box 17, 3300 AA Dordrecht, The Netherlands.

Kluwer Academic Publishers incorporates
the publishing programmes of
D. Reidel, Martinus Nijhoff, Dr W. Junk and MTP Press.

Sold and distributed in the U.S.A. and Canada
by Kluwer Academic Publishers,
101 Philip Drive, Norwell, MA 02061, U.S.A.

In all other countries, sold and distributed
by Kluwer Academic Publishers Group,
P.O. Box 322, 3300 AH Dordrecht, The Netherlands.

Printed on acid-free paper

Softcover reprint of the hardcover 1st edition 1996

Contents

* Chapters indicated with an asterisk are reprinted from *Fertilizer Research*, Volume 43, Nos. 1–3 (1995/1996).

Session IVa: MINERAL FERTILIZERS MANAGEMENT
Chairman: M.A.C. Fragoso, Estaçao Agronomica Nacional, Oeiras, Portugal

The Organizing Committee

Prof. Dr. C. Rodriguez-Barrueco. IRNA-CSIC, Salamanca, Spain
Chairman

Prof. Dr. A. Amberger. Inst. Plant Nutrition, Technical Univ., Munich, Germany
Deputy President of CIEC, Co-Chairman

Prof. Dr. G. Nicolás Rodrigo. Universidad de Salamanca, Spain

Prof. Dr. J. Barceló-Coll. Universidad Autónoma de Barcelona, Spain

Dr. J.A. Díez-López. C. Ciencias Medio Ambientales; CSIC, Madrid, Spain

Prof. Dr. Ch. Hera. Joint FAO/IAEA Division, Wien, Austria

Local Arrangements Committee

C. Rodriguez-Barrueco
N. Martin Diego, Secretary
P. Hernandez Cartes
L. F. Rodriguez Burrieza
M. Tapia Cid
J. Garcia Talegon
A. Gonzalez Nuñez
M. Igual Arroyo
A. Garcia Sanchez
E. Iglesias Jimenez
N. A. Vicente Hernandez
Viajes Y Congresos, Salamanca

The Organizing Committee Acknowledges the Sponsorship of:

European Commission
Consejo Superiór de Investigaciones Científicas (CSIC)
Ministerio de Agricultura, Pesca Y Alimentación
Ministerio de Industría y Energía
DGICYT. Ministerio de Educación y Ciencia
Universidád de Salamanca
Diputación Provinciál de Salamanca
Cámara de Comercio e Industría de Salamanca
Asociación Naciónál de Fabricantes de Fertilizantes
Mirat, S.A., Salamanca
Fertiberia
International Fertilizer Industry Association (IFA)
European Fertilizer Manufacturers Association
El Consejo Británico
Iberdrola
Caja Salamanca y Soria
Patronato Provinciál de Turismo, Salamanca

Preface

Consejo Superior de Investigaciones Científicas (CSIC), International Centre of Fertilizers (CIEC), Universidad de Salalamanca, three famous and traditional scientific organizations have sponsored the VIII International Symposium CIEC from 26–29th September 1994 in Salamanca, Spain, under the theme:

Fertilizers and Environment

The goals of this Symposium were both to emphasize, that
– food production is still the main problem in many – especially developing – countries of the world and fertilizers are an important instrument to produce more and better food for the still growing world population
– protection of the soils from detrimental impacts of environment or unproper application of fertilizers in sustainable agricultural systems maintaining or even increasing soil fertility
The Nobel laureate Dr. Norman E. Borlang recently has made the following important statement:

"For those of us on the food production front, let us all remember that world peace will not – cannot – be built on empty stomachs. Deny farmers access to modern factors of production – such as improved varieties, fertilizers and crop protection chemicals – and the world will be doomed – not from poisoning, as some say, but from starvation and social chaos".

The highlighted topics of this Symposium covered:

– importance and benefits of fertilizer application for food and forestal production

– modern management practices to increasing efficiency of fertilizers and to minimizing losses

– amounts of mineral and organic manures applied in a nutrient concept to sustaining or increasing soil fertility

– fertilizer impact on environment (air, water, soil) by nitrate leaching, denitrification, ammonia volatilization, heavy metal pollution, etc.

– fertilization and legislation

– concepts for environmental balanced fertilizer recommendations on farms.

The great number of participants of about 200 from over 28 countries have demonstrated the great interest and responsibility to combine both food production and environmental protection based on scientific fertilizer management concepts.

Prof. Dr. C. Rodriguez-Barrueco (CSIC)
Prof. Dr. A. Amberger, Deputy President
Prof. Dr. G. Nicolás Rodrigo, Univesity of Salamanca

Opening Speech

Speech delivered at the opening ceremony of the international symposium on 'Fertilizers and Environment'

Excmo. Sr. D. Francisco Jambrina Sastre
Consejero de Medio Ambiente y Ordenación del Territorio Junta de Castilla y León Spain

The concept of sustainable development has become one of the main reflections of the end of the present century. The achievement of a model for a sustainable development in Castilla y León is our aim, within which the conservation of environment is receiving careful attention.

Development has traditionally been assimilated to growth, and the latter has been associated to accelerated production and consumption. Nowadays, man is conscious that those models can not be maintained for long, since they do not respond to present market regulations, neither they make an efficient use of the natural resources nor assign to them their real value. At the same time, practices associated to those models are leading to an accelerated destruction of natural resources thereby negatively affecting the production processes. Therefore, it is imperative to search for new development models addressed to improve the quality of human life without overloading the ecosystems.

We must stand as an example for the less developed countries in the efforts to produce new methodologies and to put them at work under the Common Agriculture Policy while conserving environment and regional diversity.

The new Agricultural Policy of the European Union shall be based on the analysis of its potential impact on the environment, as well as it should contribute to the restoration of soil, water and wildlife. In short, the new European Policy must correct the territorial and environmental upset caused by CAP until now.

If we want a sustainable development for Castilla y León our own regional programmes must envisage the conservation of the ecosystems diversity and vitality, and the preservation of habitats and wildlife protection. It will be the only way to make progress in our aim to improve the quality of life of our people.

The correction of territorial desequilibria and the approximation of development levels of Castilla y León to those of other European regions must be made within a solid concept of protection of environment and not at any price. We are given the opportunity to build up a development upon new standards which will, no doubt, allow us to achieve the goal of a better quality of life.

Other considerations, of a more specific character, shall determine the sustainability parameters for each of the sectors under study, eventually outlining a framework for regional development as a whole. One of those sectors, most important for nature conservation is agriculture. The intensive exploitation and degradation of natural resources, on which agriculture depends (water, soil, air) stand up as irreversible processes requiring an immediate change of man behaviour. Soil erosion due to inadequate practices, overexplotation of aquifers, soil and water contamination, salinization and nutrient losses, lowering of animal and plant diversity, are just a few of the consequences of the model of agriculture practiced along this XXth century now ending.

A new model of development and, as a consequence a new model for agriculture requires the introduction of practices that are compatible with nature conservation, able to increase rather than reduce soil capacity, allow the recovery of wildlife and game species, and to ensure the conservation of genetic diversity through the use of crop species and livestock better adapted to our regional characteristics. In few words, a new design for our agriculture must take quality into account at a greater extent.

The role of the Public Administration shall contemplate programming, incentivating and educational aspects. Castilla y León ranks among the regions with programmes for sustainable agriculture practices.

Thus, a Program for the cerealist estepa of over 1.600.000 Has in the central part of our region has been implemented and approved by the European Commission on 16th September 1993. It is our policy that it will be but the first of a series of issues to achieve a more rational and sustainable agriculture.

Those Programmes shall pay attention to the so called ecological agriculture, as well as to reduce consumption of irrigation waters, to transform marginal lands into extensive grasslands, to regenerate a shrubby vegetation and autochtonous tree cover, to control unwanted fire-risk bushes and to maintain endangered autochtonous species.

It is also important to investigate and to keep the farmer informed on rational irrigation practices as well as on an adequate use of fertilizers in irrigated crop lands. Livestock farming, a sector of greatest interest in our region must also receive attention at the time of implementing the above criteria of sustainability. On the one hand extensive livestock, nowadays endangered by present CAP and international commerce, so important for conservation of habitats of high ecological value i.e. "dehesas" or firebreaks control in forests ecosystems. On the other hand, the serious problems produced by an intensive livestock exploitation leading to a high concentration of residues and to the nitrate contamination of soils, waters, and food chain. For the purpose, the Junta de Castilla y León is about to implement an specific issue to provide with solutions in the treatment of farm residues. Its final success shall come however, through the collaboration and full willingness of farmers.

As you may gather, environment and agriculture interactions hold a great complexity. Still, even more so in a world of contrasts where some regions may encourage a crop yield reduction due to the high costs involved in finding a way out to a surplus production, while the opposite situation in other areas with an under production unable to feed the existing population is also true.

Within this dilemma, research has to found the ways to provide with satisfactory methodologies to secure sufficient yields to warrant the continuity of human life. An integrated concept of fertilization practices and criteria so that crop yields increase through a maintenance of soil productivity with no hazards for environment will be required.

Fertilization, or let me say "sustainable nutrition" of crops, as defined by the World Conference of Environment and Development of Rio de Janeiro, stands up as one of the main objectives along next century in all and every human community of our planet.

That essentially implies a protection of the regional diversity with recommendations specific to the geographic and cultural characteristics of the area. In our geographic area of Castilla-León, the maintenance of land productivity, and the preservation of traditions are to be taken into account both for the protection of environment and rural development. The updating of traditional methodologies to improve fertilizer rational use and to lower down both the harmful effects and the levels needed. must receive the attention of experts.

Research must also be undertaken on the use of farm and solid urban residues and sludges from water depurating stations as fertilizers for agriculture and forestry. Besides, efforts must be addressed towards providing the farmer with all sort of information on how to increase efficiency and on how to secure soil productivity and environment conservation at a long term. That will be the key for success in sustainable agriculture.

On the whole, planning efforts from the Administration will not contribute significantly to sustainable development of Castilla y León without the conscious collaboration from farmers and experts. The scientific research and practical application of knowledge as well as to seek rentability of new agricultural management practices are the milestone towards the whole process of sustainable development, and for the purpose the expertise of scientists participating in this meeting is of a fundamental value.

Symposia like the one now hosted by Salamanca should help to find adequate recommendations adapted to each regional requirements with the final aim of designing a model of sustainable agriculture, that is to say, a model of compatible agriculture consequently respectful with environment. Within that context so important for our future development I encourage all of you to continue doing the necessary efforts.

Salamanca, 26th September 1994

C. Rodriguez-Barrueco (ed.), Fertilizers and Environment, 5–11.

Heavy metal contaminants in inorganic and organic fertilizers

J.J. Mortvedt
Department of Soil and Crop Sciences, Colorado State University, Fort Collins, CO 80523, USA

Key words: biosolids, cadmium, heavy metal limits, lead, phosphate fertilizers, regulations, sewage sludge

Abstract

Commercial phosphate (P) fertilizers contain small amounts of heavy-metal contaminants which were minor constituents in phosphate rock (PR). Animal manures and sewage sludges (biosolids) are the main organic fertilizers and the latter also may contain heavy-metal contaminants. Heavy metals in biosolids may be found in the inorganic form or may be organically complexed, which could affect their chemical reactions in soil. These heavy metals may accumulate in soil with repeated fertilizer applications. Cadmium (Cd) is the heavy metal of most concern because it may affect human health. Other heavy metals of possible significance are arsenic (As), chromium (Cr), lead (Pb), mercury (Hg), nickel (Ni), and vanadium (V). Some countries have set tolerance limits on heavy-metal additions to soil because their long-term effects are unknown. These limits usually are set for the tillage layer (surface 20–30 cm) of soil where most root activity occurs. Controls on heavy-metal concentrations in sewage biosolids and their maximum total and annual loading rates to soil have been imposed in some countries. Regulations also have been proposed for phased-in limits on maximum heavy metal concentrations permitted in P fertilizers, or they are already in effect. Most of the fertilizer regulations relate Cd limits to P concentrations, so P application rates dictate Cd inputs to soil. Regulations affecting sewage biosolids include a number of heavy metals, while those concerning P fertilizers only include limits on Cd at this time.

Increased concern about the entry of various heavy metals into the human food chain has occurred in recent years, and all aspects of agricultural production and food processing are being examined. Cadmium (Cd) is the heavy metal of most interest because potentially it may be harmful to human health. Therefore, we are giving more attention to its avenues of entry into the human food chain. Among these avenues is the Cd component of P fertilizers and sewage biosolids applied to agricultural lands. While animal manures also contain Cd and other heavy metals, these concentrations are low. Therefore, no buildup of heavy metals would be expected from moderate manure application rates on agricultural lands.

Phosphate rock (PR) contains various metals as minor constituents in the ores. Varying amounts of these elements are transferred to P fertilizers in production processes, and later are applied to soils with these fertilizers. Sewage sludge also contains heavy metals which vary considerably in concentration, depending on the composition of the treatment plant waste streams.

There are also numerous other heavy metal contaminants in P fertilizers and sewage biosolids. Concentrations of these metals vary considerably, depending on the PR source. Some other heavy metals of possible significance are: arsenic (As), chromium (Cr), lead (Pb), mercury (Hg), nickel (Ni), and vanadium (V). However, these metals are of less concern than Cd, either because they are not as readily absorbed by plants from P-fertilized soils or their apparent relative effects on human health are less than that of Cd.

Heavy metal concentrations in phosphate rock deposits

Most known PR deposits have been assayed for heavy metal content. A recent report (Kongshaug *et al.*, 1992) gives an excellent summary of some heavy metal concentrations found in various PR deposits. Table 1 shows the average heavy metal concentrations in deposits comprising 91% of the known world PR reserves. Concentrations vary considerably with metal (Cr is highest and Hg is lowest) and with region.

Table 1. Average heavy metal concentrations in phosphate rock (PR) deposits and estimated inputs to soil by P fertilizers (Kongshaug *et al.*, 1992)

PR deposit	Heavy metal concentration, mg/kg						
	As	Cd	Cr	Pb	Hg	Ni	V
Russia (Kola)	1	0.1	13	3	0.01	2	100
USA	12	11	109	12	0.05	37	82
South Africa	6	0.2	1	35	0.06	35	3
Morocco	11	30	225	7	0.04	26	87
North Africa	15	60	105	6	0.05	33	300
Middle East	6	9	129	4	0.05	29	122
Avg of 91% of PR reserves	11	25	188	10	0.05	29	88
mg/kg of P	71	165	1,226	66	0.29	189	578
g/ha/yr applied with 20 kg P/ha	1	3.3	25	1	0.01	4	12
Tolerance limit in soil, mg/kg[1]	2	100	100	2	50	50	300

[1]Finck (1992).

Converting the average heavy metal concentrations values in PR to a basis relative to the P concentration allows an estimation of the average heavy metal application rate to soil at a given P rate. Data in the second to the last line of Table 1 show that such rates range from 0.01 to 25 g/ha/yr for Hg and Cr, respectively, when growers apply P fertilizers from these PR deposits at a rate of 20 kg P/ha. These rates would be somewhat lower if P fertilizers, such as triple superphosphate (TSP) or diammonium phosphate (DAP), produced from the same PR sources were applied to soil.

Some countries have set tolerance limits for heavy metal additions to soil. Data in the last line of Table 1 provide the values set for German soils (Finck, 1992). Such limits generally are set for the surface plow layer (20–30 cm) of soil where most root activity exists. To estimate the number of years required for a given heavy metal contaminant in P fertilizers to reach the tolerable limit, the following example is shown: Total soil Cd generally ranges from 0.1 to 1 mg/kg. The total amount of Cd in the surface 20-cm layer is 1.5 kg/ha, assuming an average Cd concentration of 0.5 mg/kg. The tolerable limit for Cd is 2 mg/kg which is equivalent to 6 kg/ha. Thus, Cd applications could total 4.5 kg/ha to reach the tolerable limit.

From the data in Table 1, it would take 1,300 yr of P applications at a rate of 20 kg P/ha to reach the tolerable Cd limit (4.5 kg/ha divided by 3.3 g/ha). This calculation ignores other possible Cd inputs to agricultural soils, such as sewage sludges and aerial deposition, as well as Cd removal by crops. However, it illustrates the estimated time needed to reach some defined heavy metal limits for agricultural soils.

Transfer of heavy metals in phosphate rock into phosphate fertilizers

Various fractions of heavy metal contaminants in PR will be transferred into P fertilizers, depending on the manufacturing process. Ordinary (single) superphosphate is produced by reacting H_2SO_4 with PR. The resulting product will contain all of the heavy metal constituents found in the PR. However, most of today's P fertilizers is produced from wet-process H_3PO_4. By-product phosphogypsum also will contain some of the heavy metals found in the PR.

Most partitioning studies have been conducted with Cd. Wakefield (1980) reported that TSP contained 60–70% of the Cd found in PR. However, Williams and David (1973) found a close relationship between concentrations of P and Cd in superphosphates and their respective PR sources in Australia. While there are little data on the relative transfer of the other heavy metals from PR to P fertilizers, the transfer coefficients should be similar for most metals, depending on chemical reactions occurring during the production processes.

Several chemical processes to remove Cd from H_3PO_4 before it is converted to P fertilizers have been studied. Smani (1993) discussed several methods of Cd removal from wet-process H_3PO_4. While not yet tested on a commercial scale, solvent extraction appears to be a possible commercial method. Estimated costs are \$100(US)/metric t of P_2O_5 or \$46(US)/metric t of DAP, which is about 30% of the current P fertilizer prices. Thus, the price of DAP would have to increase significantly to offset the increase in production costs. Benchekroun (1992) suggested that increasing current P fertilizer prices more than 20% to pay for environmental protection could result in dramatic decreases in P fertilizer use by developing countries, where use already is low because of inadequate financial resources.

Plant uptake of heavy metal contaminants in phosphate fertilizers

Results of studies in several countries have shown that some heavy metal contaminants in P fertilizers may be available to plants. Because of the potentially adverse effects of Cd on human health, most of the studies have been focused on Cd. Williams and David (1973) reported that plant species differed considerably in their ability to take up Cd, with leafy vegetables absorbing more Cd than grasses. In addition, only 12–18% of the Cd in cereal plant tops was translocated into the grain. However, soil application of $CdCl_2$ or TSP containing Cd resulted in increased Cd concentrations in both cereal grains and the edible portions of vegetables. Topdressing pastures with TSP also resulted in increased Cd of pasture species, especially that of subterranean clover (*Trifolium subterraneum* L.).

Mortvedt and Giordano (1977) reported greater Cd uptake by maize (*Zea mays* L.) from commercial DAP fertilizers which contained from 100 to 260 mg Cd/kg P than from reagent grade DAP (5 mg Cd/kg P). Plant uptake of Cr, Ni, and Pb was quite variable and was not related directly to their concentrations in P fertilizers.

While many of these studies have been conducted in greenhouse pot experiments, others have been done under field conditions. Mortvedt *et al.* (1981) applied DAP fertilizers containing 10, 370, and 760 mg Cd/kg P to a P-deficient soil. Concentrations of Cd in both grain and straw of winter wheat (*Triticum aestivum* L.) were increased only with the highest-Cd DAP applied to an acid soil (pH 5.1) but not to the same soil limed to pH 5.9. Jaakola (1977) also reported that Cd uptake by grain and straw of spring wheat was not affected by applications of P fertilizers containing 10, 150, 285, or 405 mg Cd/kg P. Results of USDA 'market basket' surveys (Wolnik *et al.*, 1983) have shown that Cd levels in several important USA food crops were lower than previously reported. The average weekly Cd intake per person in the USA was estimated at about 100 μg, as compared with the maximum weekly Cd intake of 400–500 μg recommended by the World Health Organization (Anon., 1989). Estimated per capita weekly Cd intake in Australia was 125–225 μg, based on a 1990 market basket survey (Anon., 1990). While Cd uptake by crops might be somewhat higher on P-fertilized acid soils, it seems doubtful that weekly Cd intake by humans will approach the maximum recommended levels listed above.

Long-term effects of heavy metals applied in phosphate fertilizers

Accumulations of Cd and other heavy metals applied to agricultural soils with P fertilizers are difficult to estimate because we can not easily assess the mechanisms for addition and removal. Avenues for addition, other than with P fertilizers, are atmospheric deposition, return of crop residues, and application of animal manures, sewage biosolids, and other non-P fertilizers. After application to soils, Cd is very immobile so it tends to accumulate in the surface soil. Removal is mainly through grain, forage, and livestock products, with some removal possible from soil erosion.

One method used to assess the net effects of heavy metal contaminants in P fertilizers applied to soils is to analyze crops and soils from long-term soil fertility experiments. While heavy metal analyses may not have been completed on many P fertilizers, we can estimate such concentrations if the source of PR used to produce these P fertilizers is known. Isermann (1982) analyzed soils from 20 west European, long-term (26- to 138-yr) experiments and calculated that Cd inputs from P fertilizers were of the same magnitude as those from farmyard manure or crop residues applied to soils.

Smilde and van Luit (1983) compared trends in soil Cd in P-treated and control plots in some long-term field experiments in The Netherlands. Estimated totals of 135 to 450 g Cd/ha applied over a period of time (up to 64 yr) in five experiments did not affect Cd concentrations in wheat and barley (*Hordeum vulgare* L.) grain, potato (*Solanum tuberosum* L.) tubers, sugar beet (*Beta vulgaris* L.) leaves, or onion (*Allium cepa* L.) bulbs.

Rothbaum *et al.* (1986) analyzed soils from three long-term experiments at Rothamsted, England, and one in New Zealand. Annual rates were equivalent to 33 kg P/ha and 5 g Cd/ha for 95 yr in England and to 37 kg P/ha and 20 g Cd/ha for 30 yr in New Zealand. Very little Cd accumulated in the surface soil of two

Table 2. Proposed or implemented Cd limits for P fertilizers in several countries

Country	Cd limit, mg/kg P	Effective year
Australia	450	In effect
	350	1995
	300	2000
Austria	120	In effect
Belgium	200	Voluntary
Denmark	150	In effect
	110	1995
Finland	50	In effect
Germany	200	Voluntary
Japan	343	In effect
Norway	100	In effect
	50	1995
Sweden	100	In effect
Switzerland	50	In effect
The Netherlands	35	

[1] Adapted from Anon. (1989).

experiments, but they found about half of the applied Cd in the surface layer of soil.

Mortvedt (1987) conducted a similar study of nine long-term (> 50 yr) experiments in the USA. Annual Cd applications were estimated to range from 0.3 to 1.2 g/ha (assuming 5 mg Cd/kg of TSP). Results from analysis of maize, soybean (*Glycine max* L. Merr.) and wheat leaves or grain, and timothy (*Phleum pratense* L.) forage indicated that uptake of Cd contaminants in P fertilizers was negligible.

Regulations on heavy metals in phosphate fertilizers

In the past decade, attention has been directed towards the regulation of maximum Cd concentrations permitted in fertilizers (specifically P fertilizers). It was estimated in 1987 that the annual Cd input to soil at average fertilizer application rates was 3.5 g/ha in the European Community (EC) countries, as compared with Cd inputs from sewage biosolids of 60 to 167 g/ha in EC countries and 500 g/ha in the USA (Anon., 1989).

The Dutch Government proposed regulations in 1987 limiting the maximum Cd concentration in P fertilizers to 35 mg Cd/kg P (15 mg Cd/kg P_2O_5) by the mid-1990s (Table 2). The Danish Government announced proposals to gradually reduce the maximum Cd concentration in P fertilizers in four phases from 1987 to 1997 (Anon., 1989). This proposal was later revised to slightly higher allowable Cd levels by 1995. (O. Kaarstad, pers. comm., 1994). The proposed progression in guidelines for maximum Cd concentrations in imported and manufactured P fertilizers in Australia is voluntary. Norway, Sweden and Switzerland also have proposed limits on Cd concentrations in P fertilizers, while Germany has a Cd limit on a voluntary basis. In addition to their Cd limit of 100 mg/kg P, Sweden has imposed a tax of 30 SEK/kg Cd on fertilizers with Cd concentrations between 5 and 100 mg Cd/kg P (I. Steen, pers. comm., 1995). Austria, Finland and Japan have statutory requirements regarding maximum Cd concentrations in P fertilizers. These countries base their regulations on Cd concentrations related to P concentrations in fertilizers, rather than on a unit weight basis. Therefore, P application rates will dictate Cd inputs to soil, regardless of the P concentration in a fertilizer.

One reason for proposing regulations to limit Cd concentrations in P fertilizers is that some surveys have shown that the average Cd content of surface soils and in some plant tissues is increasing. In a comprehensive study using archived samples collected at the Rothamsted Experiment Station in England since about 1850, Johnston and Jones (1992) reported that the net annual Cd input by atmospheric deposition, which had averaged 3.2 g/ha over a 100-yr period, increased to 14 g/ha during the past 20 yr. By comparison, annual Cd inputs would be 6.9 g/ha with superphosphate containing 345 mg Cd/kg P applied at a rate of 20 kg P/ha. Cadmium accumulated in acid soils containing about 5% organic matter but not on neutral soils low in organic matter. While Cd did not accumulate in subsoils low in organic matter, it was not confirmed if Cd losses by leaching had occurred. In addition to controls on industrial Cd emissions, limits on Cd inputs from P fertilizers have been considered in England and the UK.

In contrast, the amounts of Cd added to soils via aerial deposition in rural agricultural soils in the USA are on the order of several g/ha, which are similar to those added to soil from normal P fertilization programs (Page *et al.*, 1987). No limits on industrial Cd emissions or Cd concentrations in P fertilizers have been proposed in the USA at this time.

Once such limits on Cd in P fertilizers are in place in some countries, others may consider similar legislation. While there is less interest in placing limits on other heavy metals in P fertilizers at this time, other metals may be included even though there is much less data on which to base any regulations. Australia has a limit of 500 mg of Pb/kg and 5 mg of Hg/kg of product

Table 3. Median concentrations and ranges of heavy metals in dry digested sewage biosolids in the USA

Element	Dry sludge (mg/kg)	
	Median	Range
As	10	1–230
Cd	10	1–3,410
Cr	500	10–99,000
Cu	800	84–17,000
Pb	50	13–26,000
Hg	6	1–56
Mo	4	1–214
Ni	80	2–5,300
Se	5	2–17
Zn	1700	101–49,000

Adapted from Chaney (1983).

for all fertilizers at present. At this time, it appears that there may be some justification to consider limiting Cd concentrations in P fertilizers. However, there is very little evidence to justify placing limits on other heavy metals in P fertilizers.

Heavy metal concentrations in sewage biosolids

Disposal of sewage biosolids is a problem facing municipalities around the world. Quantities of biosolids continue to increase with increases in population density in cities. One alternative to disposal is recycling by agricultural utilization. Application rates of biosolids to agricultural lands should be such that the plant nutrients contained in the biosolids partially or entirely meet the needs of the crop. Concentrations of N and P in most biosolids usually are not in the same ratio as those of crop needs. Therefore, application rates should be chosen to avoid accumulation of P in soil when using biosolids mainly to supply N to the crop.

In addition to valuable plant nutrients, biosolids also contain many organic and inorganic trace constituents which are potentially harmful to plants, animals, or humans. Large variations of metal concentrations have been found in sewage biosolids monitored in the United States, Canada and Europe. The distributions were skewed towards high concentrations, and arithmetic means overestimated the average metal concentrations of most biosolids. Median concentrations generally are used to describe the average metal contents of biosolids (Sommers, 1977). These values also can be used as reasonably attainable background levels in biosolids after pretreatment to remove known sources of metals. Table 3 lists median values and ranges of various metals in dry digested sewage biosolids.

Bioavailability of metals in sewage biosolids

Numerous studies have been conducted to determine the bioavailability of metal contaminants as well as plant nutrients in sewage biosolids. Results show that bioavailability of metals generally decreases with increasing pH, organic matter content, and clay content of soil.

Metal uptake also varies with the chemical form of metals in biosolids. Fractionation schemes have been developed to characterize chemical forms. Processes which control solution activity include solid-phase equilibria, adsorption-desorption, chelation and oxidation-reduction (Logan and Chaney, 1983). Recent use of computer speciation models has helped to better predict the chemical forms in biosolids.

Transfers of heavy metals in sewage biosolids to the food chain begin with plant uptake from land-applied biosolids. Relative uptake of some of these metals, especially Cd, by various crops is well documented. Annual application rates and loading limits of the heavy metals should integrate knowledge of characteristics of the biosolids and soil, as well as crops to be grown. Computer modelling is being used to predict effects of the above variables on plant uptake. Chaney *et al.* (1987) estimated maximum allowable soil Cd load levels on the basis of increases in dietary Cd intake by humans.

Regulations on heavy metal concentrations in sewage biosolids

Several countries have introduced controls on heavy metal concentrations in sewage biosolids because of the above concerns regarding heavy metal effects in the human food chain. During the early 1980s, a commission of the EC recommended maximum annual and total loading rates of various metals to agricultural lands. For Cd, these were 0.1 and 0.15 kg Cd/ha/yr for recommended and mandatory maximum annual loadings, respectively, and 2.4 and 8.4 kg/ha for recommended maximum and mandatory cumulative limits, respectively (Webber *et al.*, 1984).

After The Netherlands introduced controls on Cd concentrations in biosolids used as fertilizers, the com-

Table 4. Heavy metal concentration limits and annual load limits of sewage biosolids in the USA (USEPA, 1993)

Element	Ceiling limit	APL[1]	Maximum annual load
	mg/kg dry weight		kg/ha/yr
As	75	41	2.0
Cd	85	39	1.9
Cr	3,000	1,200	60.0
Cu	4,300	1,500	75.0
Pb	840	300	15.0
Hg	57	17	0.85
Mo	75	—[2]	3.8
Ni	420	420	21.0
Se	100		4.9
Zn	7,500	2,800	140.0

[1]Alternate Pollutant Limit;
[2]Concentration not yet determined.

mission of the EC issued a directive specifying mandatory and recommended limits of 40 and 20 mg Cd/kg (dry-weight basis), respectively, for sewage biosolids applied to agricultural soils (Anon., 1989). Another EC decision limited Cd concentrations in industrial effluents discharged into the Rhine River. Hutton (1982) prepared a comprehensive report on sources, human exposure, and environmental impact of Cd in EC countries.

The 1972 USA Federal Water Pollution Control Act Amendments (Public Law 92-500) encouraged the recycling of plant nutrients in sewage biosolids through land application and consequent growth of agricultural and forest products. One difficulty facing regulatory agencies and municipalities interested in applying biosolids to agricultural lands is that of knowing the maximum application rates which would not result in harmful accumulations of heavy metals in crops and soils. The 1977 and 1981 Amendments to the Clean Water Act approved utilization of sewage biosolids on agricultural lands if it was done in an acceptable manner. Some of the requirements were chemical analysis of the biosolids, using recommended rates of application, record keeping, and monitoring of heavy metals in crops, soils and groundwaters.

When it was determined that the heavy metal concentrations were very high in some sewage biosolids, a federal industrial pretreatment program was implemented in the USA about 1976. This was to control the discharge of these heavy metals into publicly owned treatment facilities. Thereafter, the metal loads to municipal wastewater treatment facilities decreased, and heavy metal concentrations in biosolids decreased with time.

The USA Clean Water Act, Section 503 (Sludge Rule) regulations concerning utilization and disposal of sewage biosolids are given as Ceiling Limits and Alternate Pollutant Limits (APL) (Table 4). The Ceiling Limits are maximum allowable heavy metal concentrations when biosolids are pasteurized and stabilized. If a biosolid meets the APL and is applied at an agronomic rate (based on N and/or P needs), then cumulative load limits do not pertain to these materials. These regulations were first published in 1989, revised after public hearings by 1992, and took effect in February 1994.

References

Anon. (1989) Cadmium in phosphates: One part of a wider environmental problem. Phosphorus and Potassium 162 (4): 23–30

Anon. (1990) The 1990 Australian market basket survey report. National Health and Medical Council, National Food Authority. Australian Government Publishing Company, Canberra

Benchekroun A (1992) Identifying and managing the environmental issues facing the phosphate fertilizer production sector – the viewpoint of the World Phosphate Institute. In: Schultz JJ (ed) Phosphate Fertilizers and the Environment, pp 143–146. Spec. Pub. IFDC-SP-18, Int. Fert. Dev. Ctr., Muscle Shoals, AL

Chaney RL (1983) Potential effects of waste constituents on the food chain. In: Parr JF (ed) Land Treatment of Hazardous Wastes, pp 50–76. Noyes Data Corp. Park Ridge, New Jersey

Chaney RL, Bruins RJF, Baker DE, Korcak RF, Smith JE and Cole D (1987) Transfer of sludge applied trace elements to the food chain. In: Page AL *et al.* (eds) Land Application of Sludge, pp 67–99. Lewis Publishers, Chelsea, MI

Finck A (1992) Dünger und Düngung: Grundlagen und Anleitung zur Düngung der Kulturpflanzer (2nd Ed), pp 389–395. VCH Verlagsgesellschaft mbH, Weinheim, Germany

Hutton M (1982) Cadmium in the European community: A prospective assessment of sources, human exposure and environmental impact. Comm. of the European Communities. Tech. Rept. 33-ENV U.K., Chelsea College, Univ. of London, England

Isermann K (1982) Effect of phosphate fertilization on the cadmium content of soil determined on the basis of numerous long-term trials in Western Europe. Landwirtsch Forsch Sonderh 39: 283–301

Jaakola A (1977) Effect of fertilizers, lime and cadmium added to soil on the cadmium content of spring wheat. J Sci Agric Soc Finl 49: 406–414

Johnston AE and Jones KC (1992) The cadmium issue – long-term changes in the cadmium content of soils and the crops grown on them. In: Schultz JJ (ed) Phosphate Fertilizers and the Environment, pp 255–269. Spec Pub. IFDC-SP-18, Int. Fert. Develop. Ctr., Muscle Shoals, AL

Kongshaug G, Bockman OC, Kaarstad O and Morka H (1992) Inputs of trace elements to soils and plants. Proc. Chemical Climatology and Geomedical Problems, Norsk Hydro, Oslo, Norway

Logan TJ and Chaney RL (1983) Utilization of municipal wastewater and sludge on land-metals. In: Page AL *et al.* (eds) Utilization of Municipal Wastewater and Sludge on Land, pp 253–326. University of California, Riverside, CA

Mortvedt JJ (1987) Cadmium levels in soils and plants from some long-term soil fertility experiments in the United States of America. J Environ Qual 16: 137–142

Mortvedt JJ and Giordano PM (1977) Crop uptake of heavy-metal contaminants in fertilizers. In: Wildung RA and Drucker H (eds) Biological Implications of Heavy Metals in the Environment, pp 402–416. ERDA Rep. Conf. 750929, Oak Ridge, TN

Mortvedt JJ, Mays DA and Osborn G (1981) Uptake by wheat of cadmium and other heavy metal contaminants in phosphate fertilizers. J Environ Qual 10: 193–197

Page AL, Chang AC and El-Amamy M (1987) Cadmium levels in soil and crops in the United States. In: Hutchinson TC and Meema KM (eds) Lead, Mercury, Cadmium and Arsenic in the Environment, SCOPE 31, pp 119–146. John Wiley and Sons, New York

Rothbaum HP, Goguel RL, Johnston AE and Mattingly GEG (1986) Cadmium accumulation in soils from long-continued applications of superphospate. J Soil Sci 37: 99–107

Smani MS (1993) Removal of cadmium from phosphoric acid. *In* Proc. Fourth Int. Conf., Phosphorus, Life, and the Environment Sept. 8–11, 1992, Ghent, Belgium. World Phosphate Institute, Casablanca, Morocco

Smilde KW and van Luit B (1983) The effect of phosphate fertilizer cadmium on cadmium in soils and crops. Rapport Institut Bodemvruchtbaarheid no. 6–83. Haren, The Netherlands

Sommers LE (1977) Chemical composition of sewage sludges and analysis of their potential use as fertilizers. J Environ Qual 6: 225–232

United States Environmental Protection Agency (1993) Clean Water Act 40 CFR 503, Sludge Rule. U.S. Govt. Print. Off., Washington, DC

Wakefield ZT (1980) Distribution of cadmium and selected heavy metals in phosphate fertilizer processing. Bull. Y-159. National Fertilizer Development Center, Tennessee Valley Authority, Muscle Shoals, AL

Webber MD, Kloke A and Tjell J (1984) A review of current sludge use guidelines for the control of heavy metal contamination in soils. In: Hermite PL and Ott H (eds) Processing and Use of Sewage Sludge. Reidel Publishing Co., Dordrecht, The Netherlands

Williams CH and David DJ (1973) The effect of superphosphate on the cadmium content of soils and plants. Aust J Soil Res 11: 43–56

Wolnik K, Fricks FL, Capar SG, Braude GL, Meyer MW, Satzger RD and Bonner E (1983) Elements in major raw agricultural crops in the United States. 1. Cadmium and lead in lettuce, peanuts, potatoes, soybeans, sweet corn, and wheat. J Agr Food Chem 31: 1240–1244

C. Rodriguez-Barrueco (ed.), Fertilizers and Environment, 13–18.

The present situation of fertilizer production and use in the world

K.F. Isherwood
Information Service, International Fertilizer Industry Association (IFA), Paris, France

Abstract

In 1960 88% of the world fertilizer usage was accounted for by the developed countries. Today the proportion is 45%. Fertilizer consumption in the developed countries remains large but since 1989 it has fallen in absolute as well as proportional terms. This is due particularly to the spectacular falls in fertilizer demand in the countries of the FSU and Central Europe, and to a lesser extent in Western Europe. Conversely, fertilizer consumption in developing countries, now accounting for 55% of total world consumption has continued to increase. On the whole, the outlook for the fertilizer industry is more promising than in recent years. Fertilizer demand has already started to recover in some Central European countries and the rate of fall in Western Europe has slowed greatly. The prospects in Latin America and in Asia are good. World grain stocks are low and commodity prices have improved. As regards supply, the fertilizer industry has suffered in recent years from substantial over-capacity, with resulting low prices and profitability. As a result there is considerable hesitation about investing in new projects and the supply surpluses are shrinking, particularly for ammonia and urea but also for phosphoric acid. The world's population continues to increase, and certain, although unfortunately not all, developing regions are becoming much more prosperous. Increased agricultural production will be required from a land base which is not only finite, it is decreasing due to soils degradation, urbanization etc. World cereal stocks are near the minimum tolerable level. At the same time the use of mineral fertilizers is under attack from environmentalists, often without scientific foundation. Fertilizers are regarded as responsible for the pollution of water supplies, harming the soil structure, leading to intensive agriculture which disfigure the countryside, and in general leading to a type of agriculture which is unsustainable. Ideally, all the sectors concerned, including the environmentalists, should work together, not blindly to reduce mineral fertilizer use, but to ensure future food supplies by promoting their correct use. This paper is based on statistics and reports provided by IFA members in the different countries of the world. The objective is to make assessments of the world fertilizer supply and demand situation, which fertilizer manufacturers are then free to use as they think best to guide their investment policies.

Fertilizer consumption, 1960 to 1993

World

Between 1930 and 1960, world nitrogen, phosphate and potash consumption developed in line with each other, in similar quantities for each nutrient. From 1960 onwards the consumption of nitrogen increased faster than that of phosphate and potash but the consumption of all three nutrients grew substantially. There was a fall in world fertilizer consumption following the 1973/74 oil crisis but the growth of nitrogen consumption quickly resumed. The consumption of phosphate and potash fertilizers also recovered but to a lesser extent. There was a further pause in world fertilizer consumption following the second oil crisis, in 1979/80. The previous growth pattern then resumed until 1989, when world consumption began to fall.

The results of the most recent IFA survey indicate that world fertilizer consumption, in terms of total nutrients, fell by 4% in 1993/94. Between 1989/90 and 1991/92 world fertilizer consumption, in terms of total nutrients, fell by almost 9 million tonnes. Between 1991/92 and 1993/94 it is estimated to have fallen even more, by a further 14 million tonnes.

Developed countries

In 1963/64, 86% of the world fertilizer consumption was accounted for by the developed countries: Western Europe, North America, Oceania, Eastern Europe, the former USSR (FSU), South Africa and Japan. Today, these regions account for 45% of total world fertilizer consumption. Their population growth has tended to level off, almost everyone is adequately fed, and world agricultural exports have stagnated due to economic problems in the importing countries.

The phosphate and potash consumption of developed countries has tended to decline since 1979. In spite of this, crop yields have continued to increase and costly surpluses have developed in Western Europe. In general, soils had been well fertilized with phosphate and potash for many years and some reduction of application could be permitted during periods of economic difficulties for the farmer.

In the developed countries as a whole, including the formerly centrally-planned economy countries, fertilizer consumption fell by 23 million tonnes nutrient between 1989/90 and 1992/93. Essentially, the reduction is due to a decline fertilizer use in the countries of Central Europe and of the Former Soviet Union, and, to a lesser extent, of Western Europe. We estimate a further, but lesser fall, in 1993/94, due essentially to the FSU.

Developing countries

While the fertilizer consumption of developed countries has faltered, that of developing countries continued to increase steadily. In 1990/91, fertilizer consumption in developing countries for the first time exceeded that of the developed countries.

With their rapidly increasing populations, many developing countries have been compelled to give agricultural production and the development of fertilizer use a high priority. Their substantial expansion of fertilizer use started in the early 1960's. By 1993/94 the share of world consumption accounted for by developing countries had increased to 55%. By 1998/99 we estimate that it will be 58%.

The consumption of nitrogen in developing countries has developed more rapidly than that of phosphate and potash. Fertilizer application has in fact become unbalanced in many countries i.e. too much nitrogen in relation to the other nutrients, and the present developments are aggravating the situation. Many developing countries have soils which are highly deficient in phosphate and their levels of fertilizer application are low. The soils of developing countries, particularly in the drier regions, tend to be richer in potash but large quantities of this nutrient are removed in harvested crops.

To 1998

The outlook is more promising. Between 1992/93 and 1998/99, world fertilizer consumption is expected to increase again, by 9 million tonnes nutrient. At this rate, nitrogen consumption will recover to its 1988/89 level by 1998/99 but phosphate and potash consumption will still be well below the levels of the peak year.

We expect that fertilizer consumption will begin to recover in the developed countries in 1995. A gradual recovery is expected in Central Europe from 1994 onwards and, hopefully, in the FSU from 1995. Consumption in these regions in 1998, and for some time after that, will be well below the 1989 level.

Concerning the relative importance of the different regions as regards fertilizer consumption, in 1993/94 Socialist Asia accounted for 23% of the world total, North America for 18%, Western Europe for 14%, South Asia for 13%, Central Europe and the FSU for 9% (compared with 16% in 1990). The sum of consumption in these regions amounts to 77% of the global total. Fluctuations in these five regions have a strong influence on the fertilizer market.

The situation in these regions is as follows:

Socialist asia

The largest fertilizer-consuming country in the world is China. Apparent fertilizer consumption (production plus imports) in China fell in both 1992 and 1993. In 1992, Chinese imports of urea and DAP fell by 9% and 13%, respectively, compared with 1991. This trend continued in 1993, due to reduced subsidies, decentralized purchases and uncertainty concerning foreign exchange allocations.

The net result of these financial measures is a substantial increase in the cost of fertilizers and the prices of the agricultural products are not compensating for the increased costs. There is a rural exodus due to the better living standards available in towns, especially in the coastal regions, and some farmers have given up cultivating their land.

China's problem will be to meet increases in demand during the coming years, from the same (or reduced) agricultural area. Growth of fertilizer consumption must inevitably resume.

North America

A significant increase of fertilizer consumption in the USA is reported for the 1993/94 season in the USA. The poor corn and soybean harvests of 1993 resulted in low stocks and zero set-aside i.e. a larger planted area.

As regards this year's harvest in the USA, bumper corn and soyabean harvests are expected, although wheat production will probably be down.

The development of fertilizer consumption in the USA depends largely on the sown area, i.e. the set-aside area, which in turn depends on the weather and harvests in the U.S. and elsewhere, and these are not predictable. We assume a medium-term growth rate of 1% p.a.

Western Europe

Nitrogen and potash consumption in Western Europe has been falling since 1985/86, phosphate since 1980. There was a particularly strong fall in 1992/93, due to the combined effect of falling prices and incomes, set-aside, fears concerning the impact of the GATT negotiations etc. A continuing decline of fertilizer consumption is anticipated, due to measures designed to reduce the agricultural production surpluses, but one which is much less severe than that which occurred between 1989/90 and 1992/93 i.e. the situation is stabilizing.

South Asia

India accounts for more than three quarters of fertilizer consumption in South Asia, the Indian sub-continent. In India, fertilizer consumption fell in 1993/94, for the first time in 18 years, due to changes in pricing and subsidy policies. N consumption in fact increased, but P and K consumption declined. Fortunately weather conditions were favourable in both 1992/93 and 1993/94 and foodgrain production was reasonable, although it would have been higher with proper fertilizer application – and it was well below target.

The countries of this region have very little safety margin as regards agricultural production and have no alternative to more and better fertilizer use.

Central Europe and the FSU

Between 1989 and 1992 fertilizer consumption fell in the countries of Central Europe by 70%. During the same period FSU consumption declined by 50%. However, whereas the fall in the countries of Central Europe appears to have stopped, fertilizer consumption continues to fall sharply in the FSU.

The factors accounting for the sharp decline of fertilizer consumption in these regions include the removal of high subsidies, high interest rates, shortage of cash and a lack of credit facilities, the need to pay for imports in hard currencies, increased input prices, higher oil costs, uncertainties about land ownership, a poor distribution system and the collapse of trade in agricultural products between the former Comecon countries. In some areas a heavy, perhaps excessive, use of fertilizers in the past, when fertilizers were allocated rather than marketed, resulted in high soil nutrient levels.

Other regions

Latin America is a region of the developing world with considerable potential for increased fertilizer consumption. Several Latin American countries are showing signs of growth in fertilizer consumption. Brazil and Mexico account for two thirds of total fertilizer consumption in this region. Fertilizer consumption in Brazil increased by about 13% in 1993, following an increase of 12% in 1992, and it is expected that growth will continue in 1994, largely due to favourable economic conditions for Brazilian agriculture.

A gradual increase of fertilizer consumption in the countries of South-east Asia, with their increasing populations, limited land area and substantial economic growth, is expected. Fertilizer consumption in Africa is expected to increase but from a very low base level. In the Middle East, Egypt, Iran and Turkey are the major fertilizer consuming countries. In each of these countries the market has been disturbed recently by the removal of subsidies, privatization of distribution etc. but in the longer term steady growth is expected.

Production and supply/demand, 1994 to 1998

Nitrogen

In 1980/81 the developing countries accounted for 31% of nitrogen fertilizer production. By 1992/93 their share had increased to 45%. The main raw materials for the manufacture of nitrogen fertilizers are air and energy and in consequence their production is more evenly distributed throughout the world, although there has been a trend towards production in locations where cheap natural gas is available.

The number of new projects for ammonia and nitrogenous fertilizers is limited. For the past few years, our surveys show shrinking supply surpluses. A tight supply situation has in fact developed recently. We cannot exclude the possibility that more frequent temporary tight supplies could occur in the coming years.

Considerable uncertainty surrounds the effective nitrogen supply capability of Central Europe and the FSU. Energy prices have increased very substantially and a number of plants are no longer competitive. In the FSU, there is a large potential supply surplus over domestic demand but exports are limited due to problems with ammonia pipelines, port congestion etc. In recent months there has been a sharp increase in ammonia prices. Urea prices also have firmed due largely to restricted export availability from the FSU.

Our estimates show a surplus of ammonia and urea capacity during the next five years, but one which is small. The same applies to ammonia.

Phosphate

The main producers of phosphate rock and phosphate fertilizers are the USA, the former USSR, China, Africa, the Maghreb countries, Egypt, Senegal, Togo and South Africa, and the Middle East. Several of these countries are developing countries and the phosphate industry makes an important contribution to their economies.

Over the past two decades there has been a distinct trend towards the processing of phosphate rock close to its source and this is expected to continue. However, even in these regions there is little new capacity and it is not impossible that phosphate fertilizers will be in short supply in a few years' time.

There is also a tendency for processed phosphates to substitute for phosphate rock in international trade. From the mid-1970's to the early 80's, most of the increase in processed phosphate trade was in the form of phosphoric acid. More recently, it is diammonium phosphate which has accounted for most of the increase.

There are few new projects in the world for additional phosphoric acid and down-stream phosphate capacity. Furthermore, most of the former phosphoric acid production capacity of Western Europe is now closed and two more plants could close during the forecast period.

Our calculations show the persistence of a supply surplus, amounting to 2.6 million t P_2O_5 in 1994, and falling to 1.9 million t P_2O_5 in 1998. The firm new phosphoric acid projects do not compensate the recent plant closures and those already announced. If the import demand of China, India and Central Europe proves higher than anticipated, and/or if the implementation of some new capacities is delayed, the supply/demand balance for phosphoric acid and the derived fertilizers could tighten.

The supplies of phosphate rock should be sufficient to supply demand during the next five years, at least on a global basis.

Potash

Potash is produced largely where the ores are located. In 1992/93 the former USSR counted for 30% of the world's production, Canada for 31%, Germany for 15%. Together France, Spain and the UK accounted for 10% and Israel and Jordan for 9%, the USA for 4%. Production elsewhere is negligible.

1993 was the fifth consecutive year showing a decline in potash production. Between 1988 and 1993, world production decreased by 11.5 million t K_2O.

The potash market has improved significantly in the past few months thanks to a high demand for US agriculture and to substantial purchasing by China. However, our calculations show that a significant potash supply surplus exists and will persist over the 5 year period considered.

Sulphur

1992 was marked by a shift from a supply deficit to a supply surplus situation. That year saw the end of a long period of world inventory reduction. In 1993, world deliveries declined by at least 2.3 million t while world production decreased significantly less. In North America and elsewhere sulphur is being put into stock. A very significant world supply surplus exists.

Population

According to the United Nation's estimates, between 1950 and 1990, the population of the industrialized countries increased at a rate of 0.9% p.a., whereas population growth rate in the developing countries was 2.2%. They forecast that the rate of growth between 1990 and 2025 will be 0.4% p.a. for the industrialized countries and 1.6% for the developing countries. In 1950, the developing countries accounted for 67% of the world's population; by 2025 the figure could be 83%. The world's population increased by 2.8 billion people between 1950 and 1990. Between 1990 and 2025, an extra 3.2 billion people are expected, an increase of 60%.

The agricultural situation

Food production

In general, food production has kept up with the growth of population, although there are hundreds of millions of people in the world who are under-nourished.

Until about 1950, the increase in agricultural production was due to the expansion of the area of cultivated land and mechanization. Since that date, there have been substantial advances in agricultural production, associated with the increased use of fertilizers and agro-chemicals, more irrigation and the introduction and adoption of hybrid varieties of maize and high-yielding wheat and rice.

What is the global food situation today? It is not all that favourable if we take as a yard-stick the cereal stocks available in the world. Indeed, cereal stocks declined from 456 million tonnes in 1986/87 to 326 million tonnes in 1993/94. The latter level represents about 18% of world cereal consumption; FAO considers that a range of 17 to 18% is the minimum necessary to safeguard world food security. I should not dramatize this situation and supplies are still sufficient to meet global effective demand but stocks are at a low level.

As mentioned above, a near-record maize crop is expected in the USA this year, but wheat production could be slightly down. Substantial falls of wheat production are expected in Australia, Canada and the FSU; world wheat stocks could fall to a 14-year low in 1994/95. Rising wheat prices coupled with reduced food aid could affect the food security situation of low-income food-deficit countries.

Past experience shows that major droughts or floods in some major agricultural regions can reduce world food production by 4 to 5% in one year. When this happens again, the impact on world food stocks, and hence on prices, can be imagined.

The agricultural situation is very different in regions with agricultural surpluses and those with agricultural deficits.

Countries with an Agricultural Surplus

Overproduction

Countries with an agricultural surplus are generally located in the industrialized regions of North America, Western Europe, Oceania and Japan.

The basic tendency at the aggregate level has been for agricultural production growth to outstrip that of demand, both domestic and foreign. This situation reached crisis proportions in recent years, as is demonstrated by the fairly drastic changes being made in the EU's Common Agricultural Policy, whose main aim is to curb over-production.

In the early 1960s, there was room for expansion of food demand. Populations and incomes were growing. However, in the 1970s this situation changed drastically. Overall, the economic growth was slower and population growth also started to slow down. As a consequence, the growth rate of domestic demand for agricultural products dropped considerably in the industrialized world. However, during the 1970s developments in the foreign trade sector compensated for the deceleration of growth in domestic demand.

Since the beginning of the 1980s, the state of international agricultural trade has in many respects deteriorated. There was the slow growth, and even decline, of import demand, stemming partly from the effects of slow economic growth, debt-servicing problems, the decline in petroleum prices and, on the positive side, a stronger growth in domestic agricultural production in some developing countries.

Environment

The mineral fertilizer industry is vulnerable in two respects – its production facilities and its products. Until four or five years ago, most attention was paid to the environmental problems related to the production of fertilizers, and the problems were largely resolved. There are still some highly polluting plants in the countries of Central Europe and the FSU but the technology for clean production exists.

Today, the major problems concern the use of fertilizers. Up to the mid-1980s, the use of mineral fertilizers was largely spared environmental considerations. In fact, fertilizers were regarded as the 'spearhead of agricultural development'. There were some concerns, for example about eutrophication in the 1970s, but the subject had become less topical. Furthermore, it was generally accepted that mineral fertilizers had an important role to play in ensuring that the world's population, quite a large proportion of which was and which remains under-nourished should be adequately fed. Also agricultural development was regarded as an important contributor to the general economic development.

Then, towards the end of the 1980s a combination of events put mineral fertilizers into the limelight as regards the environment. In the developed countries, particularly in Western Europe, the agricultural sector had became unpopular with the public, being regarded as over-subsidized, producing unwanted surpluses and, in the process, destroying much of the countryside. There was concern about the quality of drinking water, particularly as regards the levels of nitrates. The environmental pressure groups became an extremely strong force and the fertilizer industry was not prepared for the intensity of the attack on its products arising from this combination of circumstances.

Food-deficit countries

To take the example of Asia, the total population increased from 1.4 billion in 1950 to 3.1 billion in 1990. Thanks to the 'green revolution', food production kept up with this increase in population. By 2025 there could be 4.9 billion people in Asia. The food requirements will increase accordingly (without taking account of dietary improvements).

A FAO study indicates that, among the Asian countries, there will be cereal deficits by the year 2000 in Pakistan, Japan, the Republic of Korea, Taiwan Province, Sri Lanka and Bhutan, if immediate measures of plant nutrient management are not implemented. By the year 2010, India, Bangladesh, the Philippines, Nepal and Cambodia could be in the same situation. The situation would deteriorate progressively thereafter. There is little unused, productive agricultural land and yields per ha must be increased, substantially. This can only be achieved by the correct use of fertilizers and other inputs, accompanied by correct agricultural practices.

In Africa, the annual population growth rate between 1990 and 2025 is estimated at 2.6% p.a. During the 1980s, unlike the situation in Asia, agricultural production per caput fell. The food deficit, i.e. the food import requirement, is deteriorating year after year. The growth of agricultural production in sub-Saharan Africa is likely to continue to be well below that of population, implying rapid growth of import requirements.

The area of agricultural land is not only limited, it is declining due to factors such as urbanization and soil degradation. There is also a widespread reduction in soil fertility. This is less evident than a complete loss, but is more insidious. In many regions, particularly in Africa, soils are being 'mined' of their nutrients i.e. the quantities of nutrients being supplied are insufficient to compensate for those being removed by crops and lost in various ways, is not sustainable.

In recent years, the synergistic effect of the adoption of improved agricultural technologies has improved fertilizer use efficiency, i.e. the quantity of agricultural produce per unit quantity of nutrients has tended to increase. It is to be hoped that this trend will continue.

C. Rodriguez-Barrueco (ed.), Fertilizers and Environment, 19–22.

Higher and better yields with less environmental pollution in Egypt through balanced fertilizer use

Mohamed M. El-Fouly & A.F.A. Fawzi
Botany Department, National Research Centre, Cairo - Dokki, Egypt

Key words: Egypt, fertilizers use, optimization, N-losses, micronutrients, soil testing, plant analysis

Abstract

Nitrogen use per unit area in Egyptian agriculture is over 300 kg ha^{-1}. An annual average increase of about 2% till year 2000 was estimated. Fertilizers use is characterised by excessive N application, moderate P use and neglection of K and micronutrients. Consequently, nutrient imbalance occurs in crop plants. Subsidising N fertilizers resulted in their misuse and increasing losses. Efficiency of N use in different areas and crops is low. High nitrate concentrations were detected in drainage as well as ground water. Subsidy was removed and fertilizers handling was left to the free market in 1992/93, which resulted in drastic price increases. It could be demonstrated for many years and in farmers fields that yields can be considerably increased (20%) through balanced fertilizers use based on soil testing and plant analysis and adjusted according to the prevailing farming system, leading to more return and less N-use. Use of micronutrients foliar fertilizers leads to increase in root growth and higher uptake of macronutrients. Use of relatively high rates of fertilizers is still a must, however it should be optimised to obtain the highest possible efficiency. Making crop and location specific fertilizer recommendations available to farmers helps in increasing high quality yields, which results in high economic benefits, keeping agricultural production sustainable and decreasing pollution.

Introduction

Use of chemical fertilizers started in Egypt early this century (El-Fouly, 1993). The N followed by P fertilizers are still the most commonly known to farmers. Fertilizer consumption grew since 1970 and is prognosed to increase by 10% within 1990–2000 (Table 1). Both the expansion of desert reclamation and the need for higher yields implied more use of the traditionally known chemical fertilizers.

Table 1. Amounts of NPK fertilizers distributed during 1970–1990 and prognosed for year 2000 (El-Fouly, 1993)

N	Amount (×1000 T)			
	1970	1980	1990	2000*
N	317	490	793	1586
P_2O_5	55	99	165	330
K_2O	1.4	11.5	21.0	42.0

*Prognosed

Pattern of fertilizer use

Fawzi (1992) mentioned that since the sixties the Egyptian fertilization pattern was characterized - even in orchards - by excessive N, moderate P, rare K and no micronutrients, leading to low yield, as compared with other countries (Table 2). For crops, more than 350 kg N ha^{-1} is applied to 2.2 crops/year (Table 3). Low cost of subsidised fertilizers formed one of the reasons for excessive N-application. Insufficient information on the balanced plant nutrition concept and its positive environmental dimension formed another reason for neglecting K and micronutrients.

Impacts on yield, economy and environment

Data of PBDAC (1991) and CAPMS (1992) reflect insufficiently high yields with rations of subsidised fertilizers (Table 4). According to data on most of

Table 2. Amounts of nutrients applied to produce one ton of oranges, and average yields in different countries (Fawzi, 1992)

Country	kg per ton			fruit yield
	N	P_2O_5	K_2O	(t ha^{-1})
USA	2.3	1.5	2.5	>48
Morocco	4.6	3	4.5	36-48
Egypt	19.5	4	0.5	14-20

Table 3. Nitrogen amounts applied/crop rotation based on data from PBDAC (1991)

Crop rotation	total N-dose (kg N ha^{-1})
Clover/cotton/vegetables	372
Clover/maize/vegetables	387
Wheat/maize/potato	400

the alluvial soils of Egypt, crop plants on one ha are provided with 50 kg N from soil, 30 kg N from the preceding legume crop (clover) and 10 kg N if adding 20 m^3 farmyard manure (Eid *et al.*, 1974; Fawzi and El-Fouly, 1993; Hammisa *et al.*, 1974). Thus, 90 kg N ha^{-1} are usually available to summer crops grown after clover and manured.

Calculating N-balance for both crops on national level (Table 5), it could be seen that N-losses in summer crops are very high and efficiency of N-fertilizers used is very low. Consequently, both drainage water (Fig. 1) and ground shallow water used for village drinking (Fig. 2) are heavily polluted with nitrate in summer (Fawzi, 1992).

Role of micronutrients

It is known and well documented that most of crops in Egypt suffer from micronutrient deficiencies (El-Fouly, 1983). It was mentioned by Abdalla and Mobarak (1992 a,b) that micronutrient sprays increased NPK and Mg uptake in shoot, which might be attributed to enhancing root growth (Table 6). Similar micronutrients effects were reported on soyabean and maize as well as on sour orange seedlings (Boutros *et al.*, 1988; Mobarak and Abdalla, 1992; Mobarak *et al.*, 1992). El-Sayed *et al.* (1992) indicated that micronutrients treatment strengthened the physiological sink in youg leaves and pods of fababean.

Table 4. Fertilizer rations distributed to farmers and average crop yields (mean of 1987-1991) (PBDAC 1991 - CAPMS 1992)

Crop	Average Yield (t)	Fertilizer rations (kg/ha^{-1})		
		N	P_2O_5	K_2O
Wheat(grains)	4.9	184	36	0
Rice (grains)	6.4	98	36	0
Maize (grains)	5.4	221	36	0
Clover (fodder)	32.00(/cut)		36	0
Faba bean (seeds)	2.7	37	71	0
Soyabean (seeds)	2.7	184	57	0
Cotton (seed+hair)	2.1	166	36	0
Potato (tubers)	22.6	295	107	57*
Citrus (fruits)	17.2	443	71	114*

*Only if available

Table 5. Nitrogen balance in maize and cotton sown after clover as winter crop

Parameter	Maize 000 ha 869	Cotton 000 ha 358
Yield in t ha^{-1} (National average 1991)	5.89	2.22
kg N ha^{-1}		
1) Total plant uptake	147	130
2) Provided from soil + previous crop + manure	90	90
3) To be provided by chemical fertilizer (1–2)	57	40
4) Provided by chemical fertilizer	220	166
5) Loss in chemical fertilizer (4–3)	163 (74%)	126 (76%)
6) % Use of total available N (1÷(2+4) %)	47%	51%
7) National N loss/crop (1000 t)	142	145

Based on data from: Hammissa *et al.* (1974), Eid *et al.* (1974) Halliday and Trenkel (1992) Fawzi and El-Fouly (1993)

Table 6. Effect of micronutrients foliar application (MN) on dry weight of root and macronutrients uptake by shoot in fababean (Abdalla and Mobarak, 1992a,b)

Character	Control = (100%)	+ MN* (%)
Root d.w. (mg/pot)	6.04	7.33 (122%)
Uptake (mg/pot)		
N	403	521 (129%)
P	23	29 (125%)
K	366	488 (133%)
Mg	65	78 (134%)

*MN = Micronutrients.

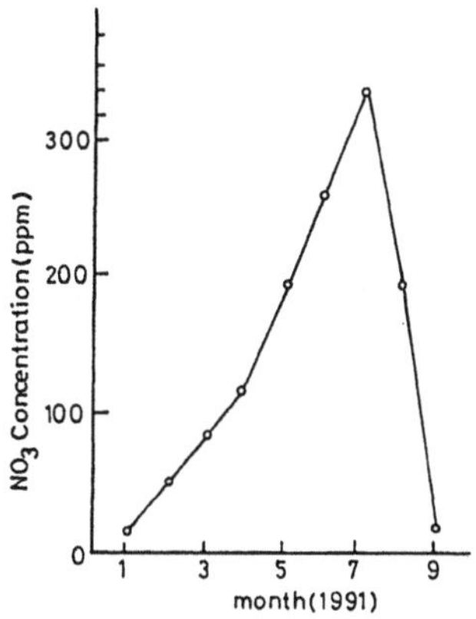

Fig. 1. NO_3 concentration in drainage water of Sharkia governorate (Delta).

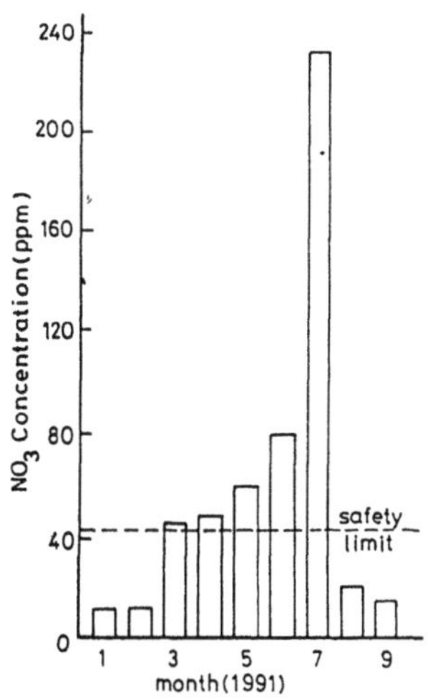

Fig. 2. NO_3 concentration in ground-drinking water of Sharkia governorate (Delta).

Table 7. Response of potato to optimised fertilization (less N + K + micronutrients)

Dose (kg/ha)			Micronutrient foliar	Tuber yield
N	P_2O_5	K_2O	application	($t\ ha^{-1}$)
910	190	0	0	23 (100%)
740	190	115	0	27 (120%)
580	190	115	+	32 (139%)

Based on data from (Nour and El-Sayed, 1991)

Table 8. Effect of micronutrients foliar application on yield of major field crops (1985-1993) on growers fields

Crop	Area sprayed (ha)	Yield ($t\ ha^{-1}$) -MN[a] (100%)	+MN(%)	Net return (L.E.)[b]
Wheat	5800	4	5.0 (125%)	3.2:1
Cotton	1000	2.8	3.1 (111%)	5.8:1
Faba bean	500	2.5	3.1 (124%)	9.1:1
Maize	800	6	7.5 (124%)	5.0:1
Rice	450	6.5	8.00 (123%)	5.4:1
Soybean	150	2.4	3.6 (50%)	9.0:1

Based on data from NRC/GTZ internal reports.
[a]MN = micronutrients.
[b]L.E. : Egyptian pound = 0.296 US $.

Means for optimizing fertilizers use under free market conditions

Subsidy was removed in 1992/93 and fertilizer handling was left to free market, leading to drastic increases in price. Although farmers are aware of the fertilizers benefit, they are unable to take the proper decision on their use.

The NRC/GTZ project on "Optimising Fertilizer Use" oriented its results to advisory services and to demonstrate the benefit of balanced fertilizer use. Fertilizer recommendations are worked-out on farm level upon soil tests, plant analysis and adjusted to the prevailing farming system (NRC-GTZ, 1991, 1992, 1993). Farmers were often advised to improve NPK balance. Fawzi *et al.* (1983) El-Fouly *et al.* (1991), Nour and El-Sayed (1991) reported significant yield increases due to less N-dose and to K-application. Micronutrients (Zn + Mn + Fe) foliar application added further increment to that gained by balancing NPK (Table 7). In practice, farmers who followed the advice were able to increase their yields (NRC - GTZ,

1991, 1993, 1994). An average of 20% more yield was achieved on farmers fields with 3–9 Egyptian pounds (L.E.) net return for every L.E. spent on implementing fertilizer recommendation (Table 8).

Conclusion

Use of relatively high rates of fertilizers is still a must, since the agricultural land of Egypt is very limited and the demand for higher production is pressing. However, fertilizers use should be balanced and optimized to obtain the highest possible efficiency. Researchers and extentionists should give more attention to work out crop and location specific fertilizer recommendations based on soil testing and plant analysis and make them available to farmers. This will help in increasing high quality yields which results in high economic benefit for the farmers, keeping agricultural prouction sustainable and decrease pollution.

References

Abdalla FE and Mobarak Zeinab M (1992a) Shoot intake of nutrients from different micronutrient fertilizer formulations in fababean. Af J Agric Sci 19: 147–160

Abdalla FE and Mobarak Zeinab M (1992b) Uptake of NPK and Mg by fababean after foliar treatment with chelated and non-chelated micronutrient fertilizers. Af J Agric Sci 19: 161–172

Boutros BN, Mandour MS and Fawzi AFA (1988) Growth and nutritive status of sour orange rootstock to micronutrients application. Egypt J Soil Sci 28: 139–151

Central Authority for Public Mobilization and Statistics (CAPMS) (1992) Statistical Yearbook, pp 62–65

Eid MT, Hammissa MR, Serry A, Abd El-Samie M, El-Banna E and Omar S (1974) Efficiency of nitrogen fertilizer applied to wheat as affected by time of application. Agric Res Rev 52: 125–133

Eid RAH (1994) Developing a simulation model for identifiying optimum fertilizer treatments for a crop varieties M. Sc. Thesis, Cairo, University

El-Fouly MM (1983) Micronutrients in arid and semi-arid areas : Level in soil and plants and the needs for fertilizers with reference to Egypt. In: Proc. 15th Coll. Inter. Potash Inst., Bern. p163

El-Fouly MM (1993) Fertilizers. In: Craig GM (ed) Agriculture of Egypt, pp 363–382. Oxford Univ. Press, Oxford, UK

El-Fouly MM, Fawzi AF, Firgany AH and El-Baz FK (1991) Uptake and removal of potassium by maize and effect of potassium sulphate on yield. Tropen-landwirt Z Trop Subtrop 92: 73–97

El-Sayed AA, Abdalla FE and Fawzi AFA (1992) Uptake and partitioning of micronutrients in fababean grown on alluvial soil. Af. J Agric Sci 19: 173–180

Fawzi AFA (1992) Fate of nitrogen fertilizer in soil to major crops. Proc. Egypto - German Seminar on Environmental and Cultural Aspects of Fertilizer Use, 25–28 Nov. 1991, Cairo. El-Fouly (ed) pp 23-38 (*In Arabic*)

Fawzi AFA and El-Fouly MM (1993) Inputs for computerized expert system for fertilizer recommendations in Fayoum governorate. NRC-GTZ report 94p (Unpubl.)

Fawzi AFA, Firgany AH, Rezk AI, Kishk MA and Shaaban MM (1983) Response of *Vicia faba* bean to K and micronutrient fertilizers. Egypt J Bot 26: 113-121

Hammissa MR, Abdel Samie ME, El-Banna E and Khadr MS (1974) Corn fertilization programme in A.R.E : Utilization of nitrogen by maize as affected by time and method of application. Agric Res Rev 52: 25–48

Halliday DJ and Trenkel ME (Eds) (1992) IFA World Fertilizer Use Manual. Publ. International Fertilizer Industry Association, Paris. 632p

Mobarak Zeinab M and Abdalla FE (1992) Nutrients uptake by maize plants as affected by foliar micronutrients application. Afr J Agric Sci 19: 153–205

Mobarak Zeinab M, El-Sayed AA, Abdalla FE and El-Bendary AA (1992) Differential responses of soybean varieties to micronutrients foliar application. Afr J Agric Sci 19: 123–136

NRC-GTZ (1991) Multiptlier Program for Optimizing Fertilizer Use. Report 15. Fawzi AFA (ed) 254p. (Unpubl.) (*In Arabic*)

NRC-GTZ (1993) Report on serving farmers on ARO lands season 1991/92. Ed. Fawzi MM and El-Sayed AA (eds) 99p. (Unpubl.) (*In Arabic*)

NRC-GTZ (1994) Report on serving farmers on ARO lands, Season 1993/94. Rezk AI (ed) 72p. (*In Arabic*)

Nour UM and El-Syed AA (1991) Multiplier Development Extension program for Potato Growers in Etai El-Baroud, Beheira Governorate, Egypt. In: El-Fouly MM (ed) Report. 87p. (Unpubl.) (*In Arabic with English Abstract*)

Principal Bank for Development and Agricultural Credit-PBDAC (1991) PBDAC - Circulation No. 136 (Unpubl.)

C. Rodriguez-Barrueco (ed.), Fertilizers and Environment, 23–25.

Fertigation for minimizing environmental pollution by fertilizers

J. Hagin[1] & Anat Lowengart[2]
[1]Faculty of Agricultural Engineering, Technion - Israel Institute of Technology, Haifa 32000; [2]Ministry of Agriculture Extension Service, Israel

Key words: fertigation, fertilizer application, irrigation, nutrients requirements

Abstract

Intensification of agriculture by irrigation and enhanced use of fertilizers may generate pollution by increased levels of nutrients in underground and surface waters. Most of the irrigation is by open systems having a relatively low efficiency of water application. A higher efficiency may be gained by pressurized irrigation systems. Drip irrigation generates a restricted root system requiring frequent nutrient supply that may be satisfied by applying fertilizers in irrigation water, i.e. by fertigation. Maximization of crop yield and quality and minimization of leaching below the rooting volume may be achieved by managing fertilizers concentrations in measured quantities of irrigation water, according to crop requirements.

Water pollution hazard by fertilizers

Irrigation with enhanced use of fertilizers generates very often pollution, by increasing levels of nutrients in underground and surface waters. In an intensively cropped valley in Israel the use of fertilizer nitrogen increased from 42 to 153 kg ha^{-1} over the period of 50 years since 1936. An increase in nitrate concentrations from about 40 to 105 mg l^{-1} in water wells was observed in the same valley in the 30 years since 1953. There are indications that most of the increase in nitrate concentration originates from the increased use of fertilizers, although some other sources contribute their share (Salinger, 1985).

Other measurements done in very different conditions present similar results. It is stated (WRFW, 1990) that

> New technology and the demand for greater agricultural productivity have led to an exponential increase in the use of fertilizers and pesticides. The runoff of these chemicals particularly nitrogenous fertilizers, creates one of the most widespread and serious of all water quality problems, particularly in industrialized countries. Evidence from long term monitoring in the United Kingdom suggests that nitrate levels have been building up for years. For example the nitrate concentration in the river Thames increased from about 2 mg l^{-1} in the year 1938 to about 10 mg l^{-1} in 1984 (WRFW, 1990).

An example of a possible polluting effect of applying high levels of nitrogen is given from a fertilization experiment in avocado (Lahav and Kalmar, 1993). Four levels of N were applied over four years.

Increasing the N applied from 80 kg ha^{-1} to 640 kg ha^{-1} increased the NO_3-N concentration in soil from 4.2 to 427.2 mg kg^{-1} in the 0–30 cm layer and from 0.5 to 232.0 mg kg^{-1} in the 60–90 cm layer.

The above examples illustrate the gap between the need for crop yields increases by irrigation and fertilization and between the demand for minimization of water pollution by plant nutrients, specifically by nitrates.

Irrigation methods

Most of the irrigation in the world is open. The efficiency of water use in such systems is low and one half to one third of the applied water may be lost carrying with it considerable amounts of nutrients. Pressurized irrigation systems have a higher water use efficiency, ranging from 70% to 95% and water and nutrient losses

Table 1. Nutrients in two liquid fertilizers, weight percentage

Fertilizer	N	P	K
A	7.3	1.4	2.7
B	5.0	0.9	3.1

Table 2. Nutrient input and uptake per week, of tomatoes in detached culture

Nutrient	Applied		Uptake
	$(mg\ l^{-1})$	$(kg\ ha^{-1})$	$(kg\ ha^{-1})$
N	250	87	85
P	100	35	19
K	350	122	190
Ca	175	61	43
Mg	40	14	11

may be well controlled. Constraints on use of pressurized irrigation are the initial and maintenance costs and expertise. Drip irrigation is probably the most effective method of water application. It generates a restricted root system requiring frequent nutrient supply. This may be satisfied by applying fertilizers with irrigation water, i.e., by fertigation.

Fertigation

Nutrient supply to crops in fertigation is determined by their concentration in irrigation water, nutrient uptake by the plant, rate of evapotranspiration and reactions such as precipitation or fixation by the growth medium.

Optimization of nutrient and water supply for maximizing crop yield and quality and minimizing leaching below the rooting volume is achieved by managing fertilizers concentrations in measured quantities of irrigation water, according to crop requirements.

Detailed knowledge of nutrient uptake by plants is a prerequisite for optimizing nutrient application, both, for satisfying plant requirements and minimizing losses to the environment.

Nutrition of greenhouse tomatoes was studied intensively (Gambash and Nitzani, 1989; Hagin and Segelman, 1990; Sonneveld and Welles, 1988; Zaiddan, 1989) and it is taken here as an example for balancing crop nutrient requirements with loss prevention.

Controlling drip fertigation necessitates frequent measurement of growth medium moisture by tensiometers, by direct water measurement or by estimating water losses according to open pan evaporation. Further, nutrient concentration in the irrigation water and in the growth medium solution has to be monitored. Measured amounts of fertilizer solutions have to be injected into the irrigation water through fertilizer tanks or pumps (Shani and Sapir, 1986; Sne, 1988).

Nutrients ratios and concentrations in irrigation water for greenhouse tomatoes vary with the season according to plant requirements. The variation in nutrients ratios is obtained by using different fertilizer solutions (Table 1) at the beginning of the growing season (Fertilizer A) and at the peak season (Fertilizer B). Both liquid fertilizers contain NH_4NO_3, KNO_3, H_3PO_4 and micronutrients and have a pH of 3.5. N concentrations vary from about 150 to 250 g m^{-3} and parallel to it the concentrations of other nutrients.

A nutrient balance, at the peak growing season (Table 2), was calculated for 25,000 tomato plants per ha grown in a soilless culture and fertigated with a B fertilizer solution (Table 1). Calcium and magnesium nitrates solutions are added separately and as necessary. The rate of irrigation is 35 m^{-3} ha^{-1} $week^{-1}$. The nutrient uptake data are taken from literature (Huett and Dettmann, 1988) and the recommended nutrient concentrations are according to the Israel Ministry of Agriculture Extension Service.

The data in Table 2 indicate that fertilizer management according to recommendations does not leave appreciable surpluses of N, K, and Mg for leaching. Amounts of P applied are higher than the calculated uptake. It may be assumed that this is due to possible precipitation of applied P in the rooting medium. A review of literature (Bar-Yosef *et al.,* 1980; Huett and Dettmann, 1988; Tsikalas and Manios, 1985) reveals that fertigation recommendations are within a similar order of magnitude irrelevant of the growth medium.

In addition to nutrients requirements of the crop some other factors have to be accounted for. As an example, in tomato culture, the osmotic pressure of the nutrient solution is an important factor in quality control of the fruit. The osmotic pressure is closely related to the level of mineral nutrients applied and usually measured by electrical conductivity (EC) of the solution. Increasing the EC of soil extracts (1:1) above 0.5 dS m^{-1} decreased somewhat tomato yield

Table 3. Nutrients uptake rates and, application recommendations (kg ha^{-1} day^{-1}) Field grown hybrid corn

Days after emergence	N		P		K	
	Uptake	Appl.	Uptake	Appl.	Uptake	Appl.
18–38	7.0	9.3	0.5	0.24	10	5.0
38–45	9.0	12.0	1.0	0.48	16	8.0
45–52	10.0	13.3	1.0	0.48	11	5.5
52–60	8.0	10.7	0.5	0.24	8	4.0
60–68	3.0	4.0	0.5	0.24	–	–
68–85	–	–	1.0	0.48	–	–

(Bar-Yosef *et al.,* 1980). On the other hand it is reported that higher EC values induced little loss in fruit yield and had a beneficial effect on fruit flavor (Gough and Hobson, 1990; Sonneveld and Welles, 1988). Recommendations aimed at flavor development are based on increasing the EC of the water applied at the late stages of growth. One possibility to achieve this is by an increased concentration of KCl in the irrigation water, above the strict nutritional requirement (Mizrahi, 1988).

Fertigation of a hybrid field grown corn is another example for balancing plant nutrients requirements and application levels (Berkovitz, 1985). Daily uptake of nutrients was measured. Recommendations for nitrogen are based on a 75% utilization of the applied and nil contribution from the soil. P and K requirements are assumed to be covered approximately 50% by residual in the soil and basic dressing. Nutrient uptake and the resulting recommendations are presented in Table 3.

Conclusions

The presented examples illustrate that for both, satisfying the crop demand for nutrients and minimizing surface and underground water pollution, a detailed knowledge of nutrient uptake rates is necessary. Further, a well controlled irrigation and fertilization system is required. This may be achieved by fertigation when applied by pressurized irrigation, specifically by drip irrigation.

References

Bar-Yosef B, Sagiv B and Eliah E (1980) Fertilization and irrigation of winter tomatoes grown in glasshouse in the Besor area. Prel Rep, Div Sci Publ Bet Dagan (775): pp 90

Berkovitz S (1985) Corn growing. Extension Service, Ministry of Agriculture, Israel. 106p

Gambash S and Nitzani Y (1989) Fertilization Guide. Tomato. Haifa Chemicals Ltd. 18p

Gough C and Hobson G E (1990) A comparison of the productivity, quality, shelf life characteristics and consumer reaction to the crop from cherry tomato plants grown at different levels of salinity. J Hortic Sci 65: 431–439

Hagin J and Segelman G (1990) Trends in fertilizers and fertigation. The S Neaman Institute, Technion - Israel, Institute of Technology. 77p

Huett DO and Dettmann EB (1988) Effect of nitrogen on growth, fruit quality and nutrient uptake of tomatoes grown in sand culture. Aust J Exp Agric 28: 391–399

Lahav E and Kalmar D (1993) Comparison of nitrogen sources and fertilizer levels in Avocado. Akko Exp Sta Rep 75–79

Mizrahi Y (1988) Saline irrigation regime for improving tomato fruit quality without reducing yield. J Am Soc Hort Sci 113: 202–205

Salinger Y (1985) The changes in the chemical quality of the water resources in the Yavneel valley. Thesis, Hebrew. University, Jerusalem, p 50–56, 74

Shani M and Sapir E (1986) Fertilization in irrigation. Extension Service, Ministry of Agriculture, Israel. 44P

Sne M (1988) Fertigation. Extension Service, Ministry of Agriculture, Israel. 12p

Sonneveld C and Welles GWH (1988) Yield and quality of rockwool-grown tomatoes as affected by variation in EC-value and climatic conditions. Plant Soil 111: 37–42

Sonneveld C and Voogt S J (1981) Nitrogen, potash and magnesium nutrition of some vegetable fruit crops under glass. Neth J Agric Sci 29: 129–139

Tsikalas PE and Manios VI (1985) Nutrition of tomatoes in growing-bags in greenhouse. I.Effect on productivity. Georgike Ereuna 9: 279–289

World Resources (1990–91) Fresh Water. World Resources Institute, United Nations Environment Program, United Nations Development Program. pp 164–166

Zaiddan O (1989) Growing tomatoes in greenhouses. Extension Service, Ministry of Agriculture, Israel. 36p

C. Rodriguez-Barrueco (ed.), Fertilizers and Environment, 27–30.

The agronomical and physiological efficiency of nitrogen applied for arable crops in Poland

E. Fotyma & M. Fotyma
Institute of Soil Science and Plant Cultivation, Osada Palacowa IUNG, 24–100 Pulawy, Poland

Key words: nitrogen efficiency, cereals, potatoes, beets

Abstract

The common relation yield / nitrogen dose can be broken into two relations: nitrogen uptake / nitrogen dose and yield / nitrogen uptake. Such procedure narrows the original variation and separates it in two different and easier to recognise categories, external and internal. The results of permanent field experiments conducted in Poland with all the most important arable crops has been re-examined according to this idea. The model for fertilizer recommendations in respect to nitrogen was proposed.

Introduction

The recommendation system for nitrogen fertilization in Poland is based on response curves. The most common is quadratic polynomial model, though spline model in the form of two broken lines was used as well (Fotyma, 1988). The coefficients of the response equations are calculated from the results of field experiments with several nitrogen doses. By pooling the results of many experiments with the same crop the poor fit of experimental date with the response curve (low correlation coefficient) is usually recorded. It can be partly explained by the broad scatter of yields in the control treatments due to the differences in natural soil fertility and partly by climatic and agronomic factors influencing the slope of the response curves in individual experiments. Instead of describing the relation between nitrogen dose and yield in one step Janssen (1990) proposed to break it in two partial relations corresponding to steps II and III.

step I: nitrogen dose – yield
step II: nitrogen uptake – yield
step III: nitrogen dose – nitrogen uptake

In step III the problems that cause a low efficiency of the crop in taking up the nitrogen from fertilizers are reflected (external or agronomical nitrogen efficiency). Step II gives insight into the problems that cause low efficiency of the crop to utilize nitrogen already taken up from fetilizers (internal or physiological nitrogen efficiency). Janssen's idea was tested in reference to the results of numerous field experiments in Poland.

Methods

Permanent field experiments in four-crop rotations with 6 consecutive doses of nitrogen were carried on since 1985 in 4 Experimental Farms belonging to the Institute of Soil Science and Plant Cultivation in Pulawy. The description of the experimental sites and the preliminary results for the years 1985–1989 were presented in former publications (Fotyma, 1992, 1992a). This paper covers the experimental period 1985–1992 and all results have been re-examined in respect to Janssen's formula. Three basic equations (quadratic polynomials) were calculated for each crop separately:

$$Y = a + bN + cN^2 \tag{1}$$

$$U = a' + b'N + c'N^2 \tag{2}$$

$$Y = a'' + b''U + c''U^2 \tag{3}$$

where:

Y – yield of main crop product in t·ha^{-1} (kg $\times 10^3$·ha $^{-1}$)
U – total uptake of nitrogen by crop in kg N. ha $^{-1}$
N – nitrogen dose in kg N·ha^{-1}

The respective response curves are presented in the form of four quadrants diagrams (Janssen, 1990)

Table 1. The statistics of variation of nitrogen uptake (V kg N $\cdot$ha^{-1}) in control treatment

crop	average	coeff. of variation	minimum	maximum
winter wheat	86	29	49	136
triticale	65	37	33	128
winter rye	62	29	36	102
spring barley	70	36	32	118
spring wheat	70	37	23	130
oats	79	31	37	118
potatoes	67	57	6	157
sugar beets	149	49	73	407
fodder beets	161	49	64	338

leaving one quadrant empty. Agronomical (external) and physiological (internal) nitrogen efficiencies were calculated from the formulas:

$$AE = \mathrm{d}Y/\mathrm{d}N \qquad (4)$$

$$AF = \mathrm{d}Y/\mathrm{d}U \qquad (5)$$

where:
AE – agronomical efficiency in kg of product for 1 kg N in fertilizers
AF – physiological efficiency in kg of product for 1 kg N taken up by the crop
$\mathrm{d}Y$ – increase of the yield
$\mathrm{d}N$ – range of nitrogen doses
$\mathrm{d}U$ – increase of nitrogen uptake

Utilization coefficient for nitrogen was calculated according to difference method

$$UC = (\mathrm{d}U/\mathrm{d}N)\times 100 \qquad (6)$$

Results and discussion

The response curves for 9 most important arable crops in Poland are presented on Fig. 1a–i. The best fit of experimental to theoretical date was found for the relation yield/uptake, which confirms Janssen's (1990) idea. Potential (maximal) crop yields of about 4.7 t·ha^{-1} spring cereals, 5.0–5.9 t·ha^{-1} winter cereals, 37 t·ha^{-1} potatoes and 50–73 t·ha^{-1} beets were recorded. With the potential yield cereals and potatoes took up 120–170 kg N·ha^{-1} and beets about 300 kg N·ha^{-1}. The uptake of nitrogen with the potential yield calculated for yield unit was quite similar for all cereals and fell in the range 2.7–2.9 kg N per 100 kg of grain. This unit uptake amounted 0.3 kg N, 0.4 kg N and 0.5 kg N per 100 kg tubers and/or roots of potatoes, fodder beets and sugar beets, respectively. From the total amount of nitrogen taken up by cereals and potatoes 60–80 kg N originates from the soil (in control treatment) and the rest from fertilizers. The share of soil nitrogen was much higher in case of beets and accounted for 140–160 kg N·ha^{-1}. For a particular crop the uptake of nitrogen in control treatment, called by Janssen (1990) potential nitrogen supply showed a very broad variation from year to year and between experimental fields. The statistical characteristic of this variable is presented in Table 1.

The identification and quantification of the sources of this variation is now examined. The utilization coefficient of nitrogen depended on crop species and fertilizer dose. For optimal nitrogen dose this coefficient was about 80% for beets, 60% for all cereals but rye and 50% for rye and potatoes. The agronomical (external) efficiency of nitrogen decreased with increasing doses of fertilizers which is a quite common phenomena. The average agronomical efficiency for optimal nitrogen dose fell in the range 11.5–12.5 kg grain per 1 kg N in fertilizers for all cereals but oats (10.5 kg grain per 1 kg N) and triticale (16.5 kg grain per 1 kg N) and 85–98 kg tubers and/or roots per 1 kg N for row crops. The physiological (internal) efficiency decreased as well with increasing fertilizer doses. It can be explained by widening the straw/grain or roots/leaves ratio and accumulation of nitrogen in crop products.

Conclusions

Splitting the common relation between yield and nitrogen dose in two partial relations: nitrogen dose/nitrogen uptake and nitrogen uptake/yield offers a new approach to model of fertilizer recommendations. The model can be presented in the form of simple equation:

$$Nopt = (Up - Ns)/UC \qquad (7)$$

where:
$N\,opt$ – recommended dose of nitrogen fertilizers
Up – uptake of nitrogen with the potential crop yield
Ns – soil nitrogen supply
UC – utilization coefficient of nitrogen from fertilizers

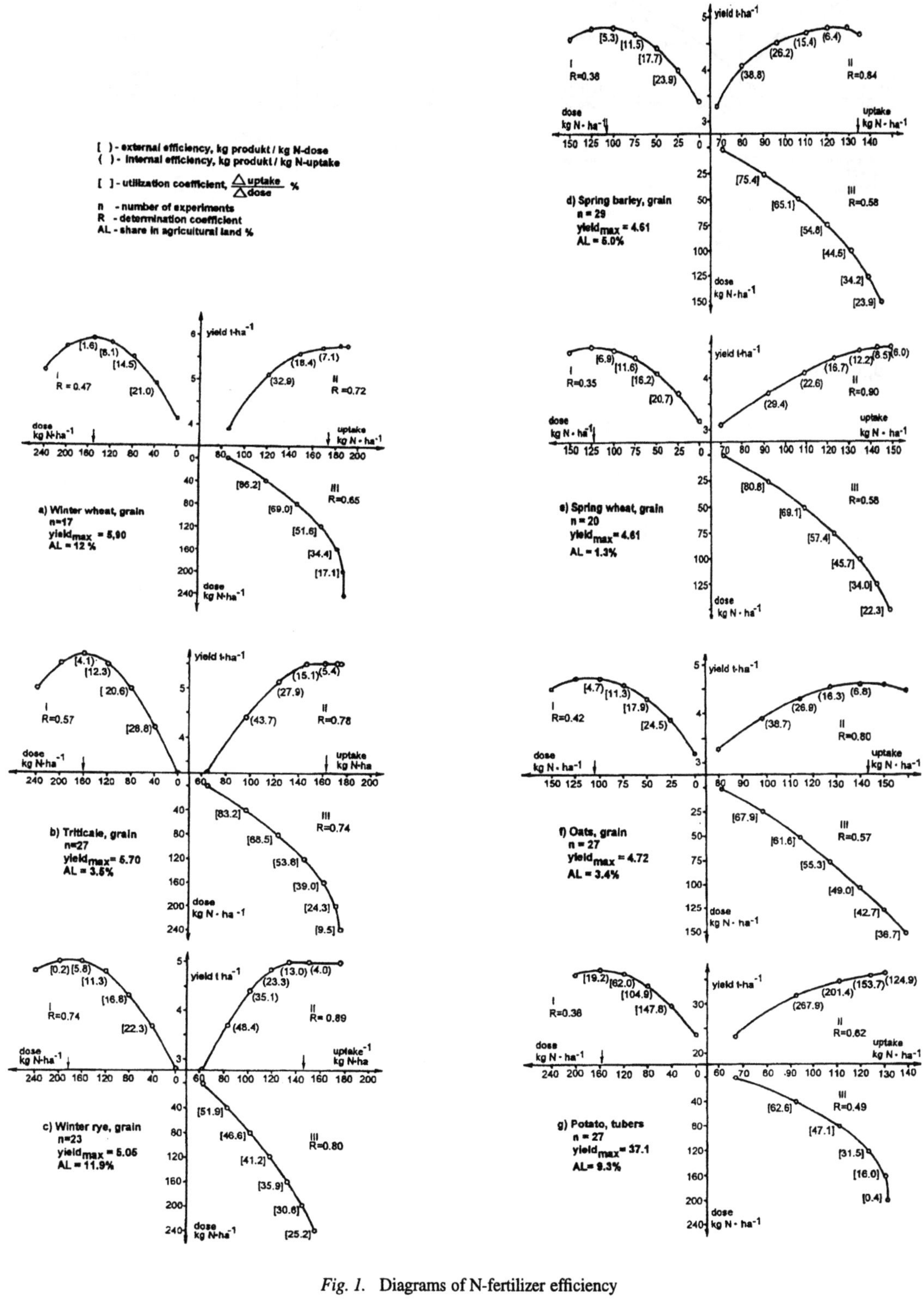

Fig. 1. Diagrams of N-fertilizer efficiency

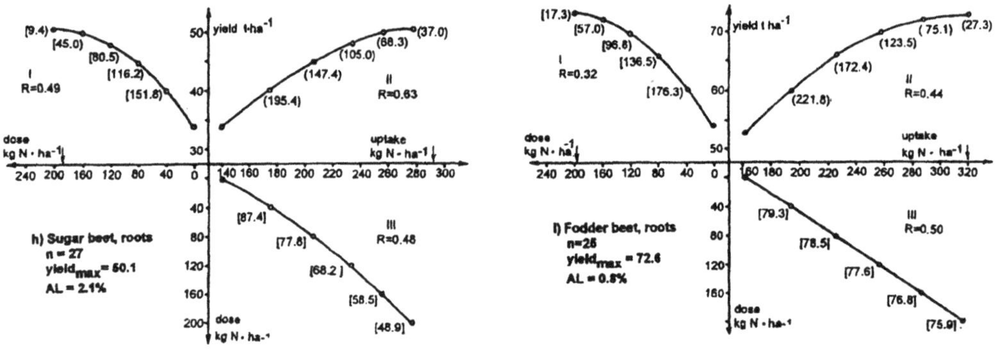

Fig. 2.

Acknowledgements

This work was sponsored by Polish State Committee for Scientific Research in form of the grant no 5S30205004 " Principles of nitrogen fertilization in integrated system of crop production ".

The authors are very much indebted for this helpful and kind support.

References

Fotyma E (1988) The response of arable crops to nitrogen fertilization (in Polish). Pamiêtnik Pulawski 101: 7–34

Fotyma E, Fotyma M (1992) The direct and after-effect of nitrogen in two-course crop rotations (in Polish). Pamiêtnik Pulawski 100 pp 37–60

Fotyma E, Fotyma M (1992a) Utilization of nitrogen from fertilizers by arable crops (in Polish). Pamiêtnik Pulawski 101: 7–34

Janssen BH *et al.* (1990) A system for quantitative evaluation of the fertility of tropical soils QUEFTS. Geoderma 46: 299–318

C. Rodriguez-Barrueco (ed.), Fertilizers and Environment, 31–37.

Nitrogen balances in long-term field experiments

T. Németh
Research Institute for Soil Science and Agricultural Chemistry of HAS (RISSAC), H-1022 Budapest, Herman O. u. 15., Hungary

Key words: deep-drilling, environmental conditions, N-balances, N-fertilization, nitrate accumulation.

Abstract

The influence of fertilization with different N doses on nitrate-N content of the deeper soil layers were investigated in soils of long-term experiments at the experimental stations of the National Fertilizer Experimental Network in Hungary. Distribution of nitrate-N with depth was also studied in plots fertilized with different N doses. Although these experiments were set up according to an identical experimental plan at every location, after 20 years of experimentation substantial differences could be detected in the nitrate-N contents, nitrate-N distribution profiles on plots of the same treatments, as well as in the yield and the nitrogen uptake of the cultivated plants at the various sites. These differences are attributed to differences in soil properties and in environmental conditions existing at the individual experimental stations. The conclusions drawn from the data are very important when evaluating the role of nitrogen fertilization in the nitrate-contamination of soils, surface water and groundwater, and for a more precise determination of the N doses to be applied under specific soil and environmental conditions.

Introduction

Agricultural land use is always expected to supply the nation (or the world population) with enough and good quality food through rational soil management. It means, that the farmers need to preserve and maintain, even increase the productivity of the cultivated lands, giving high priority for environmental protection and biodiversity. The productivity of the land can be characterized with different physical, chemical and biological (microbiological) factors, i.e. soil moisture regime, nutrients supply, microbiological activity. The preservation and maintenance of fertility necessitate following and investigating the nutritive element regime of the soils, which represent the "life media" for microbiological activities as well as for natural vegetation and cultivated crops. These facts show that the soils are the most important factors of production of agriculture and the most important – renewable – natural resources (Várallyay, 1990). Also the soil is one of the most important compartments in the element cycles, where the different processes intersect each other. The soil is able to retain the nutrient compounds, this property of the soil brings about a relative accumulation of available forms of those elements which are most essential for plants.

Nitrogen is one of the most special among the essential elements, because about 96% of it is organically bounded. Plants are able to take up the nitrogen from mineral forms. The proportion of nitrogen in these forms (exchangeable ammonium and nitrate) is greatly affected by certain climatic and soil conditions. The mineral nitrogen content of the soil can increase through N-mineralization from the organic matter of the soil as well as via fertilization. Overfertilization with nitrogen can result in contamination of the environment.

For understanding the role of the different processes a budgetary approach appears to offer a good tool through analyzing the turnover of the main elements in the soil-plant system. To avoid the mistakes caused by using the data of only one growing season a long-term analysis is more adequate. To develop comprehensive nitrogen management (fertilization) system one needs to consider the nitrogen cycle of the certain area. The full nitrogen budget contains all the nitrogen transformation processes, inputs and outputs, but for practical purposes a simplified version is also convenient.

The inputs are applied fertilizers (mineral and organic), extent of N-fixation (symbiotic and non-symbiotic), atmospheric deposition etc., while the main outputs are removal with harvested yield, leaching, denitrification, volatilization, run-off, etc. There are also some internal processes as mineralization-immobilization. For an agricultural system the fluxes of these processes (mineralization-immobilization) are usually in equilibrium, during a longer period. For non-leguminous arable cropping systems a simplified nitrogen balance is often constructed (Goss *et al.*, 1994; Hill, 1986; Lund, 1982; Tanji *et al.*, 1977). In this case for calculating the simplified nitrogen balance usually only the fertilization (organic and mineral) and the removal with harvested yield are taken into account.

The simplified nitrogen, phosphorus and potassium balances of the Hungarian agriculture between 1930 and 1992 were summarized by Kádár (1987) and Csathó (1994). When analyzing the NPK balances of Hungarian agriculture it can be concluded that the surplus appearing in the nutrient balance sheet is a consequence of intensive fertilization. Their data also show, that the nitrogen balance of the Hungarian agricultural soils became positive in the early seventies (with 7 kg/ha), remained positive till the end of the eighties (between 16–27 kg/ha), and became negative again in the last few years (−3 kg/ha in 1990, −55 kg/ha in 1991).

The long-term experiments give a good basis to investigate more adequately the nitrogen balance of different fertilizer treatments in crop rotation under field conditions. The long-term field experiments are suitable for measuring the main components of the budget according to the separate (different) treatments. If the balance is becoming positive, any excess of N-inputs over output of nitrogen in agricultural products (surplus nitrogen) represents a potential loss to the environment. This surplus nitrogen can contaminate the environment under certain conditions through leaching, surface runoff and in gaseous form. Under such environmental circumstances when the groundwater table is deep and the precipitation is likely not enough to wash out the soil layer down to the groundwater table, the surplus of the nitrogen can accumulate in the form of nitrate in the deeper soil layers (Németh, 1993; Németh, 1994a; Németh *et al.*, 1987).

To investigate the effects of the N-fertilization on the nitrogen balance and to follow the fate and behaviour of nitrogen in the soil-plant system a long-term fertilizer experimental network was chosen.

Materials and methods

The experimental series chosen for our measurements started in 1968 on 9 experimental stations in different parts of the country, on different soil types, under diverse environmental conditions (supervised by PANNON Agricultural University, Keszthely). The plant order and the fertilization treatments were identical in all sites.

In the first four cycles (1968/69 – 1983/84) winter wheat, maize, maize and winter wheat were grown, and in the following cycle winter wheat, maize, sunflower and winter wheat represented the plant order. The basic soil analysis data of this network were published by Debreczeni and Debreczeni (1994).

Deep-drilling was carried out on 8 experimental stations from the above mentioned 9, in July 1988 following the harvest of winter wheat. Soil samples were taken on the unfertilized plots and on the plots which received 50, 150 and 250 kg N/ha in each year, in four replications. During the twenty experimental years the rate of N-fertilization was changed twice (keeping the appropriate steps between the treatments) – according to the increasing farm application of the nitrogen – this is why the sum of the 20 years application in Tables 3 and 4 is not equal to 20 × 50, 20 × 150 and 20 × 250. The deep-drilling was done down to 3 meters and samples were taken after every 20 cm. The soil moisture contents were determined immediately from the samples, while the mineral-N and other chemical analyses were done after air-drying the soils. The results of the nitrate accumulations in the soil profiles were discussed earlier by Németh (1994a, 1994b).

For the budgetary approach the nitrogen contents of the harvested crops were determined in each treatments. These analyses gave the basis of the calculations. The calculation of the simplified nitrogen budget was the following:

$$N_{budget} = -(N_i - N_0) + N_f \qquad (1)$$

where

N_{budget}= N budget in the i-th treatment (kg/ha) (in our case i = 50, 150, 250 kg N/ha/y),

N_i= amount of nitrogen taken up by the crop in the same treatment (kg/ha),

N_0= amount of nitrogen taken up by the crop of the unfertilized plots (kg/ha),

N_f= amount of nitrogen given in the fertilizer in the selected treatment (kg/ha).

In the soil samples from deep-drilling the nitrate-N content of the soil profiles were determined (in the average of each 20 cm soil layer). These results made possible to investigate the N-balances together with the nitrate-N accumulations and distribution in the soil profiles. In this case the following equation was applied:

$$N_{budget/nitrate} = -(N_i - N_0) + N_f - N_{nitrate} \quad (2)$$

where

$N_{budget/nitrate}$= N budget in the i-th treatment (kg/ha) (in our case i = 50, 150, 250 kg N/ha/y),

N_i= amount of nitrogen taken up by the crop in the same treatment (kg/ha),

N_0= amount of nitrogen taken up by the crop of the unfertilized plots (kg/ha),

N_f= amount of nitrogen given in the fertilizer in the selected treatment (kg/ha).

$N_{nitrate}$= amount of the nitrate-N in the soil profile (kg/ha)

Results and discussion

The investigated 8 soils belong to 3 groups according to the Hungarian soil classification system: 4 chernozem soils (Hajduböszörmény, Karcag, Iregszemcse, Nagyhörcsök), 3 brown forest soils (Bicsérd, Kompolt, Putnok) and 1 alluvial soil (Mosonmagyaróvár). In Table 1 the specific nitrogen content of the cultivated plants, while in Table 2 the amounts of the nitrogen uptaken by the crops after 20 experimental years can be seen.

The results of the nitrogen uptake of the cultivated crops show, that in the two most fertile soils located at Hajduböszörmény and at Bicsérd, N uptake of the plants were yearly 50 kg/ha higher in the average of the treatments. The unfertilized plots of the experimental sites it was 25–50 kg/ha/y higher than at the other 6 experimental sites.

On the unfertilized plots the yearly nitrogen uptake of the harvested crops was 110 kg in the average of the eight experimental sites, while the same value for the treated plots were 157 kg, 179 kg and 202 kg, respectively.

The sum of fertilizer applied nitrogen can be compared to the sum of the nitrogen taken up by the crops during the same period. In Table 3 the differences between the added fertilizer and the crop nitrogen uptake in the three nitrogen treatments are summarized. On the two most fertile soils (Hajduböszörmény and Bicsérd) the crops have taken up more nitrogen in all N-treatments than what was applied in the form of fertilizer, so the balances are negative. (This means that under these circumstances the crops took more nitrogen from other sources. On the remaining 6 experimental sites the balances became positive after the yearly application of 250 kg N/ha, i.e. the yearly applied 250 kg N/ha dose was higher than the nitrogen demand of the crops. One of the other sources – the most important among them – is the nitrogen originating from the organic matter content of the soil through mineralization. Having the crop nitrogen data from the unfertilized plots, there is a possibility to take into account also the rate of the nitrogen originated from the other sources. The results of this calculation can be seen in Table 4.

Balances in Table 4 were calculated according to Equation (1), i.e. the nitrogen content of the crops grown on the unfertilized plots were subtracted from the nitrogen content of the fertilized plots for the 50, 150 and 250 kg N/ha/year treatments, respectively. From this Table it can be seen, that the nitrogen balances are still negative at all experimental sites after the yearly application of 50 kg N/ha. The nitrogen balance depended on both, the nitrogen uptake of the crops (which varied in this treatment between 2761 and 3924 kg N/ha/20 years, 138 and 196 kg N/ha/y) and the original fertility of the land (characterized by the nitrogen content of the plants grown up on unfertilized plots, which varied between 1790 and 2757 kg N/ha/20 years, 90 and 138 kg N/ha/y). This means that nitrogen originated from the soil – or from other sources – varied between 90 and 138 kg N/ha/year depending on environmental circumstances (i.e. soil physical, chemical and biological properties and climate).

After the yearly application of 150 and 250 kg N/ha the nitrogen balances became positive. This positive balance varied between 27 and 92 kg N/ha/y when 150 kg N/ha was applied yearly and between 88 and 137 kg N/ha/y after than 250 kg N/ha was the yearly application rate. On the most fertile soils the balances are less positive, than on the others. The highly positive balances after the overfertilization appear also in this calculation method, based on the nitrogen uptake of the harvested crops on the unfertilized crops. The positive

Table 1. Specific nitrogen content of the crops calculated for the selected experimental sites (Nitrogen content of 1 ton harvested crops together with crop aftermath)

Crops	N-treatments (kg/ha)	Experimental sites 1	2	3	4	5	6	7	8	Average
Winter wheat	0	26	26	23	19	25	21	24	26	*24*
	50	30	29	27	24	30	26	28	30	*28*
	150	33	32	30	28	35	30	31	33	*33*
	250	40	37	36	34	47	34	37	39	*37*
Maize	0	24	18	20	21	29	26	32	22	*23*
	50	27	21	21	22	34	29	31	23	*26*
	150	29	24	22	23	38	32	29	23	*28*
	250	29	24	25	27	42	37	34	23	*30*
Sunflower	0	46	43	60	54	54	74	68	95	*61*
	50	49	60	63	57	54	78	68	96	*65*
	150	53	80	67	59	54	81	68	97	*68*
	250	66	120	81	65	59	111	72	97	*83*

Experimental sites: 1. Hajdubözörmény, 2. Karcag, 3. Iregszemcse, 4. Nagyhörcsök, 5. Bicsérd, 6. Kompolt, 7. Putnok, 8. Mosonmagyaróvár

Table 2. Nitrogen taken up by the cultivated crops (N kg/ha/20 years)

Experimental sites	Added N in fertilizers (kg N/ha/y) 0	50	150	250	Average
Added during 20 years N (kg/ha/20 years)	0	715	2425	4135	
Chernozem soils					
Hajduböszörmény	2756.7	3923.5	4353.2	4893.8	3981.8
Karcag	1957.8	2761.4	3242.1	3462.2	2855.9
Iregszemcse	2076.1	2796.7	3114.0	3470.8	2864.4
Nagyhörcsök	1790.7	2816.7	3085.9	3654.0	2836.8
Average	*2145.3*	*3074.6*	*3448.8*	*3870.2*	*3134.8*
Brown forest soils					
Bicsérd	2711.8	3776.9	4588.2	5087.8	4041.2
Kompolt	2080.7	2920.0	3445.0	3943.0	3097.2
Putnok	2072.1	2864.4	3328.3	3976.2	3060.3
Average	*2288.2*	*3187.1*	*3787.2*	*4335.7*	*3399.6*
Alluvial soil					
Mosonmagyaróvár	2271.4	3297.7	3512.9	3829.0	3227.8
Average	2214.7	3144.7	3583.7	4039.6	3245.7

Table 3. Nitrogen calculation comparing the added N-fertilizer to the nitrogen taken up by the cultivated crops (kg N/ha/20 years)

Experimental sites	Treatments (N kg/ha/y)		
	50	150	250
Chernozem soils			
Hajduböszörmény	−3208.5	−1928.2	−758.8
Karcag	−2046.4	−817.1	672.8
Iregszemcse	−2081.7	−689.0	664.2
Nagyhörcsök	−2101.7	− 660.9	481.0
Brown forest soils			
Bicsérd	−3061.9	−2163.2	− 952.8
Kompolt	−2205.0	−1020.0	192.0
Putnok	−2149.4	− 4903.3	158.8
Alluvial soil			
Mosonmagyaróvár	−2582.7	−1087.9	306.0

Table 4. Nitrogen balances in the long-term fertilizer experiments (kg/ha) (netto nitrogen balances[x])

Experimental sites	Added N in fertilizers (kg N/ha/y)		
	50	150	250
Added during 20 years N (kg/ha/20 years)	715	2425	4135
Chernozem soils			
Hajduböszörmény	−451.8	828.5	1997.9
Karcag	−88.5	1140.8	2630.6
Iregszemcse	− 5.6	1837.2	2740.3
Nagyhörcsök	− 311.0	1129.8	2271.8
Brown forest soils			
Bicsérd	− 350.1	548.6	1758.9
Kompolt	−124.3	1060.7	2272.7
Putnok	−77.2	1168.8	2230.9
Alluvial soil			
Mosonmagyaróvár	−311.3	1183.5	2577.4

[x]The nitrogen taken up by the crops of the unfertilized plots was subtracted

balance is shown not only by this calculation but also in the specific nitrogen content of the differently fertilized plots (Table 1). From these results it can be concluded, that after the yearly application of 150 and 250 kg N/ha, the specific nitrogen content of the harvested yields exceeded (specially in the case of sunflower) the usual value in most experimental sites. The usual values measured on a medium supplied soil in Hungary are 27 kg N/t for winter wheat, 25 kg N/t for maize and 41 kg N/t for sunflower.

In the plots of the fertilized treatments the real ratio of the other – non fertilizer – sources taking part in the crop nitrogen nutrition lies somewhere between the values obtained in the two balance calculations shown in Tables 3 and 4. E.g. on Hajduböszörmény soil it does not exceed 2756.7 kg N/ha/20 years and it is not less

Table 5. Nitrogen balances in the long-term fertilizer experiments calculating with the residual nitrate-N content of the soil profiles (kg/ha) (netto nitrogen balances[x])

Experimental sites	Added N in fertilizers (kg N/ha/y)			
	0	50	150	250
Added during 20 years N (kg/ha/20 years)	0	715	2425	4135
Chernozem soils				
Hajdúböszörmény*	− 2893.9	−599.2	547.6	1473.0
Karcag	−2116.7	−348.4	540.4	1883.8
Iregszemcse	−2257.1	−200.2	874.3	1958.8
Nagyhörcsök	−1990.4	−522.0	703.2	1863.1
Brown forest soils				
Bicsérd	−2883.7	−558.0	254.7	1070.0
Kompolt**	−2212.3	−219.2	918.0	2043.7
Putnok**	−2159.1	−197.5	1058.0	2062.6
Alluvial soil				
Mosonmagyaróvár	−2554.6	−536.8	660.4	1745.1

[x]The nitrogen taken up by the crops of the unfertilized plots was subtracted.
*Nitrate-N in the 0–200 cm soil layer.
**Nitrate-N in the 0–100 cm soil layer.

than 758.8 kg N/ha/20 year. It is also clear from the experimental data, that calculating with the nitrogen content of the control (unfertilized) plants as a constant causes an overestimation of the mineralization in the soils of nitrogen fertilized treatments.

The results also show, that the optimum rate of the nitrogen fertilization is between 50–150 kg N/ha/year.

Having the nitrogen balance results for the eight experimental sites we were looking for how and where this extra nitrogen could be detected in the soil-plant system. For further calculations measuring the nitrate-N content of the soil profiles seemed to be a good tool.

In Table 5 the nitrogen budget (taking account also the nitrate-N content of the soil profiles) of the experimental sites can be seen according to *Equation (2).*

These calculations were made on the basis of data shown in Table 4 where the nitrogen taken up by the crops of the unfertilized plots were subtracted. The amounts of nitrogen found in the form of nitrate in the soil profiles were approximately the same in the first two treatments, i.e. in unfertilized plots and in yearly 50 kg N/ha N-application treatments at all locations. On chernozem soil after the yearly application of the two higher nitrogen doses higher nitrate-N accumulation was found in the soil profiles as compared to the brown forest soils. This means that if we have positive nitrogen balances using this type (*Eq.1*) of budgetary calculation, a greater part of the surplus nitrogen can be found in the profiles of chernozem soils, than in the leaching type brown forest soils.

On the basis of the balance studies according to *Equation 2* it can be stated, that the nitrogen balances seemed to be too high after the yearly application of 150 and 250 kg N/ha. This positive balance varied between 12.7–52.9 kg N/ha/y in the 150 kg N/ha/y treatments, between 53.5–103.1 kg N/ha/y in the 250 kg N/ha/y treatments under diverse soil and environmental conditions.

Acknowledgement

This research is supported under Grant No. T006463, National Scientific Research Found (OTKA).

References

Csathó P (1994) NPK balances of the Hungarian soils in 1990 and in 1991. (in Hungarian). Növénytermelés 43: 551–561

Debreczeni B and Debreczeni B-né (Eds.) (1994) Fertilization Research (1960–1990) (in Hungarian). Akadémiai Kiadó, Budapest, Hungary

Goss MJ, Beauchamp EG and Miller MH (1994) A farming system approach to minimizing nitrogen losses to the environment. *In* Transactions 15th World Congress of Soil Science, July 10–16, 1994, Acapulco, Mexico. Volume 5a, 123–137. International Society of Soil Science

Hill AR (1986) Nitrate and chloride distribution and balance under continuous potato cropping. Agric. Ecosys. Environ. 15: 267–280

Kádár I (1987) Nutrient regime of arable land use. (in Hungarian) Növénytermelés 36: 517–526

Lund LJ (1982) Variations in nitrate and chloride concentrations below selected agricultural fields. Soil Sci. Soc. Am. J. 46: 1062–1066

Németh T (1993) Fertilizer recommendations – Environmental aspects. Zesz. Probl. Post. Nauk Roln. 400: 95–104

Németh T (1994a) Nitrate-N accumulations in the soil profiles of long-term fertilizer experiments. *In* Transactions 15th World Congress of Soil Science, July 10–16, (1994, Acapulco, Mexico. Volume 5b: 46–47. International Society of Soil Science

Németh T. (1994b) Studying the nitrate leaching and accumulation in long-term fertilizer experiments. *In* Fertilization Research 1960–1990. Eds. B Debreczeni and B-né Debreczeni. pp 124–130. (in Hungarian) Akadémiai Kiadó, Budapest, Hungary

Németh T, Kovács G and Kádár I (1987–1988) Nitrate, sulphate and "water soluble salt" accumulation in the soil profiles of long-term fertilization experiment. (in Hungarian) Agrokémia és Talajtan. 36–37, 110–126.

Tanji KK, Fried M and Van De Pol RM (1977) A steady-state conceptual nitrogen model for estimating nitrogen emissions from cropped lands. J. Environ. Qual. 6: 155–159.

Várallyay Gy (1990) Soil quality and land use. Chapter 3. *In* State of the Hungarian Environment. Eds. D Hinrickson and Gy Enyedy. pp 91–123. Central Stat. Office. Budapest, Hungary

C. Rodriguez-Barrueco (ed.), Fertilizers and Environment, 39–41.

The effect of fertilizers on maize yield under two tillage systems

Z. Dumanovic, Z. Videnovic & G. Vasic
Maize Research Institute Zemun Polje, S. Bajica 1, 11080 Belgrade-Zemun, Yugoslavia

Key words: Maize yield, fertilizers, tillage system, irrigation

Abstract

In order to investigate the possibilities of maize production under lower intensity tillage systems, the 3 factorial experiment was set up on chernozem type of soil in Zemun Polje-Belgrade, with the following factors: tillage systems (reduced tillage and no tillage), amount of fertilizers (without fertilizers and with 660 kg/ha of NPK fertilizers) and irrigation (without irrigation and with irrigation). Each treatment was replicated four times in split-plot design, using plots of 28 m^2 size. With the exception of 1989, higher yields were recorded when 660 kg/ha of fertilizers were used, in comparison with yields obtained without fertilizers. They averaged 9.79 t/ha, meaning an increase of 24.24% (1.91 t/ha). The effect of fertilizing on maize yield was better under no tillage in comparison with reduced tillage system. In no tillage system, yield increase due to use of fertilizers averaged to 30.60% (1.94 t/ha) in dry land farming and to 32.58% (2.59 t/ha) under irrigation. Under reduced tillage system, it averaged to 14.34% (1.04 t/ha) in dry land farming and to 20.42% (2.04 t/ha) under irrigation. The effect of fertilizing was better when irrigation was applied. When no fertilizers were used, the yield averaged to 6.80 t/ha in dry farming and to 8.97 t/ha under irrigation. Yield increase, due to application of fertilizers, averaged to 21.93% (1.49 t/ha) in dry land farming and to 25.81% (2.32 t/ha) under irrigation.

Introduction

Minimum tillage in Yugoslavia has been studied by Kosovac and Tadic [2], Konstantinovic [1], Videnovic and Vasic [3] and others. They found it possible to use minimum tillage on lighter soils, but didn't find conspicious advantages over other tillage systems in our conditions.

The aim of this study was to investigate the possibilities of maize production under lower tillage intensity and high dose of mineral fertilizers.

Materials and methods

During the period 1988–1993, maize (*Zea mays*) was grown on chernozem type of soil in Zemun Polje. The maize hybrid ZP-704 (FAO maturity group 700) was planted in the experiment at the density of 60600 plants/ha.

Three factors were investigated: tillage (no tillage (NT) and reduced tillage (RT)), fertilizing (0 and 660 kg/ha of NPK fertilizers) and irrigation (with (I) and without irrigation (NI)).

In both tillage systems maize was planted with John Deere Max Emerge II planter in spring. It was planted directly into the wheat stubble in no tillage plot. In reduced tillage plot soil was cultivated with rotary tiller 15 cm deep.

In fertilized plots, the whole amount of fertilizer was spread over the surface in the autumn. NPK fertilizer was used, containing: 300 kg/ha of N, 211 kg/ha of P_2O_5 and 150 kg/ha of K_2O.

Irrigation schedule was based on gravimetric determination of the water content in the 0–90 cm soil layer at 7 day intervals. The following quantities of water were used in irrigated plots: 377.4 mm in 1988 (from 4 irrigations), 177.7 mm in 1989 (applied in one irri-

Table 1. Total annual precipitation (mm) and mean temperatures (°C) at experimental site in the period 1988–1993

Period	1988	1989	1990	1991	1992	1993	1953–1984
Precipitation							
I–XII	499.7	558.9	438.7	713.5	445.9	446.4	602.9
IV–IX	246.8	392.6	228.0	381.3	236.6	207.7	345.8
Temperatures							
I–XII	11.6	11.8	12.3	11.1	13.0	12.2	11.1
IV–IX	18.6	18.0	18.2	18.0	20.6	20.3	17.7

Table 2. Maize yield at Zemun Polje

	kg/ha of NPK	1988	1989	1990	1991	1992	1993	Average (t/ha)
NT-NI	0	4.45	5.80	6.82	7.93	5.37	7.70	6.34
	660	10.02	5.75	8.26	9.25	7.70	8.69	8.28
Increase (t/ha)		5.57	−0.05	1.44	1.32	2.33	0.99	1.94
Increase (%)		57.17	−0.86	21.11	16.65	43.39	12.86	30.60
RT-NI	0	6.57	6.54	7.82	9.27	6.76	6.55	7.25
	660	8.98	6.35	9.64	10.79	7.10	6.86	8.29
Increase (t/ha)		2.41	−0.19	1.82	1.52	0.34	0.31	1.04
Increase (%)		36.68	−2.90	23.27	16.40	5.05	4.73	14.34
NT-I	0	6.37	6.03	9.99	6.36	9.48	9.49	7.95
	660	11.88	6.68	11.39	9.41	10.48	13.43	10.54
Increase (t/ha)		5.51	0.65	1.40	3.05	1.00	3.94	2.59
Increase (%)		86.50	10.78	14.01	47.96	10.55	41.52	32.58
RT-I	0	9.68	6.92	10.06	8.96	12.95	11.35	9.99
	660	13.40	7.08	13.00	10.81	15.27	12.60	12.03
Increase (t/ha)		3.72	0.16	2.94	1.85	2.32	1.25	2.04
Increase (%)		38.43	2.31	29.22	20.65	17.91	11.01	20.42

gation), 355.2 mm in 1990 (from 5 irrigations), 266.4 mm in 1991 (from 3 irrigations), 255.0 mm in 1992 (from 6 irrigations) and 337.5 mm in 1993 (from 9 irrigations).

Basic meteorological data are presented in Table 1.

Results and discussion

In five out of six years, maize yield increased when high dose of fertilizer was used. Under irrigation, yield increased in all six years (Table 2).

The maize yield averaged to 8.27 t/ha in no tillage farming and to 9.39 t/h in reduced tillage. The yield increase averaged to 31.79% (2.27 t/ha) in no tillage and to 17.87% (1.54 t/ha) in reduced tillage.

The yield averaged to 6.80 t/ha in dry land farming and to 8.97 t/ha when maize was irrigated, that is the increase of 2.17 t/ha (31.91%). Maize yield increase due to application of high dose of NPK fertilizer averaged to 1.49 t/ha (21.9%) in dry land farming and to 2.32 t/ha (25.9%) under irrigation.

Yield increase between years varied more in non irrigated than in irrigated plots. Also, it varied more in no tillage than in reduced tillage plots. For non irrigated plots, coefficient of variance averaged to 125.91% for no tillage and to 104.83% for reduced tillage. In irrigated plots, it averaged to 85.06% for no tillage and to 64.41% for reduced tillage.

Conclusions

According to our results it can be concluded that:

- Maize can be grown without fertilizers.
- Maize yield increase due to application of high dose of NPK fertilizer averaged to:
  30. 60% (1.94 t/ha) in non irrigated no tillage;
  14. 34% (1.04 t/ha) in non irrigated reduced tillage;
  32. 58% (2.59 t/ha) in irrigated no tillage;
  20. 42% (2.04 t/ha) in irrigated reduced tillage.
- Maize yield varied more when not irrigated than under irrigation.
- Maize yield varied more in no tillage than in reduced tillage.

References

1. Konstantinovic J (1982) Reduced tillage – present state and prospectives of use in Vojvodina. Proceedings 9th Conf. of the Int. Soil Tillage Res. Org. at Osijek, pp 103–110.
2. Kosovac Z and Tadic L (1973) Proizvodnja postrnog silaznog kukuruza bez oranja na cernozemu (Production of silage corn as a second crop on chernozem without tillage). Zbornik radova Instituta za poljoprivredna istrazivanja, Novi Sad, September 1973, pp 99–102.
3. Videnovic Z and Vasic G (1982) Study of minimum tillage in dry farming and irrigation conditions on chernozem type of soil. Proceedings of the 9th Conf. of the Int. Soil Tillage Res. Org. at Osijek, pp 148–155.

C. Rodriguez-Barrueco (ed.), Fertilizers and Environment, 43–45.

Effect of some nutrients on yield and composition of sugar beet grown in highly calcareous soil

B.I.M. Moussa[1], S. El-Demerdashe[1] & H.H. El-Mashhadi[2]
[1]*Desert Research Center, Cairo, Egypt;* [2]*Soils & Water Research Institute, Agricultural Research Center, Giza, Egypt*

Key words: Calcareous soil, sugar beet, P, K, Zn

Abstract

This work is undertaken on sugar beets grown in a field experiment of calcareous soil at El-Nobaria area, Egypt. Three rates of P, K and Zn were used, where P was added before cultivation and K was added in two equal doses after 30 and 45 days from cultivation, while zinc was added foliarly after 30, 45 and 60 days from cultivation. The results showed that the yield of leaves, roots and total sugars of sugar beets increased significantly with the application of Zn, P and K, whether added solely or in combination. The most adequate combination treatment was Zn2 P1 K2, i.e. 0.5% Zn SO_4 + 100 kg triplephosphate + 200 kg potassium sulphate/fed. The concentration of the studied elements in roots and leaves varied with treatments depending on the type and combination of nutrients. Considering the soil chemically extractable elements, data revealed that their contents increased with increasing application, with the exception of Zn which is nearly unchanged due to its foliar application.

Introduction

Very recently, several projects for sugar beet production have been set out in Egypt. Since the calcareous soils comprise one of the most promising desert lands for future expansion, attention should be given to improve its nutritional status and productivity.

Previous work proved that sugar beet can be successfully grown in calcareous soils where sugar-cane fails completely to grow. Therefore, this investigation is carried out to study and evaluate the response of sugar beets to P, K and Zn as well as their effect on the yield of sugars. The supplying power of such calcareous soils for available elements is also considered.

Materials and methods

A field experiment was conducted at El-Nobaria area on calcareous (32% $CaCO_3$), non saline (2.2 dS m^{-1}), mildly alkaline soils (pH = 7.8). To evaluate the effect of soil application of P and K as well as foliar application of Zn on the yield and total sugars content of beets, the experiment included 27 treatments and arranged in a split-split block design with 5 replicates. Phosphorus was added as triple phosphate, before cultivation at the rates of 0, 30, 60 kg P_2O_5/fed, while K was added as K_2SO_4 in two equal portions at the rates of 0, 50 and 100 kg/fed. Also, Zn was added as zinc sulphate foliarly after 30, 45 and 60 days of cultivation at the rates of 0, 0.25 and 0.5%. Nitrogen was added as ammonium sulphate (100 kg/fed.) in two equal portions to all plots. At the end of the experiment, representative samples of leaves and roots were collected. Total sugars content was determined as described by Dubois *et al.* (1965). The plant samples were oven dried and digested according to Thomas *et al.* (1967). Soil samples were taken to represent all treatments and subjected to the ordinary preparation. Soil P, K and Zn were extracted by Olsen *et al.* (1954), Black (1983) and Lindsay and Norvell (1978), respectively. Plant digest and soil extractable P and Zn were determined spectrophotometrically while K was determined by flamephotometry.

Results and discussion

Data in Table (1) revealed that the yield for both roots and leaves of sugar beet increased significantly with

Table 1. Effect of P, K and Zn on yield and concentration of P, K, Zn and sugar content of sugar beet and chemically extractable P, K and Zn in soil

Treatments			Roots					Leaves				Soil		
P rate	K rate	Zn rate	Yield (t /fed)	P (%)	K (%)	Zn (ppm)	Sugar content (%)	Yield (t/fed)	P (%)	K (%)	Zn (ppm)	P (ppm)	K (ppm)	Zn (ppm)
		0	24.8	0.135	0.59	15	13.5	4.62	0.370	1.64	21	6.8	35	0.40
	0	0.25	29.1	0.130	0.70	18	17.3	5.51	0.300	1.95	22	6.7	30	0.50
		0.50	30.7	0.130	0.78	25	20.0	5.81	0.255	2.01	26	6.8	28	0.59
		0	27.4	0.120	0.71	22	13.9	5.26	0.300	2.85	33	5.2	47	0.40
0	50	0.25	31.4	0.150	0.83	35	17.5	5.58	0.330	3.01	29	5.3	41	0.53
		0.50	32.9	0.168	1.09	45	20.2	7.14	0.345	3.20	33	5.4	34	0.63
		0	27.6	0.115	1.45	28	15.0	5.85	0.270	3.59	36	4.0	63	0.45
	100	0.25	32.3	0.200	1.83	43	17.8	6.19	0.350	3.76	38	4.1	52	0.58
		0.50	34.1	0.230	1.98	56	20.5	7.94	0.380	4.20	39	4.4	43	0.63
		0	28.7	0.155	0.91	14	14.6	6.57	0.390	1.80	18	9.8	31	0.35
	0	0.25	30.8	0.150	1.05	16	18.8	6.92	0.380	2.05	20	10.8	28	0.47
		0.50	32.5	0.145	1.25	18	19.9	8.06	0.320	2.13	29	10.9	25	0.51
		0	28.0	0.150	1.15	16	14.8	7.56	0.310	3.15	19	10.6	42	0.38
30	50	0.25	34.3	0.200	1.35	29	19.0	9.79	0.430	3.12	23	9.7	30	0.50
		0.50	36.1	0.235	1.45	39	20.1	9.96	0.475	2.33	30	9.9	26	0.59
		0	31.8	0.145	1.65	21	15.2	8.40	0.280	3.85	21	8.7	60	0.39
	100	0.25	34.2	0.250	1.93	36	19.7	10.87	0.480	3.04	23	8.5	51	0.53
		0.50	36.5	0.268	1.70	48	20.8	11.06	0.495	2.79	32	8.9	46	0.60
		0	26.3	0.160	0.81	13	15.8	6.08	0.430	1.70	17	15.5	25	0.30
	0	0.25	30.3	0.160	0.74	15	19.1	5.93	0.380	1.25	19	18.1	21	0.35
		0.50	31.3	0.155	0.66	17	20.4	6.50	0.330	1.00	25	18.0	18	0.39
		0	28.9	0.159	0.98	15	16.0	6.63	0.350	1.95	17	13.6	36	0.32
60	50	0.25	33.5	0.220	0.86	30	19.3	7.59	0.440	1.73	20	13.4	30	0.35
		0.50	34.4	0.270	0.78	38	20.6	8.42	0.480	1.60	28	13.0	25	0.39
		0	29.3	0.145	1.35	24	16.4	7.37	0.300	2.34	19	12.9	53	0.32
	100	0.25	33.6	0.266	1.15	30	20.5	8.44	0.490	2.00	21	12.9	42	0.38
		0.50	35.2	0.282	1.00	40	21.5	9.35	0.512	1.85	30	12.8	36	0.40

L.S.D. (0.05):		P	K	Zn	P x K	P x Zn	KxZn	P x K x Zn
	Yield	0.81	0.63	1.03	0.98	1.12	1.06	1.23

adding P, K and Zn. The percent increases in roots yield were (8–47%), (9–21%) and (13–15%) for P, K and Zn additions, respectively. The corresponding increases in leaves yield were (5–23%), (11–35%) and (15–27%). These findings are in harmony with many investigators; P, Shirinyan and Ponomarenko (1981) and Dahdoh *et al.* (1988); K, Bucher *et al.* (1982) and Dahdoh *et al.* (1988); and Zn, Dahdoh *et al.* (1988).

With regard to the plant elements content, it is noticed that adding each element to soil or foliarly increased its content in plant in the presence of the other two elements. Phosphorus content decreased with increasing K addition to soil at Zn0, while follows an opposite trend with Zn addition. The same behaviour was observed when comparison was made between plant P and Zn, where P content decreased with increasing foliar Zn application, while increased under K addition to soil. These data verify the antagonistic relationship between P and Zn; and synergistic effect between P and K; Puente *et al.* (1982), and Dahdoh *et al.* (1988).

For plant Zn, data showed that adding K to soil increased Zn content, while adding P led to an opposite trend.

For plant K, data showed that adding Zn foliarly increased K content at any level of soil P application, while P increased plant K content at P30 and decreased it at P60 treatment.

The interactions between the studied elements showed positive effect on the yield of sugar beet and the magnitude of increase is shown to be dependent on the nature of each pair or the combination of all the studied elements together. In short the combined treatments P30 K100, P30 Zn0.25%, K100 Zn0.5% and P30 K100 Zn0.5% gave the highest yields reachimg about 90, 85, 48 and 139% respectively for leaves compared to the control treatment. The corresponding increases for roots were 21, 32, 30 and 47% respectively. The sugars content increased with increasing application of P, K and Zn with a more pronounced increase associated with Zn application. However, the increase of sugars content relative to control is marked at Zn 0.25% while being of less magnitude at higher Zn levels. Regarding the soil chemically extractable elements, data indicate that P increased with increasing its application while decreased with increasing K application, confirming the antagonistic relationship between P and K El-Zainy et al. (1991). Addition of Zn did not affect extractable P.

The soil chemically extractable K increased with increasing its application while decreased with increasing P and Zn additions. This may be explained on basis of the effect of Zn on sugar beet growth which stimulate more absorption of K, since sugar crops have higher affinity for K uptake. This is reflected by the depletion of K level in soil. The soil chemically extractable Zn increased with increasing Zn or slightly with K application, while decreased with increasing P levels. Again. this adds further confirmation on the antagonistic relationship between P and Zn.

References

1. Black CA (1983) Methods of Soil Analysis. Soil Sci. Soc. Am. Inc. Publ. Madison, Wisconsin, USA.
2. Bucher R, Diez T and Bihler E (1982) Results of two long-term potassium fertilizer experiments in the loess region of northern Bavaria. Part I: The influence of increasing rates of potassium on plant yield, quality and K uptake. Landwirstschafttiche Forschung, Son Derheft 38: 162–177. Cf Soil & Fert 46: 10443 (1983).
3. Dahdoh MSA, Kamh RN, El-Kadi MA and El-Hassanin AS (1988) The productivity of sugar beets grown in calcareous soils with regard to Zn, P and K fertilization. Desert Inst Bull, ARE 38(2): 229–237.
4. Dubois MA, Gilles TK, Hamilton PA and Smith PA (1965) A colorimetric method for determination of sugar and related substances. Anal Chem 28: 350.
5. El-Zainy HH, Abd El-Halim AK and El-Noemani AA (1991) Effect of foliar spraying of urea and superphosphate on growth and yield of (*Vicia faba*) plant grown under different levels of water supply. Egypt J Agron 15(1–2): 125–137.
6. Lindsay WL and Norvell WA (1948) Development of a DTPA soil test for zinc, iron, manganese and copper. Soil Sci Soc Am J 42: 421–428.
7. Olsen SR, Cole CV, Watanabe FS and Dean LA (1954) Estimation of available phosphorus in soils by extraction with sodium bicarbonate. USDA Circ 939. US Government Printing Office, Washington, DC, USA.
8. Puente LS de la, Molino IMM del, Prat Perez L and Crisanto Herrero T (1982) A study of the quality of sugar-beet on some farms in Salamanca. Anuario IRNA-CSIC. Salamanca 8: 243–270.
9. Sanchez de la PL; Martin del HGIH; Prat PL and Grisanto HT (1982). Study on the quantity of sugar beet on some farms in the province of Salamanca. Aplicada de SCSIS, Salamanca, 8; 243–270 (c.f. Information Service, INC. Print 22/5/1–186; Item 19 of 186; 441).
10. Shirinyan MKh and Ponomarenko GT (1981) Response of sugar beet to fertilizer application in relation to phosphorus and potassium contents of soil. Agroch 4: 66–70.
11. Thomas RL, Sheward RW and Mayer JR (1967) Comparison of conventional and automated procedures for nitrogen, phosphorus and potassium analysis of plant material using a single digestion. Agron J 59: 240–243.

C. Rodriguez-Barrueco (ed.), Fertilizers and Environment, 47–49.

Nitrogen losses and yield of cauliflower

Effects of the reduction of mineral nitrogen fertilization

J. Navarro Pedreño[1], R. Moral, I. Gomez, J. Mataix & C. Ramos[2]
Div. de Agroquímica, Facultad de Ciencias, Universidad de Alicante, P.O. Box 99, 03080-Alicante, Spain;
[1] Universidad Autónoma de Madrid, Madrid, Spain; [2] IVIA-Moncada, Valencia, Spain

Key words: Cauliflower, nitrate lixiviation, nitrogen fertilization

Abstract

The traditional nitrogen fertilization (+N, about 500 kg of N/ha) for cauliflower plant was compared with a reduced fertilization treatment (−N, less than 300 kg of N/ha) in a calcareous soil of Spain. Nitrogen losses as nitrate were diminished in (−N) treatment and the yield was the same for both treatments.

Introduction

Nitrogen addition in mineral forms is the most important fertilization in agriculture. Traditional nitrogen fertilization is based in the use of important amounts of nitrate salts leading to nitrogen losses and causing environmental damage to water systems [4] and economical damage to farmers because of the use of excessive amounts of fertilizers added without positive plant response [2]. The mineral nitrogen not used by plants is an important source of pollution [3].

Cauliflower is an important horticultural plant in Mediterranean areas with a period of cultivation about three to five months. This plant requires important inputs of nitrogen for its development. The present work has the objective to demonstrate that an important reduction of nitrogen addition to the soil (about 300 kg/ha or less for horticultural plants) could prevent nitrogen losses without diminishing yield and could favor the farmer economy.

Methods

An experiment using this plant (var. nautilus) was done in a field of 1562 m^2, dividing it in two plots of 781 m^2 to study the effects of diminishing nitrogen fertilization on yield and nitrogen losses. The soil was a calcareous cultivated soil of the Mediterranean area that could be considered as a calciorthip [6].

Two treatments were established. The first one was the traditional fertilization applied and called +N treatment. This consists in an addition of NPK fertilizer (15:!5:15) to the soil a month before transplanting of plants (48 kg of N/ha) and ammonium nitrate three times along the period of cultivation (422 kg of N/ha), with a total input of nitrogen of 470 kg/ha (close to 500 kg/ha). The other treatment was denominated −N and the same initial fertilization was added, but less ammonium nitrate was applied along the cultivation (212 kg of N/ha), total nitrogen was 260 kg/ha (less than 300 kg/ha).

Two cycles of cultivation were studied: 1° (August to November of 1993), and 2° (January to April of 1994) applying the same fertilization in both periods and the plant density was kept in 3 plants/m^2. Yield and nitrogen concentration in aerial part of the plants were determined for each period. In the soil, four horizons were studied: 0–30 cm (arable layer), 30–60 cm (limit for roots length), 60–90 cm and 90–120 cm. Each soil sample was composed of five subsamples taken in zig-zag for each treatment in the plots. In all of them N as nitrate [5], ammonium and organic forms [1] were measured as well as organic matter. Sampling periods were: before the first period of cultivation (I), in the middle of them (II) and after the second cycle of cultivation (III). Irrigation was basically with recycled

Table 1. Characteristics of irrigation water

pH	(dS/m) EC	mg/l Cl^-	HCO_3^-	SO_4^{2-}	PO_4^{3-}	NO_3^-	NH_4^+	BO_3^{3-}	K	Na	Ca	Mg
7.7	2.11	309.6	300.0	380.0	2.6	42.3	23.1	1.9	27.5	27.8	76.0	65.2

Table 2. Results for N and organic matter of dry soil in samples I, II and III

Horizon		NO_3^- (mg/kg)			NH_4^+(mg/kg)			N org. (g/kg)			O.M. (%)		
	(cm)	I	II	III	I	II	III	I	II	III	I	II	III
+N	0–30	37.6	213	261	11.7	14.9	15.5	1.04	1.14	1.58	1.14	1.68	1.88
	30–60	45.2	153	234	12.0	14.1	14.3	0.98	0.91	1.27	0.83	1.34	1.67
	60–90	38.7	126	146	12.3	13.7	13.9	0.85	0.65	0.97	0.66	0.86	1.07
	90–120	29.1	74.0	76.0	10.3	13.7	13.4	0.63	0.52	0.70	0.53	0.62	0.70
−N	0–30	37.6	160	148	11.7	12.4	13.8	1.04	1.12	1.14	1.14	1.64	1.67
	30–60	45.2	93.5	74.5	12.0	10.1	11.8	0.98	0.93	1.08	0.83	1.17	1.52
	60–90	38.7	59.5	66.0	12.3	10.8	12.6	0.85	0.68	0.80	0.66	0.68	1.21
	90–120	29.1	54.0	47.5	10.3	12.5	14.2	0.63	0.57	0.73	0.53	0.51	0.91

Table 3. N content in plant tissue and yield of cauliflower

Treatments →	+N 1°	+N 2°	−N 1°	−N 2°
N (%)	4.5	5.8	4.3	5.3
Yield (kg/ha)	33231	46086	38550	46020

waste water and the total amount of water applied was about 2250 L/m^2 along the period of cultivation. Its characteristics are shown in Table 1.

Results

Table 2 showed the results obtained in the soil analysis. It was observed that a reduction of mineral nitrogen fertilization produced an important diminution of nitrate losses and poor effect was observed for ammonium and organic nitrogen (N org.) when comparing both treatments for each soil layer. The organic matter (O.M.) content could be affected by roots and plant residues as well as biological activity explaining this fact the increment observed in the top layers (0–30 and 30–60).

The effects of treatments in N content of aerial part and yield were not important when comparing treatments. However, an increment of N concentration in plant tissue was observed in the second cycle of cultivation. +N and −N treatments produce similar values as Table 3 demonstrates for both parameters. Yield of −N treatment in the first period was higher than +N. Highest yield was found in the second cycle of cultivation for both treatments (spring period) as it happened with nitrogen concentration in plant.

Discussion

The results showed that no negative effect was observed in the yield of the cauliflower plants as well as an important diminution of nitrogen lixiviation as nitrate was produced when a reduction of nitrogen fertilization was applied. These results suggest the possibility of recommending a diminution of the mineral nitrogen addition to soil for the cultivation of cauliflower in the conditions of the area studied (south-east of Spain).

References

1. Bremmer JM and Breitenbeck GA (1983) A simple method for determination of ammonium in semimicro-Kjeldahl analysis of soils and plant materials using a block digester. Commun Soil Sci Plant Anal 14(10): 905–913.
2. Cayuela JA, Fernández JE, Moreno F, Murillo JM and Cabrera F (1993) Estimación de las pérdidas de nitrato en un suelo con cultivo de maíz y riego. Proc. IX Congreso Nacional de Química (Química Agrícola y Alimentaria) 3: 91–98.
3. MacDonald AJ, Powlson DS, Poulton PR and Jankinson DS (1989) Unused fertilizer nitrogen in arable soils – Its contribution to nitrate leaching. J Sci Food Agric 46: 407–419.
4. Rossi N, Ciavatta C and Vittori Antisari L (1991) Seasonal pattern of nitrate losses from cultivated soil with subsurface drainage. Water Air Soil Poll 60: 1–10.
5. Sempere A, Oliver J and Ramos C (1993) Simple determination of nitrate in soils by second-derivative spectroscopy. J Soil Sci 44: 633–639.
6. Soil Survey Staff (1992) Keys to Soil Taxonomy, Blacksburg (VA, USA), Pocahontas Press Inc.

C. Rodriguez-Barrueco (ed.), Fertilizers and Environment, 51–54.

The effect of calcium fertilization and liming on the soil improvement and grain yield of winter wheat, corn and soybean

R. Ognjanovic, S. Lomovic, D. Djokic & M. Jelic
Agricultural Research Institute 'Serbia', Centre for Small Grains, Kragujevac, Yugoslavia

Key words: Calcium fertilization, liming, grain yield, wheat, maize, soybean

Abstract

Investigations were conducted on the smonitza (vertisol) soil type and on a stationary three-field crop rotation (soybean–wheat–maize) over 1991–1993. Application of calcium fertilizer (Njival Ca) was to amend common properties of acid soil, as well as its productive ability. Its application into the soil increased soil pH by 0.69 in water and by 1.05 in KCl. Humus content increased relatively by 8 and 16%, and total nitrogen by 7.5 and 27% for soybean and corn trial, respectively. Content of available phosphorus was significantly enhanced dependent on crops, treatments and soil layers from 1.77 to 5.57 mg/100 g soil or 20–40% respectively. Available potassium was negligibly increased for all three crops. Content of calcium was slightly increased from 0.92 to 1.39%. Statistically higher grain yield for wheat and soybean and fresh matter for maize in the second and third years of investigation for all treatments were obtained compared to control.

Introduction

Sixty per cent of Serbia soils are acid [11,14], that being a major limiting factor in plant production. Permanent increase of acidity is caused by stronger leaching of calcium in the deeper layers of soil. Intensive practice management, application of mineral fertilizers, 'acid rain' as well as omission of organic matter application (fertilizers and residue) also cause soil acidity. That has led to disturbances in chemical, biological and physical water traits in soils. Increase of soil acidity causes leaching and changes in dynamics of calcium and immobility of some biogenic and trace elements. In that manner, soils become deficient particularly in biogenic elements, especially available phosphorus, and mobility of iron, aluminium and toxic elements is increased [1, 3, 4, 8, 9].

In the aim of decreasing calcium leaching, new paths are sought to slow down the above-mentioned processes of degradation. One way is application of lime and other liming materials into the soil (slaked and quick lime, saturated mud, dolomite, ash, gypsum etc.). The mentioned materials are different sources, therefore their effect in the soil is different. It can be used with or without organic and mineral fertilizers. Fertilization calcium and liming slightly affected soil pH, increasing available forms of nutrients, especially phosphorus, as well as the microflora and microbiological processes in the soil and the content in certain macro and trace elements [4, 5, 6, 10, 13, 16, 17].

Methods

Studies were conducted in the stationary field trial over 1991–1993 years, on the experimental field of the Institute for small grains in Kragujevac. Three-fields crop rotation (soybean, wheat and corn) was used. latin squared design with four repetitions was applied. Following variants were used:

1. NPK 120:90:60 without calcium;
2. NPK 120:90:60 + 0.8 t/ha Njival Ca;
3. NPK 120:90:60 + 1.6 t/ha Njival Ca;
4. NPK 120:90:60 + 4.0 t/ha Njival Ca;
5. NPK 120:90:60 + 8.0 t/ha Njival Ca.

Experimental plot was 100 m^2 (10 × 10 m). Calcium fertilizer (Njival Ca) was applied as ground lime. Njival Ca as a product from Serbian glasswork in Paracin from the mine 'Plana' on the Baba mountain. It consists of 98.5% $CaCO_3$.

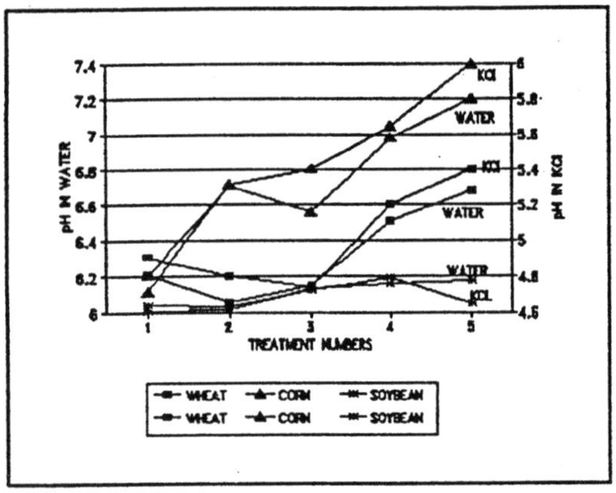

Fig. 1. Effect of 'Njival Ca' on the soil pH in water and KCl.

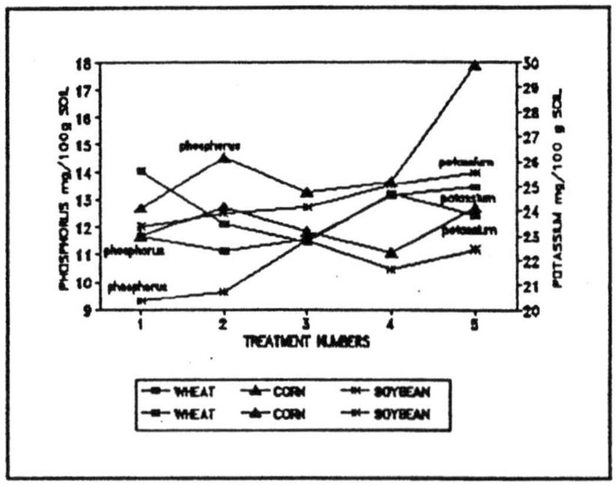

Fig. 3. Effect of 'Njival Ca' on the soil phosphorus and potassium.

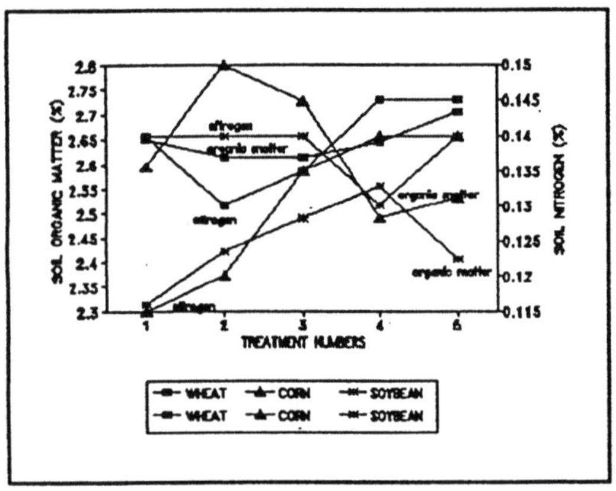

Fig. 2. Effect of 'Njival Ca' in the humus content and total soil nitrogen.

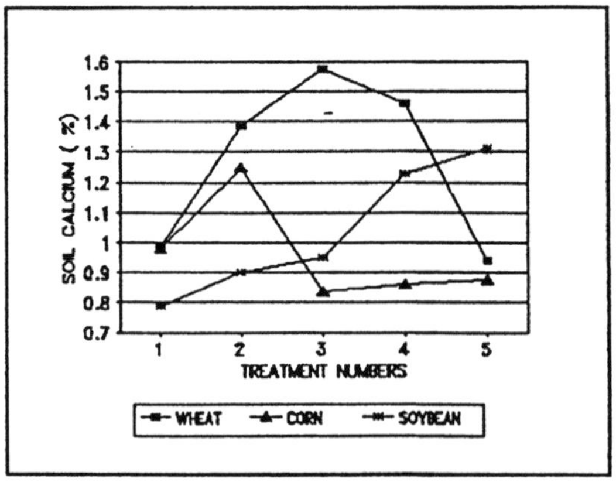

Fig. 4. Effect of liming on the soil calcium content.

Each variant was fertilized with NPK fertilizer 10:30:20 at the rate of 300 kg/ha before sowing. The rest 3/4 nitrogen was applied in the top dressing. nitrogen was not applied in the top dressing for silage corn and soybean.

Variants no. 2 and 3 are fertilization variants used every year, while variants no 4 and 5 are variants of liming added periodically – for five years once. Variant 1 is a control without added calcium.

Soil samples for chemical analysis were taken from 0–20 cm and 20–40 cm depth for each crop, on average. Humus content, total nitrogen, available phosphorus, potassium and calcium as well as pH values of soil in water and KCl were determined. grain yield of wheat and soybean and fresh matter of maize at the wax maturity stage were obtained per hectare.

Results

pH values in water raised by 0.62, 1.00 and 0.16 for wheat, maize and soybean, respectively. Soil pH values (in KCl) raised by 0.80 and 1.28 for wheat and corn, respectively. No considerable changes of pH value as affected by liming in soybean trial were detected. A reason for this could be due to the late time of liming.

Humus content slightly changed in the wheat trial, while it was enhanced by 13 and 12% for maize and soybean respectively.

Content of total nitrogen increased by 0.02, 0.03 and < 0.01% as affected by the applied lime on wheat, corn and soybean fields, respectively.

Content of available phosphorus was considerably increased by adding 'Njival Ca'. The highest enhancement was obtained in the corn field 5.20 mg/100 g soil. Increases amounted to 2.33 and 2.31 for wheat and

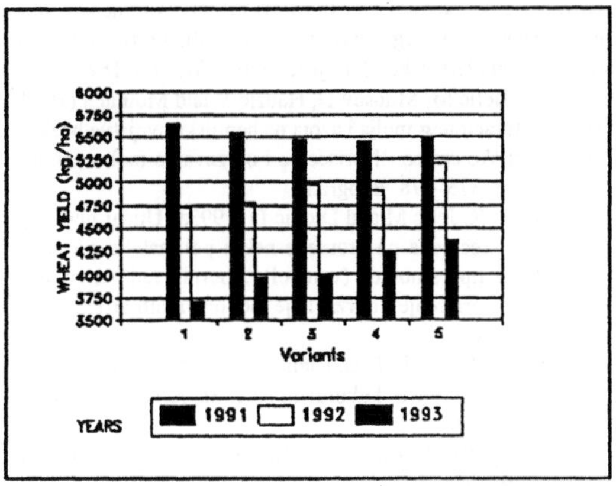

Fig. 5. Effect of liming on the grain yield of winter wheat.

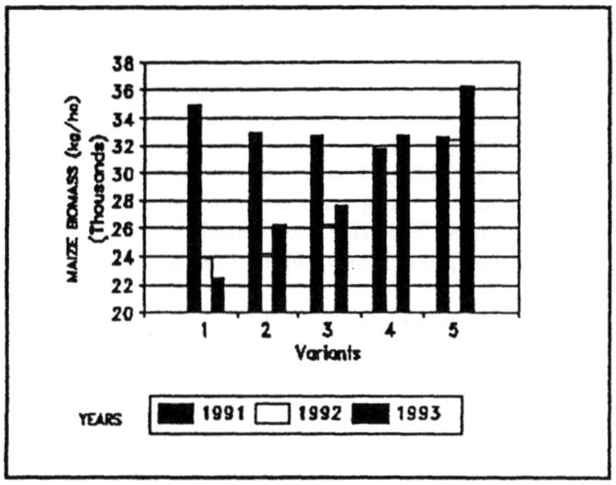

Fig. 6. Influence of liming on the fresh biomass of maize.

soybean, respectively, accounting for 20–40%, over the control. Quantitatively, for whole soil layer (0–40 cm, about 5×10^6 kg soil(above mentioned increases

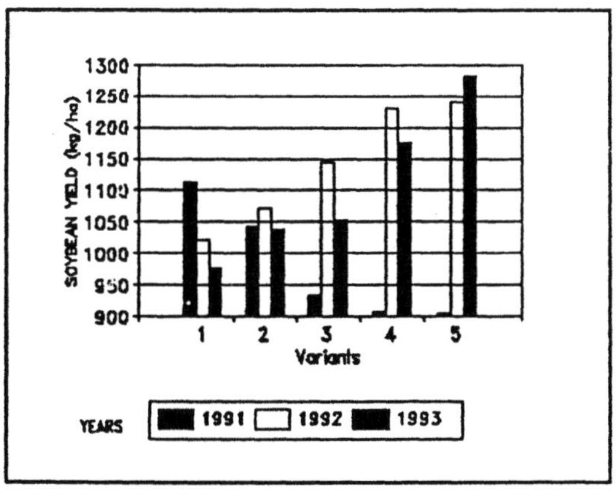

Fig. 7. Impact of liming on the grain yield of soybean.

correspond to 120 kg/ha for wheat and 260 kg/ha for maize.

Content of available potassium in the soil was negligibly changed when the calcium fertilizer 'Njival Ca' was added to all three crops.

Application of lime increased water soluble calcium in the soil for all three fields, but especially for wheat and soybean. Mean content of calcium in the fertilized variants of wheat was 1.34 and 0.99 in the control. Calcium content was 1.10 in the fertilized variants and 0.79 in the control for soybean.

Statistically significant differences among variants in the first year of studying were not found. A positive effect of application of 'Njival Ca' on grain yield for all three crops in second and third years was observed. Grain yield was positively correlated to increases of added lime. These enhancements were statistically highly significant (0.01) for wheat in the variant 5 and significant (0.05 probability) in treatments 2 and 3 in 1992, as well as it very significant by increased for 4 and 5 treatments in 1993, compared to control. Fresh matter of maize slightly increased in treatments 3, 4 and 5 in 1992 and 1993. It was also obtained increased fresh matter for variant 2 in 1993, compared to control. Grain yield of soybean very significantly increased for treatments 3, 4 and 5 in 1992, very significant for 4 and 5, and significant (0.05 probability) for treatment 3 compared to control.

Discussion

Liming materials used to reduce soil acidity, add calcium to the soil, and increase available phosphorus. Calcium positively affected hydrogen ion release forming an adsorptive complex and reducing substitutional soil acidity. It positively affected soil phosphorus changes from unavailable into available forms. Decreasing acidity reduces solubility of aluminum and iron colloid complexes and their ability to bind unavailable phosphorus. In this way, crops susceptibility to unfavourable conditions of phosphorus fertilizations are reduced. It create conditions for advanced early root formation and development of young plants [7]. Calcium application (Njival Ca) positively affected improvement of efficient uptake of other mineral nutrients.

Changes in the soil as a consequence of using 'Njival Ca' positively affected yield and another traits of wheat, maize and soybean [18].

It is seen that negligible lower grain yield of 2–5 variants compared to control occurred, because investigated soil type had not high soil acidity (in the control 6.35 and 5.00 in water and in KCl respectively). It is suggested that changes in the content of certain macro and trace elements, not investigated here, might positively affect grain yield of above mentioned crops [2, 10]. It is emphasized that the used fertilizer slightly decreased acidity of very acid soils.

References

1. Basic F, Butorac A, Vajnberger A and Malbasic D (1987) Komparativna istrazivanja efikasnosti razlicitih vapnenih materijala u kalcifikaciji kiselih tala. Poljoprivredne aktuelnosti 28: 153–170.
2. Bertic B, Vukadinovic V, Juric I, Zugec I and Kovacevic V (1985) Tjecaj kalcizacije i fosfatizacije na prisutnost mikroelemenata u tlu. Agrohemija 6: 391–397.
3. Butorac A and Tomic F (1987) Sadasnje stanje hidro i agromelioracija u SR Hrvatskoj i smernice za njihovo koriscenje u praksi. Poljoprivredne aktuelnosti 28: 85–105.
4. Glazunova NM (1990) Provedenija fosfotov dernovo – podzolastih pocvah pri izvestkovanii. Agrohimiya 4: 31–38.
5. Grincenko TA and Yacuk UL (1973) Effektivnost udobrenij izvesti i gipsa na opozolenih pocvah Rovenskoj oblasti USSR. Agrohimiya 8: 57–63.
6. Ivojlov AV (1991) Vlijanie izvestkovanija i meneralnih udobrenij pri dlitelnom ih primenenii v sevooborote na urozaj polevih kuljtur i ego kacestvo. Agrochimiceskie svojstva cernozema vjazelocennogo tjazelosuglinasto. Agrohimiya 10: 85–93.
7. Jelic M, Lomovic S and Ognanovic R (1994) Produkciija psenice na zemljistu vertisol posle upotrebe 'Njivala Ca'. SMIS '94, 'Proizvodnja hrane i energije', pp 218–222, Beograd.
8. Jovandic P (1975) Dinamika reakcije karbonata, humusa i zamenljivih katjona u karbonatnom aluvijumu Sarajevskog polja. Agrohemija 9–10: 391–400.
9. Kovacevic V, Zugec I and Bertic B (1988) Poremecaji mineralne ishrane kao organicavajuci faktor biljne proizvodnje na tlima Istocne Hrvatske. Zemljiste i biljka 37: 183–189.
10. Nesic Lj, Belic M, Milosev D, Hadzic V and Molnar I (1993) Uticaj saturacionog mulja i azota na hemijska svojstva bezkarbonatne ritske crnice. 'Koriscenje i odrzavanje melioracionih sistema', pp 375–378, Beograd.
11. Ognjanovic R, Jelic M and Djokic D (1993a) Uticaj unosenja mlevenog krecnjaka na promene nekih pokazatelja plodnosti zemljista tipa smonica (vertisol). Zbornik radova sa simpozijuma 'Koriscenje i odrzavanje melioracionih sistema', pp 375–378, Beograd.
12. Ognjanovic R, Djokic D and Jelic M (1993b) Uticaj 'Njivala Ca' na prinos psenice, kukuruza i soje. Savremena poljoprivreda 1: 368–369.
13. Popovic Z (1960) Kalcifikacija i djubrenje. Strucna biblioteka, knjiga 3, Beograd.
14. Stevanovic D, Jakovljevic M and Brkovic M (1992) Problematika fertilizacije kiselih zemljista Srbije. Zbornik radova 'Fertilizacija na prelazu dva veka', pp 61–72. Poljoprivredni fakultet, Institut za ratarstvo i povrtarstvo, Novi Sad.
15. Stevanovic D, Antonovic G and Brkovic M (1993) Kisela zemljista Srbije i njihova zastita primenom hemijskih melioracija. Zbornik radova sa Simpozijuma 'Koriscenje i odrzavanje melioracionih sistema', pp 319–324.
16. Vetcinina EM (1990) Vlijanie vozrastajuscih doz izvesti na agrotechniceskie svojstva pocvi i rozhranost seljskohozjastvennij kuljtur. Vnedrenie dostizh. nauki i peredov opita v agroprom. komplekse Ivanov. obl. Taz dokl. nauc. prak. konf., pp 66.
17. Zugec I, Juric I and Kovacevic V (1987) Neke agromelioracijske mjere kao faktor povecanja plodnosti tla Istocne Hrvatske. Poljoprivredne aktuelnosti 38: 171–181.
18. Zdravkovic L, Ognjanovic R and Djokic D (1994) Uticaj krecnog djubriva na prinos psenice, kukuruza, soje i popravku zemljista. SMIS'94, 'Proizvodnja hrane i energija', Zbornik radova, pp 364–370, Beograd.

C. Rodriguez-Barrueco (ed.), Fertilizers and Environment, 55–61.

Use of fertilizers in Russia is a factor of soil productivity

M. Ovcharenko
The State Agrochemical Service of MINSELHOZ of Russia, Orlikov per. 1/11, Moscow, 107139, Russia

Key words: Agrochemical service, soil productivity, soil sampling, fertilizers, nutrient balance, croplands, market

Abstract

One of the main reasons of a low and unstable development of agricultural production in Russia, together with unfavorable weather conditions, is an unsatisfactory state of croplands resulting in a constant decrease of soil productivity, which can be lost completely in the nearest future in vast areas of Russia. Total area of croplands in the country amounts to 222.1 mln. ha or 13% of land resources: 132.5 mln. ha – arable soils, 8 mln.ha – pastures and haylands. The State pursues a policy of soil protecting and keeping up soil fertility. However these measures are not enough. Agricultural producers and farmers themselves must make efforts to increase efficiency in agriculture.

Introduction

Soil fertilizer application provides for a stable crop development and growth, a formation of reproductive organs and high quality yields. The use of macro and micro-nutrients in sufficient rates for plant production allows to keep up and to increase soil productivity. Many scientific research centers and agrochemical service stations of the Russian Department of Agriculture and Food are constantly controlling the soil nutrient balance. There are 113 regional, territory and zonal centers and stations within the Russian agrochemical service system. The staff of these centers exceeds 11 thousand people.

Methods

Agrochemical service investigates annually about 30 mln. ha of croplands. Soil sampling is carried out both on the elemental sites (1–3 ha) of the irrigated lands and on the large sites (5–10 ha) in the zones of high humidity (non-chernozem zone).

Soil sampling on the dry land sites equal to 10–25 ha is practised as well. 15–30 probes (20–30 grams each) are taken from the depth of top soil. They form a joint mixed sample which is sent to the agrochemical laboratory. It is analyzed there for determination of total humus content, pH, exchangeable Ca, Mg, K, acid-soluble plant available P_2O_5, $N\text{-}NO_3$ + $N\text{-}NH_4$, micronutrient, trace metals level and pesticide residues. Using an IBM program, fertilizer application to certain fields is worked out. The program estimates for crop productivity and nutritive requirements. It shows a fertilizer assortment, time and methods of fertilizer use, applicators, effectiveness and economic return. There are some data about the ecological situation too.

On receipt of the program agricultural producers and farmers can order the needed fertilizers through the regional dealing system. On the basis of the chemical analysis results laboratory experts can make a plan of acid soil liming and saline soil gypsuming which are fulfilled by special mechanized teams. All these works are paid from the State budget.

During the vegetative period agrochemists control both a soil nitrogen content (soil diagnostics) and a total nitrogen, nitrate nitrogen content in plants (plant diagnostics). The level of total phosphorus and potassium are also the objects of analysis. On the basis of analysis results a correction of nutrient rates is carried out during the phase of development. Yield analysis on special sites makes it possible for experts to determine crop removal (kg/ha), to make nutrient balance and to estimate the further sources of nutrient supply. All these measures promote to keep up and increase soil fertility. All the results are brought to a data bank.

Table 1. Agrochemical characteristics of arable soils in Russia

Items	Years 1966–1970	1976–1980	1986–1990	1992	1994 expected	2000 program
Acid soils						
total:mln. ha involving soils	44	43	41	41	42	40
with pH < 5.0	26	22	18	17	13	16
Soils with low phosphorus content						
mln. ha	72	52	36	31	33	30
%	55	40	28	24	26	23
Soils with low potassium content						
mln. ha	15	13	11	11	12	10
%	12	10	8	8	9	8

Table 2. Use of nitrogenous, phosphoric and potassic fertilizers is Russia

Items	Years 1966–1970	1971–1975	1976–1980	1981–1985	1986–1990	1994 expected	2000 programmed
Total NPK mln.t	3.5	6.1	8.5	10.9	12.9	1.4	16.5
kg/ha arable soil	28	48	65	83	99	11	127
consists of:							
– nitrogenous							
mln.t	1.4	2.7	3.3	4.5	5.2	0.9	6.9
kg/ha	11	21	25	34	40	6	53
– phosphoric							
mln.t	1.1	1.7	2.7	3.5	4.6	0.4	6.4
kg/ha	9	14	21	27	35	3	49
– potassium							
mln.t	1.0	1.7	2.5	2.9	3.1	0.2	3.2
kg/ha	8	13	19	22	24	1	25
including:							
– phosphorite meal							
mln.t	0.5	0.6	0.7	0.6	0.7	0.1	0.7
kg/ha	4.0	5.0	5.0	5.0	5.0	1.0	5.0

N:P:K ratio =
1966–70 1:0.8:0.7;
1985–90 1:0.9:0.6 (optimal);
1994–1:0.4:0.1;
2000–1:0.9:0.5.

Results

Every third hectare of Russian croplands has an increased pH value and a low content of available phosphorus (Table 1). That is why soils in Russia are considered as low-productive and require a compulsory application of mineral and organic fertilizers as well as liming. Acid soils need the use of high lime rates as an ameliorant. Investigations show that the acidity of soddy-podzolic soils has not decreased even at rates 6–7 t/ha for 25–28 years. The total area of such soils in Russia amounts to 41–43 mln.ha.

Table 3. Use of organic fertilizers in arable soils in Russia

Mean per year	1966–1970	1971–1975	1976–1980	1981–1985	1986–1990	1991	1992	1993 expected	1994 programmed	2000
Applied organic fertilizers										
mln.t	208	325	352	460	482	374	288	229	150	450
t/ha	1.6	2.2	2.7	3.6	3.7	2.7	2.1	1.8	1.2	3.6
Including: delivered for composts										
mln.t					92	65	57	66	20	80

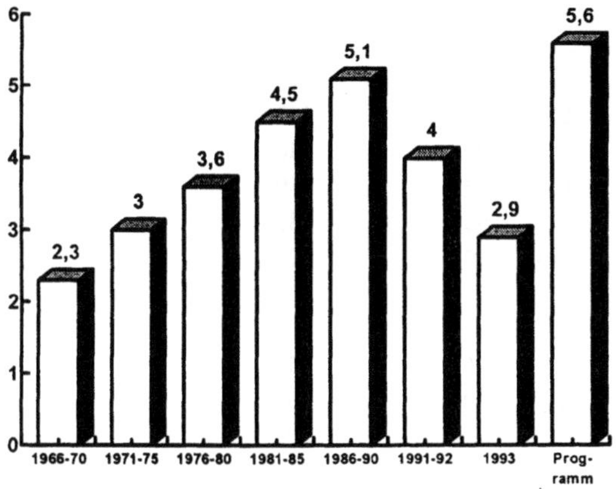

Fig. 1. Liming of acid soils m/n (average per year).

A large amount of arable soils is of a chernozem type, the acidity of which has increased significantly the last quarter of the century (pH KCL = 5.0–5.2). Nowadays these soils require the application of lime fertilizers.

The use of fertilizers in Russia has increased from 28 kg/ha to 99 kg/ha for 1990–93; but in 1994 it is estimated to apply 10–11 kg active ingredient/ha. According to the program 'Soil Productivity' it is required to use 16.5 mln.t. or 127 kg/ha to keep up optimal fertility. In 1985–90 nutrient ratio was close to optimal 1:0.9:0.6 (Table 2).

The use of organic fertilizers is of great significance in Russia and the last 25 years it has reached 482 mln.t. About 20% of organic fertilizers is peat used with liquid manure as composted peat (Table 3).

In 1991–94 the mineral fertilizer supply decreased significantly and now it is 9–10 times as little as in the previous years, that is not more than 1.5 mln.t.

The use of mineral fertilizers is 3–5 times less in Russia in the whole but in some farms the mineral fertilizers have not been applied at all the last two–three years. The total grain yield as seen is 20 mln.t. less than in 1992 (Table 4).

The cereal yield has not changed greatly for 30 years and was stable in 1986–93, namely 1.6 c/ha on an average.

Nutrient balance in Russian croplands was 64 kg/ha in 1966–70 and 70 kg/ha in 1991–95. During 1980–90 the balance was positive including phosphorus. It varied from +8 to +23 kg/ha (Table 5).

In 1994 the negative nutrient balance in Russian croplands was equal to −90 kg in the whole but in some regions and farms it was 150–180 kg/ha. A large amount of nutrients was removed with weeds (Table 6).

In 1975–95 State Programs on Chemization 'Soil Productivity' were adopted to decrease the area of acid and saline soils as well as the soils with low phosphorus content. The Programs also provided for the increase of soil fertility, liming, phosphatization of acid soils, gypsuming and ameliorative treatment of alkali soils. All these works were State-paid (Table 7, Fig. 1, Table 8).

Since entering the market the financing system in Russia has changed. It concerns goods producers, trade-mediation network, raw material resources, power engineering, fuel and transport systems as well as Russian industry (Table 9).

Industrial monopolization, price and income increase, insolvency and unreliable partnership encouraged inflation.

Goods producers are now both without State-order and without trade-mediation order. As a result there is a false production over age in all branches of industry. There is a need for goods but there is nothing to pay for them, that is why there is no demand.

Table 4. Mineral fertilizers

	supply 1986–90 average	1991	1992	1993	1994 expected	2000 programmed
mln.t NPK	12.9	10.1	5.5	3.7	1.4	16.5
kg/ha	99	78	43	29	11	127
	Application					
Total amount applied.mln.t	11.3	10.1	8.6	4.4	1.6	16.5
kg/ha seeding	95	85	73	46	25	135
out of them to:						
cereals	85	80	68	44	25	134
fodders	88	80	67	37	20	121
sugar beets	423	360	360	247	100	420
sunflowers	88	90	90	27	20	100
soya	80	80	80	37	20	120
flax	190	170	170	86	60	180
potatoes	284	265	265	176	90	285
vegetables	225	160	160	143	90	225
pastures and grasses	5	5	3	1	–	13
Area fertilized (%)						
cereals	70	63	54	47	25	85
fodders	65	59	49	41	20	80
pastures and grasses	4	3	2	1	–	10
Total yield of cereals mln.t	104	89	107	99	85–90	125

Table 5. Mineral fertilizer use and cereal yield in Russia

1966–70	1971–75	1976–80	1981–85	1986–90	1991–95	2000 expected	
mineral fertilizers applied, mln.t	3.5	6.1	8.5	10.8	12.9	4.4	16.5
kg/ha cropland	28	48	65	83	99	34	127
cereal yield, c/ga	13	13	14	13	16	16	19
nutrient balance in cropland kg/ha	−64	−34	−20	+5	+9	−70	+40
including: phosphorus kg/ha	−5	–	+8	+17	+23	−13	+35

Table 6. Nutrient balance in Russian croplands, (kg/ha)

	1986–90	1992	1993	1994 expected
Total amount of fertilizers applied	147	92	58	40
mineral fertilizers	100	65	34	22
organic fertilizers	47	27	24	18
Total removal	138	135	139	130
yield removal	113	110	106	90
weed removal	25	25	33	40
Balance	+9	−43	−81	−90

Table 7. Chemical reclamation of soils

	1986–1990	1991	1992	1993	1995 programmed
Liming, mln.ha	5.4	4.4	3.8	2.9	6.8
Lime fertilizers supplied, mln.t	39	28	21	15	45
Phosphatization, mln.ha	1.9	2.0	1.5	1.0	2.5
Gypsuming, mln.ha	0.2	0.2	0.1	0.03	0.2
Reclamation of saline soils, mln.ha	0.3	0.2	0.1	0.05	0.5
Complex agrochemical reclamation of saline soils mln.ha	2.0	1.0	0.5	0.2	4.0

Table 8. State support on soil productivity increase

*	A new system of payment from regional budgets for liming, phosphatization of acid soils, gypsuming and reclamation of saline soils has been adopted since 1992. (This work had been paid on federal level before)
*	Agricultural producers and farmers have got 30% supplements for fertilizers bought since 1993 (Plants get 50% supplements)
	Financing system
*	Agricultural producers and farmers get a credit at 150–213% interest rate.
*	Fertilizer plants get credits at 213% interest rate in commercial banks

Table 9. Mineral fertilizers prices in Russia

	1991	1992	1993	Thousand rubles per MT 1994
March	0.4	7.9	53	374
August	0.4	21	140	490
				US dollars per MT
March	228	143	77	215
August	220	103	142	229
	Grain prices in Russia			
Russian ruble	487	2630	36000	150000
US dollars	270	12.8	36.5	70

Table 10. Change of mineral fertilizer use ratio in Russia (mln.t NPK)

Items	1990	1992	1993	1994 (half year)
Production				
mln.t NPK	16.0	12.0	9.9	4.25
%	100	100	100	100
Domestic supplies				
mln.t NPK	10.7	5.5	3.7	1.05
%	67	46	37	25
Export				
mln.t NPK	5.3	6.5	6.2	3.2
%	33	54	63	75

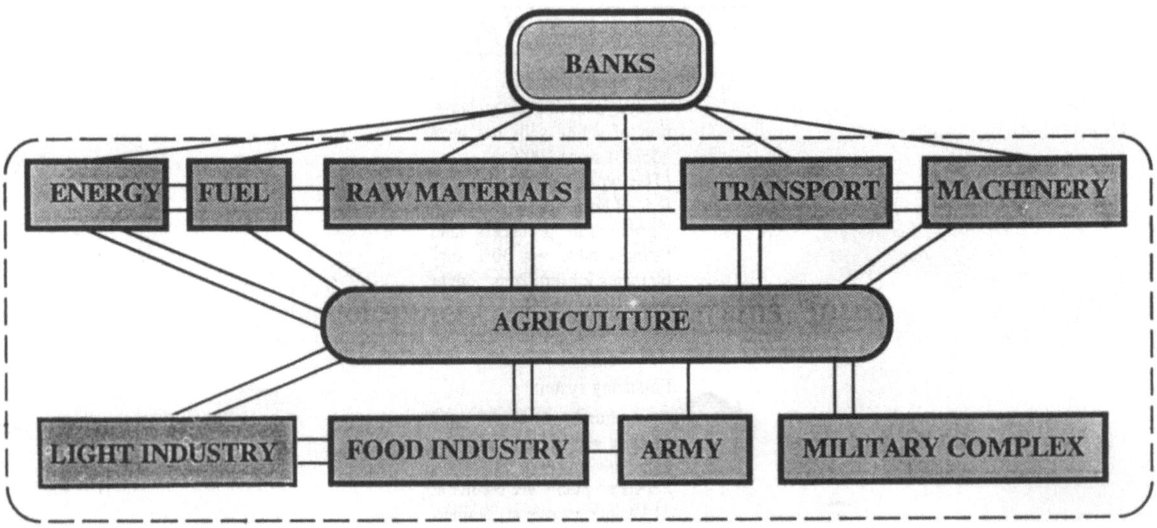

Fig. 2. Economic relations of agriculture.

It is just the same situation with fertilizer plants, agricultural producers, mechanical engineering and other interrelated branches. Fertilizer prices rose quicker than prices of agricultural production. There is no finances available to purchase fertilizers now. As a result the domestic plants began to offer fertilizers to the international market (Table 10).

Discussion

The analysis of fertilizer needs, production and use in Russia shows a certain interdependence. It is impossible to increase fertilizer production and their use in agriculture without solving complex problems in the Russian market: economical, financial and political.

In a free market environment it is required the following:

– a guaranteed bank credit to agricultural producers and a state order for agricultural products;
– a raise of output quality, establishment of wholesale market and stock exchange for healthy competition of products;
– guaranteed credits for all kinds of industry, development of economic capacity;
– investment of commercial banks by International Bank and Eurobank on Reconstruction and Devel-

opment for industry reconstruction program. For instance: plant + marketing service + distributor + dealer;

– establishment of pricing policy control, adoption of energy – fuel – grain equivalent, fuel – machine building – grain equivalent, etc., to wipe out price disparities between agricultural and industry products.

The main tendencies of modern agricultural production are the following:

– increase of organic fertilizer use including phosphorus for composts;
– adoption of beans (cereal and feed) in crop rotation;
– establishment of favorable terms for the use of green manure in crop rotation;
– expansion of production and use of organic–mineral mixtures;
– use of fertilizer blending.

Suggestions

A vast system of agrochemical service, scientific research institutes, educational agricultural institutions and universities gives everything necessary for the development of new programs, introduction of advanced know-how and adoption of scientific achievements in agricultural production.

The establishment of Stock-Exchange Bank will make it possible for farmers to get short-term credits bearing no interest as well as long term credits at low interest rate.

Conclusion

The main goal of the whole mankind is to keep up and increase soil productivity.

C. Rodriguez-Barrueco (ed.), Fertilizers and Environment, 63–64.

The fertilizer productive action and soil nutrient balance determination

Anatole V. Postnikov
All-Russian Institute for Agricultural Chemistry, Moscow

Key words: Soil nutrient balance, fertilizer productive action

The analysis of nutrient balance in agriculture enables to evaluate efficiency upon application of fertilizers, to determine changes of soil agricultural properties and to outline their characteristics depending on conditions and targets.

The balance of nutrient elements is defined as the difference between their entry and consumption. The general and productive balance is determined. In determining the general balance, the general quality of soil incoming nutrients is compared with their consumption by the crop without considering nonproductive waste.

The general balance is determined in those cases when it is necessary to establish the value of the fertilizer consumption and the increase of the content of different soil nutrients, taking into account that not all nutrient amounts are spent effectively and a certain part is wasted from the soil or changes into a state unavailable for the plant. The productive balance by A.V. Postnikov method (1980) is determined together with the general one forming a more complete idea concerning the importance of different sources in the rotation of biogenic elements and their influence on harvest and soil fertility.

This is estimated comparing the general nutrient income into the soil, with consumption at harvest and all other sources of consumption (washout, erosion, denitrification, etc.).

Table 1 is the example of computation of nitrogen, phosphorus and potassium balance in the agriculture of the Central Economic region of Russia.

In determining nutrient balance besides the accepted ratio of the nutrient application a new agrochemical idea is introduced – the productive action of fertilizers and the ratio of the fertilizer productive action reflects the total nutrient element action, and the increased content of this element in the soil in a state accessible for the plants.

Table 1. The balance of nutrients in the agriculture of the Central Economic region of Russia, thousand tons, 1990

Index	N	P_2O_5	K_2O
Entry			
Applying with mineral fertilizers	784.1	618.3	692.4
Applying with organic fertilizers	384.1	192.1	461.0
Applying with seeds	58.3	18.6	32.0
Symbiotic fixation	83.4	–	–
Nonsymbiotic fixation	69.2	–	–
Fallout with precipitation	56.5	–	–
Total	1435.6	829.0	1185.4
Discharge			
Harvest consumption	467.9	217.9	639.8
Washout	60.8	–	25.4
Gaseous losses	290.1	–	–
Losses due to erosion	36.7	9.9	88.9
Fixation into a state difficult to attain	–	269.4	–
Total	855.5	497.2	754.1
General balance			
Difference between entry and harvest consumption	+967.7	+609.1	+545.6
Productive balance			
Difference between entry and total discharge	+580.1	+331.8	+431.3

The fertilizer productivity action ratio is determined according to the following formula:

$$R' = \frac{B + na}{D}, \text{ where}$$

R' – fertilizer productive action ratio;

B – the consumption of nutrients by harvest, kg/ha;

na – nutrient amount accumulated in soil, kg/ha;

D – nutrient dosage, kg/ha.

Table 2. Norms of nutrients for the increase of phosphorus and potassium content by 10 mg per 1 kg of soil, kg/ha

Soils	Soil texture	Norms		Method of determination
		P_2O_5	K_2O	
Soddy-podzol soils	1*	50–60	40–60	after Kirsanov
	2	70–90	60–90	
	3	100–120	80–100	
Soddy-podzol-gley soils	on avr.	150–160	–	after Kirsanov
Grey forest soils	1	70–80	60–70	after Kirsanov
	2	90–100	70–80	
	3	120–140	80–90	
Chernozem, podzolized	1	80–90	80–90	after Chirikov
bleached	2	90–100	80–90	
	3	100–120	80–90	
Thick chernozems,	1	90–100	–	after Chirikov
ordinary and typical	2	100–110	–	
	3	120–130	–	
Southern and carbonated chernozems	on avr.	110–130	–	after Machigin
Chestnut soils	on avr.	90–100	–	after Machigin

*1 – sandy and sandy-loamy soils;
2 – loamy soils;
3 – clay and heavy-loamy soils.

For example, in the Central region during 1971–1990, P_2O_5 content increased by 72 mg per 1 kg of soil or by 216 kg/ha. The productive action ratio of the applied phosphorus fertilizers was 50%. This index is one of the main criteria to evaluate fertilizer application efficiency.

In organizing the work on soil fertility control one must know up to what level fertility should be increased in the different soils and how many fertilizers will be required.

The dosage of phosphorus and potassium fertilizers estimated for the planned harvest and increase of soil nutrients are determined by A.V. Postnikov according to the formula:

$$D = \frac{C}{R''} + \frac{(Pa - Rc) - NC0,1}{T}, \text{where}$$

D – nutrient dosage, kg/ha;

C – nutrient consumption with planned harvest, kg/ha;

R'' – ratio of nutrient consumption from fertilizers;

Pa – planned amount of soil nutrients, mg/kg;

Rc – real content of soil nutrients, mg/kg;

NC – norm of nutrient consumption for the increase of its content by 10 mg/kg of soil;

T – time, during which it is planned to increase nutrient content to the projected level.

Table 2 represent the norms of nutrient consumption for the increase of mobile phosphorus and exchange potassium for 10 mg per 1 kg of soil. They are estimated for the main types of soils and differentiated according to their mechanical content. The above mentioned values are estimated for the 20 centimetre soil layer. If the field which is planned to be fertilized in a complex agrochemical way, has a different pasture layer, then the dosage of nutrient element is correspondingly decreased or increased. For example, if the pasture layer is 30 centimetre, its dose increases by 1.5 times, etc.

C. Rodriguez-Barrueco (ed.), Fertilizers and Environment, 65–73.

Principles of forest fertilisation – illustrated by New Zealand experience

I.R. Hunter[1] & W. Smith[2]
[1]*Natural Resources Institute, Chatham, Kent ME4 4TB, UK;* [2]*Tasman Forestry Limited, Ngahere House, Vaughan Road, Rotorua, New Zealand*

Key words: forests, fertiliser, application, fertiliser-types, plant nutrition

Abstract

While part of the world is still being deforested, in another part agricultural land is being relinquished and forests are being replanted. The common factor that links these divergent trends is that foresters have had very little choice over which land changes use and tended to get land agriculturalists do not want. Some has been poor, infertile land. Thus forestry today is practised on land some of which is inherently susceptible to nutrient deficiency. Although natural forest is rarely nutrient deficient there are fewer and fewer places where truly natural forest exists. Past nutrient removals in timber, branches, leaves and even forest floor litter have interrupted the nutrient cycle. The land coming back into forestry from agriculture can suffer deficiencies caused by past agricultural history. Thus nutrient deficiency can be expected, and fertiliser use will be an important component of managing these forests. There are certain factors relating to forest fertilisation that distinguish it from agricultural fertiliser practice.

1. Foresters generally are reluctant to use fertiliser. This is partly because of the uncertainty of obtaining an effective and economic response and partly because of the environmental overlay associated with forests.
2. Foresters are more ready to use fertilizers to protect their crop from gross damage (because that is more easily defined) than they are just to increase yield (because they have been unable to justify this in economic terms).
3. Forest land can be rough or steep and the crop itself is dense. Land-based applications of fertiliser are difficult. Aerial applications predominate. Compared to agricultural use, the cost of application is high relative to the cost of the fertiliser, however the total cost over the life of the crop is likely to be less than in a succession of annual crops.
4. There is only limited sideways nutrient transfer within the crop, hence accurate application is essential.
5. There are considerable advantages to using slow release fertilisers to lengthen the duration of response.
6. The nutritional management of plantations on poor soils can be very complex because the fertiliser need varies both in quantity and quality over time.

Introduction

In the temperate world deforestation has slowed down and in many cases reversed. In the past however, many temperate latitude countries have undergone massive deforestation. Britain went through several periods of deforestation beginning with 80% forest cover 7000 years ago and culminating in less than 5% of the country being covered by trees in the early years of this century. Much of the land deforested went, sometimes indirectly, into agriculture. Some that had been cleared proved suitable for early agriculture but unsuitable for modern agriculture and languished under heath vegetation for centuries until reforestation took place in this century. The land allowed to go back into forest was almost exclusively land that was too poor for farming on the high moors of north and west Britain. In New Zealand the history of deforestation is much more recent (Nichols, 1980) and the motivation for deforestation was more directly for agricultural use. As elsewhere, mistakes in choice of land were made, but generally the early pioneers showed good judgement in leaving the steep, infertile soils under podocarp and southern beech forest. When reforestation started with an exotic pine in the 1920s one of the largest single areas of land planted up was the area of pumice soils in the centre of the North Island (Hunter, 1991a). This area had been abandoned for farming because livestock developed undiagnosed trace element deficiency. Three-quarters of the original estate was sited on reverting or abandoned agricultural land. One of the main causes of abandonment was that fertility maintenance was unprofitable. When Hunter *et al.* (1991) summarised foliar nutrient concentrations in radiata pine in New Zealand they found that a large proportion of the forest estate had various nutrient deficiencies. Thus until recently temperate foresters have had the choice of land on which to practice forestry dictated to them by other, more dominant users and this land tended to be the more infertile.

The temperate latitudes now produce more food than they need for their own uses although many in the Third World starve. Paradoxically, the response of the Developed Countries has been to 'set-aside' agricultural land. Temperate-latitude foresters have been able to acquire some of this land for afforestation. Past, productive, use for agriculture does not, unfortunately, always imply fertility for tree crops. The pattern of nutrient demand for trees and crops differs. Farmers, particularly livestock farmers, have sometimes chosen to fertilise their crop (or stock) directly rather than maintain the fertility of the land. Foresters sometimes experience nutritional problems on such land. Kucuba (1983) found that Scots pine planted on abandoned agricultural land in Poland responded strongly to NPK fertiliser which response he attributed to the presence of severely exhausted bands of soil resulting from earlier heavy cropping. Hunter (1994) showed that Pinus radiata planted on old pasture was more likely to suffer Mg deficiency than trees planted on old-forest sites. This was thought to be partly due to out-competition by grass and partly due to the farmers habit of supplying stock needs for Mg by direct feeding, thereby progressively depleting soil Mg. The prevalence of nutritional problems on old-field sites in the United States is well known.

In most of the tropical latitudes rising population coupled with low productivity agriculture maintains continual land pressure and hence deforestation for agricultural use. Local people are often aware of the agricultural potential of land in their neighbourhood and select the better for conversion. Forests are being increasingly restricted to land that is too infertile, steep, cold or wet for the main agricultural practice.

The factors that influence a forester in deciding to use fertiliser

Active forest fertilisation started in many countries in the 1960's and the activity has now become integrated into the routine management of forests. There has been a large literature on the opportunities to fertilise yet the ongoing programme operates at a level less than that indicated to be the optimum for maximum growth. This has come about as a result of detailed consideration by foresters and it is appropriate now to review some of the factors that appear to have influenced them in choosing the scale of the programme.

It is generally believed that natural forests do not suffer from nutrient deficiency in the sense that they do not display gross abnormalities of growth which can be attributed to a nutritional shortage. However many natural forests are almost certainly restricted in potential growth by nutrient deficiencies. The outstanding performance of Australian eucalypts on fertilised prepared sites relative to that in their natural stands is evidence of that (e.g. Hopmans *et al.*, 1990; Barros *et al.*, 1991). However, the discussion may be largely academic because truly natural undisturbed forest is rarely found. Areas in the Black Forest of Germany that showed the new-type of forest decline were found to have a history of charcoal burning from the 13th century, followed by intensive use for several centuries (Hüttl *et al.*, 1990; Ende and Zottl, 1990). Recently the mixed conifer-hardwood forest in Baden-Wurtemburg has been replaced by faster-growing lowland provenances of spruce in pure stands. Thus forest management of these sites has reduced the supply of nutrients by intensive management but increased the demand for nutrients by using more demanding faster growing trees. Nevertheless foresters are constrained in their nutritional management of forests by the public supposition that these forests are natural ecosystems.

Although normal stem-only harvesting of forests removes relatively little nutrient compared to many agricultural crops (Steele *et al.*, 1981), the removal can still be significant relative to the site capital of the nutrient. Dyck *et al.* (1988) thought that a high percentage of New Zealand's forest estate might be susceptible to decline in productivity over time through developing nutrient deficiency as a result of harvesting. Ballard and Will (1981) found that litter-raking and wood removal could rapidly deplete the availability of two nutrients with which the site was poorly supplied. Evans (1986) found that the productivity of plantations in Swaziland decreased over the rotations

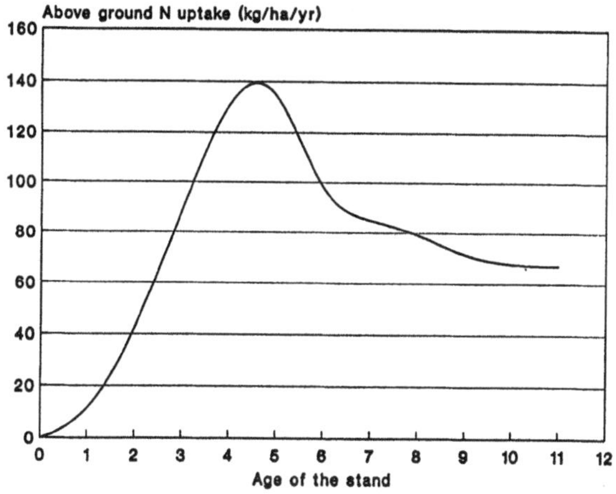

Fig. 1. Annual above ground nitrogen uptake in young radiata pine stands in New Zealand

on soils that were susceptible to rapid nutrient depletion. Foresters are very aware of these trends and are actively concerned to manage soil organic matter so as to conserve it and avoid the need for fertilising.

Tree seedlings to be used in plantations are generally grown in a specialised nursery for up to three years before being transplanted onto the site. Growing in and removing repeated crops of young seedlings from such a nursery can place a severe nutrient drain on the soil. It is perhaps the one operation in forestry where nutrient management should be similar to that practised by every arable farmer. Benzian (1965) demonstrated that in U.K. conditions Sitka Spruce seedling crops removed as much nutrients as wheat, hay or potatoes. Knight (1978b) showed that 1 year old transplants removed a maximum of 76 kg/ha of N; 9 kg of P and 49 kg of K. However 1.5 year old seedlings removed a maximum of 232 kg of N; 32 kg of P and 140 kg of K; almost as much as a normal rotation length stem harvest of radiata pine (Webber and Madgwick, 1983). Annual testing of the soil and replacement of nutrients is imperative (Knight, 1978a). Flinn *et al.* (1980) followed the nutrition of successive radiata pine seedling crops in three nurseries in Victoria. They showed that without the information yielded by repeated monitoring, fertiliser regimes led to a widening of nutrient ratios and less than optimal nutrition. The need for fertilisers in forest nurseries is widely recognised and in New Zealand was the main forestry use of the small quantities of fertiliser used annually before 1970 (Fig. 3). There will always be a market for fertilisers for such uses.

Certainty and magnitude of response and the nature of the risk avoided

As a young tree grows it enters a period of a few years where its size is increasing very rapidly and it is creating a large array of foliage which is proportionately high in nutrients. Madgwick *et al.* (1988) found 42 grammes of N per tree in closely grown radiate pine at age 5 and Webber and Madgwick (1983) found 1205 grammes of N per tree at age 29. These N contents imply a 350 fold increase in N content in the first five years but only a 28 fold increase over the next 24 years. Net nutrient demand is high in the early years but is followed by a stage where photosynthate allocation shifts in favour of wood creation and nutrient recycling reduces the net demand for nutrients. Consequently trees tend to show a nutrient uptake pattern over time similar to that shown in Fig. 1. In a plantation all the trees are going through these stages simultaneously which led Miller (1981) to divide the stand into three ages corresponding to those three stages. In an uneven-aged natural forest the trees go through these stages asynchronously and hence the stand's pattern of nutrient uptake over time cannot be so categorised. Miller (1981) thought that maximum fertiliser responses would occur in the first stage of stand growth, when nutrient demand is very high. At an advanced age accumulation of nutrients in undecomposed litter on the forest floor may restrict nutrient uptake (Miller, 1981).

Experience in New Zealand would tend to validate the hypothesis that the greatest responses occur in young stands – responses to boron and nitrogen fertiliser have been greatest in young open stands (Hunter *et al.*, 1985a ; Hunter *et al.*, 1990) – with the exception that phosphorus deficiency can first occur when that phase is past and large responses to P fertiliser can be achieved in crops in the second stage of growth (Hunter *et al.*, 1985b).

In intervening, by fertilisation, to assist the seedlings through that early stage of exponential growth we are somewhat hampered by the lack of readily available tools. The fertiliser industry is dominated by the needs of agriculturists who typically require immediate availability of nutrients (Ignatieff and Page, 1960). Empirical trials with such fertilisers applied close to seedlings soon after planting often show tree growth gains (Hunter, 1981). However, if the site is strongly deficient in a nutrient, the fertiliser application does not prevent the further onset of deficiency by the time of canopy closure (Hunter and Skinner, 1986). It

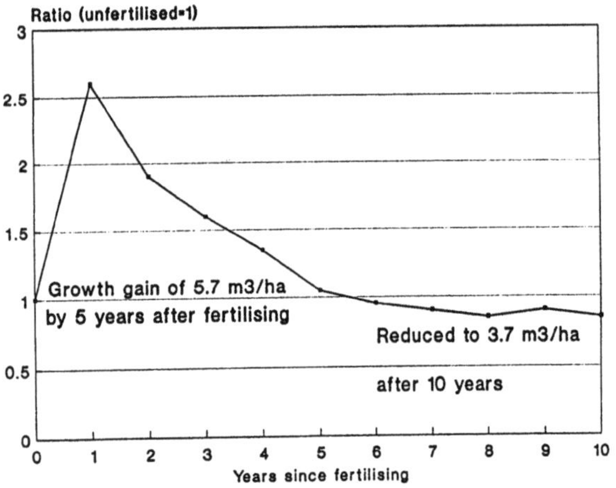

Fig. 2. Ratio of growth in N fertilised 10 year old radiata pine to growth in the unfertilized control

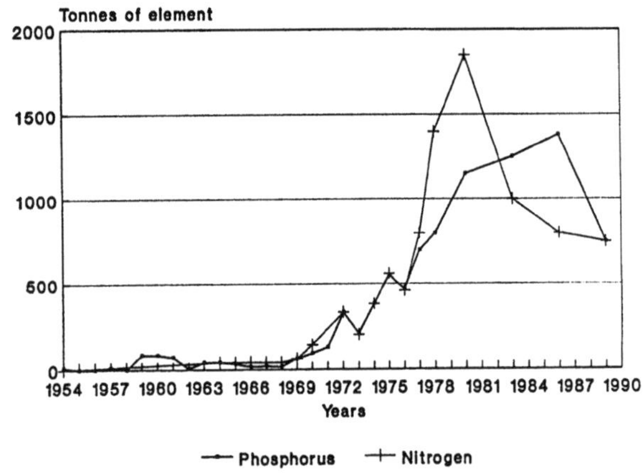

Fig. 3. Annual usage of N and P fertiliser in forests in New Zealand. (elemental tonnes)

is found moreover that the utilisation of applied fertiliser from such an early application is very low (Ballard, 1978). Amongst the reasons that contribute to the poor uptake are the placement of the fertiliser, close to the roots at time of planting but not close to the active roots at later ages and fertiliser interactions with the soil such that much of the applied fertiliser is either unavailable or leached by the time the exponential growth phase starts. It seems to be difficult to satisfy the demands of trees in this phase of growth.

There is a considerable difference in the nature and pattern of response to N and P fertiliser which affects their use by foresters. Sites in New Zealand differ in their natural fertility for both P and N. Some sand-dune forest sites are grossly deficient for N. However the strategy of raising the N status of the site by the use of underplanted lupin has been successful in preventing N deficiency in the tree stand except in a minority of situations (e.g. Hunter and Hoy, 1983). Many forest sites have marginal N concentrations in their foliage. However, this marginal status does not seem to prevent the crop from closing canopy. In such stands N fertiliser is largely ineffective in increasing growth (Hunter *et al.* 1985a), whereas if the stands are thinned or pruned a nitrogen fertiliser response occurs. McIntosh (1984) reported similar findings for pine in the UK. Several studies (e.g. Brix, 1981) have shown that the main process by which nitrogen fertiliser brings about an increase in tree growth seems to be through increased needle mass on the trees. The gain in growth through increased needle efficiency is small. Hence, a silvicultural treatment which makes it possible for the trees to increase needle mass is necessary before a response can be expected. The longevity of response to N fertiliser is short. In most stands foliar nitrogen concentrations are only detectably different from the unfertilised state for one or two years after fertilising. At the very deficient site described by Hunter and Hoy (1983) differences were detectable for 5 years. Typically growth is accelerated for up to 5 years before returning to the same rate as in the unfertilised state or even possibly to below that rate. Fig. 2, which is redrawn from Hunter *et al.* (1985a) shows a situation where after an initial increase in growth lasting 5 years, growth was subsequently poorer than in the unfertilised state and 30% of the gain in growth which was predicted at four years after fertilising was lost.

By contrast sites can become P deficient at any stage during the rotation from time of planting onwards. The differing pattern of response to P may be partly due to uptake and progressive retention of P within the above ground ecosystem and partly due to an interruption of the recycling of P from fallen needles due to the activity of soils with high P retention. Stands that are P deficient have lower quantities of less-efficient foliage than those that are not (Hunter and Hunter, 1991) and respond to P fertiliser at all stages of their development. By comparison with N fertiliser both foliar and growth responses to P fertiliser are long lived (Hunter *et al.*, 1985b). Results of several studies (Hunter and Hoy, 1983; Hunter *et al.*, 1986; Hunter, *et al.*, 1985c; Mead *et al.*, 1984; Hunter and Graham, 1982) can be summarised as in Table 1.

Ballard and Will (1978) suggested that the opportunity existed to expand N fertilisation to 14,000 tones per hectare per year and phosphorus fertilisation to 3,000 tonnes per hectare per year by 1985. Hunter

Table 1. Probable response by radiata pine to N and P fertiliser in m^3/ha at clearfelling

Degree of deficiency	Not thinned	Thinned or pruned
Slight deficiency (Foliar N < 15 > 12 mg/g; Foliar P < 1.1 > 0.8 mg/g)	No response to N fertiliser, 100–300 m3/ha to P	20–40 m3/ha to N fertiliser, 100–300 m3/ha to P
Strong deficiency (Foliar N < 12 mg/g; Foliar P < 0.8 mg/g)	40–70 m3/ha to N fertiliser, 300–400 m3/ha to P	40–70 m3/ha to N fertiliser, 300–400 m3/ha to P

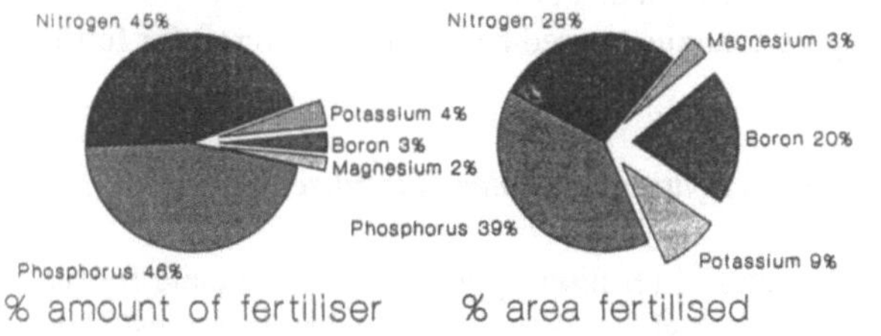

Fig. 4. Proportional importance of various fertiliser elements in NZ forestry in 1990

(1984) discussed how such fertiliser use could be integrated with forest management. Boardman and Simpson (1981) presented a similar view from an Australian perspective. However, as Fig. 3 shows far from increasing, N fertiliser use particularly and P fertiliser use to a lesser extent have never approached the projected peak and have recently tended to decrease. The reasons given by New Zealand foresters for this decline in use illustrate their prime research needs. The lack of certainty of the magnitude of an N response; the difficulty of combining fertilizer operations and silvicultural operations and the small size of the N response relative to other management options are given as reasons for the lack of take up of N fertiliser opportunities. On the other hand, while P fertiliser rates have halved from those used by Ballard and Will (1978) as a result of greater efficiency in application; fertilising frequency has reduced due to better formulations and fertility carry-over between rotations has been better than expected, leading to less weight of fertiliser applied, the programme area is closer to Ballard and Will's (1978) projection for P than it is for N.

Fig. 4 shows that in 1990 forest fertilising in New Zealand included a substantial programme of boron fertilising. Foresters in New Zealand therefore have been reluctant to use fertiliser to simply increase growth but they have been prepared to use fertiliser to prevent the kind of gross malformation that occurs with boron deficiency.

Experience elsewhere would tend to both support and contradict New Zealand foresters' judgement. In Sweden in the 1980s and more recently in British Colombia, the certainty of an impending log shortage led those foresters to apply fertiliser to existing stands in order to increase future harvests. On the other hand there has been only a sporadic uptake of the cultivation and fertiliser technology that has been shown to greatly boost pine growth in the SE USA. Perhaps the generalisable lesson is that foresters are unwilling to add to the cost of growing the stand unless there is a clearly demonstrated gain and a clear need for the extra wood which has wide political support.

Environmental considerations

Forests are often spread over the headwaters of major catchments or in areas of high recreation use. Increasingly the public is suspicious of 'unnatural' activities in forested areas. The importance placed on the environmental aspects of forest fertilization can be judged from the fact that the world-wide literature is almost equally divided between articles concerning the techniques of fertiliser application combined with growth responses and the environmental impacts of forest fertilization (e.g. Neary and Leonard, 1978; McLaughlin *et al.*, 1985; Gonzalez and Plamondon, 1978; Meehan *et al.*, 1975).

New Zealand is not impacted by air pollution but forests in America and Europe are. Much of the justification for forest fertilization currently lies with the need to rehabilitate damaged forests (Nilsson *et al.*, 1988). One of the main agents of damage is nitrate (Skeffington and Wilson, 1988) such that N fertilization is of much less interest in such forests, while excess nitrate and sulphate leach cations (Rosen, 1989) and cause nutrient imbalance. Substantial fertilizer programmes exist in these countries aimed at redressing the cation imbalance and reducing acidity (Huttl *et al.*, 1990). While some of the principles elucidated in this paper would be relevant to such a fertilization programme, many would not be. Foresters implementing such programmes would not generally be as concerned about the economic rate of return to the operation as those involved in commercial fertilisation.

Forest site constraints on fertiliser application technology

Fertiliser may need to be applied to forest stands at some time after they have been planted. Trees are typically planted fairly close together at distances between 1.5 and 4 metres. Forests are frequently on steep slopes. Consequently the ground-based systems used in agriculture cannot be used as effectively. For example, in the 1960s one New Zealand company used modified skidders to spread fertiliser from trails that had been used when thinning the stand. They found that many parts of the stand received no fertiliser while other parts were fertilised several times. Drakenberg and Sennblad (1976) report on the use of tractors with spreaders for fertiliser application in Sweden but Swedish forests are usually sited on rough but level terrain. However Schmidt (1983) reports that in Germany helicopters are three times as expensive as ground-based applicators and can only be justified for inaccessible terrain. Even application of fertiliser is very important in a forest stand (Hunter and Siviter, 1990). The response curve to fertiliser application has the usual Miterslich shape from which it follows that areas overfertilised will not compensate in growth for areas underfertilised. Intervals between fertiliser applications tend to be greater than 5 years so any uneven application will not be corrected by chance variations in the next annual dose as might occur in most agricultural systems. In pastoral systems, stock will transfer fertility. Forest fertilising in New Zealand began to use the aerial topdressing systems developed for agriculture in the early 1960's (Conway, 1962) but as a minor customer of the industry had to accept the standards set in agriculture. Visible 'striping' effects of growth across fertilised forests led to an investigation (Ballard and Will, 1971) which showed that the fixed-wing aircraft with gravity fed hoppers were missing large parts of the stand. Eilert (1967) reported similarly disappointing results from Australia. Later applications did not necessarily correct the problem because the possible flight-paths were often dictated by topography. Research in New Zealand in the late 1970s and early 1980s led to the development of an integrated system using helicopters (which with their slower flight speed can be more accurately positioned); motor-driven hoppers suspended under the helicopter (which are capable of great accuracy in fertiliser metering) and radar guidance of the helicopter to enable the pilot to monitor his position and check that the whole compartment is fertilised (Hedderwick and Will, 1982). Latterly satellite-sensing global positioning systems have been used for guidance. Systems similar to this were used in Sweden, and in Germany for fertilising the Black Forest. However, the system is expensive and under current economic conditions in New Zealand, not always affordable. Some managers continue to used fixed-winged aircraft for fertilising.

Factors influencing choice of fertiliser type

The rates of fertiliser applied in New Zealand forests are fairly high in relation to annual rates in most New Zealand agricultural crops – 60–80 kg/ha of P; 200 kg/ha of N. The expenditure is significant in the overall growing cost of a forest. Although there is an inherent conflict between the cumulative action of interest rates (which tend to penalise early action) and the need to ensure a healthy crop, most foresters would chose a type of fertiliser which gave a long lasting response from an early application made once, over one which did not. As Knight (1983) showed, foliar responses to nitrogen fertiliser and boron fertiliser (when applied as sodium borate) tend to be short-lived (less than 5 years). Foliar responses to very soluble P fertiliser last somewhat longer. Later research showed that foliar responses to K fertiliser were also very short lived. Recent research has identified more slowly soluble sources of B (colemanite and ulexite) and P (rock phosphates) which give much longer foliar responses (Hunter *et al.*, 1990; Hunter and Hunter, 1991). These fertilisers have the additional benefit that foliar concentrations do not increase so sharply immediately following fertilising. Boron can be toxic in excess and increasing the concentration of any nutrient out of relation to other nutrients can cause secondary problems. Slow release forms of N fertiliser such as sulphur coated urea and IBDU have often been shown to bring improved growth but, because of their greatly increased cost, never economically. Foresters would however be interested in cost-effective slow release forms of N and K.

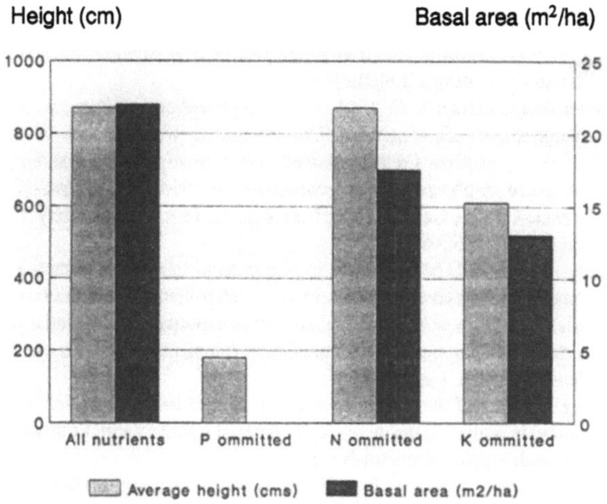

Fig. 5. Response of young radiata pine growing on a clay podzol in New Zealand to varying combinations of fertiliser

Monitoring the response

Most nutrient-poor forest soils can be identified as responsive to only one nutrient. However on many sites other nutrients interact to impact on expected growth response. The nutritional management of plantations on poor soils can be complex and require constant monitoring. Thus the upland heaths of Britain and the claylands of northern New Zealand respond to P. The welded-pumice soils of southern Kaingaroa respond to Mg (Hunter *et al.*, 1986). However in the 1970s considerable success was achieved in establishing radiata pine on the extremely poor podzolised soils of northern New Zealand (Hunter and Skinner, 1986). These soils proved to be predominantly deficient in phosphorus but also required K, responded to N and in some areas showed copper (Hunter *et al.*, 1990b) and zinc deficiencies (Thorn and Robertson, 1987). Trials with varying rates and types of fertiliser showed that the stands were sensitive to the amount and type of fertiliser applied – omitting a crucial element or applying fertiliser in the wrong ratio of elements could actually depress yields. Fig. 5 shows the variable response of one such stand to varying combinations of fertiliser from which one element was omitted (Hunter, 1991b). The primary deficiency was clearly P but omitting N or K while supplying P also caused a loss of increment. Knowledge of the nutritional status of the stands is therefore very important. Foresters in New Zealand have used foliar analysis for nearly 30 years to build up a detailed knowledge of the nutrition of their main species and have latterly reinforced that knowledge with the use of DRIS (Svenson and Kimberley, 1988) and vector analysis (Hunter *et al.*, 1990c). Elsewhere new techniques using root bioassays have been developed (Harrison *et al.*, 1990).

References

Ballard R and Will GM (1971) Distribution of aerially applied fertiliser in New Zealand forests. New Zealand Journal of Forestry Science 1(1): 22–36.

Ballard R and Will GM (1978) Past and projected use of fertilizers in New Zealand Forests. New Zealand Journal of Forestry Science 8: 15–27.

Ballard R and Will GM (1981) Removal of logging waste, thinning debris, and litter from a *Pinus radiata* pumice soil stand. New Zealand Journal of Forestry Science 11(2): 152–63.

Barros NF, Novais RF, Neves JCL and Leal PGL, (1991) Fertilizing eucalyptus plantations on the Brazilian Savannah Soils. pp 610–621. In: Schonau, APG (ed) Proceedings of the Symposium on Intensive Forestry: The Role of Eucalypts. Pretoria, South African Institute of Forestry

Benzian B (1965) Experiments on Nutrition Problems in Forest Nurseries. Vol 1. Forestry Commission Bulletin No 37, 251 pp. H.M.S.O. London

Brix H (1981) Effects of thinning and nitrogen fertilization on branch and foliage production in Douglas Fir. Canadian Journal of Forest Research 11: 502–11

Conway MJ (1962) Aerial application of phosphate fertiliser to radiata pine forests in New Zealand. Commonwealth Forestry Review 41: 234–245

Drakenberg, K and Sennblad, G, (1976) Skogsgodsling for sjalvverksamma skogsgare. Rapporter och Uppsater, Institutionen for skogsteknik, No. 103, 64pp

Dyck WJ, Hunter IR and Mees CA (1988) Forest Classification For Predicting site productivity decline in New Zealand. In: Williams TM and Gresham CA (eds) "Predicting Consequences of Intensive Forest Harvesting on Long-Term Productivity by Site Classification". Baruch Forest Science Institute of Clemson University, Georgetown, S.C. IEA/BE Project A3 (CPC-10) Report No 6

Eilert AG (1967) Variability in aerial application of fertilizer. Australian Forestry Research 2: 37–9

Ende H-P. and Zöttl HW (1990) Effects of magnesium fertilizer on the vitality and nutrition of a European beech forest (*Fagus sylvatica*) stand in the southern Black Forest of West Germany. Water, Air and Soil Pollution 54: 561–566

Evans J (1986) Productivity of second and third rotations of pine in the Usutu Forest, Swaziland. Commonwealth Forestry Review 65(3): 205–214

Flinn DW, Hopmans P and Craig FG (1980) Survey of the nutrient status of *Pinus radiata* seedlings and of soil properties in three Victorian nurseries. Australian Forestry 43: 58–66

Frederick DJ, Madgwick HAI, Jurgensen MF and Oliver GR (1985) Dry matter, energy, and nutrient contents of 8-year-old stands of *Eucalyptus regnans, Acacia dealbata* and *Pinus radiata* in New Zealand. New Zealand Journal of Forestry Science 15(2): 142–157

Gonzalez A and Plamondon AP (1978) Urea fertilization of natural forests: effects on water quality. Forest Ecology and Management 1: 213–221

Harrison AF, Dighton J, Taylor K, McDonald M and Jones HE (1990) Phosphorus nutrition of trees and grasses using a root bioassay. Pp

373–378. In: Harrison, Ineson and Heal (eds), Nutrient Cycling in Terrestrial Ecosystems. Elsevier

Hedderwick GW and Will GM (1982) Advances in the aerial application of fertiliser to New Zealand Forests: Use of an electronic guidance system and dust-free fertiliser. Forest Research Institute Bulletin no. 34, 16p

Hopmans P, Stewart HTL, Flinn DW and Hillman TJ (1990) Growth, biomass production and nutrient accumulation by seven tree species irrigated with municipal effluent at Wodonga, Australia. Forest Ecology and Management 30: 203–211

Hunter, IR and Graham, JD, (1982) Growth response of phosphorus deficient Pinus radiata to various rates of superphosphate fertilizer. New Zealand Journal of Forestry Science 12: 49–61

Hunter IR (1984) The Management Significance of Fertilising. New Zealand Journal of Forestry 29: 51–59

Hunter IR, Graham JD, Prince JM and Nicholson GM (1985a) What Site Factors Determine the Fourth Year Basal Area Response of Radiata Pine to Nitrogen Fertiliser? New Zealand Journal of Forestry Science 16(1): 30–40

Hunter IR, Graham JD, Gallagher SS and Calvert KT (1985b) Long Term Foliar Phosphorus Response of Pinus Radiata to Superphosphate Fertilizer. New Zealand Journal of Forestry Science 15: 89–100

Hunter IR, Graham JD and Calvert KT (1985c) Effects of nitrogen fertiliser on radiata pine growing on pumice soils. New Zealand Journal of Forestry 30: 102–14

Hunter IR, Prince JM, Graham JD and Nicholson G (1986) Growth and Nutrition of radiata pine on rhyolitic tephra as affected by magnesium fertiliser New Zealand Journal of Forestry Science 16(2): 152–165

Hunter, IR and Skinner, MF, (1986) Establishing radiata pine on the North Auckland Podzol. New Zealand Journal of Forestry (1986): 17–22. New Zealand Forest Service Reprint No. 1944

Hunter IR, Will GM and Skinner MF (1990a) A strategy for the correction of boron deficiency in radiata pine plantations in New Zealand. Forest Ecology and Management 37(1–4): 77–82

Hunter IR, Hunter JAC and Nicholson G (1990b) Current problems in the copper nutrition of radiata pine in New Zealand: A review. Forest Ecology and Management 37 (1–4): 143–150

Hunter IR, Truax T and Prince J (1990c) Vector Analysis of radiata pine foliage from a nitrogen and phosphorus fertiliser trial, pp 121–130. In: "Forest soil and Modern Forest Management". Proceedings of the 1st International Symposium on Forest Soils. Bo-Qun Lin (ed), Harbin, China. July 22–27 (1990). Publishing House of the NE Forest University, Harbin, China

Hunter IR and Siviter JB (1990) Why do foresters care so much about even spreading of fertiliser? pp 320–325. In: The challenges for fertiliser research in the 1990s. Proceedings of the New Zealand Fertiliser Manufacturers Research Association Conference 1990. Waipuna, New Zealand, 20–22 November 1990. Also: New Zealand Ministry of Forestry Forest Research Institute Reprint No 2330. NZFRI, Private Bag 3020, Rotorua, New Zealand

Hunter IR (1991a) Reforestation in New Zealand, pp 29–40. In: 'Large Scale Reforestation'. Winjum JK and Schroeder PE (eds), Proceedings of the International Workshop on large scale reforestation. Corvallis, Oregon, May 9–10 1990. USEPA Office of Research and Development, Washington, USA

Hunter IR and Siviter B (1991) Why do foresters care so much about even spreading of fertiliser? pp 320–335. In: Proceedings of New Zealand Fertiliser Manufacturers' Research Association Conference. Auckland, November 1990. Also New Zealand Ministry of Forestry Reprint No. 2330

Hunter, IR, Rodgers, BE, Dunningham, Prince, JM and Thorn, A, (1991) A Nutritional Atlas for radiata pine in New Zealand. Ministry of Forestry Bulletin No. 165

Hunter IR and Hunter JAC (1991) Apparent phosphorus uptake and change in nitrogen content of *Pinus radiata* growing on soils of different phosphorus retention treated with superphosphate and A-grade rock phosphate. Proceedings Xth World Fertiliser Conference, Cyprus, October 1990. New Zealand Journal of Forestry Science 21(1): 50–61

Hunter IR (1991b) The nutritional implications of the establishment phase in plantation forestry. In: Menzies, Parrot and Whitehouse (eds), Efficiency of Stand Establishment Operations, Proceedings of the IUFRO Symposium on plantation establishment. Rotorua New Zealand August 1989

Hunter IR (1994) The occurrence and treatment of magnesium deficiency in radiata pine in New Zealand. New Zealand Forest Research Institute Bulletin No. 172

Hüttl RF, Fink S, Lutz H-J, Poth M and Wisniewski J (1990) Forest decline, nutrient supply and diagnostic fertilisation in southwestern Germany and in Southern California. Forest Ecology and Management 30: 341–350

Ignatieff V and Page HG (1960) Efficient use of fertiliser. FAO Agricultural Studies No 43. FAO. Rome

Knight PJ (1978a) Fertiliser practice in New Zealand Forest Nurseries. New Zealand Journal of Forestry 8: 27–53

Knight PJ (1978b) The nutrient content of Pinus radiata seedlings: A survey of planting stock from 17 New Zealand forest nurseries. New Zealand Journal of Forestry Science 8: 54–69

Knight PJ, Jacks H and Fitzgerald RE (1983) Longevity of response in *Pinus radiata* foliar concentrations to nitrogen, phosphorus and boron fertilisers. New Zealand Journal of Forestry Science 13(3): 305–24

Kucuba S (1983) Wplyw zabiegow agromelioracyjnych na wzrost uprawy sosnowej na gruntach porolnych. Sylwan 127: 61–70

McLaughlin RA, Pope PE and Hansen EA (1985) Nitrogen fertilization and ground cover in a hybrid poplar plantation: effects on nitrate leaching. Journal of Environmental Quality 14: 214–245

Madgwick HAI, Sims A and Oliver GR (1988) Nutrient content and uptake of close-spaced *Pinus radiata*. New Zealand Journal of Forestry Science 18: 65–76

McIntosh R (1984) Fertiliser experiments in established conifer stands. Forest Record of the Forestry Commission No. 127, 23pp

Mead DJ, Draper D and Madgwick HAI (1984) Dry matter production of a young stand of Pinus radiata: some effects of nitrogen fertiliser and thinning. New Zealand Journal of Forestry Science 14: 97–108

Meehan, WR, Lotspeich, FB and Mueller, EW, (1975) Effect of forest fertilization on two southeastern Alaska streams. Journal of Environmental Quality 4: 50–55

Miller HG (1981) Forest fertilisation: Some guiding concepts. Forestry (1981) 54: 157–167

Neary DJ and Leonard JH (1978) Effects of forest fertilization on nutrient losses in streamflow in new Zealand. New Zealand Journal of Forestry Science 8: 189–198

Nichols JL (1980) The Past and Present Extent of New Zealand's Indigenous Forests. Environmental Conservation 7 (4): 309–10

Nilsson SI, Berden M and Popovic B (1988) Experimental work related to nitrogen deposition, nitrification and soil acidification. Environmental Pollution 54: 233–248

Rosen K (1989) Observed and predicted changes in exchangeable calcium, magnesium and potassium in Sedish forest soils, pp 115–122. In: Szabolcs (ed), Ecological Impact of Acidification. Hungarian Academy of Sciences

Schmidt M (1983) Walddugung: Erfolgsaussichten und Technik. Landtechnische Zeitschrift 34: 1533–1534

Skeffington RA and Wilson EJ (1988) Excess nitrogen deposition: issues for consideration. Environmental Pollution 54: 159–184

Steele KW, McCormick SJ, Percival N and Brown NS (1981) Nitrogen, phosphorus, potassium, magnesium and sulphur requirements for maize grain production. New Zealand Journal of Experimental Agriculture 9: 243–249

Svenson GA and Kimberley MO (1988) Can DRIS improve diagnosis of nutrient deficiency in *Pinus radiata*? New Zealand Journal of Forestry Science 18: 33–42

Thorn AJ and Robertson ED (1987) Zinc deficiency in *Pinus radiata* at Cape Karikari, New Zealand. New Zealand Journal of Forestry Science 17: 129–132

Webber B and Madgwick HAI (1983) Biomass and nutrient content of a 29 year-old *Pinusradiata* stand. New Zealand Journal of Forestry Science 13: 222–8

C. Rodriguez-Barrueco (ed.), *Fertilizers and Environment*, 75–81.

The effect of Al and Mn on growth and mineral composition of *Casuarina equisetifolia* Forst.

R. Kasraei[1], C. Rodríguez-Barrueco[2] & M. Igual Arroyo[2]
[1]*College of Agriculture, University of Tabriz, Iran;* [2]*IRNA-CSIC, Salamanca, Spain*

Key words: Casuarina, aluminium, manganese, toxicity

Abstract

In pot experiments (perlite) with *Casuarina equisetifolia* a strong growth depression and clear Al and Mn toxicity symptoms were found with higher treatment levels of Al and Mn. In nutrient solution dry matter yields at the highest Al treatment (2000 μM) were 37,5% for shoot and 49,5% for root biomass compared to the control. With the highest Mn treatment (3200 μM) the yield was 49% of control for both shoot and root dry matter. The lowest levels of Al and Mn treatments (250 μM Al and 400 μM Mn) resulted in a dry weight production of 78,6% for shoots and 99,3% for roots in the case of Al and 80,7% for shoots and 89,8% for roots in the case of Mn compared with the dry weight produced by control plants. The Al accumulation was relatively higher in roots than in shoots and increased with increasing Al concetrations in the nutrient solution. Also, Mn content in roots under low Mn treatments was as high as in shoots. High Al and Mn concentrations had a negative effect on the uptake of cations (Ca, Mg and Na). Aluminium had no effect on K uptake. A high decrease in the phosphorus concentrations in shoots and roots occurred in the higher treatments of Mn and especially Al. The uptake of NO_3-N from the nutrient solution was positively influenced only by the lower Al and Mn treatments. Higher concentrations of Al and Mn had no effect on the N uptake. The principal toxicity symptoms shown by the plants were the blue-green colour with brown-necrotic spots on their shoots for Al and white-chlorotic spots in the upper parts of the shoots for Mn.

Introduction

Casuarinas possess many attributes which characterize successful exotic species. They are multi-purpose plants with uses other than for fuelwood: they are adaptable to a range of environments including infertile, saline and arid sites; and can be easily established and managed. They have the ability to fix atmospheric nitrogen and some grow well under a range of enviromental stresses. Research into the physiological attributes will enable species to be manipulated to improve their productivity.

Among the known phytotoxic effects of certain elements Al toxicity is a serious problem to crop production in acid soils [22]. The toxic effects of Al on crop plants have been reviewed by Foy [7]. Aluminium toxicity is prevalent in acid soils since Al is more soluble under acid condition and eventually becomes biologically active.

Two to six billion ha of strongly acid solis, mostly in humid regions, have Al toxicity problems. Inhibition of shoot growth and root elongation, physiological and biochemical changes, and changes in nutrient and water uptake are characteristic symptoms of Al stress. Manganese toxicity has been recognised as an important factor limiting plant growth on acid and poorly drained soils [8]. Manganese toxicity produces visible symptoms that can vary between plant species.

Material and methods

Three months old seedlings of *Casuarina equisetifolia* were transplanted into pots containing perlite. One week after transplantation the different treatments commenced. There were four replicates and the experiment was arranged in a completely randomized design. The nutrient solution contained (μM:): 400 $CaCl_2$, 150 $MgSO_4 \cdot 7H_2O$, 40 $Na_2HPO_4 \cdot 12H_2O$, 94 NaH_2 $PO_4 \cdot 2H_2O$, 10 FeNaEDTA, 1010 NO_3K, 0.5 $MnSO_4 \cdot H_2O$, 0.1 $CuSO_4 \cdot 5H_2O$, 0.1 $ZnSO_4 \cdot 7H_2O$, 5 H_3BO_3, 10 NaCl, 0.05 $Na_2MoO_4 \cdot 2H_2O$. The seedlings

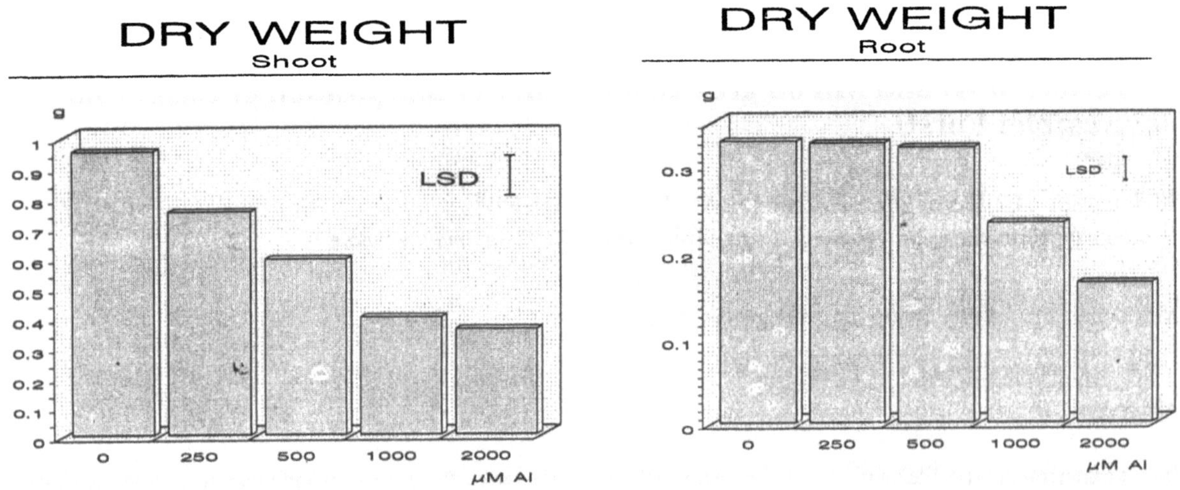

Fig. 1. Dry weight of shoots (left) and roots (right) at different levels of aluminium in the nutrient solution. Media of four plants. (LSD at p¾0.05).

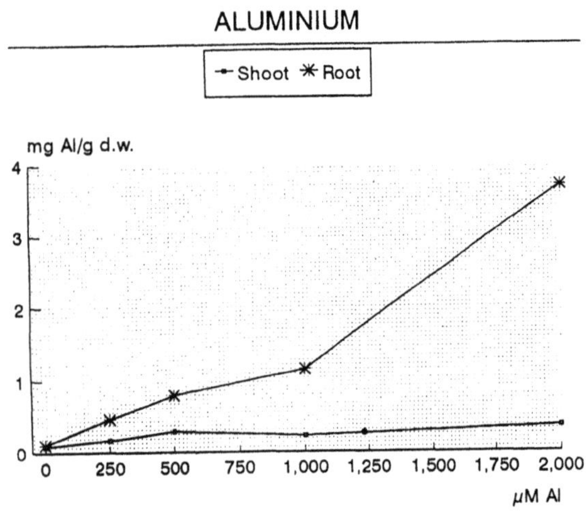

Fig. 2. Aluminium content of shoots and roots at different levels of aluminium in the nutrient solution. Values are media of four plants.

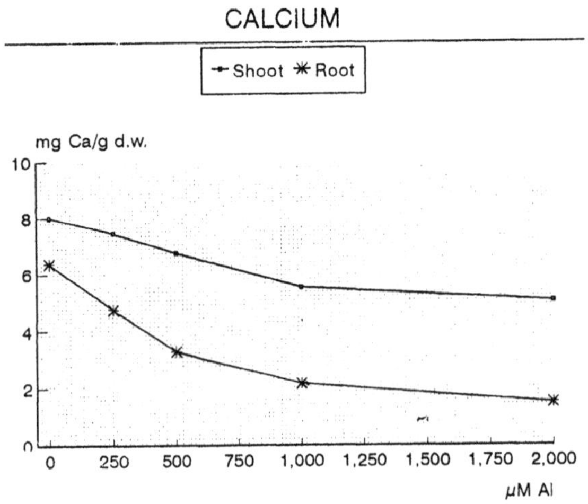

Fig. 3. Calcium content of shoots and roots at different levels of aluminium in the nutrient solution. Values are media of four plants.

were exposed to five levels of Al (0, 250, 500. 1000 and 2000 μM) supplied as $AlCl_3$ or to five levels of Mn (5, 400, 800, 1600 and 3200 μM) supplied as $MnSO_4$. The pH in all treatments was adjusted to 4.0 with HCl or NaOH. The appropriate solutions were applied to field capacity twice weekly.

Twelve weeks after transplantation the pots were harvested. Dry weights were determined for shoots and roots. Samples were analyzed for N by micro Kjeldahl procedure; for K, Ca, Mg, and Mn by atomic absorption spectrophotometry; for P by colorimetry using the vanado-molybdophosphoric yellow method and for Al by inductively coupled plasma emission spectrometry.

Results and discussion

In this experiment we examined the effect of different Al and Mn concentrations in nutrient solutions on dry weight production, cation uptake and mineral composition of roots and shoots and the occurrence of toxicity symptoms in *Casuarina equisetifolia*.

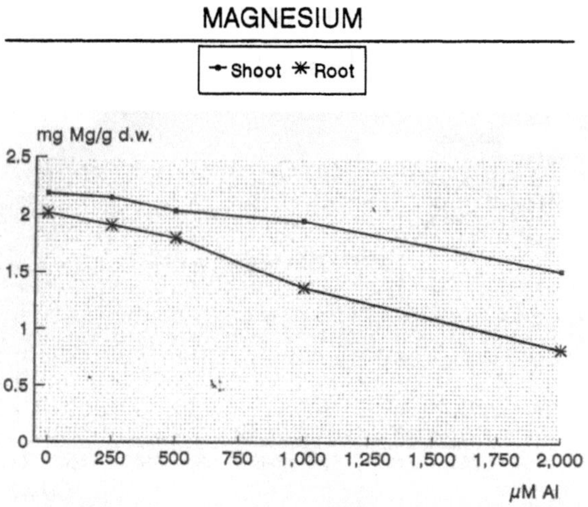

Fig. 4. Magnesium content of shoots and roots at different levels of aluminium in the nutrient solution. Values are media of four plants.

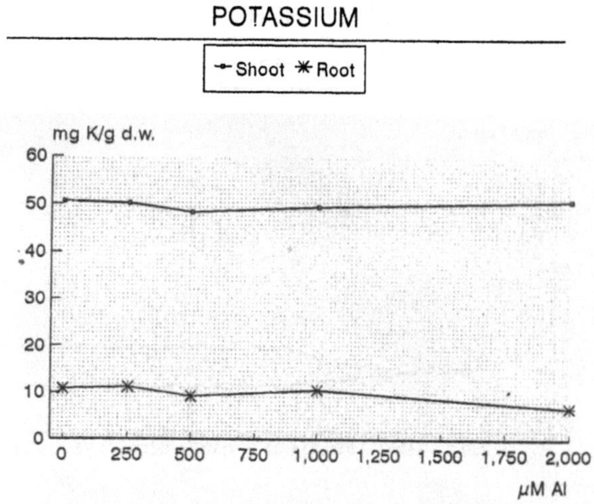

Fig. 6. Potassium content of shoots and roots at different levels of aluminium in the nutrient solution. Values are media of four plants.

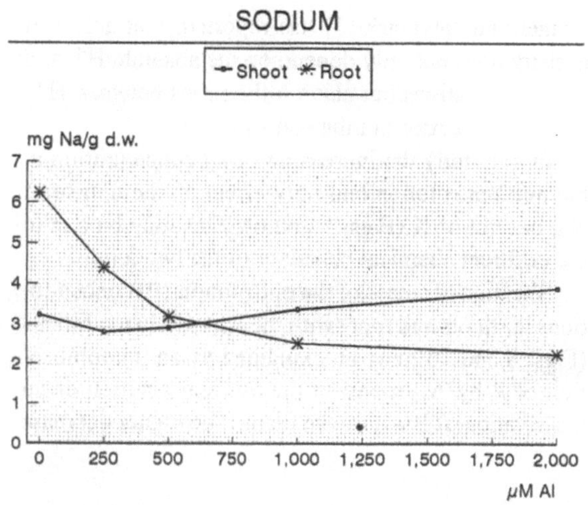

Fig. 5. Sodium content of shoots and roots at different levels of aluminium in the nutrient solution. Values are media of four plants.

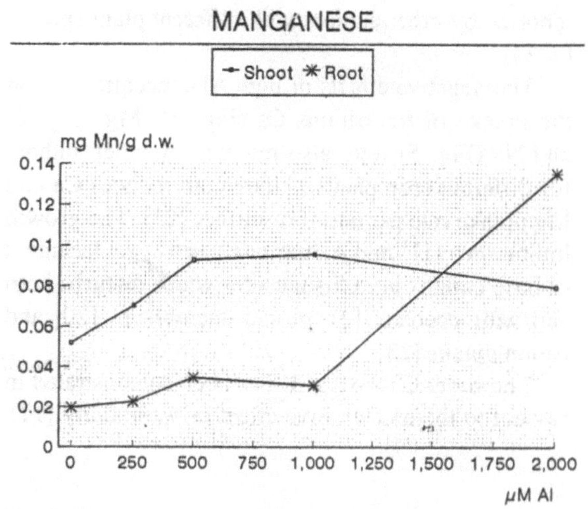

Fig. 7. Manganese content of shoots and roots at different levels of aluminium in the nutrient solution. Values are media of four plants.

Figures 1–17 show the data for the total dry matter production and the mineral composition of shoots and roots.

The relative difference of dry weight production between treatments and control were 78,6% in shoot and 99,3% in root for the 250 μM Al treatment. With the highest Al treatment level (2000 μM) dry weight production for shoots reached only 37,5% and for roots 49,5% of dry matter produced in the control (Fig. 1). The first symptoms of Al toxicity and a clear reduction of shoot growth occurred about 5 weeks after the Al treatment in the higher Al levels. Blue-green colour and brown-necrotic spots in the shoots and dark brownish roots may be interpreted as special Al toxicity symptoms for *Casuarina equisetifolia*.

The Al concentrations of the roots increased parallely with Al treatment levels but Al concentrations of the shoots maintained more or less stable throughout all treatment levels and were considerably lower than those found in roots (Fig. 2). Frederique *et al.* [10] found also higher accumulations of Al and Fe both in roots and in shoots in the pea mutant E.107. Furthermore Al toxicity symptoms were more pronunced in roots than in shoots. In our experiment damages occurred principally in shoots (Fig. 1). The observed Al toxicity symptoms were not comparable to those

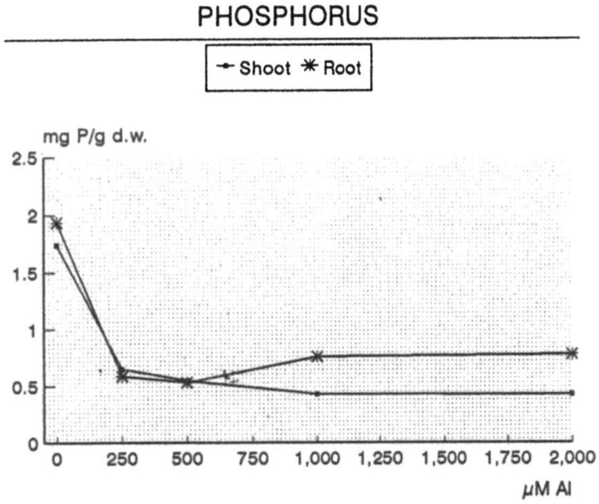

Fig. 8. Phosphorus content of shoots and roots at different levels of aluminium in the nutrient solution. Values are media of four plants.

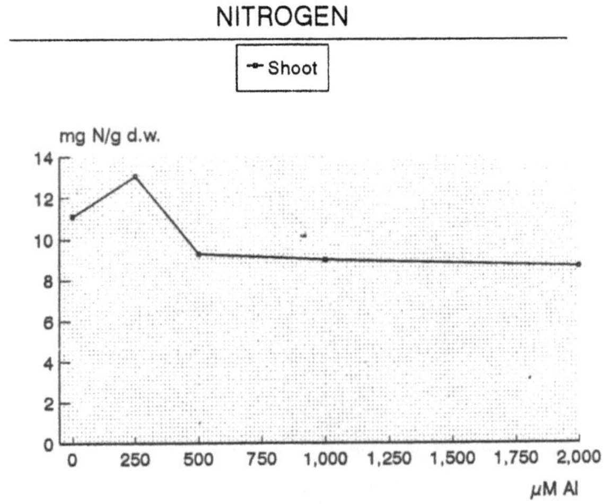

Fig. 9. Nitrogen content of shoots at different levels of aluminium in the nutrient solution. Values are media of four plants.

reported by other authors with different plant species [7, 24].

The negative effects of high Al concentrations on the uptake of the cations Ca (Fig. 3), Mg (Fig. 4) and Na (Fig. 5) were also reported by other authors for different crop plants. Aluminium reduces Ca and Mg uptake rate per unit root surface [23]. The growth limitation by H^+ and Al may be caused by replacement of base cations in exchange sites which disturbs both cell wall structure [2], plasma membranes [20] and cation uptake [23].

The order Ca> Mg> K has been demonstrated in respect to the ameliorative effect on Al toxicity [17]. Brunet and Neymark [3] has reported, that acid soil toxicity does not only depend on the absolute H^+ and Al concentrations but also on the ratios between H^+, Al and base cations in the soil solution.

In our study the increasing of Al concentration in the nutrient solution had only effect on the uptake of Na, but not of K (Figs. 5 and 6). This may be due to the different chemical character of the two ions.

The high decrease of the phosphorus (P) concentrations in shoots and root swith the Al and Mn treatments (Figs. 8 and 16) may be explained by an immobilisation of P by Al or Mn in the nutrient solution or an inactivation of P within the plant. The higher decrease

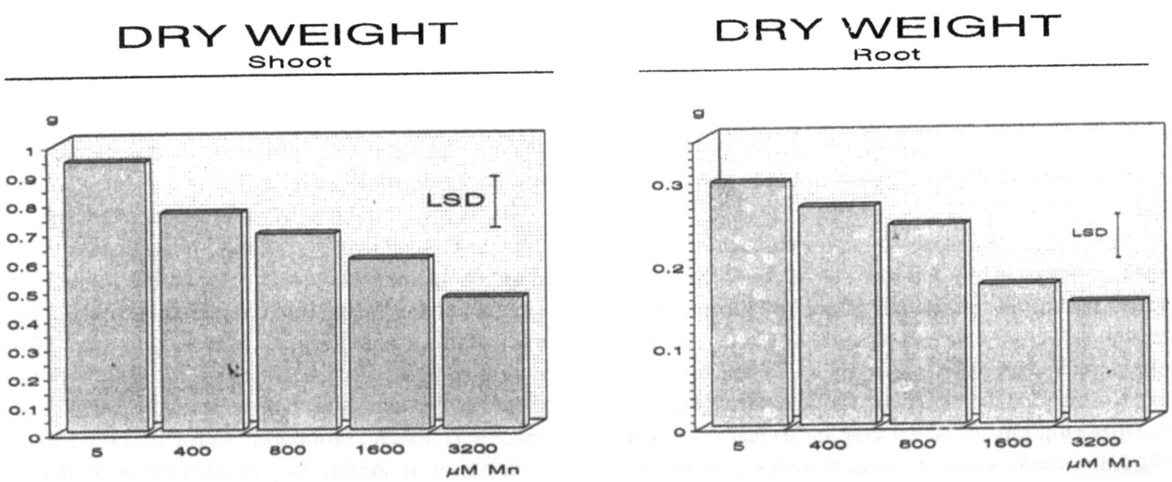

Fig. 10. Dry weight of shoots (left) and roots (right) at different levels of manganese in the nutrient solution. Media of four plants. (LSD at $p \leq 0.05$).

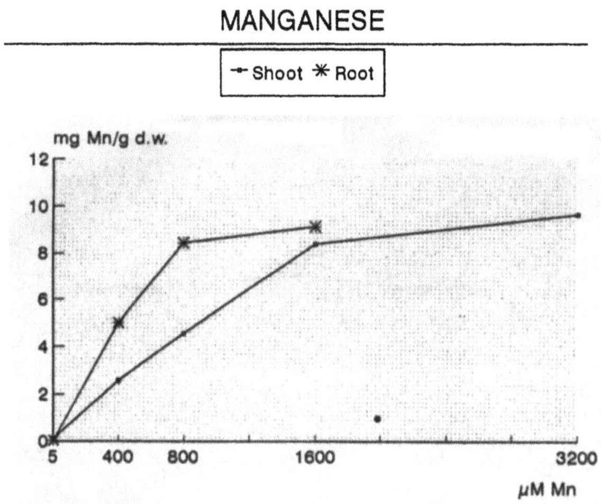

Fig. 11. Manganese content of shoots and roots at different levels of manganese in the nutrient solution. Values are media of four plants.

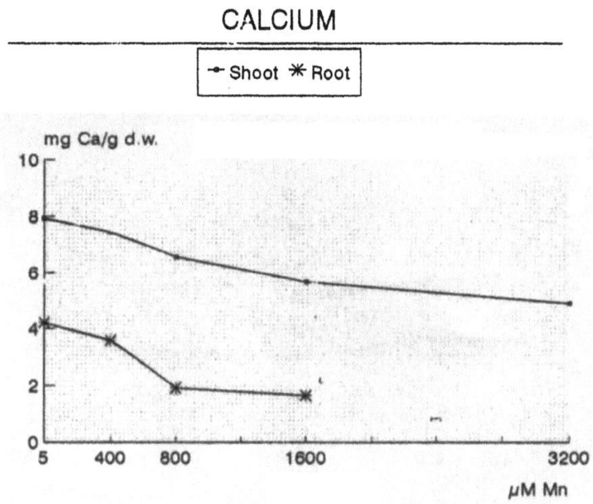

Fig. 12. Calcium content of shoots and roots at different levels of manganese in the nutrient solution. Values are media of four plants.

of the plant P concentration in the Al treatments compared to those of the Mn treatments may be due to a stronger precipitation of P by Al than by Mn in the nutrient solution.

The N uptake by the plants is characterized by an increase of the N content under low treatment levels of Al and Mn, while the plant N content is maintained constant at higher levels of Al and Mn in the nutrient solution (Figs. 9 and 17).

A highly significant depression of dry weight production of shoots and roots (51% compared to control) was also found with the highest Mn treatment (3200 μM) (Fig. 10). In the range of 400–800 μM Mn the Mn concentrations were higher in the roots than in the shoots (Fig. 11). The observed Mn toxicity symptoms were not comparable with those found by many other authors on different crop plants [8, 14, 15, 21]. About 6 weeks after the Mn treatment toxicity symptoms occurred simultaneous with a growth depression in the higher Mn treatments. The principal symptoms were white-chlorotic spots in the upper part of the shoots and dark brownblack roots. Mn toxicity can be influenced by genotypic differences between plants [14], temperature [25] and light intensity [26].

The negative effect of elevated concentrations of Mn on the concentration of cations mainly Ca and Mg in shoots and roots (Figs. 12 and 13), Na only in roots (Fig. 14) and K only in shoots (Fig. 15) were also found by other authors for different crop species. Heenan and Campbell [13] reported that Mn depresses the uptake of other cations. High Mn supply depressed the uptake of Ca and Mg in tomato but had relatively little influ-

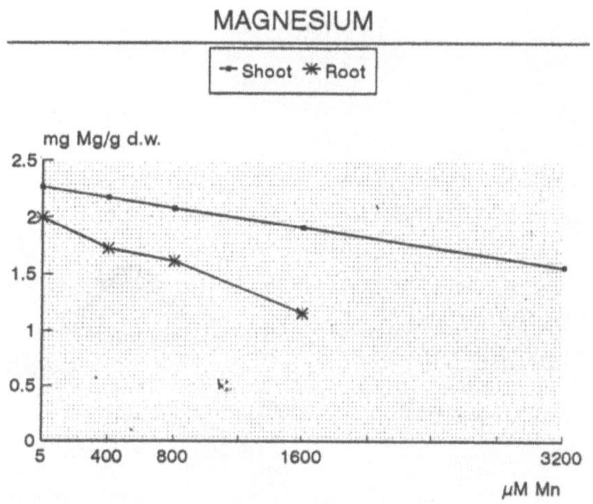

Fig. 13. Magnesium content of shoots and roots at different levels of manganese in the nutrient solution. Values are media of four plants.

ence on the K uptake [18]. Mn toxicity induces Ca deficiency particulary in younger tissues both in cotton [9] and in bean [15]. In our experiment the negative effect of manganese on the plant uptake of K and Na could be detected since the lower levels (400–800 μM of Mn). Higher levels of Mn (800–3200 μM) in the nutrient solutions showed no additional effect on the uptake of K and Na (Figs. 15 and 14).

Chinnery and Harding [4] reported that the presence of Fe, Ca and Mg can modify the uptake of manganese from solution. Maas *et al.* [19] showed that cations stimulate the uptake of manganese ions into excised roots. Uptake of Mn has been shown to retard

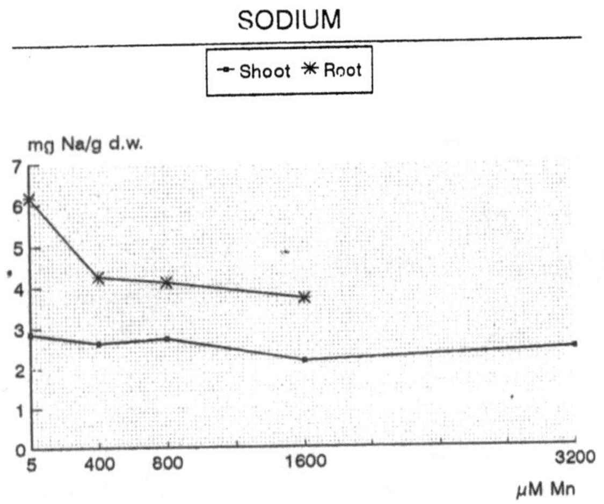

Fig. 14. Sodium content of shoots and roots at different levels of manganese in the nutrient solution. Values are media of four plants.

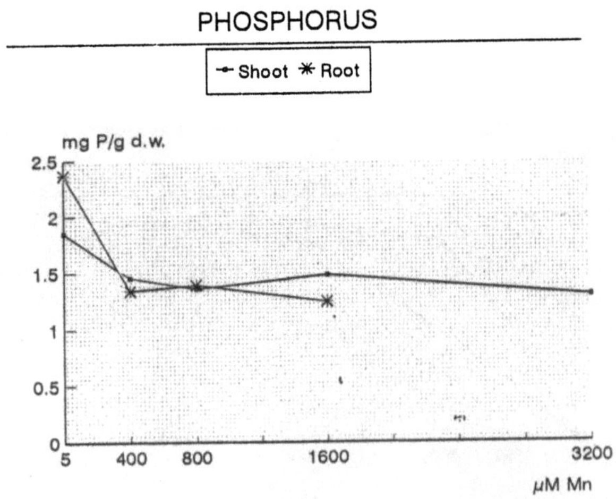

Fig. 16. Phosphorus content of shoots and roots at different levels of manganese in the nutrient solution. Values are media of four plants.

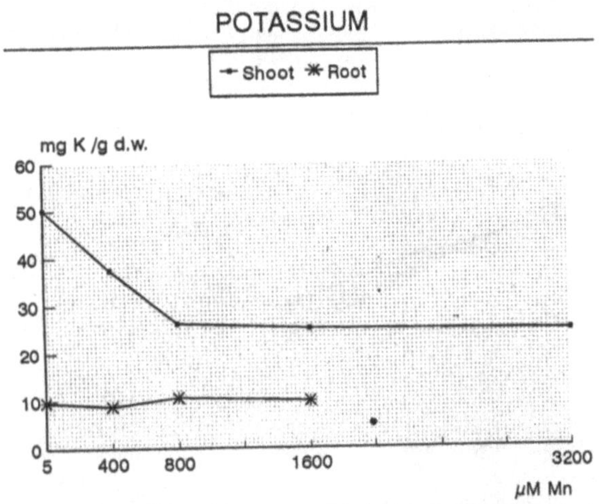

Fig. 15. Potassium content of shoots and roots at different levels of manganese in the nutrient solution. Values are media of four plants.

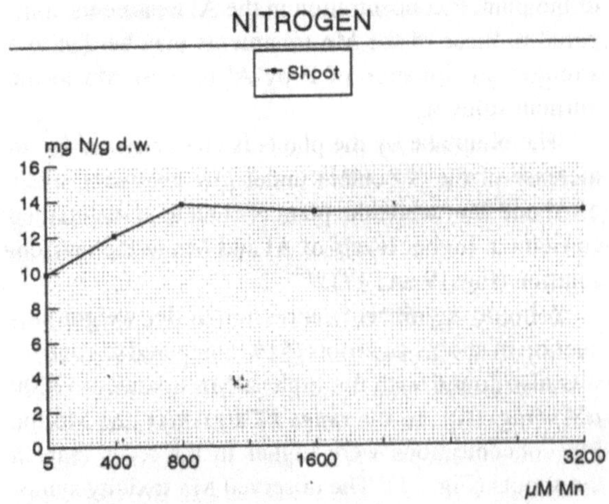

Fig. 17. Nitrogen content of shoots at different levels of manganese in the nutrient solution. Values are media of four plants.

the uptake of Ca and Mg [12]. Allen and Robinson [1] reports that the Mn toxicity is the inhibition of cations uptake such as Mg and ferrous iron (Fe^{+2}) and ferric iron (Fe^{+3}) by Mn adsorption. That was the case of our experiment, since white chlorosis spots is shoots in not only a symptom of Mn toxicity but also a symptom of Mg deficiency.

Lime treatment is a recommended method for alleviating Al toxicity in acid soils [6], although many investigators have shown that Al toxicity is not always economically correctable with conventional methods of soil liming [5].

Ganesan and Sankaranaraanan [11] have reported that Ca is more likely to bring out the beneficial effect of ameliorating Al toxicity through some other mechanisms other than Al uptake. Therefore the alternative approach to alleviate Al toxicity would be to select breed plant genotypes that have greater tolerance to Al.

References

1. Allen VG and Robinson DL (1980) Occurrence of Al and Mn^{2+} in grass tetany cases and their effects on the solubility of Ca^{2+} and Mg^{2+} in vitro. Agron J 72: 957–960.

2. Blamey FPC, Asher CJ, Kerven GL and Edwards DG (1993) Factors affecting aluminium sorption by calcium pectate. Plant and Soil 149: 87–94.
3. Brunet J and Neymark M (1992) Importance of soil acidity to the distribution of rare forest grasses in south Sweden. Flora 187: 317–326.
4. Chinnery LE and Harding CP (1980) The effect of ferrous ion on the uptake of manganese by *Juncus effusus*. Ann Bot 26: 409–412.
5. Evans CE and Kamprath EH (1970) Lime reponses as related to per cent Al saturation solution Al, and organic matter content. Soil Sci Soc Amer Proc 34: 893–896.
6. Foy CD (1984) Physiological effects of hydrogen, aluminium and manganese toxicities in acid soil. In: Adams F (ed) Soil Acidity and Liming. Agronomy Monograph n° 12 (2nd ed) American Society of Agronomy, Madison, WI, USA, pp 57–97.
7. Foy CD (1988) Plant adaptation to acid, aluminium-toxic soils. Commun Soil Sci Plant Anal 19: 959–987.
8. Foy CD, Chaney RL and White MC (1978) The physiology of metal toxicity in plants. Ann Rev Plant Physiol 29: 511–566.
9. Foy CD, Webb HW and Jones JE (1981) Adaptation of cotton genotypes to an acid, manganese toxic soil. Agron 73: 107–111.
10. Guinel FC and LaRue TA (1993) Excessive aluminium accumulation in the pea mutant E107 (brz). Plant and Soil 157: 75–82.
11. Ganesan K, Sankaranarayanan C and Balakumar T (1993). Physiological basis of differential aluminium tolerance in rice genotypes. Commun. Soil Sci. Plant Anal. 24: 2179–2191.
12. Goss MJ and Carvalho MJ (1992) Manganese toxicity: The significance of magnesium for the sensitivity of wheat plants. Plant and Soil 139: 91–8.
13. Heenan DP and Campbell LC (1981). Influence of potassium and manganese on growth and uptake of magnesium by soybean (*Glycine max* (L.) Merr. cv. Bragg). Plant Soil 61: 447–456.
14. Horst WJ (1983) Factors responsible for genotypic manganese tolerance in Cowpea (Vigna unguiculata). Plant and Soil 72: 213–218.
15. Horst WJ and Marschner H (1977) Effect of excessive manganese supply on uptake and translocation of calcium in bean plants (*Phaseolus vulgaris* L.) Z Pflanzenphysiol Bd 87: 137–148.
16. Horst WJ (1988) The physiology of manganese toxicity. In: Graham RD, Hannam RJ and Uren NC (eds) Manganese in Soils and Plants. Kluwer Academic Publishers, Dordrecht, The Netherlands, pp 175–188.
17. Kinraide TB and Parker DR (1987) Cation amelioration of aluminium toxicity in wheat. Plant Physiol 83: 546–551.
18. Le Bot J, Kirkby EA and Beusichem ML (1990) Manganese toxicity in tomato plants: effects on cation uptake and distribution. J Plant Nutr 13: 513–25.
19. Maas EV, Moore DP and Mason BJ (1969) Influence of calcium and magnesium on manganese absorption. Plant Physiol 44: 796–800.
20. Matsumoto H, Yamamoto Y and Kassai M (1992) Changes of some properties of the plasma membrane-enriched fraction of barley roots related to aluminium stress: membrane associated AT Pase. aluminium and calcium. Soil Sci Plant Nutr 38: 411–419.
21. Ohki K (1984) Manganese deficiency and toxicity effects on growth, development, and nutrient composition in wheat. Agron J 76: 213–218.
22. Parker DR, Jinraide TB and Zelanzy LH (1988) Aluminium speciation and phytotoxicity in dilute hydroxy-aluminum solutions. Soil Sci Soc Amer J 52: 438–444.
23. Rengel Z (1992) Role of calcium in aluminium toxicity. New Phytol 121: 499–4513.
24. Roy AK, Sharma A and Talukder G (1988) Some aspects of aluminum toxicity in plants. Bot Rev 54: 145–178.
25. Rufty TW, Miner GS and Raper CD Jr (1979) Temperature effects on growth and manganese tolerance in tobacco. Agron J 71: 638–644.
26. Wissemeier AH and Horst WJ (1992) Effect of light intensity on manganese toxicity symptoms and callose formation in cowpea (*Vigna unguiculata* (L.) Walp.). Plant and Soil 143: 299–309.

C. Rodriguez-Barrueco (ed.), Fertilizers and Environment, 83–87.

Nutrient cycling and sustainability

I. Láng
Hungarian Academy of Sciences, H-1051 Budapest, Roosevelt tér 9, Hungary

Key words: sustainable development, sustainability in agriculture, nutrient cycling

Abstract

A historical survey of the changes in the concept of sustainable development is given. Clarification of the interrelations between food security strategy and sustainability in agriculture are outlined and the possible role of nitrogen, phosphorus and potassium in environmental pollution examined. Conclusions as to the more important steps to be taken towards realizing sustainable agriculture are drawn.

Sustainable development

Let me first say a few words on the concept of 'sustainable development'.

This concept was first formulated in the early 1980s. Lester R. Brown's book "Building a Sustainable Society" was published in 1981. It summarized views suggesting that it would be possible to harmonize the material needs of society, the growth of the population and the rational utilization of the natural resources so that environmental pollution would also be minimized. This new economic development model is called 'sustainable development'. In the following years scientists have increasingly dealt with the various aspects of sustainable development.

In December 1983, the UN General Assembly established the World Commission on Environment and Development, also known as the Brundtland Commission, after the name of the Norwegian politician Mrs. Gro Harlem Brundtland, president of the commission.

In February 1987, the Commission reported on its work in the book *Our Common Future* which the UN General Assembly discussed later that year and expressed its general agreement to it. The report focussed on the concept of sustainable development. It did not suggest, as did the report of the Club of Rome *Limits to Growth*, that economic growth should be restrained. Rather, *Our Common Future* concluded that people in the Third World would be able to satisfy and surpass their basic needs if economic development continued in their countries. With restrained development, the Commission argued, poverty would be conserved on a world scale in the interest of environmental protection and such a situation would be unjust and unacceptable.

The concept of sustainable development, as presented in the Brundtland Commission's report, integrated economic and environmental policies, so that in cases of conflicts between the two, ecological interests are given preference.

Sustainable development will require a modification in the lifestyle of developed countries and should include a voluntary decrease of wasteful consumption, the development of a new value system and a new relation to nature.

The technological aspects of sustainable development include radical decreases in energy and material use, and increases of energy efficiency, recycling, and the production of environmentally sound products.

According to the definition of the Brundtland Report (*Our Common Future*), "Sustainable development is a development that meets the needs of the present without compromising the ability of future generations to meet their own needs".

According to the definition formulated by IUCN, UNEP, WWF, sustainable development is "improving the quality of human life while living within the carrying capacity of supporting ecosystems".

Sustainable nutrition security was defined by the famous agricultural scientist, M.S. Swaminathon as "providing physical and economic access to balanced diets and safe drinking water to people at all times".

The US National Research Council identified in 1989 the principal goals of sustainable agriculture. These are:

- incorporating natural processes, such as nutrient cycles, nitrogen fixation and pest-predator relationship into the agricultural production process;
- reducing the use of off-farm inputs, which have a harmful effect on the environment or on the health of farmers and consumers;
- making greater use of the biological and genetic potential of plant and animal species;
- improving the correspondence between cropping patterns and the productive potential and physical limitations of agricultural lands to ensure long-term susceptibility of current production levels;
- emphasizing improved farm management and conservation of soil, water, energy and biological resources.

Strategy on food security

The sustainability of agriculture in any country largely depends on the given country's strategy on food security. This strategy, in turn, is determined in each country by such factors as the area of arable land per capita, the fertility of soils, the length of the vegetation period, water reserves, level of the country's economic development, demand on agricultural export, magnitude of funds available for agricultural import, the purchasing power of the population, and the like. That is, the strategy on food security is determined by natural and social factors alike.

Essentially, three types of strategies can be discerned:

- no attempt is made at self-sufficiency in food; the missing produces or products are acquired by imports;
- self-sufficiency is the basic goal to be attained;
- over and above self-sufficiency, a considerable amount of agricultural export is also needed to keep the balance of the economy.

These strategy types are, of course, simplified ones, each permitting several variations. In one country, for example, self-sufficiency may be attained on the model of low-input agriculture, because large land resources are available and the climate is also advantageous, moreover, water resources are sufficient to meet the needs. In another country, in turn, self-sufficiency can only be realized on the model of high-input agriculture, because the limited natural resources would offer no other chance to achieve the strategic goal.

Unfortunately, there is no global strategy on food security, which would co-ordinate and optimize the national strategies. It would be much easier to realize sustainable agriculture on the global level if the national strategies were deduced from a global one. However, such a global strategy is not in sight yet, though some new approaches and regional co-operations are taking shape. Such are, for example, the agricultural policy of the European Union and the free trade area of the North-American continent (including the United States, Canada and Mexico).

Consequently, the national programme of sustainable agriculture is greatly influenced by the national strategy on food security. Sharp contradictions may arise when one-sided demands for revenues in hard currency lead to such agricultural production which excludes the attainment of sustainability. In such cases, the exaggerated export of agricultural products as a strategic goal has to be changed.

European programme towards sustainability

In 1992 the Commission of the European Communities (now already called European Union) approved the Programme of Policy and Action in relation to the Environment and Sustainable Development called "Towards Sustainability".

The document stated: changes in farming practices in many regions of the Community have led to overexploitation and degradation of the natural resources on which agriculture itself ultimately depends: soil, water and air.

This approach is clearly reflected in recent deliberations on the reform of the Common Agricultural Policy which, inter alia, have indicated that "producing more" cannot be any longer considered as the central point of Community policy.

The following targets are proposed in the Programme of the European Union up to 2000:

- standstill or reduction of nitrate levels in groundwaters;
- reduced incidence of surface waters with a nitrate content exceeding 50 mg/l, or giving rise to eutrophication of lakes and seas;
- stabilization or increase of organic material levels in the soil;
- significant reduction of pesticide use per unit of land under production and conversion of farmers

to methods of integrated pest control, at least in all areas of importance for nature conservation;

– increase of forest plantation, also on agricultural land.

It is important to state that the European Union–when setting the long-term goals of sustainable agriculture in its agricultural policy–emphasized the decrease of environmental damage (nitrification, eutrophication) arising from the excessive use of chemical fertilizers, and not the decrease of the amount of chemical fertilizers to be used. There is, namely, a marked difference between the two. Most probably, the amounts to be used should also be decreased in many places to lessen the damages, but further steps must as well be taken in respect of the use of chemicals, soil cultivation, soil conservation, plant rotation, land use, and so on.

A short history

In his book entitled *Sustainable Agricultural Systems* (edited by Clive A. Edwards *et al.*, published by US Soil and Water Conservation Society in 1990), Richard R. Harward summarizes the history of sustainable agriculture.

The *model of industrial agriculture* was taking shape gradually, in line with the technological development of the 20th century. This model included such things as intensive mechanization, new crop varieties, and new chemicals (mineral fertilizers, pesticides). Developed in parallel with this was the *model of alternative agriculture*, which aimed at a more efficient agricultural production by other ways and means, namely by adopting methods standing nearer to nature. It was within this mainstream that the *theory of biodynamic agriculture* was conceived in the 1930s. It included the stimulation and regulation of complex life processes by biodynamic preparations for soils, plants and manures. It also included the consideration of the effect of cosmic and terrestrial forces on biological organisms.

The *conception of biological agriculture*, which was scientifically more exacting than biodynamic agriculture, meant a further step forward in this evolutionary process. *Humus farming* reached its peak in the early 1950s. The principles of composting and compost use were well articulated by this time, and considerable research has since been done on the handling of municipal waste, with emphasis on methodologies.

In the 1960s, the *industrial model* was widespread in the developed countries. Chemical fertilizers and pesticides were comparatively inexpensive. Heavy-duty tractors also became widespread. Capital for investment was readily available, perhaps generated in large part by undervalued energy costs.

It was also in the same period that the *Green Revolution* took shape in the developing countries. The first results were really spectacular: many countries of the Third World succeeded in attaining self-sufficiency in food. Later, however, unfavourable effects, both social and natural, emerged: the problems of equity, of rural income, of product diversity, of environmental impact and of huge neglected areas of poor soil and water resources.

During the 1970s, the adverse effects of high-input agriculture were already felt in both the advanced and the developing countries. The increasing awareness of the impact of modern (industrial) technologies on the environment became clear as the pesticides in the food chains were traced. Crop nutrients began to accumulate in streams and in underground aquifers. The energy 'shortage' of the early 1970s was an event that shook the consciousness of scientists. For the first time it became painfully aware that earth resources were limited. It was also at this time that the conception of *sustainable agriculture* began to take shape. The principles underlying this conception can be summed up in the following three points:

– the interrelatedness of all parts of the farming system, including the farmer and his family;

– the importance of many biological balances in the system;

– the need to maximize desired biological relationships in the system and to minimize the use of materials and practices that disrupt those relationships.

Definitions are disputed by many experts; however, it seems that consensus is possible on three major points:

– agriculture must be increasingly productive and efficient in resource use;

– biological processes within agricultural systems must be much more controlled from within (rather than by external inputs of pesticides);

– nutrient cycles within farms must be much more closed.

The use of fertilizers

In the course of debates, opposers to the use of chemicals often fail to make a distinction between the individual types of chemical fertilizers, emphasizing only their adverse effects in a generalized way. Obviously, this is not correct, and the individual kinds of chemical fertilizers must be considered separately.

In the case of *nitrogen*, leaching causes environmental pollution, as it increases the nitrate content of groundwaters. While this effect is really a widespread one, it should also be emphasized that the nitrate content of underground-waters does not exclusively derive from chemical fertilizers. Settlements without proper drainage and animal farms contribute at least as much to pollution as the nitrogenous fertilizers.

Hence it follows that it would be desirable to apply the nitrogenous fertilizers in such an amount which is adjusted to the actual need of the given plant. Unfortunately, generally a larger amount of nitrogen must be applied than would be strictly needed for the planned yield and quality, because even with the utmost precaution, part of the nitrogen will be washed out. It is, therefore, essential to determine the specific nitrogen need of the individual species of cultivated plants, on the one hand, and to develop the method of applying the substance, on the other.

Indeed, it may be the case that a high nitrogen concentration may occur in the plant, which would cause disease in the case of human consumption. A widely known example for this is the decline in the oxygen conveying capacity of the blood under the effect of nitrates, after reduction to nitrites, which get into the organism of infants. This may be a rather frequent occurrence in the case of consuming certain vegetables (sorrel, spinach). This kind of concentration may be eliminated by applying organic manures and by decreasing the ratio of nitrogenous chemical fertilizers.

In the case of *phosphorus*, an oversupply leads to its fixation in the soil. This is, however, an economic loss and not an environmental damage. In later years, this fixed phosphorus might again become absorbable, so fixation is only a temporary loss. Phosphorus fixed in the soil may get into rivers and lakes through erosion. With this, it can contribute to the process of eutrophication, which has been noticed in several places. Thus, in such cases the reduction of erosion is the primary task.

The Scientific Committee on Problems of the Environment of ICSU published a summary of a survey on the global issues of phosphates in 1994. (*Phosphates – Global Issues. Problems, Options and Opportunities.* Edited by H. Tiessen.)

I am quoting here a passage of this report:

"A critical issue in the management of agricultural lands which contribute to a diffuse input of P to surface waters, is the extreme sensitivity of waters to increased P loads. Quantities of P sufficient to cause significant eutrophication in waters are almost invariably much too small to be measurable in a P balance of agricultural lands, and would be entirely lost in the natural variability of measurable P contents of a soil. It is therefore impossible to predict the impact on waters of land-based P balances, beyond the very general statement that increased P applications and P losses are a potential cause for water pollution. On the other hand, water bodies are very sensitive indicators for P pollution. River transport of P and the trophic status of lakes is an integrated measure of P use and misuse on land. River transport of P may therefore be a diagnostic for monitoring land use practices in the watershed, and it provides an indicator for the need to improve P management on land."

Potassium has not yet raised anxieties about pollution. According to widely held views, it is not likely that the potassium level prescribed by the directives for the quality of drinking-water would significantly rise owing to potassium of agricultural origin, because the potassium fixing capacity of soils is very high.

According to figures published in *Potassium in Ecosystems* (Proceedings of the 23rd Colloquium of the International Potash Institute, held in Prague, 1992), the minimum need for man has been established as 2000 mg per day in order to maintain normal body reserves and a normal potassium concentration in the plasma. Additional potassium intake does not seem to present any disadvantages. The amounts which are not required are eliminated mainly by the kidneys, but also by gastro-intestinal secretion.

Organic manures

Scientific literature on the models of sustainable agriculture unequivocally underline the necessity of returning the largest possible part of organic matters of agricultural origin into the soil. It is one of the important principles that nutrient cycles within the farm must be much more closed.

Similar is the situation on the global level too; namely, the largest part of the biomass of plant ori-

Table 1. Yearly primary production

Dry material	10^{15} gr/year
Total primary production on the land area of the Earth	
117.5	= 100 per cent
of which:	
Plant cultivation	
9.1	= 7.7 per cent
Grassland	
20.0	= 17.0 per cent
Forests	
79.9	= 68.0 per cent

Source: The Global Carbon Cycle, 1979.

gin gets back into the soil. The related data are shown in Table 1.

If the yearly total primary production on the land area of the earth is taken as 100, then not more than 7.7% of the total originates from plant cultivation using chemical fertilizers. Grasslands account for 17 and forests for 68%. (Accounting for the remaining 7.3% are mainly ecosystems of waters throughout the mainlands.) Thus, a round 92% of the total primary production remains on the spot and the nutrients get back into the natural cycle.

In the case of agriculture, however, part of the nutrients will always be removed, as a considerable part of the produces is sold or transported from the farm. On this account, chemical fertilizers have to be used to make up for this loss.

The nitrogen fixing ability of *papilionaceae*, i.e. the biological fixation of nitrogen, can also be used for reducing the application of nitrogen produced by chemical processes. It seems possible too that the new biotechnological procedures related to gene transfer will lead to a breakthrough in this field in one or two decades, permitting the biological nitrogen fixation of non-legume plants. At present, however, this is only a theoretical possibility, and it would be illusory to expect it to become a reality before long.

The large-scale growing of papilionaceae is generally limited by the extent of the available arable land. Therefore, it would also be an illusion to think that the 'biological nitrogen' could virtually replace the use of 'chemical nitrogen'. Consequently, biological nitrogen is, for the time being, only a supplementary source, but its utilization is very important even so.

Conclusions

To realize sustainable agriculture, it is necessary to give equal attention to the natural, economic and technological factors, so that they might be treated as a large complex system.

Particularly important actions to be taken are:

– to maintain soil organic matter and crop diversity, to practice rotations and nutrient cycling;

– to minimize soil loss, reduce energy use, develop better technologies for biological nitrogen fixation, to maintain existing genetic diversity;

– to protect groundwater from contamination, reduce or eliminate the use of pesticides, to reduce the use of fertilizers, and to improve the methods of their application;

– to provide food of acceptable quality and diversity, and to develop market channels that respond to market and social needs.

References

Our Common Future. Report of the World Commission on Environment and Development. Oxford University Press, 1987

UN Conference on Environment and Development. AGENDA–21. Rio de Janeiro, 1992

World Resources. Oxford University Press, 1988, 1990, 1992, 1994

Sustainable Agricultural Systems. Edwards, CA *et al.* (eds) Soil and Water Conservation Society, 1990

Phosphates – Global Issues. Problems, Options and Opportunities. Tiessen, H (ed) SCOPE, 1994

Potassium in Ecosystems. Proceedings of the 23rd Colloquium of the International Potash Institute held in Prague (Czechoslovakia), 1992

C. Rodriguez-Barrueco (ed.), Fertilizers and Environment, 89–93.

Effect of different rates of N-fertilizers on nodulation, nodule activities and growth of two field grown cvs. of soybean

A.M. Abdel Wahab * & M.H. Abd-Alla
Department of Botany, Faculty of Science, Assiut University, 71516, Assiut, Egypt

Key words: Nodulation, N-fertilizers, N_2-fixation, nitrate reductase, soybean

Abstract

The effect of N fertilizers on nodulation, nitrogenase, nitrate reductase activities and growth of two cultivars of soybean, Clark and Crauford was evaluated in a field experiment. KNO_3 or NH_4Cl were applied 27 days after planting at 0,16, 32, 64 and 128 kg N/ha. Nodulation and growth of both cultivars significantly increased when N was applied at low levels whereas specific N_2-ase activity (SNA) slightly and insignificantly increased. Cv Crauford showed a greater positive response than cv. Clark. Higher rates of KNO_3 and NH_4Cl (128 kg N/ha) significantly depressed nodulation and SNA but slightly decreased the plant dry matter. Cv. Crauford was more tolerant to N fertilizers than cv. Clark. The decline in SNA was ascribed to increased nitrate reductase activity (NRA) and higher accumulation of nitrites in nodule cytosol. NRA and nitrate contents in nodules of cv. Clark were greater than that in cv. Crauford. Results showed that NH_4^+ is the preferred N source with occasional increases in nodule number and weight. This study provides an evidence for the nodulation and growth variability of soybean cultivars fertilized with different levels of N. The results also suggest that diminishing NRA could contribute to increased N_2 fixation and the interaction between NO_3^- assimilation and N_2 fixation is strongly dependent on the plant cultivar.

Introduction

Recent literature on environmental problems related to agriculture reflects international concern on intensive use of N fertilizers to increase food production. There are increases of NO_3^- levels in ground and surface water resources, creating health hazards to humans and animals through NO_3^- enrichment.

One of the widely cultivated grain legumes is soybean. It is considered to have an average potential of symbiotic N_2 fixation compared to other legumes (Hardarson, 1993). A limited initial fixation and substantial N_2 fixation occurred only some weeks after planting soybean, which was just maintained for short periods (Zapata *et al.*, 1987). Field studies show that N fertilization is needed to achieve substantial yield, indicating that the symbiotic N_2 fixation is unable to provide enough N for maximum yield of soybean (Deibert *et al.*, 1979; Buttery and Dirks, 1987). The high variability of fixation among soybean cultivars and the need to apply N fertilizers to obtain high yields are important reasons to study the fixation potential of soybeans. Accordingly, cultivars which maintain high rates of N_2 fixation in fertile N rich soils can be identified (Senaratne *et al.*, 1987). Selection of cultivars supporting a symbiotic system with a high tolerance against N fertilizers would be thus desirable. On the other hand, fertilizer rates exceeding those exerting a "starter-N" effect, generally reduce nodulation and N_2 fixation (Afza *et al.*, 1987).

In the present study two cultivars of *Glycine max* L.; Clark and Crauford were compared for their symbiotic performance aiming at the selection of cultivars with superior nodule activity in response to different levels of mineral nitrogen applied 27 days after planting.

Materials and methods

The experiment was carried out at the farm of the University. The soil had the following characteristics: pH 7.5; 1.68% organic matter; 0.014% NO_3^-; 0.8% NH_4^+ and clayey in texture.

Cultivars Clark and Crauford, were used. Seeds were pelleted with a peat-based mixed inoculum of

* Author for correspondence

Table 1. Nodule number and mass (mg/plant) of two soybean cultivars fertilized with different levels of combined nitrogen applied at 27 days after planting.[a] Nodule mass values in parentheses

Treatment (kg N/ha) cv.	Days after treatments							
	14		28		42		56	
	Cl	Cr	Cl	Cr	Cl	Cr	Cl	Cr
0 (control)	16 (170)	25 (260)	28 (320)	46 (560)	48 (530)	71 (1040)	47 (580)	75 (870)
KNO_3								
16	24 (220)	39 (330)	34 (360)	59 (670)	55 (670)	84 (1310)	50 (650)	85 (960)
32	21 (210)	29 (300)	30 (350)	50 (630)	51 (630)	82 (1240)	49 (620)	81 (960)
64	10 (100)	25 (280)	15 (170)	48 (600)	28 (270)	76 (1140)	26 (290)	73 (280)
128	9 (80)	20 (210)	11 (140)	39 (480)	25 (210)	52 (860)	24 (240)	50 (650)
NH_4Cl								
16	26 (240)	34 (370)	37 (400)	56 (700)	57 (690)	87 (1350)	52 (655)	86 (900)
32	25 (260)	32 (330)	32 (380)	53 (650)	54 (640)	84 (1260)	48 (630)	82 (890)
64	13 (120)	28 (300)	18 (190)	50 (620)	31 (280)	79 (110)	27 (292)	7 (800)
128	18 (90)	23 (230)	14 (150)	40 (500)	27 (240)	53 (920)	24 (250)	49 (680)
L.S.D.								
P = 5%	5 (15)	4 (53)	4 (30)	6 (68)	6 (23)	12 (110)	5 (28)	7 (90)

[a]Each value represents the mean of three replicates. cv. Cl = Clark; Cr = Crauford

Bradyrhizobium japonicum strains (RCR 3407 and 3442). Seeds were drilled in holes 4 cm deep and 20 cm apart on two sides of the soil ridges. The temperature ranged between 38/24 °C (day/night).

The experimental design was split-plot with three replications, main plots were two cultivars and subplots were treatments of two N sources (KNO_3 and NH_4Cl) applied at rates of 0, 16, 32, 64 and 128 kg N/ha 27 days after planting. The whole experimental area was fertilized just before sowing with 40 kg superphosphate/ha. The plots were irrigated at 12- and 15-day intervals. Samples in replicates were taken at 14,28,42 and 56 days after treatments (DAT) (41, 55, 69 and 83 days after planting). Parameters estimated were nodule number, nodule fresh mass, shoot and root dry weight, N_2-ase activity, (C_2H_2 reduction, Hardy *et al.*, 1968), and *in vivo* nitrate reductase of whole nodules (Deroche and Babalar, 1987); nitrite content of nodule cytosol (Griess-Ilosvay method, Lalisse-Grundmann *et al.*, 1988). Analysis of variance was done by PC-state programme and least significant difference (LSD = 0.05) was used to test the significance between treatments.

Results and discussion

Nodulation

Although the mechanisms by which N_2 fixation is inhibited by combined N are not well understood, a number of hypotheses have been put forward for the direct and indirect effects on N fertilizer on crop species. Nodule number and nodule air-dry weight increased at all sampling dates when N was applied at low levels (16 and 32 kg N/ha, Table 1). Soybean cv. Crauford showed a greater response to low rates of combined N than cv. Clark. The 64 and 128 kg N/ha significantly lowered nodule population and mass of cv. Clark which continued throughout until the experiment terminated. However, 64 kg N had insignificant effect on nodulation of cv. Crauford, whereas 128 kg N significantly reduced nodule number and weight (Table 1). At the last harvest, higher levels of combined N (64 and 128 kg N/ha) reduced nodule number by 3 and 33% for cv. Crauford and 45% and 50% for cv. Clark, respectively.

Tables 1 and 2 indicate that NH_4^+ is the preferred N source, occasional increases in nodule number and mass being observed. This finding is confirmed by a number of studies with N_2-fixing crops and trees and suggest that some species prefer NH_4^+ (Goi *et al.*, 1992). Our data confirm the findings of several authors (Park and Buttery, 1989; Muller *et al.*, 1993).

Table 2. Nitrogenase (A) (μmol C_2H_4 $h^{-1}g^{-1}$ fr. wt. nodules), nitrate reductase (B) (μmol NO_2 $h^{-1}g^{-1}$ fr. wt. nodules), activities and nitrite content of nodule cytosol (C) (μg g^{-1} fr. wt. nodules) of two soybean cultivars fertilized with different levels of combined N applied at 27 days after planting[a]

Treatment	Days after treatments																							
(kg N/ha)	14						28						42						56					
cv.	Cl			Cr			Cl			Cr			Cl			Cr			Cl			Cr		
	A	B	C	A	B	C	A	B	C	A	B	C	A	B	C	A	B	C	A	B	C	A	B	C
0 (Control)	17.5	0.47	0.14	22.6	0.35	0.12	17.0	0.43	0.11	23.1	0.28	0.08	10.4	0.17	0.07	15.9	0.07	0.05	3.1	0.08	0.07	4.8	0.12	0.03
KNO_3																								
16	21.5	0.60	0.25	23.1	0.42	0.20	19.9	0.63	0.15	24.5	0.42	0.12	9.8	0.39	0.10	16.4	0.22	0.05	3.3	0.13	0.08	3.5	0.15	0.03
32	20.8	0.65	0.28	22.3	0.53	0.25	19.8	0.81	0.16	22.9	0.61	0.14	10.1	0.60	0.10	15.4	0.36	0.05	3.3	0.21	0.10	3.1	0.12	0.04
64	17.9	1.13	1.02	21.9	0.64	0.88	15.1	2.07	0.85	19.9	1.71	0.62	7.1	2.73	0.70	15.7	2.60	0.32	3.02	2.20	0.62	3.6	1.56	0.10
128	10.2	1.35	1.32	15.2	0.95	1.04	11.3	2.61	1.59	16.3	1.97	1.28	8.6	4.60	1.83	15.2	3.00	1.34	3.0	2.19	1.15	4.3	2.07	0.82
NH_4Cl																								
16	21.7	0.56	0.24	19.1	0.40	0.20	19.9	0.53	0.13	25.5	0.36	0.09	10.3	0.32	0.09	16.9	0.28	0.04	3.5	0.16	0.07	4.0	0.08	0.03
32	17.9	0.61	0.26	21.3	0.47	0.23	19.3	0.72	0.15	23.8	0.45	0.11	10.5	0.42	0.10	16.3	0.35	0.05	3.2	0.28	0.07	3.7	0.12	0.05
64	16.3	1.02	0.91	20.7	0.58	0.84	14.7	1.89	0.79	19.9	1.66	0.57	10.2	3.20	0.61	14.2	2.33	0.23	3.2	2.60	0.44	3.8	1.37	0.07
128	9.7	1.21	1.18	14.8	0.86	1.06	11.9	2.33	1.28	17.2	1.85	1.18	9.6	4.80	1.44	13.2	2.90	1.18	3.4	2.30	1.08	3.3	1.55	0.63
L.S.D. P=5%	3.9	0.21	0.09	4.5	0.12	0.11	1.54	0.17	0.10	3.7	0.15	0.18	N.S	0.23	0.14	N.S	0.41	0.19	N.S	0.26	0.16	N.S	0.32	0.18

[a]Each value represents the mean of three replicates; N.S.= Not significant.

Low and moderate levels of N fertilizer applied as a "starter" at early growth stages stimulate the formation and development of nodules once seedlings consumed the reserve N of cotyledons and if active N_2 fixation is delayed (Sprent and Thomas, 1984).

Nodule activity

Low levels of N fertilizer (16 and 32 Kg N/ha) slightly elevated SNA up to 42 DAT. However, high rates of NO_3^- or NH_4^+ significantly reduced SNA of both cultivars. A reduction of 41, 33 and 20% of activity was caused by 128 kg N for cv. Clark at sampling dates of 14, 28 and 42 DAT, whereas the inhibition was 32, 29 and only 4% of cv. Crauford SNA at the same sampling dates. The decrease in nodule activity could be ascribed to nodule senescence and destruction of bacteroids induced by combined N (Abdel Wahab and Abd-Alla, 1988). Thus, despite of the large number of nodules at 42 and 56 DAT (Table 1). SNA was almost suppressed. This is confirmed by the results of Novak *et al.* (1993) that suppression of nodule activity is independent of nodule growth. The depression in SNA may also have resulted from increased NRA and high accumulation of nitrite in nodule cytosol (Table 2).

The results indicate that soybean cv. Clark had fewer nodules and mass whereas cv. Crauford was more tolerant with pronounced nodulation and higher N_2-fixing activity when subjected to relatively higher levels of N fertilizers. Absolute N_2-ase activity (per plant) as calculated from values in Tables 1 and 2 (by multiplying total nodule fresh mass by SNA, not shown in the text) also suggests the greater N_2-fixing potential of cv. Crauford.

Some studies with soybean have indicated that the sensitivity of N_2-fixation to combined N also depends on rhizobial strains (Senaratne *et al.*, 1987). However, results presented here are supported by the conclusions of Gibson and Harper (1985), and Muller *et al.* (1993), that judicious use of mineral nitrogen to improve nitrogen fixation by legumes would be most profitable towards host cultivars rather than rhizobial strains.

Results of Table 2 closely agree with those of Nelson and Edie (1988) who concluded that pea cv. Juneau A 317 and A 334 differ in their ability to assimilate mineral nitrogen (NO_3^-) according to nitrate reductase activity in each cultivar. Thus pea cv. Juneau A 317 grew very poorly with NO_3^- as the sole N source compared to cv. A 334 (Vigue and Warner, 1987). Fur-

Table 3. Plant dry weight (g/plant) of two cultivars of soybean fertilized with diferent levels of combined N 27 days after planting[a]

Treatment (kg N/ha) cv.	Days after treatments							
	14		28		42		56	
	Cl	Cr	Cl	Cr	Cl	Cr	Cl	Cr
0 (Control)	3.2	4.2	6.3	7.9	11.5	13.9	16.3	18.7
KNO_3								
16	4.4	5.5	8.18	10.6	14.3	16.6	20.6	18.7
32	4.6	6.0	9.30	11.6	14.9	18.2	19.6	25.9
64	3.3	5.0	6.60	9.8	11.1	14.5	15.3	21.7
128	3.2	3.8	5.10	6.9	8.3	11.2	12.9	15.3
NH_4Cl								
16	4.7	5.9	8.40	11.3	15.9	17.9	21.6	25.8
32	5.0	6.6	8.90	12.3	16.3	20.1	21.7	26.9
64	3.5	5.1	5.90	9.3	11.8	15.6	16.2	22.2
128	3.3	4.4	5.40	6.7	9.3	12.1	13.6	16.7
L.S.D.								
P=5%	1.1	1.2	2.10	1.8	1.9	2.3	2.4	3.3

[a] Each value represents the mean of three replicates

ther studies by Taylor *et al.* (1988) indicated greater accumulation of NO_2^- by cv. Juneau A 317 than cv. Juneau WT nodules. Eaglesham *et al.* (1983) reported that the relationship between NO_3^--reduction and N_2-fixation is important to maximize the use of both N sources (N_2 and NO_3^-) by nodulated legumes. In addition, Lluch *et al.* (1988) found that diminishing NRA could contribute to increased N_2 fixation and the interaction between NO_3^- assimilation and N_2 fixation is strongly dependent on the plant cultivar.

Plant growth

Dry matter of both cultivars increased in the presence of N fertilizer up to 64 Kg N/ha. However, cv. Crauford showed a more positive response than cv. Clark at all harvesting stages. Data of Table 3 indicate that dry matter accumulation slightly increased in plants fertilized with NH_4^+ rather than NO_3^-. Hine and Sprent (1988) similarly showed that *Vicia faba* and *Phaseolus vulgaris* grown on urea or ammonium salts exhibited greater weights than on NO_3^-. Similar observations have been also reported (Rys and Phung, 1984). The higher yield and greater growth rates were suggested to result from a reduced energy requirement in protein synthesis and increased photosynthetic capacity when using NH_4^+ instead NO_3^-. Higher rates of combined N resulted in decrease of plant dry weight. This is partly due to the faster respiration rates of plants, which presumably reflect the very high energy cost (Day *et al.*, 1986) or may be attributed to a possible release of compensatory amounts of H^+ ions associated with NH_4^+ uptake which might lead to acidification of soil solution to a point that limits somewhat plant growth or to NH_4^+ toxicity (Haynes and Goh, 1978).

Results of the present study suggest that an approach for achieving maximum growth and greater potentiality of N_2 fixation and fertilizer N assimilation by soybean is to select cultivars which produce symbioses less affected by external combined N or which recover more rapidly from such effects.

References

Abdel Wahab AM and Abd-Alla MH (1988) Effect of combined nitrogen on the structure of N_2-fixing nodules in two legumes. In: Bothe H, de Bruijn FJ and Newton WE (eds) Nitrogen Fixation: Hundred Years After, p 535. Gustav Fischer, Stuttgart/New York

Afza R, Hardarson G and Zapata F (1987) Effect of delayed soil and foliar N fertilization on yield and N_2 fixation of soybean. Plant and Soil 97: 361–368

Buttery BR and Dirks VA (1987) The effect of soybean cultivar, Rhizobium strain and nitrate on plant growth, nodule mass and acetylene reduction rate. Plant and Soil 98: 285–293

Day DA, Lambers H, Bateman J, Carrol BJ and Gresshoff PMA (1986) Growth comparisons of a supernodulating soybean (*Glycine max*) mutants and its wild-type parent. Physiol Plant 68: 375–382

Deibert EJ, Bigeriego M and Olson RA (1979) Utilization of ^{15}N fertilizer by nodulating and non-nodulating soybean isolines. Agron J 71: 717–723

Deroche ME and Babalar M (1987) *In vivo* nitrate reductase activities in different organs of lucerne (*Medicago sativa*) plants: Effect of combined nitrogen during first vegetation growth and after shoot harvest. Physiol Plant 70: 90–98

Eaglesham ARJ, Hassouna S and Seeger R (1983) Fertilizer-N effects on N_2 fixation by cowpea and soybean. Agron J 75: 61–66

Gibson AH and Harper JE (1985) Nitrate effect on nodulation of soybean *Bradyrhizobium* japonicum. Crop Sci 25: 497–501

Goi SR, Sprent JI, James EK and Jacob-Neto J (1992) Influence of nitrogen form and concentration on the nitrogen fixation of *Acacia auriculiformis*. Symbiosis 14: 115–122

Hardarson G (1993) Methods for enhancing symbiotic nitrogen fixation. Plant and Soil 152: 1–17

Hardy RWF, Holsten RD, Jackson EK and Burns RC (1968) The acetylene-ethylene assay for N_2 fixation: Laboratory and field evaluation. Plant Physiol 43: 1185–1207

Haynes EW and Goh KM (1978) Ammonium and nitrate nutrition of plants. Biological Rev 53: 465–510

Hine JC and Sprent JI (1988) Growth of *Phaseolus vulgaris* on various nitrogen sources. The importance of urea. J Exp Bot 67: 779–784

Lalisse-Grundmann G, Brunel B and Chalamet A (1988) Denitrification in a cultivated soil: optimal glucose and nitrate concentrations. Soil Biol Biochem 20: 839–844

Lluch C, Caba JM, Hervas A and Ligero F (1988) Effect of nitrate and cultivar genotype on nitrogen metabolism in nodules of *Vicia-faba* minor 1. In: Bothe H, de Bruijn FJ and Newton WE (eds) Nitrogen Fixation: Hundred Years After, p 572. Gustav Fischer, Stuttgart/New York

Muller S, Pereira PAA and Martin P (1993) Effect of different levels of mineral nitrogen on nodulation and N_2 fixation of two cultivars of common bean (*Phaseolus vulgaris* L.) Plant and Soil 152: 139–143

Nelson LM and Edie SA (1988) Effect of nitrate on nitrogen fixation and nodule carbohydrate and organic acid concentrations in pea mutants deficient in nitrate reductase. Physiol Plant 73: 534–540

Novak K, Skrdleta V, Nemcova M and Lisa L (1993) Behaviour of pea nodulation mutants as affected by increasing nitrate level. Symbiosis 15: 145–206

Rys GJ and Phung T (1984) Effect of nitrogen form and counterion on establishment of *Rhizobium trifolii-Trifolium repens* symbiosis. J Exp Bot 35: 1811–1819

Senaratne R, Amornpimol C and Hardarson G (1987) Effect of combined nitrogen on nitrogen fixation of soybean (*Glycine max* L. Merrill) as affected by cultivar and rhizobial strain. Plant and Soil 12: 45–50

Sprent JT and Thomas RJ (1984) Nitrogen nutrition of seedling grain legumes: Some taxonomic, morphological and physiological characterization. Plant Cell and Environ 1: 637–645

Taylor DC, Shelp BJ, Nelson LM and Grodzinski B (1988) Carbon and nitrogen partitioning in young nodulated pea (Wild type) and nitrate reductase deficient mutants plants exposed to NH_4NO_3. Physiol Plant 74: 593–601

Vigue JT and Warner RL (1987) Nitrate utilization and dinitrogen fixation (acetylene reduction) by nitrate reductase-deficient mutants of pea. Crop Sci 27: 548–552

Zapata F, Danso SKA, Hardarson G and Fried M (1987) Time course of nitrogen fixation in field grown soybean using nitrogen -15 methodology. Agron J 79: 172–176

C. Rodriguez-Barrueco (ed.), Fertilizers and Environment, 95–100.

Impact of vesicular arbuscular mycorrhizal fungi and *Rhizobium* on the growth and P, N and Fe uptake by faba-bean

I.A. El-Ghandour[1], M.A.O. El-Sharawy[2] & E.M. Abdel-Moniem[2]
[1]*Soil and Water Department, Soil Microbiology Unit, Atomic Energy Authority, Cairo, Egypt;* [2]*Soil Science Department, Faculty of Agriculture, Ain Shams University, Shobra El-Khima, Egypt*

Key words: Phosphorus, nitrogen, iron, VAM, Rhizobium, faba-bean

Abstract

The effect of rhizobial inoculation, vesicular arbuscular mycorrhizal (VAM) fungi on the growth, P, N and Fe uptake by faba-bean plants (*Vicia faba* L.V. Giza 2) grown in virgin sandy soil, treated with super or rock-P were studied under green-house conditions. The earthern pots received a half of the recommended rate of P either as single super phosphate or rock-P in the rate of 20 mg P/kg soil, and calcium ammonium nitrate was added in the rate of 10 mg N/kg soil. Iron was applied in two levels 0 and 5 mg Fe/kg soil, in the form of iron sulphate ($FeSO_4 \cdot 7H_2O$). Dry matter yield, as well as P, N and Fe-uptake were determined. Nodule numbers and their dry weights, spore numbers and mycorrhizal root infection were determined. Results indicated that, fungal infection and rhizobial inoculation either alone or in combination increased dry matter yield as compared to uninoculated plants, whereas the percentages increase in dry matter were 34, 26 and 57% in case of super-P application, while they were 56, 47 and 89% in case of rock-P when inoculated with *Rhizobium*, VAM and dual inoculation respectively. Also P, N and Fe uptake were significantly increased due to inoculation, and dual inoculation resulted the highest effect. Generally, inoculation with *Rhizobium* and/or mycorrhizae can remove the deficient effect of P and Fe on N_2-fixation and plant growth in the soil of low nutrients content.

Introduction

Faba-bean (*Vicia faba* L.) is the most important grain legume in Egypt. It is produced on more than 50% of the total area under food legume, including soybean. It serves as an important source of protein for humans, especially for those with low income. Productivity of such crops is known to be poor in sandy soil mainly due to the scarcity of macro and micronutrients, soil organic matter and absence of the root-nodule bacteria (Hallsworth, 1972). In arid and semiarid regions the future expansion of agriculture depends on the reclamation of the sandy soil. Improvement of faba-bean production in such soil using the minimum of fertilizer input, requires alternative resources for nutritional elements such as nitrogen fixers as well as mycorrhizal fungi, that can affect phosphorus and other micronutrients availability. Certain limitations are inherent in soil properties which may reduce the possibility of yield increase resulting from a single rhizobial treatment. Among these limitations is nutrients deficiency both for plant and rhizobia. Mycorrhizal associations have been shown to be beneficial for plant growth and nutrients absorption particularly where nutrients, are limiting factor for plant growth (El-Ghandour, 1992). Nitrogen fixation depends on the balanced nutrition of leguminous plants which have high P, Mo, Zn and Fe requirements, and VA mycorrhizae can satisfy this demand, particularly on P-deficient soil (Suman Bala and Singh, 1985; Kucey and Janzen, 1987).

The present work was conducted to study the effect of VAM and rhizobial inoculation on growth as well as P, N and Fe uptake by faba-bean plants grown in sandy soil, and also their contribution to N_2-fixation by plant.

Material and methods

Fertilization treatments

A factorial greenhouse experiment was conducted in 20 cm diameter plastic pots filled with 5 kg of virgin sandy soil taken from Atomic Energy Authority, Inshas area (Sharkeia Governorate). The soil has 90.5% sand, 4.7% silt and 4.8% clay with pH 7.9, organic matter 0.07%, total N 4.0 ppm, P-Olsen 4.0 ppm and available Fe 4.3 ppm.

Each pot was filled with 5 kg of soil. The pots were divided into two main groups, one of which has received P in the form of single superphosphate (7.2% P) at rate 20 mg P/kg soil; the other received a fine rock-phosphate (11.7% P) at the same rate, every main group was treated with two levels of iron separately, 0 and 5 mg Fe/kg Soil in the form of ($FeSO_4{\cdot}7H_2O$). All pots received 10 mg N/kg soil added as calcium ammonium nitrate.

Fungal and bacterial culture

Spores were extracted from an alluvial soil by means of the wet sieving technique described by Gerdemann and Nicolson (1963). The spores suspension were recorded using × 35 dissecting microscope. Approximately 75 spores was placed 6–8 cm below the seeds to ensure that all growing roots passed through the inoculum layer.

An effective strain of *Rhizobium leguminosarum* biovarviceae ARC 201 was used. The number of rhizobial cells per gram inoculant ranged from 30–35 × 10^8. Gum arabic was used as sticker to ensure viable rhizobia per seed, before sowing. Microbial treatments included: uninoculated (control), rhizobia, VAM and rhizobia + VAM.

Four seeds/pot were sown, then thinned to two seedlings after growing. Each treatment was replicated three times.

Sampling

Plant samples were taken at the time of preflowering (65 days). Plants were uprooted along with soil core containing the entire root system. The roots were washed with water to eliminate soil particles. Nodules were counted and the average number/plant were recorded. Nodules were detected from the root system, then oven dry weight was recorded. A sub-sample of washed roots was cleared in 10% KOH and fungus stained with trypanblue (Philips and Hayman, 1970), percentage mycorrhizal infection was determined. Numbers of VA mycorrhizal spores in the soil surrounding the roots were determined according to (Gerdemann and Nicolson, 1963). Shoots and root part of faba-bean were air dried, then oven dried at 70 °C and dry weight values were recorded. Representative samples of shoots were ground for N analysis by microkjeldahl method. Phosphorus as well as Fe content in plant material were analyzed according to John (1970) and Black (1965). Iron was determined using atomic absorption spectrophotometer.

Analysis of variance for the final data was carried out and significant differences among the means of various treatment were distinguished by the L.S.D. according to (Snedecor and Cochran, 1980).

Results and discussions

Dry matter yield

Effect of inoculation with *Rhizobium* and/or mycorrhizae on shoot and root dry weights are presented in Table 1. Data indicated that inoculation with *Rhizobium* significantly increased the dry matter yield of aerial and ground parts of faba-bean plants, where the mean value were 32.5 and 16.5% for shoot and root in case of super-phosphate, while it was 35.3 and 7.2% when rock-P was applied.

Hamdi *et al.* (1966) and Abdel-Ghaffar (1988) reported that, inoculating faba-bean with *Rhizobium* increased the yield and the quality of grain compared to uninoculated treatments. Also data showed that single inoculation with mycorrhizae in the presence of P-sources increased dry weight of shoot faba-bean by about, 27.7 and 47.8% over the uninoculated when super and rock-P amended, respectively. Barea and Gonzalez (1986) reported that VAM enhance plant growth and N_2-fixation by legumes. Also Pacovsky *et al.* (1986) mentioned that VAM fungi is more effective in soils amended with sparingly soluble P. These results are argeement with those obtained by El-Ghandour (1992). The maximum yield was recorded with combined inoculation (*Rhizobium* + VAM) even in the presence of super or rock-phosphate. This may be due to synergistic beneficial effect between the two symbionts on plant growth (Harris *et al.*, 1985).

Concerning the effect of Fe-application, data show that increasing Fe-level lead to increase in different plant parts in the presence of P-sources, but the increase was not significant. Papastylianou (1993) observed that in peanut, the moderate chlorosis of the plants not treated with Fe did not significantly affect the final yield. On the other hand significant increase in dry matter yield with Fe-fertilization was observed when plants inoculated with *Rhizobium* and/or mycorrhizae under super or rock-P application. In this concern, increasing shoot

Table 1. Effect of inoculation with *Rhizobium* (RH) and/or mycorrhizae (VAM) on shoot and root dry weights (g) under P-sources and Fe-Levels

Treatments	Shoot			Root		
	−Fe	+Fe	Mean	−Fe	+Fe	Mean
Super-P						
Control	22.7	25.9	24.2	13.5	14.7	14.1
RH	30.5	34.5	32.5	15.3	17.6	16.5
VAM	29.4	32.3	30.9	14.9	16.1	15.5
RH+VAM	36.8	39.1	38.0	18.5	20.0	19.3
Mean	29.6	33.0			15.6	17.1
Rock-P						
Control	21.2	23.9	22.6	13.8	15.6	14.7
RH	33.8	36.8	35.3	15.6	18.8	17.2
VAM	31.8	34.9	33.4	14.9	17.0	16.0
RH+VAM	40.5	45.1	42.8	19.8	21.5	20.7
Mean	31.8	35.2			16.0	18.2
L.S.D. 0.05						
Fe-rates		2.53			1.23	
Inoculations		5.24			0.95	
P-sources		2.00			N.S.	
Fe × P × I		6.14			N.S.	

Table 2. Effect of inoculation with *Rhizobium* (RH) and/or mycorrhizae (VAM) on nodule number and dry weights (D.W.) under P-sources and Fe-levels

Treatments	Nodule No/pot			Nodule D.W. mg/pot		
	−Fe	+Fe	Mean	−Fe	+Fe	Mean
Super-P						
Control	13.3	16.7	15.0	66.7	80.5	73.6
RH	45.5	50.2	46.4	178.5	227.3	202.3
VAM	25.7	30.1	27.9	101.6	120.4	111.0
RH+VAM	49.9	55.7	52.8	242.0	278.7	260.4
Mean	32.9	38.2		147.2	176.7	
Rock-P						
Control	16.9	18.9	17.9	79.9	94.5	85.9
RH	50.3	56.3	53.3	251.5	275.8	263.7
VAM	31.6	35.4	33.5	154.8	177.0	165.9
RH+VAM	60.0	69.8	64.9	282.0	342.0	312.0
Mean	39.7	45.1		191.9	222.3	
L.S.D. 0.05						
Fe-rates		4.67			19.94	
Inoculations		8.71			23.82	
P-sources		6.85			24.22	
Fe × P × I		11.91			55.40	

dry weights were 33.2, 24.7 and 51% in case of super-P; 54, 46 and 88.7% when rock-P was applied for inoculation with *Rhizobium*, mycorrhizae and combined inoculation respectively. Baraka and Chen (1982) reported that Fe-deficiency may have effect on nodule initiation and effect on host plant growth.

Regardless inoculation and Fe treatments, shoots and roots dry weight was higher in case of rock than super-P whereas the mean of shoots were 33.5 when rock-P was applied while it was 31.4 in case of super phosphate. In spite of this, increases were not significant although superior performance for rock-P was shown. Chien *et al.* (1993) marked that dry matter yield of different soybean plant parts showed an increase in dry matter yield with different P sources as compared with check.

Spore numbers, mycorrhizal infection, nodule numbers and dry weights

Nodule numbers and nodule dry weights as affected by P-source are presented in Table 2a. When superphosphate was applied nodule No. and dry weight were 35.6 nodule/pot and 162.0 mg/pot respectively, while 42.4 nodule/pot and 207.1 mg/pot where found for the previous parameters in case of rock-P application. These findings are in agreement with results obtained by Saad (1990) and El-Ghandour (1992). Also, data show that increasing Fe-level lead to improvement effect on nodule No. and dry weights either with super or rock-P dressings. O'Hara *et al.* (1988a, b) mentioned that, Fe deficiency induced N deficiency due to inhibition of the development of nodules. Also, the effect of iron on nitrogen fixation is direct, where the symbiosis requirement for iron is greater than that of host plant growth (Robson, 1978, 1983).

Concerning inoculation, the data in the same table indicated that the, studied inoculants resulted in significant increase in nodule No. and dry weight. In this connection the effectiveness of the inoculants was of the following order: dual inoculation with *Rhizobium* + VAM > single inoculation, in the presence of P-sources and Fe-levels. The gained results were the same as those obtained by Pacovsky *et al.* (1986) and Mosse (1977). They explained that in alkaline soil and low P-deficient, rock-phosphate enhanced nodulation and nitrogen fixation of mycorrhizal plants. Also nodulated roots can enhance the production of root exudates which may affect the growth of soil microflora. Rhizo-

Table 3. Effect of inoculation with *Rhizobium* (RH) and/or mycorrhizae (VAM) on mycorrhizal root infection and spore no under P-sources and Fe-levels

Treatments	VAM infection%			Spore No/pot × 10^3		
	−Fe	+Fe	Mean	−Fe	+Fe	Mean
Super-P						
Control	18	20	19	5.1	10.5	7.8
RH	23	28	25.5	30.2	35.3	32.8
VAM	50	55	52.5	65.0	63.2	64.1
RH+VAM	65	68	66.5	68	70	69
Mean	39	42.8		42.1	44.8	
Rock-P						
Control	22	25	23.5	12	18	15
RH	25	33	29	35	45	40
VAM	45	59	52	70	72	71
RH+VAM	70	72	71	80	85	82.5
Mean	40.5	47.3		49.3	55	
L.S.D. 0.05						
Fe-rates		3.2			2.1	
Inoculations		4.4			22.5	
P-sources		2.9			7.2	
Fe × P × I		31.7			30.5	

bia may also increase the permeability of root cell to the fungus.

Regarding spore number and root infection, data presented in Table 2b showed that in the presence of rock-phosphate mycorrhizal parameters were found to be more beneficial than when superphosphate was used. Mycorrhizal infection and number of spores were 40.9%, 43.9% and 43.4%, 52.2% in case of super and rock-P, respectively.

Also, data show that, sporulation was affected favourably by rhizobial inoculation in the presence of super or rock-P and Fe levels. The same trend was observed with single inoculation by VAM, but the most pronounced results were obtained in case of a combined inoculation under rock-P application. These results are explained by Fairchild and Miller (1990); Koide and Li (1990) and Thomson *et al.* (1990).

P, N and Fe uptake

Results of Table 3 showed that under different P-sources and Fe applications, inoculation with *Rhizobium*, mycorrhizae and combined inoculation significantly increased P-uptake in faba-bean shoots. In this

Table 4. Effect of inoculation with *Rhizobium* (RH) and/or mycorrhizae (VAM) P, N and Fe uptake under P-sources and Fe-levels

Treatments	P mg/pot			N mg/pot			Fe mg/pot		
	−Fe	+Fe	Mean	−Fe	+Fe	Mean	−Fe	+Fe	Mean
Super-P									
Uninoculated	45.4	54.4	49.9	391.3	492.1	444.7	5.01	7.49	6.25
Rhizobium	64.1	75.9	70.0	661.5	759.0	685.3	8.86	10.48	9.67
Mycorrhizae	67.6	77.5	72.6	558.6	646.0	602.3	7.37	9.55	8.46
Rhizobium + mycorrhizae	86.5	97.8	92.2	828.0	958.0	393.0	12.16	13.71	112.94
Mean	65.9	76.4		598.9	713.8		8.35	8.65	
Rock-P									
Uninoculated	31.8	40.6	36.2	339.2	418.3	378.8	5.73	7.17	6.54
Rhizobium	55.8	66.2	61.0	659.1	717.6	688.4	10.16	11.96	11.06
Mycorrhizae	57.2	69.8	63.5	540.6	610.8	575.7	9.25	11.02	10.14
Rhizobium + mycorrhizae	81.0	94.7	87.9	804.0	1059.9	931.9	14.91	17.15	16.03
Mean	56.5	67.8		585.7	701.7			10.01	11.83
L.S.D. 0.05									
Fe rates		8.67			95.84			0.226	
Inoculation		11.84			105.79			1.115	
P-sources		4.35			N.S.			1.021	
Fe × P × I		16.05			111.15			1.99	

respect, P-uptake increased by about 41.2, 48.9 and 90.50% in case of super-P, as compared with 75.5, 79.9 and 154.7% whit rock-P dressing. These results confirm the observation of Saad (1990) and El-Ghandour (1992). Also, Murdoch *et al.* (1967), and Jackson *et al.*, (1972) reported that, in soil when relatively insoluble phosphate sources are added, such rock-phosphate, VA mycorrhizae seemed to hydrolyse rock-P. That result could be due to a more efficient uptake of chemically dissociated P-ions by fungal hyphae.

Regardless of inoculation, P-uptake significantly increased from both sources of phosphate. Also, the same trend was observed at two Fe-level.

Concerning the effect of inoculation in the presence of P and Fe fertilization, results indicated that N-content of the plants followed the order : dual inoculation > *Rhizobium* > mycorrhizae > uninoculated. Mukhtar and Abu Naib (1988) mentioned that in low N-soil, inoculation of faba bean with effective strains of rhizobia increased N-content. Also, Barea and Gonzalez (1986) reported that, VAM are able to increase N-concentration in plant shoots by one or more mechanism such as, indirect enhancement of symbiotic N_2-fixation as a consequence of increased P-supply by VAM or direct uptake of N compounds by VAM hyphae.

Also, data showed that fertilization with P-sources and Fe increased N-content compared to control where N-uptake, in case of iron application either when super or rock-P were dressed, were 713.8 and 701.7 mg/plant, while in iron deficient treatments were 598.9 and 585.5 mg/plant, respectively. Caixian Tang *et al.* (1992) reported that iron deficiency, generally, decreases nodule formation, leghaemogobin production, and nitrogenase activity. This leading to low nitrogen concentration in the shoots of some legumes. Chien *et al.* (1993) reported that N-uptake was much lower in the non-nodulating soybean than in the nodulating soybean at the 100 mg P/kg rate when P was added in the form of rock-P.

Results of Fe-uptake indicated that inoculation increased Fe-content where super or rock-P applied. In case of super-P, inoculation increased Fe-uptake by about 1.5, 1.2 and 2.1 fold of control when inoculated with *Rhizobium*, mycorrhizae and dual inoculation, respectively, while in rock-P amendments 1.7, 1.6 and 2.5 fold increases over the control where obtained. These results are in the same line of those obtained by Raman Raj (1988) and Kucey and Janzen (1987); they observed that VAM increased Fe-concentration and uptake in field beans. Also, plants inoculated with *Glomus albidus* are able to convert more root tissue into mycorrhizal than uninoculated and this reflected on the nutrient uptake (P and Fe) and symbiotic N_2-fixation. Wallace (1982) reported decreased in T203 soybean while the nodule were active, this also suggests some interaction between N_2-fixation and available Fe.

References

Abdel-Ghaffar AS (1988) Effect of edaphic factors n biological nitrogen fixation in *Vicia faba* under Egyptian field conditions. In: Beck DP and Materon LA (eds) Nitrogen Fixation by Legumes in Mediterranean Agriculture. ICARDA 1988. MNP, Dordecht, Boston and Lancaster

Barea JM and Gonzalez SB (1986) VA mycorrhizae as modifiers of soil fertility. Transactions of the XIII Congresses of the International Society of Soil Science. Hamburg, 13–20 August, 1986, p 588

Baraka P and Chen Y (1982) Iron Nutrition of plants in calcareous soils. Adv Agron 35: 217–240

Black CA (1965) Methods of analysis part 1, 2. Amer. Soc. of Agron., Inc. Pub. Madison Wisconsin, USA

Caixian Tang, Robson AD and Dilworth MJ (1992) The role of iron in the (Brady) *Rhizobium* Legume symbiosis. J Plant Nutr 15(10): 2235–2252

Chien SH, Carmona G, Menon RG and Hellums DT (1993) Effect of phosphate-rock sources on biological nitrogen fixation by soybean. Fert Res 34: 153–159

El-Ghandour IA (1992) Effect of biofertilizers on the availability of nutrients to plants. PhD Thesis, Fac. Agric., Ain Shams University

Fairchild MH and Miller GL (1990) Vesicular arbuscular mycorrhizas and the soil-disturbance induced reduction of nutrient absorption in maize III. Influence of P amendments to soil. New Phytol 114: 641–650

Gerdemann JW and Nicolson TH (1963) Spores of mycorrhizal endogone species extracted from soil by wet sieving and decanting. Trans Brit Mycol Soc 46: 235

Hallsworth EG (1972) Factors affecting the response of grain legumes to the application of fertilizers. Intern. Atomic Energy Agency Vienna: 1–15

Hamdi H, Taha S, El-Damaty A and El-Sherif A (1966) The effect of bacterial inoculation and phosphatic fertilization on some legume in sandy loam soil of UAR. J Soil Sci UAR 6: 1–10

Harris D, Pacovsky RS and Paul EA (1985) Carbon economy of soybean *Rhizobium* – *Glomus* associations. New Phytol 101: 427–440

Jackson NE, Franklin RE and Miller RH (1972) Effect of VA mycorrhizae on growth and phosphorus content of three agronomic crops. Soil Sci Soc Amer Proc 36: 64–67

John M (1970) Colorimetric determination of phosphorus in soil and plant materials with science. Vol 109, No 4

Koide RT and Li M (1990) On host regulation of the vesicular arbuscular mycorrhizal symbiosis. New Phytol 114: 59–74

Kucey RM and Janzen HH (1987) Effect of VAM and reduced nutrient availability on growth and phosphorus and micronutrient uptake of wheat and field beans under greenhouse conditions. Plant and Soil 104: 71–78

Mosse B (1977) the role of mycorrhiza in legume nutrition on marginal soils. In: Vincent JM, Whithney AS and Bose J (eds)

Expbiting the Legume – *Rhizobium* Symbiosis in Tropical Agriculture, pp 275–292. College of Tropical Agriculture, University of Hawaii, USA (Miscellaneous Publications, No 145)

Mukhtar NO and Abu Naib SA (1988) Biological nitrogen fixation by Faba-bean, lentil and chickpea in the traditional versus the new areas of sudan. In: Beck DP and Materan LA (eds) Nitrogen Fixation by Legume in Mediterranean Agriculture. ICARDA. MNP, Dordrecht, Boston and Lancaster

Murdoch CL, Jackobs JA and Gerdemann JW (1967) Utilization of phosphorus sources of different availability by mycorrhizal and non-mycorrhizal maize. Plant and Soil 27: 329–334

O'Hara GW, Boonkerd N and Dilworth MJ (1988a) Iron deficiency specifically nodule development in peanut inoculated with *Bradyrhizobium* Sp New Phytol 108: 51–57

O'Hara GW, Hartzook A, Bell RW and Loneragan JF (1988b) Response to *Bradyrhizobium* strains of peanut cultivars grow under iron stress. J Plant Nutr 11: 843–852

Pacovsky RS, Fuller G, Standdorf AR and Paul EA (1986) Nutrient and growth interaction in soybean colonized with *Glomus fasciculatum* and *Rhizobium japonicum*. Plant and Soil 92: 37–45

Papastylianou (1993) Timing and rate of iron chelate application to correct chlorosis of Peanut. J Plant Nutr 16(7): 1193–1203

Phillips JN and Hayman DS (1970) Improved procedures for clearing roots and staining parasitic and vesicular arbuscular mycorrhizal fungi for rapid assessment of infection. Trans Br Mycal Soc 55: 158–161

Raman Rai (1988b) Interaction response of *Glomus albidus* and *Cicer Rhizobium* strains on iron uptake and symbiotic N_2-fixation in calcareous soil. J Plant Nutr 11(6–11): 863–869

Robson AD (1978) Mineral nutrition limiting nitrogen fixation in legumes. In: Andrew CA and Kamprath EJ (eds) The Mineral Nutrition of Legumes on Tropical and Subtropical Soils, pp 277–293. CSIRO, Melbourne

Robson AD (1983) Mineral nutrition. In: Broughton WJ (ed) Nitrogen Fixation, Vol 3, pp 36–55. Oxford University Press, New York

Ross JP and JW Gilliam, 1973. Effect of endogone mycorrhizae on phosphorus uptake by soybeans from inorganic phosphate. Soil Sci. Soc. Am. Proc. 37, 237–239

Saad RN (1990) Studies on legumes mycorrhiza-rhizobia symbiotic system in newely reclaimed soil. PhD Thesis, Fac. Agric., Ain Xhams University

Snedecor GW and Cochran WG (1982) Statistical methods 7th ed. The Iowa State University Press, Ames Iowa, USA

Soerensen KU, RE Terry and VD Jolley, 1988. The interaction of iron-stress response and root nodules in iron efficient and inefficient soybeans. J. Plant Nutr., 11(6-110), 853–862

Suman Bala and Singh OS (1985) Response of lentil to VA mycorrhizal inoculation and plant available P levels of unsterile soils. Plant and Soil 87: 445–447

Thomson BD, Robson AD and Abbott BO (1990) Mycorrhizas formed by *Gigaspora calospora* and *Glomus fasciculatum* on subterianean clover in relation to soluble carbohydrate concentrations in roots. New Phytol 114: 217–225

Young CC, TC Juang and HY Guo, 1986. The effect of inoculation with vesicular arbusular mycorrhizal fungi on soybean yield and mineral phosphorus utilization in subtropical – tropical soils. Plant and Soil 95, 245–253.

C. Rodriguez-Barrueco (ed.), Fertilizers and Environment, 101–108.

Fertilisation balances in two agriculturally used water catchment areas

M. Kücke & P. Kleeberg
Institute of Plant Nutrition and Soil Science, Federal Agricultural Research Centre Braunschweig, Bundesallee 50, 38116 Braunschweig, Germany

Key words: mineral N fertilisation, N fertilisation balances, organic N fertilisation, surplus, water catchments

Abstract

N fertilisation balances have been calculated since 1987 in two different water catchments in Northern Germany with different soil (sand soils and loam soils) and cropping conditions (different crop rotations) for individual fields. The fertilisation excess was similar between both catchments on average (50 to 60 kg N ha^{-1}). Enormous differences occur on field scale: N balances of fields with regular organic manuring usually exceed those of unmanured fields by at least 100 kg N ha^{-1} because farmers do not account for the nutrients in the organic fertilisers for their mineral fertilisation. Crops with low N surplusses were potato and brewery barley (mainly cultivated in the sand soil region) while for winter cereals the N fertilisation surpluses were 40 to 60 kg N ha^{-1} on the loamy sites and up to 100 kg N ha^{-1} on the sandy soils. It is concluded that individual field balances are good instruments to indicate nutrient excess situations and can serve as a basis for fertilisation recommendations. However average balances on larger scales (catchment balances, national balances) are not suitable for solving the nitrogen excess problem in agriculture or even to estimate the contribution of agriculture to ground and surface water pollution. Differentiation between N balances of different soils and cropping systems is necessary.

Introduction

Fertilisation balances are suitable tools to quantify the excess nutrient input on a field, farm or regional scale. Since 1985, several papers have reported the mineral excess situation in Germany (Bach, 1989) and European agriculture (Isermann and Sturm, 1990). As a consequence, nutrient balances have been focused in policy discussions concerning the influence of agronomic practice on the nutrient input to ground and surface water. These balances have been used as a basis to calculate the contribution of fertiliser use in agriculture to the eutrophication of the fresh surface and sea water (Werner *et al.*, 1991). In most cases, national statistical data are used for the calculations of nutrient excess balances. They are therefore unsuitable for the identification of regional differences due to the local cropping systems and site conditions. However, fertilisation recommendations are based on the specific crop situations.

This study evaluates the nutrient balance in two Northern German water catchments which have been analysed for their nutrient and water dynamics since 1986 by the interdisciplinary research group "Water and Matter Dynamics in Agroecosystems". The data for the calculation of the nutrient excess situation in these areas come from annual interviews of each farmer. These interviews provided data on crop type, yield and fertilisation practice for each field in both

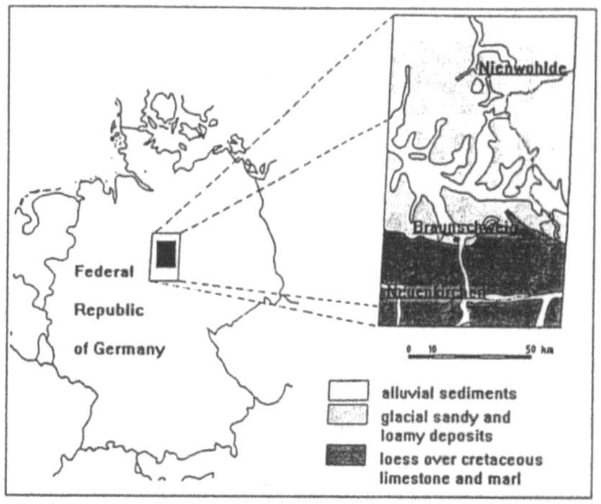

Fig. 1. Location of the investigation areas and their geological situation (Othmer and Bork, 1989).

Table 1. Characterisation of the investigation areas

	Krummbach area	Eisenbach area
Arable area	1100 ha	500 ha
Dominant soil types (FAO)	Gleyic luvisols	Luvic areonosols
Drained (% of arable land)	About 90%	About 30%
Precipitation mm (aver. 10 years)	600 mm	600 mm + 80 to 150 mm irrigation
Plant available water capacity of soils mm	140-200	50-90
Seepage mm	180	350

Table 2. Crops in the water catchments Krummbach (Loam soils) and Eisenbach (Sand soils, between 1987-1992 (% of the arable land)

	1987	1988	1989	1990	1991	1992	ϕ
Krummbach area							
W-wheat	59.8	48.7	57.6	54.3	52.8	54.4	54.5
W-barley	10.0	10.4	6.3	13.4	9.1	3.1	8.7
S-beets	28.0	35.9	33.8	28.6	33.1	32.8	32.0
Others	3.1	5.3	2.3	3.8	4.9	9.7	4.9
Eisenbach area							
W-barley	15.1	13.2	4.4	5.5	12.8	7.6	9.8
S-barley	16.0	22.0	27.4	18.5	30.1	21.7	22.6
W-rye	12.6	13.1	13.3	13.1	7.1	4.6	10.6
Triticale	5.8	1.1	-	7.7	3.2	3.3	4.2
W-wheat	2.4	6.3	8.5	4.0	1.0	2.6	4.1
Potatoes	25.2	20.0	18.2	22.2	23.5	23.7	22.1
S-beets	19.0	18.9	20.9	23.7	14.3	15.1	18.7
Maize	-	-	-	-	3.6	2.6	3.1
Others	3.9	5.4	7.3	5.4	4.5	18.8	7.6

catchments. These individual field data allow the calculation of nutrient balances for each individual field and can be used for site and crops specific analysis of the nutrient excess situations.

Materials and methods

The investigations were carried out between 1990 and 1993 in two water catchment areas intensively investigated in the Special Collaborative Program 179 "Water and Matter Dynamics in Agro-Ecosystems" (TU Braunschweig). The 'Krummbach' water catchment covers about 12 km^2 and is located in the northern foothills of the Harz mountains in Lower Saxony, Germany (loam soils derived from loess, predominantly gleyic luvisols (FAO), Fig. 1, Table 1). The 'Eisenbach' water catchment is situated farther north in the heathland region about 20 km south of Uelzen (sand-derived soils, predominantly luvic arenosols (FAO)) and extends over about 5 km^2. The crop distribution in the loess loam catchment is dominated by sugar beet and winter wheat. The importance of winter barley has decreased from about 10% of the area in 1991 to 3% in 1992. In the sand soil catchment, sugar beets, potatoes, summer barley and winter rye dominate the

Table 3. N balances of crops in the Krummbach watershed on fields with and without organic manuring (1991 and 1992) % = % area of the crop in the catchment; Y = yield, dt FM ha^{-1}; I = input (kg N ha^{-1}); min = mineral fertilisers; E = Export by harvest products (kg N ha^{-1}); org. = farmyard manure + slurry; B = saldo ($\sum I$ - E, kg N ha^{-1})

		Only mineral fertilisation					Mineral and organic fertilisation						
Crop	ha	%	Y	I_{min}	E	B	%	Y	I_{min}	I_{org}	$\sum I$	E	B
1991													
Sugar beet	382.3	81.1	551	155	102	+53	18.9	607	129	93	222	112	+110
W-wheat	592.1	99.6	92	215	177	+38	0.4	69	200	136	336	135	+201
Potato	4.0	0	-	-	-	-	100	410	80	180	260	144	+116
W-rape	4.1	0	-	-	-	-	100	35	156	100	256	83	+173
1992													
Sugar beet	380.4	83.7	458	144	79	+65	16.3	562	128	127	255	97	158
W-wheat	643.6	100	82	213	154	+59	0	-	-	-	-	-	-
Potato	1.8	0	-	-	-	-							
Oats	1	0	-	-	-	-							

crop rotation (Table 2). Because of the low water storage capacity of the sandy soils, economic crop production is possible only with irrigation. More details on the dominant soil types and geology are provided by Othmer and Bork (1989) and on the hydrological situations by Kücke and Hasenpusch (1995). Nutrient balances were calculated from the data of annual interviews of farmers, carried out since 1987 in both catchments. Nutrient inputs were calculated from the farmers fertiliser use. Nutrient input by organic fertilisers were calculated from the data on the amount of manures or slurry applied by the farmer and the average nutrient concentration (total nitrogen) in typical manures sampled in both catchments. Yield data were supplied by the farmers. These were separately checked by large control plot harvests by uses of a combine harvester and a special weighting construction on several sites of each area. Numerous harvest samples were collected for each important crop in both watersheds and analysed for their nutrient content. Nutrient excess situations on the field scale were calculated by the formula

$$\text{fertilisation excess (balance)} = I_{\text{mineral fertilisers}} + I_{\text{organic fertilisers}} - E_{\text{harvest products}}$$

where I means input and E export. Atmospheric nutrient inputs or N_2 fixation and losses to the environment are not accounted in these calculations. Catchment balances have been calculated from the field balances. The major problem with this methodology is the quantification of organic fertilisation. The amounts of farmyard manure or slurry applied can rarely be reported with accuracy by the farmers. N losses during storage and application can only very roughly be estimated. Therefore those balances calculated for 1987 to 1990 by Hersemann and Sauerbeck (1990) include no field specific data on organic fertilisation. For their calculations, only the mineral fertilisation input was considered. Since 1991, data on organic fertilisation have been examined for each individual field.

Results

In Table 3 the N balances of the crops cultivated 1991 and 1992 on the loam soils in the Krummbach watershed are listed separately for fields without and with organic fertiliser use. In this catchment, potatoes and winter rape were regularly treated with organic fertilisers but the area cultivated with these crops was negligible low. Most organic fertiliser was given to sugar beets. On organically manured fields sugar beet yields as well as N exports by harvest products were slightly higher in both years compared with fields without manuring.

Although the farmers have applied slightly less mineral fertilisers on organically fertilised fields, this reduction was not equivalent to the additional N input by manure application. N balances of the manured crops were therefore at least double of those of the unmanured fields and exceeded 100 kg N ha^{-1} in most cases.

Table 4. N balances of crops in the Eisenbach watershed on fields with and without organic manuring (1991 and 1992) % = % area of the crop in the catchment; Y = yield, dt FM ha^{-1}; I = input (kg N ha^{-1}; min. = mineral fertilisers; E = Export by harvest products (kg N ha^{-1}); org. = farmyard manure + slurry; B = saldo ($\sum$I - E, kg N ha^{-1})

		Only mineral fertilisation					Mineral and organic fertilisation						
Crop	ha	%	Y	I_{min}	E	B	%	Y	I_{min}	I_{org}	$\sum$I	E	B
1991													
Sugar beet	78.2	68.0	405	201	73	+128	32.0	531	92	94	186	99	+87
Maize	18.7	0	-	-	-	-	100		114	123	237	237	0
Potato	123.6	80.5	398	119	139	-20	10.5	409	83	181	264	143	+121
W-rape	9.7	0	-	-	-	-	100	32	190	91	281	83	+198
S-barley	158.2	93.7	56	112	85	+27	6.3	60	115	95	210	90	+120
W-barley	67.7	78.5	57	190	87	+103	21.5	56	143	188	331	83	+248
W-rye	39.8	87.4	56	193	86	+107	12.6	60	197	209	406	92	+314
Triticale	17.2	27.3	61	192	108	+84	72.7	72	200	89	289	128	+161
1992													
Sugar beet	128.4	62.9	458	144	79	+65	37.1	471	145	112	257	82	+175
Maize	14	0	-	-	-	-	100	591	83	174	257	210	+47
Potato	128.4	84.7	384	133	134	-1	15.3	400	76	95	171	140	+31
W-rape	8.4	0	-	-	-	-	100	32	74	92	166	99	+67
S-barley	115.2	94.8	48	111	70	+41	5.2	52	87	91	178	75	+103
W-barley	41.3	87.9	61	161	119	+42	12.1	66	96	192	288	129	+159
W-rye	8.4	80.2	54	157	85	+72	19.8	48	96	196	292	76	+216
W-wheat	13.7	100	47	181	83	+98	0	-	-	-	-	-	
Triticale	17.5	71.4	52	166	106	+60	28.6	62	96	196	292	126	+166

On the sandy soils of the Eisenbach catchment (Table 4) preferentially manured crops were sugar beet, maize, winter rape and triticale. As in the Krummbach area, yields and nutrient exports were in most cases slightly higher after organic manuring. Nevertheless, organic fertilisation caused a doubling of the nutrient balance. In 1991, the mineral fertiliser input to sugar beet and potatoes was lower on organically manured fields. For cereal crops the mineral input was similar regardless of the level of organic fertilisation. In contrast, farmers clearly reduced the mineral N inputs to winter cereals in 1992.

Development of the mineral fertilisation since 1987

On average, between 1987 to 1992, the mineral N fertilisers use was 30 kg N ha^{-1} higher on the loam soils of the Krummbach area compared with the sandy soils of the Eisenbach catchment (Table 5). Corresponding to this, 36 kg N ha^{-1} more was exported by the

Table 5. N surplus by mineral N fertilisation in two Northern German Water Catchments

	Mineral N-Fertilisation	N Export	Saldo
Krummbach area			
1987	188	135	+53
1988	188	123	+65
1989	190	126	+64
1990	186	149	+37
1991	190	148	+42
1992	191	131	+60
average	189	135	+54
Eisenbach area			
1987	176	104	+72
1988	170	91	+79
1989	165	94	+71
1990	170	105	+65
1991	145	101	+44
1992	130	99	+31
average	159	99	+60

Data between 1987 to 1990 reported by Sauerbeck and Hersemann (1990).

Table 6. N surplus by mineral fertilisation of the most important crops in the Krummbach watershed between 1987 and 1992 (organic manuring not accounted for)

Crop/ year	% of arable land	Miner. fertiliser kg N ha^{-1}	Crop. yield dt/ha^{-1}	N content kg N ha^{-1}	Removal kg N ha^{-1}	Excess = saldo
Sugar beets						
1987	28	143	487	0,20	97	+46
1988	36	151	450	0,20	90	+61
1989	34	153	549	0,20	110	+43
1990	29	136	603	0,20	121	+15
1991	33	153	563	0,19	104	+49
1992	33	156	536	0,17	92	+64
Average	32	149	531	0,19	102	+46
W-wheat						
1987	59	210	86	2,13	156	+54
1988	48	216	82	2,13	149	+67
1989	58	212	61	2,57	133	+79
1990	54	215	91	2,20	170	+45
1991	53	214	92	2,30	177	+37
1992	54	213	82	2,20	154	+59
Average	54	213	82	2,26	157	+57
W-barley						
1987	10	187	79	1,82	122	+65
1988	10	185	76	1,83	118	+67
1989	6	175	80	1,97	134	+41
1990	13	182	88	1,72	129	+53
1991	9	185	91	1,79	138	+47
1992	3	172	92	1,86	145	+26
Average	9	181	84	1,83	131	50

Data between 1987 to 1990 reported by Sauerbeck and Hersemann (1990).

harvest products. This results in a mineral N fertilisation excess of 54 kg N ha^{-1} and 60 kg N ha^{-1} in the Krummbach and the Eisenbach areas respectively. There is no obvious trend detectable for decreasing mineral N fertiliser use in the Krummbach area. However, in the Eisenbach catchment, the mineral N fertilisation decreased from 170 kg N ha^{-1} (average 1987 to 1990) to 138 kg N ha^{-1} (average 1991 and 1992) in the last 2 years. Because there was no corresponding decrease in N export, the N surplus decreased from 72 kg N ha^{-1} (average 1987-1990) to 38 kg N ha^{-1} (average 1991/1992). For the most important crops of the Krummbach territory (sugar beets, winter wheat and winter barley) mineral N-fertilisations and N balances were rather stable over the last 7 years (Table 6). There is no evidence that fertilisation practices have changed.

Differences in the mineral nitrogen surpluses between individual crops are greater in the Eisenbach catchment. (Table 7). Highest mineral N surpluses are recorded for winter cereals (w-wheat, w-barley and w-rye) and sugar beet while those for summer barley and potatoes were low or even negative. The major reasons for the decreasing trend of the mineral N fertilisation excess on area basis since 1991 (Table 5) are: (1) The area of w-rye and w-wheat (crops with high surplus-

Table 7. N excess by mineral fertilisation of the most important crops in the Eisenbach watershed between 1987 and 1992 (organic manuring not accounted)

Crop/ Year	% of arable land	Yield dt ha^{-1}	Min.fertil. kg N ha^{-1}	Removal kg N ha^{-1}	Balance kg N ha^{-1}
S-barley					
1987	16	53	108	70	+38
1988	22	50	110	66	+44
1989	27	46	108	70	+38
1990	29	53	112	78	+35
1991	30	56	113	85	+28
1992	22	48	109	71	+39
Average	24	51	110	73	+37
W-rye					
1987	13	50	197	70	+127
1988	13	46	208	64	+144
1989	13	51	193	81	+112
1990	13	51	193	77	+116
1991	7	57	194	87	+107
1992	5	53	145	83	+61
Average	11	51	188	77	+111
Potatoes					
1987	25	416	148	146	+2
1988	20	403	127	141	-14
1989	18	431	149	151	-2
1990	22	399	154	140	+14
1991	24	399	115	140	-25
1992	24	387	124	135	-11
Average	22	406	136	142	-36
Sugar-beets					
1987	19	478	212	96	+116
1988	19	410	224	82	+142
1989	21	432	196	86	+110
1990	24	507	199	101	+98
1991	14	431	192	80	+112
1992	15	456	140	79	+61
Average	19	452	194	87	107
W-barley					
1987	15	57	210	100	+110
1988	13	53	195	93	+102
1989	4	63	200	106	+94
1990	6	59	182	99	+83
1991	13	57	180	85	+95
1992	8	62	153	121	+32
Average	10	59	187	101	+86
W-wheat					
1988	6	57	226	94	+132
1989	9	43	247	90	+157
1990	4	65	184	122	+62
1991	1	78	198	137	+61
1992	3	47	181	83	+99
Average	5	58	207	105	+102

[a]No winter wheat in 1987; data between 1987 to 1990 reported by Sauerbeck and Hersemann (1990).

es) was reduced. (2) Winter rye and especially winter wheat showed clearly lower N surpluses in the last two (w-rye) and three years (w-wheat). (3) The potato area (crop with low mineral N surplus) was lower than average in the last two years. Moreover the potato surpluses where lower in these two years compared to the previous years.

Discussion and conclusions

The national N balances reported by Isermann and Sturm (1990) or Bach (1989) indicate an average nitrogen excess of 167 kg N ha^{-1} for Germany. These calculations are based on statistical data and on assumptions for atmospheric N inputs or biological N_2 fixation. Therefore these calculations cannot point out regional differences due to local cropping and soil conditions. The data presented here clearly demonstrate that differentiation at crop and field scale is necessary if potentials for N fertilisation reductions are to be detected and used for more effective fertilisation recommendations. Even though the average N excess in the both catchments of this investigation is similar, the cropping situations are incomparable because of differences in soil type, yield potential, crop distribution, irrigation and manure use. The average mineral fertilisation excess of the two catchments ranged between 50 to 60 kg N ha^{-1}. This is based on average mineral fertiliser inputs of 189 (Krummbach) resp. 159 kg N ha^{-1} (Eisenbach). Taking into account the specific N inputs by organic manures (17 kg N ha^{-1} in the Eisenbach and 5 kg N ha^{-1} in the Krummbach catchment) and estimations on average atmospheric N input (30 kg N ha^{-1} and biological N_2 fixation, the inputs for these areas increase to 218 (Krummbach area) and 236 kg N ha^{-1} (Eisenbach area). The average N excess increases to 91 and 119 kg N ha^{-1} respectively. In these calculations about 20% of the N input and about 45% of the N excess is due to assumptions which cannot be affected by fertilisation management.

The crop specific N balances as well as the comparison of organically manured and unmanured fields demonstrate potentials for effective N fertilisation reductions. In both catchments, N fertilisation can be decreased especially to sugar beets and winter cereals while the potentials to decrease the mineral N inputs to potatoes and summer barley appear to be low. Moreover it is concluded that organic manure should be applied in lower doses over more crops and fields. This would not decrease the average amount of manure N input per catchment but should result in a more ecological use and increase the efficiency of N use by the crops. If the nutrients in the manures are taken into account for the calculation of the fertilisation demand there are huge potentials for reducing the tremendous N excess found on manured fields. The values for 1992 in the Eisenbach catchment indicate that this has a large effect on the average N balance of the area.

Fertilisation experiments in both catchments (Kücke and Kleeberg, in press) where N fertilisation was reduced according to the expected N offtakes by the crops showed that such an adaptation on loess loam soils had negligible effect on the yields (-2 to - 4%) but effectively decreased NO_3 content in the soil solution and NO_3 leaching. In contrast, on sandy soils, yield decreases of sugar beet, winter rape and summer barley were more severe (- 5 to- 17%) if the mineral N fertilisation was reduced while the effect of omitted manure application was low. On these soils, NO_3 content of the soil solution and N leaching was less effected or even increased by fertilisation reduction.

Acknowledgement

These investigations have been funded by the German Research Organisation in the interdisciplinary research group "Water and Matter Dynamics in Agroecosystems" running since 1986. The authors thank Burghard Bosse for interviewing the farmers, Rosi Rietz, Ursula Krüger and Tanja Römmer for assisting the plant sampling and plot harvests and for analysing the plant samples.

References

Kücke M and Hasenpusch K (1995) Dynamic of nutrient losses by drainage and brook waters in two northern Germany water catchments with different soil and cropping conditions. In: Rodriguez-Barrueco C (ed) Fertilizers and Environment, pp 383–386 (this volume).

Kücke M and Kleeberg P Nitrogen balances and soil nitrogen dynamics in two areas with different soil, climatic and cropping conditions. European J. Agron.

Bach M (1987) Die potentielle Nitrat-Belastung des Sickerwassers durch die Landwirtschaft in der Bundesrepublik Deutschland. Göttinger Bodenkundliche Ber 93: 1–186.

Isermann K und Sturm H (1990) Stickstoff- und Phosphorbilanzierung der Landwirtschaft im Vergleich westeropäischer Länder. VDLUFA-Schriften 32/90: 229–235.

Othmer H and Bork HR (1989) Characterisation of the investigation sites. Landschaftsgenese Landschaftsökologie 16: 73–86.

Sauerbeck D and Hersemann H 1990 The nitrogen balance of agriculturally used water catchment areas. In: Merck R, Verecken H and Vlassak K (eds) Fertilisation and the Environment, pp 330–339, 1990. Intern. Symp. 28.-30.8.1989, Leuven, Belgie. Presses Universitaires de Louvain, Leuven, Belgie.

Werner W, Olfs H.W., Auerswald K., Isermann K (1991) Stickstoff- und Phosphateintrag in Oberflächengewässer über "diffuse Quellen". In: Hamm A (ed) Studie über Wirkungen und Qualitätsziele von Nahrstoffen in Fließgewässern. Academia-Verlag St. Augustin. pp 665–764.

C. Rodriguez-Barrueco (ed.), Fertilizers and Environment, 109–113.

Tillage (plowing, subsoiling and no-tillage) effect on soil erodibility

I. Runoff and infiltration

F.C. Barreiros, F.P. Pires & E.M. Sequeira
Estação Agronómica Nacional, Quinta do Marquês, 2780 Oeiras, Portugal

Abstract

The effect of three different soil tillage systems (plow, subsoiling and no-tillage) on runoff and infiltration was determined *in situ*, using a rain simulator. These determinations were carried out in a field trial under the project "Soil tillage in dry and irrigated farming" of the EU Program AGRIMED, laid out in a granite soil Inland Central Portugal (Fundão).

The effect of treatments in one hour mean of runoff and infiltration rates were statistically highly significant. No-tillage treatment significantly decreased the runoff and increased infiltration rates in relation to plowing, but increased also significantly the soil bulk density, and decreased total soil porosity. Sub-soiling had an intermediate behaviour in all observed characteristics.

After discussion it is concluded that no-tillage is the best soil treatment to improve the observed soil physical characteristics, and it seemed there was no reason to apply sub-soiling, at least for this 2-year experiment, in this soil.

Introduction

It is proved that soil technology such as plowing, subsoiling, minimum tillage or no-tillage strongly affect soil chemical and physical characteristics (Azevedo, 1973; Davidson and Santelman, 1973; Philips and Young, 1973; Elliot, 1975; Cordier *et al.*, 1979; Agboola, 1981). However, these effects are dependent on soil susceptibility, specially to physical degradation (Balloni and Favilli, 1987; Fedorof, 1987; Steenvorden and Bouma, 1987).

No-tillage or minimum tillage decrease energy consumption, reduce soil compaction by machinery traffic, increase soil organic matter content but also reduce soil erosion risk (Philips and Young, 1973; Yassouglou, 1987; Yassoglou, 1987; Morgan, 1992).

This lower erosion risk results from no disruption of soil surface structure (Hewitt and Dexter, 1980), and also by the maintenance of a protective organic mulch (litter, straw and rests of previous crops) or inorganic mulch (surface gravel and stones) (Philips and Young, 1973; Yassouglou, 1987; Sequeira, 1988, 1992).

A1-a	A1-b	A1-c
A2-a	A2-b	A2-c
A3-a	A3-b	A3-c
A4-a	A4-b	A4-c

Fig. 1. Crop rotation type design in the tillage field trial. Crops: A1 – forage; A2 – triticale; A3 – lupine; A4 – wheat. Tillage treatment: a – plow; b – no-tillage; c – subsoiling.

This mulch dissipate the rain drop energy, decreasing it dispersive effect (Ferreira *et al.*, 1984), well proved by one of the authors (Barreiros, 1989).

We intend to determine the effect of three different soil technologies (plow, subsoiling and no-tillage) on soil erodibility, describing its effect on runoff and infiltration rates.

With the use of one rain simulator, especially designed to exactly reproduce the rain drop effect (calibrated droppers), using the same standard rainfall on plots submitted to three soil technologies applied to a standard crop rotation of four years, we determined the variation of runoff and infiltration rates.

Material and methods

This field trial was run under EU AGRIMED Program (1988–1990), "Soil Tillage in Dry and Irrigated Farming" Project.

The aims of the Project were: 1 – to compare crop yields in two typical crop rotations, one of them – wheat, lupine, tritical, forage crop, and another one similar to this one after 4 years of temporary improved pasture, submitted to three different soil technologies (plow, subsoiling, no-tillage); 2 – to assess the soil chemical and physical behaviour.

Soil morphology

The studied soil was a Dystric Cambisol from granite (Lithic Xerochrets) (Cardoso, 1965, 1971).

0–19 cm pale brown 10YR6/3 (dry), brown 10YR5/3 (moist), gravelly sandy loam, with some quartz stones, weak crumb structure, very friable to friable, few fine and medium pores, many fine and medium roots (mechanical impedance 1.7 to 2.2 kg cm^{-2}). Clear boundary to 19–46 cm very pale brown 10YR7/3 (dry), brown 10YR5/3 (moist), gravelly sandy loam, with few quartz stones, very weak crumb structure, very friable to friable, few fine and medium pores, few fine roots (mechanical impedance 2.0 to 4.0 kg cm^{-2}). Clear boundary to 46–70 cm very pale brown 10YR8/3 (dry), pale brown 10YR6/3 (moist), gravelly loam, weak coarse sub angular blocky structure, friable, few fine pores, without roots, some weathered granite stones. Very weathered soft granite rock with many small muscuvitic plates (mechanical impedance 2.5 to 4.0 kg cm^{-2}).

Rotation

Fig. 1 presents the rotation type modality used for erodibility tests. The statistical design was in split plot randomized blocks, with annual crop and tillage as split plot.

At autumn of the 2nd year of the rotation, in each of the 12 plots of the field trial (4 culture of the rotation × 3 tillage treatments), with the rain simulator two rainfalls were applied (replicates), determining the runoff, the infiltration, the soil material carried out by runoff and an index of "splash".

The rain simulator is an adaptation of Adams *et al.* (1957) apparatus, made by one of the authors (Barreiros, 1989).

The rain simulator works as follows:

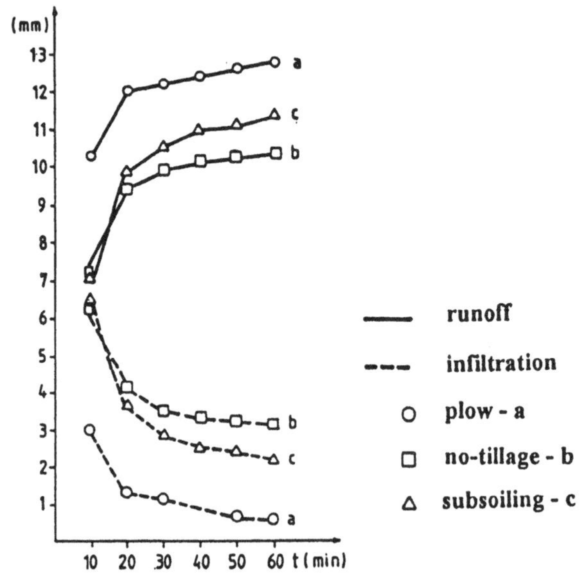

Fig. 2. Mean runoff and infiltration mm, for each 10 minutes interval, by tillage treatments.

- artificial rain application (distilled water) of 1375 cm^3 (< > 79.49 ≈ 80 mm) over areas of 172 cm^2 delimited by cylindrical infiltrometers, with an intensity of 80 mm $hour^{-1}$;
- each raindrop (from individual calibrated trickler) is about 5.56 mm diameter, the height of fall is 1 m, so the energy of each drop is 8.82 10^{-4} Joules; this means a rainfall with about 784.9 10^{-5} Joules per hectare;
- slope of the infiltrometer surface, 0%;
- the soil surface was clean of all vegetation, and about 0.5 cm of soil surface was cut out;
- the soil was moistened at field capacity before the test; 24 hours before the beginning of the test the rain simulator infiltrometer (diameter, 0.5 m) was saturated with water, and covered with a plastic;
- each test ran for about 60 minutes; the time of initial runoff was noted, and at 10 minute intervals runoff water and soil material carried out were collected; the infiltration was obtained by difference between rainfall and runoff at each 10 minute intervals.

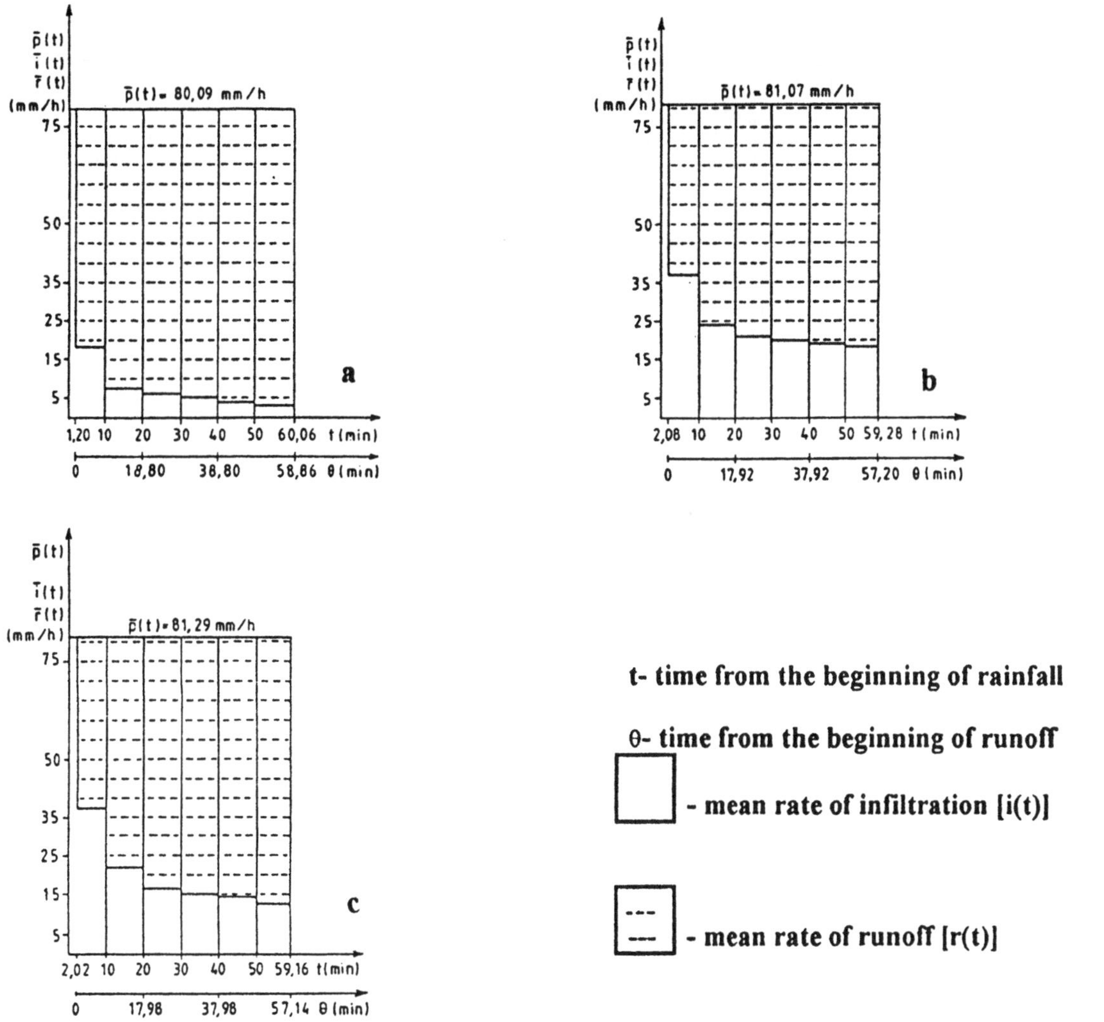

Fig. 3. Means of runoff mean rates [rt], infiltration mean rates [it], and rainfall intensity [pt], obtained for the rainfall simulations for each tillage treatments: a – plow; b – no-tillage; c – subsoiling. Time of runoff start – t, and time from the beginning of runoff – θ.

Table 1. Final mean data of soil material lost by runoff, "splash" and runoff + "splash", at the end of the simulated rainfall 80 mm rainfall in 1 hour), in the tillage treatments: plow, subsoiling and no-tillage, expressed in tones ha^{-1} h^{-1}.

soil losses	tillage treatments		
	plow	subsoiling	no-tillage
carried out by runoff	2.32 ± 0.44	1.88 ± 0.32	1.51 ± 0.45
'splash'	22.07 ± 1.11	19.57 ± 1.82	13.82 ± 1.61
runoff + 'splash' total)	24.35 ± 1.48	21.38 ± 2.02	15.28 ± 1.99

Results and discussion

Fig. 2 shows the mean time of initial runoff, mean runoff volume (in mm of rain), and infiltration volume (in mm of rain), for each 10 minute interval.

In Fig. 3, a, b and c, the mean infiltration and runoff rates (mm h^{-1}), for each soil tillage treatment, are shown.

Table 2. Synopsis for analysis of variance of soil losses by runoff, by "splash", total soil loss, and solids concentration in runoff

Source of variation	kind of soil loss	Solids in runoff							
runoff	'splash'	total							
	d.f.	F values	signif.	F values	signif.	F values	signif.	F values	signif.
tillage treatment	2	16.31	**	127.10	***	101.70	***	8.84	*
crop	3	0.10	n.s.	2.17	n.s.	1.21	n.s.	0.20	n.s.
replicates	1	0.22	n.s.	0.62	n.s.	1.00	n.s.	6.30	*

*** significant at 0.1% level, ** significant at 1% level, * significant at 5% level, n.s. not significant

In all treatments, runoff increased quickly up to 20 minutes, and afterwards tended to an asitotic equilibrium rate, and obviously the infiltration has a complemental behaviour.

Until the beginning of runoff, all rainfall infiltrated. The rate of infiltration and of runoff was corrected to cover 10 minutes at the first interval.

In plow treatment runoff started at 1.2 minutes; in the subsoiling plots at 2.02 minutes and in the no-tillage plots at 2.08 minutes.

In Table 1 the final mean value of runoff and infiltration rates by tillage treatments are presented. In Table 2 the results of analysis of variance are shown.

Only soil tillage treatment was highly significant (p 0.1%); replications and crops did not show any significant effect after 2 years of field experiment.

Runoff in plow plots was significantly greater than in the other two treatments, in the beginning of rainfall, as well as at the equilibrium stage, at the end of the test. At this final equilibrium phase, plow plots showed runoff about four times higher than under no-tillage (Fig. 2 and 3 a, b, c).

Subsoiling had an intermediate behaviour, starting with an infiltration rate greater than under no-tillage, but after some minutes (< 20 minutes), it decreased to an intermediate position closer to that under no-tillage.

Santos and Oliveira (1991) under the same project and at the same field trial, found that the soil bulk density was greater with no-tillage treatment, by the reduction on total soil porosity. They found the following: for no-tillage a bulk density of 1.59, with a total porosity of 37%; for subsoiling a bulk density of 1.58, with a total porosity of 37.9%; for plowing a bulk density of 1.55, with a total porosity of 39.4%.

They also found that, using undisturbed soil samples (cylindrical cores with 10 cm height and 5 cm diameter) the hydraulic conductivity had a behaviour similar to that found in our results: no-tillage, K = 7.59 cm h^{-1}; plow, K = 6.85 cm h^{-1}.

However, for subsoiling plots they found a lower result (K = 4.38 cm h^{-1}), as the subsoiling plot soil cores were collected far from the places where the subsoiler lamina disturbed the soil. This result represents the compacted fraction of the plot and our results represents the intermediate with the disturbed places where subsoiler lamina crossed.

Such places, that showed a greater infiltration rate at the beginning of the test, were quick stopped by sediments, as shown in Fig. 2, so, the infiltration rate dropped to a lower position.

The greater infiltration rates under no-tillage, in spite of a greater soil bulk density and a lower total soil porosity, may be explained by there having been no destruction of vertical pores (dead roots, etc.), which significantly increases the hydraulic conductivity, as was found by Blevins *et al.* (1984) and Goss *et al.* (1984).

The subsoiling treatment, as explained earlier, strongly increased not only the bulk density of most part of the soil, by machinery wheels tramping, but also disrupt the vertical channels in the disturbed and undisturbed fraction by lateral pressure caused by subsoiler lamina.

This is partially compensated by the increase on hydraulic conductivity at subsoiler rips. This fact was shown by the higher results of the bigger infiltrometer of the rain simulator.

References

Adams JE, Kirkham D and Nielsen DR (1957) A portable rainfall-simulator infiltrometer and physical measurements of soil in place. Soil Sci Soc Am Proc 21: 473–477.

Azevedo AL (1973) Evolução do teorem matéria orgânica dos solos sujeitos a diferentes tratamentos. Anais do Instituto Superior de Agronomia 34: 65–114.

Balloni W and Favilli F (1987) Effects of agricultural practices on the physical, chemical and biological properties of soils: Part I – Effect of some agricultural practices on the biological soil fertility. In Barth H and L'Hermite P (eds) Scientific Basis for Soil Protection in the European Community. Elsevier Applied Science, London, 161–180.

Barreiros FC (1989) Estudo da erodibilidade de algumas unidades-solo. Dissertação para acesso à categoria de Investigador Auxiliar. Equivalent to a PhD dissertation. EAN, Oeiras.

Blevins RL, Smith MS and Thomas GW (1984) Changes in soil properties under no tillage. In Phillips RE and Phillips SH (eds) No Tillage Agriculture. Principles and Practices.

Cardoso JVC (1965) Os solos de Portugal. Sua Classificação, Caracterização e Génese. I- A Sul do Rio Tejo. Secretaria de Estado da Agricultura. DGSA, Lisboa.

Cardoso JVC, Bessa MT and Marado MB (1971) Carta dos solos de Portugal (1:1000000). Agronomia lusit 33: 481–602.

Cordier Y, Vitlox O and Frankinet M (1979) Labour, demi-labour, or semis directe en continue; consequences geotechniques. Bull Rech Agron, Gembloux 14: 121–142.

Davidson JM and Santelman PW (1973) An evaluation of various tillage systems for wheat. Agric Exp Sta Oklahoma Sta Univ 3711.

Elliot JG (1975) Reduced cultivation and direct drilling in farming systems. Outlook on Agriculture 8 250–254.

Fedorof N (1987) The production potential of soils. Part I – Sensitivity of principal soil types to the intensive agriculture of North-Western Europe. In Barth H and L'Hermite P (eds) Scientific Basis for Soil Protection in the European Community. Elsevier Applied Science, London, 65–86.

Ferreira IMM, Ferreira AJR and Sims DA (1984) Preliminary analysis of runoff plot data from Vale Formoso Research Station for the years 1962/63–1979/80 in terms of Universal Soil Loss Equation. Drainage and Soil Conservation Project for Alentejo Region. DGHEA, Lisboa.

Goss MJ, Ellers W, Boon FR and Hoose KR (1984) Effects of soil management pratices on soil physical conditions affecting root growth. J Agric Eng Res 30 131–141.

Hewit JS and Dexter AR (1980) Effects of tillage and stubble management on the structure of a swelling soil. J Soil Sci 31 203–217.

Morgan RPC (1992) Soil erosion in the northern countries of the European Community. Workshop on the elaboration of a code of good agricultural pratices 21–22/05/92. Final Report. European Institute for Water, Brussels, 61–67.

Phillips SH and Young Jr HM (1973) No-Tillage Farming. Ed Reiman Associates, Milwaukee, Wisconsin.

Santos MJ and Oliveira AV (1991) Caracterização física do solo ao fim de dois anos de ensaio de mobilização em regime de sequeiro. Pedologia, Oeiras 25: 44–50.

Steenvoorden JH and Bouma J (1987) Optimizing the use of soils: New agricultural and water management aspects. In Barth H and L'Hermite P (eds) Scientific Basis for Soil Protection in the European Community. Elsevier Applied Science, London, 389–408.

Sequeira EM (1988) Protecção do solo no Alentejo. Programa de investigação na área da Pedologia, para concurso de acesso a Investigador Coordenador. EAN, Oeiras.

Sequeira EM (1992) Soil erosion risks in Southern European Countries. Workshop on the elaboration of a code of good agricultural pratices 21–22/05/92. Final Report. European Institute for Water, Brussels, 75–82.

Yassoglou NJ (1987) The production potential of soils: Part II – Sensitivity systems in Souther Europe to degrading influxes. In Barth H and L'Hermite P (eds) Scientific Basis for Soil Protection in the European Community. Elsevier Applied Science, London, 87–122.

C. Rodriguez-Barrueco (ed.), Fertilizers and Environment, 115–119.

Tillage (plowing, subsoiling and no-tillage) effects on soil erodibility

II. Soil losses by runoff and "splash"

F. C. Barreiros, F. P. Pires & E. M. Sequeira
Estação Agronómica Nacional, Quinta do Marquês, 2780 Oeiras, Portugal

Abstract

The effect of three different soil tillage systems (plow, subsoiling and no-tillage) on soil losses by runoff and "splash", was determined in situ using a rain simulator. These determinations were carried out in a field experiment under the Project "Soil Tillage in Dry and Irrigated Farming" (EU Program AGRIMED), installed in a Cambisol from granite, Inland Central Portugal (Fundão). The tests were carried out after removing of mulch and 5 mm of the surface layer. The effect of soil tillage treatments on one hour mean rates of soil loss by runoff, soil loss by "splash", and total soil loss were highly significant.

No-tillage treatment significantly decrease the soil loss by runoff, the soil loss by "splash" and consequently the total soil loss in relation to plow (60%), as previously verified in relation to runoff. Subsoiling had an intermediate behaviour in all cases.

In conclusion, no-tillage seemed to be the best treatment to protect soil from erosion. As it was previously concluded for runoff and infiltration, at least in the initial situation (2-years), with a sandy loam soil with low content of organic matter, subsoiling was not justified.

Introduction

Following a previous paper (Barreiros *et al.*, 1994) where we presented the results of runoff and infiltration of a in situ "typical" rainfall (with a rain simulator), it was still necessary to quantify the soil loss by runoff, as well as to characterize the "splash" effect.

In the previous paper, as well as it was pointed out by other authors (Blevins *et al.*, 1984; Goss *et al.*, 1984; Santos and Oliveira, 1991) it was referred that no-tillage increased soil bulk density at the upper soil layer, but simultaneously increased the infiltration rate.

It was necessary to assess the effect of these modifications on destacability and transportability of the soil particles.

So, no-tillage maintains organic and inorganic protecting surface mulch, as previously referred by many authors (Philips & Young, 1973; Yassouglou, 1987; Sequeira, 1988, 1992), reduces soil erosion by the decrease of runoff as it was previously shown (Barreiros *et al.*, 1994), but certainly increases soil structure stability, soil cohesion, and reduces soil particle detachment and transportability, so it decreases soil erodibility.

We intend to quantify these effects on soil physical properties under three soil tillage treatments.

Material and Methods

In the same trial previously described (Barreiros *et al.*, 1994), and using the same rain-simulator and identical methodology, we collected and quantified, at 10 minutes interval the solid material carried out by runoff. This gave a good index of transportability evolution of soil particles during the rainfall period.

By these data it was possible to assess the effect of the runoff increase on the quantity of carried soil particles, as well as the effect of treatments on soil particle destacability when assessed by the sediment carried out by unity of runoff.

The presence of a water layer above the soil surface changed all conditions. It increased the drop energy available to loose soil particles (Ferreira, 1984) especially at a layer depth between one and three rain drops diameter.

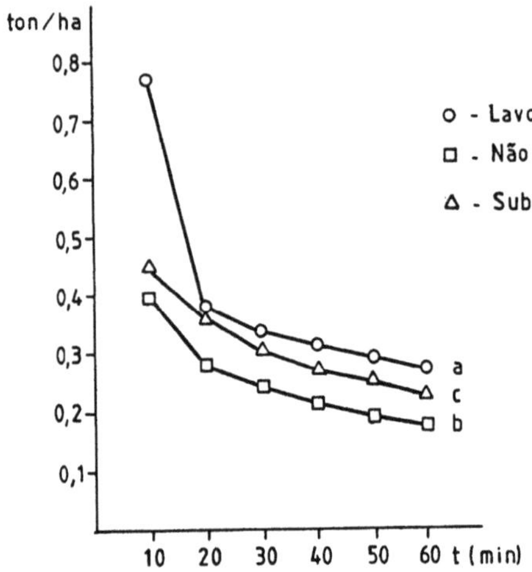

Fig. 1. Soil loss material, carried by runoff at each 10 minutes of applied rainfall, by each tillage modality - plow a - no-tillage b - sub-soiling c

So, the sediments carried out by runoff indicate the destacability and transportability, of soil particles, and they are normally associated with finer fraction of soil particles.

The method used to determine the "splash" carried material was, to collect, by washing to a bottle, the material that remains in the shelter and in the gutter, at the end of each rainfall test, and characterize that material.

Although this soil material do not represent an effective soil loss by "splash" as the finer material was carried out with runoff, it is a good index of the "splashed" material, loosed and projected by direct impact of the drop.

Results and Discussion

Mean soil loss by runoff in each tillage treatments, at 10 minutes interval, (expressed in tones per hectare is shown in Fig. 1. In Fig. 2, a, b and c, the mean rates of soil loss by runoff, (expressed in tones per hectare and per hour), for 10 minutes intervals, in each soil tillage treatments, are shown (a- plow; b- no-tillage; c-subsoiling).

In all treatments, the carried material decreased sharply until 20 minutes, and afterwards tended to an assintotic equilibrium. However, it is clear that:

- plow treatment leads to a very high initial soil loss, probably caused by the loosening of fine soil particles which are transported to the surface by plowing;

Table 1. Final mean data of soil material lost by runoff, "splash" and runoff+"splash", at the end of the simulated rainfall (80 mm rainfall in 1 hour), in the tillage treatments: plow, subsoiling and no-tillage, expressed in tones ha^{-1} h^{-1}.

soil losses	tillage treatments		
	plow	subsoiling	no-tillage
carried out			
by runoff	2.32 ± 0.44	1.88 ± 0.32	1.51 ± 0.45
by "splash"	22.07 ± 1.11	19.57 ± 1.82	13.82 ± 1.61
runoff+"splash"			
(total)	24.35 ± 1.48	21.38 ± 2.02	15.28 ± 1.99

- the increase of fine soil particles at surface (silt and clay), detached by drops impact in plow plots, tend to obstruct macropores, so to decrease permeability and increase transported soil material;

- under other tillage treatments there are no fine soil particles to be loosen at soil surface, with the exception of subsoiled plots, in disturbed places.

- The different behaviour at the equilibrium (Fig 1) certainly resulted from the different runoff rates found under the treatments (Barreiros *et al.*, 1994).

To assess the effect of treatments on soil destacability, we tried to estimate the soil particles carried out by unite of volume of runoff (g l^{-1}). This concentration was greater in plow plots, 3.19 ± 0.52 g l^{-1}, than in subsoiling 3.03 ± 0.51 g l^{-1}, and than in no-tillage 2.56 ± 0.60 g l^{-1}. So, as it was expected, notillage treatment decreased not only the total amount of water runoff, but also decrease the particle destacability.

This difference is greater until 10 minutes (plow - 7.56 ± 1.02 g l^{-1}, subsoiling - 6.17 ± 1.05 g l^{-1}, no-tillage - 5.51 ± 0.83 g l^{-1}), and diminish after, probably due to the decreasing of the fine soil particles quantity that could be loosen by drop impact, as it is clearly observed through the final values (plow 2.11 ± 0.42 g l^{-1}; subsoiling 1.99 ± 0.43 g l^{-1}; no-tillage 1.74 ± 0.30 l^{-1}).

However, there is a great difference in the final soil loss by "splash" (Table 1) in the no-tillage plots and in the other two treatments (no-tillage 13.8 + 1.6, subsoiling 19.6 ± 1.8 and plow 22.1 ± 1.1 tones per hectare per hour).

It must be take in account that this is a sandy loam Cambisol (Santos and Oliveira, 1989), with 20% of gravel in the upper layer, and 60% of coarse sand and 20% of fine sand in the fine in the fine soil fraction. So,

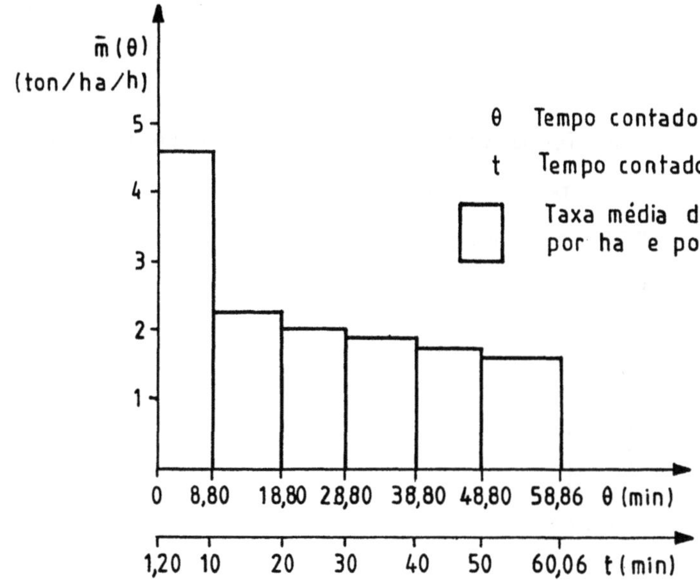

Fig. 2. Mean rate of soil material carried by runoff [m (θ)] expressed in t ha^{-1}, in each 10 minute interval of rainfall applications, for each soil tillage modality: a- plow; b- no-tillage; c- sub-soiling. t - time from the beginning of rainfall θ - time from the biginning of runoff - mean rate of soil matterial carried by runoff in ton ha^{-1}h^{-1} - [m(θ)]

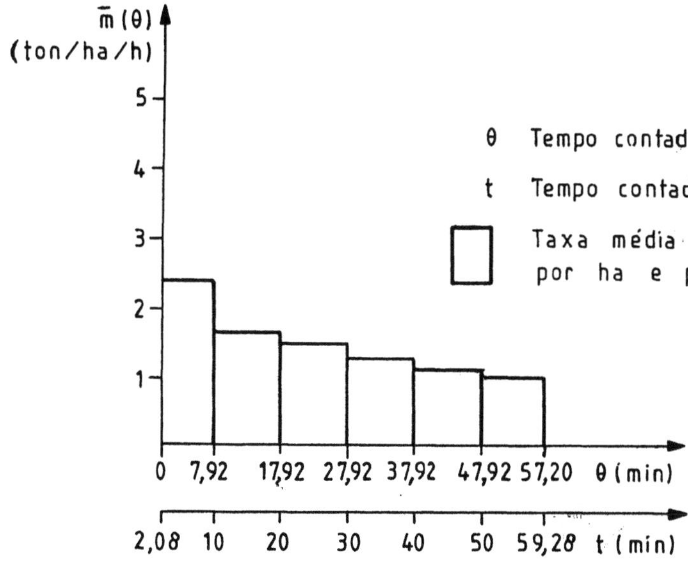

Fig. 2. continued

the fine material that could be carried out in the test by runoff is less than 10% of the total.

In this soil, the "splash" contribution represents about 90% of the total soil loss, probably because the soil coarser fractions rest in the "splash" fraction ; in fact, the collected soil peds, gravel and sand remain in the gutter and in the shelter, i.e., they were not carried out by runoff.

Table 1 shows the final mean data, of lost material, by runoff, by "splash" and by the sum of runoff and "splash" for each tillage treatment. In Table 2, the analysis of variance can be found, where the kind of

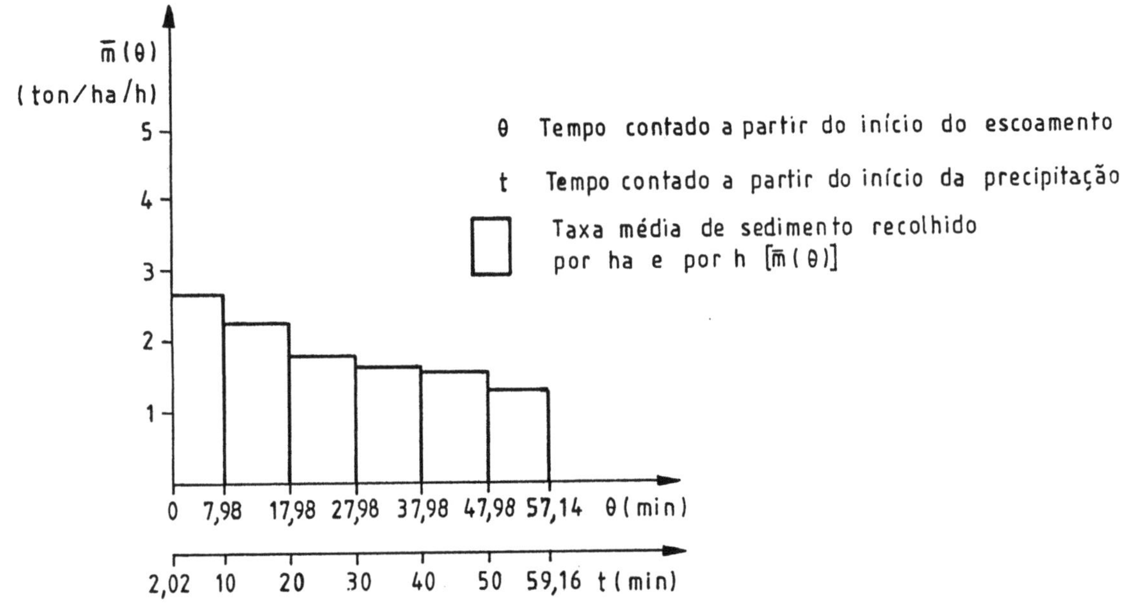

Fig. 2. continued

Table 2. Synopsis for analysis of variance of soil losses by runoff, by "splash", total soil loss, and solids concentration in runoff

Source of variation	kind of soil loss							Solids in runoff	
		runoff		"splash"		tota l			
	d.f.	F values	signif.	F values	signif.	F. values	signif.	F values	signif.
tillage treatment	2	16.31	**	127.10	***	101.70	***	8.84	*
crop	3	0.10	n.s.	2.17	n.s.	1.21	n.s.	0.20	n.s.
replicates	1	0.22	n.s.	0.62	n.s.	1.00	n.s.	6.30	*

*** significant at 0.1% level. ** significant at 1% level. * significant at 5% level, n.s. not significant.

soil loss is related with the tillage treatment., the used crop and the replications.

In Table 2, it can be observed that the effect of crop (leguminous or cereal) has no significative effect. The replicates have a low significant effect (P5%) in solid concentration carried out by runoff. However, all the tillage treatments have a significant effect, highly significant for soil losses carried out by runoff, by "splash" and by the sum of runoff and "splash ".

In these kind of soils, Barreiros (1989) has evaluated the effect of the protective "mulch", when using the rain. simulator; in fact, total soil loss was reduced from 24 ton ha^{-1} h^{-1} to about 1.3 ton ha^{-1} h^{-1}.

These sandy loam soils, with weak structure, with loosening fine and coarse sand materials, are soils where protective mulch is rather important to decrease soil "splash" by high energy rain drops (Barreiros, 1989). However, no-tillage improved soil structure and coherence, thus decreasing soil detachment and particle transport.

References

Barreiros F C (1989) Estudo da erodibilidade de algumas unidades-solo. Dissertação para acesso à categoria de Investigador Auxiliar. Equivalent to a PhD dissertation EAN, Oeiras.

Barreiros F C (1994) Tillage (plowing, subsoiling and no-tillage) effect on soil erodibility. I. Runoff and infiltration. In publication

Blevins R L, Smith M S and Thomas G W (1984) Changes in soil properties under no tillage. In: Phillips R E and Phillips S H (eds) No Tillage Agriculture. Principles and Practices.

Ferreira A G (1984) Influence of a shallow water layer over the soil in the erosion by raindrop impact. PhD dissertation, University of California, Davis, USA.

Goss M J, Ellers W, Boon F R and Hoose K R (1984) Effects of soil management pratices on soil physical conditions affecting root growth. J. Agric. Eng. Res. 30: 131–141.

Phillips S H and Young Jr H M (1973) No-Tillage Farming. Ed. Reiman Associates, Milwaukee, Wisconsin.

Santos M J and Oliverira A V (1991) Caracterização em regime de sequeiro. Pedologia, Oeiras 25: 44–50.

Sequeira E M (1988) Protecção do solo no Alentejo. Programa de investigaçào na área da Pedologia, para concurso de acessso a Investigador Coordenador. Ean, Oeiras.

Sequeira E M (1992) Soil erosion risks in Southern European Countries. Workshop on the elaboration of a code of good agricultural practices (21–22/05/92). Final Report. European Institute for Water, Brussels. pp 75–82.

Yassoglou N J (1987) The production potential of soils: Part II-Sensitivity systems in Southern Europe to degrading influxes. In: Barth H and L'Hermite P (eds.) Scientific Basis for Soil Protection in the European Community. Elsevier Applied Science, London. pp 87–122.

C. Rodriguez-Barrueco (ed.), Fertilizers and Environment, 121–125.

Effect of time and rate of nitrogen application on cotton

H.B. Setatou & A.D. Simonis
Soil Science Institute, 541 10 Thesssaloniki, Greece

Abstract

Field studies were conducted in nine cotton production areas of Greece, for 15 successive years, to detective the optimum time and rate of N fertilizer application for maximum yields and better fibre characteristics of cotton (*Gossypium hirsutum*). Two to five rates of N fertilizer (60–180 kg N/ha) were applied: before sowing, when the plants had 3–4 leaves and 20 days later. The yield of cotton increased in most of the cases by N fertilization, but not significantly to both N rates and time of application. Under Greek conditions – limited growing season and low precipitation in spring – the split application of N fertilizers, compared to a single presowing application did not differ, as regards its effect on yield and fibre quality of cotton. Split application of fertilizers, supplementary to that applied at sowing could be of significance, when N plant demands are high and in addition there is a risk for excessive plant growth or for leaching losses. In the case of high N fertilizers rates, the split application must be preferred.

Introduction

Cotton (*Gossypium hirsutum* L.) requires proper nitrogen (N) fertilization to achieve maximum yields (Constable and Rochester, 1988). Increased yield in response to N is due to more and larger fruit. However, a point is reached on the N response curve, where lint yield is maximized, and in some environments a reduction in yield or fibre quality can occur if excessive N is applied (Maples, 1990). Components of these reductions include delayed maturity, reduced ginning percentage, greater incidence of disease and greater attractiveness to insect pest (Hearn, 1975; Constable and Hearn, 1981).

Greek cotton production has increased rapidly in the last decade (Setatou and Simonis, 1993). This expansion has been due mostly to an increased area of cropping, but there has also been emphasis on increasing yield particularly in view of increasing costs of production. One ingredient of high yield is sound nutrition, and there has been a gradual trend for N rates to increase over this time, and for fertilizer application to be used. Research elsewhere (Maples and Frizzell, 1985a) has shown that applying N closer to the stage of peak demand can improve N utilization, particularly on lighter textured soils where leaching is more probable. Later applications may allow the crop to be determined later in the season, giving more precise total fertilizer rates (Maples, 1990). However, wet conditions following late fertilizer applications could lead to severe production problems by further delaying maturity and by producing rank vegetative growth (Maples and Frizzell, 1985b).

The maturity of the cotton crop as evidence by first and second picking lint yield, was found to depend on the rate of N application (Massagni *et al.*, 1993). In general, cotton growth is delayed by N fertilization (Setatou and Simonis, 1993).

Although extensive studies have been conducted to determine the effects of fertilizer application on cotton lint yields, less attention has been given to their influence on fibre characteristics (Murray *et al.*, 1965). Increases in lint length from N fertilizer treatments have been reported (Gardner and Tucker, 1967). However, other investigators stated reductions in both fibre length and strength (Jackson and Tilt, 1968) or no effect of fertilizers upon fibre properties (Hardy and Garret, 1965).

The objective of these studies was mostly to determine if split applications could improve the utilization of N fertilizer and lead to improved yield or profitability.

Table 1. Description of N experiments with cotton

Years											
1965–1971			1972–1975			1976–1977			1978–1980		
Time of N application											
A	B	C	A	B	C	A	B	C	A	B	C
kg N/ha											
0	0	0	0	0	0	0	0	0	0	0	0
80	0	0	60	0	0	60	0	0	90	0	0
40	40	0	30	30	0	60	30	0	120	0	0
0	40	40	30	30	0	60	30	0	120	0	0
120	0	0	90	0	0	60	30	30	90	30	0
80	40	0	60	0	30	90	0	0	90	0	30
40	40	40	60	0	30	90	30	0	150	0	0
			120	0	0	90	0	30	90	30	30
			90	30	0	120	0	0	120	30	0
						90	0	30	120	0	30
									180	0	0
									120	30	30

A= at sowing; B= when the plants had 3-4 leaves; C= 20 days later

Methods

Sixty-three experiments were conducted at nine sites of cotton production areas of Northern Greece for 15 successive years. The soils used were fine to coarse textured, with low organic matter (1.0–1.5%) and pH ranged from 5.3–7.2.

There were four series of experiments differing in the rate and time of N application (Table 1).

The experimental design was a randomized complete block with five replications. Individual plots included four rows that were 10 m in length and spaced 1 m apart. Greek cotton variety 4S was planted on March–April each year and the final plant populations were approximately 165,000 plants per ha. The fields were irrigated by furrows and cultural practices for weed and insect control were conducted on an as-needed basis for all experiments.

The fertilizers used were ammonium sulphate for the application at sowing and ammonium nitrate for the split applications. The various rates, as indicated in Table 1, were applied a) before sowing; b) when the plants had 3–4 leaves; and c) 20 days later. A standard dressing of 60 kg P/ha/year was applied to all plants, as superphosphate at planting.

Yield was measured by machine harvesting the two center rows of each plot and weighing the seed cotton. Boll weight and lint percentage were determined by picking 50 mature cotton bolls from each plot, one day before the first picking at random. Fibre samples were also taken from each plot, blended and otherwise prepared for analyses in the fibre laboratory. Fibre length was measured on a digital fibrograph, fineness with the micronaire and strength with the stelometer.

Data were analysed statistically according to the standard methods (Steel and Torry, 1983).

Results

Seed cotton yield at maturity as influenced by fertilizer rate and time of application, for the four series of experiments, are shown in Table 2.

The yield of cotton was increased in most of the experiments by N application, but not significantly to both N rates and time of application. Seed cotton yield ranged on an average from 2470 to 3380 kg/ha in the four series of experiments. Yields obtained with high N rates approached the maximum yielding ability of the variety used, indicating that factors other than N nutrition were not limiting. These results are in agreement with the data of other investigators (Amer and Abuanin, 1969), but also they are opposed to other findings (Amer *et al.*, 1964).

Split applications increased yields in some experiments, but only at high rates of applied N fertilizer. In all other experiments, there was no yield difference with time of application. Other researchers also found no increase in seed yields due to time of N application (Constable and Rochester, 1988).

Table 3 shows the effect of fertilization on the maturity of cotton. It was found that N fertilization caused a delay in the maturity of cotton plants, ranging from 0.2 to 2.5 days in comparison to the control in most of the experiments. However, this delay is important only when the harvesting conditions are unfavourable.

Among the fibre characteristics only lint percentage was affected by fertilization (Table 4). A decrease in lint percentage (0.5 to 1.0) unit) was observed. The data found in relation to the effect of N fertilization (rate-time) on the fibre length and the fibre strength were contradictory, while there was an indication that the fibre fineness was related to the delay of cotton maturity.

Table 2. Effect of time and rate of N application on the yield of cotton.(Kg/ha)

Years:1965 - 71

time of application * A B C	Larissa (3)	Karditsa (6)	Phthiotis (4)	Arta (7)	Average (20)	%
0 0 0	2720	3270	1390	2300	2470	100
80 0 0	3300	3440	1860	2350	2720	110
40 40 0	3030	3470	1770	2450	2700	109
0 40 40	3160	3410	1680	2380	2670	108
120 0 0	3120	3510	1920	2340	2730	111
80 40 0	3130	3420	1890	2460	2740	111
40 40 40	3150	3360	1890	2430	2710	110

*numbers in parenthesis= number of experiments.

Years:1972 - 75

time of application A B C	Thes/ki (1)	Kilkis (1)	Veria (5)	Serres (4)	Larissa (4)	Karditsa (5)	Viotia (5)	Average (25)	%
0 0 0	3110	2980	2650	3240	2550	2410	2070	2600	100
60 0 0	3490	3420	3260	3670	2650	3430	2470	3120	120
30 30 0	3520	3540	3290	3750	2750	3400	2470	3150	121
30 0 30	3390	3420	3250	3600	2650	3360	2420	3080	118
90 0 0	3420	3220	3370	3630	2780	3570	2640	3210	123
60 30 0	3000	3450	3350	3580	2830	3690	2510	3190	123
60 0 30	3570	3440	3330	3590	2710	3710	2570	3210	123
120 0 0	3490	3390	3350	3630	2860	3770	2700	3280	126
90 30 0	3830	3380	3370	3520	2980	3800	2730	3310	127
90 0 30	3530	3370	3520	3600	2740	3830	2670	3290	127

Years:1976 - 1977

time of application A B C	Veria (1)	Larissa (2)	Karditsa (2)	Viotia (2)	Average (7)	%
0 0 0	3640	2970	3720	2400	3110	100
60 0 0	3610	3230	4040	2640	3340	107
60 30 0	3700	3340	3690	2630	3290	106
60 0 30	3830	3400	3890	2570	3360	108
60 30 30	3750	3380	3780	2500	3300	106
90 0 0	3520	3330	3960	2650	3340	107
90 30 0	3690	3420	3690	2540	3280	105
90 0 30	3690	3410	3930	2580	3360	108
120 0 0	4000	3200	3970	2680	3380	109

Years:1978 - 80

time of application A B C	Veria (2)	Larissa (3)	Karditsa (3)	Viotia (3)	Average (1)	%
0 0 0	2430	3220	2830	2960	2870	100
90 0 0	2420	3350	3030	2860	2960	103
120 0 0	2800	3610	2910	3010	3110	187
60 30 30	2470	3650	3190	2890	3110	108
90 30 0	2280	3690	3150	2840	3050	106
90 0 30	2600	3550	2970	2900	3040	106
150 0 0	2170	3560	3250	2950	3050	106
90 30 30	2200	3600	3130	2980	3050	106
120 30 0	2440	3630	3110	2970	3090	108
120 0 30	2260	3540	3140	2980	3040	106
180 0 0	2330	3370	2970	3000	2970	103
120 30 30	2210	3460	3200	2860	3000	105

A=at sowing, B-when the plants had 3-4 leaves, C-20 days later

Discussion

The data presented indicate that split application of N fertilizer had no effect on the yield and fibre properties of cotton.

Cotton requires large amounts of N, particularly under irrigation. The amount of N applied to cotton should be sufficient to achieve maximum returns of the crop, while avoiding problems, such as excessive growth and delayed maturity which in turn could lead to reduced fibre quality, or cause defoliation, or harvest difficulties (Constable and Rochester, 1988). The extent of rain precipitations is very important during the first three months of the cotton vegetative cycle (Brown, 1984). In the cotton producing areas of Greece, rain precipitations during the first three months after planting are only 170 mm (average of 30 years), quantity that is very low for some benefit to be expected. Moreover, the vegetative growth period for cotton in Greece is very limited and as a consequence suffi-

Table 3- Effect of time and rate of N application on the earliness of cotton (days of the control)

Years : 1965 -

Time of application A	B	C	Larissa (3)	Karditsa (6)	Phthiotis (4)	Arta (7)	Average (20)
0	0	0	25/9	1/10	21/9	2/10	27/9
80	0	0	1.4	-1.3	0.9	-0.2	0.2
40	40	0	0.6	-0.9	-0.1	1.7	0.7
0	40	40	2.0	-1.4	1.1	0.7	0.6
120	0	0	0.9	-0.8	1.0	-0.2	0.2
80	40	0	1.7	0.5	1.3	1.0	1.1
40	40	40	1.3	0.5	1.2	0.7	0.9

Years : 1972 - 75

Time of application A	B	C	Thessaloniki (1)	Kilkis (1)	Veria (5)	Serres (4)	Larissa (4)	Karditsa (5)	Viotia (5)	Average (25)
0	0	0	29/9	30/9	19/9	3/10	27/9	21/9	14/9	25/9
60	0	0	1.2	2.8	1.4	1.6	-0.8	0.7	0.2	0.7
30	30	0	0.8	2.4	1.8	1.1	-0.1	1.5	0.9	1.1
30	0	30	1.5	2.1	2.1	1.0	0.7	3.0	1.5	1.7
90	0	0	0.4	3.4	2.4	1.2	1.4	1.5	0.9	1.5
60	30	0	2.6	5.0	2.9	1.9	0.7	2.5	0.9	2.0
60	0	30	0.9	4.5	1.7	2.6	0.3	2.3	1.7	1.8
120	0	0	2.2	3.9	4.0	4.5	0.6	1.8	1.3	2.5
90	30	0	2.2	1.8	2.5	2.5	0.8	2.5	0.8	1.9
90	0	30	3.2	3.7	3.1	3.5	1.8	2.6	1.2	2.5

Years : 1976 - 77

Time of application A	B	C	Veria (1)	Larissa (3)	Karditsa (3)	Viotia (3)	Average (10)
0	0	0	26/9	30/9	27/9	14/9	24/9
60	0	0	-2.3	-0.9	0.3	0.6	-0.3
60	30	0	-0.7	-0.5	2.9	1.0	0.9
60	0	30	-3.0	1.4	1.8	0.2	0.5
60	30	30	0.9	0.4	2.0	0.5	0.9
90	0	0	0.8	0.3	1.6	-0.6	0.5
90	30	0	-1.2	0.5	0.4	0.0	0.1
90	0	30	-1.6	0.7	-0.7	0.0	-0.2
120	0	0	0.1	0.6	1.0	-0.2	0.4

Years: 1978 - 80

Time of application A	B	C	Veria (1)	Larissa (2)	Karditsa (2)	Viotia (2)	Average (7)
0	0	0	24/9	16/9	21/10	26/9	29/9
90	0	0	-1.1	0.3	0.6	0.1	0
120	0	0	-1.2	0.7	0.7	1.1	0.7
60	30	30	1.2	0.9	0.9	1.1	1.0
90	30	0	-0.5	1.5	0.1	0.6	0.4
90	0	30	-0.9	1.5	0.9	1.5	0.8
150	0	0	-2.0	0.7	0.6	0.6	0
90	30	30	-1.3	1.5	1.0	1.2	0.6
120	30	0	-2.1	0.1	1.0	1.5	-0.8
120	0	30	-0.5	0.5	1.0	1.3	0.6
180	0	0	-1.6	1.5	0.8	0.6	0.3
120	30	30	-1.8	1.4	0.8	1.7	1.1

* numbers in parenthesis = number of experiments

A = at sowing , B = when the plants had 3 - 4 leaves , C = 20 days later

cient amount of N at the initial growth phase of plants is more important.

Although certain environmental conditions such as moisture and temperature greatly affect both yield and fibre properties, the data presented suggest that fertilizer exerts its influence primarily on yield components and not on lint quality. Apparently, the cotton plant requires a certain level of nutrients to produce a given amount of seed, which controls lint yield (Murray *et al.*, 1965).

Cotton yield response to fertilization depended on the location (Table 2). In areas with favourable environmental conditions (Viotia, Karditsa) the effect of fertilization on cotton yield was higher. On the contrary in sites located in a marginal for cotton growth environment, the effect was generally limited.

Table 4. Effect of time and rate of N application on cotton lint percentage (average %)

1965-70

	A	B	C	
0				37.5
80	36.9	37.0	37.1	37.0
120	36.9	36.7	36.8	36.8
	36.9	36.9	37.0	

1971-75

	A	B	C	
0	38.1	37.8	37.7	38.6
60	37.4	37.5	37.7	37.9
90	37.6	37.6	37.5	37.5
120	37.7	37.6	37.6	37.6

The effect of fertilization in most of the cases was the same among years, although yields were quite variable. The average increase resulting from fertilization, for each year, was almost constant. It seems that variability in the effect of fertilization might be expected when the deviation from the environmental conditions, as determined by the soil, the climate and the cultural practices, is very high.

In conclusion, the data suggest that, under Greek conditions – limited growing season and low precipitation in spring – the split application of N fertilizers, compared to a single presowing application did not differ, as regards its effect on yield and fibre quality of cotton. Split application of fertilizers, supplementary to that applied at sowing could be of significance, when N plant demands are high and in addition there is a risk for excessive plant growth or for leaching losses. In the case of high N fertilizers rates, the split application must be prefered.

References

Amer F and Abuamin H (1969) Evaluation of cotton response to rates, sources, and timing of nitrogen application by petiole analysis. Agr J 61: 635–639

Amer F, Elgabaly M and Balba M (1964) Cotton response to fertilization on two soils differing in salinity. Agron J 57: 208–211

Brown H (1984) A study of the effect of fertilizers on various characters of the cotton plant. Louisiana Agric Exp Stn Bull 406

Constable GA and Hearn AB (1981) Irrigation for crops in a subhumid environment. VI. Effect of irrigation and nitrogen fertilizer on growth, yield and quality of cotton. Irrig Sci 3: 17–28

Constable GA and Rochester IJ (1988) Nitrogen application to cotton on clay soil: Timing and soil testing. Agr J 80: 498–502

Hardy GW and Garrett JD (1965) Nitrogen sources, levels and timing for cotton on clay soils in N Arkansas. Univ of Arcansas, Agric Exp Stn Bull 140

Hearn AB (1975) An economic assessment of the response of cotton to nitrogen and water in the Ord Valley. North Western Australia. Cotton Grow Rev 52: 165–188

Jackson EB and Tilt PA (1968) Effects of irrigation intensity and nitrogen level on the performance of eight varieties of uplant cotton, *Gossypium hirsutum* L. Agron J 60: 13–17

Maples RL (1989–90) Nitrogen can increase cotton yield and soil organic matter. Better Crops, Winter, 16–17

Maples R and Frizzell M (1985a) Timing of nitrogen fertilizer applications for cotton. In: Nelson TC (ed) p 307. Proc Beltwide Cotton Prod Res Conf, New Orleans, LA, 6–11 Jan National Cotton Council, Memphis, TN

Maples R and Frizzell M (1985b) Effects of varying rates of nitrogen on three cotton cultivars. Univ of Arcansas Agric Exp Stn Bull 882

Massagni HJ, Keissling TC and Maples RL (1993) Response of fast fruiting cotton cultivars to N on a clay soil. J Prod Agr 6: 104–111

Murray JC, Reed RM and Oswald ES (1965) Effect of fertilizer treatments on the fiber properties of cotton. Agr J 57: 227

Setatou HB and Simonis AD (1993) Response of cotton to NPK fertilization – The Greek experience. World Cotton Res. Conf. 1, 13–17 Feb. 1994, Brisbane, Australia (in press)

Steel, R and Tozzie, S (1960). Principles and procedures of statistics, McGrow (ed.) pp 479.

C. Rodriguez-Barrueco (ed.), Fertilizers and Environment, 127–130.

Stimulation of nodulation, nitrogen fixation and plant growth of faba bean by cobalt and copper additions

A.M. Abdel Wahab*, M.H. Abd-Alla & A.E. El-Enany
Department of Botany, Faculty of Science, Assiut University, 71516, Assiut, Egypt

Key words: Cobalt, copper, faba bean, leghaemoglobin, N_2-fixation, nodulation

Abstract

The effect of different concentrations of cobalt or copper on nodulation nitrogenase activity and growth of faba bean was studied in pot experiments. Cobalt at 100 ppm increased nodule number and weight by 63.4 and 81.3%, respectively. Cobalt at this level induced 114.8% increases in absolute N_2-ase activity. Copper promoted nodule mass by 44.7% and absolute N_2-ase by 61.6%. Leghaemoglobin content, dry matter and total nitrogen content of shoots and roots were also significantly increased. Higher levels up to 200 ppm of both cobalt and copper had no deleterious effects on *Rhizobium-Vicia faba* symbiosis.

Introduction

Cobalt and copper are essential trace elements for host legumes and their associated bacteria (Riley and Dilworth, 1985; Seliga, 1993). Their absence may restrict development of free living rhizobia in rhizophere, growth and nodulation of host plant and impair the nodule functions. Deficiency of cobalt and copper occurs in many organic, sandy and calcareous soils throughout the world.

Cobalt is essential for the growth of legumes under symbiotic conditions (Bollard, 1983). Thus, cobalt limitation has been shown to reduce nodule initiation growth and nitrogen fixation in *Lupinus angustifolius* dependent on symbiotically fixed N_2 (Chatel *et al.*, 1978; Dilworth *et al.*, 1979). However, it has not been demonstrated whether cobalt may be limiting for nodule initiation or processes prior to it.

Copper deficiency was found to have an indirect effect on nodule formation. Nodules from Cu-deficient plants had fewer bacteroids, more starch and less cytochrome c-oxidase than nodules from Cu-sufficient plants (Hallsworth, 1972). Increasing the soil level of nitrogen should be accompanied by an increase in the copper supply to legume plants in order to obtain an effective population of rhizobia, good nodulation and efficient nitrogen fixation. Under conditions of copper deficiency, symptoms of nitrogen deficiency are produced in subterranean clover despite the development of effective nodules (Hallsworth *et al.*, 1964).

The effect of copper and cobalt on nodulation and nodule physiology of *Vicia faba* plants, as far as the authors are aware, has not been examined. The purpose of this investigation was to determine the effect of cobalt and copper nutrition on nodulation, nodule activity and growth of *Vicia faba* plants growing in an agricultural soil.

Materials and methods

Plant culture

Surface-sterilized faba bean seeds (*Vicia faba* cv. Giza 3) were inoculated with 5 ml *Rhizobium leguminosarum* biovar viceae strain RCR 1001 per five seeds (inoculum was 10^6 cells/ml), and planted into plastic pots containing 5 Kg autoclaved clay soil. The soil analysis was as follows: pH, 7.1; total soluble salts 0.26%, organic matter content 2.6% and total nitrogen 0.15%. The pots were given nutrient solution (Riley and Dilworth, 1985) only once with the addition of increasing levels of cobalt and copper (25, 50, 100 and 200 ppm) as cobalt or copper sulphate solution per pot. Control pots were given cobalt or Cu-free nutrient solution. The pots were thereafter irrigated with

* Author for correspondence.

tap water at 15- and 18-days intervals. Seedlings were thinned to two per pot after 7 days. The temperature ranged between 23–9 °C (day/night).

Treatments were layed out in a randomised block design with three replicates each. Plants were harvested 67 days after planting. Parameters measured were nodule number, nodule air dry mass, N_2-ase activity (colorimetrically, La Rue and Kurz, 1973), Leghaemoglobin of nodule cytosol (Johnson and Hume, 1973) and protein contents of nodule cytosol and bacteroids (Lowry *et al.*, 1951). Dry weights of shoots and roots, and N content of plant tissues were determined by the Kjeldahl method.

Analysis of variance was done by PC-state programme and the least significant difference (LSD) was used to test the significance between treatments.

Results and discussion

Nodulation and nodule activity

Cobalt up to 100 ppm and Cu up to 50 ppm significantly increased nodule number and their mass (Table 1). The nodule number of the 100 ppm Co-treated plants showed 63.4% increase over the controls and their weight increased by 81.3%. The nodule number of 50 ppm Cu-treated plants showed 13.3% increases over the Cu-deficient plants and their mass increased by 44.7%. Calculations of nodule number/g dry mass root indicate that Co and Cu applications had a more positive effect on nodule formation rather than on root growth (Table 1). Moreover, absolute N_2-ase activity was remarkably enchanced though Specific N_2-ase activity (SNA), was slightly increased. Thus 100 ppm Co and 50 ppm Copper induced a 114.8 and 61.6% increases in absolute N_2-ase activity, whereas nodule activity (SNA) increased only by 18 and 9%, respectively. The enhancement in total N_2-ase activity could be attributed to the increased nodule number and mass per plant (Table 1) as a result of Co and Cu treatments. Higher rates upto 200 ppm of both Co and Cu had no adverse effects on nodulation and N_2-fixing activity. Results of the present study indicate that leghaemoglobin (Lgb) and protein contents of nodule cytosol and bacteroids were also significantly increased.

Results of Table 1 show significant increases of Lgb content. This finding supports previous suggestions concerning the possible role of Co in Lgb synthesis which is essential for active N_2-fixing sites. Results reported here indicate that soil with suboptimal Co can limit nodulation and consequently the total N_2-ase activity and growth of faba bean plants.

Although legumes require very small amounts of cobalt, field situations have been identified where there was a significant response to applied cobalt e.g. with Lucerne (Powrie, 1964) and with *Lupinus* (Chatel *et al.*, 1978). Responses in nodule weight, growth and N content of plants were also observed (Bollard, 1983).

Robson *et al.* (1979) and Dilworth *et al.* (1979) suggested that the essential role of Co in N_2-fixing legumes may be to allow synthesis of adequate vitamin B_{12}, which in turn is probably required for the synthesis of Lgb.

Inhibition of nodule initiation and/or functions could be ascribed to Co-deficiency. It is therefore recommended to add small amounts of Co salts when fertilizing legumes with e.g. N, P and K.

The results of Cu treatments (Tables 1 and 2) indicate that, in faba bean, copper influenced two separate processes. Firstly, it took part in the development of *Vicia* plants through increases in shoot and root dry weight and N content. The second process affected was symbiotic N_2-fixation. This was deduced not only on the basis of increase in nodulation and N_2-ase as an effect of Cu treatment, but above all, of higher levels of Lgb in nodules and higher N accumulation in whole plants. These results closely agree with the findings of Seliga (1993). Who reported that application of Cu to soil increased N_2 fixation (C_2H_2 reduction), nodule number and weight as well as the yield and N content of lupin. In other experiments, a low level of N_2 fixation in nodules of Cu deficient plants was accompanied by starch accumulation in nodules (Weryszko, 1992).

Apparently, the activation in SNA (Table 1) was due to Cu-treatments increasing the Lgb and protein contents of nodule cytosol and bacteroids. Seliga (1993) reported that nodules of Cu-deficient plants contained less than half the Lgb concentration of copper-adequate plants. Yates and Hallsworth (1963) found that soluble amino acids of the nodule increaseed within 24 hours of increasing the copper supply to the plant. Sandmann and Boger (1983) reported that Cu plays an important role in photosynthetic electron transport enzymes and several other metalloenzymes of plants. This in turn is translated to a better and healthy plant growth with adequate photosynthate supply to nodules.

The influence of Co and Cu application on dry matter production and nitrogen index (Table 2) show that

Table 1. Effect of cobalt and copper additions on nodulation. N-ase activity, leghaemoglobin and protein contents of *Vicia faba* nodules[a]

Treatment	Nodules/plant		Nodule number/ g root	Nitrogenase activity μ mol C_2H_4/h		Leghaemoglobin mg/ g nodule fresh wt	Protein mg/g nodule fresh wt	
ppm	No.	Fresh wt(g)		per plant	per g nodule fresh wt	Cytosol	Cytosol	Bacteroid
Control (0)	123	1.23	42.40	7.4	6.01	3.4	5.3	7.4
Cobalt								
25	170	1.79	48.60	10.9	6.08	3.8	6.5	8.2
50	190	2.17	52.7	16.2	7.50	4.3	6.7	8.4
100	201	2.23	52.9	15.9	7.10	4.1	6.8	8.9
200	118	1.22	43.1	7.3	6.0	3.5	5.2	7.3
Copper								
25	150	1.55	46.9	9.3	6.3	3.5	6.2	7.2
50	163	1.78	52.3	11.8	6.6	3.9	6.1	8.0
100	132	1.45	47.3	8.9	6.2	3.6	6.4	7.5
200	121	1.26	48.3	7.8	6.1	3.4	6.8	7.1
L.S.D.(P=5%)	32.3	0.46	–	3.2	0.35	0.29	0.39	2.6

Table 2. Effect of cobalt and copper treatments on dry matter production and N accumulation of *Vicia faba* plants[a]

Treatment ppm	Dry matter g/plant		Total nitrogen mg/plant	
	Shoot	Root	Shoot	Root
Control (0)	6.36	2.9	445	116
Cobalt				
25	8.25	3.5	602	158
50	8.31	3.6	631	156
100	9.42	3.8	668	168
200	6.22	2.7	450	112
Copper				
25	8.10	3.4	575	138
50	7.90	3.1	577	141
100	6.40	2.8	491	113
200	6.23	2.5	434	110
LSD (*p*=5%)	1.22	0.4	89	40

[a] Each value represents the mean of three replicates.

Co up to 100 ppm and Cu up to 50 ppm significantly promoted plant growth and significanty increased the accumulation of N in whole plants. Higher levels of Co and Cu (200 ppm) had insignificant effect on dry matter accumulation and nitrogen content of shoots and roots. Ahmed and Evans (1961) suggested that Co requirements for N_2-fixation are greater than those for host-plant growth. Earlier, Greenwood and Hallsworth (1960) reported that higher levels of Cu increased dry matter yields and N concentration of plant tops. confirming the results obtained here.

Results of the present work and those obtained by other authors are enough to suggest that Co and Cu application is essential for the enhancement of nodulation and N_2 fixation of faba bean and other legumes.

References

Ahmed S and Evans HJ (1961) The essentiality of cobalt for soybean plants grown under symbiotic conditions. Proc Nat Acad of Sci, USA 47: 24–35.

Bollard EG (1983) Involvement of unusual elemnts in plant growth and nutrition. Encyclopedia of Plant physiol 15B: 695–744.

Chatel DL, Robson AD, Gartrell JW and Dilworth MJ (1978) The effect of inoculation and cobalt application on the growth of and nitrogen fixation by sweet lupins. Austral J Agric Sci 29: 1191–1202.

Dilworth MJ, Robson AD, and Chatel DL (1979) Cobalt and nitrogen fixation in *Lupinus angustifolius* L. II Nodule formation and function. New Phytol 83: 63–79.

Greenwood EAN and Hallsworth EG (1960) Studies on the nutrition of forage legumes. II. Some interactions of calcium, phosphorus copper and molybdenum on the growth and chemical composition of *Trifolium subterranean* L. Plant Soil 12: 97–127.

Hallsworth EG (1972) Factors affecting the response of grain legumes to the application of fertilizers. In use of isotopes for study of fertilizer utilization by some crops. IAEA, Vienna.

Hallsworth EG, Greenwood EAN and Yates MG (1964) Studies on the nutrition of forage legumes. Plant Soil 20: 17–33.

Johnson HS and Hume DJ (1973) Comparison of nitrogen fixation estimates in soybeans by nodule weight, leghaemoglobin content and acetylene reduction, Cand J Microbiol 19: 1164–1168.

LaRue TA and Kurz WGW (1973) Estimation of nitrogenase using a colorimetric determinations for ethylene. Plant Physiol 51: 1074–1075.

Lowry OH, Rosbrough NJ, Farr AL and Randall RJ (1951) Protein neasurement with Folin phenol reagent. J Biol Chem 183: 265–275.

Powrie JK (1964) The effect of cobalt on the growth of young lucerne on a siliceous sand. Plant Soil 21: 81–93.

Riley IT and Dilworth MJ (1985) Cobalt status and its effects on soil populations of *Rhizobium lupini*, Rhizosphere colonization and nodule initiation. Soil Biol Biochem 17: 81–85.

Robertson WK, Thompson LG and Martin FG (1973) Manganese and copper equirements for soybean. Agron J 65: 641–644.

Robson AD, Ditworth MJ and Chatel DL (1979) Cobalt and nitrogen fixation in Lupinus angustifolius L.I. Growth, nitrogen oncentrations and cobalt distribution. New Phytologist 83: 52–62.

Sandmann G and Boger P (1983) The enzymological function of heavy metals and their role in electron transfer processes of plants. Encyclopedia of Plant Physiol 15B: 564–596.

Seliga H (1993) The role of copper in nitrogen fixation in Lupinus luteus L. Plant Soil 155/156: 349–352.

Weryszko-Chmielewska E (1992) Anatomical and morphological changes in organs of some plant species at copper deficiency (*in Polish*). Ed. Agricultural University, Lublin, Poland.

Yates MG and Hallsworth EG (1963) Some effects of copper in the metabolism of nodulated subterranean clover. Plant Soil 19: 265–284.

C. Rodriguez-Barrueco (ed.), Fertilizers and Environment, 131–149.

The role of inorganic fertilizers and their management practices

Christian Hera
Soil Fertility, Irrigation on Crop Production Section, Joint FAO/IAEA Division, Vienna, Austria

Key words: soil fertility, fertilizer, plant nutrition, crop response, nutrient mining, isotopes

Abstract

Nutrient management is the key issue in sustainable soil fertility. N, P, K fertilization aims not only for a high economic return of the investment through optimized yield and quality, but also for minimum environmental hazards. The basic concept underlying integrated plant nutrition systems, is the maintenance and possible increase of soil fertility for sustaining enhanced crop productivity through optimal use of all sources of plant nutrients, particularly inorganic fertilizer, in an integrated manner and as appropriate to each specific ecological, social and economic situation. Much research has established the importance of fertilizers in increasing the fertility of soil and in influencing its productivity. It has been observed that applying fertilizers causes many changes in the soil, including chemical changes, that can positively or negatively influence its productiveness. Only a fraction of the fertilizer applied to the soil is taken up by the crop, the rest either remains in the soil or is lost through leaching, physical wash-off, fixation by the soil, or release to the atmosphere through chemical and microbiological processes. The critical information on the relative merits of different fertilization practices such as method of fertilizer placement, time and rate of application and type of fertilizers, is essential. Results from different field and laboratory experiments which helped to achieve maximum efficiency, in the most economical and sustainable way of fertilizer use to reduce the nutrient losses and production costs to the farmers and prevent environmental pollution are presented in the paper.

Introduction

Food requirements for the ever increasing world population, which now stands at 5.5 billion and will increase to nine billion people by the year 2030, especially in developing countries, can be satisfied only by applying improved technologies, included rational use of fertilizers, adequate to different soil and climatical conditions [2, 7, 15–17]. We all depend on plants for our food, and plants depend on mineral nutrients and water for their growth and development.

Loss of soil fertility in many developing countries due to continual nutrient mining by crop removal, without adequate replenishment could result in a deterioration not less dangerous than other forms of environmental degradation. The human activities led to the degradation of nearly 2000 million hectares of land in the last 40 years. The per capita arable land area fell by 25 percent between 1950 and 1970 and will decline by a further 15 percent by the year 2000 [2, 7].

To meet the challenge of food security for present and future populations, to conserve natural resources and to protect the environment, it is imperative that agricultural production be conducted within an overall framework of sustainable development. An important factor in this respect is the maintenance and enhancement of soil fertility through an appropriate application of plant nutrients in order to replenish the nutrients removed by the harvest of produce and to build up the nutrient status of soils that are inherently infertile or have been depleted.

Over the last ten years FAO has been actively supporting the development of an Integrated Plant Nutrition System (IPNS), an approach through which the management of plant nutrition and soil fertility in cropping and farming systems is adapted to site characteristics and to locally available resources.

Though the efficient use of mineral fertilizers is recognized to be a quick and sure way of boosting crop production, it must be realized that their cost and other constraints frequently deter farmers, especially from

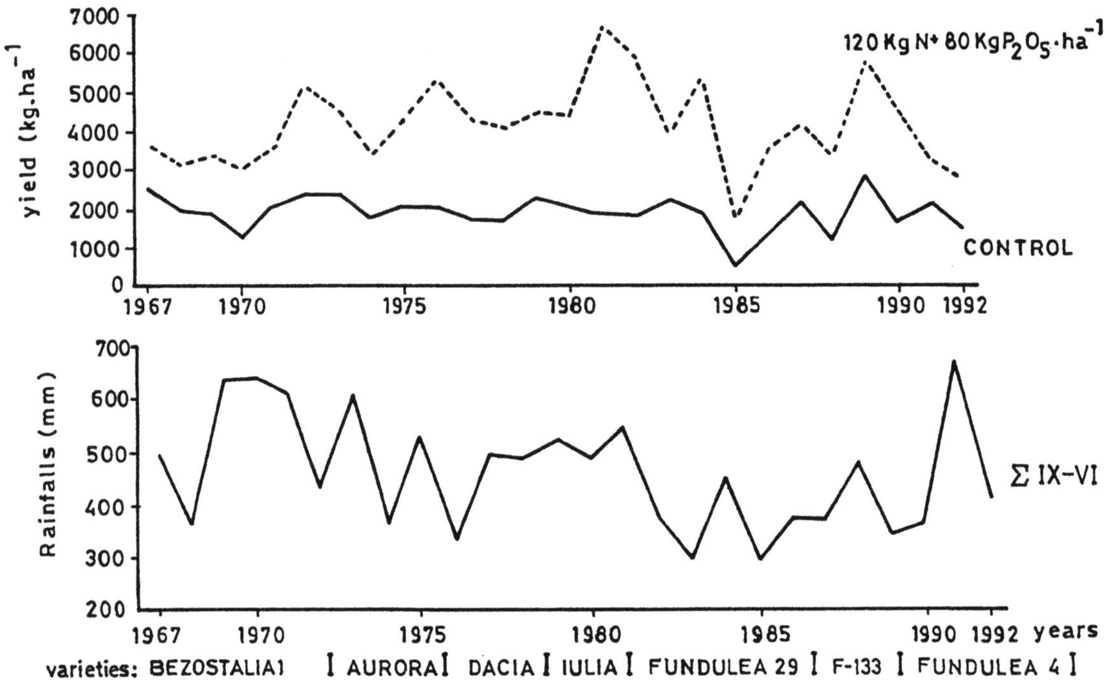

Fig. 1a. Winter wheat yield and rainfalls Fundulea – Romania (1967–1992).

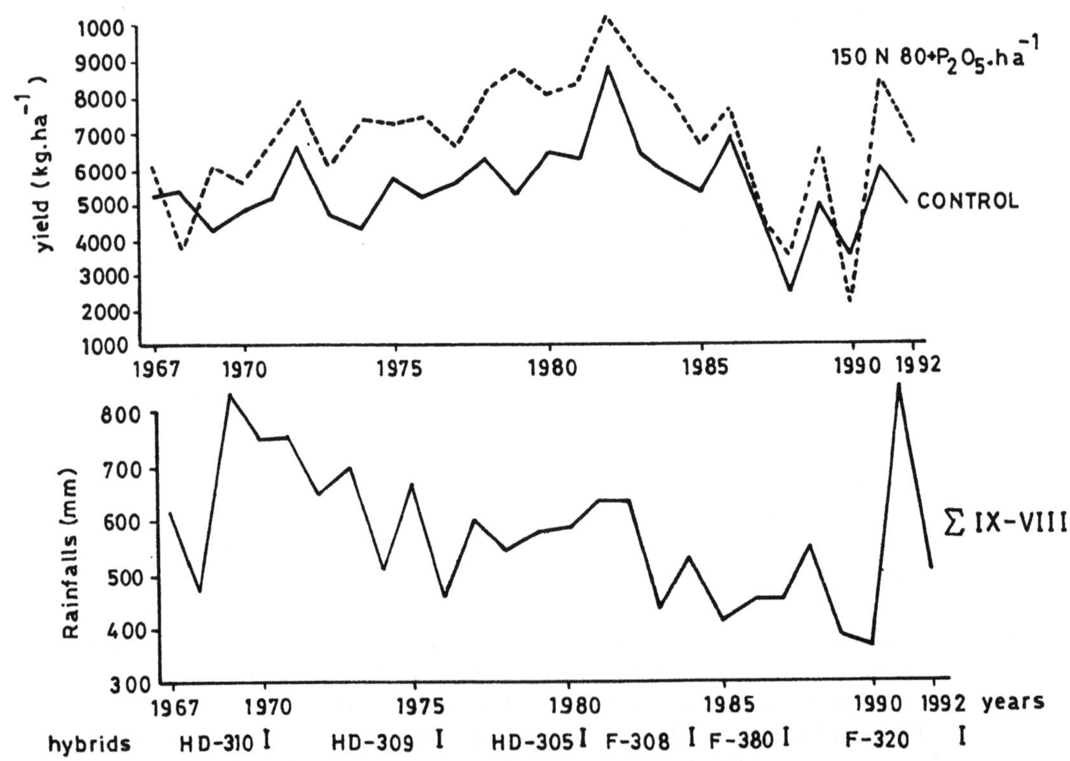

Fig. 1b. Maize for grain yield and rainfalls Fundulea – Romania (1967–1992).

developing countries, from using them in recommended quantities and in balanced proportions [8, 19, 23, 24]. Hence a judicious combination of mineral fertilizers with organic sources of plant nutrients is being promoted. Minimization of losses and replenishment of nutrients from both internal and external sources are major issues. This approach is the one which is advocated in FAO's IPNS. This system concurrently pays attention to the interaction with other management practices and addresses farming systems rather than individual crops or fields [7, 8, 16].

The use of organic sources cannot replace the use of mineral fertilizers. Even though the effects of organic inputs go beyond the nutritional aspects, by contributing to improving physical soil properties and to a better fertilizer use efficiency, the recycling of organic materials does not suffice to fully replenish the nutrients that are removed by crop harvests [16]. An increased and more efficient use of mineral fertilizers in developing countries must be envisaged in the medium term.

Special attention has to be given to the agronomic and economic evolution of plant nutrition practices and plant nutrition sources. The integration of biological processes in the plant nutrition system should be evaluated. The role of biologically fixed N may have been exaggerated, and there is a need to examine if and when these benefits are likely to occur, and how to maximize their effect [4].

The goal must be a better utilization of the all nutrients sources, to prevent avoidable losses, to minimize expensive inputs, to protect the environment, and to assure food security for an ever increasing world population.

Results and discussions

If water is not a limiting factor, application of fertilizer normally significantly increases yield. The beneficial effect of fertilization depends on the overall interaction of chemical and physical soil properties, climatic conditions, crop characteristics and agricultural practices.

Results from long term field experiments with fertilizers

The effects of inorganic fertilizers on yields of wheat and maize grown in a 2-year rotation since 1967 on a cambic chernosiom at Research Institute for Cereals and Industrial Crops, Fundulea, Romania, are briefly

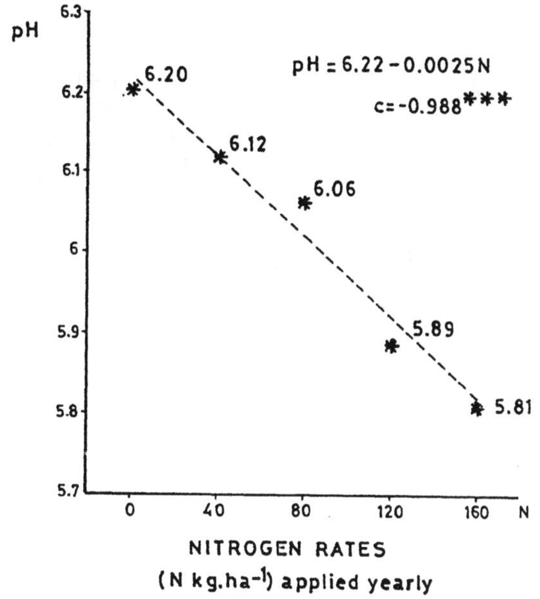

Fig. 2. pH evolution after 26 years in relation to the quantity of fertilizer applied on the cambic chernozem at Fundulea – Romania.

described. The particle size distribution on the soil is 35% fine sand, 30% silt and 35% clay with 250 me kg^{-1} total exchangeable bases. Initially, soil pH was about 6.5 and total N 0.155% [1, 14, 20].

Yields

Fig. 1a shows yields of wheat and rainfall (November–June) and Fig. 1b those of maize and rainfall (November–August) between 1967 and 1992. On the control plot, yields of wheat grain have remained reasonably constant at about 1.5 t ha^{-1} whereas those of maize grain after remaining at about 5.5 t ha^{-1} until 1985, have decreased in recent years, perhaps due to a run of seasons with below average November–August rainfall.

On soils given 80 kg P_2O_5 ha^{-1} each year and 120 kg N for wheat and 150 kg N for maize, yields have also tended to be less in recent years compared to the first 20 years. Applying fertilizer gave a larger proportional increase in wheat yields than in maize yields.

Changes in some soil properties

Soil samples from the tilled layer were taken in 1992, after 26 years, and analyzed for total nitrogen (by the Kjeldhal method), pH in water (soil: solution ratio 1: 2.5) and humus content (by the modified Schölemberger method). Available phosphorus and exchangeable potassium were determined after

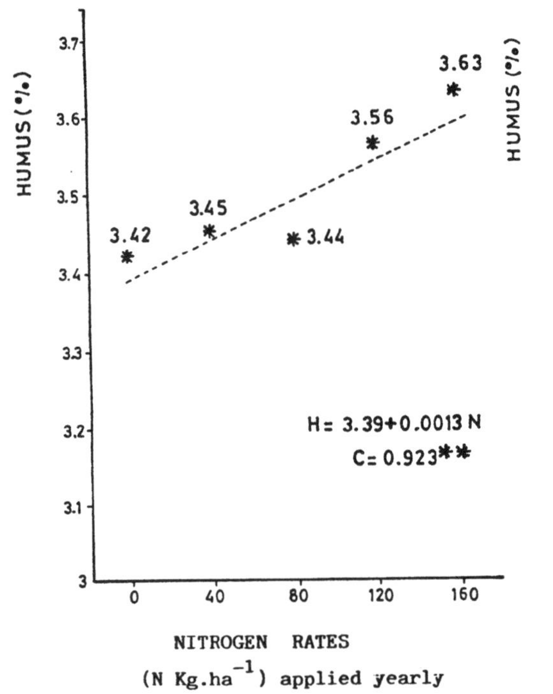

Fig. 3. Humus evolution after 26 years in relation to the quantity of fertilizer applied on the cambic chernozem at Fundulea – Romania.

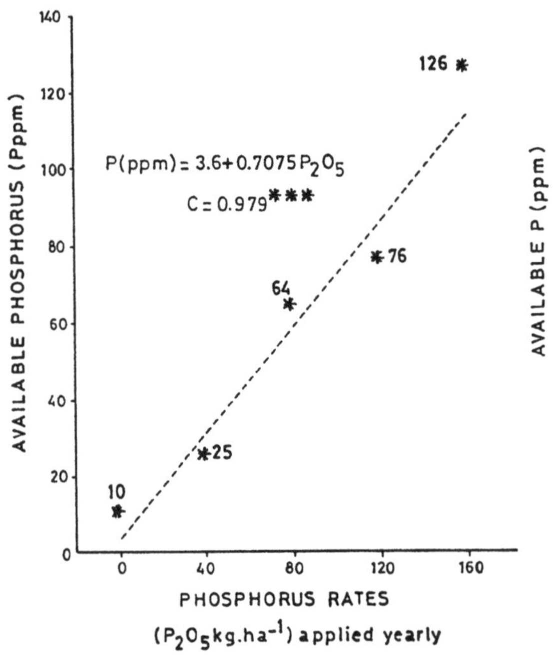

Fig. 4. Available phosphorus evolution after 26 years in relation to the quantity of fertilizer applied on the cambic chernozem at Fundulea – Romania.

extraction with calcium lactate and ammonium acetate respectively. Changes in soil agrochemical indices, after 26 years of applying different amounts of nitrogen, phosphorus and potassium fertilizers are significant and there is a linear relationship between the agrochemical index and amount of nutrient applied.

Soil pH Soil pH (Fig. 2) has declined, depending on the amount of N applied. It now ranges from 6.2 on the plot without N to 5.8 where 160 kg N ha^{-1} has been applied annually for 26 years. The decrease can be described by the following equation:

$$pH = 6.22 - 0.0025N \quad (1)$$

where N = amount of N (kg ha^{-1}) applied annually. If the total amount of N is taken into account, instead of yearly application, then equation (1) becomes:

$$pH = 6.22 - 0.000096\,QN \quad (2)$$

where QN is the total amount of nitrogen (kg ha^{-1}) used.

Equation (2) can be used to predict the change in pH of the cambic chernozem at Fundulea when a certain quantity of nitrogen is applied as ammonium nitrate to arable crops. For example, by applying 120 kg N;ha^{-1} for 9 years, the pH of this soil will decrease by 0.1 pH unit. Phosphorus and potassium fertilizers had less influence on soil pH.

Soil organic matter Nitrogen fertilizers have affected both the humus and total nitrogen content of the soil. The relationship between the quantity of nitrogen applied each year and the humus content of the soil is shown in Fig. 3. Soil humus now ranges from 3.42% without nitrogen to 3.63% with 160 kg N ha^{-1} each year for 26 years.

The relationship between humus content and applied fertilizer N is described by equation (3):

$$H(\%) = 3.39 + 0.0013\,N \quad (3)$$

where H = percentage humus, N = nitrogen (kg ha^{-1}) applied annually.

If the total amount of nitrogen applied is used, then equation (4) is obtained:

$$H(\%)\ 3.39 + 0.00005\,QN \quad (4)$$

where QN = the total quantity of applied nitrogen (kg ha^{-1}).

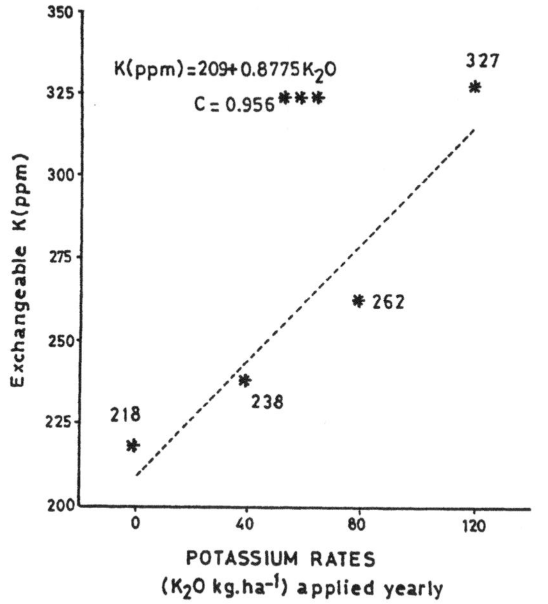

Fig. 5. Exchangeable potassium evolution after 26 years in relation to the quantity of fertilizer applied on the cambic chernozem at Fundulea – Romania.

From equation (4), it can be shown that applying 2000 kg N ha^{-1} to crops over a number of years increases soil humus content by 0.1%. This could be achieved by applying 125 kg N ha^{-1} $year^{-1}$, for a period of 16 years.

Soil nitrogen Previous research has shown that the total nitrogen content of the soil has been significantly increased where nitrogen fertilizers have been applied. Soil nitrogen now ranges from 0.148% without fertilizer nitrogen to 0.167% with 160 kg N ha^{-1} annually to maize and wheat grown in rotation for a period of 26 years. Equation (5) describes the change:

$$N\% = 0.145 + 0.00011\,N \qquad (5)$$

where N = the quantity of nitrogen (kg ha^{-1}) applied annually for 26 years. The average annual application of N needed to maintain the initial content of total soil nitrogen (Ni = 0.158%) was 118 kg N ha^{-1}. If less than this quantity was applied then soil N declined and where no nitrogen fertilizer was applied, the total nitrogen content of the soil at Fundulea decreased by 0.0005%, yearly.

Readily soluble phosphorus The readily soluble phosphorus content of soil is strongly influenced by the use of phosphorus fertilizers. After 26 years at Fundulea there was 126 mg kg^{-1} readily soluble P in the fertilized plot given 160 kg P_2O_5 ha^{-1} $year^{-1}$ compared to 10 mg P kg^{-1} in the unfertilized plot. There was a strong linear correlation between the applied P fertilizer and the available phosphorus in soil (Fig. 4)

$$P = 3.6 + 0.7075 P_2O_5 \qquad (6)$$

where P is readily soluble P (mg kg^{-1}) and P_2O_5 (kg P_2O_5 ha^{-1} $year^{-1}$) is the phosphorus applied for 26 years to maize and wheat grown in rotation. Equation (6) shows that approximately 44 kg P_2O_5 ha^{-1} $year^{-1}$ must be applied to maintain the initial readily soluble phosphorus content of the soil at 31 mg P kg^{-1}.

By introducing into equation (6) the proportion of the added P that remains readily soluble, the initial content of available phosphorus (Pi, mg kg^{-1}), the total phosphorus applied (QP_2O_5 – kg ha^{-1}) in *n* years and the annual maintenance quantity of phosphorus (44 kg P_2O_5 ha^{-1}) the change in the readily soluble phosphorus (P, mg kg^{-1}) after *n* years would be:

$$P(\text{mg kg}^{-1}) = Pi\,(\text{mg kg}^{-1}) + 0.0272\,(QP_2O_5 - n \cdot 44) \qquad (7)$$

From equation (7) it can be shown that to increase the readily soluble P content of the soil at Fundulea by 1 mg kg^{-1}, requires 81 kg P_2O_5 $ha^{-1}year^{-1}$, i.e. 37 kg P_2O_5 ha^{-1} must be applied annually in addition to the phosphorus required for maintenance (44 kg P_2O_5 ha^{-1} $year^{-1}$).

Readily soluble potassium The initial content of exchangeable K (K_i) was 234 mg kg^{-1} and after 26 years exchangeable K had declined to 218 mg kg^{-1}, in the unfertilized plot and increased to 327 mg kg^{-1}, where 120 kg K_2O ha^{-1} had been applied annually (Fig. 5). The linear regression gives:

$$K = 209 + 0.8775\,K_2O \qquad (8)$$

where K is exchangeable K (mg kg^{-1}) and K_2O (kg K_2O ha^{-1}) is the amount of K applied annually for 26 year. From equation (8) it can be calculated that to maintain the initial soil K content (234 mg kg^{-1}) would have required 28.5 kg K_2O ha^{-1} $year^{-1}$. Taking into account the initial content (K_i) and the amount of K (28 kg) to maintain the initial level, the proportion of added K that remains exchangeable and the total quantity of potassium used (QK_2O) in a period of *n* years, the change in exchangeable soil K at Fundulea is:

$$K(\text{mg kg}^{-1}) = K_i(\text{mg kg}^{-1}) + 0.03375\,(QK_2O - n28) \quad (9)$$

From equation (9) to increase readily soluble K by 1 mg kg^{-1} requires an annual addition of 31 kg K ha^{-1} in addition to the 28 kg K ha^{-1} needed to maintain the initial level of soil K. This quantity is small and may be explained by the richness in potassium of the cambic chernozem at Fundulea and by the capacity of the soil to maintain the pool of exchangeable K from the non-exchangeable pool and mineral reserves.

From the 26 years results with inorganic fertilizers in the long term field experiments, the following conclusions are:

- The application of inorganic fertilizers has had both direct and indirect effects on some agrochemical indices of the soil. Whereas nitrogen fertilizers have changed significantly the soil pH, the total nitrogen and humus content of the soil, the phosphorus and potassium fertilizers have influenced significantly only the readily soluble P and K content of the soil, respectively.
- The indirect effect of the application of a nutrient through an increase in the consumption of the other elements, has been much weaker than the direct effect. Thus, the application of a larger quantity of nitrogen increased the amount of phosphorus taken from the soil compared to where a smaller amount was given and this decreased the amount of this element in the soil. This decrease, however, was much less than when phosphorus fertilizers were not applied.
- A decrease of 0.1 pH unit in the cambric chernozem at Fundulea occurred after 9 years when 120 kg N $ha^{-1}year^{-1}$ was applied as ammonium nitrate.
- An annual application of ammonium nitrate, supplying 118 kg N ha^{-1}, together with P maintained the total nitrogen content of the soil. Where nitrogen fertilizers were not applied total soil nitrogen decreased by 0.0005% N annually. An increase of 0.1% N was obtained after a period of 16 years by applying 125 kg N $ha^{-1}year^{-1}$.
- To increase the readily soluble P content of the cambic chernozem at Fundulea by 1 mg kg^{-1} P required 37 kg P_2O_5 ha^{-1} in addition to the quantity (44 kg P_2O_5 $ha^{-1}year^{-1}$) required to maintain the initial level of readily soluble P.
- To maintain the initial exchangeable K content of the cambic chernozem at Fundulea required only 28 kg K_2O $ha^{-1}year^{-1}$ because of the richness of this soil in potassium and the ability of the non exchangeable and mineral K to support the pool of exchangeable K.

Aspects of fertilizer efficiency

General considerations

The extent to which crop growth is limited by insufficient presence of nutrients depends on the crop demand and the capacity of the soil to supply them.

With ample supply of water and nutrients, 20 to 30 t of total dry matter ha^{-1} y^{-1} can be obtained [14, 20, 26]. If water is not limited, warmer climates are more favourable because of longer growing periods and more efficient photosynthesis [6]. The dry matter of crops grown with ample supply of nutrients seldom contains less than 1.5% N, 0.3% P and 1.5% K [14, 20]. However, if maximum yield is to be achieved, crops must contain at least 300 kg ha^{-1} of N, 60 kg ha^{-1} of P and 300 kg ha^{-1} of K. These quantities can only be obtained through application of fertilizers, because the amount of nutrients provided by the soil itself is generally small. In addition, the inherent capacity of soils to supply nutrients generally diminishes from temperate regions towards to the equator [6]. In tropical and subtropical regions, heavy rainfall and a faster breakdown of organic matter lowers the reserve of nutrients.

Efficiency studies

It has been shown that the efficiency of fertilizers is often lower than 50% [1, 9, 12–14, 18, 21, 25]. During more than 25 years, fertilizer N efficiency studies have been conducted throughout the world [7, 14, 15, 26]. Factors such as fertilizer placement, timing, type of fertilizer, cultivation practices (irrigation, plant density, cropping sequence), identification of N efficient genotypes, competition in mixed agricultural and natural ecosystems, etc. have been studied and progress has been made in selecting the best conditions for increasing the fertilizer efficiency [9–11, 14].

Fertilizer placement Among the factors affecting crop production, fertilizer placement plays an important role. A strong argument in favour of this affirmation are the results received by five countries participated in the first Research Coordination Programme organized by the Agriculture Unit of the International

Table 1. Effect placement on the % P in Rice plants that was derived from fertilizers

Location	Treatments						LS 5%
	Surface	Hoeing	Hill 10 (a)	Hill 20	Row 10 (b)	Row 20	
Philippines (Los Baños)	17	17	6	4	4	3	2
Thailand (Bankhen)	68	68	50	34	51	36	9
Thailand (Surin)	37	40	22	15	26	23	2
Burma (Cyogon)	11	17	6	4	4	3	2
Burma (Mandalay)	25	25	6	6	6	4	5
Pakistan (Tandojam)	48	50	5	4	4	4	5
Egypt (Sakha)	64	60	37	38	38	37	10

[a] Placement at 10 cm depth in the planting hill.
[b] Placement at 10 cm depth between the rice rows.

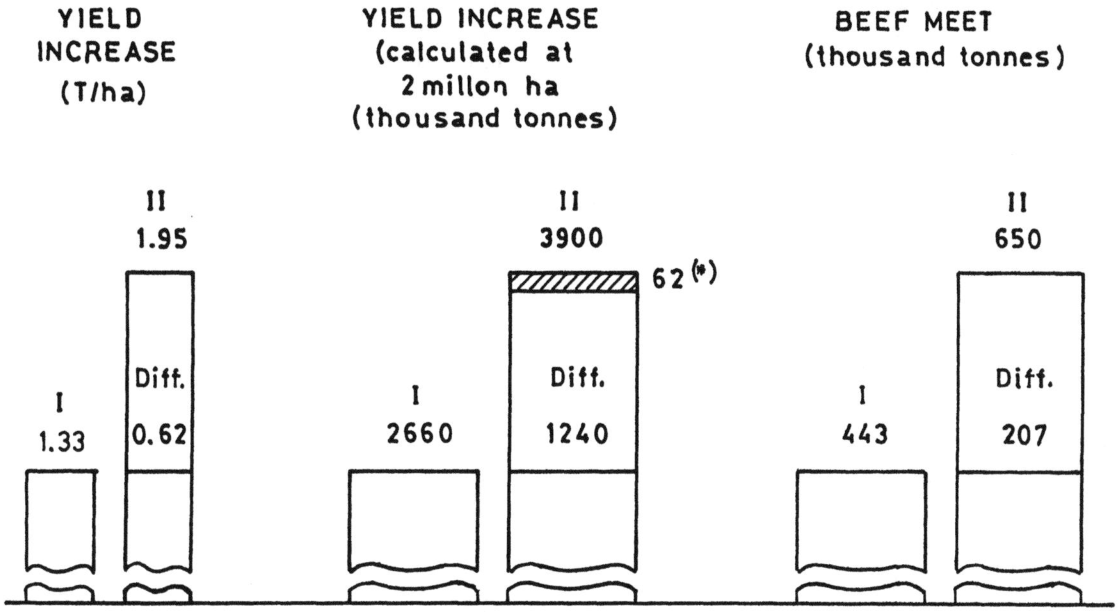

Fig. 6. Advantage of using the method of application of N and P fertilizer at seeding in bands (II) as compared with the broadcast application (I). The figure of 62 thousand tonnes (marked with an asterisk) represents the value of the investment made in bringing in the new method.

Atomic Energy Agency on Rice Fertilization (Table 1).

Surface application and hoeing into the surface were equally effective. At all locations, all other treatments were less effective in supplying fertilizer phosphorus to the rice plant. This was in spite of the fact that placement at 10 cm depth in the planting hill actually involved placing the fertilizer in the hole in which the rice plant was transplanted, a treatment that most of the participants and observers anticipated would be the most effective treatment. Before the field experiment with P-32 labelled superphosphate were performed, the five year programme with the non-labelled fertilizer had given inconclusive results and the debate concerned the desirability of extending the programme for another five years. In only a one year experiment with radioactive labelled fertilizer, the participating countries in the programme had answered the question [9, 18].

The placement of nitrogen fertilizer also influenced to a great extent the yield and nitrogen uptake by the plants (Table 2). By applying 80 kg N/ha in bands at seeding and 80 kg N/ha side-dressing, (I) during the vegetation period it was observed that a yield increase with 620 kg/ha of maize kernels and an increase of coefficient of utilization of N coming from fertilizer

Table 2. Nitrogen utilization by maize according to fertilizer placement

	Fertilizer (as NH_4NO_3 placement)		
	I	II	LSD 5%
*Average yield**			
Grain (quintals/ha)	92.6	86.4	5.9
Stalks (quintals/ha)	82.0	80.6	9.8
N consumption from soil and fertilizer (kg/ha)			
Grain	135.2	127.9	7.0
Stalk	44.9	48.1	3.5
Total	180.1	176.0	
N consumption from the fertilizer (kg/ha)			
Grain	46.9	28.8	5.1
Stalk	14.3	13.0	2.7
Total	61.1	41.8	
Fertilizer utilization rate** (%)	38.2	26.2	
Distribution of N consumption from the fertilizer			
Plough-down (broadcasting)			
In grain (kg/ha)	–	16.0	
In stalk (kg/ha)	–	6.3	
Total (kg/ha)	–	22.3	
Fertilizer utilization rate*** (%)	–	27.8	
Band application (at seeding)			
In grain (kg/ha)	12.8	12.8	
In stalk (kg/ha)	6.7	6.7	
Total (kg/ha)	19.5	19.5	
Fertilizer utilization rate*** (%)	24.3	24.3	
Side-dressing (during the vegetative period)			
In grain (kg/ha)	34.0		
In stalk (kg/ha)	7.6		
Total (kg/ha)	41.5		
Fertilizer utilization rate*** (%)	52.0		

*Calculated at 15.5% moisture content in grains and 0 (zero) in stalks.
**Reported for 160 kg N/ha.
***Reported for 80 kg N/ha.
I – 80 kg N/ha and 80 kg P_2O_5/ha applied in bands at seeding + 80 kg N/ha side dressing during the vegetation period.
II – 80 kg N/ha and 80 kg P_2O_5/ha broadcasted and ploughed in the soil + 80 kg N/ha side dressing during the vegetation period.

from 41.8% to 61.1% in comparison with the application of 80 kg N/ha broadcasting and plough-down and 80 kg N/ha in bands at seeding (II). Placement of 80 kg N/ha in bands at seeding and 80 kg/ha side-dressing at vegetation, in comparison with 80 kg N/ha broadcasting and plowed-down and 80 kg N/ha side-dressing at vegetation, conduct also in the cost reduction, as the work is done simultaneously with seeding and weeding, by adding to the seeders and cultivator machines the required equipment for fertilizer application [16].

Table 3. Influence of split application on labelled fertilizer recovery and loss (after [16])

Timing	Application	% N. recov. plant	% N. recovery soil (0–30 cm)	Unrecovered N-15 (% of applied N)
Basal	Incorporation	5.1	8.2	85
12 DT	Broadcast	20.6	26.0	53
5–7 DBPI	Broadcast	46.3	26.2	27
PI	Broadcast	34.4	16.4	49

DT: days after planting.
DBPI: days before panicle initiation.
PI: panicle initiation.

Table 4. Uptake of phosphorus from different sources at different stages of growth (Datta and Vyas, 1967)

Treatment		Uptake of total P in mg/g plant material			Uptake of fertilizer phosphorus(%)			Utilization of fertilizer (%)		
Stage	Application	I	II	III	I	II	III	I	II	III
Super-phosphate	Spray	0.89	1.04	1.33	40.0	40.7	13.8	0.54	0.76	1.07
	Soil	0.90	1.00	1.08	4.6	15.2	10.2	0.08	0.34	1.85
Mono-calcium phosphate	Spray	1.06	1.19	1.49	27.5	30.9	17.3	1.61	1.84	4.20
	Soil	1.07	1.19	1.24	3.2	15.9	12.2	0.27	2.20	5.50
Dicalcium phosphate	Spray	0.88	0.94	1.42	29.2	35.6	24.7	0.87	1.00	2.10
	Soil	0.45	0.50	1.30	2.6	14.1	12.5	0.03	0.90	3.60
Ammonium phosphate	Spray	1.02	0.90	1.34	32.6	56.0	22.2	1.83	2.70	4.80
	Soil	0.45	1.40	1.10	3.4	18.3	10.4	0.40	2.50	5.80
Average	Spray	0.94	1.3	1.4	30.6	40.8	19.5	1.20	1.6	3.0
	Soil	0.74	1.0	1.2	3.5	15.9	11.3	0.19	1.5	4.2
F-test (application)		Sig.	Sig.	Sig.	Sig.	Sig.	Sig.	Sig.	Not Sig.	Sig.
$S.E_m.\pm$		0.03	0.06	0.05	1.8	1.03	1.07	0.06	1.15	0.26
C.D. @ 5%		0.09	0.24	0.16	5.2	2.96	3.08	0.19	–	0.82

Sig = Significant.

Regarding good results obtained in the field experiment with N-15 labelled fertilizers in Romania, the new method of fertilizer placement was extended on an area of 2000000 ha planted with maize. The results presented in Fig. 6 highlight the 6 years average yield increase of 620 kg/ha kernels obtained in research and production as compared to those obtained through the classical method of broadcast application of fertilizers. The investment made for adapting the seeders and the cultivators to the new methods of fertilizer application represents only 5% of the total value of the yield increase that can be obtained every year. The advantage of the new method is obvious if we also take into consideration the elimination of the supplementary work in applying the fertilizer in the classical method, and as a result, the diminution of the fuel consumption and the elimination of the possibilities of worsening the physical soil characteristics by a supplementary soil farming. Translating the obtained maize yield increase into tones of meat, we can say that 207000 additional tones of meat can be produced every year.

It has been demonstrated that the proper method of fertilizer placement can result in lower amounts of residual N and consequently in lower leaching. Split application may further increase the fertilizer efficiency [22]. This is illustrated in Table 3.

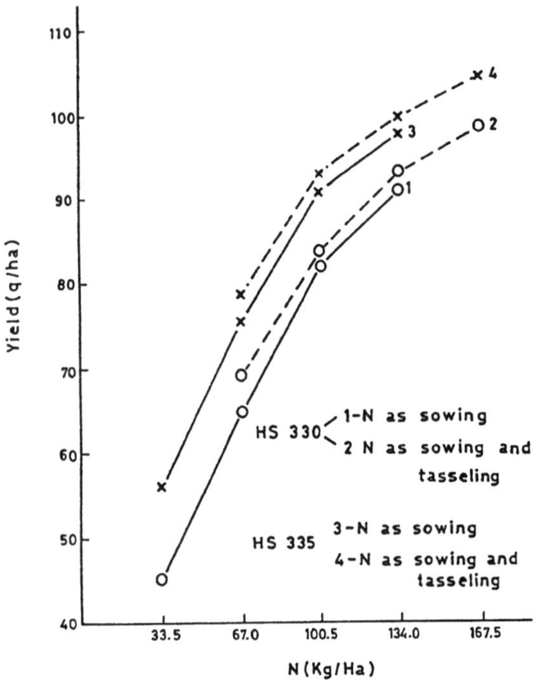

Fig. 7. Influence of nitrogen fertilizer application on maize with irrigation water by sprinkling.

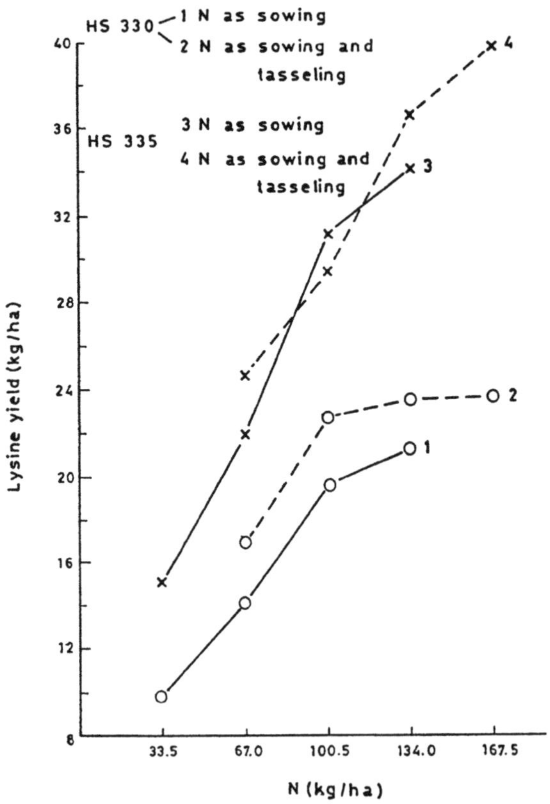

Fig. 8. Influence of sprinkling fertilizer application on lysine production.

Foliar application of fertilizers A clear distinction between root and non-root absorbed nutrients occurring simultaneously in the plants can be made by the use of isotopes. Datta and Vyas [5] carried out an evaluation of the magnitude of the nutritional contribution of a number of foliar-applied phosphate fertilizer labelled with P-32. Results received in New Delhi, India, on the uptake by maize of total and fertilizer phosphorus and percentage utilization of the applied fertilizer at the equivalent doses of 5 lb. P_2O_5/acre of superphosphate, monocalcium phosphate, dicalcium phosphate and ammonium phosphate through foliar and soil application at three stages of growth (24 (I), 39 (II) and 60 (III) days after germination) are presented in Table 4.

From the results presented in Table 4, it can be seen that the total uptake of phosphorus from foliar application was significantly superior compared with soil application at all stages of growth. In the data on percent fertilizer phosphorus uptake it is observed that foliar application was eight and two times as efficient as soil application at the first and second stage respectively. It was also superior at the third stage. Percent utilization of the fertilizer phosphorus from foliar application was nearly six times as efficient as soil application in the early stage. In the later stages of growth, however, there was not much difference in the utilization.

Foliar application increased the percentage uptake of fertilizer phosphorus [5]. It was observed that at early stages of growth, foliar application was not only superior to the equivalent rate of soil application but also higher than 20 and 80 lb. P_2O_5/acre soil application. Phosphorus is needed by the plants at the early stages of growth and if they do not get sufficient phosphorus at the early stages, then they cannot extract phosphorus efficiently in the later stages. Thus, higher uptake of phosphorus through foliar application at early stages could be attributed firstly to the young and expanding leaves which are active and secondly to root growth being less.

Nitrogen fertilizer can be also successfully used as foliar application. The results received in Romania [17] applying 33.5 kg N/ha at tasseling stage of growth by sprinkler irrigation, using a normal maize hybrid, HS 330, and an opaque hybrid, HS 335, showed not

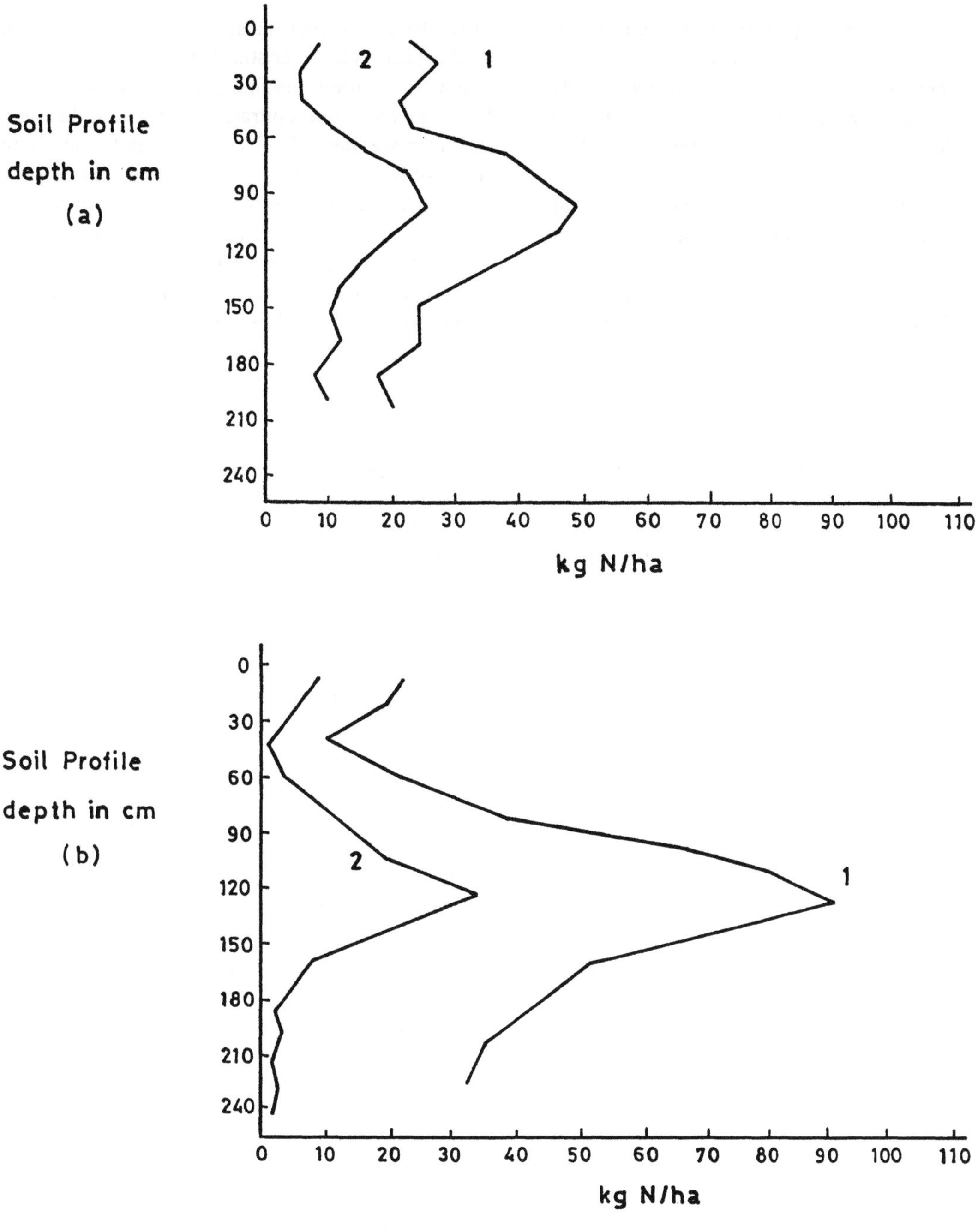

Fig. 9. Distribution of total NO_3 (1), fertilizer NO_3 (2) on chernozium soil profile at Fundulea, in the treatments with 120 (a) and 240 (b) kg N/ha.

only yield increases but also and increase in the lysine yield per hectare (Figs. 7 and 8).

Forms and time of fertilizer application In general fertilizer usage, the expected efficiencies are approximately 30 to 70 percent of added nitrogen, 5 to 25 percent of added phosphorus, and 40 to 80 percent of added potassium. These values can be improved by using special care or can also be even lower because of carelessness or bad climatical conditions [6, 26].

Looking at the fertility problem from the growers' point of view, the addition of any amount of fertilizer is of interest only if it profitably enhances yields, either larger or better quality. A number of factors affect fertilizer efficiency and crop responses to added fertilizer.

The effectiveness of fertilizer depends on having it in the forms and where the plant can get it when it needs it. Usually a deficiency, however temporary, will affect some growth. Some of the results received, by using labelled fertilizers with stable and radioactive isotopes, are of real benefit for the practical fertilizer use.

To determine with high accuracy the quantity of nutrients absorbed by wheat plants and to study the rhythm of nitrogen absorption during different phenophases, field experiments with N-15 labelled fertilizers in the framework of the Research Coordinated Programme by the Joint FAO/IAEA Division of Nuclear Techniques in Food and Agriculture were organized. Both nitrogen fertilizers – ammonium nitrate and urea, in rates of 120 kg N/ha were applied either all in autumn at seeding (A-treatment) or in rates of 60 kg N/ha in autumn and 60 kg N/ha in spring, a case in which N-15 labelled fertilizer was applied on different plots (B-treatment, Bl, B2). In treatment C, to study the rhythm of N absorption during the vegetation period of wheat, besides 20 kg N/ha applied in autumn, 100 kg N/ha was applied, namely:

a – early in spring, at the third node;
b – at the boot stage;
c – at 50% heading;
d – at 50% flowering;
e – grain at milk stage.

This C-treatment was divided into 5 sub-plots, each one receiving 20 kg N/ha, labelled with N-15, only at one phenophase, the rest of the phenophases receiving unlabelled nitrogen fertilizer. In this way, it can be determined exactly, how nitrogen derived from fertilizer is used in each phenophase, the rest of the factors remaining constant.

Of the two factors studied, namely the form of fertilizer – NH_4NO_3 or $(NH_2)_2CO$ – and the time of application, at the same quantity of active ingredient, it was the time of application that had the strongest influence on kernel yield (Table 5). If, at kernel yield, the time of application influenced the efficiency of ammonium nitrate, at straw yield the differentiation, depending on the time of application, was stronger in the case of urea. The different effects in the case of the two forms of fertilizer, on kernel and straw yield may be affected by the specific characteristics of the fertilizer-soil-plant interaction. During the first period of vegetation (till heading), ammonium nitrate is better used than urea, the latter needing to be decomposed in the soil in the forms of NH_4^+ and NO_3^-. Owing to the formation of a larger vegetative mass, the lodging phenomenon was stressed, and as a result, kernel yield obtained from fertilizer applied at a rate of 120 kg N/ha, as ammonium nitrate, in autumn, was lower than the one obtained by the urea treatment. It is well understood that NO_3^- leaching deeper than the active area of the root system must also not be neglected [14, 16].

The lower yield obtained with the treatment C, in which a rate of 120 kg N/ha was applied in six stages, proves that plant needs were not satisfied with the low nitrogen rates distributed at different times. Nitrogen fertilizer influenced the yield quality in a positive way, increasing the protein content in the kernel. As a result, an increase in the total nitrogen output by the yield was noted, as a function of form and time of fertilizer application of the amount of N taken by the kernels, 33–34% is derived from fertilizer. The highest % Ndff values were obtained when half of the total nitrogen rate was applied in autumn and the other half early in spring (Table 6). The coefficient of utilization of N coming from fertilizer in kernel and straw yields range from 47.2% to 62.4% in the case of ammonium nitrate and from 44.9% to 54.1% in the case of urea, depending on the time of application. The highest utilization coefficient was noticed in the case of a split application – 60 kg N/ha in autumn and 60 kg N/ha in spring with an increase of 15.2% in the case of NH_4NO_3 and 9.5% in the case of $(NH_2)_2CO$ [17].

For good kernel formation, wheat needs a good nitrogen supply when 50% of the plants are at flowering. From the nitrogen applied as fertilizer, up to this phenophase over 50% is used, with a maximum of 69.8%, when the fertilizer is applied at 50% heading stage of the plants (Table 7). It follows that, to obtain wheat yields that are superior quantitatively and qualitatively, a good nitrogen supply for the plants is needed during almost the whole vegetation period, the first phases being decisive as regards the quantity, and the heading-flowering stages as regards the quality.

Of the two forms of application studied, a small difference was noticed in favour of NH_4N0_3, especially at fertilization at the third node stage and at the boot stage [17].

These results show a preferential absorption function of the chemical composition of the fertilizers and

Table 5. Influence of different forms and times of application of the N fertilizer on wheat yield

Treatments	Yield (q/ha)					
	Kernel			Straw		
	NH_4NO_3	$(NH_2)_2CO$	x	NH_4NO_3	$(NH_2)_2CO$	x
A	33.5	39.4	36.5	108.5	94.5	101.5
B	40.2**	40.7	40.4*	113.4	118.8*	116.1*
C	38.3	36.7	37.5	116.3	110.0*	113.1*
LSD 5%	3.8	3.8	3.7	11.0	11.0	7.8

* Significant at 5% level.
** Significant at 1% level.
q= metric quintal= 100 Kg.

Table 6. Influence of different forms and times of application of N fertilizer on the utilization coefficient from fertilizer

Treatment	N content (%)	Absorbed N (kg/ha)	Ndff (%)	Absorbed N (kg/ha)		Utilization coefficient (%)
				From soil	From fertilizer	
In Kernel						
I. A	2.29	76.7	39.6	46.3	30.4	25.3
B_1	–	–	18.6	–	17.6	29.3
B_2	–	–	24.5	–	23.1	38.5
B	2.35	94.5	43.1	53.8	40.7	33.9
C	2.35	90.0	36.4	57.2	32.8	32.8
II. A	2.29	90.2	36.1	57.6	32.6	27.3
B_1	–	–	15.3	–	14.6	24.3
B_2	–	–	22.0	–	20.9	34.8
B	2.34	95.2	37.3	59.7	35.5	29.6
C	2.39	87.7	33.2	58.6	29.1	29.1
LSD 5%	0.125		3.5			
In straw						
I. A	0.66	71.6	36.8	45.3	26.3	21.9
B_1	–	–	23.2	–	18.1	30.2
B_2	–	–	20.5	–	16.0	26.7
B	0.69	78.2	43.7	44.0	34.2	28.5
C	0.79	91.9	25.5	68.5	23.4	23.4
II. A	0.60	56.7	37.2	35.6	21.1	17.6
B_1	–	–	15.9	–	12.8	21.3
B_2	–	–	20.6	–	16.6	27.8
B	0.68	80.8	36.5	51.3	29.5	24.5
C	0.80	88.0	25.4	65.7	22.3	22.3
LSD 5%	0.14		3.0			

I. NH_4NO_3.
II. $(NH_2)_2CO$.

Table 7. Influence of different times of application on the N uptake from fertilizer

Fertilizer application, at stage: (C treatment)	Kernel Ndff (%)			Kernel Utilization coefficient			Straw Ndff (%)			Straw Utilization coefficient			Total Utilization coefficient		
	I	II	x	I	II	x	I	II	x	I	II	x	I	II	x
a– In spring, at the 3-rd node	7.2	5.4	6.3	32.4	23.7	28.0	6.5	5.0	5.7	29.9	22.0	25.9	62.3	45.7	50.4
b– Boot stage	7.0	5.4	6.2	31.5	23.7	27.6	4.9	6.1	5.5	22.5	26.8	24.7	54.0	50.5	52.2
c– At 50% heading	9.6	9.9	9.7	43.2	43.4	43.3	5.9	5.9	5.9	27.1	26.0	26.5	70.3	69.4	69.8
d– At 50% flowering	8.6	9.4	9.0	38.7	41.2	39.9	4.3	3.9	4.1	19.7	17.2	18.4	58.5	58.4	58.4
e– Kernel milk stage	4.0	3.1	3.5	18.0	13.6	15.8	3.8	4.6	4.2	17.5	20.2	18.8	35.5	33.8	34.6

I. NH_4NO_3.
II.–$(NH_2)_2CO$.

Table 8. Nitrogen uptake from $NO^-{}_3$ and $NH^+{}_4$ from ammonium nitrate

Time and rate of N application (kg/ha) Autumn	Spring	Kernel % Ndff	Kernel N absorbed from fertilizer (kg/ha)	Kernel Utilization coefficient (%)	Straw % Ndff	Straw N absorbed from fertilizer (kg/ha)	Straw Utilization coefficient (%)	Total % Ndff	Total N absorbed from fertilizer (kg/ha)	Total Utilization coefficient (%)
$^{15}NH_4NO_3$										
60*	60	8.0	7.6	25.3	7.9	6.1	20.3	15.9	13.7	45.6
60	60*	10.4	9.82	32.7	9.0	7.0	23.3	19.4	16.8	56.0
Total		18.4	17.4	29.0	16.9	13.1	21.8	35.3	30.5	50.8
$NH_4{}^{15}NO_3$										
60*	60	10.2	9.6	32.0	10.6	8.3	27.7	20.8	17.9	59.7
60	60*	12.0	11.3	37.7	13.0	10.2	34.0	25.0	21.5	71.7
Total		22.2	20.9	34.8	23.6	18.5	30.8	45.8	39.4	65.6

*Labelled with ^{15}N.

of the ions which play a role in nitrogen plant nutrition. To determine the $NH_4{}^+$ and $NO_3{}^-$ ion contribution in nutrition, depending on the time of application and on the soil-ion-plant interaction, by means of NH_4NO_3 labelled with N-15, the contribution of the two ions $NO_3{}^-$ and $NH_4{}^+$ applied as fertilizers at the time of grain formation was studied.

Thus, from the total of 120 kg N/ha applied half in autumn and half in spring, $NH_4{}^+$ accounts for 29% at kernel formation, and 21.8% at straw formation, the general utilization coefficient being 50.8%.

In the same conditions, the nitrate form was more efficient, the utilization coefficient being 34.8% for kernel, 30.8% for straw, the total being 65.6% (Table 8).

The higher contribution of the $NO_3{}^-$ at the time of grain formation indicated by % Ndff, the utilization coefficient being higher in all cases, may be caused by the higher mobility of this ion. It is known that, in contrast to the $NH_4{}^+$ ion, the $NO_3{}^-$ ion is retained in a much lesser extent by the soil adsorbtive complex, circulating in the soil solution by mass flow. The fact that the nitrate moves with the water, allows the root system to exploit the nitrate ion better. This can explain the superior values of the utilization coefficient of $NO_3{}^-$.

Nitrogen × phosphorus interaction Hera *et al.* reported many results that were received from field experiments, when ammonium salt, especially $(NH_4)_2SO_4$, was mixed with superphosphate, leading to a better utilization of the fertilizer phosphate than when the two fertilizers were applied separately. This did not occur where nitrate salt was the source

Table 9. The effect of mixing of $(^{15}NH_4)_2SO_4$ and ^{32}P Superphosphate on the uptake of fertilizer nitrogen and phosphorus by rice (60 day harvest)

Location	% P derived from the fertilizer		% N derived from the fertilizer	
	separated	mixed	separated	mixed
Burma	4.4	5.0	23	28
Ceylon	38	50	32	36
Rep. of China	5.8	6.5	27	27
Hungary	7.5	13.5	48	52
India I	10	17	44	47
India II	44	60	46	36
Italy	22	36	15	20
Korea I	7.8	7.2	50	49
Korea II	9.8	11.5	26	22
Madagascar	67	75	40	37
Pakistan	40	40	33	36
Dacca	56	53	20	22
Phillippines	8	15	31	28
Thailand	72	82	28	30
Egypt I	18	34	25	25
Egypt II	20	27	33	33
Romania I	12	19	34	39
Romania II	28	47	43	52

*applied in rows at 5 cm depth.

Table 10. N Absorption and utilization coefficient of the N fertilizer by sunflower, as a function of the NP application rate

N and P_2O_5 (kg/ha)*	Yield (q/ha)	Absorbed N (kg/ha)		Utilization coefficient (%)
		Total	From fertilizer	
P_0N_0	28.9	–	–	–
$P_{30}N_0$	30.4	–	–	–
N_{40}	33.3	75.2	15.3	38.2
N_{80}	34.6	81.7	29.1	36.3
N_{120}	36.8	87.9	46.1	38.4
$P_{60}N_0$	29.7	–	–	–
N_{40}	34.6	75.4	18.3	45.7
N_{30}	36.8	88.3	33.9	42.3
N_{120}	36.6	95.9	51.3	42.7
LSD 5%	2.3	5.4		
			3.7	3.9

*As NH_4NO_3 and superphosphate.

of nitrogen. This observation was further extended in an international coordinated program on the efficiency of fertilizer use by maize where the question was examined at eighteen different locations adding the additional question of the effect of mixing the fertilizers on the uptake of fertilizer nitrogen [12, 25]. These questions could be quantitatively answered in a field experiment by labelling the superphosphate with P-32 and the nitrogen source with N-15. Some of the results from this experiments are presented in Table 9. The results confirmed the observation that when applied $(NH_4)_2SO_4$ was mixed with superphosphate,

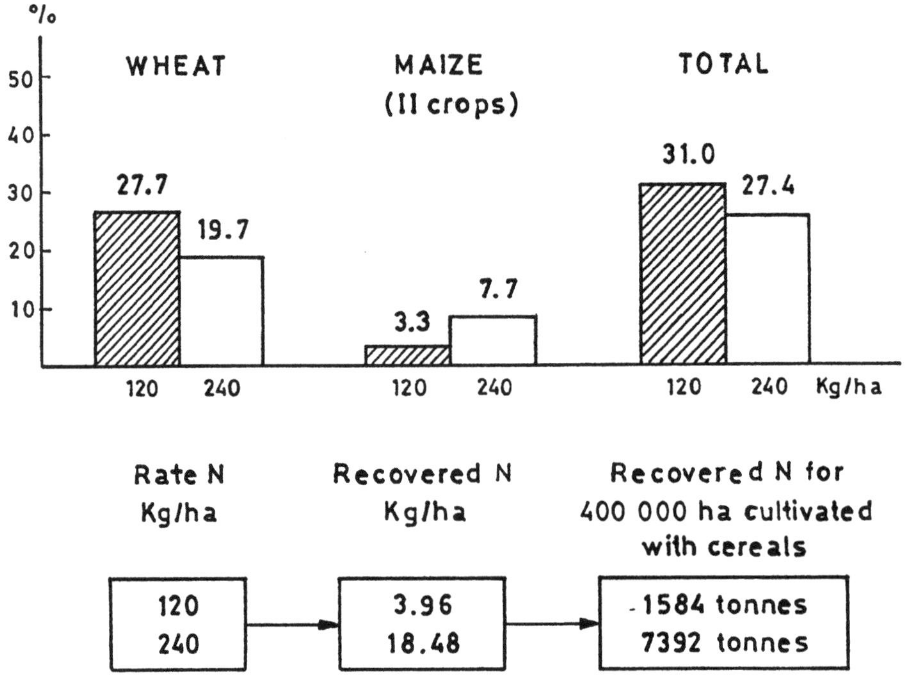

Fig. 10. Nitrogen utilization coefficient in wheat as a main crop and maize as a second crop.

the plants took up appreciably more of the phosphorus from the superphosphate. Further however, the same mixing had no continuous effect on the uptake of fertilizer nitrogen from the $(NH_4)_2SO_4$ source.

The results of earlier research on sunflowers had underlined the reduced effect of fertilizers on this crop, as on other crops, although the sunflower is a big consumer of nutrients. This is due mainly to the high capacity of its root system to absorb nutritive elements from less soluble forms from the soil [13, 14, 16]. In order to determine with precision the utilization coefficient of fertilizers with this crop, a number of experiments was conducted at the Research Institute for Cereals and Industrial Crops, Fundulea, Romania, using labelled fertilizers. Some of the results are presented in Table 10. The results show that by a mixed application of N and P fertilizers, significant yield increases, which reached 790 kg kernels per hectare, were obtained, as against the non-fertilized control.

The nitrogen fertilizer and their interaction with phosphorus fertilizer made an important contribution to the yield increases. The amount of nitrogen absorbed by the sunflower seeds from the fertilizer determined with N-15, increased as the rate of N fertilizer increased. The utilization coefficient of nitrogen derived from the fertilizer ranged between 36.3% to 45.7%, this being negligibly influenced by the growth of nitrogen fertilizer rate. The utilization coefficient of the nitrogen from fertilizer is significantly influenced by the levels of P fertilizer.

Leaching and nutrient recovery Under intensive agriculture, with high rates of nitrogen fertilizer, there is a major concern for a potential contamination of groundwater with NO_3^- not taken up by the crops, which may accumulate at different depths in the soil depending on rainfall or irrigation conditions and various soil properties.

In order to study what would happen to the N unused by the main and secondary crop yield, an experiment with labelled ammonium nitrate was undertaken on irrigation conditions at Fundulea Research Institute for Cereals and Industrial Crops to determine the quantity of the N leached in the soil and that recovered by silo maize crop, seeded immediately after harvesting the wheat. For a 120 kg N/ha rate, about 16 kg N/ha are found at 80 cm depth, and for a 240 kg N/ha rate, about 28 kg N/ha at 110 cm depth, the quantity of leached N decreasing with depth of profile, though N derived from the fertilizer can still be found at the depth of 200 cm (Fig. 9). A part of the N leached in the soil was

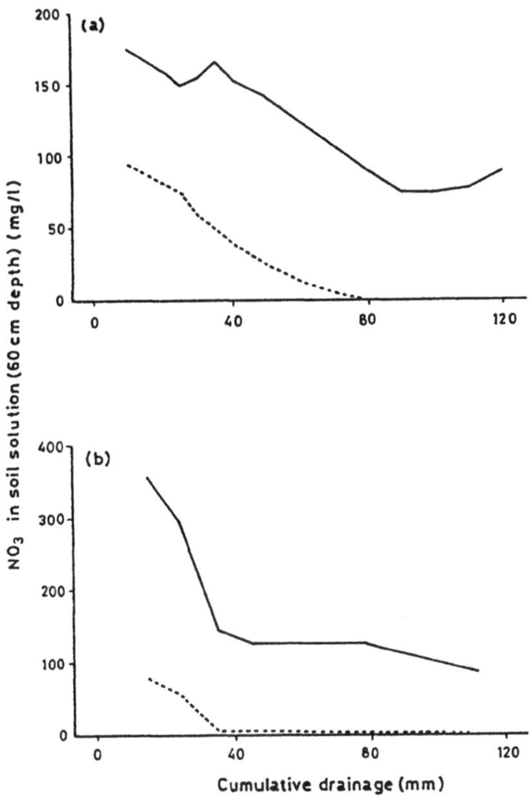

Fig. 11. Examples of the decrease in soil water nitrate achieved by winter cover crops sown before spring crops: (a) shows the benefit on a soil over limestone (winter 1990/91), and (b) shows the effect on a chalk soil (winter 1989/90). Continuous lines, falow; dotted lines, rye cover crops.

recovered by the maize crop seeded immediately after harvesting the wheat (Fig. 10).

Fig. 11 provides examples of the effect of rye cover crops grown in experiments on limestone and chalk soils [23]. Compared with winter fallow they decreased nitrogen losses from 32 to 7 kg N/ha (limestone) and from 27 to 2 kg N/ha (chalk). On many arable farms, maintaining a green crop cover over winter will be the most effective way of decreasing the risk of nitrate leaching. In function of the climatic conditions, green cover could be either a winter cereal or rape crop if established early enough, or if a spring crop is planned, a cover crop established in the early autumn and plowed down in the spring.

The amount of N applied to the major arable crops in England and Wales increased greatly during the early 1980s, but since then have remained stable or slightly decreased [3]. The significant decreases in N use on cereals and winter oilseed rape mainly result from the more restricted use of N in the autumn (Fig. 12).

The effects of a range of nitrogen application rates on post-harvest nitrate leaching measured by porous crops are presented in Fig. 13 [3].

In order to reduce N losses from urea, urease inhibitor were used successfully (Table 11).

Table 11. Cumulative loss from urea as controlled by the concentration of the urease inhibitor, nBTPT, upon 14 days in a sandy loam soil (after [26])

Treatment	% fertilizer loss
Urea alone	52
Urea + 0.5% nBTPT	5
Urea + 0.1% nBTPT	8
Urea + 0.05% nBTPT	14
Yrea + 0.01% nBTPT	23
Urea + 0.005% nBTPT	32

Conclusions

The inorganic fertilizers and water are the most important factors for plant growth and the easiest to manipulate for increased crop yields. For the above mentioned reason the fertilizer use has increased rapidly, but environmental concern has also increased. As we all depend on plants for our food, and plants depend on mineral nutrients and water for their growth and development, the rational use of fertilizers in different soils and climatical conditions is of great importance for both the developing and developed world. It is imperative that fertilizer management be conducted within an overall framework of sustainable development and environmental concern.

An important factor for the conservation of natural resources and for environmental preservation is the maintenance and enhancement of soil fertility through appropriate application of plant nutrients. The integrated plant nutrition system is a realistic approach through which the fertilizer and soil fertility management in cropping and farming systems is adapted to site characteristics and to locally available resources. Loss of soil fertility, especially in developing countries, due to continual nutrient mining by crop removal, without adequate replenishment could result in a deterioration not less dangerous than other forms of environmental

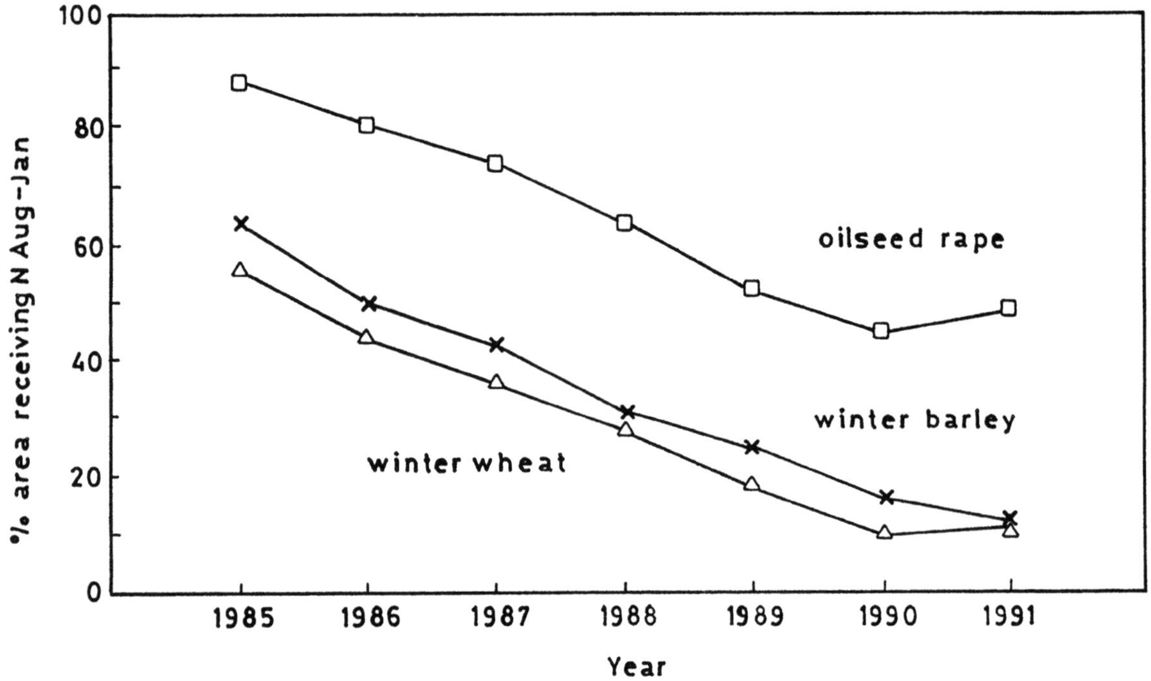

Fig. 12. Trends in N applied to arable crops in autumn from Chalmers *et al.*

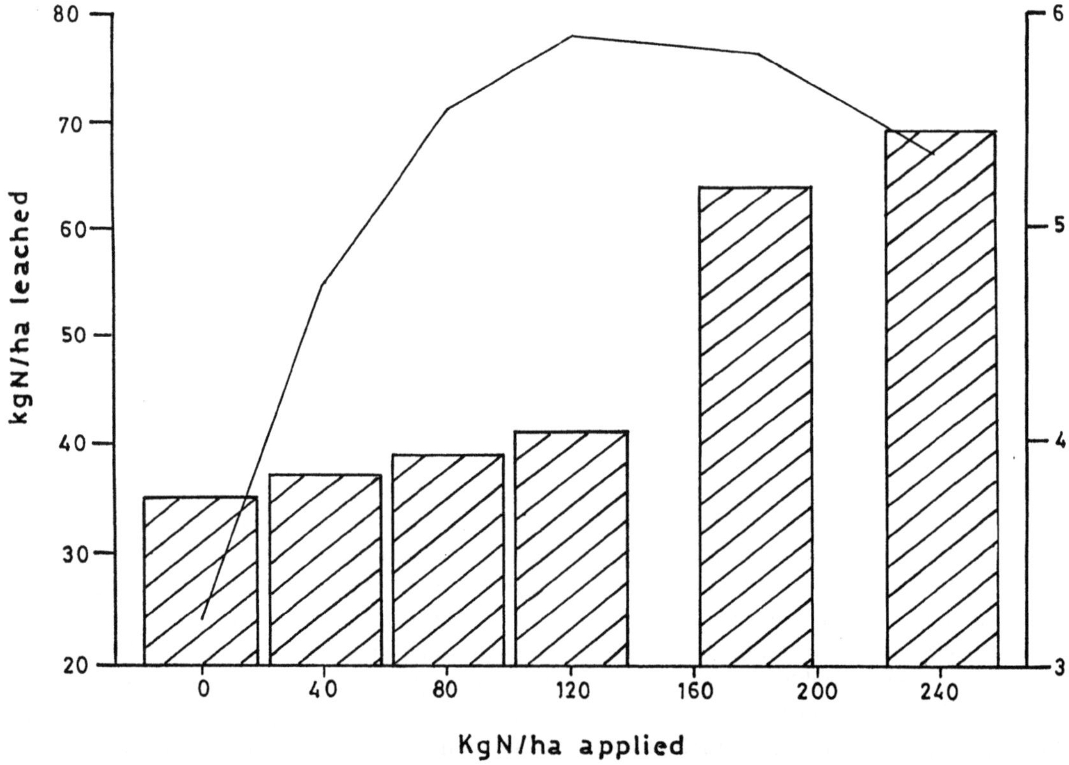

Fig. 13. The effect of nitrogen rate on yield of spring wheat, harvested 1990, and nitrate leached below 90 cm during winter 1990/91 at ADAS Gleadthorpe (sandy loam soil). The solid line represents yield and the bars show the amount of N leached.

degradation. There is our duty, to avoid soil deterioration for the benefit of future generations.

References

1. Borlan Z and Hera, C (1984) Optimizarea agrochimica a sistemului sol-planta. Editura Academiei. Bucharest, 51–157
2. Borlaug NE and Dowswell (1994) Feeding a human population that increasingly crowd a fragile planet. Suplement to Transations 15th World Congress of Soil Science – Mexico 1–15
3. Chalmers AG, Church BM, Dyer CJ and Leech PK (1991) Survey of fertilizer practice. Fertilizer use on farm crops, England and Wales 1990 ADAS/FMA. London
4. Danso SKA (1994) The role of biological nitrogen fixation in soil/plant system. 8th Nitrogen Workshop. Book of abstracts, p 6
5. Datta NP and Vyas KK (1967) Uptake and Utilization of Phosphorus – by Maize from Foliar Spray, In Isotopes in Plant Nutrition and Pysiology. IAEA, Vienna, 371–376
6. Davidescu D (1968) Ambiance, théorie et pratique de la fertilization. Zeszyty problemoiwe postepow nauk rolniczych. Zeszyt 84. Warsaw, 149–192
7. Fao (1994) Expert consultation on integrated plant nutrition systems, Rome
8. Finck A (1994) From the fertilization of crops to the management of plant nutrients in crop rotation and farming systems. Expert consultation on integrated plant nutrition systems, Rome
9. Fried M (1978) Direct Quantitative Assessment in the Field of Fertilizer Management Practices. In: Proc. XIth International Congress of Soil Science, Alberta University Press, 103–129
10. Greenwood DJ, Cleaver TJ, Turner MK, Hunt J, Niendorf KB and Loquens SMH (1980) Comparisons of the effects of phosphate fertilizer on the yield, nitrogen content and quality of 22 different vegetables and agricultural crops. J Agric Sci Camb 95, 457–469
11. Greenwood DJ, Cleaver TJ, Turner MK, Hunt J, Niendorf KB and Loquens SMH (1980) Comparisons of the effects of potassium fertilizer on the yield, nitrogen content and quality of 22 different vegetables and agricultural crops. J Agric Sci Camb 95, 441–456
12. Hera C (1979) The contribution of isotopes and radiation techniques to the development of agricultural production in Romania. In: IAEA (ed) Isotopes and radiation in research on soil-plant relationships. IAEA-SM-235/53. Vienna, 149–162
13. Hera C (1979) Some Aspects of Nitrogen Fertilization using N-15. Proc. IAEA-SM-235/36, 107–123
14. Hera C and Borlan Z (1980) Ghid pentru aclatuirea planurilar de fertilizare. Editura Ceres. Bucharest, 63–107
15. Hera C, Idriceanu A and Bologa M (1984) Technici Nucleare in Agricultura. Bucuresti – Editura Stiintifica si Enciclopedica, 102–237
16. Hera C (1993) Contribution of nuclear techniques to the assessment of nutrient availability for crops-Expert consultation on integrated plant nutrition systems. Rome, 1–25
17. Hera C (1994) Role of Isotopes in fertilizer studies. Proceedings of the International Conference on Fertilizer Usage in the Tropics. Kuala Lampur, 300–314
18. IAEA (1978) Isotope studies in rice fertilization. Technical Reports, Series No 181, IAEA Vienna 1–132
19. Loomis RS and Gerakes PA (1975) Productivity of agricultural ecosystems. In: Cooper, JP (ed) Photosynthesis and productivity in different environments. Cambridge, Cambridge University Press, 145–172
20. Mihaila V and Hera C (1994) Some recent results from the longterm experiments at Fundulea, Romania. Long term Experiments in Agricultural and Ecological Sciences, UK, 220–227
21. Powlson DS (1993) Understanding the soil nitrogen cycle. Soil Use and Management, Vol 9, Nr 3, 86–94
22. Satrusajang A, Snitwongse P, Buresch RJ and Friesen DK (1991) Nitrogen-15 and sulfur-35 balances for fertilizers applied to transplanted rainfed lowland rice. Fert Res 28, Kheves Ac Publisher 55–65
23. Shepherd MA, Davis DB and Johnson PA (1993) Minimizing nitrate losses from arable soils. Soil use and management, Vol 9, Nr 3, 96–99
24. Sylvester-Bradley R (1993) Scope for more efficient use of fertilizer nitrogen. Soil use and mangement, Vol 9 Nr 3, 112–117
25. Triboi E (1974) Etallissement d'un systeme rationnel de fertilite Phosphatee sur deux types de chernoziums du SE de la Roumanie. These de Doctorat en Agronomie. IANB. Bucharest
26. Van Cleemput O and Hera C (1994) Links between fertilizer nitrogen efficiency and loss of nitrogen out of the rooting zone (unpublished)

C. Rodriguez-Barrueco (ed.), Fertilizers and Environment, 151–154.

Suitability of North Carolina natural phosphate to Polish agriculture

M. Fotyma[1], L. Hammond[2] and K. Kesik[1]
[1]*Institute of Soil Science and Plant Cultivation, Osada Palacowa IUNG, 24–100 Pulawy, Poland;* [2]*Texasgulf Inc., Raleigh NC 27622–0321, USA*

Key words: phosphorus fertilizers, phosphate rock, maize, winter rape

Abstract

North Carolina Natural Phosphate (NCNP) is a natural marine sediment, excavated, washed and sold in unprocessed form as a phosphate fertilizer. In vegetation and field experiments carried on in Poland in 1992–1994 with potatoes, rye, maize and winter rape the efficiency of phosphorus in NCNP was in the range of 75–100% of phosphorus in superphosphate. Blending and granulating of NCNP with urea and sulphur proved to be very effective. Further works on blending technology and the usability of NPS or NPKS blends are recommended.

Introduction

During the period 1975–1989, the consumption of phosphorus fertilizers in Poland averaged about 50 kg P_2O_5 per ha of agricultural area. Due to fertilizer production subsidies during these years the cost to the farmer was equivalent to only 1 kg grain per 1 kg of phosphorus. The fertilizers consumed were single superphosphate – 50%, triple superphosphate – 14% and ammonium phosphate/potassium-ammonium phosphate – 36%. The use of finely ground phosphate rock from Morocco and Tunisia did not exceed 1%. Since the introduction of market rules in 1990 and the lifting of the fertilizer subsidies, the consumption of phosphorus in Poland dropped to about 15 kg P_2O_5 per ha. The grain equivalent for fertilizers has increased to about 4–5 kg grain per 1 kg of phosphorus. The phosphate fertilizer factories, mainly single superphosphate, are now operating at only about 30% of production capacity. To avoid bankruptcy, the factories are enlarging the assortment of fertilizers by bulk-blending and offering NP, NPK, NPMg and PK blends. The primary components of these blends are single superphosphate to supply P_2O_5 and ammonium sulphate to supply N. These are the least expensive but most difficult products to sell as straight fertilizers in Poland.

Under these circumstances, the use of a highly reactive natural phosphate would be expected to show potential as a more economic source for either direct application or mixture with other fertilizers. Seventy percent of the agricultural soils in Poland are acidic and would be particularly suitable for the use of highly reactive natural phosphate. In 1991 the investigations on the suitability of North Carolina Natural Phosphate (NCNP) has started in this respect in Poland.

Materials and methods

NCNP is a natural marine sediment lying about 100 feet below the surface near the Atlantic coast in North Carolina, USA. Excavating equipment is used to remove this overburden and the 10–15 m thick vein of phosphate ore is extracted using huge draglines. The phosphate ore is then screened, put through a flotation process (washed) to remove impurities, and dried. No chemical processing is involved, and since it naturally occurs in a friable, sandy form rather than an aggregated rock, no grinding is required. Despite the fact that it has not been ground, it exposes a high surface area, 22 $m^2 \cdot g^{-1}$, a factor which contributes to its rapid rate of dissolution in the soil. Screen analysis shows that, even without grinding, there is no pebble present (i.e. > 1.19 mm) and that 97% of the rock passes a 35 mesh (0.42 mm) screen and 49% passes a 65 mesh (0.21) screen. NCNP is not soluble in water but 21% of phosphorus it contains is soluble in neutral ammonium citrate (AOAC method). Solubility measurements in neutral ammonium citrate, 2% citric acid, 2% formic

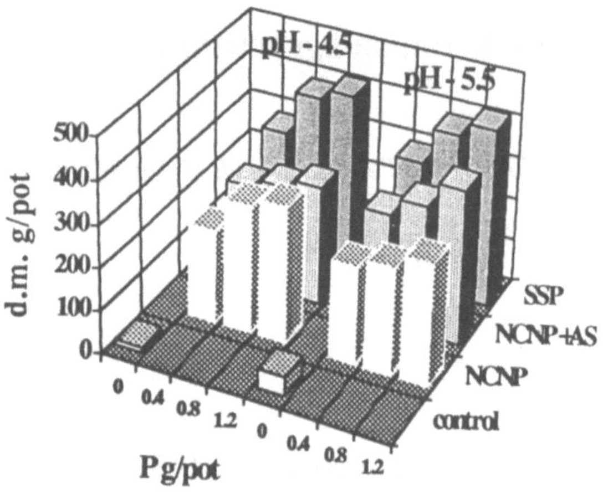

Fig. 1. Field experiments 1992–1993.

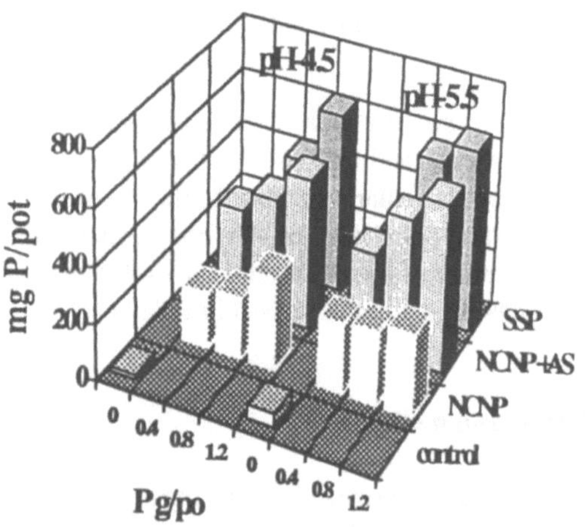

Fig. 2. Field experiments 1992–1993.

acid etc. indicate that NCNP is one of the most reactive natural phosphate and could anticipate the fast rate of dissolution in acid soils as well.

In 1992 and 1993 the vegetation and field experiments on the efficiency of NCNP has been carried on in Experimental Farm Grabów belonging to the Institute of Soil Science and Plant Cultivation in Pulawy – the largest agricultural Institute in Poland. In the vegetation experiments the absorption rate of phosphorus from NCNP and the accumulation of maize dry matter was measured depending on the phosphorus dose and soil pH. In the field experiments the efficiency of several phosphorus fertilizers was compared in two-course crop rotation with market and fodder crops. The soil was sandy loam, acid (pH 4.5–5.5) and showed a very low content of available phosphorus (30 mg $P_2O_5 \cdot kg^{-1}$

Table 1. The yield of dry matter of the crops grown in two-course rotations

Treatment	Yield of dry matter t. ha^{-1} for 2 years in the crop rotation	
	potatoes – rye	maize – maize
Without phosphorus + ammonium nitrate	15.1	15.0
Single superphosphate[1] + ammonium nitrate	18.3	18.6
NCNP[1] + ammonium nitrate	17.7	17.2
NCNP[1] + ammonium sulphate	18.0	17.0
LSD 0.05	0.5	0.8

[1] 120 kg $P_2O_5 \cdot ha^{-1}$ for 2 years.

soil acc. to Egner method). In all phosphorus treatments nitrogen was supplied as ammonium nitrate. In the treatment with NCNP ammonium sulphate was applied as well. In 1994 the preliminary experiments with NPS blends were carried on with winter rape as test crop. Two granulated blends were prepared by cold compressing and crushing the products:

NPS 1–9.4: 16.5: 13, NCNP + ammonium sulphate

NPS 2–12.4: 18: 14, NCNP + urea + elemental sulphur

The experiments has been conducted both in the vegetation house and on the fields belonging to private farmers on typical for Poland light and acid soils.

Results

Vegetation experiment with maize 1992

In the vegetation experiment all factors significantly influenced the dry matter yield of maize which was grown as a test crop (Fig. 1). The yields of maize grown on NCNP were in the range of 75–90% in relation to the yield on single superphosphate (SSP). On the very acid soil, pH 4.5, higher yield was recorded when NCNP was supplemented with ammonium nitrate as a nitrogen source. On moderately acid soil, pH 5.5, supplementation with ammonium sulphate (NCNP + AS) was more efficient. Ammonium sulphate activated obviously slightly soluble in water NCNP, though both fertilizers were applied separately and not in single granule. This activation was reflected more dramatically in absorption rate and in the final uptake of phosphorus by maize (Fig. 2). The final phosphorus uptake was almost doubled in the treatment with nat-

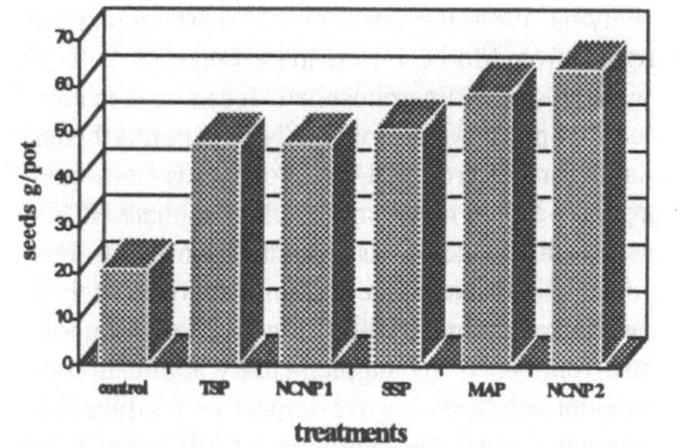

Fig. 3. TSP = Triple superphosphate; NCNP 1 = NCNP + amonium sulphate; SSP = Single superphosphate; MAP = ammonium phosphate; NCNP 2 = NCNP + elementar sulphur.

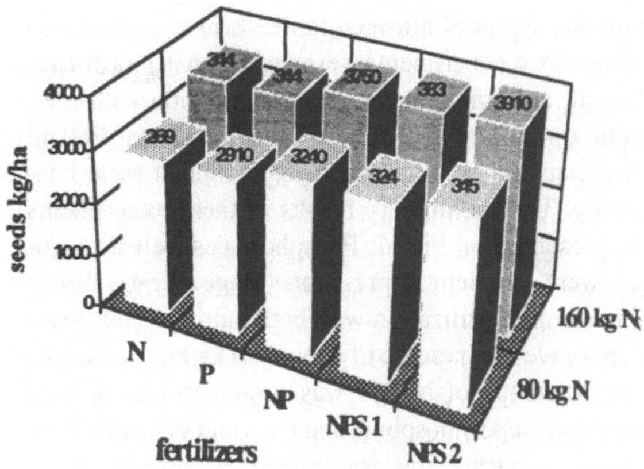

Fig. 4. N = urea; P = triple superphosphate; NP = urea + triple superphosphate (single fertilizers); NPS 1 = NCNP + ammonium sulphate (blended and granulated); NPS 2 = NCNP + urea + sulphur (blended and granulated). All fertilizers applied in the autumn with uniform dose of potassium. 80 kg N-ammonium nitrate applied at the beginning of vegetation, 160 kg N-ammonium nitrate applied at the beginning of vegetation (50%) and at the budding stage (50%).

ural phosphate supplemented by ammonium sulphate (NCNP + AS) in comparison to phosphate supplemented with ammonium nitrate NCNP.

Field experiments 1992–1993

Field experiments were carried on in two independent, two-course, rotations: potatoes – winter rye and maize for silage – maize for silage. In Table 1 the sum of dry matter yields of main crop products for 2 years are presented. In the rotation with market crops the yield of dry matter of potatoes and winter rye grown on NCNP was about 90% of the yield on single superphosphate. In the rotation with fodder crop the yield of maize grown on both fertilizers was practically the same. The supplementation of NCNP with ammonium sulphate, in cotradiction to vegetation experiment, did not increase the efficiency of phosphorous.

It can be explained by much looser contact of both fertilizers in the larger volume of soil in field conditions. In both, vegetation and field, experiments there was no difference in cadmium content in crops grown on single superphosphate and NCNP. Cadmium in NCNP appears obviously in unavailable form.

Vegetation experiment with winter rape 1994

Winter rape in Poland is sown at the end of August and develops a strong canopy before winter time. Consequently the crop shows high demand for nutrient, particularly for phosphorus and nitrogen, already in the fall. Winter rape is also high demanding for sulphur fertilization. In vegetation experiment carried on soil poor in phosphorus several N, P, NP and NPS fertilizers applied before sowing of rape were tested. In the following spring all treatments were uniformly and liberally fertilized with nitrogen in the form of ammonium nitrate + calcium nitrate. The preliminary results of this experiment are presented in Fig. 3. The most limiting factor was phosphorus. In the control treatment and the treatment with nitrogen alone a very low yield of rape seeds were recorded. Application of phosphorus in single or compound fertilizers increased dramatically the yield of seeds. The best fertilizer for fall application proved to be NCNP 2, e.g. the blend of NCNP, urea and elemental sulphur. Only the second was ammonium phosphate which contains phosphorus in the form fully soluble in water.

One of the reasons of the superiority of NCNP 2 over MAP could be the content of sulphur in the former fertilizer. This assumption is confirmed by slightly higher yield obtained on SSP (containing calcium sulphate) in comparison to TSP. Ammonium sulphate in NCNP 1 seems to be the unsuitable component of nitrogen-phosphorus-sulphur blends presumably due to acidifying effect on the already acid soil.

Field experiments with winter rape 1994

The idea underlying the field experiments was generally the same as in vegetation experiments. Several N, P, NP and NPS fertilizers were applied before sowing of winter rape together with the uniform dose of potassium and in the spring the crop was top-dressed with two doses of nitrogen in the form of ammonium nitrate. Two experiments were carried on in North-East Poland, environmentally very clean region with low-input agriculture and one experiment in West Poland characterized by intensive highly efficient rural husbandry. The preliminary results of these experiments are presented on Fig. 4. Phosphorus as well as nitrogen were deficient in pre-winter stage of rape development and fertilization with both nutrients (NP treatment) gave the increase of about 300 kg·ha^{-1} of seeds. The efficiency of NPS 1 was similar to that of urea and triple superphosphate and the yield of winter rape grown on NPS 2 was insignificantly higher in comparison to straight fertilizers. It can be concluded that NCNP activated by either ammonium sulphate (NPS 1) or sulphur (NPS 2) is as good source of phosphorus as fully soluble triple superphosphate.

Discussion and conclusions

In the vegetation and field experiments with potatoes, winter rye and maize for silage NCNP applied as a straight fertilizer proved to be slightly and in most cases insignificantly less efficient as fully soluble phosphorus fertilizers (SSP, TSP ora MAP). In the older experiments carried out in Poland in early 80 ties (Kêsik and Fotyma, 1988) the yields of maize grown on rock phosphate from Morocco were in the range 83–93% of the yield on single superphosphate. It can be then concluded that phosphorus from NCNP is slightly better available for this crop that the phosphorus from finely ground (< 0.06 mm) 'soft' rock phosphate.NCNP offers the advantage of easier application due to less fine dust. The results of vegetation experiments with maize showed the possibility for activating the phosphorus from NCNP by supplementary application of ammonium sulphate. It gave impact to produce (on experimental scale) the blends of NCNP either with ammonium sulphate or with urea and sulphur as a compound NPS fertilizers. In the vegetation and field experiments these blends proved to be very promising in fertilization of winter rape. The Polish farmers are nowadays interested in buying the cheap blended fertilizers containing all basic nutrients or at least nitrogen and phosphorus.

References

Kêsik K and Fotyma M (1988) Assessment of fertilizer value of superphosphate-ground phosphate rock mixture (*in Polish*) Pamiêtnik Pulawski 91, pp 185–203

C. Rodriguez-Barrueco (ed.), Fertilizers and Environment, 155–159.

Improvement of leguminous vegetables production in calcareous soil by addition of some acidifying materials

I. Effects on yield and yield components

N.N. Michail, F.S. Faris, M.W.A. Hassan & R.G. Kerlous
Soils and Water Research Institute, Horticulture Research Institute, Agric. Res. Center, Giza, Misr (Egypt).

Key words: calcareous soil, acidifying, leguminous, yields, yield components

Abstract

Two field experiments were carried out in two sites of Noubaria Agricultural Research Station Farm. The investigated acidifying materials were sulphur, superphosphate and Phosphorine. The materials were mixed with cowpea seeds just before sowing in a rate 1 kg for seeds required to 1 fed (4200 m^2) in the first experiment. Additions were 100, 30 and 1 kg S, P_2O_5 and Phosphorine/fed, respectively to soils before common bean planting in the second experiment where the two varieties Giza 3 and Giza 6 were put under investigation. The residual effect of the soil additions was investigated using bean as a test crop.

The obtained results showed that acidification materials were effective in such soil for seed yields of cowpea and common bean - while the residual effect on bean was unconsiderable. The variety Giza 6 of common bean was significantly superior to Giza 3. The harvesting index was higher by using sulphur or superphosphate than Phosphorine in the case of cowpea while in common bean superphosphate and Phosphorine were more active in increasing this index than sulphur. Acidification process resulted in some increases in harvest index of bean regardless the used material. As for 100 seed weight, it was raised in significant level by using .sulphur or superphosphate than Phosphorine or control in the case of cowpea while all treatments were statistically equal in the case of common bean and bean.

Introduction

Calcareous soils represent a vast area in Egypt. In these soils, $CaCO_3$ plays a considerable role in phosphorus nutrition of plants.

The effect of acidifying materials applied to calcareous soils on the availability of soil phosphorus has been reported by many investigators. Badr El-Din *et al.* (1981) reported that, sulphur application to calcareous soils affected phosphorus nutrition and may increase the availability of other plant nutrients particularly the minor elements by lowering the soil pH. El-Shall *et al.*, (1987) found a marked increase in the yield of barley due to sulfur application in calcareous soil at Ras Sidre in South Sinai. The response to sulfur differed according to variety, yield being more than doubled for cc 163 variety but only inereased by 30% for Griza 1221 variety. Wassif *et al.* (1991) showed that the solubility and leachability of certain plant nutrient cations (Ca^{2+}, Mg^{2+}, Na^+ and K^+) and anions (Cl^- and $SO_4^=$) in two calcareous soils increased significantly as the elemental sulphur application rate and salinity level of irrigation water increased.

Also, calcium superphosphate was widely considered as an acidifying material in addition to its importance as phosphorus source. The majority of researchers studied it for both purposes. Mack *et al.* (1964) found increases in dry matter and P contents of french bean and pea by P application where the response of bean was greater than peas. Coertze (1977) reported an excellent production for snap bean was found in medium loamy soils and adequate fertilization. Singh *et al.* (1981) found that P applied at 60 and 90 kg/ha significantly increased yields in comparison with 30 kg P, when 4 levels of P (0, 30, 60 and 90 kg/ha) were applied. El-Gharably and Abdel-Razek (1982) studied the effect of various levels of P being added to a sandy calcareous soil on snap bean grown in a greenhouse for 35 days. They found that the maximum growth was obtained by 150 ppm P application. They also found a positive effect of P, Fe, Ca, Mg, Mn and Zn by P application. Negm *et al.* (1992) observed that phosphorus application increased the number of branches and flowers per plant. The uptake of N, P and K in different plant parts was increased gradually by the increase in P concentration in the fertilizer solution.

Phosphorine, the phosphate solubilizing biofertilizer was studied by Taha *et al.* (1969) and Osman *et al.* (1992). They mentioned that the compound is produced and practiced now in different countries where an inverse relationship was recorded between pH values and the amounts of released phosphorus, and highly positive correlation (R = 0.95) has been obtained between total acidity and released amounts of soluble phosphorus. Moreover, they recommended the application of that compound which, as they reported contained starter, inoculation and carried, for different crops. They added that Phosphorine rates in soil and water decreases the production costs and minimizes pollution. The effect of Actinomyces, Fungi and Bacteria solubilise rock phosphate and tricalcium phosphate enhancing more available P in calcareous soils was reported by Molla *et al.* (1986), Sattar and Gaur (1986), Thomas and Shantaram (1987) and Khalafallah *et al.* (1988). Moreover, Sham Sei Fan and Morard (1993) observed that phosphorus uptake was slight during the development cycle in a paralleled trend of dry matter production and so, continuous providing of P should be released.

From otherwise common bean varieties, Giza 3 and Giza 6 were compared by Faris *et al.* (1992) who reported that Giza 6 variety exhibited the characters of heavy and large seeds, high productivity, resistant to mosaic virus and rust infection. Giza 3 is one of Giza 6 parents associating with Swiss Blanc.

The current work is therefore, conducted to evaluate sulphur, superphosphate and Phosphorine as acidifying materials and their effects on legumes yields and yield parameters under calcareous soil conditions.

Materials and Methods

Two field experiments were carried out in two sites of Noubaria Agricultural Research Station Farm. A soil sample was taken before sowing and prepared for the determination of physical and chemical analyses as show in Table 1 according to Black (1982).

The investigated treatments were elemental sulphur B superphosphate C and Phoshorine D in addition to control A. Each treatment was replicated 4 times in 16 plots for cowpea experiment and in a complete randomized block design: 32 plots for two common bean varieties namely Giza 3 and Giza 6 in a split block design.

Seeds of cowpea and common bean were planted on 26th of May 1992 and 1st of March, res-pectively where the common agricultural practices were done till harvesting on 29th August and 11th June 1992 for cowpea and common bean, respectively. The addition of acidifying materials were added mixing with cowpea (*Vigna sinensis* savi) seeds in a uni-rate of 1 kg substance/seeds required to one fed while in common bean (*Phaseolus vulgaris*) experiment, soil addition of 100 kg S, 30 kg P_2O_5 or 1 kg Phosphorine/fed. was conducted just before planting. Faba bean (*Vicia faba* L.) seeds were planted after common bean in the same plots, to study the residual effect of these treatments through a complete growing season starting from 15 November 1992 till harvesting on 29th of April 1993. A plant sample of 3 plants was taken on the 7th of July 1992, after 45 days of planting representing the vegetative growth stage as well as another one at maturity from the plots of each experiment. The plants were washed, oven dried, weighed where the mature plants were divided into seeds and straw. The plant materials were ground and prepared for chemical analysis. The seed yield was collected for each crop, air dried and weighed. Some yield components as harvest index (seeds/whole plant ratio) and 100 seed weight were estimated.

Results and Discussion

Seed yields

Table 2 shows that cowpea seed yields of B, C and D treatments were statistically the same while each of them was significantly higher than control treatment A indicating that cowpea responded to acidification process regardless the used material.

The behaviour of acidifying materials for common bean was similar to that for cowpea. As for varieties, it was noticed that Giza 6 was significantly superior to Giza 3 regardless acidifcation. The interaction of treatments and varieties showed no significant difference between each other. Concerning bean as a test crop to investigate the residual effect of these additions, there

Table 1. The soil components and chemical analyses for the investigated soil sample.

Depth (cm)	Particle size %				Texture class	$CaCO_3$ fractions %			
	Clay	Silt	F. sand	C. sand		Clay	Silt	F. sand	C. sand
0–25	33.32	11.40	40.04	14.10	Sandy clay loam	11.55	12.21	4.40	28.16
25–50	32.47	18.87	36.04	13.12	Sandy clay loam	17.51	4.24	8.76	25.51

Depth (cm)	T.S.S. %	Anions me/100 g soil				Cations me/100 g soil				O.M %
		$CO_3^=$	HCO_3^-	Cl^-	$CO_4^=$	Ca^{++}	Mg^{++}	Na^+	K^+	
0–25	0.29	–	0.55	6.00	0.31	1.75	1.25	3.54	0.32	0.63
25–50	0.38	–	0.70	4.50	1.18	1.60	1.40	3.00	0.38	0.61

Table 2. Seed yield of legumes as affected by acidifying materials in kg/fed

Crop	A	B	C	D	Mean	L.S.D.
Cowpea	244.8	321.9	333.1	313.4	–	21.3
Common Bean Giza 3	224.6	395.5	306.2	331.9	314.5	treat. 96.8
Giza 6	302.9	440.9	420.0	396.4	390.0	vari 68.4
Mean	263.7	418.2	363.9	364.1	–	t.v.:n.s
Faba Bean	699.5	920.6	866.5	782.9	–	n.s.

was no significant difference between treatments A, B, C and D indicating that common bean almost consumed the added materials and therefore, there was little residual effect of these additions. These results were in agreement with El-Shall *et al.* (1987) for sulphur, Singh *et al.* (1981) for superphosphate, and Taha *et al.* (1969) and Osman *et al.* (1992) for Phosphorine.

Harvesting index

Harvesting index, the ratio of seeds to whole plant was calculated to detect the produ-ctivity unit of each crop as affected by acidifying treatments. Table 3 clarifies these indexes. Cowpea index was grouped into two statistical groups. The lower one including treatments B and C while the other including A, and D was significantly higher than the 1st group. It was noticed also that there was no significant differences within the treatments of each group. It could be concluded that B and C raised both of the seeds and straw in the same rate but in A and D treatments seed production was more superior to straw. The index decreased in common bean, as Table 3 shows, according to the descending order D > C = A > B with small differences not more than 2.2. Also the varieties gave some differences where Giza 3 harvest index exceeded than of Giza 6 in 2.5. The highest ratio appeared with treatment C in variety Giza 3 while the lowest value was obtained in the same treatment when added to Giza 6. Concerning bean, Table 3, shows that the highest ratio with treatment D and C where statistically each of them was significantly higher than control; the ratio in treatment C was similar to that in control treatment A with small differences with B and D, indicating that bean responded to acidifcation process regardless the used material. In this connection Negm *et al.* (1992) attributed that due to the increases in number of branches and flowers per plant.

Table 3. Harvesting index (seeds/whole plant ratio) of legumes as affected by acidifying materials.

Crop	A	B	C	D	Mean	L.S.D.
Cowpea	27.9	25.7	25.5	27.3	–	0.7
Common Bean Giza 3	24.3	23.1	25.2	23.2	24.0	Not
Giza 6	20.7	20.3	20.8	24.6	21.40	deter-
Mean	22.5	21.7	22.6	23.9	–	mined
Faba Bean	30.4	31.1	30.9	29.1	–	1.6

Table 4. Weights of 100 seeds of legumes as affected by acidifying materials in gm.

Crop	A	B	C	D	Mean	L.S.D.
Cowpea	20.7	19.8	19.0	20.1	–	0.9
Common Bean Giza 3	24.9	26.5	27.7	21.4	25.1	treat : n.s.
Giza 6	28.7	27.3	29.8	25.0	27.7	vari : n.s.
Mean	26.8	26.9	28.7	23.2	–	t.v. : n.s.
Faba Bean	68.5	68.7	70.8	68.5	–	n.s.

Specific seed weights:

Regarding the specific seed weights (100 seed weight). Data in Table 4 indicate that weight of 100 cowpea seeds of B, and C were significantly higher than that of control treatment A. The treatment D was similar to control treatment A. It is clear that Phosphorine application produced small size seeds when it was compared with sulphur or superphosphate treatments. Concerning common bean, factors of treatments and varieties were not significant among each other. Generally, the varieties gave some differences where Giza 6 was riased than Giza 3 with about 2. 5 g and treatment C was the highest among acidification treatments. In case of bean. It is clear that all of treatments were without significant differences because the common bean almost consumed the added materials in addition to many other factors and therefore there was little residual effect of these additions.

References

Badr El-Din MM, Negm MA and Abd-Elnaim EM (1981) Effect of sulphur addition of the nutritional status of phosphorus in calcarcous soil; Agricultural Research Review 59.

Black CA (1982) Methods of Soil Analysis Parts 1 and 11. Am. Soc. Agron. Series (9). Madison, Wiscon, U.S.A.

Coertze AF (1977) Climate and soil requirements for green beans. Farming in South Africa Series: Vegetable Cultivation in South Africa Green beans and green peas No. B. 1/1977

El-Gharably GA and Abd El-Razek M (1982), Responses of snap beans to applications of P and Fe in calcareous soil. Journal of Plant Nutrition 5: 721–728.

El-Shall A, Wassif M, Hilal M and El-Bagouri I (1987) Response of barley to sulfur application in a calcareous soil under saline irrigation water. Desert Instit. Bull., A.R.E. 37. No. 1: 177–130.

Faris FS, Nassar SH and El-Dessouky EM (1992) Giza 6, a new variety of common bean, Agric. Res. Rev., Cairo, [in press].

Khalafallah MA, Saleh EA, Mahmoud/SAZ and Abas MH (1988) Phosphate dissolving Actinomycetes in some Egyptian calcareous soils. Egypt. J. Microbiol. 23: pp. 413–427.

Mach HJ, Fang SC and Apple SB (1964) Effect of soil temperature and phosphorus fertilization on snap beans and peas. Proceedings of the American Society for Horitcultural Science 84: 332–338.

Molla MAZ, Chowdhury AA and Islam, AS (1986) Microbial mineralization of organic phosphate in soil. Microbiology 19: 308–316.

Negm MA, Abdel-Reheem MA and Montasser SY (1992) Effect of different rates and frequencies of, foliar applications of phosphorus on sunflower growing on calcareous soils. Egypt. J. Soil Sci. 32: 513–521.

Nicholaides III JJ, Chancy HR, Nelson LA and Shelter JE (1985) Snap bean grade and yield response to N rate and time of application and P and K rate. Communications in Soil Science and Plant Analysis, 16: 741–757.

Osman YM, Ibrahim AN and Khirey AM (1992) Phosphorine, phosphate solubilizing biofertilizer, its production, application and effect on crop yield. 2nd International Scientific Conference, the Application and Utilization of the Agricultural Scientific Results in Developing Countries, 31st Aug. - 3rd Sept. 1992, Godollo, Hungary.

Sattar MA, Gaur AC (1986) Dissolution of rock phosphate by rhizosphere microorganisms isolated from Bangladesh soils, Bangladesh J. of Agriculture 11: 27–43.

Sham Sei Fan A and Morard P 1993 Nutrient uptake by Toulouse violet (*Viola odorata* var. parmensis) during its developmental cycle. M.A.C. Fragoso and. M.L. Van Beusichem (eds). Optimization of plant nutrition, 269–275.

Singh KN, Prasad RD, Tomar VPS (1981) Response of french bean to different levels of nitrogen and phosphorus in Nilgiri-Hills under rainfed condition. Indian Journal of Agronomy, 26: 101–102.

Taha SM, Mahmoud SAZ, El-Damaty AH and Abdel-Hafez AM (1969) Activity of phosphate dissolving bacteria in Egyptian soils. Plant and soil 31: 149–160.

Thomas GV, Shantaram MV (1987) Solubilization of inorganic phosphates by bacteria from coconut plantation soils. J. of Plantation Crops 14: 42–48.

Wassif MM, Elgala AM, Mostafa MA, and El-Maghraby SE (1991) Effect of elemental sulfur and water salinity levels on ion solubility in two calcareous soils. 2nd African Soil Sci. Soc. Conf, Cairo, Egypt.

C. Rodriguez-Barrueco (ed.), Fertilizers and Environment, 161–170.

Nitrogen fertilization management in orchards to reconcile productivity and environmental aspects

M. Tagliavini[1], D. Scudellazi[2], B. Marangoni[1] & M. Toselli[1]
[1]*Department of Horticulture and Forestry, University of Bologna, 40126, Bologna, Italy;* [2]*ERSO, Viale Bovio 600, 47023 Cesena (Forli), Italy*

Key words: N requirements, N status, N use efficiency, root distribution, soil nitrates

Abstract

Nitrogen fertilization in orchards of Emilia-Romagna Region, (Italy) was based in the past on excessive, not split, applications often supplied late in winter; the NUE (Nitrogen Use Efficiency) was therefore low and the risk of nitrate leaching was high. This paper summarizes the studies conducted in the last 10 years at the Department of Horticulture and Forestry of the University of Bologna aimed to develop a more rational use of nitrogen in orchards and vineyards. Root escavation of mature trees revealed that the use of localized irrigation (drip or microjet) causes a concentration of roots in the area wetted by the emitters. In such a situation, band applications of N to the tree row may allow a reduction of amounts of N fertilizer, while widespread applications, especially if the orchard soil is tilled, lead to an accumulation of nitrates in the alley. Results of several field trials where increasing N rates were applied indicate that the kind of response to N supply depends on the presence in soil of natural sources of nitrogen. This fact clearly stresses the necessity of evaluate the N status of an orchard before N fertilization. Rapid estimation of leaf chlorophyll by portable instruments is a promising index of leaf N concentration, only provided that calibration is made for each cultivar. A method, currently under testing in orchards and vineyards of Emilia-Romagna, is proposed here to adjust N fertilizer rates to the demand of the crop and to the level of available N in soil as determined in soil or soil solution samples.

Introduction

For decades, fruit growers in Italy have relied on nitrogen fertilization to increase yield, and application rates in fruit orchards of Emilia-Romagna, one of the most intensive agricultural area of Italy, ranged approximately from 150 kg ha^{-1} for apple to 300 kg ha^{-1} for peach (Cobianchi and Turci, 1980; Lalatta, 1980). Such amounts are clearly in excess with respect to plant demand (Weinbaum *et al.*, 1992) and yearly applications have tremendously increased the N fertility of some soils and tree N reserves to the point that interrupting the N supply for long periods might not result in a decrease of yields (Scudellari *et al.*, 1993). These outdated recommendations have also included non-split and broadcast applications early in the season, often before bud burst, when tree uptake is low and deciduous trees entirely rely on N reserves stored the previous year (Millard and Thompson, 1989; Muñoz *et al.*, 1993).

In more recent years, characterized by an over-production of many fruit crops, growers are also requested to produce fruits with high quality standards and storage potential. Excessive N supply has been often associated with a depression of fruit quality (Scudellari *et al.*, 1993) and poor storage life (Marcelle, 1984). Many causes have been evoked to explain these adverse effects, including the competition that nitrogen exerts on other nutrients as calcium, and the stimulation of vegetative growth caused by high soluble N in the soil solution.

Environmental concerns call now for more than ever urgent no-polluting strategies of fruit orchard management. The entry of NO_3^- into ground and drinking water has recently gained major attention. This risk is of great concern in well drained sandy soils, but should not be underestimated in loamy and clay-rich soils of the eastern Po Valley (Rossi *et al.*, 1991) to which the plain area of the Emilia-Romagna Region geographically belongs. These soils have an alluvial origin and are characterized by a raise of shallow water table at the end of the winter up to 80 cm (Patruno *et al.*, 1986). These shallow water tables sustain superficial waters and are connected with ground waters (Baratti, 1991), the latter often used as drinking sources. Moreover, it should be considered that well fissured clay soils allow a high rate of vertical movement of water (Shaw, 1962). For these reasons, nitrates, the main fraction of soluble mineral N in alkaline and calcareous soils of the Po Valley, will likely be subjected to leaching towards deep soil layers down to the water table as a result of winter and spring rains (Rossi *et al.*, 1991). The pathway followed by nitrates then includes (i) transportation via the superficial waters to the Adriatic sea, where it contributes to the eutrophication phenomenon (Getti and Pantani, 1991), and (ii) contamination of ground waters that may induce methamoglobinemia in humans if used for drinking purposes.

High leaching losses of nitrate occur when the N fertilizer rate is not adjusted to the demand of the crop (Weller, 1983) and to the level of available nitrogen in soil (Ziegler *et al.*, 1992), and when low recovery of N from fertilizers occurs; a low nitrogen use efficiency (NUE) also results from incorrect timing of supply and distribution practices of the fertilizers. Moreover, fruit trees are generally expected to have low recovery capacity of fertilizer N as typically they have lower root density as compared to many field crops (Atkinson and Wilson, 1979). This paper summarizes the research activity recently conducted at the 'Department of Horticulture and Forestry' of the University of Bologna with the aim of developing a more rational usage of N fertilizers in fruit orchards. A method is presented by which fertilization rates are calculated taking into account the goal of minimizing the risk of nitrate leaching.

Water management, root distribution and soil nitrates

The knowledge of root development in soils is of great importance for improving the recovery of fertilizer-N. Plant genotype and soil characteristics as well as the orchard management are the main responsibles of root growth pattern. Under dry climates it has been proved that root distribution depends mainly on the irrigated soil volume: therefore a considerable reduction of root explored soil volume occurs with drip irrigation (Levin *et al.*, 1980). As localized irrigation methods (microjet and drip) have become frequently adopted

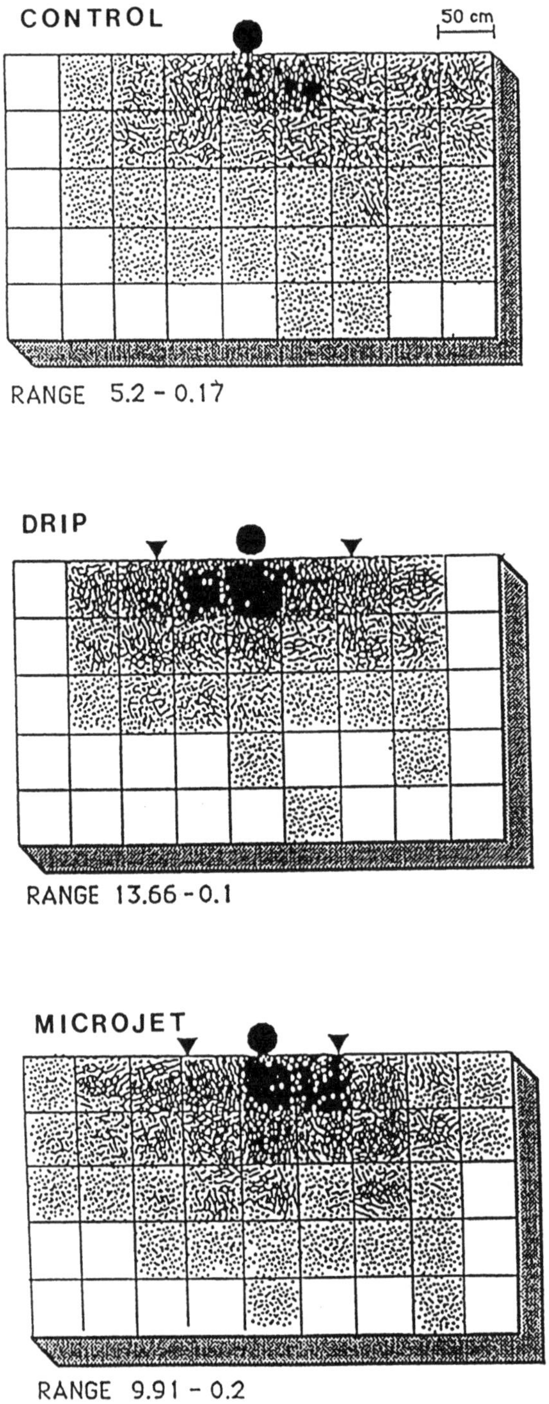

Fig. 1. Horizontal distribution of roots of Myrobalan B into a frame of 50 cm mesh, as affected by irrigation system. Within each square, root density is linearly and positively related to the number of points; the maximum and minimum root density (g l^{-1}) inside meshes is reported below each diagram. Circles and triangles indicate trunk and emitter positions, respectively.

in recently planted orchards of Emilia-Romagna, the question arose as to whether such a way to supply water could have had an effect on root growth distribution, even in areas with yearly rainfall averaging 700–800 mm. For such a reason, root excavations were performed in 1990 in two mature orchards located in Emilia-Romagna and subjected to different irrigation treatments. Both orchards, one with peach and the other with apricot trees, had a floor management with grassed alleys and herbicide row strips, approximately two meters wide.

The peach orchard was established eight years before roots were excavated, with planting distances of 5 × 5 m, on a deep alluvial clay-loam soil. Trees of the cv Flavorcrest were grafted either on seedling and on the hybrid peach × almond, GF 677. Treatments included unirrigated controls, drip and overhead sprinkler irrigation with a restitution of 100% ET. Two trees per graft combination were considered and, for each tree, half of the root system was excavated by the aid of pressured water (Böhm, 1979). Total roots and roots with size less than 5 mm were divided according to their horizontal distribution along the row and between the rows, into a frame of 50 cm mesh.

Roots in the apricot orchard (cv S. Castrese on Myrobalan B), with planting distances of 4.5 × 3.5 m, were excavated nine years from planting. The orchard soil was sandy clay loam and trees were subjected to drip and microjet irrigation. Not irrigated trees were present as control. Excavation methodology and number of trees considered were as in the peach orchard.

In both orchards, the deepest roots were found at approximately 70–80 cm depth. Table 1 reports the percentage of total and fine (less than 5 mm diameter) roots present under a weed-free row strip of 2 m width. In both orchards, drip irrigation resulted in a higher concentration of total and fine roots close to the tree rows as compared to control. Distribution of peach rootstock roots was similar in non-irrigated trees and in those subjected to overhead irrigation (Table 1). Peach seedlings and GF 677 roots showed a similar soil distribution. As shown in Fig. 1, root density of drip irrigated and microjet irrigated trees was highest under the emitters. Myrobalan root system intensively developed under the grassed interrow if no irrigation was applied (Table 1).

These data support the assumption that under soil and climate conditions of Emilia-Romagna, the method of supplying water to the soil affects the root distribution. Localized irrigation systems, in particular, cause a concentration of roots in the area wetted

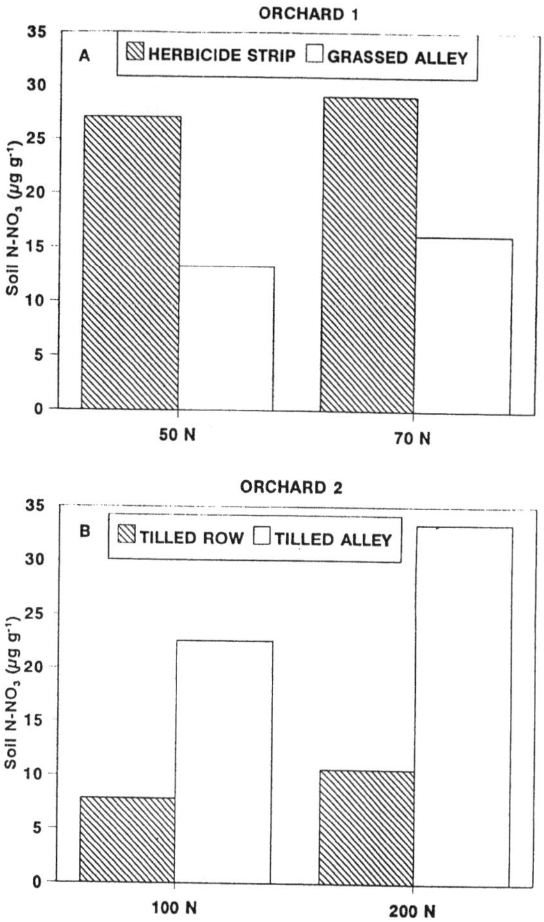

Fig. 2. Soil N-NO_3 as determined in July at a depth of 0–40 cm in correspondence to tree row and alley of an orchard with grassed alley and herbicide row strip (A) and of another with soil completely tilled (B). N fertilization rates (kg ha^{-1}) are reported in the horizontal axis.

by the emitters. In such a situation, band application of nitrogen in a strip along the tree row should improve the fertilizer-N recovery and reduce the concentration of nitrates in soil as compared to broadcast application of the fertilizers. This fact is especially true in the orchards whose soils are completely tilled despite the presence of drip irrigation, as it appears in the following example that refers to recently conducted field studies (Marangoni *et al.*, 1994; Scudellari, unpublished results). Two orchards were considered, the first (orchard 1) having a grassed and regularly mowed interrow (Fig. 2A) and the second (orchard 2) having soil completely tilled (Fig. 2B). Fertilizers were broadcast distributed to the whole soil surface and despite the different N rates applied, strong differences arose on soil distribution of nitrates in the two orchards

Table 1. Percentage (in weight) of total and fine (diameter size less than 5 mm) roots found in a 2 m wide row strip of soil, as affected by irrigation system

Treatments	Total roots			Fine roots		
	Peach		Apricot	Peach		Apricot
	Seedling	GF677	Myrobal	Seedling	GF677	Myrobal
Control	86 ± 7[1]	84 ± 5	71 ± 1	70 ± 14	78 ± 1	49 ± 4
Drip	94 ± 6	94 ± 2	86 ± 6	82 ± 8	87 ± 6	69 ± 7
Microjet	–	–	80 ± 2	–	–	65 ± 7
Overhead sprinkler	78 ± 12	81 ± 1	–	69 ± 3	79 ± 6	–

[1]Mean and standard deviation.

Table 2. Concentrations of ammonium and nitrate in the solution of a soil fertigated with three rates of N as ammonium nitrate[1]

N supplied by fertigation		Soil solution	
		$N\text{-}NH_4$	$N\text{-}NO_3$
(mg tree^{-1})	(mg kg^{-1} soil)	(mg l^{-1})	(mg l^{-1})
0	0	3 ± 2	42 ± 10
660	3	2 ± 1	51 ± 10
1320	6	2 ± 1	66 ± 8

[1]Unpublished results obtained in a newly planted intensive apple orchard;
[2]Averages of 4 values and standard deviations.

in July; in the orchard 1, the presence of grasses lowered the nitrate concentration in the interrow (Fig. 2A), while nitrates accumulated in the tilled interrow soil of orchard 2, as a result of the absence of tree and grass N uptake. Our results fit with findings by Shaller (1991) and confirm that the presence of grasses in the interrow plays an important and often underestimated role for reducing the environmental risks associated with leaching of mineral ions.

The need of estimating the N status in the orchard

Several sources of plant available N exist in soils (Mengel, 1991): ammonium fixed in 2:1 clay minerals or clay-adsorbed, nitrogen from irrigation water, precipitations, mineralization of organic matter and atmospheric sources. The relative and overall importance of these sources varies according to soil type, climate and previous crop management, hence their contribution to the tree nutrition is variable also. Since plants have access to nitrogen in the soil solution, the contribution of all the above mentioned natural sources of N to tree nutrition depends on their transformation into soil solution N. Although ammonium and nitrate are the two main forms of N suitable for plant uptake, several evidences obtained in calcareous and alkaline soils of the Emilia-Romagna indicate that the contribution of ammonium is negligible as compared to nitrate. Table 2 reports the nitrate and ammonium concentration in the solution (extracted by soil solution access tubes) of a sandy loamy soil (pH = 8.1) after fertigation with three rates of ammonium nitrate, providing a total of 0, 660 and 1320 mg N/tree. Possible explanations of the increase of soil solution nitrate but not ammonium (Table 2) are the ammonia volatilization, known to be considerably higher in lime and alkaline soils as compared to acidic soils (Oenema and Velthof, 1993), the high nitrification activity and ammonium fixation.

Several field trials in orchards of Emilia-Romagna region have been undertaken with the aim of evaluating the response to increasing N rates; Table 3 summarizes the nitrogen status of the unfertilized trees. The two studies on nectarines (Table 3) revealed a complete different response of trees to withdrawal of N-supply: neither a yield decline (data not shown) nor a depression of leaf N was observed in the orchard with cv. Stark Redgold, while interrupting the N supply for two years resulted in a severe N deficiency in the other nec-

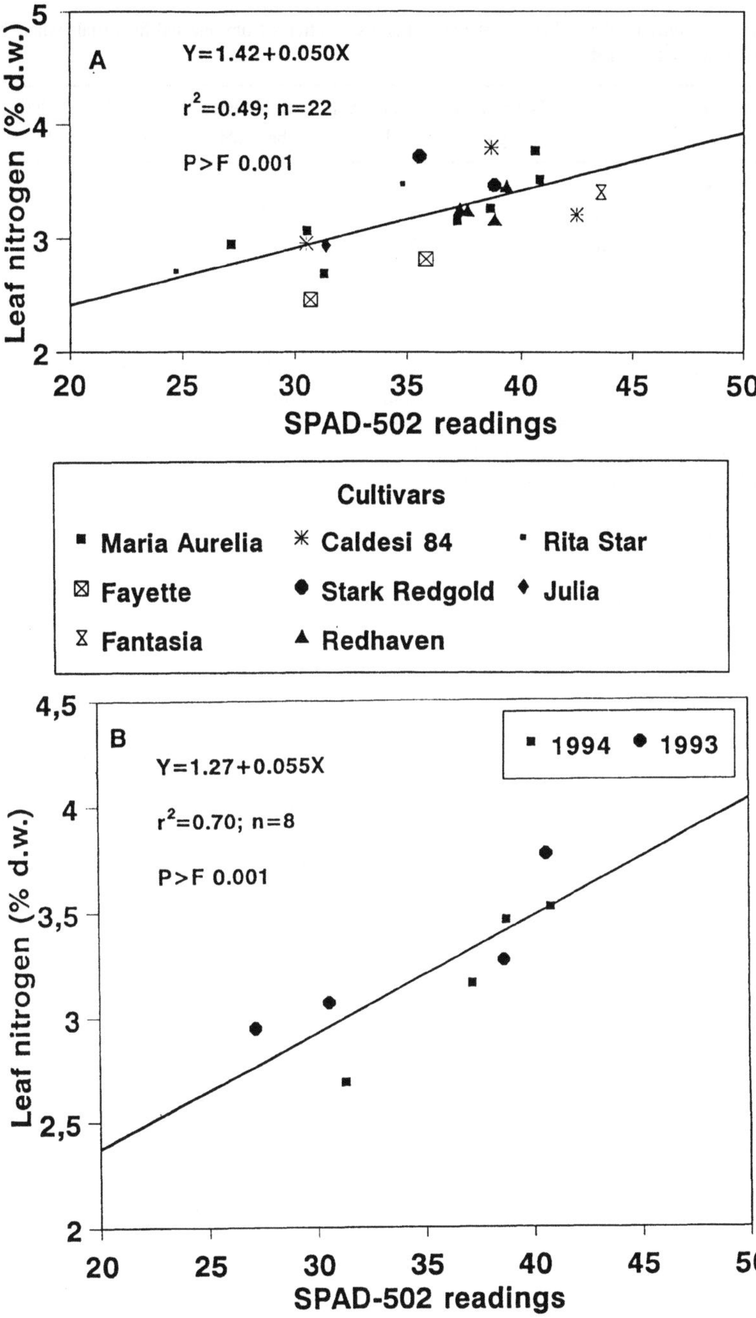

Fig. 3. Linear regressions of leaf-N concentration and SPAD reading. Data refer to determinations and analyses conducted on the same leaves, always collected in July and August. Pooled data from 8 peach cultivars are shown in (A), while data in (B) refer to cv. Maria Aurelia only, and each point is the average of 6 replications.

tarine orchard (Table 3). When comparing the available nitrogen in the soil of the two nectarine orchards, a complete different picture arose (Table 3), with N-deficient trees located in a soil with very low nitrate concentration.

The need of estimating the N status of the orchard before recommending N rates clearly arises from the previous example. Leaf chlorophyll meter readings (SPAD-502, Minolta Corp.) have been used to estimate leaf N concentration in field crops (Peng *et al.*,

Table 3. Concentration values of total leaf N and soil nitrates from selected field trials where N supply was interrupted

Tree crop	Years of no N-supply	Total leaf N (%d.w.)	Sufficiency N threshold (% d.w.)	Soil[7] N-NO_3 ($\mu g\ g^{-1}$)
Nectarine:				
cv. Stark Redgold[1]	4	3.47	3.0[5]	21
cv. Maria Aurelia[2]	2	2.68	3.0[5]	1
Pear:				
cv. Abbé Fetel[3]	4	1.85	1.8[6]	9
Apple:				
cv. Hi Early[4]	9	2.29	1.9[5]	–
cv. H. Stripe[4]	9	2.36	1.9[5]	–

[1]Adapted from Marangoni *et al.* (1993);
[2]Adapted from Tagliavini *et al.* (1994);
[3]Scudellari (unpublished results);
[4]Casalicchio *et al.* (1992);
[5]Jones *et al.* (1991);
[6]Shear and Faust (1980);
[7]Soil nitrates determined at the same time as leaf N.

Table 4. Annual removal of nitrogen by a mature peach orchard

Tree part	Amounts (t ha^{-1})	N Concentration (%)	N removed (kg ha^{-1})
Fruits	15–30	0.09–0.12[1]	13–36
Leaves	1.1–1.7	3.1[2]	34–53
Pruning wood	1.5–2.1	1.0[2]	15–21
Permanent framework:			
Trunk and branches	3.5	1.0[2]	35
Roots	1.2	1.0[2]	12
Total	–	–	109–157

[1]On a fresh weight basis; [2]On a dry weight basis.

1993; Schepers *et al.*, 1992) and few studies (Neilsen *et al.*, 1995) have suggested their use for fruit trees under specific circumstances. The most important advantage related to the use of the portable instrument SPAD-502 is the rapidity of its use. We have been recently used the portable chlorophyll meter on several trials involving mineral nutrition of many fruit crops (Tagliavini *et al.*, 1994). Figure 3A reports the relationship between SPAD-502 readings and leaf nitrogen concentration of peach leaves, pooling data from eight peach and nectarine varieties. As it appears from the figure, the possibility of predicting the N concentration of peach leaves by a single regression equation based on SPAD readings (suitable for several cv) is not very promising, as about 50% of the variability of N data is not explained by the linear model. However, when we considered only data from cv Maria Aurelia (Fig. 3B) r2 were 0.79 and 0.97 for 1993 and 1994 respectively, and pooling data from the two years we obtained a r2 of 0.70. In this respect, our results fit with conclusions of Khemira *et al.* (1994) on pear and Neilsen at al. (1995) on apple who found that relationships between leaf color and total N is not constant for variety and year. The possible use of this diagnostic tool of N status has to be also necessarily based on the absence of any limitation of other nutrients involved in the chlorophyll formation (e.g. iron, magnesium).

Table 5. N-uptake in grape, peach and plum trees in selected periods

Crop	Reference period	% of N removal on total absorbed
Grape (Riesling vine)[1]		
	by full bloom	25
	between full bloom and veraison	50
	between veraison and harvest	25
Peach (cv. Flavorcrest)[2]		
	by mid-May (fruit thinning)	10
	between mid-May and end of August	65
	later	25
Prune (cv. Agen)[3]		
	by end of April	15
	between May and end of August	60
	later	25

[1]Calculated from data of Lohnertz (1991), Germany;
[2]Calculated from data by Soing and Mandrin (1993), South-France;
[3]Calculated from data by Weinbaum *et al.* (1978), California (USA).

The proposed method

To reconcile productivity and environmental aspects, fertilizer recommendations have to be based on plant demand and nitrogen availability in soil. From above it clearly appears that the same amount of fertilizer can be not sufficient to meet plant needs in one soil or can be even in excess in another soil. Under soil and climatic conditions of Emilia-Romagna, the major risk of contamination of waters with leaching ions takes place late in winter-early spring (Rossi *et al.*, 1991) and reducing the presence of NO_3^- ions at the end of vegetative period (late autumn) would be a helpful strategy for minimizing it. The method proposed here is mainly based on 1. total nitrogen uptake by the orchard; 2. kinetics of N removal from the soil to the tree and 3. determination of nitrate-N content (per hectare) before spring and post-harvest fertilizations.

Total nitrogen removal and its kinetic
Many attempts of calculating total seasonal N removals from fruit crops have been made: e.g. Greenham (1980) for apple; Smith *et al.* (1988) for kiwifruit; Soing and Mandrin (1993) for peach; Lohnertz (1991) for grape. Total N removed by an orchard or a vineyard is distributed in several organs, namely fruits, leaves, pruning wood and perennial framework (above and underground). Net removals are easy to calculate once known the average N concentration in fruits and pruning wood (Table 4) and amounts yielded and pruned. Removals in leaves and in the permanent tree framework are hardly estimated with accuracy as they depend on vegetative growth, which in turn is affected by the cultivar, the rootstock, the soil fertility and the climate.

In the example reported in Table 4, which refers to a mature peach orchard, we have estimated the leaf N removal by (i) calculating the number of leaves, on the assumption that average ratio between number of leaves and fruits ranges between 20 and 30 (Sansavini *et al.*, 1985); 2. assuming a fruit production of 25 t/ha and an average fruit weight of 180 g; 3. assuming an average dry weight of 450 mg/leaf (Poni *et al.*, 1992) and 4. assuming a concentration of 3.1% N in leaves. Calculations were made considering a peach orchard with 500 to 700 tree/ha. Kinetic of N removal was estimated on the basis of published literature. Table 5 summarizes it for peach, prune and grape.

Determination of nitrogen availability in soils
The determination of nitrate availability is proposed either from soil or soil solution samples. Soil samples for nitrate extraction need to be collected from the rooting depth in three periods, just before the timings scheduled for possible N fertilization. In the case of a peach orchard, the first soil collection corresponds to fruit set, before tree N reserves from storage organs

are completely mobilized and the trees rely entirely on root N uptake (Muñoz *et al.*, 1993). The second soil collection and nitrate extraction is recommended in the peach orchard at fruit thinning (mid-May under our conditions) and the third is suggested before post-harvest N fertilization (e.g. early in September). The quantity of N-NO_3 extracted is calculated on a hectare basis provided that the soil volume explored by roots, the soil density and the water content are known.

Alternatively to soil collection, in fertigated or drip irrigated orchards, nitrate availability might more easily be estimated on soil solution samples, extracted by suction lysimeters. The use of such lysimeters provides a rapid, non-destructive tool for controlling nutrient supply in fertigated orchards in north-west United States and south-west Canada (Neilsen *et al.*, 1995). The conversion formula from soil solution nitrates to amounts of available N per hectare is described by Thicoïpé *et al.* (1993).

Ammonium is not considered as its presence in the soil solution is considered negligible due to high pH of soil solution and high nitrification activity.

A practical example

Recommended timings for possible N supply in peach orchards in Emilia-Romagna are fruit set, fruit thinning and post-harvest (early September). The formula for calculating the fertilization rate in the three timings is the following:

Recommended fertilization rate (kg/ha)=[(total amount of N removed in the year) * (percentage of removal in the reference period/100)] – [(soil N-NO_3, in ppm) * (soil volume, in $m^3/1000$) * (soil bulk density, kg l^{-1})].

Soil volume considers a soil depth of 90 cm and an horizontal distribution of roots variable for each crop, rootstock, orchard floor management and irrigation system. In the case of a peach orchard with drip irrigation we may consider that the majority of roots are confined in a strip of 1 m each side of the row; considering an interrow distance of 4 m, the total soil volumes per hectare to be used in the formula is 4,500 cubic meters. In the formula the soil volume is divided by 1000 to have recommended rates as kg/ha. The percentages of N removals in different periods were as in Table 5, in the assumption that data obtained in Southern France can be applied to Northern Italy conditions.

The practical application of the method reported here refers to the case of a peach orchard removing yearly 150 kg N per hectare, in which N-NO_3 concentrations in soil were 3, 8 and 4 ppm (on soil dry weight) before fruit set, fruit thinning and in post-harvest, respectively.

At fruit set the recommended rate is:

$$[150 * (10/100)] - (3 * 4.5 * 1.4) = -3.9 \text{ kg/ha}$$

that means that no N is required.

At fruit thinning the recommended rate is:

$$[150 * (65/100)] - (8 * 4,5 * 1.4) = 47 \text{ kg/ha}$$

In post-harvest (early September), the recommended rate is:

$$[150 * (25/100)] - (4 * 4.5 * 1.4) = 12 \text{ kg/ha}$$

The total recommended N rate to the peach orchard is therefore 59 kg/ha.

Since 1993 the method has been used in 15 fruit orchards and vineyards located in the province of Ravenna. In 1993, the amounts of N applied to orchards fertilized with the present method were less then 50% of those supplied to the average of that area and no visible differences in tree yield and growth were noticed; it should be considered, however, that tree response to fertilizer management can only be assessed in the long period, therefore a long-term testing program is undertaking in several areas of the Emilia-Romagna region.

As the method has been especially designed to be applied under field conditions, we have taken into consideration rapidity and feasibility aspects; therefore in its application, nitrate determinations are recommended in water saturated extracts from soil samples or, in drip irrigated orchards, in soil solution extracted by suction lysimeters (Irrometer Co. Inc., Riverside, CA, USA). The latter extraction method has been successfully coupled with the use of portable instruments for detecting nitrate concentration (e.g. the Merck Reflectoquant RQflex 16970) directly in the field. For a correct application of this method, it would be essential that data on N uptake rate and kinetic, and root distribution pattern are available for the different orchard conditions.

The contribution of the mineralization of organic matter to the buildup of available nitrogen in the soil is only indirectly and not directly estimated as in the method described for N recommendations in cereals by Mengel (1991). In the latter method, called the 'Giessen Method' the soil hydrolizable organic N compounds are measured and assumed to be easily mineralizable (Ziegler *et al.*, 1992); the resulting amounts of mineral nitrogen are then detracted from the crop N needs to provide N recommendation rates in spring. This method implies that all the hydrolizable organic N is mineralized by May-June. By our method, that assumes a NUE = 1, N recommendations at fruit set, and partially at fruit thinning, do not consider the contribution of mineral N from organic sources and therefore N rates may be slightly overestimated; it should be considered however that possible mineralization of organic matter should raise the nitrate content at times of the second and third soil sampling, therefore causing a reduction of successive recommended N rates. In this respect our method differs from the Nmin method (Nöfe *et al.*, 1990) that gives recommendations only on the basis of mineral N found early in spring.

To the author's knowledge, the method presented herein represents the first attempt, done under Italian conditions, to guide the N fertilization in orchards by adjusting the N rates to tree demand and soil N availability. After testing the method, some improvements could be necessary (e.g. reducing the number of collection times and the depth of sample collection) and plant indicators of N status (e.g. by SPAD readings) will likely complement soil analysis.

Acknowledgements

This study has been supported by the Regione Emilia-Romagna as part of a programme managed by ERSO of Cesena (Forli). The authors are in debt with F. Pelliconi of the "Consorzio Agrario Provinciale, Ravenna" for cooperating in the evaluation of rapid indexes of N status in orchards. We also wish to acknowledge the contribution of the Agronomy Department of Bologna University and of the 'Canale Emiliano-Romagnolo, CER' in the root excavation study.

References

Atkinson D and Wilson SA (1979) The root soil interface and its significance for fruit tree roots of different ages. In: Harley JL and Russell RS (eds) The Root Soil Interface, pp 259–281. Academic Press, London

Baratti S (1991) I problemi delle acque sotterranee nei grandi comprensori irrigui. In: Agricoltura e Ambiente, pp 741–752. Edagricole, Bologna, Italy

Böhm W (1979) Methods of studying root system. Springer-Verlag, Berlin

Casalicchio G, Marangoni B, Cobianchi D, Scudellari D, Faedi W, Maltoni ML and Corrado C (1992) Esperienze sperimentali sulla fertilizzazione del melo in Emilia-Romagna. Notiziario Tecnico ERSO

Cobianchi D and Turci E (1980) Concimazione, produzione e condizioni nutritive di pescheti nel forlivese. Atti XV Convegno Peschicolo Ravenna Italy, 191–206

Getti PF and Pantani C (1991) Contributo delle concimazioni e degli allevamenti zootecnici all'inquinamento dei corpi idrici. In: Agricoltura e Ambiente, pp 311–348. Edagricole, Bologna, Italy

Greenham DWP (1980) Nutrient cycling: the estimation of orchard nutrient uptake. Acta Horticulturae 92: 345–352

Jones BJ Jr, Wolf B and Mills HA (1991) Plant Analysis Handbook. Micro-Macro Publishing, Inc. Athens, GA, USA

Lalatta F (1980) La fertilizzazione nell'arboricultura da frutto. Edagricole, Bologna, Italy

Levin I, Assaf R and Bravdo B (1980) Irrigation, water status and nutrient uptake in an apple orchard. Acta Horticulturae 192: 255–263

Lohnertz O (1991) Soil nitrogen and the uptake of nitrogen in grapevine. In: Proceedings of the International Symposium on Nitrogen in Grapes and Wine, Seattle 1991, pp 1–11. The American Society for Enology and Viticulture

Khemira H, Sugar D and Righetti TL (1994) Relationship between chromaticity values and nitrogen content of pear leaves. Acta Horticulturae 367: 305–308

Marangoni B, Scudellari D, Brigati S, Neri F, Spada GL and Toselli M (1994) Fertilizzazione del pesco e qualita' dei frutti. Atti del XXX Convegno del Pesco, Lugo (Ra) Italy (*in press*)

Marcelle R (1984) Mineral analysis and storage properties in fruit. In: Martin-Prevel P (ed) Proceedings of VIth International Colloquium for the Optimization of Plant Nutrition, pp 365–371

Mengel K (1991) Mineralization de l'azote organique et optimization de l'apport d'engrais azoté. C. R: Acad Agric Fr 76 (8): 3–12

Millard P and Thompson CM (1989) The effect of the autumn senescence of leaves on the internal cycling of nitrogen for the spring growth of apple trees. J Exp Bot 40: 1285–1289

Muñoz N, Guerri J, Legaz F and Primo-Millo E (1993) Seasonal uptake of 15N-nitrate and distribution of absorbed nitrogen in peach trees. Plant and Soil 150: 263–269

Nöfe D, Lerche K and Schönberg G (1990) The content of inorganic nitrogen in soils of orchards in different plantation areas. Acta Horticulturae 274: 339–345

Neilsen D, Hogue EJ, Herbert LC, Parchomchuuk P and Neilsen GH (1995) Use of rapid techniques for estimating the status of fertigated apple trees. Acta Horticulturae 383: 211–218

Oenema O and Velthof GL (1993) Ammonia volatilization from compound nitrogen-sulfur fertilizer. In: Fragoso MAC and Van Beusichem ML (eds) Optimization of Plant Nutrition, pp 341–350. Kluwer Academic Publishers, Dordrecht, The Netherlands

Patruno A, Cavazza L and Pistelli E (1986) Variazioni del contenuto di azoto nitrico del terreno. Influenza della lavorazione, della coltura e del decorso meteorico. Rivista di Agronomia 20 (2–3): 204–217

Peng S, Garcia FV, Laza RC and Cassaman KG (1993) Adjustment for specific leaf weight improves chlorophyll meter's estimate of rice leaf nitrogen concentration. Agron J 85: 987–990

Poni S, Tagliavini M, Neri D, Scudellari D and Toselli M (1992) Influence of root pruning and water stress on growth and physiological factors of potted apple, grape, peach and pear trees. Scientia Horticulturae 52: 223–236

Rossi N, Ciavatta C and Vittori Antisari L (1991) Seasonal pattern of nitrate losses from cultivated soil with subsurface drainage. Water, Air, and Soil Pollution 60: 1–10

Sansavini S, Corelli L and Giunchi L (1985) Peach yield efficiency as related to tree shape. Acta Horticulturae 173: 139–158

Schepers JS, Francis DD, Vigil M and Below FE (1992) Comparison of corn leaf nitrogen concentration and chlorophyll meter readings. Commun Soil Sci Plant Anal 23 (17–10): 2173–2187

Scudellari D, Marangoni B, Cobianchi D, Faedi W and Maltoni ML (1993) Effect of fertilization on apple tree development, yield and fruit quality. In: Fragoso MAC and Van Beusichem ML (eds) Optimization of Plant Nutrition, pp 357–362. Kluwer Academic Publishers, Dordrecht, The Netherlands

Shaller K (1991) Ground water pollution by nitrate in viticultural areas. In: Proceedings of the International Symposium on Nitrogen in Grapes and Wine, Seattle 1991, pp 12–22. The American Society for Enology and Viticulture

Shaw K (1962) Loss of mineral nitrogen from soil. J Agric Sci 58: 145–151

Shear CB and Faust M (1980) Nutritional ranges in deciduous tree fruits and nuts. Hortic. Reviews 2: 143–163

Smith GS, Buwalda JG and Clark CJ (1988) Nutrient dynamics of a kiwifruit ecosystem. Scientia Horticulturae 37: 87–109

Soing P and Mandrin JF (1993) Nutrition du pecher. Cinetiques des consommations. Infos-Ctifl 92: 33–36

Tagliavini M, Scudellari D, Marangoni B and Pelliconi F (1994) Valutazione di indici rapidi per la determinazione del livello azotato nel frutteto. Atti Giornate Scientifiche Societé Orticola Italiana, S. Benedetto del Tronto, 24 giugno 1994: 487–488

Thicoïpé JP, Dapoigy L, Estrade JR and Fleury A (1993) Measure de nitrates: prelevement de la solution du sol avec le lysimetre a succion. Infos-Ctifl 92: 42–46

Weinbaum SA, Merwin ML and Muraoka TT (1978) Seasonal variation in nitrate uptake efficiency and distribution of absorbed nitrogen in non-bearing prune trees. J Amer Soc Hort Sci 103(4): 516–519

Weinbaum SA, Scott Johnson R and DeJong TM (1992) Causes and consequences of overfertilization in orchards. HortTecnology 2(1): 112–121

Weller F (1983) Nitrogen turnover in some orchard soils of South West Germany. Z Pflanzenernahr Bodenk 146: 261–270

Ziegler K, Nemeth K and Mengel K (1992) Relationship between electroultrafiltration (EUF) extractable nitrogen, grain yield, and optimum nitrogen fertilizer rates for winter wheat. Fert Res 32: 37–42

C. Rodriguez-Barrueco (ed.), Fertilizers and Environment, 171–175.

Natural radioactivity in phosphate fertilizers

L.C.Scholten & C.W.M.Timmermans
KEMA Nederland B.V., P.O. Box 9035, 6800 ET Arnhem, The Netherlands

Key words: radioactivity, phosphate, fertilizer, regulations, radiation doses

Abstract

Radioactive materials, like uranium and radium, are normal constituents of the earth's crust. The radioactivity of phosphate rock is enhanced by geological processes. Exposure of workers and the public to radiation from phosphate rock and fertilizer is therefore not unlikely. The European Commission has issued a draft proposal for revision of the Basic Safety Standards for the protection of workers and the general public against the dangers of ionizing radiation. In this proposal the exposure to natural radiation sources is also regulated. The radioactivity present in some of the phosphate ores is such that the production and use of phosphate fertilizer should not be allowed without prior notification, and it is essential that a system of control be set up. The radiation dose to which workers in a fertilizer plant are subjected is not negligible, although depending on the inhalation of dust. Stocks of fertilizer form a serious radiation problem. At the retrieval of phosphate from the rock, gypsum is formed whose release causes environmental problems. The contamination of land and food with radioactivity is compared with current legislation in the Netherlands.

Introduction

There is a tendency in the modern western world towards decreasing the risks to man from anthropogenic sources. Also, the acceptability of the harmful effects of ionizing radiation is decreasing. The knowledge about the effects of radiation has improved significantly over the past decades. Both issues together have led to proposals for lower regulatory limits for ionizing radiation and as a consequence to lower limits for radioactivity content of raw materials and products [3].

Radioactivity is not only a residuary product of nuclear energy. It is also a normal constituent of the earth's crust. As with all elements, the distribution of radioactive elements in the crust is not even. Geological processes have enhanced the radioactivity of sedimentary phosphate rock. By processing phosphate rock to fertilizer the radioactivity of the ore is transferred to the product and to the waste products.

The current regulations already impose that workers in the fertilizer industry must be protected against radioactivity in phosphate. Under the proposed regulations the use of phosphate fertilizer will not even be allowed without prior notification or a government licence. This of course, could influence the use of phosphate fertilizer.

The by-product gypsum is normally released into the environment or used as soil amendment. However, some consequence analyses show that in future this might no longer be acceptable [6,11]. Also, the use of phosphate fertilizer on farmland becomes questionable.

Natural radioactivity

The stellar material from which the earth was formed about 4.5 billion years ago contained many unstable nuclides. The majority of these unstable nuclides have long since decayed into stable elements. However, some of the original (primordial) nuclides, whose half-lives are about as long as the earth's age, are still present. The main primordial nuclides are listed in Table 1.

Fig. 1 shows the decay scheme of the ^{238}U-series. The radionuclide ^{238}U decays into ^{234}Th by emitting an α-particle. The newly formed nuclide is also unstable and decays further. Finally, after a total of 14 such steps, emitting 8 α-particles and 6 β-particles, accom-

Table 1. Main primordial nuclides and mean occurrence in earth crust

Nuclide	Half-life in billions of years	Mean concentration in earth crust [13] (Bq/kg)
^{238}U	4.47	25
^{235}U	0.70	1
^{232}Th	14.1	25
^{40}K	1.28	370

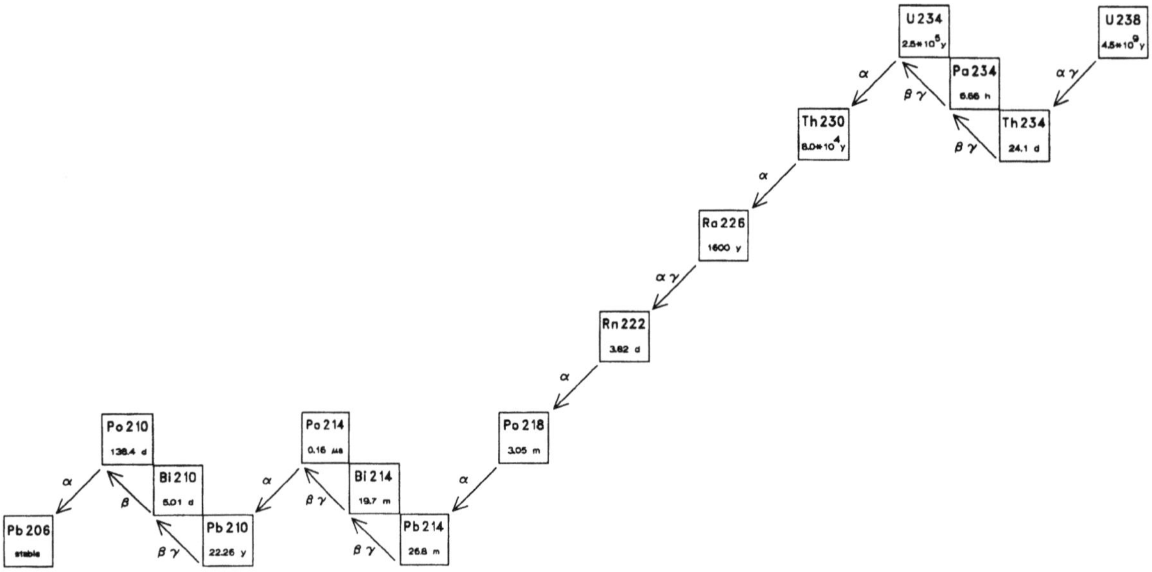

Fig. 1. Decay scheme of the ^{238}U series

panied by γ-radiation, stable lead is formed. Since all radionuclides following ^{238}U have shorter half-lives than the mother nuclide, the undisturbed decay series is said to be in secular equilibrium, i.e. all daughter nuclides have the same activity as ^{238}U.

The decay series includes the nuclide ^{226}Ra which has a half-life of 1600 years. This element has chemical properties clearly different from those of uranium; solubility of radium in water, for instance, is much better. Since ^{226}Ra has a fairly long half-life, natural leaching processes can change the concentration of this nuclide in rock. If leaching occurs, the decay series of uranium is no longer in secular equilibrium.

^{226}Ra decays into ^{222}Rn, an inert noble gas that does not form any chemical bonds and can escape into the atmosphere. The escape of radon also disturbs the equilibrium in the decay series. The risks related to radon itself are limited. However, gaseous radon decays further into short lived radioactive solid products. These ionic products attach rapidly to aerosols and dust particles in the air, and can thus be inhaled and deposited in the lungs. Once deposited, the radiation emitted at the decay of these products, can cause damage to the deep lungs.

^{232}Th decays in 10 steps, emitting 6 α-particles and 4 β-particles. The half-lives of the radium isotopes and the radon isotope in the ^{232}Th series are short, so that in natural thorium the series is always in equilibrium.

^{40}K decays to stable argon or to stable calcium. It gives no extra hazard in the body, because the potassium content in the body is always in equilibrium. However, the emitted γ-radiation at the decay to argon can cause external irradiation.

The natural radioactivity in phosphate rock depends on its origin. In sedimentary rock it is much higher than in volcanic rock. The main sources are in Florida, Morocco and Kola. Some typical concentrations are given in Table 2.

Table 2. Some typical values of activity concentrations in phosphate rock [13]

Origin	Activity concentration (Bq/kg)		
	^{238}U	^{226}Ra	^{232}Th
Florida	1500	1600	16
S. Carolina	4800	4800	78
Morocco	1700	1700	30
Kola	90	40	90
China	150	150	25

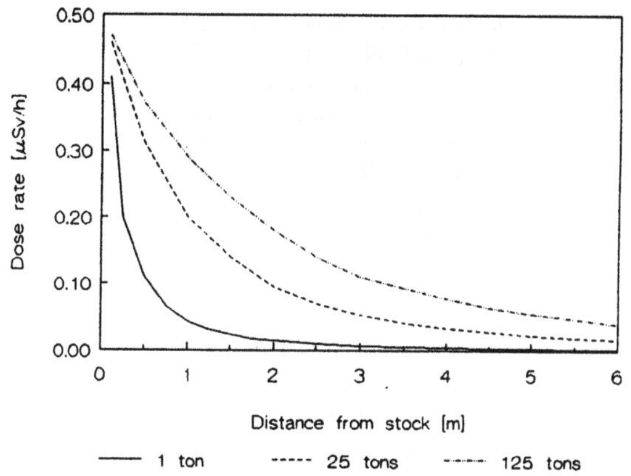

Fig. 2. Dose rates as function of distance from a stock of PK fertilizer

Exemption levels

As all materials contain in principle at least small quantities of natural radioactive elements, levels are specified below which materials are exempted from regulatory control. The legal exemption levels for the countries of the European Union are based on two Council Directives of the European Commission [3,4]. Exemption levels for specific activity as given in these directives are:

- 500 Bq/g for solid natural radioactive substances,
- 100 Bq/g for all other radionuclides.

In addition to the levels for specific activity, exemption levels are given for the total amount of radioactivity. Both levels must be superseded before regulatory control becomes compulsory. In practice with fertilizers, however, the level for the amount of activity is always exceeded, so that the level of the specific activity is determinant.

With a view to reducing the risks to the public of detrimental effects of ionising radiation, the EU has issued a draft for a new Council Directive [9]. In this draft the exemption levels for moderate amounts of material containing the uranium, radium or thorium decay series are decreased to only 1 Bq per gramme. Although these levels are only valid for moderate amounts, they may be extended to larger volumes since Member States are obliged to investigate and identify practices where workers might be exposed to significant quantities of natural radionuclides. If the exemption levels are used for such quantities, it can be seen from Table 2, that normal commercial phosphates exceed these levels. Consequently, phosphate rock, fertilizer and waste products might be considered as 'radioactive substances'.

Radiation protection

The basic principles in respect of radiation protection are [5]:

- the use of radioactivity must be justified,
- doses must be kept as low as reasonably achievable (ALARA),
- no individual may be exposed beyond stated limits.

The use of fertilizer for growing crops gives sufficient grounds for the justification. The ALARA-principle implies that reasonable measures must be taken to reduce radiation doses, and that costs have to be weighed against the gain in reducing the radiation doses. The individual limits for effective radiation doses are [5]:

- workers under supervision 50 (20) mSv y^{-1},
- members of the public 5 (1) mSv y^{-1}.

In the near future the dose limits will be lowered to the figures in the brackets.

The general trend for reducing technological risks to the public will eventually lead to a further reduction to the insignificant individual radiation dose from a single source down to 0.01 mSv y^{-1} and of the integrated collective dose to 1 Sv y^{-1} [12]. This corresponds to an individual mortality risk of about 10^{-6} per year. In this aspect stock piles and the use of fertilizer are to be regarded as separate sources.

Man can be exposed to radiation through: –direct radiation from a source, –internal contamination as a result of inhalation or ingestion of radioactive matter. The dose by direct radiation is dependent on the time spent in the neighbourhood of the source and the dis-

Table 3. Effective doses for users of phosphate fertilizer in mSv y^{-1} [10]

	External	Internal
stocks	0.05	2.25
strewing	≤ 0.02	≤ 1.0

tance from the source. In general, the radiation dose by external radiation is distributed equally over the body.

By internal contamination only a specific organ can be damaged, e.g. by inhalation the lungs. The exposure lasts as long as the radioactive matter remains in the body. The rate of uptake into and removal from the body is strongly dependent on the chemical and physical characteristics of the material. To estimate the radiological burden, models are developed which calculate the total dose, integrated over the time span the matter remains in the body. In order to compare the doses from internal irradiation with those of external irradiation, the internal doses are expressed in effective doses, i.e. in external doses, giving the same amount of total risk. The internal doses are committed to the point of time of intake.

Exposure of workers

Workers in fertilizer production plants are exposed to direct radiation from stocks of raw phosphate ore and products. Moreover they are exposed to 'hot-spots', i.e. parts of the installations where radioactivity has accumulated. We measured at plants general radiation levels up to 0.0015 mSv h^{-1}. Extrapolated to full employment these levels give an annual dose of more than 2 mSv. Some hot-spots were found with radiation levels of up to 0.025 mSv h^{-1}.

The plants are dusty, so there is also a risk of internal exposure by inhalation. From measurements at several plants we estimate potential inhalation doses for workers up to 2 mSv y^{-1}.

Radiation doses near stocks are calculated with the computer code MICROSHIELD [8]. Results are given in Fig. 2 for some sizes of stocks with PK fertilizer containing normal quantities of activity.

In [10] an estimate is given of the exposure of farm workers to radiation from stocks near farms and the strewing of fertilizer, taking into account the time spent in the neighbourhood of the fertilizer (Table 3).

These doses are not insignificant, but still below the current limits for the general public. However, when the limit for the public will be lowered, there exists a reasonable chance that it will be exceeded then. This may mean that workers have to be set under supervision as 'radiological workers' and that precautionary measures will be prescribed by the Inspectorates.

Environment

Most fertilizer plants are situated near estuaries. The by-product gypsum is released via surface water into the sea or disposed in ponds where it settles. Due to co-precipitation in the process the gypsum is contaminated with ^{226}Ra. In the European Union each year about 6–7 MT of gypsum is released to the sea and 8–11 MT is deposited in piles [1]. An extensive study on all environmental impacts of phosphogypsum is given in [11]. Here we restrict us to the radiological aspects.

The activity of ^{226}Ra is up to 1250 Bq/kg. This leads to contamination of the environment as for example is described for the North Sea [6] and the Huelva region in Spain [7].

The general criterium in the Netherlands for environmental pollution is the mortality risk for an individual member of the public. This may not be higher than 10^{-6} per year from a single source, corresponding to an annual radiation dose of only 0.04 mSv (risk factor 2.5% Sv^{-1}). The dilution in the North Sea is high, but dose estimates show that for some high risk groups the limit can be exceeded [6]. Dumping on land, as in the Huelva region [2,7], is prohibited in the Netherlands, as doses, even after the restoration, are too high.

Generally speaking, the transfer of natural radionuclides to vegetation is low. Therefore the risk to the public from consumption of crops with enhanced natural activity derived from the use of fertilizer is negligible [11,13]. Extended use of phosphate fertilizer raises the dose rate in air above the ground by about 0.8 nGy h^{-1} [13]. This may give rise to an additional exposure of man at continuous residence on such previously contaminated farm land of 0.007 mSv y^{-1}. If phosphogypsum is used as a soil amendment, the doses may be no longer insignificant, mainly due to the exhalation of radon [11].

Discussion

The occurrence of primordial radionuclides in phosphate ore and fertilizer has been recognised for a long time. The radiation hazards have long been ignored. The acceptability of risks from ionizing radiation is decreasing. This tendency leads to more stringent regulation. In the forthcoming years lower regulatory limits are likely to become obligatory, especially in the EU.

Precautionary measures and restrictions in order to protect workers in the factories and on the farm-yards may become compulsory.

Disposal of the by-product phosphogypsum is already a problem. Other environmental effects are only minor.

All this might result in a price increase for fertilizers, entailing additional expense for users and society in general.

References

1. Beatslé LH (1991) Study of the radionuclides contained in wastes produced by the phosphate industry and their impact on the environment: Report EUR13262 EN, Luxembourg: Publication Office of the European Communities.
2. Cancio D *et al.* (1993) Radiological considerations related with the restoration of a phosphogypsum disposal site in Spain. Proc. Int. Symp. on Remediation and Restoration of radioactive contaminated sites in Europe. XI-5027/94, Luxembourg: Publication Office of the European Communities.
3. Councel Directive 80/836/Euratom dated 15 July 1980 pertaining to modification of the regulations for the establishment of basic standards for health protection of the population and workers against the hazards related to ionizing radiation (PB no. L 246 dated 17–9–1980).
4. Councel Directive 84/467/Euratom dated 3 September 1984 pertaining to modification of the regulations for the establishment of basic standards for health protection of the population and workers against the hazards related to ionizing radiation (PB no. L 265 dated 05–10–1984).
5. ICRP (1977) Publication no.26, Recommendations of the Int. Comm. on Radiation Protection, Oxford, UK: Pergamon Press.
6. Köster HW *et al.* (1985) Radiological model calculations for natural radionuclides released into the environment by disposal of phosphogypsum. Sci of Total Env 45: 47–53.
7. Martinez-Aguirre A and Garcia-León M (1994) The distribution of U, Th and ^{226}Ra derived from the phosphate fertilizer industries on a estuarian system in southwest Spain. J Environ Radioactivity 22: 155–177.
8. MICROSHIELD, version 3.12 (1987), Grove Engineering Inc., Rockville MD 20850.
9. Modified Proposal for a Council Directive laying down the basic standards for the protection of the health of workers and the general public against the dangers arising from ionizing radiation. Commission of the European Communities. COM (93) 349 final; Brussels, 20 July 1993.
10. Reichelt A and Lehmann KH (1993) Anthropogene Stoffe und Produkte mit natürlichen Radionukliden, München, Bavaria: Ministry for Land Development and Environment (in German).
11. Rutherford PM, Dudas MJ and Samek RA (1994) Environmental impacts of phosphogypsum. Sci of Total Env 149: 1–38.
12. Safety Series no. III-P-1.1 (1992) Application of Exemption Principles to the Recycle and Reuse of Materials from Nuclear Facilities, Vienna: IAEA.
13. Unscear (1982) Ionizing radiation: Sources and biological effects, New York: United Nations.

C. Rodriguez-Barrueco (ed.), Fertilizers and Environment, 177–183.

Coordinated phosphorus research within a network of six European institutions

Influence on soil phosphorus status and on crop response

M. Vanoverstraeten & G. Hanotiaux
Faculté des Sciences Agronomiques de Gembloux (FSAGx), Belgique, Unité des Sciences du Sol et de la Terre, 5030 Gembloux, Belgique

Key words: phosphorus, European network, maintenance fertilization, fixation capacity, comparison of methods

Abstract

After three years of a research network project on mineral phosphorus fertilization including five experimental fields located in Europe the first results are discussed. Crop response was very significant to TSP application in the alluvial calcareous polder soil of Netherlands, and in the brown silty acid soil of Scotland, both having a low level of P availability and a high fixation capacity. In the alluvial sandy loam on chalk in England, a response was observed to the first fertilization level equal to the previous crop export of phosphorus. In the brown sandy-silty soil on sand in Germany the highest rate of TSP led to a response in the third year. No effect on the final yields was observed in the brown silt loam of Belgium characterised by a textural B horizon with a high P fixation capacity. The critical values for phosphorus fertilization are discussed as the amount of P needed to maintain a target value of soil phosphorus. Concerning the supply of the different soils, no balance was reached in the Dutch and Scottish soils, a steady state was reached in the English soil with the return of the previous crop removal and the critical value for P was lower than the return of the previous crop export in the German and Belgian soils. According to the eight methods of P determination compared in the network, the P contents in the plow layer were raised in the soils of Netherlands, England and Scotland. They remained at the same level or fluctuated depending on the soil testing methods in Germany and in Belgium. High correlations exist between the different methods used in routine analysis, except for the calcium cloride and calcium acetate lactate method. Annual fluctuations in the soil P were detected at different depths depending on analytical methods and need further research.

Abbreviations: IMPHOS – Institut Mondial du Phosphate; TSP – Triple superphosphate; AFRC – Institute of Arable Crops Research; SAC – Scottish Agricultural College; IBDLO – Instituut voor Bodemvruchtbaarheid; FAL – Federal Agricultural Research Center of Braunschweig-Völkenrode; SPAA – Station Provinciale d'Analyses Agricoles de la province de Liège; MLURI – Macauley Land Use Research Institute; DPVE CEN – Département Physiologie Végétale et Ecosystèmes Centre d'Etudes Nucléaires

Introduction

The IMPHOS-FSAGx phosphorus network in Western Europe supports scientific research on fertilizer efficiency: it is carried out in six West European countries and is coordinated by the Soil and Earth Sciences Unit of the University of Gembloux. Each of the research centers contribute to the results of the project and monitors a standardized experimental field. Three research objectives were defined:

1. Evaluation of the effect of phosphate application on crop yield and soil phosphorus content. The fertilizer rate is adjusted to the phosphate export by the crop. Phosphorus distribution in soil profiles is monitored in order to assess the risk of phosphorus leaching.

Table 1. Methods used in the network for available P determination

Method	Type of extractant	Units of results	Country of use
Cottenie (1979)	Ammonium acetate (0.5 N) EDTA	mg P/100 g	Belgium Wallonie
Modified Morgan (1991)	Acetic acid/ammonium acetate (0.5 M)	mg P/l	Scotland
Schüller (1969) calcium acetate lactate	Calcium lactate (0.05 M) calcium acetate (0.05 M) acetic acid (0.3)	mg P_2O_5/100 g	Germany Hanover
Olsen (1954)	Sodium bicarbonate (0.5 M)	mg P/l–mg P/kg	England-Spain
Sissingh (1971)	Water	mg P_2O_5/l	The Netherlands
Houba (1990)	Calcium chloride (0.01 M)	ppm P	Germany Hanover
Fardeau (1988)	Isotopic dilution	ppm P	France
Somasiri, Birnie and Edwards (1991)	Ion exchanging resins	mg P/l	Scotland

2. Comparison of routine soil tests for phosphorus run by the countries of the network to establish target values for maintenance fertilization.
3. Establishment of scientific bases for phosphate maintenance fertility in Western European soils.

The partners of the network are: Dr. M. Allison (AFRC, Great Britain), Dr. P. Dyson (SAC, Great Britain), Ir. P. Ehlert (IBDLO, The Netherlands), Dr. M. Kücke (FAL, Germany), Prof. J. Torrent (Univ. Cordoba, Spain), Ir. D. Van Vyve (SPAA, Belgium).

Collaborators: Dr. T. Edwards (MLURI, Great Britain), Dr. JC. Fardeau (CEN, France).

Coordination and results analysis: Dr. M. Vanoverstraeten (FSAGx, Belgium).

Scientific supervision: Prof. G. Hanotiaux (FSAGx, Belgium) and the Scientific Committee of Imphos.

Materials and methods

The standardized experimental protocol consists of four treatments (4 rates of triple superphosphate): P0 = control with no mineral phosphate; P1 = P input equals the previous crop removal; P2 = twice the crop removal; and P3 = three times the crop removal. Nitrogen and potassium are applied according to the fertilizer recommendations of the partner. Soil samples from each plot of the experimental field are taken after harvest and are circulated within the network. Each soil sample is characterised according to the routine analytical method for phosphorus availability and to the phosphorus fertility assesment standards currently in use in the countries participating in the network. This is to allow comparison of analytical methods. The activities of each partner consists in the organisation and monitoring of the experimental field, the sampling of soil and plants, their analyses as well as those of the soil samples of the partners. The collaborators conduct their own routine method on available phosphorus. The activities of the coordinator consist of sampling, subsampling, shipment, distribution and analysis of some five thousand soil samples and the interpretation of the collected data per year and per rotation.

Results and discussion

Three aspects of the results are discussed: the effect of TSP on crop yield and on soil phosphorus status, the comparison of methods and the concept of maintenance fertilization. Table 2 illustrates the yield results for the first three years and Table 3 shows the mean values for P balance and the fluctuations in available P values in the soils for each field own method.

The effect of TSP on crop yield and on soil P status

After the first three years, mineral phosphorus fertilization influenced the yields and the soil phosphorus status to the following degree. The yield effect was very significant in the Netherlands (alluvial calcareous soil of polder) and in Scotland (brown silty acid soil) where soils have a low level of available phosphorus. In the Dutch soil, an additional crop response to the

Table 2. Comparison of yields for the different treatments

		P0	P1	P2	P3
The Netherlands					
Year 1 – Spring wheat	Level of fertilization kg P_2O_5	0	45	90	135
	Grain yield t/ha	5.60	5.88	5.92	5.95
	P content %	0.47	0.46	0.48	0.45
	P_2O_5 uptake grain kg/ha	59.86	62.52	64.65	60.85
	P_2O_5 total uptake grain + straw	72.43	74.63	79.53	74.10
Year 2 – Sugar beet	Level of fertilization kg P_2O_5	0	75	150	225
	Sugar yield t/ha	14.60	16.30	16.90	17.20
	P_2O_5 total uptake kg/ha	68.91	91.62	100.35	103
Year 3 – Spring barley	Level of fertilization kg P_2O_5/ha	0	92	183	275
	Grain yield t/ha	3.50	4.10	4.30	4.30
	P content %	0.42	0.45	0.45	0.44
	P_2O_5 uptake grain kg/ha	33.43	41.45	43.97	42.82
	P_2O_5 total uptake grain + straw	38.24	46.72	49.24	48.55
Belgium					
Year 1 – Sugar beet	Level of fertilization kg P_2O_5/ha	0	46	93	139
	Sugar yield t/ha	11.63	12.41	11.59	11.56
	P_2O_5 total uptake kg/ha	46.69	55.51	46.56	48.39
Year 2 – Winter wheat	Level fertilization kg P_2O_5/ha	0	72	145	217
	Grain yield t/ha	7.40	7.20	7.00	7.00
	P content %	0.34	0.34	0.36	0.34
	P_2O_5 uptake grain kg/ha	57.94	55.65	57.71	54.04
	P_2O_5 uptake grain kg/ha	75.34	70.76	75.34	69.85
Year 3 – Winter barley	Level of fertilization kg P_2O_5/ha	0	62	125	187
	Grain yield t/ha	6.20	6.12	6.10	6.10
	P content %	0.96	0.96	1.01	1.01
	P_2O_5 uptake grain kg/ha	59.31	58.81	61.37	61.60
	P_2O_5 total uptake grain + straw	66.55	65.22	68.22	68.49
Germany					
Year 1 – Sugar beet	Level of fertilization kg P_2O_5/ha	0	65	130	195
	Sugar yield t/ha	8.40	8.90	9.00	8.70
	P_2O_5 total uptake kg/ha	70.99	73.28	73.28	68.70
Year 2 – Winter wheat	Level of fertilization kg P_2O_5/ha	0	72	146	204
	Grain yield t/ha	4.80	5.00	5.00	5.10
	P content %	0.25	0.27	0.28	0.28
	P_2O_5 uptake grain kg/ha	28.36	32.58	32.87	33.24
	P_2O_5 total uptake grain + straw	33.90	37.80	38.56	39.31
Year 3 – Winter barley	Level of fertilization kg P_2O_5/ha	0	38	78	117
	Grain yield t/ha	5.80	5.90	6.10	6.20
	P content %	0.36	0.37	0.39	0.40
	P_2O_5 uptake grain kg/ha	49.00	55.00	55.00	58.00
	P_2O_5 total uptake grain + straw	62.00	64.00	72.00	75.00

Table 2. Cont.

Scotland					
Year 1 – Winter wheat	Level of fertilization kg P_2O_5/ha	0	70	140	210
	Grain yield t/ha	7.64	7.94	8.00	7.61
	P content %	0.33	0.34	0.35	0.38
	P_2O_5 uptake grain kg/ha	50.38	53.59	56.79	56.33
	P_2O_5 total uptake grain + straw	55.20	59.10	62.50	62.10
Year 2 – Winter barley	Level of fertilization kg P_2O_5/ha	0	65	130	195
	Grain yield t/ha	4.82	5.12	5.37	5.22
	P content %	0.32	0.35	0.36	0.36
	P_2O_5 uptake grain kg/ha	41.22	48.09	52.67	50.38
	P_2O_5 total uptake grain + straw	45.30	54.00	57.20	56.10
Year 3 – Oil seed rape	Level of fertilization kg P_2O_5/ha	0	54	108	162
	Grain yield t/ha	3.11	3.01	2.97	2.98
	P content %	0.72	0.74	0.74	0.79
	P_2O_5 uptake grain kg/ha	51.07	50.61	50.15	53.82
	P_2O_5 total uptake grain + straw	61.60	63.43	64.35	68.70
England					
Year 1 – Sugar beet	Level of fertilization kg P_2O_5/ha	0	50	100	
	Sugar yield t/ha	5.09	7.07	6.66	
	P_2O_5 total uptake kg/ha	32.80	36.07	38.00	
Year 2 – Oats	Level of fertilization kg P_2O_5/ha	0	50	100	
	Grain yield t/ha	5.50	5.18	5.63	
	P content %	0.36	0.37	0.39	
	P_2O_5 uptake kg/ha	45.36	43.28	51.98	
	P_2O_5 total uptake grain + straw	48.55	54.27	62.47	
Year 3 – Winter wheat	Level of fertilization kg P_2O_5/ha	0	50	100	
	Grain yield t/ha	4.79	6.35	6.09	
	P content %	0.33	0.37	0.34	
	P_2O_5 uptake grain kg/ha	36.64	53.36	47.40	
	P_2O_5 total uptake grain + straw	48.32	65.49	61.83	

highest P level was observed. In the plow layer the phosphorus contents were raised in the Netherlands under treatments P2 and P3 and in Scotland at P3 fertilization level. In England (alluvial sandy loam on chalk) the effect of P1 treatment relative to control P0 was observed with sugarbeet and winter wheat. In the plow layer the phosphorus contents were raised from P2 fertilization level. In Germany (brown sandy-silty soil on sand), the highest rate of superphosphate led to a yield response on winter wheat in the third year. In the plow layer, phosphate fertilization had variable effects on the available phosphorus contents. In Belgium (brown silt loam with a textural B horizon), the responses are significant only at intermediate stages of cereal growth. As was the case in Germany, the phosphorus contents remained either at the same level or increased or decreased depending on soil testing methods.

Furthermore, it appears that the phosphorus contents are subject to fluctuations from year to year. These fluctuations were not of the same magnitude for the different methods of soil analysis. Some methods show

Table 3. Mean values for P balance and fluctuations in available P values in the soils for each field own method – 3 years interval (1990–1993)

	The Netherlands			Belgium			Germany			Scotland		England		
Units	kg	mg P_2O_5/ 100 g		kg	mg P_2O_5/ 100 g		kg	mg P_2O_5/ 100 g		kg	mg P_2O_5/ 100 g	kg	mg P_2O_5/ 100 g	
Method	P_2O_5/ ha	water		P_2O_5/ ha	NH_4Ac		P_2O_5/ ha	CAL		P_2O_5/ ha	acetic acid	P_2O_5/ ha	Sodium bicarbonate	
Treatment	P balance	Δ P topsoil	Δ P subsoil	P balance	Δ P topsoil	Δ P subsoil	P balance	Δ P topsoil	Δ P subsoil	P balance	Δ P topsoil	P balance	Δ P topsoil	Δ P subsoil
P0	−179,6	−0.2	−0.1	−201.6	0.1	−2.5	−164.9	0.3	−1.5	−162.7	−0.7	−129.7	0.2	0.3
P1	−1,0	0	−0.1	−19.4	1.5	−1.6	1.1	−0.7	−1	12.7	0.1	−4.3	1	−2.6
P2	193,9	0.1	−0.3	161.1	2	−0.2	170.4	0.4	−1	193.9	−0.1	137.7	3.2	−1.2
P3	409,4	0.8	0	342	2	0.7	333.7	4.2	−0.9	380.1	0.5	×	×	×

Table 4. Correlation table between the eight extraction methods for available phosphorus – Year 1 – Plough layer. n = 64; P = 0.001***

	M1	M2	M3	M4	M5	M6	M7
M2	0.93						
M3	0.79	0.82					
M4	0.91	0.92	0.81				
M5	0.93	0.93	0.88	0.96			
M6	0.72	0.67	0.68	0.73	0.78		
M7	0.90	0.88	0.88	0.94	0.98	0.71	
M8	0.92	0.86	0.86	0.94	0.98	0.78	0.97

M1 = NH_4Ac; M2 = acetic acid/NH_4Ac.; M3 = CAL; M4 = $NaHCO_3$; M5 = water; M6 = $CaCl_2$; M7 = Isotopic exchange kinetic; M8 = Exchange resins.

a decline in phosphorus levels whereas others show an increase. To verify and explain these variations, one or more repetition of the triennial crop rotation is essential.

Phosphorus migration in the root zone of soil profiles is measured by successive sampling of deep horizons underlying the Ap horizon (plough layer). The absorption of phosphorus in the soil layer below the plough layer was high. This absorption limits the risk of phosphorus migration below the root zone.

The comparison of methods

Statistical analyses showed that high correlations exist between the different methods used in routine analysis, except for the $CaCl_2$ and Calcium acetate lactate methods (Table 4). Regression analyses indicate that the best correlation exist between the Olsen and isotopic dilution methods, respectively, which may therefore be considered as reference methods. The comparison of soil fertility classes according to the standards used in participating countries did not allow us to check the validity of these standards in terms of predicting yield response to phosphate fertilization. Difficulties of comparison among classes appear to be mainly due to the soil types and target yield.

The concept of maintenance fertilization

This concept refers to P fertilizer applications to compensate for crop removal of P from soils considered sufficiently high in phosphorus to the extent that they do not require a P fertility build-up: the maintenance fertilization is defined as the amount of phosphorus needed to maintain a target value of soil phosphorus.

Four trends were observed:

1. The Netherlands and Scotland: no balance was reached in the supply of the soil which remain at a low fertility level. Negative variations in the P input-output balance, and the low level of phosphorus in soils receiving P1 calls for careful consideration. The sites do not show stable high P level in the soil until higher P fertilizer rates are applied.
2. England: a steady-state of the phosphorus status of the soil was reached when the export of phosphorus by the crop was compensated. Efficiency here only requires the return of the amount of P exported by the plants.
3. Germany: though the sandy loamy soil here presents a low phosphorus fixation capacity, the site provided mineral phosphate for the crop. After three years the critical value for P fertilization appeared to be between P0 and P1.
4. Belgium: with a high P absorption capacity in the deep horizon, the soil testing methods reveal puzzling fluctuations in the soil phosphorus content which could not be explained. All the analytical methods demonstrate a high supply of available P. As no final yield effect appeared in this soil fertilization seems to serve in stocking P under the plough layer in the highly fixating B horizon.

In all cases, negative variations in available P in the subsoil profile account for fertility maintenance evaluation.

Conclusion

The heterogeneity of crop responses to phosphoric fertilizer applications reflects typically the diversity in the European pedo-climatic situations and provides ample justification for undertaking this European experimental work. The three years of experimentation are not enough to understand and to measure crop responses and changes in the phosphorus fractions extracted by the different methods of soil testing and in the P dynamics. Further research would contribute to a more detailed identification of the measurable fraction of soil phosphorus based on the analysis of data collected from each trial in relation to phosphorus dynamics in the soil (top-subsoil) and with the crop type. The concept of maintenance fertilizer should be evaluated in the context of the cropping system and the prevailing notion of sustainability. Therefore the designing of referential grids would greatly assist crop system management. Better than the interpretation of simple punctual analyses, the guidelines generated and validated for a soil type identified within its geographical and agronomical referential grid will help solve the fertilization efficiency problem. The soil maps based on objective criteria remain therefore a good support for constructing those referentials.

Acknowledgement

Support for this research was provided by the Administration Board of Imphos.

References

Allison M (1990–1994) Annuals reports of the Imphos network. FSAGx-Imphos

Cottenie A (ed) (1979) Workshop on standardisation of analytical methods for manure, soils, plant and water. CEE Agricultures series. EUR 6368 EN. 57 p

Ehlert P (1990–1994) Annuals reports of the Imphos network. FSAGx-Imphos

Draycott AP and Durrant MJ (1971) The relationship between soil phosphorus and the response by sugar beet to phosphate fertilizer on mineral soils. J Agric Sci CamB 77: 117–121

Dyson P (1990–1994) Annuals reports of the Imphos network. FSAGx-Imphos

Gachon L (1988) Phosphore et potassium dans les relations sol-plante. INRA, 566 p

Houba VJG, Novozamsky I, Lexmond ThM and Vander Lee JJ (1990) Applicability of 0.001 M $CaCl_2$ as a single extraction solution for the assesment of the nutrient status of soils and other diagnostic purposes. Soil Sci Plant Anal 21(19&20): 2281–2290

Kücke M (1990–1994) Annuals reports of the Imphos network. FSAGx-Imphos

Martinez J and Delas J (1990) Les essais de longue durée INRA sur la fertilisation phosphatée: recueil et synthèse des données récentes. Imphos, 101 p

McLaughlin MJ, Alston AM and Martin JK (1988) Phosphorus cycling in Wheat-Pasture Rotations. I. The source of phosphorus taken up by wheat. Aust J Soil Res 26: 323–31

Reith JWS, Inkson RHE, Scott NM, Caldwell KS, Ross Jam and Simpson (1987) Estimates of soil phosphorus for different soil series. Fert Res 11: 123–142

Ris J and van Luit B (1990) The establishment of fertilizer recommendations on the basis of soil tests. Instituut voor Bodemvruchtbaarheid, Haren-Groningen

Rohrmoser K (1985) Handbook for field trials in Technical Cooperation.Deuts. Gesellschaft für Techn. Zusammenarb (GTZ), Eschborn, Germany

Sen Tran T, Fardeau JC and Giroux M (1988) Effects of soil properties on plant-available phosphorus determined by the isotopic dilution phosphorus-32 method. Soil Sci Soc Am J 52: 1383–1390

Sissingh HA (1971) Analytical technique of the Pw method, used for the assesment of the phosphate status of arable soils in the Netherlands. Plant and Soil 34: 483–486

Somasiri LLW, Birnie A and Edwards AC (1991) Inductively coupled plasma atomic emission spectrometry for the analysis of soil extracts prepared on ion-exchange resins. Analyst 116: 601–603

Torrent J (1993) Annual report of the Imphos network. FSAGx-Imphos

Van Vyve D (1990–1994) Annuals reports of the Imphos network. FSAGx-Imphos

Vanoverstraeten M (1990–1994) Reports of the project 'Réseau phosphore en Europe occidentale'. FSAGx-Imphos

C. Rodriguez-Barrueco (ed.), Fertilizers and Environment, 185–193.

Plant nutrient sulphur–a review of nutrient balance, environmental impact and fertilizers

S.P. Ceccotti
The Sulphur Institute, 1140 Connecticut Avenue, N.W., Washington, DC 20036, USA

Key words: nutrient balance, sulphur deficiency, sulphur fertilizer

Introduction

Agriculture worldwide faces many challenges. In some regions, production must be expanded to provide food for growing populations, while in others, current production levels have to be maintained while striving for the right balance between intensive agriculture and environmental concerns.

Sulphur (S) fertilization is of growing worldwide importance as food production increases, while overall S inputs diminish. Sulphur is one of at least 16 elements essential for plant growth. Its functions within the plant are related closely to those of nitrogen (N) and the two nutrients are synergistic. Sulphur is required for plant growth in quantities equal to, and sometimes exceeding, those of phosphorus (P). Sulphur has a variety of vital functions within the plant's biochemistry. It is a major constituent of amino acids, such as cysteine and methionine–the building blocks of proteins. It also is essential in the formation of enzymes, vitamins, such as biotin and thiamine, and a variety of other important compounds in the plant, including chlorophyll. When S is deficient, both yield and quality suffer. Plants that are S-deficient characteristically are small and spindly. The younger leaves are light green to yellowish, and in the case of legumes, nodulation of the roots is reduced. The oil content of seeds is diminished and the maturity of fruits is delayed in the absence of adequate S (Buckman and Brady, 1969).

Despite the vital role of S, most of the growth in fertilizer consumption has been in S-free N and P fertilizers, even though higher yields of crops per unit area are removing greater quantities of S from the soil. Much of this S must be replaced; otherwise, S deficiencies will limit agricultural production severely. Increasing crop production, reduced sulphur dioxide (SO_2) emissions, and shifts in major fertilizer sources have led to worldwide increases of documented S deficiencies. This has caused the attention of the fertilizer industry to focus on means of introducing this important element into their fertilizer products. Many fertilizer companies are promoting S as an essential component of their products and guaranteeing its presence in specific amounts on their labels. In some cases, S has already been present, but not guaranteed. More often, it is being included because of a demonstrated need.

Major factors affecting plant nutrient sulphur balance

Over the past 20 years, the amount of S consumed in all fertilizers worldwide has fluctuated around 10 million tons per year. However, N consumption has almost doubled during this period from approximately 39 million tons in 1973 to about 75 million tons in 1991. The increased consumption of S-free, high analysis fertilizers is one of the most significant causes of S deficiency. When single superphosphate (12% S) and ammonium sulphate (24% S) were the dominant P and N sources, their incidental S content masked soil depletions. However, during the past two decades, with increased popularity of high-analysis materials that contain virtually no S, such as urea, triple superphosphate, and ammoniated phosphates, world agriculture has witnessed a drastic reduction in S additions as a percentage of the total fertilizer applied (Fig. 1).

Increased cropping intensity and higher crop yields removing greater quantities of S from the soil contribute to greater incidence of S deficiencies worldwide (Table 1). For example, from 1961 to 1991, cereal production in France increased at an annual average rate of 3.1% versus 2.3% in Germany. However, the area harvested in France grew at an average rate of only 0.1%; in Germany it even decreased by 0.1% annually. This intensification of land use, which is likely to

Table 1. Sample crop production in selected nations. Source: FAO

Year	Brazil soybeans	India rice	China rice	France cereals	Germany cereals	UK cereals
	kg ha^{-1}					
1961	1124	1542	2078	2276	2417	3176
1971	1210	1711	3314	3876	3788	3926
1981	1765	1962	4330	4729	4161	4933
1992	2035	2630	5798	6485	5336	6940

Table 2. Sulphur dioxide emission reductions in Europe (Anon. 1990a)

Country	Total SO_2 emissions 1980 (1000 tons)	Total SO_2 emissions 1987 (1000 tons)	% Change SO_2 emissions 1980–1987
Austria	177	75	−58
Denmark	219	155	−29
Finland	292	162	−44
France	1,779	923	−48
Former West Germany	1,600	1,022	−36
Hungary	817	710	−13
Italy	1,900	1,252	−34
Netherlands	244	141	−42
Sweden	232	116	−50
UK	2,335	1,840	−21

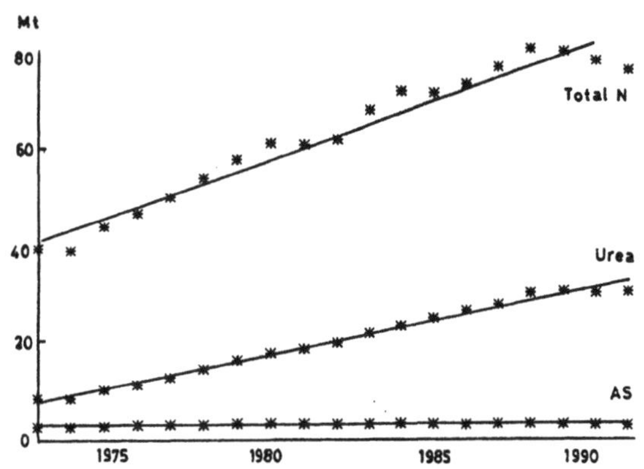

Fig. 1. World nitrogen fertilizer consumption

continue in the future, has commanded a proportionate increase in S demand.

Another factor that has affected worldwide demand for S fertilizer consumption and will become of even greater importance in the future is increased global environmental concern and compliance which has prompted reductions in SO_2 emissions that have benefitted crop growth in many countries in the past. Atmospheric SO_2 enters plants through the soil, whether through wet precipitation or dry deposition. The 1980s have been marked by a change in attitudes toward the environment; since then, many industrialized countries have made dramatic reductions in their SO_2 emissions (Table 2). Between 1980 and 1987, France, former West Germany, Finland, and Sweden achieved impressive reductions of 48%, 36%, 44%, and 50%, respectively. Similarly, the UK reduced emissions of SO_2 by 21% over the same period. During the last three decades the UK had been one of the highest SO_2 emitters in Western Europe (Anon., 1990a).

The breakdown of major emission capacities in Eastern Europe also is exposing this region to potential S deficiency. Since German reunification in 1989, most of the industrial facilities in Former East Germany closed. Prior to 1989, emissions from former German Democratic Republic brought to West Germany nearly the same amount of SO_2 as the UK, France, and Benelux combined (Schnug and Haneklaus, 1994).

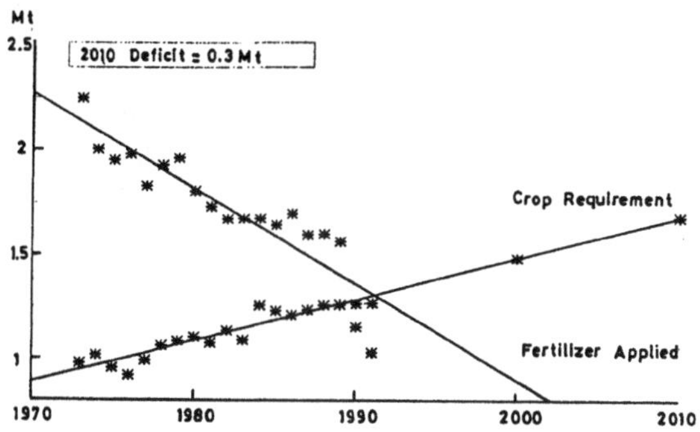

Fig. 2. Plant Nutrient Sulphur Balance in Western Europe

Emissions have slowed radically even in Poland as a result of major investments in the area of air protection. The amount of SO_2 released into the atmosphere dropped by one million tons, or almost 25% from 1989 to 1992. By 2000, a total reduction of SO_2 emissions equal to 30% of the 1989 level is expected (Pudlis, 1993).

As levels of SO_2 decrease throughout the world, more farmers will discover that S is an essential component of their fertilizer programs to maintain healthy and profitable crops. They will no longer benefit from atmospheric S feeding their crops. This is already a reality in European agriculture, where the shift from low to high-analysis fertilizers has been more rapid and complete than in developing nations. For many decades, industrialized nations could afford to neglect S fertilization because most of the crop needs were satisfied by high SO_2 depositions. However, after several years of intensive agricultural practices, including S-free fertilizers and sustained high yields, coupled with reduced SO_2 emissions, S deficiencies have become a prominent problem.

The extent of sulphur deficiency

Worldwide S fertilizer applications have shown little historic growth, since these have been primarily a component of multi-nutrient fertilizers where the S value was unrecognized. Without a change in this trend, S deficiencies and, consequently, demand for S fertilizers, will increase dramatically. The Sulphur Institute estimates that, in 1991, the total annual unrealized world market potential for plant nutrient sulphur was 6.4 million tons. The S requirement has shown steady growth since the early 1960s and is expected to continue with this trend into the 21st century. The world plant nutrient sulphur deficit is projected to reach 7.5 million tons annually by 2000 and increase to 10.4 million tons by 2010.

In Western Europe, a dramatic decrease of incidental S applications is occurring, while S requirements have continued to escalate (Fig. 2). This coincides with rapidly growing S deficiency problems throughout Western Europe. If the current trend levels off at present values, the S fertilizer deficit in Western Europe is expected to reach 0.3 million tons in 2010. Whereas, if the current trend persists, the deficit will increase by a greater margin upwards to 1.0 million tons. However, the true value depends upon future agronomic variables and market conditions. In fact, cropping patterns are changing in this region. Cultivation of oilseed rape requiring two to three times the amount of S per unit area for optimal growth as compared to wheat, has become more prevalent (Walker and Booth, 1992). Currently, rapeseed is one of the most important crops in northern Europe. In certain areas of Denmark and northern Germany, it shares more than one-third of the cropping area, making this region the most intensively cropped rapeseed area in the world (Schnug, 1989). As mentioned previously, SO_2 emissions are expected to decline even further, exacerbating the occurrence of S deficiencies. Moreover, fertilizer consumption in Western Europe has been declining since 1986–1987, falling by 24%. The fall in consumption is expected to continue during the next ten years, but more gradually. The reduction in 1992–1993 in Western Europe is estimated at 12% (Fettah and Maene, 1993); therefore, the need for S fertilizers may increase by a greater margin than the model predicts. Ultimately, changes in the EEC's Common Agricultural Policy subsidy and set aside policies, along with further developments of alternative uses, such as rapeseed oil for biodiesel production, will affect future production levels and, thus, S requirements.

Environmental impact of sulphur shortages and consequences on agriculture

Sulphur availability contributes to the overall health of a plant. The content of S-containing secondary compounds in plants is not only of importance for nutritive value or flavor, but also for resistance against pests and diseases. This is of great importance for natural resistance of plants in both agricultural and non-agricultural systems and for alternative agricultural systems where the use of pesticides is prohibited. Here, fertilization is a valuable method to enhance the natural resistance of plants against diseases and insect damage (Schnug, 1990).

During the 1980s, incidental S applications on European crops decreased dramatically (Schnug, 1991). The situation in the northern cropping areas of Europe has become so serious that Schnug (1992) reported S deficiency as the major nutritional problem in arable crops. Today in the UK, Sweden, Denmark, Germany, and France, S deficiencies are the most widespread of all nutrient deficiencies in oilseed rape. Sulphur deficiencies also affect wheat and other cereals seriously (Schnug, 1991; 1992). The average S uptake since 1980 for oilseed rape and since 1990 for wheat, is no longer satisfied by atmospheric depositions. The reduction of the S supply became dramatic during the early 1980s after the enactment of the first steps to curb emissions taken at the 1979 Geneva Convention on Long Range Transboundary Air Pollution. Sulphur dioxide emissions in the UK peaked in the late 1960s and have decreased since by approximately 40% to the current level of 1.8 million tons per year (United Kingdom Review Group on Acid Rain, 1987; 1990). Surveys conducted by the Agricultural Development and Advisory Service have shown that the area of deficiency in England grew from 24% in 1991 to 39% in 1992 (Sutton, 1994). The same report indicates that all of Scotland's oilseed rape crop and 40% of the English crop are S-deficient. On the west coast region of the German Federal State, Schleswig-Holstein, Kurmies (1957) estimated that atmospheric S input from 1955 to 1956 was 80 kg ha^{-1} per year. Thirty years later, in 1984–1985, Schnug and Holz (1987) determined a yearly S deposition of only 20 kg ha^{-1}. Most of the S deposited over Sweden originates from other countries, where the reduction in SO_2 emissions has proceeded more slowly. Emissions peaked in the early 1970s at about 0.9 million tons, and continuously decreased to about 0.2 million tons in 1990 due to environmental regulations (Simán, 1994).

Particularly noteworthy are recent crop responses to S fertilizers in soils surrounding industrial areas thought to be unresponsive only a few years ago. In Germany, S deficiencies that were first reported during the early 1980s, began spreading rapidly southward. Now, S deficiencies in oilseed rape also have been reported in southern Germany (Ofenhitzer, 1990; Wimmer, 1990). Similarly, field experiments in England prior to 1980 indicated no yield response of oilseed rape to S. Subsequently, researchers at the Agricultural Development and Advisory Service conducted field trials in England and Wales between 1981 and 1987 when responses were restricted to sites on chalk soils (Syers *et al.*, 1987). More recently, S deficiency in oilseed rape has become increasingly widespread–particularly in northern England and Scotland (Zhao *et al.*, 1991). In 1991 and 1992, research plots in Woburn, UK showed yield responses of rapeseed to S fertilization, despite SO_2 depositions of about 20–30 kg ha^{-1}, which are among the highest in the UK today (AFRC, 1992). These results are important because they occurred in an area between London and Birmingham with a history of high SO_2 depositions. Remarkably, deficiencies have been reported in the former German Democratic Republic as early as two years ago (E. Schnug, Institute of Plant Nutrition and Soil Science, March 1993, personal communication). These findings confirm that, due to the high mobility of sulphate ions in soils, areas receiving large amounts of SO_2 depositions, can become S-deficient soon after emissions are reduced.

The increasing production of oilseed rape in Germany, Denmark, France, and the UK has forced many farmers to apply S fertilizers for optimal plant nutrition. Because of substantial increases in S deficiencies of oilseed rape, the North of England Arable Centre issued an advisory to farmers in 1991, instructing them to watch for S deficiency and advising them on the use of S fertilizers to correct deficiencies (Murphy, 1991). As early as 1985, a general recommendation for S application was issued in Scotland (Anon., 1985). Similarly, in Denmark a blanket recommendation for S has been introduced (Knudsen and Pedersen, 1992). Farmers in Ireland and the UK also have seen benefits from S additions to their forages. As a result of this and other evidence, the Agricultural Development and Advisory Service (1988) in the UK recently developed a recommendation for farmers to use S fertilizers to improve forage production. In Ireland, research conducted in the late 1970s also documented grass production gains. This research and industry's promotion has increased the consumption of S fertilizers three-fold (Murphy, 1991). A recent survey conducted by The Sulphur Institute shows that current S fertilizer recommendations in the UK, Norway, Denmark, Germany, and France range from 10–25 kg S ha^{-1} for grassland, 10–80 kg S ha^{-1} for rapeseed and 5–50 kg S ha^{-1} for cereals.

In addition to reducing plant yield, S deficiency also is recognized now as adversely affecting the quality of crops grown for both human and animal consumption.

For example, in breadmaking wheat, the S nutritional status is positively correlated with yield and baking and milling quality (Schnug, 1992). A 1992 Home-Grown Cereals Authority survey found there had been a significant decline in the S status of British wheats from the main growing areas which may be responsible for some of the quality problems encountered by end users. Ten years earlier a similar survey found no evidence of S deficiency. But, in 1992, 7% of the samples had a total S concentration lower than the critical level of 0.12%. A further 34% were in the marginally deficient or 0.12–0.14% range. One in ten of the more recent samples had a N:S ratio greater than 17:1–the level regarded as critical for protein production. In 1982, no tested wheat surpassed this level (Anon., 1994). Furthermore, S fertilization has been shown to influence forage quality through increased vitamin A content of alfalfa, increased chlorophyll content of red clover, increased protein content and amino acid composition of forages, decreased N:S ratios and non-protein N and nitrate (NO_3^-) levels, and reduced hydrogen cyanide content (Tisdale, 1977). The NO_3^- concentration in vegetables and forages has become an important criterion for food quality (Schnug, 1990). Nitrogen and S are main constituents of proteins; therefore, a shortage in the S supply of crops also affects the utilization of N within plants for the synthesis of proteins. Thus, S deficiency may cause an enrichment of non-protein N compounds, including NO_3^-, in the plant tissue (Murphy, 1991). Therefore, it is important to maintain an optimal S nutritional status in order to prevent NO_3^- enrichments within plant tissue. The effect of S on forage crop quality also is important because of its ultimate impact on ruminant nutrition and performance. Increased dietary S levels in a number of studies with ruminant animals have shown increased feed uptake, dry matter digestibility, and improved N balance, all of which may result in increased meat, milk, and wool production (Morris, 1987). While both S fertilization of forages and direct dietary supplements are beneficial, several studies have shown that dry matter intake was greater where forage was fertilized with S than where the lower-quality forage was supplemented with this element (Tisdale, 1977). This accentuates the importance of supplying the right amount of S to agricultural ecosystems since where a S-deficient condition exists, dietary supplements cannot completely make up for lost production and reduced nutritional efficiency.

Fertilizing practices have been scrutinized as an important factor contributing to pollution. The contamination of ground water with NO_3^- is one of the most serious problems. Nitrogen and S are both involved in protein biosynthesis; thus, a shortage in the S supply of crops also lowers the utilization of applied fertilizer N. Besides poor efficiency for N fertilization, S deficiencies may increase the loss of N from agricultural soils through volatilization and leaching. Schnug *et al.* (1993), estimated that between 4000 to 6000 tons of N are lost annually to the environment from rapeseed croppings in northern Germany, due to insufficient S supply needed to convert N into biomass. Therefore, the diminishing supply of S to crops has far-reaching implications and consequences. Sulphur plays an important role in the primary and secondary plant metabolism as a component of proteins, glucosinolates, and other compounds that relate to several parameters determining the nutritive quality of crops. The deterioration of the S supply will have several consequences on natural ecosystems that have to be considered in future developments for improved soil fertility and environmental quality.

Economic implications of sulphur shortages

In Europe, the financial losses caused by S deficiency to farmers are well-represented by some practical examples. In northern Germany, an application of 100 kg S ha^{-1} on S-deficient oilseed rape crops, normally yields an extra 2.0 tons per ha. In economical terms, with oilseed rape selling at DM 750 (US$ 441) per ton and the price of S at DM 0.4 per kg, the S application returns DM 1500 (US$ 885) on the investment–roughly a 38:1 value-to-cost ratio (E. Schnug, Institute for Plant Nutrition and Soil Science, April 1992, personal communication). Similarly, Richards (1990) reported that in the main grass-growing areas of the UK, average yield increases of 7–10% can be expected from an application for second and third cut silage of 25 kg S ha^{-1}, resulting in an additional yield of approximately 1 ton dry-matter ha^{-1} per year. With the S application costing around UK£ 7 per ha (US$ 10) and the extra yield worth about UK£ 80 (US$ 118), it resulted in a value-to-cost ratio of 11:1. Further, Murphy (1990) estimated that annual S fertilization on deficient areas of Ireland, could result in an additional 1.5 million tons of herbage dry matter with a value of I£ 75 million (US$ 107 million). Once the farmers realize the economical implications suggested by these examples, S fertilization will become an integral part of the agronomic practices.

In many parts of the world, farmers are willing to pay for a material that they received free a few decades ago. While S in fertilizers is a value-added product for the fertilizer industry, this success would not be realized if farmers did not recognize the benefits resulting from increased yields and profits. Many manufacturers of ammonium sulphate have shifted marketing of this product from a N fertilizer to a multi-nutrient fertilizer. This product promotion has led to granular ammonium sulphate prices reaching historically high levels in the United States during the spring 1993 fertilizer season. Compared to the same period in 1992, granular ammonium sulphate FOB prices increased by 10% in the Corn Belt, while urea prices have shown only a 5% gain. Again, compared to the same period last year, during the spring 1994 fertilizer season, granular ammonium sulphate FOB prices in the Corn Belt showed an additional 5.5% increase, while urea prices decreased 2.5%. In Europe, pricing for value-added S fertilizers also has improved. In fact, in the Netherlands, DSM is reducing the volume of granular ammonium sulphate available to export markets, to supply the growing demand within the European market. In Germany, Italy, and the UK, S fertilizers currently are retailing for DM 0.40 (US$ 0.23), L 450 (US$ 0.27), and UK£ 0.3 (US$ 0.45) per kilogram of S, respectively; whereas, in the United States, S fertilizers are retailing for about US$ 0.45 per kilogram of S. This translates into a metric ton S equivalent of US$ 233 for Germany, US$ 266 for Italy, US$ 446 for the UK, and US$ 440 for the United States.

Traditional and innovative sulphur fertilizers to overcome deficiencies

The economics of fertilizer distribution and use generally dictate the manufacture of materials with high plant nutrient content. In the case of N, P, and potassium (K) carriers, this has often meant exclusion of S from the product. When high-analysis fertilizers are to be used in areas that are deficient or potentially deficient in S, this element must be applied separately or with the N-P-K materials. There are numerous materials available that may be used as sources of S. The economics of production, handling and storage, transportation and farm application will determine whether the S carrier should be added at the point of manufacture or at the point of consumption, or whether it should be applied separately. From a farmer's point of view, the most economic and convenient method of applying needed S is likely to be as an ingredient in his regular fertilizer treatment. To this extent, numerous advances have been made in formulating materials and developing innovative technologies for adding S to fertilizers (Bixby and Beaton, 1970).

Sulphur can be applied to the soil using a variety of different products, although today the most significant sources are still ammonium sulphate, single superphosphate, and potassium sulphate (Anon., 1987). Producers of byproduct material dominate the manufacture of ammonium sulphate, since synthetic production is not always advantageous economically. It is estimated that approximately 17.0 million tons of ammonium sulphate–equivalent to 4.0 million tons of S–were used as a fertilizer in 1993. About 3.0 million tons of S equivalent was used directly, the remainder of about 1.0 million tons of ammonium sulphate going into mixed grade fertilizers, and other fertilizers, such as ammonium sulphate nitrate (The Sulphur Institute, 1994). The world's leading producers are AlliedSignal with an annual capacity of 1.4 million tons, followed by BASF with 0.45 million tons and DSM with 0.25 million tons (Anon., 1993). Since ammonium sulphate is a co-product and is traded internationally, this commodity will influence the S fertilizers market. In contrast to ammonium sulphate, single superphosphate usually is consumed in the country where it is produced. Single superphosphate is a significant fertilizer in China, India, Australia, and New Zealand, but of little importance in other countries, such as those in North America and Europe. Furthermore, its production is declining as the majority of new phosphate fertilizer plants under consideration worldwide include compound fertilizers and ammoniated phosphates rather than single superphosphate–a switch that benefits the consumption of raw material S and exacerbates S deficiencies.

Ammonium sulphate and single superphosphate now are recognized as multi-nutrient fertilizers rather than solely N and P sources. Historically, the bulk of the S has been applied in these multi-nutrient fertilizers, followed by compound fertilizers. During the fertilizer year 1991–1992, of the 9.7 million tons of S fertilizers applied, 3 million tons were from ammonium sulphate and 3.7 million tons were from single superphosphate (Table 3). Therefore, these two sources alone provided nearly 70% of the S in fertilizers applied worldwide. Specific to Europe, the majority of the S has been applied through compound fertilizers followed by ammonium sulphate and single superphosphate. Ammonium sulphate consumption is prevalent in the southern European countries. In 1991–1992,

Table 3. Sulphur contributions from various fertilizer sources (1991–1992)

Fertilizer source	%S	World (million tons)	Western Europe (1000 tons)
Ammonium sulphate	24	3.0	320
Single superphosphate	12	3.7	101
Triple superphosphate	1	0.1	15
Ammonium phosphate	2	0.5	26
Compound fertilizers	Variable	1.9	476
Potassium sulphate	18	0.3	42
Others	Variable	0.2	52
Total S consumption		9.7	1032

ammonium sulphate consumption reached 93,600 tons S in Spain, 88,900 tons S in Italy, 45,600 tons S in France, and 42,900 tons S in Greece. Single superphosphate has been an important source of S, especially in Eastern Europe. The large fertilizer consumption decline that has characterized this region since the early 1990s will have significant consequences on future S demand. For example, during the fertilizer year 1990–1991 consumption of single superphosphate equalled 121,400 tons S in Poland, 58,300 tons S in former Czechoslovakia, and 44,100 tons S in Hungary. In comparison, during 1991–1992, consumption dropped to 52,100 tons S, 15,400 tons S, and 4,700 tons S, respectively. This input reduction is reflected by increased S deficits which The Sulphur Institute, in 1991–1992, estimated to equal 86,200 tons in Poland (700 tons in 1990–1991), 1,600 in former Czechoslovakia (−77,400 tons in 1990–1991), and 62,100 in Hungary (6,600 tons in 1990–1991). In Western Europe, Italy and Spain are the largest consumers of single superphosphate with consumption in 1991–1992 at around 41,600 tons S and 20,400 tons S, respectively. Even though it is costly, potassium sulphate is still used widely as a specialty fertilizer used for valuable cash crops sensitive to chloride, and where the nutritional benefits of both S and K are recognized. During the fertilizer year 1991–1992, the largest consumers of potassium sulphate were Italy, with 13,300 tons S, followed by France and Germany, each with 7,200 tons S.

While ammonium sulphate, single superphosphate, and potassium sulphate will remain important S sources, new sources are increasing their market share. In fact, as the market for S fertilizers expands, S producers worldwide are introducing new products to meet the diversified application requirements. Fertilizer producers in Oceania and North America, where S deficiencies have been recognized and corrected since the 1950s, have been at the forefront of S fertilizer research and technology. However, during the past few years, as S deficiencies spread throughout Europe, the major European fertilizer concerns also have introduced numerous S-containing fertilizers to meet the rapidly growing demand. This has been also facilitated by the introduction of the 89/284/EEC Directive published by the EEC in April 1989. According to this Directive, S may be declared in an EEC fertilizer if the content is at least 5% S trioxide (SO_3) or 2% elemental S. Prior to this regulation, there was no common practice and in some countries, such as Italy, the UK, and former West Germany, it was even prohibited to display the S content on bags; thus making S harder to market.

In response to the increasing evidence of a shortfall in S supply to arable crops and grassland, La Grande Paroisse in France has recently introduced a new variety of S products to satisfy different farming needs. These include AzophoS-1^{TM} (25–10–0–6S), AzophoS-2^{TM} (30–6–0–4S), NitrammoS^{TM} (25–0–0–5S), NitrammoSMg^{TM} (25–0–0–4S–5MgO), and Sulfertil^{TM} (20–0–0–8S). In the UK, ICI Fertilizers produces two S-containing fertilizers: Sulphur-Gold^{TM}, which contains 30% N and 19% S and is designed particularly for oilseed rape and winter cereals cropping systems, as well as Kaynitro-Gold^{TM}, which contains 24% N, 14% K, and 3.2% S. Kemira and Norsk Hydro, the two largest fertilizer concerns in Europe, along with BASF and DSM Agro, have developed S-containing fertilizers.

Sulphur Ten^{TM} (20–4–14–7S), developed for silage crops, has been marketed by Kemira for several years. More recently, Kemira Fertilisers in the UK, released DoubleTop^{TM}, a new granular product containing 27% N as well as 12% S and is formulated specially for oilseed rape and cereals. For spring applications on first cut silage, Hydro Agri has developed Sulphur Grass^{TM} (25–5–5–5S). The importance of S for second and third cut silage yields is now widely recognized and Hydro Agri's SulphurCut^{TM} with its balanced analysis (22–4–14–3S), has become increasingly popular since its launch in 1988 in the UK. The company advises that these fertilizers are formulated for grassland, but could be used on arable crops in the spring (Rush, 1994).

Eurozolfi S.r.L. of Italy, developed Acidam^{TM}, a product line of pelleted, dust-free S product (fertilizer-amendment) made by extruding and drying a mixture of finely divided S, optionally with other organic and inorganic components, in a binder. The binder breaks down in the presence of moisture releasing the very fine S particles (90–95%<100 m). This proprietary technology provides Eurozolfi with the flexibility of creating a wide range of formulations to satisfy most soils and cropping needs. The success and demand for such products is confirmed with the recent release of Biozolfo 620^{TM} by FOMET s.a.s. in northern Italy. This product contains 50% elemental S and a 50% mix of selected organic manures and, like Acidam^{TM}, is recommended for grapes, beets, vegetables, legumes, cereals, and fruits. Due to its unique characteristics, it is marketed as a fertilizer and soil amendment, included in the list of products permitted under the EEC provision 2092/1991 for organic farming. Similar products also are receiving attention in the UK where there is potential for significant leaching losses of sulphate ions. Stoller Chemicals, in the UK, has acquired the license to produce and market Tiger 90^{TM} in Europe. This material, manufactured by Tiger Resources Technology, Inc. of Canada (formerly Tiger Chemicals), contains 90% elemental S mixed with 10% bentonite clay and special additives to enhance its water degradability.

Ammonium thiosulphate has gained prominence in areas of North America and is growing in use and importance also in Europe because of its versatility and high S concentration in fluid formulations. Moreover, it offers added agronomic benefits acting as a nitrification inhibitor and as a remedy for iron chlorosis. This product is a clear liquid fertilizer suitable for direct applications or blending; therefore, it offers additional versatility to farmers and fertilizer retailers. Esseco S.p.A., is a major producer of ammonium thiosulphate in Europe, which is mostly marketed in Italy and other niche markets throughout Europe. This material also is gaining importance in the UK and in Germany where the prominent producers are William Blythe & Co. Ltd. and Th. Goldschmidt AG, respectively. Because of the increasing demand for fluid S sources, Esseco S.p.A. recently began producing potassium thiosulphate (0–0–25–17S) solution. At the same time, liquid fertilizers supplier Hydro Chafer in the UK introduced Chafer NS60^{TM} (26–0–0–3.1S)–a new liquid fertilizer formulated for early season use that is suitable for all crops, particularly cereals, oilseed rape, and grass. Hydro Chafer also offers Nuram 35S^{TM} (35–0–0–1.7S), Nufol 20+S^{TM} (20–0–0–1.7S) for foliar applications and compound fertilizers with S based on ammonium thiosulphate according to individual requirements (Rush, 1994).

Conclusions

In order to maximize and reconcile economic and environmental demands in agricultural practices, the vital role of S for agroecosystems and especially the importance of S fertilization in optimizing crop exploitation of other nutrients, in particular N, must be reconsidered. In fact, severe S deficiency in agricultural crops poses a major problem in northern Europe with strong ecological impacts. An insufficient S supply not only will reduce yields, but also will diminish quality of food and feedstuffs. Moreover, under conditions of S deficiency, efficiency of N fertilizer is reduced causing damaging N losses to the environment.

Considerable progress has occurred since the early 1980s when little data on crop responses to S were available in Europe. At that time, S recommendations were scarce and SO_2 emissions still were providing most of the S to a majority of the cropping areas. Furthermore, the fertilizer industry had little interest in promoting the S content of ammonium sulphate and single superphosphate, and relatively few companies had developed a strategy and specific efforts to promote and develop S fertilizers. Since then, The Sulphur Institute has been involved actively in promoting the use of S in European agriculture. Its educational and promotional efforts resulted in important advancements that made it possible to demonstrate the value of the S provided with fertilizers.

After years of research, the vital role of S for agriculture worldwide has been established and well-documented. As farmers realize the benefits of improved quality and yield of crops stemming from the use of S fertilizers, European agriculture is now at a stage where greenhouse and field trial results need to be transposed from theory into practical farming practices.

References

AFRC, Institute of Arable Crops Research (1992) Institute for Arable Crops Research. Rothamsted, UK. Report for 1992, 16 pp

Agricultural Development and Advisory Service (1988) Fertilizer recommendations for agricultural and horticultural crops. MAFF Reference Book 209. Her Majesty's Stationery Office. London, UK

Anonymous (1985) Fertilizer recommendations. The Scottish Agricultural College/The Macaulay Institute for Soil Research. Publication No. 160. 38 pp

Anonymous (1987) Sulphur Bentonite–A growing source of plant nutrient sulphur? Sulphur 192: 15–16. September–October, 1987

Anonymous (1990a) Elements of Uncertainty. Sulphur 207:21–27. March–April, 1990

Anonymous (1990b) More sulphur needed? Fertilizer Dealer Progress November–December 1990, 28 pp

Anonymous (1993) Chem Scope–Ammonium sulfate. Chemical Business. December 1993. 10 pp

Anonymous (1994) Sulphur on the wane. Farmers Weekly. 4 March, 1994. S11 pp

Bixby, DW, and Beaton JD (1970) Sulphur-containing fertilizers. Technical Bulletin No. 17. The Sulphur Institute, Washington, DC

Buckman, HO and N.C. Brady NC (1969) The Nature and Property of Soils. 7th ed. The Macmillan Company, Toronto, Ontario, Canada

Fettah, M, and Maene L (1993) Emerging issues in the fertilizer sector from an industry perspective. In: Proc. World Bank/International Fertilizer Industry Association 1993 Round Table. Washington, DC. 9–10 September, 1993

Knudsen, L, and Pedersen CA (1992) Sulphur Fertilization in Danish Agriculture. Sulphur in Agriculture. 17: 29–31. The Sulphur Institute, Washington, DC

Kurmies, B (1957) Uber den Schwefelhaushalt des Bodens. Die Schwefelsäure 17: 258–278

Morris, RJ (1987) The importance and need for sulphur in crop production in Asia and the Pacific region. p 4–11. In: Proc. Symposium on Fertilizer Sulphur Requirements and Sources in Developing Countries of Asia and the Pacific. Bangkok, Thailand. 26–30 January, 1987. FADINAP, FAO, TSI, and ACIAR

Murphy, MD (1990) Fifteen years of sulphur research in Ireland. Sulphur in Agriculture 14: 10–12. The Sulphur Institute, Washington, DC

Murphy, MD (1991) Sulphur deficiency warning. North of England Arable Centre Agronomy Bulletin No. 6. April 30, 1991. University of Newcastle-upon-Tyne, UK

Ofenhitzer, D (1990) Schwefelversorgung in Unterfranken. Raps 8: 12–14

Pudlis, E (1993) Air Getting Lighter. The Warsaw Voice. 3 October, 1993. 40(258): G1–G3

Richards, I.R. 1990. Sulphur as a crop nutrient in the United Kingdom. Sulphur in Agriculture 14: 8–9. The Sulphur Institute, Washington, DC

Rush, T (1994) Cereals need sulphur too. Arable Farming. February, 1994. p 4–9

Schnug, E (1989) Double low oilseed rape in West Germany: Sulphur nutrition and levels. p 84–100. In: Proc. National Conference on Double Low Oilseed Rape for the 1990s. Peterborough, UK. 8 March, 1989

Schnug, E (1990) Sulphur nutrition and quality of vegetables. Sulphur in Agriculture 14: 3–7. The Sulphur Institute, Washington, DC

Schnug, E (1991) Sulphur nutritional status of European crops and consequences for agriculture. Sulphur in Agriculture 15: 7–12. The Sulphur Institute, Washington, DC

Schnug, E (1992) Baking quality and sulphur content of wheat I. Influence of grain sulphur and protein concentrations on loaf volume. Sulphur in Agriculture. 16: 31–34. The Sulphur Institute, Washington, DC

Schnug, E, and Haneklaus S (1994) The ecological importance of sulphur. Norwegian Journal of Agricultural Sciences. Supplement No. 15: 149–156. ISSN 0802-1600

Schnug, E, Haneklaus S and Murphy D (1993) Impact of sulphur fertilization on fertilizer nitrogen efficiency. Sulphur in Agriculture 17: 8–12. The Sulphur Institute, Washington, DC

Schnug, E, and Holz F (1987) Deposition of nitrate-N and sulfate-S by precipitations in Schleswig-Holstein. Atmosph Environ 21: 1235–1241

Simán, G (1994) Sulphur in Swedish Agriculture. Norwegian Journal of Agricultural Sciences. Supplement No. 15: 31–34. ISSN 0802–1600

Sutton, CD (1994) Sulphur deficiency spreads in Britain. Fertilizer International 329: 40. January 1994

Syers, JK, Curtin D, and Skinner RJ (1987) Soil and fertilizer sulphur in UK agriculture. In: Proc. of the Fertilizer Society, No. 264

The Sulphur Institute (1994) The Sulphur Outlook. The Sulphur Institute, Washington, DC

Tisdale, SL (1977) Sulphur in forage quality and ruminant nutrition. Technical Bulletin No. 22. The Sulphur Institute, Washington, DC

United Kingdom Review Group on Acid Rain (1987) Acid deposition in the United Kingdom, 1981–1985. Second Report. Warren Spring Laboratory, Stevenage, UK

United Kingdom Review Group on Acid Rain (1990) Acid deposition in the United Kingdom, 1986–1988. Third Report. Warren Spring Laboratory, Stevenage, UK

Walker, KC, and Booth EJ (1992) Sulphur research on oilseed rape in Scotland. Sulphur in Agriculture 16: 15–19. The Sulphur Institute, Washington, DC

Wimmer, P (1990) Schwefelmangel in Koernerraps. Bayrisches Landwirtsch Wochenblatt 179: 24

Zhao Fang-jie, Syers JK, Evans EJ and Bilsborrow PE (1991) Sulphur and oilseed rape production in the United Kingdom. Sulphur in Agriculture 15: 13–16. The Sulphur Institute, Washington, DC

C. Rodriguez-Barrueco (ed.), Fertilizers and Environment, 195–198.

Optimisation of nitrogen fertilisation

Relationship between precipitation and expected yields

Mário Carvalho & Gottlieb Basch
Universidade de Évora, Departamento de Fitotecnia, 7000 Évora, Portugal

Key words: nitrogen fertilisation, precipitation, yield potential, wheat varieties

Abstract

The optimisation of nitrogen fertilisation for cereals in Mediterranean regions is very difficult, because yields and leaching losses are depending on winter rainfall. In this article five years of nitrogen fertilisation trials are analysed in order to get a model which relates winter rainfall, nitrogen fertilisation and wheat yields. A highly significant model obtained, in which soil type and the cultivar were taken into account. In order to improve nitrogen use efficiency and to avoid unnecessary nitrogen fertilisation, information on the minimum amount of nitrogen at seeding time and for how long the first top dressing can be delayed without yield losses will be necessary.

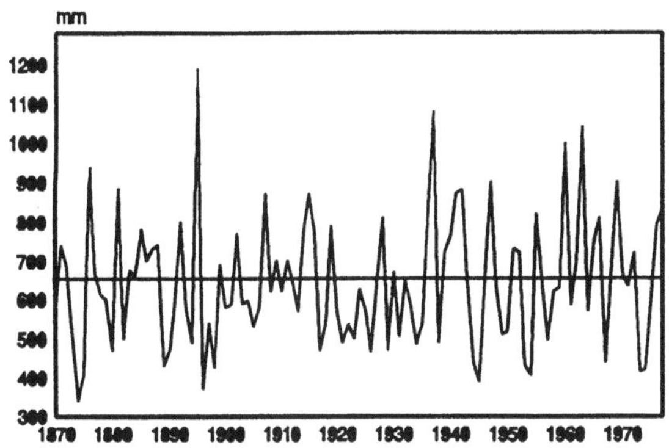

Fig. 1. Total yearly precipitation in Évora between 1870 and 1978

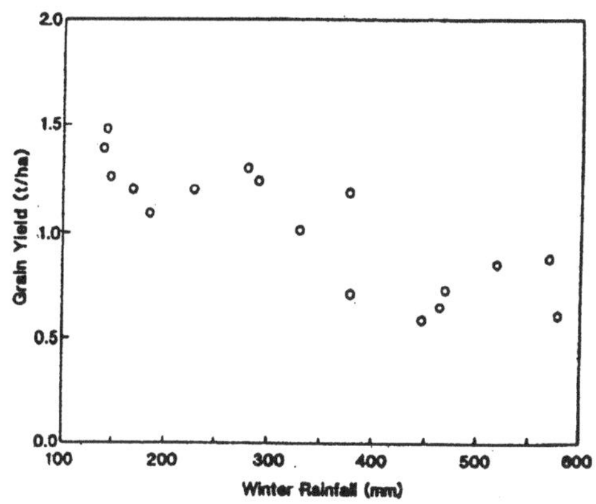

Fig. 2. Influence of winter rainfall on the yield of wheat in Évora district

Introduction

The variation of annual precipitation in Mediterranean regions is very high (Fig. 1). As rainfall can influence cereals yield potential and nitrogen losses, the optimisation of nitrogen fertilisation is therefore a very difficult task. Usually the farmers aim at the highest yield, but this behaviour leads frequently to an excessive application of nitrogen. Having in mind that winter nitrogen applications are necessary in this regions for good wheat production, the potential for N leaching losses is quite high. Therefore the development of a model relating winter rainfall, nitrogen level and expected wheat yield could help to rationalise the application of nitrogen fertiliser. Several authors have tried to establish relationships between climatic parameters, mainly rainfall, and the wheat yields in Portugal. (Carvalho, 1978; Figueiredo, 1919; Frazão, 1943; Oliveira, 1955; Oliveira and Sousa, 1969). Authors using average yields on a regional basis generally obtained negative relationships between rainfall and yields (Fig. 2). However, when the soil type is taken into account, the results can change (Fig. 3) (Goss and Carvalho, 1988). In field trials similar results were obtained for the Eutric Cambisol, however for the Pelic Vertisol

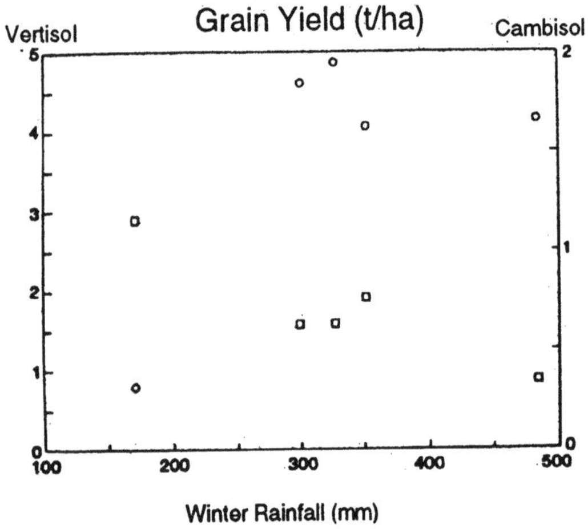

Fig. 3. Contrasting influences of winter rainfall on the yield of wheat on two different soils: circles - Pelic Vertisol; squares - Eutric Cambisol

Table 1. Precipitation (mm) - Long term averages (1941/70) and actual values during the experimental period

	1981	1982	1983	1984	1985	1986	1941-70
Jan	-	83	0	34	143	67	83
Feb	-	58	22	14	113	136	72
Mar	-	43	7	53	14	30	92
Apr	-	57	74	121	76	68	51
May	-	6	22	62	28	22	36
Jun	-	1	13	29	8	0	19
Jul	-	8	0	0	6	0	2
Aug	-	13	1	1	0	-	2
Sep	23	40	24	8	3	-	22
Oct	32	10	56	61	0	-	56
Nov	2	108	180	109	41	-	79
Dec	162	31	85	58	58	-	83

Table 2. Soil characteristics of the experimental

Horizon	Depth (cm)	Clay (%)	O.M. (%)	C.E.C. mep 100 g^{-1}	pH (water)
Ap	0–35	40.6	0.94	20.22	7.0
B	35–65	44.2	0.87	29.99	7.5

wheat grain yields were almost the same when winter rainfall exceeded 300 mm but much reduced with rainfall amounts below 200 mm.

Yield losses caused by water excess can have various reasons, but the most frequent one referred in literature is nitrogen deficiency (Fisher, 1924; Gales, 1983; Leyshon and Sheard 1974; Millington, 1961; Van Horn, 1958; Van der Paauw, 1972). The nitrogen losses caused by leaching due to excess of rainfall during the winter months in Mediterranean regions can be very high. Almeida (1965) measured losses up to 90 kg ha^{-1} $year^{-1}$ at Tapada da Ajuda (Lisbon), and this is certainly one of tbe explanations for the low nitrogen use efficiency in cereal crops in Mediterranean regions (Alves, 1979).

Materials and methods

Field trials were conducted between 1981/82 and 1985/86, on a Pelic Vertisol at the University of Évora on the Experimental Farm of Almocreva. Precipitation and soil parameters are given in the Tables 1 and 2. The levels of nitrogen used were 0, 100 and 200 kg N ha^{-1} as ammonium nitrate. Half of the fertiliser was applied at seeding time and the other half at the end of tillering. Two wheat varieties were sown: Mara and Etoile de Choisy. The first one is an alternative variety, with shorter cycle and straw; Etoile is a winter type, taller and a longer cycle. Two seeding times were used every year (Nov 20th ± 3 days; Dec 1Oth ± 3 days) and the average of the two seeding dates for each one of the cultivars was used for the model. The layout of the trial was a randomised block design with a split and 4 replications. Seeding time and nitrogen levels were the main plots and varieties the secondary ones.

In order to obtain the model only the variables presenting a significant effect were considered, which was the reason for not taking into account the linear effect of nitrogen. Only the precipitation from November to the end of February was used as rainfall variable for the elaboration of the model as the values for this period showed the highest correlation with the yield. The programme used to fit the model was MSTAT (Michigan State University, 1986).

Results and discussion

Table 3 shows the results of wheat grain yield for each one of the varieties (average of the two seeding times) and nitrogen level, and the rainfall from November to the end of February. In the models obtained from the results given in Table 3, for each one of the wheat cultivars, Y represents the grain yield (kg ha^{-1}), R the rainfall from the beginning of November to the end of

Table 3. Wheat grain yields (kg ha^{-1}) and precipitation from beginning of November to the end of February

	Nitrogen level (kg of N ha^{-1})	Mara (kg ha^{-1})	Etoile de Choisy (kg ha^{-1})	Rainfall (Nov to Feb)
1981/82	0	2618	2740	
	100	3480	3830	305
	200	4862	3070	
1982/83	0	916	999	
	100	1003	906	161
	200	880	741	
1983/84	0	1747	1712	
	100	3691	3254	313
	200	3760	3234	
1984/1985	0	1600	2015	
	100	2967	3588	422
	200	3760	3969	
1985/86	0	1342	1728	
	100	3534	3620	302
	200	4146	3074	

February (mm), and N the amount of applied nitrogen (kg N ha^{-1}).

$$MARA: \ Y = -4879 + 46.4R - 0.076R2 - 0.030N2 + 0.05NR \quad (r = 0.948 p < 0.1\%)$$

$$ETOILE: \ Y = -2621 + 29.5R - 0.045R2 - 0.057N2 + 0.05NR \quad (r = 0.956 p < 0.1\%)$$

The partial derivatives δY/δN ; δY/δR of each one of the equations are equal to:

	MARA:	ETOILE:
$(\delta Y/\delta N)$:	$N = 0.833R$	$N = 0.465R$
$(\delta Y/\delta R)$:	$R = 305 + 0.392N$	$R = 328 + 0.59N$

For both of the cultivars the models are highly signiif-icant. Therefore it is possible to calculate the nitrogen fertilisation taking into account the amount of rainfall of the winter (Nov. to Feb.), considering, however, the different lodging resistance of the varieties used. Etoile is a tall cultivar and lodging problems were clearly visible with 200 kg of N ha^{-1} while no lodging happened with Mara during the field trials, thus explaining the differences shown in Table 4. The amount of nitrogen to be applied increases with the rainfall during November to February because of an increase of the yield potential (Figures 4 and 5) and of nitrogen losses (Table 5).

To reach high wheat yields during wet winters the nitrogen fertilisation has to be increased and the potential for nitrogen leaching losses is very high. Although

Table 4. Values of rainfall and corresponding nitrogen level for maximum yield for each one of the wheat cultivars according to the models

	R (rainfall mm)	N (kg N ha^{-1})	Y max. (kg ha^{-1})
Mara	420	350	4878
Etoile	452	210	4036

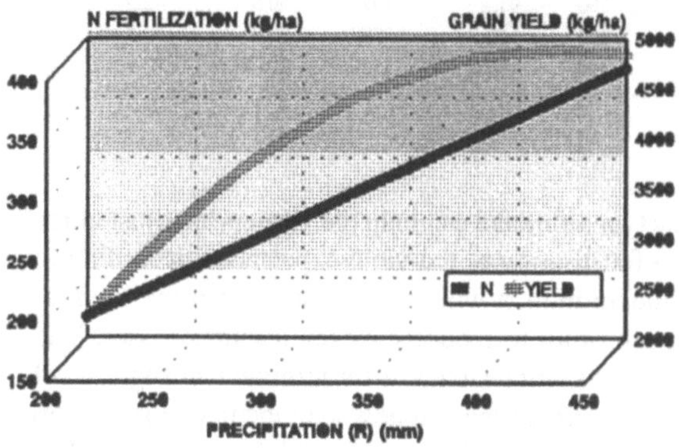

Fig. 4. Influence of winter rainfall on the nitrogen fertilisation for maximum yield and on achieved yield for Mara variety

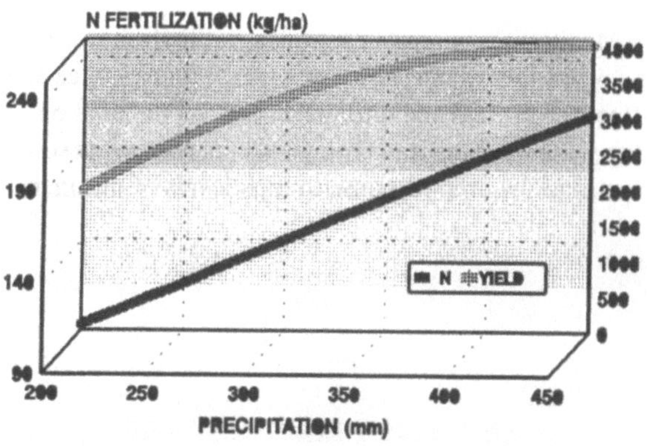

Fig. 5. Influence of winter rainfall on the nitrogen fertilisation for maximum yield and on achieved yield for Etoile variety.

Table 5. Rainfall (Nov-Feb) and calculated N fertilisation (kg N ha^{-1}) maximum yield (kg ha^{-1}), nitrogen uptake (kg N ha^{-1}) and nitrogen losses (kg N ha^{-1}) according to the models

	Rainfall (mm)	Fertilisation (kg N ha^{-1})	Grain yield (kg ha^{-1})	N uptake (kg N ha^{-1})	N losses (kg N ha^{-1})
Mara	200	167	2194	56.9	110.1
	300	250	4076	105.6	144.4
	400	333	4854	125.8	207.2
Etoile	200	93	1972	50.5	42.5
	300	140	3288	85.2	54.8
	400	186	3950	102.4	83.6

this type of model can help the farmers to adjust nitrogen fertilisation to the yield expectation according to actual rainfall, it will be necessary to improve the system in order to reduce nitrogen leaching losses. The next steps will be to know the minimum nitrogen level at the seeding time and how long the first nitrogen top dressing can be delayed without decreasing yield. The benefits of these informations would be:

- Reducing the risk of leaching losses would increase nitrogen use efficiency and reduce nitrogen use efficiency and reduce nitrogen fertilisation;
- Possible delay of the decision on the nitrogen application will reduce the amount of nitrogen applied until the end of February, and the decision on total nitrogen application can be taken based on an already secure estimation of the expected yield, thus avoiding unnecessary nitrogen fertilisation.

Conclusions

In Mediterranean regions nitrogen fertilisation can be managed taking into account winter rainfall, once it affects crop yield and nitrogen losses. As the relationship winter rainfall and crop yield depends on the soil type, and the crop response to high levels of nitrogen depends on the cultivar, both factors affect the model. To avoid high nitrogen leachinglosses during wet winter, the nitrogen fertilisation management system has still to include the information on the minimum nitrogen application at seeding time and on the period during which the first top dressing can be delayed.

References

Almeida LAV (1966) A lavagem do azoto dos adubos Pelas Aguas das Chuvas um Solo Granítico. An ISA 7: 263–288

Alves JA (1979) Efeito dos adubos azotados sobre o peso e o teor de azoto do grão, a exportação e a Recuperação do azoto aplicado na cultura do Trigo. Ministerio de Agricultura e Pescas, INIA, Oeiras

Carvalho MIGPR (1978) A produção de trigo e o clima nos distritos de Evora, Beja e Pottalegre. Instituto Superior de Agronomia, Relatório Final de Curso, Lisboa

Figueiredo FEA (1919) Observações e estudos efectuados no Laboratório de Physica agrícola e no campo experimental Meteorológico do Instituto Superior de Agronomia (1914 a 1918) - Tipogrfia Castro Irmão, Lisboa

Fisher RA (1924) The influence of rainfall on yield of wheat at Rothamsted. Phil Trans Soc, Ser B, 213: 89–142

Frazão (1943) As Chuvas de Inverno e o Rendimento Cultural do Trigo. An. ISA 14: 189–200

Gales K (1983) Yield variation of wheat and barley in Britain in relation to crop growth and soil conditions - a Review. J Sci Foog Agric 34: 1085–1104

Goss MJ and Carvalho MIGPR (1988) Causes of variation in yields of wheat under dryland farming in the Alentejo region of Portugal and some future prospects. Proceedings of the International Conference on Dryland Farming, Texas USA. pp 445–448

Leyshon AJ and Sheard RW (1974) Influence of short-term flooding on the growth and plant nutrient composition of barley. Can J Soil Sci 54: 463–473

Millington RJ (1961) Relation between yield of wheat, soil factors and rainfall. Aust J Agric Res 12: 497–508

Oliveira AJS (1955) Determinantes meteorológicas da Produção unitária de trigo. Separata da Lavoura Portuguesa, Lisboa

Oliveira AJ and Sousa ML de Barros (1969) Aspectos estatisticos da previsão da colheita. O caso da cultura do trigo no Alentejo em 1919-1967 Agron Lusitana 33 (3): 203–237

Van der Paauw F (1972) Quantification of the effects of weather conditions priori to the growing season on crop yields. Plant and Soil, 37: 375–388

Van Hoorn JN (1958) Results of a ground water level experimental field with arable crops on a clay soil. Neth J Agric Sci 6: 1–10

C. Rodriguez-Barrueco (ed.), Fertilizers and Environment, 199–204.

Plant biomass and fruit yield induction by Ti(IV) in P-stressed pepper crops

J.L. Lopez-Moreno[1], J.L. Giménez[2], A. Moreno[3], J.L. Fuentes[3] & C.F. Alcaraz[3]
[1]*Excmo. Ayuntamiento de Cehegín, Módulo de Desarrollo del Noroeste de la Región de Murcia;*
[2]*EUITA-ORIHUELA, Universidad Politécnica de Valencia;* [3]*CSIC-CEBAS, 30080-Murcia, Spain*

Key words: phosphorus, titanium, fertilizer efficiency, plant nutrition

Abstract

To study the titanium effect on P nutrition, a greenhouse experiment with *Capsicum annuum* L., cv. Bunejo plants growing under differential P fertilization was conducted. All the plants were grown under identical conditions and they only differred in the P fertilization and in Ti supply. Plant biomass production of the Ti-untreated plants was affected by the diminution of the P-feed, but the plants growing under the lowest P supply did not showed any deficiency symptom during the crop cycle. All the Ti-treated plots showed a significative increase of the plant biomass against their corresponding untreated references. The biomass enhancement was mainly caused by the increase of the fruit yield with an absolute enhancement of 62% in the plants growing under the lowest P feed, and of 45% in the plants with a complete P support.

Introduction

Growth and biomass production for a number of different plant species have been found to be enhanced by the presence of titanium, whether applied to the soil or *via* leaf spray (Pais, 1983; Ram *et al.*, 1983; Kiekens and Kamerlynck, 1987; Giménez *et al.*, 1990). In addition to this, the concentration of some essential elements will also be increased when plants are supplied with titanium. This enhanced uptake has been found to be independent of the method for the application of titanium (Feher *et al.*, 1987; Martínez-Sánchez *et al.*, 1992). The effects on the uptake enhancement of several nutrients induced by the addition of other non-essential elements were notified many years ago, as it is the case of calcium by the effect of lithium (Epstein, 1962 and 1972), nickel (Crooke, 1958), or zinc (Ernst, 1985). This mutual effect of ions is known as "Viet's effect". Up to now, there is no conclusive explanation for this phenomenon. Pais *et al.* (1977) have assumed that in the case of titanium, this element enhanced photosynthesis, thus increasing the enzyme activity and the requirement for metal ions and finally stimulated the ion uptake. In opinion of Dumon and Ernst (1988), this assumption however cannot explain an increased accumulation, because a growth-dependent demand will not give rise to excessive accumulation. They suggested other possible explanations, as the impact of an increased availability of elements, either direct or indirect *via* adsorption process, on the various forms of ATPases or in the stimulation of the synthesis of multimetal cofactors, e.g. Fe-V-S-cofactor, or of metal-binding storage proteins, as phytoferritins (Bienfait and Van der Mark, 1983). Recently other explanatory hypothesis has been proposed (Carvajal *et al.*, 1995; Alcaraz *et al.*, 1994). They proposed that tianium promotes the iron activity in cell chloroplasts and cytoplasm and consequently it increases the plant photosynthetic activity.

Whichever the pathway that titanium plays its beneficial role on plant metabolism, its promotive effects on the absorption and activity of some nutrients, have a great usefulness in the regulation of the fertilizer supplied. Therefore, the aim of this communication is to study, in a greenhouse experiment with *Capsicum annuum* L., the conjoint effects of a progressive decrease in phosphorus fertilization and the Ti(IV) leaf application, on the plant development, as well as on the phosphorus uptake.

Material and methods

Plant and experimental conditions

Seeds of *C. annuum* L., cv. Bunejo were obtained from the Centro de Investigación y Desarrollo Agrario of the Region of Murcia, Spain. Seedlings were obtained through direct sowing in 50% mixture of sand and manure (v:v) in polyurethane trays (one seed/50 cm^3 receptacle). After 95 days from sowing, the little plants were transplanted to a greenhouse equipped with a drip irrigation system and with controlled any other environmental conditions. The greenhouse was divided into three plots for the differential P fertilization, as follows: Plot 1: 33.3% P [1,000 kg/ha of 18% calcium superphosphate before the crop beginning, equivalent to 80 P fertilizer units (PFU), without any other P supply]; Plot 2: One additional P-support of 140 g/plant of the same phosphorus fertilizer, 21 days after transplanting (P-support equivalent to 160 PFU); Each plot was subdivided into two blocks for the differential Ti supplies, as follows: Block 1: without Ti, reference treatment. Block 2: one leaf spray treatment using 35 ml/plant of a 2 mg Ti/l solution [TITAVIT, Ti(IV)-ascorbate complex, 1 g Ti/l, granted by Nitrokemia, Budapest Hungary], 45 days after transplanting. In this manner, they were obtained 6 different treatments, as follows:

1. P-3: 100% P (general control, C1)
2. P-3+Ti: 100% P+Ti
3. P-2: 66.6% P (C2)
4. P-2+Ti: 66.6% P+Ti
5. P-1: 33.3% P (C3)
6. P-1: 33.3% P+Ti.

All the other crop conditions (fertilization, water doses, crop practices, phytosanitary treatments, etc.) were identical for all the experimental plants. Data of plant biomass were obtained from a sampling of ten complete plants (each plant an individual sample) performed 105 days after transplanting, and the fruit harvest was carried out one month later.

The efficiency (E) of a treatment is the ratio between the produced effect and the intensity of the treatment, and the efficiency index (EI) is the quotient of the efficiencies of two treatments.

Analytical

Plant sampling was performed (ten complete plants/treatment) 105 days after transplanting. In the laboratory, each plant was separed into roots, stems, leaves and fruits, and each one of these plant portions constituted an individual sample, which were mineralized in a mufle oven at 450 °C during 3 hours (1 g of dry powder), and then diluted with 0.6N HNO_3 upon to 25 ml. The ammonia-phospho-molybdo-vanadate complex from a suitable volume of this extract was spectrophomotrically read against a blank without sample at 460 nm.

Results

Table 1 shows the average values of the biomass production. The diminution of P-fertilization significantly affects biomass production, and this effect is mainly due to the differences in the leaf and fruit biomass (treatments 1, 3 and 5). When the plants were supplied with titanium the biomass of all the studied plant portions increases against their corresponding control treatments (2 vs 1; 4 vs 3; 6 vs 5), but it is interesting to remark that the plants growing under the lowest P supply plus titanium (treatment 6) present a biomass production higher than the corresponding plants grown under a complete P-feed (treatment 1).

The data of the Table 2 confirm the above results. Thus, the efficiency of P-fertilization on biomass production is inversely related to the intensity of the supply. On leaf biomass this effect is lower than on the other plant portions. The Ti-leaf treatment improves the values of these indexes in stems, fruits and whole plants, but scarce or negative effect is shown on the biomass production of roots and leaves.

Phosphorus concentrations in roots and leaves (Table 3) are not affected by the decrease of P supply in the fertilization (treatments 1, 3 and 5). The stem-P only decrease in the lowest fertiliser case and the levels in fruits are inverse related to the P-fertilizer intensity. These data indicate that the diminution of the P-feed intensity did not induce any P-deficiency in the plants and they permit to deduce that there is a luxury consumption of this element in the fertilizer practices on this crop. The presence of titanium increases the leaf P-concentrations, but not in the other plant organs.

According to the data on biomass and P-concentration, the efficiency of phosphorous fertilization on the real consumption of the element by the plant (Table 4) is very low in the general control plot [P3(C1)] and the efficiency index of this treatment is 3.72 times lower than that of P1+Ti, which not did

Table 1. Effect of Ti(IV) leaf spray on the biomass production in P-stressed *C. annuum* L., cv. Bunejo plants. Data are the average values of ten samples and they are expressed in g/plant

Treatment	Roots	Stems	Leaves	Fruits	Whole plant
1: 100%P(C1)	6.5 ± 0.6b	19.2 ± 1.7b	34.00 ± 2.1c	72.0 ± 12.0b	131.6 ± 13.9b
2: 100%P+Ti	7.4 ± 1.1c	22.4 ± 2.0c	31.9 ± 3.8c	104.1 ± 10.4c	165.1 ± 11.7cd
3: 66.6%P(C2)	5.8 ± 3.4a	13.8 ± 1.5ab	22.4 ± 3.2a	59.3 ± 11.3a	104.3 ± 12.7a
4: 66.6%P+Ti	7.6 ± 0.8c	30.8 ± 6.5d	32.8 ± 2.0c	60.0 ± 12.5c	167.1 ± 12.6d
5: 33.3%P(C3)	6.4 ± 0.8ab	15.0 ± 0.3a	28.2 ± 3.1b	57.8 ± 8.0a	107.3 ± 10.7a
6: 33.3%P+Ti	6.1 ± 0.3ab	19.2 ± 2.4b	33.4 ± 4.69c	95.5 ± 12.7c	154.0 ± 17.2c

For each column, means followed by a same letter are not significantly different at $p = 0.05$ by LSD test.

Table 2. Effect of Ti(IV) leaf spray on the efficiency of phosphorus fertilization on the biomass production in P-stressed *C. annuum* L., cv. Bunejo plants

Plant portion		1:P3(C1)	2:P3+Ti	3:P2(C2)	4:P2+Ti	5:P1(C3)	6:P1+Ti
Roots:	E	0.027	0.031	0.036	0.047	0.079	0.077
	EI(Ci)	1.00	1.15	1.00	1.31	1.00	0.97
	EI(C1)	1.00	1.15	1.33	1.74	2.93	2.85
Stems:	E	0.080	0.093	0.105	0.194	0.187	0.240
	EI(Ci)	1.00	1.16	1.00	1.85	1.00	1.28
	EI(C1)	1.00	1.16	1.31	2.43	2.34	3.00
Leaves:	E	0.142	0.133	0.140	0.205	0.352	0.417
	EI(Ci)	1.00	0.94	1.00	1.46	1.00	1.28
	EI(C1)	1.00	0.94	0.99	1.44	2.48	2.94
Fruits:	E	0.300	0.434	0.450	0.600	0.723	0.741
	EI(Ci)	1.00	1.45	1.00	1.33	1.00	1.02
	EI(C1)	1.00	1.45	1.50	2.00	2.41	2.47
Whole plant:	E	0.549	0.688	0.652	1.044	1.342	1.925
	EI(Ci)	1.00	1.25	1.00	1.60	1.00	1.43
	EI(C1)	1.00	1.25	1.19	1.90	2.44	3.51

E: Efficiency; EI: Efficiency Index related to each control treatment (Ci) or to the general control treatment (C1).

Table 3. Effect of Ti(IV) leaf spray on the phosphorus concentration in P-stressed *C. annuum* L., cv. Bunejo plants. Data are the average values of ten samples and they are expressed in g/kg dm

Treatment	Roots mean ± s	Stems mean ± s	Leaves mean ± s	Fruits mean ± s
1: 100%P(C1)	2.38 ± 0.26ab	2.31 ± 0.80bcd	2.67 ± 0.47a	7.07 ± 0.15bc
2: 100%P+Ti	2.64 ± 0.28bc	2.37 ± 0.24cd	3.79 ± 0.95b	6.64 ± 0.16a
3: 66.6%P(C2)	2.31 ± 0.21a	2.19 ± 0.21bc	2.55 ± 0.49a	7.25 ± 0.26cd
4: 66.6%P+Ti	2.71 ± 0.34c	2.46 ± 0.25d	3.48 ± 0.81b	6.94 ± 0.43abc
5: 33.3%P(C3)	2.31 ± 0.23a	1.65 ± 0.21a	2.41 ± 0.58a	7.31 ± 0.15d
6: 33.3%P+Ti	2.29 ± 0.19a	1.95 ± 0.28b	3.02 ± 0.80ab	6.89 ± 0.30b

For each column, means followed by a same letter are not significantly different at $p = 0.05$ by LSD test.

Table 4. Effect of Ti(IV) leaf spray on the efficiency of phosphorus fertilization on the phosphorus consumption in P-stressed *C. annuum* L., cv. Bunejo plants

Plant portion		1:P3(C1)	2:P3+Ti	3:P2(C2)	4:P2+Ti	5:P1(C3)	6:P1+Ti
Roots:	E	0.064	0.081	0.084	0.128	0.181	0.168
	EI(Ci)	1.00	1.27	1.00	1.52	1.00	0.93
	EI(C1)	1.00	1.27	1.31	2.00	2.83	2.63
Stems:	E	0.185	0.221	0.231	0.468	0.309	0.470
	EI(Ci)	1.00	1.19	1.00	2.03	1.00	1.52
	EI(C1)	1.00	1.19	1.25	2.53	1.67	2.54
Leaves:	E	0.377	0.512	0.362	0.718	0.848	1.264
	EI(Ci)	1.00	1.36	1.00	1.98	1.00	2.25
	EI(C1)	1.00	1.36	0.96	1.90	1.49	3.25
Fruits:	E	2.120	0.434	0.450	0.600	0.723	0.741
	EI(Ci)	1.00	1.45	1.00	1.33	1.00	1.02
	EI(C1)	1.00	1.45	1.50	2.00	2.41	2.47
Whole plant:	E	0.549	0.688	0.652	1.044	1.342	1.925
	EI(Ci)	1.00	1.25	1.00	1.60	1.00	1.43
	EI(C1)	1.00	1.25	1.19	1.90	2.44	3.51

E: Efficiency; EI: Efficiency Index related to each control treatment (Ci) or to the general control treatment (C1).

Table 5. Effect of Ti(IV) leaf spray on the fruit yield in P-stressed *C. annuum* L., cv. Bunejo plants

Fruit yield	1:P3(C1) mean ±s	2:P3+Ti mean ± s	3:P2(C2) mean ± s	4:P2+Ti mean ± s	5:P1(C3) mean ± s	6:P1+Ti mean ± s
g(dm)/plant	75.5 ± 7.6b	109.3 ± 8.0d	63.2 ± 8.5a	107.5 ± 10.4d	59.2 ± 8.5a	95.8 ± 8.8c
kg(fw)/Ha*	25,846	36,630	20,677	34,864	18,569	31,323

Means followed by a same letter are not significantly differents at p: 0.05 by LSD test. *Calculated on the basis of a plant density of 70,000 plants/ha.

receive P-fertilization during the crop cycle but supplied with titanium *via* leaf spray.

The beneficial effect of titanium on phosphorus utilization is clear in all the aerial portions of the plant, but it is interesting to remark that the highest efficiency indexes were obtained in the fruit.

The diminution of the P-fertilizer below to 66.6% (P1) produces a fruit yield decrease of 22%, and in the P2 treatment (2P/3 against the general control) the yield only decreased in a 16% (Table 5). All the Ti-treated subplots presented a fruit yield higher than their corresponding control references, and the highest absolute yields were offered by the P3+Ti and P2+Ti treatments (36630 and 34864 kg/ha respectively), without significative differences between them.

The efficiency of phosphorus fertilization on the fruit yield (Table 6) confirms all the exposed data. By means the addition of Ti(IV) and the diminution of P-fertilization it is possible to enhance P-efficiency by 3.72 times.

Discussion

The diminution of the P-fertilization does not promote important decrease in the whole plant biomass production. On the other hand, the data of P-concentration in the several plant portions does not show any significative differences related to the fertilizer levels. Because of this, it is possible to deduce an excessive P supply in the fertilizer doses of the general control Plot-3 and justify the great enhancement of the efficiency indexes when the P-fertilization is diminished. When Ti is supplied to the plants, the biomass production, the P-consumption, and the corresponding efficiency indexes are increased, but not the plant phosphorus con-

Table 6. Effect of Ti(IV) leaf spray on the efficiency of phosphorus fertilization on the fruit yield in P-stressed *C. annuum* L., cv. Bunejo plants

	1:P3(C1)	2:P3+Ti	3:P2(C2)	4:P2+Ti	5:P1(C3)	6:P1+Ti
E	0.314	0.455	0.395	0.672	0.740	1.198
EI(Ci)	1.00	1.45	1.00	1.70	1.00	1.58
EI(C1)	1.00	1.45	1.26	2.14	2.36	3.72

E: Efficiency; EI: Efficiency Index related to each control treatment (Ci) or to the general control treatment (C1).

centration. These results are in agreement with those communicated for the N-nutrition in experiments with N-stressed *C. annuum* L. crops (Alcaraz *et al.*, 1991; Martínez-Sánchez *et al.*, 1992), and indicate that titanium promotes the general nutrient absorption in an unspecific effect probably due to the enhancement of the activity of some micronutrients, as iron (Alcaraz *et al.*, 1994; Carvajal *et al.*, 1994b).

This higher efficiency in the absorption and use of phosphorus, must be directly related to the stimulation of the photosynthetic activity when titanium is added (Kiss *et al.*, 1985; Carvajal *et al.*, 1994 and 1994a; Alcaraz *et al.*, 1994), and in this way, it is remarkable that the plants grown under a 33.3% of phosphorus and supplied with titanium show a significative increase in their development against those of the reference without titanium, grown under 100% of phosphorus [treatment P-3(C1)].

Taking in mind this assumption, the *C. annuum* L. plants can offer a normal development with a phosphorus supply diminution of about 66.6%, if the plants receive an additional Ti(IV) leaf spray treatment, because this non-essential trace element induces phosphorus uptake and its utilization by the plants.

In the same way that the biomass production, the fruit yield was strongly increased in the Ti-sprayed plants. These results, and previous data on the improvement of fruit quality induced by titanium (Martínez-Sánchez *et al.*, 1993), show themselves as one way to reduce P fertilizer rates in the exploitation of *C. annuum* L. crops with the added advantage of a more easy control of the soil and water contamination risks.

Acknowledgements

The authors gratefully acknowledge the financial supports provided by CICYT (Spain), project AGF 93-0141 and Consejería de Agricultura de la Comunidad Autónoma de la Región de Murcia, as well as to NITROKEMIA, Budapest (Hungary) for providing TITAVIT used in the experiments.

References

Alcaraz CF, Giménez JL, Moreno A, Fuentes JL and Martínez-Sánchez F (1991) El titanio como fertilizante foliar en horticultura. Efectos sobre producción y calidad en plantas de *Capsicum annuum* L. cultivadas en fertirrigación. In: Cadahía C (ed) Proc. II congr. Nal. Fertirrigación, pp 193–198. FIAPA, Almería

Alcaraz CF, Carvajal M, Frutos MJ, Giménez JL, Martínez-Sánchez F and Pastor JJ (1994) The physiological role of titanium in *Capsicum annuum* L. plants. Proc. 6th Int. Trace Element Symp. Abs. 2. Budapest

Bienfait HF and Van der Mark F (1983) Phytoferritin and its role in iron metabolism. In: Robb DA and Pierpoint WS (eds) Metals and Micronutrients. Uptake and Utilization by Plants, pp 111–123. Academic Press, London

Carvajal M, Pastor JJ, Martínez-Sánchez F and Alcaraz CF (1993) Leaf spray with Ti(IV) ascorbate improves the iron uptake and iron activity in *Capsicum annuum* L. plants. In: Iron Nutrition in Soils and Plants. Kluwer Academic Publishers, Amsterdam: 1–5

Carvajal M, Martínez-Sánchez F and Alcaraz CF (1994) Effect of Titanium(IV) application on some enzymatic activities in several developing stages of red pepper plants. J Plant Nutr 17: 243–253

Carvajal M, Martínez-Sánchez F and Alcaraz CF (1994a) Effect of Ti(IV) on some indicators of physiological activity in *Capsicum annuum* L. J Hort Sci 69: 427–432

Dumon JC and Ernst WHO (1988) Titanium in plants. J Plant Physiol 133: 203–209

Feher M, Papp K, Fodor P and Pais I (1987) The effect of titanium on the uptake of other nutritive elements. In: Pais I (ed) New Results in the Research of Hardly Known Trace Elements and the Analytical Problem of Trace Element Research, pp 69–100. University of Horticulture, Budapest

Giménez JL, Martínez-Sánchez F, Moreno JL, Fuentes JL and Alcaraz CF (1990) Titanium in plant nutrition. III. Effect of Ti(IV) on yield of Capsicum annuum L. In: Barceló J (ed) Nutrición Mineral bajo Condiciones de Estrés, pp 123–128. SPIC-UIB, Palma de Mallorca

Kiekens L and Camerlynck R (1987) Influence of titavit on growth of maize and cowpea grown in nutrient solution. In: Pais I (ed) New Results in the Research of Hardly Known Trace Elements and the Analytical Problems of Trace Element Research, pp 133–147. University of Horticulture, Budapest

Kiss F, Deak G, Feher M, Balogh L, Szabolsci L and Pais I (1985) The effect of titanium and gallium on photosynthetic rate of algae. J Plant Nutr 8: 825–832

Martínez-Sánchez F, Giménez JL, Carvajal M, Moreno A, Fuentes JL and Alcaraz CF (1992) Efectos de tratamientos foliares con Ti(IV) sobre nutrición, producción y calidad de fruto en plantas de Capsicum annuum L. Suelo y Planta 2: 101–111

Martínez-Sánchez F, Núñez M, Amorós A, Giménez JL and Alcaraz CF (1993) Effects of titanium leaf spray treatments on ascorbic acid levels of *Capsicum annuum* L. fruits. J Plant Nutr 16: 975–981

Pais I (1983) The biological importance of titanium. J Plant Nutr 6: 3–131

Pais I, Feher M, Farkas E, Szabo Z and Cornides I (1977) Titanium as a new trace element. Comm Soil Sci Plant Anal 8: 407–410

Ram N, Verloo M and Cottenie A (1983) Response of bean to foliar spray of titanium. Plant & Soil 73: 285–290

C. Rodriguez-Barrueco (ed.), Fertilizers and Environment, 205–210.

Improvement of the N fertilizer efficiency with dicyandiamide (dcd) in citrus trees

M.D. Serna, F. Legaz & E. Primo-Millo
Instituto Valenciano de Investigaciones Agrarias, Apartado Oficial, 46113-Moncada, Valencia, Spain

Key words: dicyandiamide, nitrification inhibitor, nitrogen nutrition, citrus, yield

Abstract

Nitrification inhibitors such a dicyandiamide (DCD) help to reduce leaching losses by retaining applied N in the ammoniacal form. Research objectives were to evaluate dicyandiamide added to ammonium sulphate-nitrate (ASN) as a nitrification inhibitor in cultivated soils (Xeropsamments) and its effect on N uptake by citrus (*Citrus sinensis* (L.) Osbeck). Under field conditions, fertilization of adult trees with ASN (600 g N tree^{-1}) either with or without DCD (2% DCD-N) was compared (ASN+DCD and ASN, respectively). The NH_4^+-N concentrations in plots fertilized with ASN+DCD were significantly higher than ASN plot in the 0–15 cm layer during 5–105 day period. Nitrification started immediately after N application in both treatments (ASN and ASN+DCD). In all three soil layers analyzed, $NO^-{}_3$-N concentrations were higher in the ASN plots than in the ASN+DCD during the first 20 days. This indicates that nitrification of $NH^+{}_4$ from ASN was more rapid in the absence of DCD. On the other hand, fertilization with ASN+DCD kept higher levels of $NO^-{}_3$-N in soils than ASN during the rest of experience period (40–160 days). Addition of DCD to ASN showed a higher N concentration in the spring-flush leaves with respect to the trees fertilized with ASN, during the growth cycle. These results suggest that the use of a nitrification inhibitor permitted a more efficient utilization of fertilizer N by citrus trees. The plants treated with DCD added to ASN showed a higher yield in number of units per tree and a better fruit colour index than those treated with ASN alone.

Introduction

Most of the fertilizer nitrogen applied to soils used for citrus production, is in the form of NH_4^+ or NH_4^+-producing compounds. Applied NH_4^+ is usually oxidized quite rapidly to NO_3^- by the nitrifying microorganisms in soils (Bronson *et al.*, 1989; McCarty and Bremner, 1989). Since ammonium-N is less subject to loss, delaying the conversion of ammonium to nitrate through the use of nitrification inhibitors may reduce N losses by NO_3^- leaching and by denitrification (Reidar and Michaud, 1980). These losses account for the low efficiency with which fertilizer N is generally used by this crop (Kubota *et al.*, 1976; Feigenbaum *et al.*, 1987). In addition, nitrate-N lost through leaching is a potential groundwater pollutant (Embleton *et al.*, 1981).

Dicyandiamide (cyanoguanidine or DCD), $H_2NC(NH)NHCN$, is a dimer of cyanamide, and is an effective nitrification inhibitor (Hauck, 1980; Nommik, 1958; Reddy, 1964; Rodgers and Ashworth, 1982). Dyciandiamide contains about 67% N and is non-volatile, nonhygroscopic, relatively water soluble (23 g l^{-1} at 13 °C), and chemically and physically stable (Prasad *et al.*, 1971; Reidar and Michaud, 1980). These properties allow DCD to be effectively formulated with a wide variety of N fertilizers, including ammonium salts and urea.

Some studies have shown that addition of DCD to urea or ammonium salts increases yield and/or N uptake by crops such as rice (Norman *et al.*, 1989; Wells *et al.*, 1989), wheat (Rodgers and Ashworth, 1982; Rodgers *et al.*, 1985), and sorghum (Touchton and Reeves, 1985). In general, DCD was effective in delaying nitrification, although this was not often reflected in positive yield responses (Frye *et al.*, 1989; Malzer *et al.*, 1989). DCD completely decomposes in soil to form CO_2 and NH_4^+ as final products (Amberger and Vilsmeier, 1979; Vilsmeier, 1980). This process occurs at a faster rate at elevated soil temperatures

Table 1. N-NH_4^+ concentration at different soil depths

Depth (cm)	Days after application	Control (mg N-NH_4^+/kg soil)	ASN+DCD	ASN
	5	5.8 a	115.2 c	69.5 b
	20	5.4 a	74.5 c	22.6 b
	40	3.0 a	41.7 b	9.3 a
0–15	60	7.2 a	25.5 b	8.8 a
	85	7.4 a	16.7 b	8.6 a
	105	3.0 a	13.8 b	1.6 a
	160	4.9 a	9.8 a	2.8 a
	5	4.6 a	36.4 c	17.3 b
	20	7.4 a	19.4 b	10.5 a
	40	4.2 a	13.3 b	6.9 a
15–30	60	7.3 a	9.7 a	5.8 a
	85	7.7 a	8.4 a	3.2 a
	105	3.3 a	4.1 a	4.6 a
	160	4.0 a	1.5 a	1.9 a
	5	5.2 a	15.2 b	10.2 ab
	20	3.7 a	10.0 b	6.4 ab
	40	2.4 a	7.8 a	4.9 a
30–45	60	6.3 a	6.4 a	3.7 a
	85	5.1 a	4.9 a	3.6 a
	105	2.2 a	2.5 a	1.2 a
	160	3.0 a	4.0 a	3.2 a

The values are the means of three replications. Within a row, means followed by the same letter are not significantly different at the 5% level by Duncan's multiple range test. ASN = ammonium suphate nitrate; DCD = dicyandiamide.

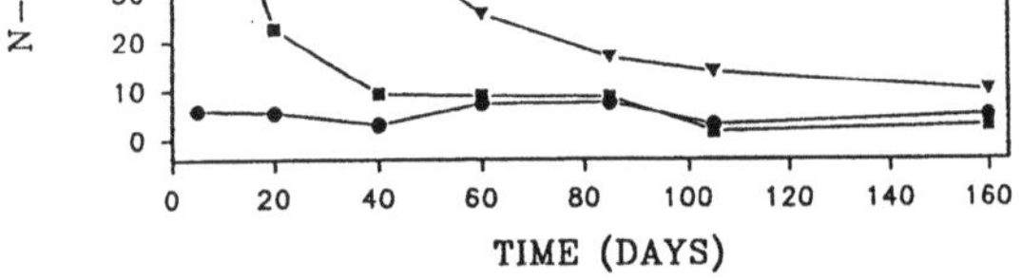

Fig. 1. Concentrations of NH_4^+-N in the upper soil layer (0–15 cm) as a function of time after fertilizer application for the three treatments. ASN = ammonium sulphate nitrate, DCD = dicyandiamide. The values are the means of three replications.

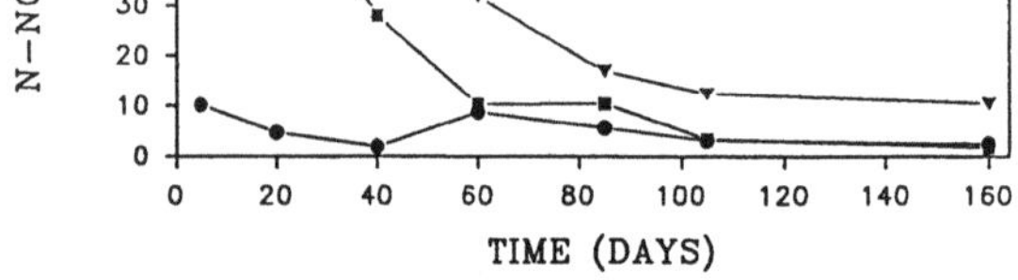

Fig. 2. Concentrations of NO_3^--N in the upper soil layer (0–15 cm) as a function of time after fertilizer application for the three treatments. ASN = ammonium sulphate nitrate, DCD = dicyandiamide. The values are the means of three replications.

Table 2. N-NO_3^- concentration at different soil depths

Depth (cm)	Days after application	Control	ASN+DCD	ASN
		(mg N-NO_3^-/kg soil)		
	5	10.2 a	41.1 b	64.4 c
	20	4.8 a	44.1 b	54.5 c
	40	2.0 a	48.0 c	18.0 b
0–15	60	8.8 a	31.8 c	10.4 b
	85	5.7 a	17.0 b	10.5 ab
	105	3.2 a	12.5 b	3.5 a
	160	2.6 a	10.5 b	1.8 a
	5	7.9 a	13.0 a	13.5 a
	20	6.6 a	22.6 b	19.9 b
	40	5.9 a	25.2 b	11.3 a
15–30	60	4.0 a	17.6 b	10.7 a
	85	5.3 a	11.8 b	4.6 a
	105	2.9 a	3.5 a	2.4 a
	160	3.1 a	1.9 a	1.8 a
	5	5.8 a	6.3 a	11.8 a
	20	1.1 a	13.5 b	14.3 b
	40	1.2 a	16.4 b	4.3 a
30–45	60	4.5 a	18.5 b	8.5 a
	85	8.9 a	8.6 a	6.8 a
	105	1.7 a	1.8 a	3.9 a
	160	2.3 a	3.5 a	3.8 a

The values are the means of three replications. Within a row, means followed by the same letter are not significantly different at the 5% level by Duncan's multiple range test. ASN = ammonium suphate nitrate; DCD = dicyandiamide.

and in the presence of metal oxides (Amberger and Vilsmeier, 1979; Vilsmeier, 1980). Additionally, soil texture, pH, moisture, organic matter content and biological activity have been shown to have an influence on the relative longevity of the inhibitor (Bundy and Bremner, 1973; Keeney, 1980; Reddy, 1964; Slangen and Kerkhoff, 1984).

The purpose of this research was to evaluate the performance of the DCD added to ammonium sulphate nitrate (ASN) as a nitrification inhibitor in soils used for citrus production and its effect on N uptake by this crop.

Material and methods

Twenty-year old Navel orange grafted on Troyer citrange rootstock, planted at 6 × 6 m distances were grown under field conditions. The soil was a deep, well-drained Xeropsamments soil with a low water holding capacity (Soil Taxonomy, 1975). This soil had a sandy-loam texture (76.4% sand, 16.6% silt, and 7% clay), basic pH (7.8), and a low level of calcium carbonate (< 1%). Olsen extractable P and 1 N ammonium acetate extractable K levels (51 and 266 mg kg^{-1}, respectively) were in the range considered as optimum (Du Plessis, 1977). The trees were irrigated by flooding every 20 days during the growth period. Approximately 50 l m^{-2} of water were used in each irrigation. This field experiment compared ammonium sulphate-nitrate with or without DCD (2% DCD-N)[1] applied at a rate of 600 g N per tree. The fertilizer was uniformly spread by hand under the tree canopy and incorporated to a 0.05 m depth by tillage, just before irrigation. Fertilizer treatments were applied one a year at the beginning of June. Unfertilized blocks were included as controls. The field experiment consisted of three treatments replicated three times in a random-

Table 3. N concentration in spring-flush leaves from non-fruiting shoots sampled at different moments of the growth cycle

Treatments	24 June	22 July	2 Sept.	6 Nov.
Control	2.10 a	2.19 a	2.33 a	2.37 a
ASN+DCD	2.64 c	2.73 c	2.87 c	2.83 c
ASN	2.55 b	2.58 b	2.68 b	2.70 b

The values are the means of three replications. Within a column, means followed by the same letter are not significantly different at the 5% level by Duncan's multiple range test. ASN = ammonium suphate nitrate; DCD = dicyandiamide.

ized complete block design. Each block consisted of 6 experimental trees and 14 guard trees. Soil samples taken to depths of 0–15, 15–30, and 30–45 cm were collected at 5, 20, 40, 60, 85, 105 and 160 days after fertilization and analyzed for NH_4^+-N and NO_3^--N. Soil samples were air-dried and ground to pass through a 2 mm sieve before analysis. Forty-eight spring-flush leaves from non-fruiting shoots were sampled from the six experimental trees of each plot, 4 times a year (June, July, September and November). The leaves were oven-dried for 48 h at 65 °C, ground in a Wiley mill to pass through a 40 mesh screen and analyzed for total N (Bremner, 1965a).

Yield in number of fruits, size (fruit weight), juice content, rind thickness, total soluble solids (TSS) and titratable acidity (TA), at the time of harvest, were measured. Sixty fruits per treatment (10 fruits per tree) were randomly picked for quality measurements. The juice was extracted with uniform pressure in a mechanical extractor. Total soluble solids (TSS, °Brix) were determined with a refractometer and titratable acidity (TA) by titration with 0.1 *N* NaOH. Color index (CI) of fruits rind was calculated by measuring the "L", "a" and "b" parameters with a Hunter lab D-25 P-2 colorimeter (Jimenez-Cuesta *et al.*, 1981). Basammon-Stabil is a nitrification inhibitor product developed by Basf Aktiengesellschaft, D-700 Ludwigshafen, Germany. It is now marketed by Basf España, S.A., Barcelona, Spain.

Analytical procedures

Total N concentration in plants was determined by the semi-micro Kjeldahl method (Bremner, 1965a). NH_4^+ and (NO_3^-+NO_2^-) were determined by steam distillation with MgO and Devarda's alloy, respectively (Bremner, 1965b).The soil was extracted with 2 *N* KCl at a soil to extractant ratio of 1:5.

Results and discussion

Table 1 and Figure 1 show that most NH_4^+-N was retained in the upper soil layer at day 5 after N application, although some NH_4^+-N moved downward, presumably as a consequence of the low cation exchange capacity in the soil. Results in Table 1 indicate that following 20 days, NH_4^+-N levels in the 0–15 cm soil layer fertilized with ASN were not significantly different from those in the control plots. By day 40, nitrification of applied NH_4^+ in the ASN treatment was essentially complete with no significant quantities of NH_4^+-N detected above those measured in the control plots. However, the NH_4^+ level in the 0–15 cm soil layer was significantly higher in the ASN+DCD fertilized plots than in the ASN plots during the 5–105 day period.

Nitrification started immediately after N application in ASN and ASN+DCD treatments (Table 2 and Fig. 2). The maximum NO_3^--N concentration in soil was found at 20 days in the ASN treatment (Fig. 2), being coincident with a reduction in NH_4^+-N concentration. This indicates that nitrification of NH_4^+ from ASN was rapid in the absence of DCD. By contrast, fertilization with ASN+DCD maintained higher levels of NO_3^--N in soil than ASN alone, during the 40–85 day period in all three soil layers studied.

Addition of DCD to ASN resulted in a significant increase in leaf N concentration with respect to the plants fertilized with ASN alone (Table 3).

These results indicate that the use of a nitrification inhibitor enabled a more efficient utilization of fertilizer N by Citrus plants. This is achieved mainly by a regulation of N supply allowing a more continuous NO_3^- release in soil and by a reduction of N losses by NO_3^- leaching and denitrification. In this experiment, NO_3^- losses by leaching were, presumably, much more important than those produced by denitrification or volatilization. The coarse texture and low organic matter content of this soil represent favourable conditions for DCD to be effective (McCarty and Bremner, 1989; Reddy 1964). Several experiments (Bundy and Bremner, 1973; Cornforth and Chesney, 1971; Prakasa Rao and Puttana, 1986; Rodgers, 1983) have shown that maintenance of high ammonium concentrations in the soil with DCD increased ammonia volatilization losses. However, these losses were considerably

Table 4. Yield and fruit characteristics of Navel oranges grown under field conditions

	Control	ASN+DCD	ASN
Fruit number/tree	501 a	721 b	685 ab
Fruit weight (g)	251	265	242
Rind (% by weight)	43.9	51.4	47.1
Rind thickness (mm)	4.7	5.6	5.3
Juice (% by weight)	41.8	41.7	44.0
Total soluble solids, TSS (%)	11.7	11.4	12.0
Total acidity, TA (%)	1.1	1.1	1.2
Maturity index, MI (TSS/TA)	10.3	10.3	10.3
Pulp (% by weight)	13.2	6.6	7.5
Colour index, CI	11.8 a	14.8 b	14.4 b

The values are the means of three replications. Within a row, means followed by the same letter are not significantly different at the 5% level by Duncan's multiple range test. ASN = ammonium suphate nitrate; DCD = dicyandiamide.

smaller when the fertilizers were incorporated into the soil compared to surface applications (Prakasa Rao and Puttana, 1986). In our experiment N losses due to ammonia volatilization should be very low since all N was incorporated to a depth of 5 cm. Table 4 shows the results obtained about plant yield and fruit characteristics. From all the parameters studied, only number of fruits and colour index showed significant differences between treatments. In these cases, the control presented a lower significantly values than the ASN+DCD and ASN treatments, However, there were not significant differences between the ASN+DCD and ASN treatments, although the values were slightly higher in the first case.

From the findings of these experiments, the addition of the nitrification inhibitor DCD to ammonium containing N sources will increase N fertilizers utilization efficiency in citrus production.

References

Amberger A and Vilsmeier K (1979) Dicyandiamideabbau in Quarzsand und Boden. Z Pflanzenernaehr Bodenkd 142: 778–785

Bremner JM (1965a) Total nitrogen. In: Black CA (ed) Methods of Soil Analysis, Part 2, pp 1149–1178. Agron Monogr 9, ASA, Madison, WI

Bremner JM (1965b) Inorganic forms of nitrogen. In: Black CA *et al.* (eds) Methods of Soil Analysis, Part 2, pp 1179–1237. Agron Monogr 9, ASA, Madison, WI

Bronson KF, Touchton JT and Hauck RD (1989) Decomposition rate of dicyandiamide and nitrification inhibition. Commun Soil Sci Plant Anal 19–20: 2067–2078

Bundy LG and Bremner JM (1973) Inhibition effect of nitrification inhibitors on the fate and efficiency of nitrogenous fertilizers under simulated humid tropical conditions. Trop Agric 60: 211–217

Cornforth IS and Chesney HAD (1971) Nitrification inhibitors and ammonia volatilization. Plant Soil 34: 497–501

Du Plessis SF (1977) Soil analysis as a necessary complement to leaf analysis for fertilizer advisory purposes. Proc Int Soc Citriculture 1: 15–19A

Embleton TW, Pallares CO, Jones WW, Summers LL and Matsumura M (1981) Nitrogen fertilization management of vigorous lemons and nitrate-pollution potential of ground water. Contrib Univ Calif Water Resources Center N° 182, 29 pp

Feigenbaum S, Bielorai H, Erner Y and Dasberg S (1987) The fate of ^{15}N labeled nitrogen applied to mature citrus trees. Plant Soil 97: 179–187

Frye WW, Graetz DA, Locascio SJ, Reeves DW and Touchton JT (1989) Dicyandiamide as a nitrification inhibitor in crop production in the southeastern USA. Commun Soil Sci Plant Anal 20 (19–20): 1969–1999

Jimenez-Cuesta M, Cuquerella J and Martinez-Jávega JM (1981) Determination of a colour index for citrus fruit degreening. Proc Int Soc Citriculture 2: 750–753

Hauck RD (1980) Mode of action of nitrification inhibitors. In: Meisinger JJ *et al.* (eds) Nitrification Inhibitors-Potentials and Limitations, pp 19–32. Spec Publ 38, ASA, CSSA, and SSSA, Madison, WI

Keeney DR (1980) Factors affecting the persistence and bioactivity of nitrification inhibitors. In: Meisinger JJ *et al.* (eds) Nitrification Inhibitors Potentials and Limitations, pp 33–46. Spec Publ 38. ASA, CSSA, and SSSA Madison, WI

Kubota S, Kato T, Akao S and Bunya C (1976) ^{15}N absorption and translocation in Satsuma trees. III. The behaviour of nitrogen supplied in early spring. Bull Shikoku Agric Exp Stn 25: 55–66

Malzer GL, Kelling KA , Schmitt MA , Hoeft RG and Randall GW (1989) Performance of dicyandiamide in the north central states. Commun Soil Sci Plant Anal 20 (19–20): 2001–2022

McCarty GW and Bremner JM (1989) Laboratory evaluation of dicyandiamide as a soil nitrification inhibitor. Commun Soil Sci Plant Anal 20 (19–20): 2049–2065

Nommik H (1958) Calcium cyanamide and dicyandiamide as sources of nitrogen for higher plants. Acta Agric Scand 9: 435–447

Norman RJ, Wells BR and Moldenhauer KAK (1989) Effect of application method and dicyandiamide on urea-nitrogen-15 recovery in rice. Soil Sci Soc Am J 53: 1269–1274

Prakasa Rao EVS and Puttana K (1986) Nitrification and ammonia volatilization losses from urea and dicyandiamide-treated urea in a sandy loam soil. Plant Soil 97: 201–206

Prasad R, Rajale GB and Lakhdive BA (1971) Nitrification retarders and slow-release nitrogen fertilizers. Adv Agron 23: 337–383

Reddy GR (1964a) Effect of mixing varying quantities of dicyandiamide with ammonium fertilizers and nitrification of ammonia in soils. Can J Soil Sci 44: 254–259

Reidar G and Michaud H (1980) Improving fertilizer efficiency. The use of a dicyandiamide nitrification inhibitor. Nitrogen 124: 31–35

Rodgers GA (1983) Effect of dicyandiamide on ammonia volatilization from urea in soil. Fert Res 4: 361–367

Rodgers GA and Ashworth J (1982) Use of nitrification inhibitors to improve recovery of mineralized by winter wheat. J Sci Food Agric 33: 1219–1226

Rodgers GA, Penny A and Hewitt MV (1985) Effect of nitrification inhibitors on uptake of mineralized nitrogen and on yields of winter cereals grown on sandy soil after ploughing old grassland. J Sci Food Agric 36: 915–924

Soil Taxonomy (1975) A basic system of soil classification for making and interpreting soil surveys. Agricultural Handbook N° 436. Soil Conservation Service, USDA

Slangen JHG and Kerkhoff P (1984) Nitrification inhibitors in agriculture and horticulture: A literature review. Fert Res 5: 1–76

Touchton JT and Reeves DW (1985) Effect of nitrification inhibitors on yield of planted and ratooned grain sorghum grown with conservation tillage. J Fert Issues 2: 32–37

Vilsmeier K (1980) Dicyandiamidabbau im Boden in Abhangigkeit von der Temperatur. Z Pflanzenernaehr Bodenkd 143: 113–118

Wells BR, Bollich PK, Ebelhar W, Mikkelsen DS, Norman RJ, Brandon DN, Helms RS, Turner FT and Westcott MP (1989) Dicyandiamide (DCD) as a nitrification inhibitor for rice culture in the United States. Commun Soil Sci Plant Anal 20 (19–20): 2023–2047

C. Rodriguez-Barrueco (ed.), Fertilizers and Environment, 211–216.

Effect of method of N-application and modified urea on N-15 recovery by rice

Soliman M.Soliman[1] & Mohemed A.S.Abdel Monem[2]
[1]*Soil and Water Dept. Atomic Energy Auth., P.O. Box 13759, Cairo, Egypt;* [2]*International Center for Agriculture Res. in Dry Areas (ICARDA), P.O.Box, 2416, Cairo, Egypt*

Key words: method of nitrogen application, modified urea, N-15 recovery, rice

Abstract

Rice is a very responsive crop to nitrogen, but the efficiency of the N-fertilizer is low. Greenhouse experiment has been conducted to evaluate several methods to improve fertilizer efficiency and reduce N-losses in rice fields. N-15 labelled urea was applied to 10 kg soils in pots, urea was applied alone, addition of two urease inhibitors N-(n-butyle) thiosphosphoric triamide (NBPT) and hydroquinone (HQ), with addition of nitrification inhibitor Dicyandimide (DCD), or with the combination of both inhibitors. The fertilizer was applied either broadcast on soil surface or at depth of 8 cm below the surface. At maturity, plants were separated into grain and straw, dried and weighted. Soil and plant samples were analyzed for total N and N-15 excess. Both fertilizer placement and inhibitor application significantly increased straw and grain yield, as well as N- uptake. Nitrogen derived from fertilizer (% Ndff) was more than doubled, when urea was applied deep and in combination with inhibitors. Total plant recovery of N-15 labelled urea ranged from 17% to 75% according to treatment. Regardless of inhibitors application, plant recovery was increased from 39% to 65% when urea was applied at depth of 8 cm. Approximately, 2/3 of the applied urea (64%) was lost, when urea was applied alone. Those losses were reduced down to 12% with deep placement and inhibitor application. The two management practices show significant effect on minimizing N-losses and increasing plant recovery.

Introduction

Approximately 40% of the world's population depend on rice (*Oriza sativa* L.) as their major caloric source (De Datta, 1981). Rice is unique in that, it is the only major food crop that is semiaquatic, that grows best in a flooded soil. This preferred habitat is a major source of difficulty in the maintenance of nitrogen added as fertilizer. While losses from N-15 labelled urea applied to upland wheat were 18% (Abdel Monem, 1986) and 17% (Abdel Monem and Ryan, 1991). Fillery *et al.* (1984) showed losses of 40 to 60% of N-15 labelled urea or ammonium sulfate broadcast to floodwater of rice. In Egypt, Abdel Monem *et al.* (1993) reported losses of 29 to 62% of N-15 labelled urea applied to rice, according to time of fertilization. They reported that the percentage of nitrogen losses is related to climatic factors and fertilizer management practices.

Several strategies exist for improving N fertilizer efficiency and reducing losses. Deep placement of applied fertilizer is the closest to farm-level acceptance. While surface placement resulted in only 23% as N-15 recovery, 63% recovery was obtained as a result to deep placement (Murayama, 1979). Khind and Datta (1975) reported increase in N-15 recovery from 50 to 66% due to deep placement as compared with surface application.

Ammonia volatilization has been highlighted as an important nitrogen losses mechanism, due to urea hydrolyses in the floodwater.Inhibiting urease enzyme, which hydrolyses urea to ammonium was another method to reduce losses. Fillery *et al.* (1986) reported the importance of phenyl phosphorodiamidate (PPD) as urease inhibitor, as it increased total N-15 recovery between 10 and 14% of the N-15 applied. A large amount of N-applied can be transformed to NO_3-N during 4–5 weeks. Brandon and Wells (1986) reported that

Table 1. Dry matter yield of rice as affected by fertilization, N-fertilizer placement and inhibitors

Treatment	Straw			Grain		
	Surface	Deep	Aver.	Surface	Deep	Aver.
	gram/pot					
Urea	28.0	30.7	29.4	27.0	34.2	30.6
Urea+N NBPT	32.3	39.5	35.9	34.6	38.0	36.3
Urea+H HQ	30.3	35.8	33.1	34.2	38.6	36.4
Urea+D DCD	33.8	44.4	39.1	34.9	41.1	38.0
Urea+NBPT+DCD	30.1	41.0	35.6	29.7	39.3	34.5
Urea+HQ+DCD	29.8	39.0	34.4	31.1	37.7	34.4
Average	30.7	38.4	34.6	31.9	38.2	35.1
LSD 5%						
N-Fertilizer	2.2	5.3				
Placement	5.4	5.9				
Fert. x placement	1.3	4.3				

application of nitrification inhibitors would increase N-recovery and reduce losses. Wilson *et al.* (1990), reported an increase of 17% in N-recovered by rice due to application of Dicyandiamid (DCD) as nitrification inhibitor to urea.

Material and methods

Greenhouse experiment was conducted in the Atomic Energy Authority and designed as randomized complete block with three replicates in plastic pots (25 cm diameter, 30 cm height) each containing 10 kg of air dry soil. This soil was collected from Shebin El-Qanater, Kaliobia Governorate at a depth of 15 cm from the surface. The soil had the following properties: pH (1:2.5), 8.1; Ec(1:5), 1.01 ms/ cm at 25 °C; clay content 58%, O.M. 0.95%, total N, 0.0125%. Labelled urea (3 atom % N-15 excess) was applied at rate of 500 mg/pot alone or in combination with DCD as a nitrification inhibitor; hydroquinone (HQ), and NBPT as urease inhibitors in solution as applied either at soil surface or at a depth of 8 cm below the soil surface. The pots were transplanted with three 21-day-old seedlings of rice, and kept under continuous flooding through the growth period. Each pot received a basal application of P and K in the form of superphosphate and potassium sulphate, respectively. At harvest, the plant parts were separated into grain and straw and the oven dry weight of each part was recorded. Both plant and soil samples were analyzed for total N by the Kjeldahl method (Bremner and Mulvaney, 1982) and ^{15}N was measured by emission spectrometry. Percentages of nitrogen derived from both fertilizers (Ndff) and soil (Ndfs) were calculated according to (Yamamuro, 1981).

Results and discussion

Straw and grain yield

Irrespective of the inhibitor, deep application of urea resulted in significant increase in straw and grain yield (Table 1). As an average, about 25% increase in straw yield and 20% in grain yield are due to deep placement, as compared with surface application. Flinn *et al.* (1984) showed that top dressing of urea raised the pH levels of the water to values at which NH_3 volatilization could have occurred, and in turn would have negative effect on the dry matter yield. Craswell and Vlek (1979) found that placing fertilizer deep in the soil greatly reduces losses and increases grain yield.

Fillery *et al.* (1986) reported increase in grain yield, as a result of application of PPD as urease inhibitor with urea. Our data presented in Table 1 show significant increase in straw and grain yield when NBPT or HQ was applied with urea. Contrary to the results

Table 2. Effect of fertilizer placement and inhibitors on total N-uptake by rice

N Placement	Urea	Urea+ NBPT	Urea+ HQ	Urea+ DCD	Urea+ NBPT+DCD	Urea+ HQ+DCD	Aver.
Surface	415	661	604	584	693	666	604
Deep	536	763	718	717	792	768	716

Table 3. Nitrogen derived from fertilizer (Ndff) and nitrogen derived from soil (Ndfs) by rice as affected by fertilizer, placement and inhibitors

Treatment	Ndff (%)			Ndfs (%)		
	Surface	Deep	Aver.	Surface	Deep	Aver.
Urea	20.3	30.8	26.2	79.7	69.2	73.7
Urea+NBPT	34.0	42.0	38.0	66.0	58.0	62.0
Urea+HQ	32.6	38.5	35.5	67.4	61.5	64.5
Urea+DCD	27.9	36.0	31.9	72.1	64.0	68.1
Urea+NBP+DCD	37.0	47.0	42.0	63.0	53.0	58.0
Urea+HQ+DCD	35.1	46.9	41.0	64.9	53.1	59.0
Average	31.2	40.2		66.9	59.8	

obtained by Abdel Monem (1991) which showed that NBPT has no significant effect on dry matter yield, when applied with labelled urea to barley. Our data (Table 1) revealed that addition of NBPT resulted in an average increase of 22 and 19% in straw and grain yield respectively. Inhibition of urease enzyme, and in turn, response as dry matter increase, are more effective in flooded system than in upland crops, due to higher urea hydrolyses in floodwater than in dry land. The results in Table 1 suggest that HQ is less effective as urease inhibitor as reflected on straw and grain yield. Mulvaney and Bremner (1978) explained that inhibition of urease activity by HQ was reversely related to clay, silt content and also to CEC of the soil in that respect that HQ will be less effective in our soil (58% clay). Studies of Wells *et al.* (1989) indicated that use of DCD delayed nitrification and tended to result in rice grain yield increase as compared with urea alone . Our results presented in Table 1 show significant increase in both straw and grain yield due to DCD application. Brandon and Wells (1986) suggested that, with a pre-plant N application, a large amount of N can be transformed to NO_3 during four to five weeks interval prior to flooding, and NO_3 subject to losses. Amberger (1989), explained that, DCD interferes with the respiration of the Nitrosomonas bacteria, which are responsible for the first step in nitrification.

N-uptake by rice

Nitrogen uptake by rice plant as a function of fertilizer placement and inhibitors application is presented in Table 2. N-uptake was higher when N-fertilizer was applied at depth of 8 cm as compared with surface application. An average of 18.5% increase in N-uptake due to deep placement is shown in the data of Table 2. N-uptake was higher when urease inhibitors were included with urea. Application of NBPT resulted in higher N-uptake than HQ. Similar to findings by Wilson *et al.* (1990), our data (Table 2) show that application of DCD resulted in higher N-uptake by rice. Possible explanation of increasing N-uptake by rice due to application of DCD is that more N accumulated in the soil organic fraction in the presence of DCD. This accumulation may be related to DCD maintaining more fertilizer N in NH_4 form, which is more readily utilized by microorganisms (Wilson *et al.*, 1990).

Table 4. N-15 balance for labelled urea applied to rice as affected by placement and inhibitors

N-balance							
Treatment	Urea	Urea + NBPT	Urea + HQ	Urea + DCD	Urea + NBPT + DC	Urea + NQ + DCD	Aver.
Surface placement							
Recovery (%)							
Plant	16.8	44.9	39.4	32.6	51.3	46.8	38.6
Soil	29.5	28.1	26.1	26.0	26.3	23.9	26.7
Total	36.3	73.0	65.5	58.5	77.6	70.7	63.6
Losses	63.7	27.0	34.5	41.4	22.4	29.3	33.1
Deep placement							
Recovery (%)							
Plant	33.0	64.6	66.3	51.6	74.7	72.0	60.4
Soil	18.7	21.6	22.3	29.5	13.2	10.8	19.4
Total	51.7	86.2	77.6	81.1	87.9	82.8	77.9
Losses	48.3	13.8	22.4	18.9	12.1	17.2	22.1

Nitrogen derived from fertilizer and soil

The nitrogen derived from fertilizer (Ndff) parameter provides a sensitive criterion by which to assess specific fertilizer practices. In our study, it was planned to evaluate two methods for fertilizer placement and urease and nitrification inhibitors. Data of Table 3 show that deep placement increased the Ndff values higher than those for surface application. Addition of inhibitors to urea resulted in an average of 11.5% increase in Ndff. The effectiveness of the urease inhibitors is higher than the nitrification inhibitor in both methods of placement that may use to the important role of urea hydrolysis and in turn ammonia volatilization as mechanism of N-losses in the floodwater. Mikkelsen and De Datta (1979) showed that ammonia volatilization losses are greatest when an ammonical N source such as urea is placed into irrigation water. Combination of both inhibitors yielded higher Ndff for the both fertilizer placement. Results of Table 3 showed that the percentage of Ndff was more than doubled when urea was applied deep and in combination with both urease and nitrification inhibitor. The percentage of Ndfs presented in Table 3, ranged from about 80 to 53% according to fertilizer placement and inhibitors application. Wilson *et al.* (1990) reported that although fertilizer N is the major pool from which the rice plant derived its N early in the season. the soil provides the remaining N for the plant. As the soil in our study is rich in organic nitrogen and organic matter in addition to high clay content, it is expected to supply rice with mineral nitrogen through the mineralization mechanism. Inhibitor application, % Ndfs was decreased to its lower value when urea was applied deep and treated with both urease and nitrification inhibitor.

Nitrogen-15 balance

Total plant recovery of N-15 labelled applied urea showed a very wide range from 16.8 to 74.7%, as showed in Table 4. Contrary to data obtained by Katyal *et al.* (1985), where plant recovery ranged from 21 to 31%, our data show much higher recovery. They explained that high percolation rate in their field trial is probably responsible for low recovery. At the same time Craswell *et al.* (1985) found that plant N-15 recovery ranged from 48 to 75% when N-15 labelled urea was applied alone either through surface or deep application. Table 4 shows that plant recovery was doubled when urea was applied deep in the soil. Regardless of the inhibitor application, plant recovery has increased as an average from 39 to 60% when fertilizer placement changed from surface to deep. Craswell *et al.* (1981) reported that deep placement of fertilizer has proved to be an effective means of reducing ammonia volatilization. Simultaneously, deep placement of urea decreases the conversion of ammonium to nitrate and thereby reduces nitrification losses. Regardless of whether ammonia volatilization or denitrification is responsible for the poor efficiency of urea in rice, the efficiency of deep placement has been confirmed both in N-15 experiments (Aleksic *et al.*, 1968), and in simple source comparison trials (Flinn at el., 1984).

Application of urease inhibitors resulted in increasing the plant recovery of labelled nitrogen applied as urea (Table 4). As it was mentioned above, as urea hydrolysis and ammonia volatilization are the main mechanisms of nitrogen losses, application of urease inhibitors is expected to increase plant recovery of applied nitrogen. Byrnes *et al.* (1984) reported that lower ammoniacal-N concentrations in floodwater in PPD as urease inhibitor amended systems reduced NH_3 loss and increased N-15 recovery from labelled urea. Also, Table 4 shows the importance of DCD as nitrification inhibitor in increasing plant recovery in both surface and deep application. Although the rate of increase in plant recovery due to DCD application is lower than that of NBPT or HQ, combined application of DCD with NBPT or with HQ resulted in even higher plant recovery. Similar results on the effect of DCD were obtained by Wilson *et al.* (1990) and Wells *et al.* (1989).

The extensive losses of applied N-15 from the soil-plant system (Table 4) are the obvious major cause for the poor recoveries of applied N by the rice. The losses from urea varied, as affected by method of application and application of inhibitors, and ranged from 64 to 12%. The high losses (64%) are comparable to losses from urea reported by Craswell *et al.* (1985). Although deep placement of urea resulted in lower losses (48%) as urea applied at the surface (64%), it is a serious cause for concern economically and environmentally. High losses of nitrogen from applied urea as ammonia volatilization would be enhanced by a pH increase in the water, which usually occurs when CO_2 in the water is largely depleted as a result of photosynthesis (Patrick, 1982). This may explain the importance of urease inhibitors which have the ability to reduce ammonia volatilization. Data in Table 4 presented that application of HQ or NBPT reduces N-losses down to 35% for surface application and to 22 and 15% for deep application. DCD applied with urea reduces N-losses and therefore enables more efficient nitrogen utilization by maintaining more of the fertilizer N and soil N in the NH_4^+ form. Combination of urease and nitrification inhibitors proved to be even better practice for minimizing N-losses (Table 4). Losses were down to 12%, as urea was applied in addition to NBPT and DCD as compared with 58%, when it was applied alone.

The use of N-15 balance technique has clearly identified N loss as a major problem in rice management. Considerable reduction in N losses would be obtained by better N-fertilizer management either by deep placement or by application of modified nitrogen fertilizers.

References

Abdel Monem, M (1986) Labelled urea fertilizer experiments on soils Mediterranean region. Ph.D. Thesis, Dept. of Agronomy, Colorado State Univ., Fort Collins, Co., USA

Abdel Monem, M (1991) Greenhouse evaluation of NBPT-urease inhibitor on urea fertilizer applied to barley using N-15. Egypt, J Appl Sci 6 (2): 186–194

Abdel Monem, M and Ryan, J (1991) Residual effect of labelled N in a vertisol of North Africa. *2nd* African Soil Science Soc. Conf. 4–10 Nov, Cairo, Egypt

Abdel Monem, M, Gadalla AM and Abbady, K (1993) Fate of urea-N applied to rice as affected by Azolla treatment. Isotope and Radiation Res. (in press)

Aleksic, Z, Broeshart, H and Middleboe, V (1968) Shallow depth placement of $(NH_4)_2SO_4$ in submerged rice soils as related to gaseous losses of fertilizer nitrogen and fertilizer efficiency. Plant and Soil, 28: 338–342.

Amberger, A (1989) Research on dicyandiamide as nitrification inhibitor and future work. Commun in Soil Sci Plant Anal 20: 1933–1953

Brandon, DM, and Wells, BR (1986) Improving nitrogen fertilization in mechanized rice culture. In De Datta SK and Patrick WH (eds) Nitrogen economy of flooded rice in soils. Martinus Nijhoff Publ. Co., The Netherlands

Bremner, JM and Mulvaney, RL (1982) Total nitrogen. In Page AL *et al.* (eds) Methods of soil analysis. Part II. Agronomy, 9: 595–624

Byrnes, BH, Savant NK and Fillery, IRP (1984) Effect of urease inhibitor, phenyl-phosphorodiamidate, on the efficiency of urea applied to rice. Soil Sci Soc Am J 74: 270–274

Craswell, ET, and Vlek, PL (1979) Fate of fertilizer nitrogen applied to wetland rice. In: Nitrogen and Rice, pp 175–192. IRRI. Los Banos, Philippins

Craswell, ET, De Datta, SK, Obcemea, WN and Hartantyp, M (1981) Time and mode of nitrogen fertilizer application to tropical wetland rice. Fert Res 2: 247–259

Craswell, ET, De Datta, SK, Weeraratne, CS and and Vlek, PL (1985) Fate and efficiency of nitrogen fertilizers applied to wetland rice. I. The Philippines. Fert Res 6 (1): 49–63

De Datta, SK (1981) Principles and practices of rice production. John Wiley and Sons. Inc., New York, pp 618

Fillery, IRP, De Datta, SK and Craswell, ET (1986) Effect of phenyl phosphorodiamidate on the fate of urea applied to wetland rice fields. Fert Res 9: 251–263

Flinn, JC, Mamaril, CP, Velascs, LE and Kaiser, K (1984) Efficiency of modified urea fertilizer for topical irrigated rice. Fert Res 5: 157–174

Katyal, JC, Singh, B, Sharma, VK and Craswell, ET (1985) Efficiency of some modified urea fertilizer for wietland rice grown on a permeable Soil. Fert Res 8: 137–146

Khind, CS, and Datta, NP (1975) Effect of method and timing of nitrogen application on yield and fertilizer nitrogen utilization by lowland rice. J Indian Soc Soil 23: 442–446

Mikkelson, DS, and De Datta, SK (1979) Ammonia volatilization from wetland rice soils. pp 135–156. In: Nitrogen and Rice. IRRI, Los Banos, Philippines

Mulvaney, RL, and Bremner, JM (1978) Use of P-benzoquinone and hydroquinone for retardation of urea hydrolysis in soils. Soil Biol Biochem 10: 297–302

Murayama, N (1979) The importance of nitrogen for rice production. Nitrogen and Rice. pp 5–23. IRRI., Los Banos, Philippines

Patrick, WH (1982) Nitrogen transformation in submerged soils. In Stevenson (ed) Nitrogen in Agriculture. Agronomy, A.S.A. Madison, Wisconsin

Wells, BR, Bollich, BK, Ebelhar, W, Mikkelsen, DS, Norman, RJ, Brandon, DM, Helms, RS, Turner, FT and Westcott, MP (1989) Dicyandiamide (DCD) as a nitrification inhibitor for rice culture in the United States. Common in Soil Sci Plant Anal 20: 2023–2047

Wilson, CF, Norman, RJ and Wells, BR (1990) Dicyandiamide influence on uptake of preplant applied fertilizer nitrogen by rice. Soil Sci Soc Am J 54: 1157–1161

Yamamuro, S (1981) The accurate determination of nitrogen-15 with an emission spectrometer. Soil Sci Plant Nutr 27: 405–419

C. Rodriguez-Barrueco (ed.), Fertilizers and Environment, 217–221.

Response to $NH_4{}^+$-N fertilization of Xeric Torriorthent soils from middle Ebro basin (Spain)

D. Badia[1] & J.Mª Alcañiz[2]
[1]*Dpto. de Agricultura, Escuela Universitaria Politécnica, Crtra, Zaragoza km. 67, 22071-Huesca, Spain;* [2]*Àrea d'Edafologia i Química Agrícola, Universitat Autònoma de Barcelona, 08193-Bellaterra, Spain*

Key words: Nitrification rate, semiarid agricultural soils

Abstract

Nitrification potential capacity of six Xeric Torriorthent agricultural soils from semiarid Ebro basin (NE Spain) was studied in an incubation experiment. A sigmoidal model was used to express the accumulation of $NO_3{}^-$-N with time. The delay period and the maximal nitrification rate were estimated to evaluate soil nitrification capacity.

Maximal rate of nitrification ranged from 1.47 to 40.34 mg $NO_3{}^-$-N mg kg^{-1} soil day^{-1}. Delay period ranged from 7.69 to 20.40 days for the experimental soils. Both parameters were negatively and significantly correlated. The lowest nitrification activity was observed in saline and highly gypsiferous soils, with a poor soil organic matter content. Maximal nitrification rate was significantly and positively correlated with total organic carbon and negatively correlated with salinity and sodicity in these semiarid soils. Delay period was significantly and positively correlated with gypsum content and negatively correlated with carbonate soil content.

Introduction

Nitrification is a process of strictly biological nature. Its agronomic importance lies in: the habitual assimilation of N by crops in the $NO_3{}^-$ form and the usual application of nitrogenous fertilizers as ammonium forms. Mineral and organic soil components can accelerate or delay nitrification process (Bundy and Bremmer, 1973; Schmidt and Belser, 1982; Gispert *et al.*, 1987; Gomah *et al.*, 1989). Possible disadvantages of quick nitrification are losses of N by denitrification, leaching and runoff, including movement of $NO_3{}^-$ out of the crop root zone with eventual $NO_3{}^-$ accumulation in ground and surface waters. On the other hand, a very low rate of nitrification under some situation should encourage losses of applied $NH_4{}^+$-N through ammonia volatilisation. In fact, it may be the major gateway of N loss in arid and semiarid conditions due to high temperature and high soil pH since nitrate leaching and denitrification are restricted by low rainfall (Praven-Kuwar and Aggarwal, 1988). Semiarid agricultural soils in the middle Ebro basin (NE Spain), because of the influence of climate, have a low organic matter content, a basic reaction and high amounts of soluble salts (Badia, 1989) which can modify nitrification process. The objectives of that research were to characterize these possible modifications in different soils.

Materials and methods

Soil characteristics

The soils were located between 0° 13′ and 0° 22′ W and between 41° 33′ and 41° 36′ N. Average annual rainfall (R) is 318 mm and evapo-transpiration (ETP) is 832 mm; so the R/ETP ratio (< 0.5) is characteristic of semiarid lands. The soils are developed on different parent material: calcareous marls, saline marls and gypsum from Oligocene period. This is reflected in the chemical composition of soils which are rich in calcium carbonate, calcium sulphate and more soluble salts, a common characteristic of soils from Ebro Basin (Herrero and Aragués, 1988). They are classified as Xeric Torriorthent (Badia, 1989). A summary of the most important physical and chemical parameters of the ploughed horizon (Ap) of every soil is shown (Table 1).

Table 1. Main physical and chemical characteristics of the soils (Ap horizon)

Soil Reference	Textural class	pH (1:2)	$CaCO_3$ (%)	Gypsum (%)	Org. C (%)	CEC cmol kg^{-1}	ECe dS m^{-1}	ESP (%)
Vilella	S-C	7.6	18.7	12.4	0.54	10.4	3.12	0.63
Valleta	C-L	8.4	31.5	1.3	1.30	10.8	1.31	0.98
Triangle	C-L	8.3	27.3	2.3	0.85	7.75	1.59	0.20
Mansaneta	S-C-L	7.9	27.1	3.2	0.47	10.3	13.8	9.49
Mont-ral	S-C-L	8.4	31.6	2.9	0.44	8.83	11.3	8.10
Miralsot	S-C	7.8	6.2	31.4	0.43	7.68	2.98	0.39

Abbreviations: Textural class (USDA): S-C, Silty-Clay; S-C-L, Silty-Clay-Loam; C-L, Clay-Loam; Org. C, organic carbon; CEC, cation exchange capacity; ECe, electrical conductivity of the saturation extract at 25 °C; ESP, exchangeable sodium percentage.

Incubation procedure and NO_3^- analysis

Nitrification potential was determined by the method of Schmidt and Belser (1982). An incubation system was carried out with soil fresh samples, within 24 h after collection. A composite sample consisting of four subsamples, collected at a depth of 0–15 cm, was obtained from each soil. Stones, roots and plant debris were first removed by sieving through 2 mm mesh. To fresh soil (about 250 g on dry weight) in a glass container, 2.5 ml of $(NH_4)_2SO_4$ solution (126 mg NH_4^+-N ml^{-1}) and sufficient water was added to bring to 50% field capacity. Containers were incubated in the dark at 25 ± 1 °C. Incubations were continued for 4 weeks. Weights were checked daily and moisture losses during incubation were compensated by adding water. Soil sampling (7 g) and nitrates measures were made each four or six days, according to the nitrification rates, by triplicate. NO_3^--N content in each soil sample was determined by extraction with $CuSO_4$ 1 N, previous elimination of chlorides with $AgSO_4$. Nitrate content was estimated colorimetrically, using a Hitachi V-3200 UV-Vis spectrophotometer, by the phenol-disulfonic acid method (Bremmer, 1965).

Data analyses and statistics

In order to express the accumulation of NO_3^--N with time (t) quantitatively, the Verhulst equation,

$$dN/dt = kN(a' - N), \quad (1)$$

which upon integration renders a sigmoidal curve, was used (Jolivet, 1983):

$$Nt = a'/1 + [a'/(N_o - 1)\exp(-a'k[t - t_o])] \quad (2)$$

where a' is the asymptotic values of nitrate, k is a constant, N_o is the initial nitrate content and t_o is the initial time. Kaleida Graph package (version 2.0.2, Abelbeck Software Inc.) was used to fit, by the least-squares method, the cumulative nitrate production *versus* time in a sigmoidal model and to obtain the k and a' parameters. K_{max} was calculated as the maximal slope of previous equation, at the inflection point (where $NO_3^- = a'/2$): $K_{max} = k[(a')^2]/4$; and the delay period (t'), according to the definition of Sabey *et al.* (1959), was calculated as the value of t when the maximal slope was extrapolated to the initial value of NO_3^-:

$$t' = (1/a'k)\ln[(a'/NO_3^-) - 1] + [NO_3^- - (a'/2)]/K_{max} \quad (3)$$

The contribution of soil organic N mineralization to NO_3^- was assumed to be very small relative to those of the nitrification; therefore, the sum of NH_4^+ and NO_3^- was assumed to be constant and equal to a'.

Significance differences between means was determined by an ANOVA test (Least Significance Differences) with the Statview statistical package (Abacus Concepts Inc., Berkeley, CA). Moreover, correlations between K_{max} and t' and soil properties were made.

Results and discussion

The evolution of NO_3^--N with time showed a sigmoidal curve (Fig. 1) with good fitting coefficients (Table 2). Estimated K_{max} ranged from 1.47 to 40.34 mg NO_3^--N mg kg^{-1} soil day^{-1} while t' ranged from 7.69 to 20.40 days. Both parameters were negatively correlated although a low level of significance ($p < 0.1$). K_{max} rates measured in this study were similar than other rates reported for mineral agricultural soils, using similar techniques (Sabey *et al.*, 1959; Malhi and McGill, 1982; Hadas *et al.*, 1986; Bhupinderpal-Singh

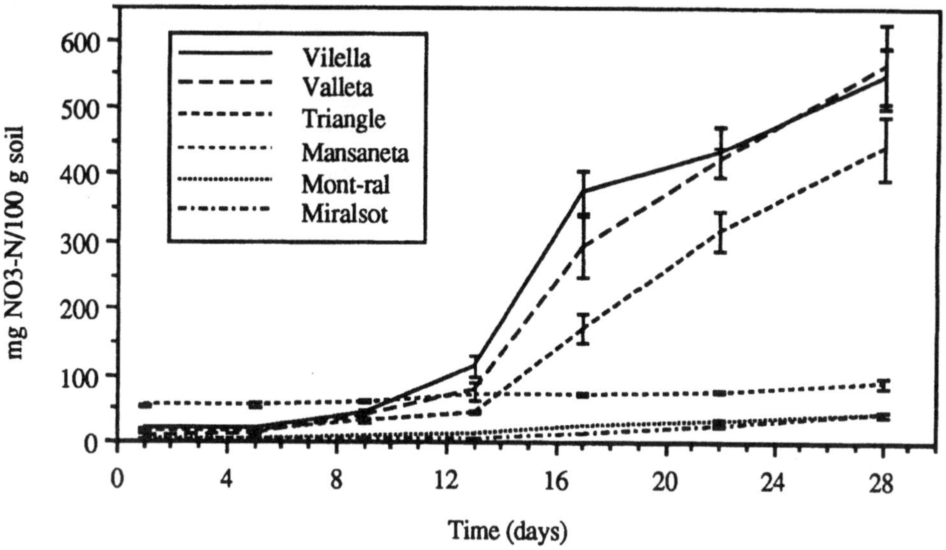

Fig. 1. Nitrification response of different semiarid agricultural soils amended with ammonium sulphate. Error bars indicate standard deviation of every value.

Table 2. Maximal nitrification rates (Kmax, mg NO_3^--N mg kg^{-1} soil day^{-1}), delay periods (day^{-1}) and correlation coefficient (R) of sigmoidal fitting. Values with the same letter, in every column, do not show significant differences ($p < 0.01$), by ANOVA

	Maximal nitrification rate, Kmax (mg NO_3^--N mg kg^{-1} soil day^{-1})	Delay period, t' (day)	R
Plot			
Vilella	40.336a	7.686a	0.982
Valleta	36.555ab	9.633ab	0.991
Triangle	29.105b	8.889a	0.991
Mansaneta	1.473c	12.012b	0.978
Mont-ral	2.306c	10.583ab	0.991
Miralsot	6.461c	20.399c	0.993

et al., 1993). Significant soil differences were proved (Table 2).

Vilella (with 40.34 mg NO_3^--N mg kg^{-1} soil day^{-1}) and Valleta (36.55) soils showed the highest K_{max} values and low t' values. The maximal rate of nitrification (K_{max}), as derived graphically by Sabey *et al.* (1959) from sigmoidal curves, when NH_4^+ concentration is not rate limiting (in this experiment, 1260 mg NH_4^+-N soil kg^{-1}), is expected to depend on soil properties (Hadas *et al.*, 1986). The lowest nitrification activity was observed in the saline soils (Mansaneta, Mont-ral). K_{max} and ECe showed a negative and significant correlation ($p < 0.01$). Previously, a negative correlation was found between microbial activity, measured as CO_2 evolved, and soil salinity (Badia and Alcañiz, 1992). Salts may interfere with one or more of the enzymes that are engaged in the process of nitrification (Frankenberger and Bingham, 1982). Analogously, McLung and Frankenberger (1985) observed that at ECe values of 20 dS m^{-1} (NaCl) decreased nitrification as high as 75% when $(NH_4)_2SO_4$ were applied at calcareous soils. These aspects agree with our results and the observations of earlier researchers on nitrification process in different soil types (Harada and Kai, 1968; Laura, 1974; McCormick and Wolf, 1980). Although the mechanisms by which the various solutes inhibit nitrification are unknown, in general terms, two effects can be recognised: cell death or inactivation and a decrease in microbial activity. In relation of this, it was observed a partial recovery of nitrification, following exposure to high osmotic pressure solutions (Darrah *et al.*, 1987; Badia and Alcañiz, 1992) which suggests that both reversible and irreversible mechanisms are involved in the inhibition of nitrification.

A slight and negative relation (r = − 0.50) was observed between K_{max} and fine particles (clay + silt) content in the soil. It was previously hypothesized that clays depressed nitrification either through reducing diffusion of NH_4^+ or restricting easy availability of this to nitrifiers through some unknown mechanism (Yadvinder-Singh and Beauchamp, 1985).

On the other hand, K_{max} in these semiarid agricultural soils were highly significant and positively correlated with total organic carbon ($p < 0.01$). In general, increasing amounts of humified organic matter enhanced microbial activity in agricultural soils (Kaiser *et al.*, 1992; Vekemans *et al.*, 1989; Gispert *et al.*, 1987; Badia and Alcañiz, 1994). Miralsot soil, less organic and with more gypsum content, showed a low K_{max} and lowest t'. In fact, t′ was positively correlated with soil gypsum content ($p < 0.01$). High gypsum content can be sufficient by large to inhibit the function of nitrifying organisms (Sindhu and Cornfield, 1967; Singh and Taneja, 1977). Unbalanced fertility and poor physical properties of gypsiferous soils can influence free space for microbes and biological activity as well as plant growth (Herrero and Porta, 1987; Badia and Alcañiz, 1994; Badia and Martí, 1994).

Summarizing, delay period and specially maximal nitrification rate, obtained from a zero-order kinetics, were good parameters to quantify soil properties effects on NO_3^- formation.

References

Badia D (1989) Los suelos en Fraga. Cartografía y Evaluación, IEA, Huesca.

Badia D and Alcañiz JMª (1992) Soil salinity effects on microbial activity of semiarid agricultural soils. Proceedings 6th. International Symposium on Microbial Ecology, Barcelona (Spain), pp 202.

Badia D and Alcañiz JMª (1994) Basal and specific microbial activity in semiarid agricultural soils: organic amendment and irrigation management effects. Geomicrobiology Journal 11: 261–274.

Badia D and Martí C (1994) Mejora del valor pastoral y medioambiental de zonas semiáridas degradadas mediante técnicas de revegetación, remicorrización y acolchado: I. Aplicación en suelos yesosos (Gypsic Regosol). Lucas Mallada 6: 17–35.

Bhupinderpal-Singh, Bijay-Singh and Yadvinder-Singh (1993) Potential and kinetics of nitrification in soils from semiarid regions of Northwestern India. Arid Soil Research and Rehabilitation 7: 39–50.

Bremmer JM (1965) Inorganic forms of nitrogen. In: Black CA (ed.) Methods of soil analysis Part 2. Am Soc Agro J Wisconsin, 1179–1237.

Bundy LG and Bremmer JM (1974) Inhibition of nitrification in soils. Soil Sci Soc Am Proc 37: 396–398.

Darrah PR, Nye PH and White RE (1987) The effect of high solute concentrations on nitrification rates in soil. Plant Soil 97: 37–45.

Frankenberger WT Jr and Bingham FT (1982) Influence of salinity on soil enzymes activities. Soil Sci Soc Am J 46: 1173–1177.

Gispert MA, Alcañiz JMª and Arcara PG (1987) Influence of two humic extracts characterized by Py-GC on soil microbial activities. The Science of the Total Environment 62: 379–385.

Gomah AHM, Al-Nahid SI and Amer HA (1989) Nitrogen mineralization in sludge-amended desert soil as affected by rate of sludge, salinity and wetting and drying cycles. Arid Soil Res Rehab 3: 417–429.

Harada T and Kai H (1968) Studies on the environmental conditions controlling nitrification in soils, I: effects of ammonium anf total salts in media on the rate nitrification. Soil Sci Plant Nut 14: 20–26.

Hadas A, Feigenbaum S, Feigin A and Portnoy R (1986) Nitrification rates in profiles of differently managed soil types. Soil Sci Soc Am J 50: 633–639.

Herrero J and Porta J (1987) Gypsiferous soils in the north-east of Spain. Soil Microbiology 5: 186–192.

Herrero J and Aragués R (1988) Suelos afectados por salinidad en Aragón. Surcos de Aragón 9: 5–10.

Jolivet E. (1983) Introduction aux Modéles Matématiques en Biologie. Actualités Scientifiques et agronomiques de l'Institut National de la Recherche Agronomique, nº11. Masson. Paris.

Kaiser EA, Mueller T, Joergensen RG, Insam H and Heinemeyer O (1992) Evaluation of methods to estimate the soil microbial biomass and the relationship with soil texture and organic matter. Soil Biol Biochem 24: 675–683.

Laura RD (1974) Effects of neutral salts on C and N mineralization of organic matter in soil. Plant Soil 41: 113–127.

Malhi SS and McGill WB (1982) Nitrification in three Alberta soils: effect of temperature, moisture and substrate concentration. Soil Biol Biochem 14: 393–399.

McCormick RW and Wolf DC (1980) Effect of sodium chloride on CO_2 evolution, ammonification and nitrification in a Sassafras sandy loam soil. Soil Biol Biochem 12: 153–157.

McLung G and Frankenberger WT Jr (1985) Soil nitrogen transformations as affected by salinity. Soil Sci 139: 405–411.

Praven-Kuwar and Aggarwal PK (1988) Reduction of ammonia volatilization from urea by rapid nitrification. Arid Soil Res Rehab 2: 131–138.

Sabey BR, Frederick LR and Bartholomew WV (1959) The formation of nitrate from ammonium nitrogen in soils: III. Influence of

temperature and initial population of nitrifying organisms on the maximum rate and delay period. Soil Sci Soc Proc 23: 462–465.
Schmidt EL and Belser LW (1982) Nitrifying bacteria. In: Page AL Miller RHS Keeney DR (eds) Methods of Soil Analysis, Part 2, 2d Am Soc Agron, Madison, WI. pp 1027–1042.
Sindhu MA and Cornfield AH (1967) Comparative effects of varying levels of chlorides and sulphates of sodium, potassium, calcium and magnesium on ammonification and nitrification during incubation of soil. Plant Soil 27: 468–472.
Singh BR and Taneja SN (1977) Effects of gypsum on mineral nitrogen status in slkaline soils. Plant Soil 48: 315–321.
Vekemans X, Godden B and Penninckx MJ (1989) Factor analysis of the relationships between several physico-chemical and microbiological characteristics of some Belgian agricultural soils. Soil Biol Biochem 21: 53–58.
Yadvinder-Singh and Beauchamp EG (1985) Alternate method for characterizing nitrifier activity in soil. Soil Sci Soc Am J 49: 1432–1436.

C. Rodriguez-Barrueco (ed.), Fertilizers and Environment, 223–225.

Effect of different fertilization systems on variation of soil fertility in long-term trials

D. Bogdanović, M. Ubavić, M. Čuvardić & M. Jarak
Faculty of Agriculture, Institute of Field and Vegetable Crops, 21000 Novi Sad, Yugoslavia

Key words: Fertilization system, mineral fertilizers, manure, liquid manure, harvest residues, soil fertility, microbiologic activity

Abstract

Soil samples were taken from long-term field trials (lasting from 14 to 28 years) after wheat harvest and studied for the effect of different fertilization systems on soil chemical properties, content of the most important biogenic elements and microbiological activity in soil. Compared with unfertilized plots, the application of different fertilization systems increased the content of biogenic elements in a chernozem. Improvement of soil fertility in biogenic elements was most affected by mineral fertilizers and manure, followed by mineral fertilizers and harvest residues, while the lowest effects were achieved by the application of mineral fertilizers alone.

Introduction

Although chernozem soil of the Pannonian plain is considered to be naturally rich and fertile and with favourable water-air and temperature properties, however, selected varieties and hybrids and agrotechnical measures and rational fertilization have to be applied with the aim to achieve high and quality yields of field crops.

Considering domestic and foreign literature, there are few data on the effect of different fertilization systems including mineral fertilizers, mineral fertilizers and manure, and mineral fertilizers and liquid manure, mineral fertilizers and harvest residues incorporated on the changes of chemical properties, content of biogenic elements and heavy metals and biological activity in soil. However, there are more reperts on many these effects on yield (Stojanović *et al.*, 1979; Broadbent, 1975; Rauhe *et al.*, 1983; Vomel and Doll, 1980).

The objective of this study was to specify the changes of a chernozem fertility which were caused by long-term application of different fertilization systems and intensive field crop cultivation.

Material and method

Soil was sampled from long-term field trials after wheat harvest and analyzed for chemical properties, content of important biogenic element and number of microorganisms. The following fertilization systems have been applied in different plots:

- The plot established in 1947/48 on a two field system - wheat/maize has not been fertilized (check);
- The plot established on maize monoculture in 1961/62 has been fertilized with increasing doses of mineral fertilizers and mineral fertilizers and manure;
- The plot established in 1964/65 has been fertilized with increasing doses of mineral fertilizers;
- The plot established in 1965/66 has been fertilized with mineral fertilizers and manure, only mineral fertilizers and mineral fertilizers and incorporated harvest residues and
- The plot established in 1980/81 has been fertilized with mineral fertilizers and mineral fertilizers and liquid manure.

The soil chemical properties and important biogenic elements were analyzed according to standard methods. The number of microorganisms was calculated by the standard microbiological methods (Pochon and Tardieux, 1962).

Table 1. Changes in chernozem fertility after long-term application of different fertilization systems

Fertilization systems	pH 1M KCl	$CaCO_3$	humus	N	P_2O_5	K_2O	Total microorganisms (10^6 g^{-1})	Azotobacter ($10^2 g^{-1}$)
		(%)			mg(100 g soil)$^{-1}$			
Control	6.3–6.9	0.0–4.6	2.0–2.8	0.10–0.14	5.0–39.0	16.0–23.0	111.33	30.45
	6.7	1.4	2.4	0.12	15.9	18.7		
NPK	6.5–7.1	0.0–2.3	2.4–3.1	0.12–0.15	30.0–40.0	14.0–29.0	131.95	44.54
	6.8	0.8	2.8	0.13	34.3	22.9		
min.fert.+ HR	6.4–7.2	0.0–2.1	2.4–2.8	0.12–0.14	26.8–47.5	25.0–27.5	152.35	55.35
	6.08	1.0	2.6	0.13	37.1	26.2		
min.fert.+ manure	6.7–6.8	0.0–1.0	2.8–2.9	0.14	22.0–26.5	19.5–38.5	153.50	43.81
	6.7	0.5	2.8		24.2	29.0		
min.fert. + liquid manure	6.8	1.8–2.4	3.2–3.3	0.15–0.17	180.0–195.0	39.0–40.0	297.77	80.36
		2.1	3.2	0.16	187.0	39.5		

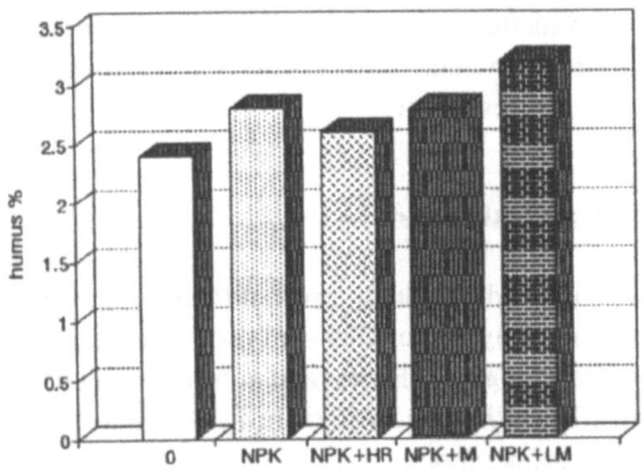

Fig. 1. Changes in humus content after long-term application of different fertilization systems.

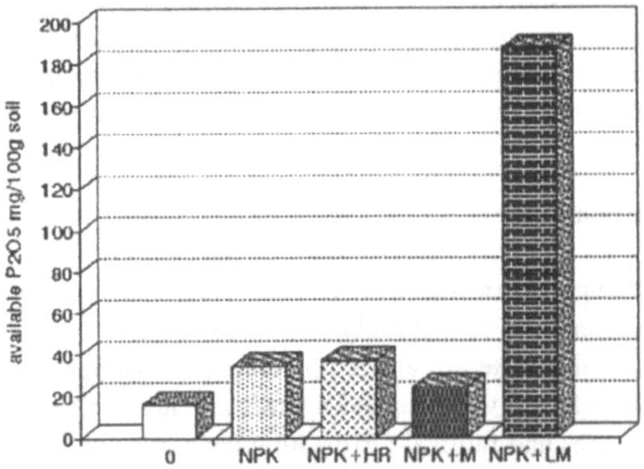

Fig. 2. Changes in phosphorus content after long-term application of different fertilization systems.

Legend

0 - Control
NPK - mineral fertilizer
NPK + HR - mineral fertilizer + harvest rezidues
NPK + M - mineral fertilizer + manure
NPK + LM - mineral fertilizer + liquid manure

Results and discussion

Depending on the duration of the trials, from 14 to 28 years, a high natural potential of chernozem soil with respect to humus is indicated by the chemical properties, content of the most important biogenic elements and the number of microorganisms in the untreated check plots (Table 1). Compared with the unfertilized plots, humus increased significantly in all fertilization systems, except in the system with mineral fertilizers and incorporated harvest residues (Fig. 1).

Significantly lower content of available phosphorus was measured in the soil fertilized with mineral fertilizers and manure (Fig. 2). Compared with others, the system in which soil was fertilized with mineral fertilizers, and mineral fertilizers and incorporated harvest residues had lower content of available potassium (Fig. 3).

Total number of bacteria in soil can be used as an indicator of general biologic activity (Acea and Carballas, 1990). The results of our study showed high number of the studied groups of microorganisms (Table 1, Fig. 4) in all systems of fertilization, particularly in the system with mineral fertilizers and liquid manure.

Fig. 5 presents wheat yield (t/ha) with respect to a chernozem production capacity in the year when the

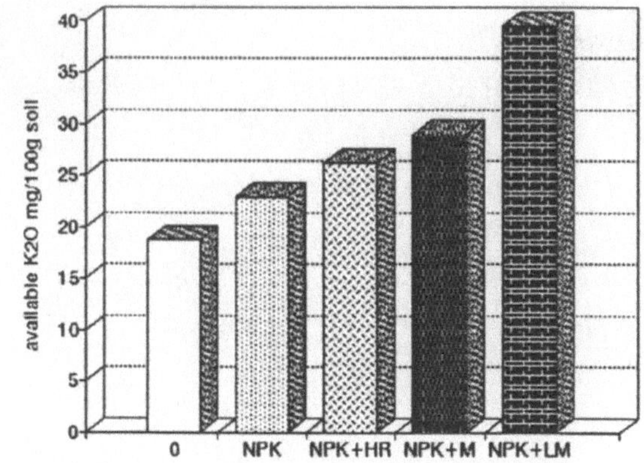

Fig. 3. Changes in potassium content after long-term application of different fertilization systems.

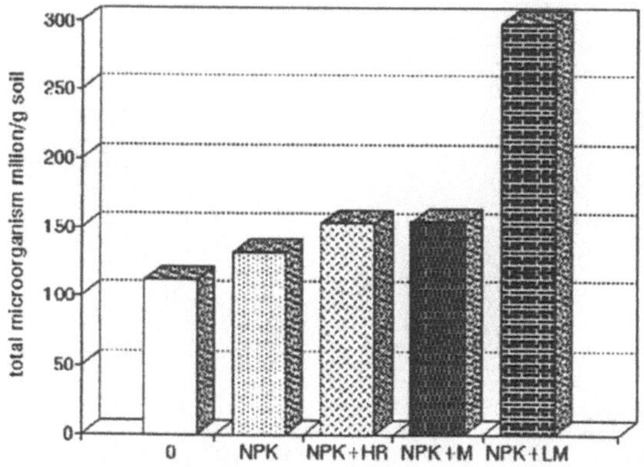

Fig. 4. Changes in total microorganism content after long-term application of different fertilization systems.

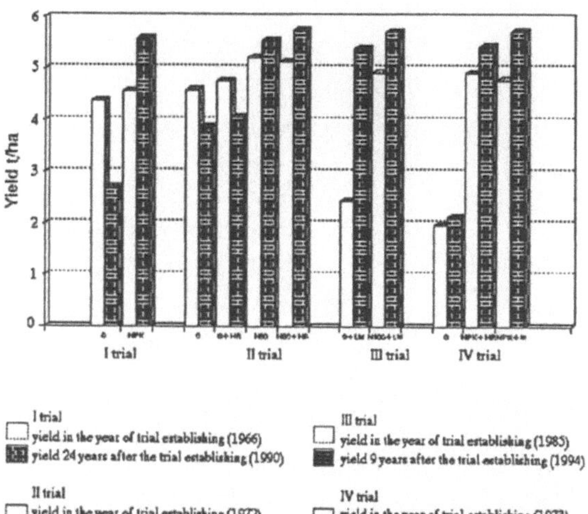

Fig. 5. Effect of different fertilization systems on production capacity of soil in wheat production (long-term trials).

trial was established and after long-term application of fertilizers. Considering unfertilized (control) variants, wheat yields were lower than on the variants with long-term application of some of the fertilization systems. This indicates the declination of natural soil fertility after long period with no fertilizer application.

References

Acea MJ and Carballas T (1990) Principal components analysis of the soil microbial population of humid zone of Galicia (Spain). Soil Biol Biochem 22,6: 749–759.

Broadbent FE (1975) Nitrogen relase and carbon los during decompozition of adbed plant residues. Soil Sci Soc Am Proc 12: 246–249.

Rauhe K, Sieger B, Hoberuck (1983) Der Wirkungsmechanismus von organischem und mineralischem Dungerstickstoff im System Boden-Pflanze-unter besonderer Berucksichtigung des 15N-Einsatzes in Daueruersuchen ouf einem Lo 4 Standart. "Zbornik radova", A periodical of scientific research on field and vegetable crops, Institute of Field and Vegetable Crops, Novi Sad 13: 115–121.

Stojanović D, Djurdjević M and Marković N (1979) The study of microelements in soils of west Serbia fertilized with mineral fertilizers. Agrohemija 5–6: 183–190.

Vomel A and Doll G (1980) Movement of nitrogen fertilizer in combination with organic fertilizers. Soil and fertilizers 7: 609.

C. Rodriguez-Barrueco (ed.), *Fertilizers and Environment*, 227–231.

Effect of soil tillage and N-fertilization on the distribution of nitrate in the soil profile in a long-term wheat–maize biculture

S. Hoffmann & T. Kismányoky
Pannon University of Agricultural Sciences, Deák F. str. 16, H-836l, Keszthely, Hungary

Key words: Nitrate-leaching, tillage, N-fertilizer

Abstract

Tillage systems need to be compared over an extended period of time to determine their impacts on different soil properties e.g. mineralization or nitrate leaching. The 20 years long-term effect of fall-ploughing, disk-harrowing and no-tillage combined with N fertilization was studied on a Cambisol.

Soil tillage did not greatly influence the total amount of nitrate in the 0–180 cm soil profile. However, we could find significant differences in the vertical distribution of nitrate. After the harvest of winter wheat there was considerable nitrate accumulation in the 0–60 cm and in the 120–180 cm layers at disc-harrowing and ploughing, respectively. At no-tillage the curve of nitrate concentrations showed an S-shape. For the next spring the accumulation maxima totally disappeared from the 0–180 cm soil profile or partly translocated. Despite of the drought, it was surprising that nitrate transport was so rapid in the deep soil layers too. Our results strongly suggest that nitrate accumulation caused by excessive N-rates or the unused rest of extremely intensive mineralization can reach relatively quickly the groundwater table at our locations. However the nitrate profils also suggest that N fertilizer applied at optimum rate for yield, does not cause greater nitrate-N accumulation in the subsoil and so provide little risk for potential contamination of groundwater at our site too.

Introduction

Agriculture is often blamed for being one ot the major source of nitrate contamination of surface and groundwater. Our field experiments in Hungary are situated at lake Balaton where this problem is especially important.

The role of N-fertilizer practice in nitrate leaching and the distribution of nitrate in the soil profile is relatively definite, but the process is not well known at all sites and climate. It is evident that N fertilizer applied at recommended rates and time causes no additional pollution. MacDonald *et al.* (1989) suggested that almost all of the nitrate at risk to leaching in winter, comes from mineralization and not from unused spring applied N fertilizer. Even unfertilized fallow plots can produce mineral -N of a respectable quantity which can cause groundwater pollution (Hoffmann, 1993). It is still debated how soil cultivation method (no-till versus conventional or reduced till) affects nitrate leaching or accumulation in the soil profile.

Recent investigations have shown different results. Some experts have found greater leaching losses or deeper nitrate infiltration with no-tilling (McMahon and Thomas, 1976; Dowdell *et al.*, 1987; Eck and Jones, 1992), while others smaller ones (Logsdon *et al.*, 1990; Hütsch and Mengel, 1992; Francis and Knight, 1993). Further others have found no difference (Lamb *et al.*, 1985; Sharply *et al.*, 1991). Hütsch and Mengel (1992) have suggested that reduced nitrate leaching at no-tillage was caused by physical conservation of nitrate in the soil matrix combined with macropore flow.

So it seems to be important to investigate the long-term effect of soil tillage and N-fertilization – especially also with excessive N-rates – on the process of nitrate accumulation in the deeper soil layers. No-till is very rarely used in Hungary, but swallow water-saving reduced tilling has its importance beside conventional tilling at our semi-continental climate.

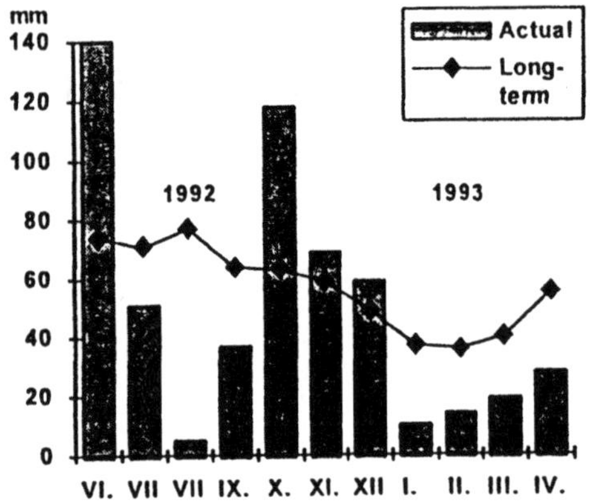

Fig. 1. Actual and long-term mean precipitation.

Materials and methods

A long-term fertilization experiment with 3 tillage systems in a wheat-maize biculture was sampled to determine accumulation of nitrate in the soil profile. The experiment was initiated in 1972 on a well drained Cambisol (clay content: 27%, humus content: 1.5%, C/N: 10.2, pH H_2O: 7.1) in a randomised split-plot design with four replications. Maize (*Zea mays* L.) and winter wheat (*Triticum aestivum* L.) alternated each other in two years rotation. Tillage systems were the following: 1. fall ploughing (fall mouldboard plough + secondary tillage); 2. fall reduced-till (disc-harrowing several times + secondary tillage); 3. no-till (chemical weed control). Sowing-time were late October and late April for wheat and maize respectively. The rate of N fertilizer treatments (average of N-doses given to winter wheat and maize) on the sampled plots was the following: N0: 0; N2: 200; N4: 300 kg/ha. There was no application of farmyard manure or slurry. Pesticides were applied to all cultivation treatments as usual. Actual and long-term monthly precipitation values are shown in Fig. 1. The long term annual mean values for precipitation and temperature are 700 mm and 10.8 °C, respectively.

10 soil cores from each plot 20 mm in diameter, was taken to a depth of 180 cm in 30 cm increments: 0–30, 30–60, 60–90, 90–120, 120–150, and 150–180 cm. Soil samples were taken from all replicates up to 90 cm and from one replication at depth of 90–180 cm. In 1992 N2 was sampled only up to 90 cm. Sampling dates were after wheat harvest in August of 1992 and before sowing of maize in April of 1993. Soil samples were dried and ground to pass a 2 mm sieve. Duplicate sample were extracted using 1% KCl and analyzed for ammonium- and nitrate-N by steam destillation. For converting the mg/kg nitrate-N values into kg/ha, bulk density of soil was measured up to 90 cm from undisturbed soil samples and below 90 cm a bulk density of 1.5 g/cm^3 was assumed.

Results and discussion

The amount of nitrate-N in late summer 1992 and the next spring, in the 0–180 cm soil profile of different tillage systems and N treatments, the averages and the decrease of nitrate-N between the two dates are shown in Table 1.

In August 1992 nitrate-N contents of all tillage systems with N0 were nearly the same. At N4 treatments the different soil cultivation methods resulted similar nitrate-N accumulation also, although under no-tillage plots there was slightly but not significantly less nitrate-N.

In April 1993 ploughing, disc-harrowing and no-till resulted in similar amounts of nitrate-N in the profile of the corresponding N treatments, too. Taken the average of all N treatments into consideration we can better see that tillage system did not much influence the nitrate-N accumulation. The 'decrease'-data of Table 1 indicate that marked amounts of nitrate-N appeared during autumn and winter from the entire soil profile of treatment N4. Becouse there was no vegetation in that period of time, loss of nitrate-N could be caused mostly by leaching and perhaps by denitrification. It could be detected some nitrate-N losses even at N0 treatments.

Figures 2, 3 and 4 illustrate how the above mentioned nitrate-N contents are distributed within the 0–180 soil profile. Following wheat harvest in August 1992 individual shapes of nitrate-N profiles could be found belonging to the different soil cultivation methods. The nitrate-N profiles of treatment N4 shows marked accumulation maxima in the 120–180 and the 0–60 cm soil layers at ploughing and disc-harrowing, respectively. At no-tillage the nitrate-N profil forms an 'S'-shape with two maxima in the 0–30 and 90–120 cm soil layers. As it was expected, profiles of N0 did not show marked nitrate-N maxima values. Nevertheless it is remarkable that their trend in the subsoil is similar to N4.

Although in 1992 in the N2 treatment the soil test was made only till 90 cm, it seems to be possible to get some information from it and draw a conclusion for the

Table 1. Nitrate-N contents (kg/ha) in late summer 1992 and the next spring 1993 in the 0–180 cm soil profile at different soil tillage systems

Sampling dates	kg nitrate-N/ha								
	Conventional till			Reduced till			No till		
	N0	N2	N4	N0	N2	N4	N0	N2	N4
Aug. 1992	238,6	–	386,1	234,4	–	383,3	234,2	–	353,7
Apr. 1993	196,4	231,5	252,3	207,7	213,7	261,4	187,3	248,6	261,4
Decrease	42,2	–	133,8	26,7	–	121,9	46,9	–	92,0
Average	261,0			260,1			257,1		

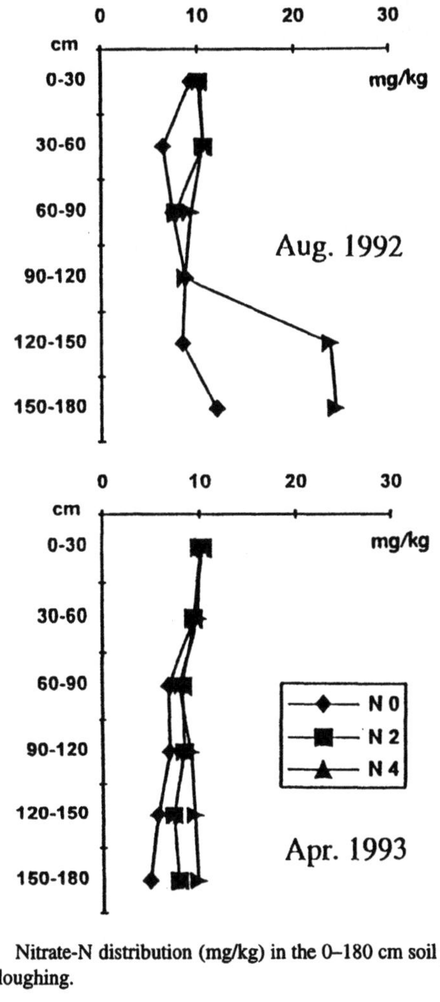

Fig. 2. Nitrate-N distribution (mg/kg) in the 0–180 cm soil profile after ploughing.

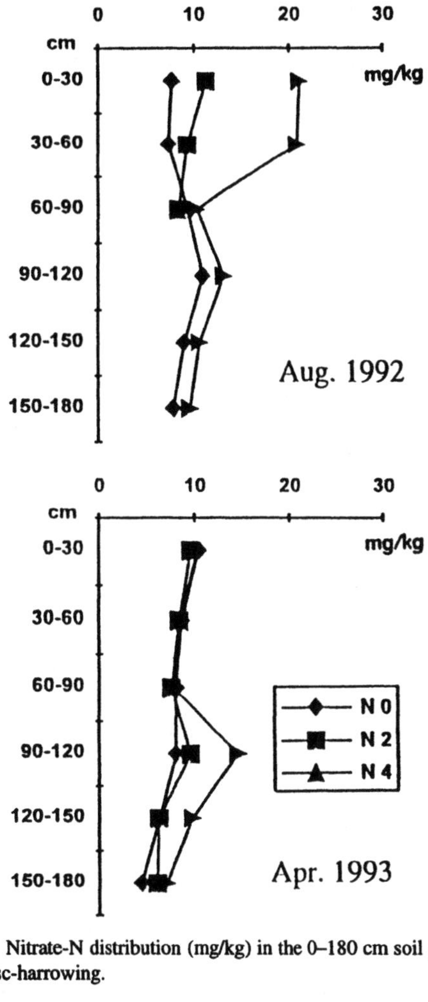

Fig. 3. Nitrate-N distribution (mg/kg) in the 0–180 cm soil profile after disc-harrowing.

entire soil profile. So the amount of nitrate-N and the shape of the profiles in the upper 90 cm suggest only a moderate accumulation in the whole 0–180 cm soil layer at cultivation methods no-till and disc-harrowing. At ploughing this estimation is uncertain.

For the spring of 1993 the marked nitrate-N accumulation maxima of N4 disappeared, although at disc-

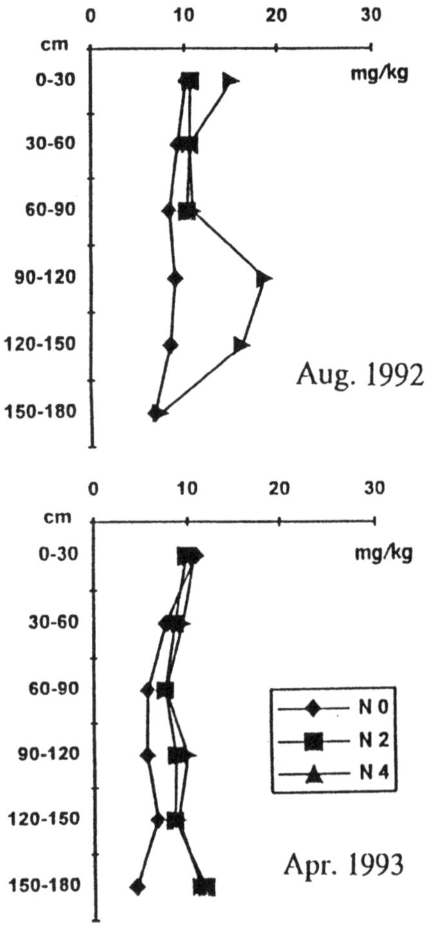

Fig. 4.

harrowing and no-till the rest of them could be found somewhat deeper. At ploughing the nitrate-N accumulation maximum migrated probably under 180 cm. There is contradiction between the nitrate decrease published in Table 1 and the shapes of the nitrate profiles. Namely, the latter suggest the highest leaching at ploughing however not proved by the data in Table 1. The nitrate profils of N2 are between N0 and N4. This does not suppose much more nitrate loss in the profile of N2 than in the N0. The relative rapid migration of nitrate-N, also in the deeper soil layers was surprising because there was a shortage of precip itation in that period of time. Namely, its amount was 172 mm smaller than the long-term precipitation. The speed of nitrate-N leaching may be about 60 cm/year, this is the double of that data reported by Kádár *et al.* (1987) in Hungary. The higher speed in our case probably is due to the sandy-texture of our soil under the 90 cm layer.

The above results strongly suggest that nitrate accumulation maximum caused by excessive N-rates or the unused rest of extremely intensive mineralization can reach relatively quickly the groundwater table at our location. Yields of w.wheat and maize (here not demonstrated) have reached their maximum at lower N-rates than N2. In most cases the recommended N-rate for maize and w.wheat in Hungary are remarkably under 200 kg/ha. The demonstrated nitrate profils also suggest that N-fertilizer applied at optimum rates for yield, do not cause greater nitrate-N accumulation in the subsoil and so provide little risk for potential contamination of groundwater at our site too.

Acknowledgement

The investigations were financially supported by the Hungarian Scientific Research Foundation (OTKA No 1868).

References

Dowdell RJ, Colbourn P and Cannell RQ (1987) A study of mole drainage with simplified cultivation for autumn-sown crops on a clay soil. 5. Losses of nitrate-N in surface runoff and drain water. Soil Till Res 9: 317–331.

Eck, HV and Jones OR (1992) Soil nitrogen status as affected by tillage, crops and crop sequences. Agron J 84: 660–668.

Francis GS and Knight TL (1993) Long-term effects of conventional and no-tillage on selected soil properties and crop yields in Canterbury, New Zealand. Soil Till Res 26: 193–210.

Hoffmann S. (1993) A N ellátás szerepe a kukoricatermesztésben barna erdőtalajokon. (The role of N-supply in maize cultivation on cambisols.) Dissertation, Pannon Univ Agric Keszthely, Hungary.

Hütsch B and Mengel K (1992) Effect of long-term ploughing and direct drilling on the distribution and leaching of mineral nitrogen in a loamy and a sandy soil. Eur J Agron 1: 213–219.

Kádár I, Németh T and Kovács G (1987) Nitrogen efficiency and nitrate leaching on a calcareous chernozem soil. 5th International Symposium of CIEC. Symposium Document Balatonfüred, Hungary 2,1: 130–133.

Lamb JA, Peterson GA and Fenster CR (1985) Fallow nitrate accumulation in a wheat-fallow rotation as affected by tillage system. Soil Sci Soc Am J 49: 1441–1446.

Logsdon SD, Allmaras RR, Wu L, Swan JB and Randall GW (1990) Macroporosity and its relation to saturated hydraulic conductivity under different tillage practices. Soil Sci Soc Am J. 54: 1096–1101.

MacDonald AJ, Powlson DS, Poulton PR and Jenkinson DR (1989) Unused fertiliser nitrogen in arable soils: Its contribution to nitrate leaching. J Sci Food Agric 46: 407–419.

McMahon MA and Thomas GW (1976) Anion leaching in two Kentucky soils under conventional tillage and a killed sod mulch. Agron J 68: 437–442.

Sharply AN, Smith SJ, Williams JR, Jones DR and Coleman GA (1991) Water quality impacts associated with sorghum culture in the southern plains. J Environ Qual 20: 239–244.

C. Rodriguez-Barrueco (ed.), Fertilizers and Environment, 233–236.

Improvement of the nitrogen uptake induced by Ti(IV) supply in nitrogen stressed pepper crops

J.J. Pastor[1], M.J. Frutos[2], M. Carvajal[2], F. Martínez-Sánchez[1] & C.F. Alcaraz[2]
[1]*EUITA-ORIHUELA, Polytechnical University of Valencia, Orihuela (Alicante), Spain;* [2]*CSIC-CEBAS, P.O. Box 4195, 30080-Murcia, Spain*

Key words: Nitrogen, titanium, fertilizer efficiency, plant nutrition

Abstract

The beneficial effect of titanium on plant metabolism can result in a more profitable use of fertilizers on many crops. An experiment was performed in a greenhouse with *Capsicum annuum* L. cv. Bunejo plants cultivated under different levels of nitrogen fertilization: 100% N – 250 N fertilizer units/ha (NFU); 75% N – 187.5 NFU; 50% N – 125 NFU; 25% N – 62.5 NFU. A third of the plants in each N support remained Ti-untreated and were used as a reference. Each of the other thirds received one and two 2 mg Ti/l, pH 6, leaf spray treatments respectively (35 ml/plant). Biomass production of the Ti-untreated references was only affected by N-supplies of 25%. Ti leaf spray treatments induced a biomass production higher than the corresponding references; the 100% N + Ti and 75% N + Ti treatments both showing the highest production. Plants under 25% N + Ti showed a level of biomass significantly higher than the 100% N without Ti reference. The efficiency index of the N-fertilization on the biomass production in the 25% N + Ti treatment increased to values five times higher than in the 100% N reference. Also, the efficiency index for the fruit yields showed a similar increase. The absolute fruit yield increased by more than 30% for the 25% N + Ti experiment against the 100% N reference (39,902 and 28,669 kg/ha respectively). The highest fruit yield was obtained for the 100% N + Ti treatment showing 60,899 kg/ha (112% of increase against the 100% N reference).

Introduction

The growth, biomass and productivity of many plant species are enhanced by titanium, whether applied as fertilizer to the soil or by leaf spray (Pais, 1983; Ram *et al.*, 1983; Kiekens & Camerlynck, 1987; Giménez *et al.*, 1990). The concentration of some essential elements has also been reported to increase when plants are supported with titanium (Feher *et al.*, 1987; Martínez-Sánchez *et al.*, 1992, 1993). Up to now, there is no conclusive explanation for this phenomenon, though it was suggested that it can be due to one of the following: the well known Viet's effect (Dumon & Ernst, 1988), a higher water uptake efficiency (Giménez *et al.*, 1990); a higher nutrient requirement induced by the increase of the photosynthetic activity (Carvajal *et al.*, 1994, 1994a; Kiss *et al.*, 1985; Pais, 1983) promoted, in its turn, by the stimulation of the synthesis of multimetal cofactors of some metal-binding storage proteins as the phytoferritins (Bienfait & Van der Mark, 1983); or by the increase in the activity of some micronutrients as iron in the leaf cell chloroplasts and cytoplasm where titanium accumulates in the leaf sprayed plants (Carvajal *et al.*, 1995; Alcaraz *et al.*, 1994). It was suggested that Ti plays an unknown role in the nitrogen fixation process in some legumes (Pais *et al.*, 1977) thus increasing the yield of these crops. In the same way, Wakamoto (1973) and Tsukamoto (1975) communicated that some of the growth promotive effects of the soil applied with Ti may be related to the stimulation of the denitrification process with further promotion of the plant N availability, but this effect does not explain the growth enhancement in both hydroponical solution culture and leaf spray experiments.

The way in which titanium plays its beneficial role on plant metabolism, and its promotive effects on the absorption and activity of some nutrients, is very useful in the regulation of the fertilizer supply. Therefore the aim of this paper is to study, in a greenhouse experiment with paprika pepper plants (*C. annuum* L.), the effects related to the progressive decrease of the nitrogen fertilization and to the Ti(IV) leaf application on the plant development, the fruit yield, and the nitrogen uptake.

Material and methods

Plant and experimental conditions

Seeds of *C. annuum* L. cv. Bunejo were obtained from the Centro de Investigaci6n y Desarrollo Agrario of the Region of Murcia, Spain. Seedlings were obtained through direct sowing in 50% mixture of sand and manure (v:v) in polyurethane trays (one seed/50 cm^3 receptacle). After 95 days from sowing, the little plants were transplanted to a greenhouse equipped with a drip irrigation system and with controlled environmental conditions. The greenhouse was divided into four plots for the different N fertilization as follows: Plot A – 100% N: 70 g/week/plant of 33.5% NH_4NO_3, equivalent to 250 nitrogen fertilizer units/ha (NFU); Plot B – 75% N: 52.5 g/week/plant, 187.5 NFU; Plot C – 50% N: 35 g/week/plant, 125 NFU; Plot D – 25% N: 17.5 g/week/plant, 62.5 NFU. Each plot was subdivided in three blocks for the different Ti supplies as follows: Block 1: Without Ti, reference treatment. Block 2: One leaf spray treatment using 35 ml/plant of a 2 mg Ti/l solution [TITAVIT, Ti(IV)-ascorbate complex, 1 g Ti/l, granted by Nitrokemia, Budapest Hungary], 45 days after transplanting. Block 3: One leaf spray as in block 2, and a second similar spray 30 days later. In this manner, 12 different treatments were obtained, as follows: 1: 100% N (R1); 2: 100% N + Ti(1); 3: 100% N + Ti(2); 4: 75% N (R2); 5: 75% N + Ti(1); 6: 75% N + Ti(2); 7: 50% N (R3), 8: 50% N + Ti(1); 9: 50% N + Ti(2); 10: 25% N (R4); 11: 25% N + Ti(1); 12: 25% N + Ti(2).

All the other crop conditions (fertilization, water doses, crop practices, phytosanitary treatments, etc.) were identical for all the experimental plants. Data for plant biomass was obtained from a sampling of ten complete plants (each plant an individual sample) performed 105 days after transplanting, and the fruit harvest was carried out one month later.

The *efficiency (E)* of a treatment is the ratio between the produced effect and the intensity of the treatment, and the *efficiency index (EI)* is the quotient of the efficiencies of two treatments.

Analytical determinations

Mineralization was performed according to the catalyzed semi-micro Kjeldahl method. Nitrogen was quantified by stream distillation of the formed NH_3 and potentiometrically determined with N/14 HCl.

Nitrate extraction from 1 g of leaf powder was performed by means of a sonical device. The nitrate of an aliquot of this extract (usually 3 ml) was stoichiometrically reduced to nitrite by native metallic cadmium in NH_4Cl-NH_4OH buffer, pH 9.6, and then 1 ml of sulfanilamide solution (1% in 25% HCl, v:v) and 1 ml of N-1 naphtyl-ethylendiamine-dihydrochloride solution (0.02% in distilled water, w:v) were added. The coloured sample was read spectrophotometrically against a reference without nitrate at 539 nm (Alcaraz *et al.*, 1982).

Results and discussion

The diminution of the nitrogen fertilization only promotes a significant decrease in the biomass production of the plants growing under the 62.5 NFU [10: 25% N (R4)] against the general control treatment [1: 100% N (R1)] (Table 1). Because of this, the efficiency of the nitrogen on the plant biomass production increases correspondingly to the diminution in the intensity of the nitrogen fertilization. Therefore, it is evident that there is an excess in the 250 NFU fertilization.

All the experimental plots treated with titanium via leaf spray, showed a significative enhancement of the biomass against their corresponding controls, and it is interesting to note that the plants growing under 25% N-fertilization plus Ti-spray supports (one or two, treatments 11 and 12) showed a biomass production significantly higher than the plants growing under the highest N-fertilizations without titanium (treatments 1 and 4). These low-N fed Ti-treated plants show an efficiency index five times higher than the general control treatment (R1).

There are no significant differences in the Total-N concentration among the four control treatments (1, 4, 7 and 10), which indicates that the decrease of the N-fertilization did not induce any N-unbalance. Similarly, the treatments with titanium did not significantly modify the total nitrogen concentrations of these plants. On the contrary, the NO_3-N levels were affected by the titanium treatments, not by the diminution of the N fertilization (Table 2). These apparent-

Table 1. Effect of Ti(IV) leaf supply on the biomass production and on the nitrogen efficiency related to the delopment of *Capsicum annuum* L. cv. Bunejo plants. Biomass data are the average values from ten samples and they are expressed as dry matter.

Treatment		Biomass, g/plant mean ± s	N efficiency on biomass yield		
			E	EI_{Ri}	EI_{R1}
1:	100% N (R1)	80.60 ± 8.34bc	0.806	1.00	1.00
2:	100% N+ Ti (1)	111.71 ± 10.74ef	1.112	1.39	1.39
3:	100% N + Ti (2)	138.14 ± 13.51g	1.381	1.71	1.71
4:	75% N (R2)	86.63 ± 8.92c	1.156	1.00	1.43
5:	75% N + Ti (1)	111.75 ± 11.25ef	1.490	1.29	1.81
6:	75% N + Ti (2)	134.33 ± 13.49g	1.791	1.55	2.22
7:	50% N (R3)	75.42 ± 7.76ab	1.508	1.00	1.87
8:	50% N + Ti (1)	111.57 ± 10.81ef	2.231	1.48	2.77
9:	50% N + Ti (2)	120.79 ± 11.73f	2.416	1.60	3.00
10:	25% N (R4)	71.05 ± 7.39a	2.842	1.00	3.53
11:	25% N + Ti (1)	101.41 ± 9.95de	4.056	1.43	5.03
12:	25% N + Ti (2)	97.25 ± 9.38d	3.890	1.37	4.83

E: Efficiency; EI: Efficiency Index related to each treatment control (Ri) or to the general reference (R1). Biomass column: means followed by a same letter are not significantly different at 5% level by LSD test.

Table 2. Total-N and NO_3-N concentrations and total-N content in *C. annuum* L. cv. Bunejo plants. Data are the average values from ten samples and they are expressed on dry matter.

Treatment		Plant concentration		Plant content
		Total-N, g/kg mean ± s	NO_3-N, mg/kg mean ± s	Total-N, g/plant mean ± s
1:	100% N (R)	23.5±2.4abcd	29±3a	2.241±0.240b
2:	100% N + Ti (1)	23.9 ± 2.0bcd	49 ± 5bcd	3.077 ± 0.279de
3:	100% N + Ti(2)	24.6 ± 2.1cd	49 ± 5bcd	4.298 ± 0.359h
4:	75% N (R2)	22.4 ± 2.0ab	29 ± 3a	1.866 ± 0.201a
5:	75% N + Ti (1)	26.9 ± 2.3e	45 ± 5b	3.179 ± 0.278ef
6:	75% N + Ti (2)	25.4 ± 2.2de	51 ± 6cd	3.680 ± 0.300g
7:	50% N (R3)	21.8 ± 1.9a	30 ± 4a	2.213 ± 0.241b
8:	50% N + Ti (1)	23.8 ± 2.0bcd	47 ± 5bc	2.769 ± 0.255cd
9:	50% N + Ti (2)	24.0 ± 2.1bcd	50 ± 6bcd	3.431 ± 0.298fg
10:	25% N (R4)	23.8 ± 2.4abcd	31 ± 3a	1.780 ± 0.199a
11:	25% N + Ti (1)	24.7 ± 2.2cde	45 ± 5b	2.593 ± 0.213c
12:	25% N + Ti (2)	22.9 ± 1.9abc	53 ± 6d	2.657 ± 0.220c

For each column, means followed by a same letter are not significantly different by LSD test at 5% level.

ly contradictory results are explained by the different units that both concentrations have (Total N: g/kg and NO_3-N: mg/kg), likewise, there was not influence of the increase of NO_3-N on total-N. Nitrate concentration in plant tissues is generally accepted as a suitable indexing parameter of its absorption rythm by plants. Therefore, from the data of Table 2 can be deduced that the higher concentration of NO_3-N, which titanium promotes, mean an enhancement of nitrate absorption. This higher concentration of nitrate could be used by the plant in the formation of new structures which gave an increase of total-N per plant. It is interesting to note that the total-N contents in the plants growing under 25% N + Ti (treatments 11 and 12) reached val-

Table 3. Effect of Ti(IV) leaf supply on the fruit yield and on the nitrogen efficiency related to fruit yield in N-stressed *C. annuum* L., cv. Bunejo plants.

Treat- ment		Fruit yield, kg/ha	E	EI_{Ri}	EI_{R1}
1:	100% N (R1)	28,669	3.58	1.00	1.00
2:	100% N + Ti (1)	44,222	5.53	1.57	1.57
3:	100% N + Ti (2)	60,899	7.71	2.15	2.15
4:	75% N (R2)	27,836	4.64	1.00	1.30
5:	75% N + Ti (1)	39,770	6.63	1.43	1.85
6:	75% N + Ti (2)	50,808	8.47	1.83	2.36
7:	50% N (R3)	26,866	6.72	1.00	1.87
8:	50% N + Ti (1)	38,259	9.56	1.42	2.67
9:	50% N + Ti (2)	45,125	11.28	1.68	3.15
10:	25% N (R4)	23,211	11.61	1.00	3.24
11:	25% N + Ti (1)	34,668	17.33	1.46	4.84
12:	25% N + Ti (2)	39,902	19.95	1.72	5.57

E: Efficiency; EI: Efficiency Index related to each reference treatment (Ri) or to the general reference (R1).

ues higher than those growing under 100% N without titanium (1, general reference).

Table 3 shows the data corresponding to the fruit yield of the different treatments. It is evident that the fruit production shows a similar trend to the total plant biomass, with a slow diminution related to the N-feed intensity and with an important increase when the plants were Ti-leaf sprayed. The highest fruit yield corresponds to the third experiment (60,899 kg/ha). When the plants growing under the lowest N-fertilization were Ti-leaf sprayed, they yielded a higher fruit production than the reference plants (treatment 1).

The beneficial effect of titanium on the nitrogen uptake and assimilation and consequently on the fruit yield, is indicated by an increase of the N-fertilizer efficiency, reaching levels up to five times higher than in plants growing under the highest N-fertilization.

Acknowledgements

The authors gratefully acknowledge the financial supports provided by CICYT (Spain), project AGF 93–0141 and Consejería de Agricultura de la Comunidad Autónoma de la Región de Murcia, as well as to NITROKEMIA, Budapest (Hungary) for providing TITAVIT used in the experiments.

References

Alcaraz CF, Romojaro F, León A & Llorente S (1982) Anionic relationships in leaf petiole sap of tomato and capsicum plants growing in a glasshouse. J Plant Nutr 5: 173–181

Alcaraz CF, Carvajal M, Frutos MJ, Giménez JL, Martínez-Sánchez F & Pastor JJ (1994) The physiological role of titanium in *Capsicum annuum* L. plants. Proc 6th Int Trace Element Symp Abs 2, Budapest

Bienfait HF & Van Der Mark F (1983) Phytoferritin and its role in iron metabolism. In: Robb DA et al. (eds) Metals and Micronutrients. Academic Press, London, 111–123 and Pierpoint WS

Carvajal M, Pastor JJ, Martínez-Sánchez F and Alcaraz CF (1995) Leaf spray with Ti(IV) ascorbate improves the iron uptake and iron activity in *Capsicum annuum* L. plants. In Abadia J (ed) Iron Nutrition in Soils and Plants. Developments in Plant and Soil Sciences. Vol 59. Kluwer Academic Publishers. The Netherlands, 1–5.

Carvajal M, Martínez-Sánchez F & Alcaraz CF (1994) Effect of Titanium(IV) application on some enzymatic activities in several developing stages of red pepper plants. J Plant Nutr 17: 243–253

Carvajal M, Martínez-Sánchez F & Alcaraz CF (1994a) Effect of Ti(IV) on some indicators of physiological activity in *Capsicum annuum* L. J Hort Sci 69: 427–432

Dumon JC & Ernst WHO (1988) Titanium in plants. J Plant Physiol 133: 203–209

Feher M, Papp K, Fodor P & Pais I (1987) The effect of titanium on the uptake of other nutritive elements. In: Pais I (ed.) New Results in the Research of Hardly Known Trace Elements and the Analytical Problem of Trace Element Research. University of Horticulture, Budapest, 69–100

Giménez JL, Martínez-Sánchez F, Moreno JL, Fuentes JL & Alcaraz CF (1990) Titanium in plant nutrition. III. Effect of Ti(IV) on yield of Capsicum annuum L. In: Barceló J (ed.) Nutrición Mineral bajo Condiciones de Estrés. SPIC-UIB, Palma de Mallorca, 123–128

Kiekens L & Camerlynck R (1987) Influence of titavit on growth of maize and cowpea grown in nutrient solution. In: Pais I (ed.) New Results in the Research of Hardly Known Trace Elements and the Analytical Problems of Trace Element Research, University of Horticulture, Budapest, 133–147

Kiss F, Deak G, Feher M, Balogh L, Szabolsci L & Pais I (1985) The effect of titanium and gallium on photosynthetic rate of algae. J Plant Nutr 8: 825–832

Martínez-Sánchez F, Giménez JL, Carvajal M, Moreno A, Fuentes JL & Alcaraz CF (1992) Efectos de tratamientos foliares con Ti(IV) sobre nutrición, producción y calidad de fruto en plantas de *Capsicum annuum* L. Suelo y Planta 2: 101–111

Martínez-Sánchez F, Núñez M, Amorós A, Giménez JL & Alcaraz CF (1993) Effects of titanium leaf spray treatments on ascorbic acid levels of *Capsicum annuum* L. fruits. J Plant Nutr 16: 975–981

Pais I (1983) The biological importance of titanium. J Plant Nutr 6: 3–131

Pais I, Feher M, Farkas E, Szabo Z & Cornides I (1977) Titanium as a new trace element. Comm Soil Sci Plant Anal 8: 407–410

Ram N, Verloo M & Cottenie A (1983) Response of bean to foliar spray of titanium. Plant & Soil 73L: 285–290

Tsukamoto F (1975) Fertilizer stabilization by titanium-iron containing substances. Jap Kokai 75–18, 242

Wakamoto K (1973) Inhibition of ammonia volatilization from soils by iron and titanium compounds. Jap Kokai 73–25, 354

C. Rodriguez-Barrueco (ed.), Fertilizers and Environment, 237–240.

Effect of slow-release N-fertilizers on yield and baking quality of winter wheat

I. Ragasits, J. Balázs & K. Berecz
Pannon University of Agricultural Sciences, Keszthely Deák F. str. 16, H-8361, Keszthely, Hungary

Key words: Wheat, slow-release N-fertilizers, N-uptake, yield, baking quality

Abstract

The aim of our study was to compare the effect of different N-forms (ammonium-nitrate, AN, ureaform /Formurin, FO/ and paraffin coated urea /Paramid, PA/) on N-leaching, N-uptake, grain yield and baking quality of winter wheat (*Triticum aestivum* L.). The experiment was carried out on Cambisol, in Keszthely (Hungary).

The favourable features of the slow-release N-fertilizers are a steady N-supply and reduced N-leaching in soil. At late autumn the FO- and PA-doses generally resulted in smaller N_{min}-amount in soil (layer of 0–90 cm) by 38% and 15%, respectively than AN-doses. The amount of N_{min} in this layer did not exceed that in control plots.

The different N-supply of the soil could be observed in N-uptake by the wheat plant also. The AN- and PA-treatments caused a similar N-uptake both at stem-extension stage and flowering, but when applying FO, the N amount in wheat plant slightly decreased. On the contrary, the yield did not considerably depend on the N-form of fertilizers.

Similarly to grain yield, the parameters of baking quality were not favourably influenced by slow-release N-fertilizers in comparison to AN-variant. 160 kg ha^{-1} N was necessary to achieve the highest wet gluten content (WG), Zeleny's number (ZN) and valorigraph value (VV). Neither FO- nor PA-treatments resulted in a significant change in the above quality parameters as compared with AN-doses. Gluten spreading (GS) and water absorption capacity (WAC) did not show any statistically proved alteration on applying different sources or doses of N-fertilizers.

The use of expensive slow-release fertilizers has proved to be reasonable from the point of view of environmental pollution. But the higher financial input will be repaid neither in the yield nor in the quality of wheat.

Introduction

The use of slow-release N-fertilizers reduce N-leaching. The favourable environmental effect of such N-forms have been proved by many specialists (Addiscott and Cox, 1976; Balázs, 1986, 1991). However, the slow-release N-fertilizers do not always have a clear positive influence on the grain yield or quality of winter wheat. Boswell (1977), and Aulakh and Rennie (1984) found these N-forms more efficient, but Balázs (1991) observed slight yield increase in some cases.

Peltonen and Virtanen (1994) reported that N-fertilizers differing in release properties did not significantly change the grain protein content and gliadin or glutenin characteristics, while loaf volume was affected in some cases. Timms *et al.* (1981) found that late application of urea increased the breadmaking quality only of flours of low protein content. Stenram *et al.* (1990) applying late urea supplement (to Ca-nitrate) concluded that "it is the amount rather than the form of N which is of relevant for total protein accumulation".

The contradictory results published and the ecological and economic circumstances, different from ours, recommend to study the use of slow-release N-fertilizers in wheat production practice under our own conditions.

Materials and methods

The field experiment was carried out on Cambisol, in Keszthely (Hungary) for four years. Three N-fertilizers differing in release characteristics were applied to winter wheat: ammonium-nitrate (AN – 34% N), paraffine coated urea (PA – 41.5% N) and ureaform (FO – 38.4% N). The fertilizers were given in autumn before sowing in doses of 0–80–120–160–200 kg ha^{-1} N, 150 kg ha^{-1} P_2O_5 and 150 kg ha^{-1} K_2O. The trial had a randomized block design in four repetitions.

Soil samples were taken in autumn, spring and after harvest up to a depth of 90 cm from layers of 0–30 cm, 30–60 cm and 60–90 cm. (NH_4 + NO_3)-N content of dry soil samples was measured using 1% KCl extractant. Plant samples (3 m of rows) were gathered at stem-extension stage, flowering and harvest. The dry matter weight was measured and N-content of plant was analysed. One year the baking quality of flour was determined according to the relating Hungarian standards: wet gluten content (WG), gluten spreading (GS), water absorption capacity (WAC), Zeleny's number (ZN) and valorigraph value (VV).

Results

N-supply in soil

Autumn and spring soil analyse data show that increasing doses of N-fertilizers raised the amount of mineral N in soil profile. The highest N_{min}-amount could be detected in AN-treatments, and the smallest amount at FO-variants. Using AN and PA, the (NH_4 + NO_3)-N amount continuously increased with the doses of fertilizers in the investigated soil profile (0–90 cm). Applying FO, a remarkable increase in soil-N appeared only in the treatment of 200 kg ha^{-1}. In autumn AN mostly caused N-accumulation in the layer of 60–90 cm (Table 1a).

In spring the paraffin of PA is not able to reduce the N-recovery any more. At that time a lower N-leaching could be found only in FO-treatments (Table 1b). Regarding N_{min}-content both in 0–90 cm and in 60–90 cm soil layers, the fertilizers have the same order, viz. AN, PA and FO. 20% less mineral N could be detected on average in the soil samples of FO-treatments.

The difference between effects of N-fertilizers on soil-N decreased at harvest (Table 1c). As our data show, unused fertilizer-N pollutes the soil. Applying FO, smaller Nmin-amount remains in the layer of 60–90 cm compared to the use of AN or PA.

Table 1. Changes in N_{min}-content of soil sampled at different times as an effect of different N-fertilizer forms.

1a. Autumn

depth cm	N-dose kg ha^{-1}	N_{min} kg ha^{-1}			
		AN	FO	PA	average
60–90	80	39.9	42.6	37.1	39.9
	120	45.2	46.6	33.9	41.9
	160	54.9	40.2	46.8	47.3
	200	65.7	36.5	47.0	49.7
	control	35.6			
	average	51.4	41.5	41.2	
LSD 5%	treatment	12.6			
	N-form	5.9			
	N-dose	6.8			

1b. Spring

depth cm	N-dose kg ha^{-1}	N_{min} kg ha^{-1}			
		AN	FO	PA	average
0–90	80	150.1	147.5	175.2	157.6
	120	171.7	159.7	212.0	181.1
	160	212.0	170.5	213.7	198.7
	200	248.1	181.0	223.3	217.5
	control	123.3			
	average	195.4	164.7	206.0	
LSD 5%	treatment	11.8			
	N-form	6.0			
	N-dose	6.9			
60–90	80	52.8	49.8	62.3	59.9
	120	66.7	55.7	72.3	64.9
	160	76.2	55.9	72.6	68.2
	200	87.5	58.5	71.0	72.3
	control	41.9			
	average	70.8	54.9	69.5	
LSD 5%	treatment	6.8			
	N-form	3.5			
	N-dose	4.0			

1c. Harvest

depth cm	N-dose kg ha^{-1}	N_{min} kg ha^{-1}			
		AN	FO	PA	average
60–90	80	33.7	32.6	42.5	36.2
	120	34.9	36.5	47.3	39.5
	160	46.1	37.0	49.0	44.0
	200	50.2	43.7	52.0	48.6
	control	30.9			
	average	41.2	37.4	47.7	
LSD 5%	treatment	5.5			
	N-form	2.7			
	N-dose	3.2			

N in wheat plant

The N-supply in soil determines the N-uptake of wheat. At treatments of AN and PA, the N-amount taken up by plant was near the same at stem-extension stage, but on plots treated with FO it was smaller on average by 10% (Table 2a, 2b). At flowering the trend of N-amount utilized by plant was similar to that of earlier growth stage. The higher AN and PA doses significantly increased the N-amount in plant, but this was not observed at FO-treatments.

The N-yield of grain showed a significant increase up to the N-doses of 120 kg ha^{-1} when AN or PA were applied. Higher N-fertilizer doses did not result in higher N-yield (Table 2c). At FO-treatments a lower N-content could be detected both in soil and plant not only at the beginning of the vegetative period, but also at harvest.

The N-fertilizers of experiment had a positive effect on grain-yield up to doses of 120–160 kg ha^{-1}, but the dose of 200 kg ha^{-1} decreased yield in most cases. The yield data show that the effect of applied fertilizers do not significantly differ from each other on this site. The relationship between grain yield and doses of N-fertilizers could be described by a quadratic equation significant at least at P=5%.

AN: $Y = 4.51 + 0.012\,X - 0.000045\,X^2$,
N-dose$_{max}$ = 131.6 kg ha^{-1}, yield$_{max}$ = 5.289 t ha^{-1}, R = 0.99

FO: $Y = 4.53 + 0.012\,X - 0.000042\,X^2$,
N-dose$_{max}$ = 143.5 kg ha^{-1}, yield$_{max}$ = 5.398 t ha^{-1}, R = 0.98

PA: $Y = 4.54 + 0.11\,X - 0.000045\,X^2$,
N-dose$_{max}$ = 130.6 kg ha^{-1}, yield$_{max}$ = 5.313 t ha^{-1}, R = 0.98

Baking quality

As it appears from Table 3, no advantage or disadvantage could be detected in the effect of the applied slow-release fertilizers on baking quality compared with AN. The WG-content, ZN and VV showed a continuous increase as a result of higher AN doses, while significant increases were recorded at a N-dose up to 160 kg ha^{-1}. The highest significant increases amounted to 51, 28 and 68% in the value of WG-content, ZN and VV, respectively.

Neither form nor dose of slow-release N-fertilizer resulted in a significantly higher value of the above baking quality parameters than those reached at 160

Table 2. Changes in N-content of wheat at different developmental stages as under different N-fertilizer forms.

2a. Stem extension

N-dose kg ha^{-1}	N kg ha^{-1} in wheat plant			
	AN	FO	PA	average
80	82.8	88.2	82.6	84.5
120	94.9	83.9	92.4	90.4
160	94.6	87.1	94.9	92.2
200	104.7	87.7	105.8	99.4
control	66.4			
average	94.3	86.7	93.9	
LSD 5%				
treatment	8.1			
N-form	4.0			
N-dose	4.6			

2b. Flowering

N-dose kg ha^{-1}	N kg ha^{-1} in wheat plant			
	AN	FO	PA	average
80	118.6	115.3	123.1	119.0
120	132.3	117.8	137.0	129.1
160	136.1	120.0	137.0	131.0
200	138.5	122.4	146.3	135.9
control	100.7			
average	131.4	118.9	135.0	
LSD 5%				
treatment	10.8			
N-form	5.5			
N-dose	6.3			

2c. Harvest

N-dose kg ha^{-1}	N kg ha^{-1} in grain			
	AN	FO	PA	average
80	89.6	94.3	104.4	96.1
120	98.7	94.8	103.9	99.1
160	100.3	102.0	104.6	102.3
200	101.0	101.5	107.6	103.4
control	75.3			
average	97.4	98.2	105.1	
LSD 5%				
treatment	7.3			
N-form	3.0			
N-dose	4.3			

Table 3. Changes in baking quality parameters as an effect of different N-fertilizer forms.

N-dose kg ha^{-1}	WG %	GS mm	VV	WAC	ZN
AN					
0	19.6	2.0	31.3	63.8	28.5
80	25.8	2.1	36.1	63.9	33.8
120	26.9	2.3	56.7	66.2	41.0
160	29.6	2.9	71.3	66.4	48.0
200	31.7	3.1	70.7	63.8	49.5
FO					
80	26.2	2.3	68.3	65.7	39.8
120	24.5	3.0	62.2	64.4	38.3
160	26.9	3.0	61.8	65.3	41.0
200	28.1	3.4	70.1	66.0	44.5
PA					
80	28.0	3.1	68.2	66.3	46.5
120	27.4	3.3	64.2	65.6	46.0
160	31.0	4.0	66.3	66.4	48.8
200	31.6	3.3	69.0	66.1	50.3
LSD 5%	3.2	1.2	11.7	2.4	7.7
P%	0.1	N.S.	0.1	N.S.	0.1

kg ha^{-1} AN. The GS and WAC did not show any statistically proved changes as an effect of the added N-doses or N-forms.

Discussion

The estimation of N-fertilizers differing in release characteristics is a complex task. Both in respects of environmental protection and from the point of view of economical aspects it is required to reduce the N-leaching to the lowest level possible in wheat production practice.

Soil analityical data proved that among the three investigated N-fertilizers the fall-applied AN caused considerable pollution in the layer of 60–90 cm. The paraffin coat of urea could decrease the N-recovery only for a short period. In spring in PA-treatments the N_{min} location in soil profile is similar to that of AN. From FO a smaller N-amount is recovered, so less N_{min} is translocated into the deeper layers. FO has favourable features in environmental respects.

Soil analyses data obtained at harvest show which N-fertilizer doses did not exceed the N-requirement of wheat. When the N-doses do not exceed the requirement of wheat, the soil pollution is insignificant. The lower N_{min} content results in a smaller N-uptake of wheat in FO-treatments. However, the grain yields were similar in the three fertilizer-treatments.

The effect of FO is positive in environmental respect, but indifferent in respect of yield and baking quality. In this case our decision depends on economical aspect. At present in wheat production the split application of traditional N-fertilizers is cheaper. Where N-fertilizers are applied in optimal doses and optimal times, the risk of N-leaching losses can be minimized.

Acknowledgement

This work was supported by the Hungarian Scientific Research Foundation, OTKA No 5009.

References

Addiscott TM and Cox D (1976) Winter leaching of nitrate from autumn-applied calcium nitrate, ammonium sulphate, urea and sulphur coated urea in bare soil. J Agric Sci, Camb 87: 381–389.
Aulakh MS and Rennie DA (1984) Transformations of fall-applied nitrogen[15]-labelled fertilizers. Soil Sci Soc Am J 48: 1184–1189.
Balázs J (1986) The effect of various N fertilizers on yield and N-leaching in experiments on winter wheat (*Triticum aestivum* L.). 9th CIEC Cong, Budapest, Proc 275–278.
Balázs J (1991) N-leaching at brown forest soils in Transdanubia. XXXIII Georgikon Days Proc 267–273.
Boswell FC (1977) Seasonal anhydrous ammonia comparison for corn with and without a nitrification inhibitor. Agron J 69: 103–106.
Peltonen J and Virtanen A (1994) Effect of nitrogen fertilizers differing in release characteristics on the quantity of storage protein in wheat. Cereal Chem 71: 1–5.
Stenram U, Heneen WK and Olered R (1990) The effect of nitrogen fertilizers on protein accumulation in wheat (*Triticum aestivum* L.) Swedish J Agric Res 20: 105–114.
Timms MF, Bottomley RC, Ellis JRS and Schofield JD (1981) The baking quality and protein characteristics of a winter wheat grown at different levels of nitrogen fertilisation. J Sci Food Agric 32: 684–698.

C. Rodriguez-Barrueco (ed.), Fertilizers and Environment, 241–244.

Effect of nitrogen fertilization and plant population on the yield of soybeans

A.D. Simonis & H.B. Setatou
Soil Science Institute, 541 10 Thessaloniki, Greece

Key words: Soybean, *Glycine max (L.) Merr.*, nitrogen fertilization, plant population

Abstract

Field experiments were conducted, over a 5-year period at 8 different sites of Greece, to determine the effects of N fertilization (0, 30, 60 kg N per ha) and plant population (10, 20, 30 plants per m) on the yield of soybeans. In general, the soybean yields were not affected by N fertilization, except in some sites with low soil fertility, where a supplementary N fertilization at the beginning of pod filling phase, resulted in a 3–20% yield increase. On an average, soybean yields were increased significantly with 20 or 30 plants per m in comparison with 10 plants per m, although the increase of plant population from 20 to 30 plants per m did not affect yields. It was concluded that, in soils with a sufficient number of rhizobia, from previous crops or from seed inoculation, N fertilization is not necessary under Greek conditions.

Introduction

The soybean (*Glycine max* [L.] Merr.) plant, as a legume plant, is expected to receive most of its nitrogen (N) supply from symbiotic N fixation. However, it was found (Allos & Bartholomew, 1959) that supplying a small amount of inorganic N to soybeans may increase the amount of N fixed and that only one-half to three-fourths of the N required for maximum yields could be supplied symbiotically. Results on soybean yield response to N applied fertilizer are often erratic (Beard and Hoover, 1971; Hanway and Weber, 1971). Yet only a few reports of measured yield increases from N fertilization are present in the literature (Bhangg and Weber, 1972) and these increases are often small. Several investigators report no effect of N fertilization on soybean yield (Weber *et al.*, 1973).

It is generally accepted that the maximum economic yield is obtained with the smaller number of plants required for their maximum biological yield (Donald, 1963). However, it was reported that soybean yields are lower with the largest number of plants (Buttery, 1969 a, b). No interaction was found between N fertilization and plant population (Anderson, 1967).

The purpose of this study was to determine the effects of N ferilization and plant population on soybean seed yield.

Methods

Thirty-two experiments were conducted during five successive years at eight sites located in Northern Greece. The soils used were fine to coarse textured, with low organic matter (0.8 to 1.2%) and pH ranging from 5.8 to 6.4.

Three levels of N (0, 30, 60 Kg N per ha) and three levels of plant population (10, 20, 30 plants per m) were tested in a factorial 3^2 design, with four replications. Soybean seeds (var. Williams), inoculated with rhizobia (10^6 rhizobia per seed, commercial inoculant Nitragin) were planted in four-row plots, 0.96 m wide and 10 m long. Nitrogen was applied as NH_4NO_3 in two split applications, one-half at the beginning of blooming and the remainder half at the pod filling period. A standard of 30 and 75 kg per ha of P and K, respectively, was added annually to each experiment. Soybeans were forrow irrigated and all other managements were made according to the practice used in the region. At maturity a 5 m length of the two centre rows was harvested for yield.

Results and discussion

Soybean yields over the five year period at the eight experimental sites are shown in Table 1. In most of

Table 1. Effects of N fertilization and plant population on the yield of soybeans (kg/ha)

Treatments	Giannitsa 1977	1978	1979	1980	1981	Aver.	Drama 1978	1979	1980	1981	Aver.	Xanthi 1978	1979	1980	1981	Aver.	Serres 1977	1978	1979	1980	Aver.
N_o = control	3830	3290	4560	4920	4080	4140	4720	3840	4560	3760	4220	3240	3830	3930	2280	3320	4960	3980	4350	2630	3980
N_1 = 30 kgN/ha	3740	3290	4480	4650	4130	4060	4700	3920	4610	3730	4240	2900	4000	3880	2620	3350	4600	4170	4180	3010	3990
N_2 = 60 kgN/ha	4010	3240	4410	4970	4330	4200	4810	3980	4560	3730	4270	3290	3580	3930	2440	3340	4800	4260	4460	2730	4060
LCD	20.2	ns	ns	28.3	ns	ns	ns	ns	ns	ns	ns	ns	ns	ns	ns	ns	ns	ns	ns	ns	ns
P_1 = 10 plants/m	3680	3250	4040	4920	4140	4010	4670	3410	4430	3470	3990	3190	3660	3900	2310	3260	4500	3910	3800	2580	3700
P_2 = 20 plants/m	3990	3160	4760	5030	4320	4250	4550	3980	4600	3830	4240	3050	4090	3800	2500	3360	5110	4260	4560	2790	4180
P_3 = 30 plants/m	3920	3410	4640	4600	4120	4140	5010	4350	4700	3930	4500	3190	3760	4040	2520	3380	4750	4240	4630	3000	4150
LCD	20.2	ns	20.3	28.3	ns	ns	ns	ns	ns	ns	21.5	ns	ns	ns	ns	22.2	33.5	ns	27.3	ns	22.2
N_oP_1	3510	3080	4050	4830	3920	3880	4640	3250	4310	3680	3970	3220	3750	4080	2230	3320	4620	3950	3770	2130	3620
P_2	4180	3140	4660	5380	4220	4320	4480	3790	4540	3550	4090	3200	4080	3970	2510	3440	5390	3880	4810	2690	4190
P_3	3810	3660	4970	4570	4090	4220	5040	4470	4820	4030	4590	3300	3670	3740	2100	3200	4860	4110	4460	3070	4130
N_1P_1	3820	3470	3940	4580	3950	3940	4510	3460	4440	3550	3990	3020	3800	3960	2560	3340	4240	3770	3620	2900	3630
P_2	3590	3160	4750	4640	4100	4110	4290	3620	4630	3940	4120	2770	4160	3740	2650	3330	4780	4410	4280	2900	4090
P_3	3810	3220	4740	4780	4040	4120	5290	4670	4770	3710	4610	2910	4040	3950	2630	3380	4780	4350	4660	3230	4260
N_2P_1	3700	3210	4140	5390	4560	4200	4840	3520	4550	3170	4020	3320	3420	3670	2140	3140	4660	4030	4020	2700	3850
P_2	4190	3170	4880	5080	4330	4330	4880	4510	4640	3990	4510	3190	4040	3700	2340	3320	5140	4490	4580	2790	4250
P_3	4140	3350	4200	4460	4250	4080	4700	3910	4510	4040	4290	3360	3590	4420	4840	3550	4600	4260	4770	2700	4080
LCD	34.9	ns	35.1	45.1	ns	ns	ns	ns	ns	ns	ns	ns	ns	ns	ns	ns	ns	ns	ns	ns	ns

Treatments	Aliartos 1978	1980	1981	Aver.	Vardates 1977	1978	1979	1980	1981	Aver.	Palamas 1978	1979	1980	1981	Aver.	Giannin 1979	1989	1981	Aver.
N_o = control	3560	5430	3460	4100	3040	4310	5070	3570	2260	3650	5090	3920	3020	3100	3780	3860	3070	3670	3490
N_1 = 30 kgN/ha	3620	5150	3190	4060	2980	4370	5520	3640	2510	3800	4990	3550	2830	3370	3680	4070	2970	3420	3440
N_2 = 60 kgN/ha	3680	5180	4180	4390	3100	4500	5240	3950	2840	3930	5200	3020	3010	3400	3660	3800	3070	3550	3550
LCD	ns	ns	47.6	ns	ns	ns	ns	32.4	24.6	18.4	ns	ns	ns	ns	ns	ns	ns	20.1	ns
P_1 = 10 plants/m	3390	5460	3400	4140	2790	4020	4980	3560	2390	3550	5000	3200	3060	3300	3640	3730	2820	3380	3330
P_2 = 20 plants/m	3650	5130	3610	4120	3060	4550	5600	3840	2600	3930	5240	3810	2950	3250	3810	3930	3050	3590	3570
P_3 = 30 plants/m	3810	5370	3770	4290	3280	4600	5240	3760	2620	3900	5040	3480	2840	3320	3670	4010	3220	3670	3570
LCD	20.1	ns	ns	ns	51.3	56.9	41.1	ns	ns	18.4	ns	33.0	ns	27.6	ns	ns	ns	ns	ns
N_oP_1	3180	5660	3560	4130	2640	3730	4740	3480	1810	3280	4920	4020	3100	2960	3750	3580	2940	3450	3320
P_2	3400	5230	3330	3990	3270	4380	5740	3680	2430	3900	5000	3890	3320	3150	3840	3990	3190	3760	3680
P_3	3600	5400	3540	4180	3220	4820	4720	3550	2520	3770	5360	3860	2630	3180	3760	3620	3070	3810	3500
N_1P_1	3650	5340	3070	4020	2660	4290	5240	3320	2370	3580	5100	2980	3150	3360	3650	3630	2580	3310	3170
P_2	3680	5100	3200	3990	3040	4550	5820	3780	2620	3960	5080	3810	2430	3470	3700	4310	3060	3360	3580
P_3	3630	5600	3300	4180	3250	4260	5500	3820	2530	3870	4780	3850	2920	3280	3710	3850	3260	3610	3570
N_2P_1	3850	5380	3560	4260	3070	4060	4980	3890	2960	3790	4980	2610	2930	3580	3530	4180	2940	3390	3500
P_2	3760	5060	4320	4380	2860	4720	5230	4040	2760	3920	5660	3720	3110	3140	3910	3910	2900	3670	3490
P_3	3820	5100	4650	4520	3380	4720	5510	3920	2800	4070	4980	2720	2980	3490	3540	3930	3390	3600	3640
LCD	ns	ns	ns	ns	ns	32.8	ns	ns	ns	ns	ns	ns	ns	ns	ns	ns	ns	ns	ns

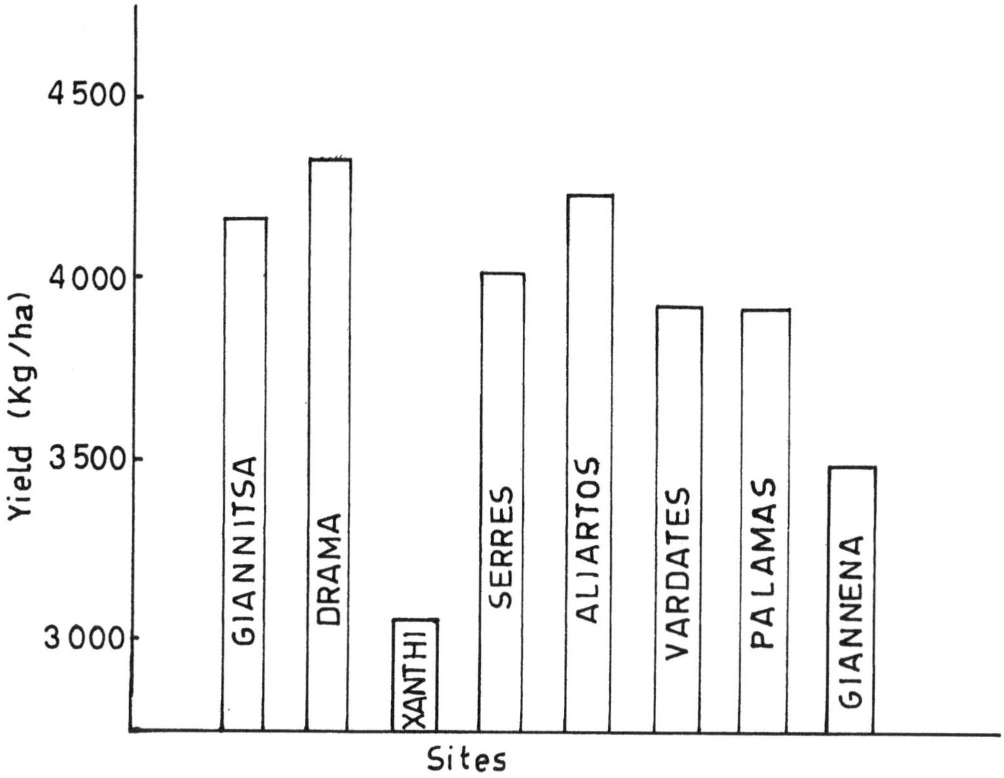

Fig. 1.

the experiments, soybean yields were not affected by N fertilization. However, in some sites (Giannitsa, Aliartos, Vardates) fertilization at the pod filling period increased significantly yields (3–20%). These results are in agreement with the data reported by other investigators (Galdwell & Vest, 1970). The increased seed yields observed at these sites, where N addition during the plant growth period and before the rhizobia begin to supply them with N, resulted in increased yields, may be due to the low soil fertility.

Fertilization with 30 kg N per ha had no effect on the yields (Table 1). However, an increase of 70 kg per ha (2%), although not significant, was observed with the addition of 60 kg N per ha (Table 1).

The effect of sites on the yield is shown in Fig. 1. Yields varied from 3340 kg per ha (site of Xanthi) to 4240 kg per ha (site of Drama). The relative low yields at the site of Xanthi are due to the high coefficient of variance (20.8%).

Plant population affected significantly soybean yields in most of the experiments. On an average, an increase up to 250 kg per ha (16%) was observed with stands of 20 or 30 plants per m. This increase was significant compared to 10 plants per m (Table 1). No relationship was observed between N fertilization and plant population.

In summary, under Greek conditions, in soils with a sufficient number of rhizobia from previous crops or with seed inoculation, N fertilization of soybean plants is not necessary.

Conclusions

In general, the soybean yields were not affected by N fertilization, except in some sites with low soil fertility, where a supplementary N fertilization at the beginning of pod filling phase, resulted in a 3–20% yield increase. On an average, soybean yields were increased significantly with 20 or 30 plants per m in comparison with 10 plants per m, although the increase of plant population from 20 to 30 plants per m did not affect yields. It was concluded that, in soils with a sufficient number of rhizobia, from previous crops or from seed inoculation, N fertilization is not necessary under Greek conditions.

References

Allos HF and Bartholomew WV (1959) Replacement of symbiotic fixation by available nitrogen. Soil Sci 87: 61–66.

Anderson IC (1967) TIBA-treated plants produce larger yields of soybeans. Soybean Digest 24: 29–30.

Beard BH and Hoover RM (1971) Effect of nitrogen on nodulation and yield of irrigated soybeans. Agron J 63: 815– 816.

Bhangoo MS and Weber DF (1972) Effect of fertilizer nitrogen, phosphorus, and potassium on yield and nutrient content of Lee soybeans. Agron J 64: 743–746.

Buttery BR (1969) Effects of plant population and fertilizer on the growth and yield of soybeans Can J Plant Sci 49: 659–673.

Buttery BR (1969) Analysis of the growth of soybeans as affected by plant population and fertilizer. Can J Plant Sci 49: 675–673.

Caldwell BE and Vest G (1970) Effects of *Rhizobium japonicum* strains on soybean yields. Crop Sci 10: 19–21.

Donald CM (1963) Competition among crop and pasture plants. Advances in Agronomy 15: 1–118.

Hanway JJ and Weber (1971) Dry matter accumulation in soybean plants as influenced by N, P, and K fertilization. Agron J 63: 263–266.

Weber LF, Boone LV, Chambliss CG, Cristiansen AT, Mulvaney DL, Oldham MG and Pendelton JW (1973) Soybean yields with direct and residual nitrogen fertilization. Agron J 65: 547–550.

C. Rodriguez-Barrueco (ed.), Fertilizers and Environment, 245–247.

The effect of fertilizers on incidence of the fungus

Microdochium bolleyi causing root rot of maize

S. Stojkov & V. Pencic
Maize Research Institute Zemun Polje S. Bajica 1, 11080 Belgrade-Zemun, Yugoslavia

Key words: Microdochium bolleyi, maize, fertilizing

Abstract

Previous investigations have shown that soil fertilizing could affect development and content of pathogen micoflora in maize root. During two years the effect of fertilizers under different conditions of irrigation, crop rotation and tillage on incidence of the fungus *Microdochium bolleyi* (Sprague) De Hoog and Hermandies-Nijof on maize root was studied. Fertilizing increased incidence of the fungus on maize root, and it ranged, depending on the growing system, from 6.5% to 18.1%, except in no-tillage system where it led to decrease of incidence of fungus by 4.1%. Fertilizing resulted in increase of the fungus incidence on maize root in all crop rotation systems. The increase was the highest in the crop rotation with wheat, and lowest in the crop rotation with soyebean.

Introduction

One of the most widespread and economically most important diseases is the parasitic lodging of maize, which is caused by the stalk and root rot. The disease is of a complex character. Various kinds of fungi cause the disease more often than bacteria. The fungus *M. bolleyi* pathogenicity in maize was detected by Krüger, 1976, and in Yugoslavia by Stojkov, 1993.

There are numerous different data on effects of fertilizers on incidence intensity of pathogen causing root and stalk rot Maric *et al.*, 1989, Balaz, 1986, Krüger and Rodgaki-Papadaki, 1980). Liljeroth and Baath (1989) point out that fertilizing with an increased nitrogen ratio decreases root infection by *M. bolleyi.* Krüger and Rodgaki-Papadaki (1980) noticed that irrigation decreased, while fertilizing with a higher nitrogen ratio increased intensity of maize root rot caused by this fungus.

The purpose of this study was to determine the effect of cropping practices, especially fertilizing, on incidence of *M. bolleyi* in maize under agroecological conditions of Zemun Polje.

Material and methods

This study was conducted during two years (1990–1991) in Zemun Polje on the chernozem type of soil. The trial was set up as a three-factorial, using the split-split plot design in four replications. In each year of the investigation the preceding crop was wheat. The effects of fertilizers were studied under different conditions of irrigation (dry land farming/irrigation), tillage (minimum and conventional one). Treatments with and without fertilizing were used in all variants. The commercial fertilizer (NPK) was applied at the ratio of 330 kg ha^{-1} in the autumn.

The investigation of crop rotation effects on *M. bolleyi* incidence was done in the two factorial block design experiment with four replications. Maize was grown in continuous cropping with maize-wheat, maize-soybean and maize-wheat-soybean rotations. There were variants in each growing system without and with fertilizers (355 kg ha^{-1} NPK for maize and wheat and 280 kg ha^{-1} for soybean). Fungi isolation from the maize root was done according to Krüger (1989).

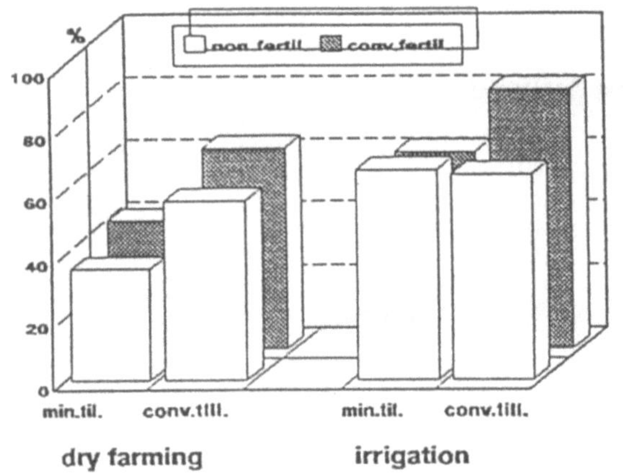

Fig. 1. Incidence of *M. bolleyi* on maize root under different growing system.

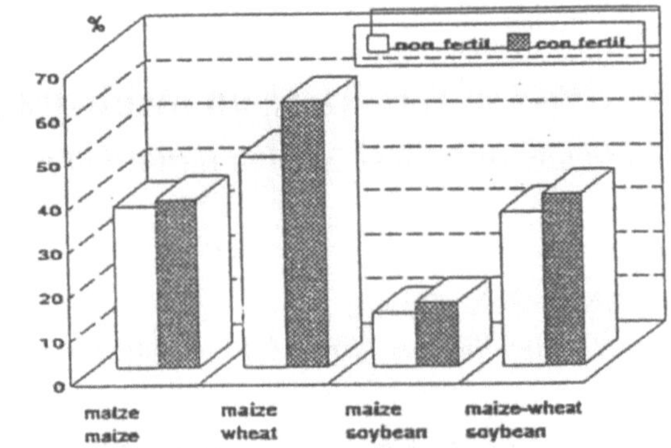

Fig. 2. Intensity of maize root infection by *M. bolleyi* depending on fertilizing in crop rotation (1990–1991).

Results and discussion

As the year of investigation had a low effect on fungus frequency, the further analysis was done on the basis of means for both years.

Fertilizing increased incidence of the fungus in all variants on the maize root, and it ranged, depending on the growing system, from 6.5% to 18.1%, except irrigation-minimum tillage variant where it led to decrease of incidence of *M. bolleyi* by 4.1% (Fig. 1).

The significant differences were determined between irrigation and dry land farming variants. The intensity of maize root infection by *M. bolleyi* under irrigation was by 8.0% to 31.3% higher than under dry land farming conditions over all variants. The effect of fertilizing on increase of this pathogen isolation was higher under irrigation on the average by 10.3%

The highest fungus frequency on maize root was detected under standard cropping practices, i.e. under conventional tillage and fertilizing of soil, in both dry land farming and irrigation conditions. Fertilizing affected, to a greater extent, more intensive fungus development on maize root in the variants of conventional tillage, where the intensity of pathogen incidence on fertilized areas was higher by 18.5%.

In relation to crop rotation it was, also, noticeable that the year of investigation did not affect *M. bolleyi* frequency on maize root. Wheat as a preceeding crop, either in maize-wheat rotation or maize-wheat-soybean rotation had the highest effect on the intensive development of *M. bolleyi* on maize root.

A bit different results on incidence intensity of *M. bolleyi* were obtained when different amounts of fertilizers had been applied in different crop rotation systems. Fertilizers increased intensity of this fungus incidence on maize root in all growing systems from 2.3 % to 12.3% (Fig. 2). The increase was the highest in two crop rotation with wheat during the first year, while it was the lowest in the two crop rotation with soybean during the second year of investigation.

The percent of isolation from wheat root ranged from 28% to 45% in the three crop and the two crop rotation, respectively. In this case, also, fertilizing increased this fungus incidence intensity by 10 % and 5% in the two crop and the three crop rotation, respectively. The presence of the pathogen was not detected on the studied soybean plants.

Conclusion

Obtained results point out that the effect of fertilizing on *M. bolleyi* incidence is significant as it was proved for other rot pathogens in previous studies. This effect depends on applied measures of maize growing and it is mostly in positive correlation. Considering that fertilizing, crop rotation with wheat and irrigation cause the most profound *M. bolleyi* incidence, a more signifcant percentage of this fungus incidence could be expected on maize under intensive condition of production.

References

Balaz, F. (1986) Uticaj plodoreda i djubrenja na vodni ražim zemljišta, biljaka i razvoj fuzariozne truleži stabla kukuruza. Savremena poljoprivreda 34: 197–288.

Krüger, W. (1976) On the occurrence of root and stalk rot of maize in the German Federal Republic 1969–1976. Mitt. aus der Biol. fur Land und Forst. 172: 49.

Krüger, W., Rogdaki-Papadaki C. (1980) On the effect of temperature, soil type, soil compression and fertilization on root rot and the fungus spectrum of maize. Zeit. fur Pflan. und Pflanzenschutz 87: 298–316.

Krüger, W. (1989) Final report of the European Cooperative Research Network on Maize, Martonvasar, August 21–22.

Liljeroth, E., Baath E. (1989) The influence of variety and nitrogen amendments on abundance of *Microdochium bolleyi* on barley roots. Zentralbatt fur Mikrobiologie 144: 181–185.

Maric, A., Jociè B., Stevanoviè M., Stojšin V. (1989) Uticaj agroekoloških uslova na pojavu oboljenja i razvoj kukuruza tokom 1989. godine u Vojvodini. Polj. fak. IRP N.Sad 16: 65–79.

Stojkov, S. (1993) Proucavanje *M. bolleyi* -prouzrokovaca trulezi korena kukuruza. Mag. rad. Polj. fakultet Novi Sad.

C. Rodriguez-Barrueco (ed.), Fertilizers and Environment, 249–251.

Effect of irrigation systems and type of nitrogenous fertilizers in the grain yield and N-uptake on a maize crop

A. Vallejo[1], M.C. Cartagena[1], R. Caballero[2] & J.A. Díez[2]
[1]*ETSI Agrónomos, UPM, Madrid, Spain;* [2]*Centro de Ciencias Medioambientales, Finca Experimental La Poveda, CSIC*

Key words: Nitrogenous fertilizers, maize, irrigation systems

Abstract

A field experiment was performed with maize, cv Juanita (Pioneer), in order to determine the influence of nitrogenous fertilization and type of irrigation system on maize grain and biomass yield, as well as on plant N use. The fertilizer treatments were: urea, a controlled release fertilizer (IBDU), a municipal refuse compost and a control (without nitrogenous fertilizer). The irrigation systems were a conventional one (with drainage) and an optimized one (without drainage).

Results obtained under these experimental conditions showed that grain yield and residual biomass production were higher in plots irrigated with the conventional system. No significant differences were observed between fertilizer treatments in either of the two parameters. However, the plant's N-uptake depended only on the fertilizer treatments.

The N lost during the cultivation period was evaluated by a N-balance in the top-soil.

Introduction

In a maize growing experiment carried out in 1991 in the Jarama river plain, keeping to the same nitrogenous fertilizer (300 kg ha^{-1}) and irrigation water doses as used by local farmers, it was seen that drainage water losses were over 20% of that provided (Diez *et al.*, 1994) and that N lost in the top soil was over 250 kg ha^{-1} (Cartagena *et al.*, 1995).

Diez *et al.* (1994), evaluated nitrate lost through leaching and provided to the aquifer. In some cases it exceeded 100 kg N ha^{-1} during growing time. It was likewise seen that when the fertilizer applied was urea, losses were even higher than those occurring on plots treated with slow release fertilizers.

One way proposed for lessening N losses through leaching consists in adjusting the irrigation dose to crop water requirements (optimized irrigation). The purpose of this paper, then, is to study the influence of the irrigation system, conventional (with drainage) and optimized (without drainage), and the type of nitrogenous fertilizer when applied in doses less than that used by local farmers, have on grain and plant production and on plant N uptake.

It is also intended to evaluate N lost in the top soil in each of the treatments.

Material and methods

The experiment was carried out at La Poveda Field station (30 km SE Madrid) in the Jarama river basin. Twenty four (11 m × 9 m) single plots with a 3 m separation between each other were laid out for irrigation management (two levels) and sources of N fertilizer (four levels). The experimental design was in complete random block with three replications. Soil samples were taken on all the single plots prior to seedbed preparation with the purpose of evaluating nutrient dynamics and especially available N, P and K in the top-soil by using the electroultrafiltration technique (Wicklicky & Nemeth, 1981).

Maize, cv Juanita (Pioneer) was sown in April 1993 in rows with 75 cm spacing end density of 70,000 plants ha^{-1} and harvested in December. Prior to seedbed preparation, a compound (0:14:7) ($N:P_2O_5:K_2O$) and K_2SO_4 were applied at the rates of 714 and 100 Kg ha^{-1}, respectively.

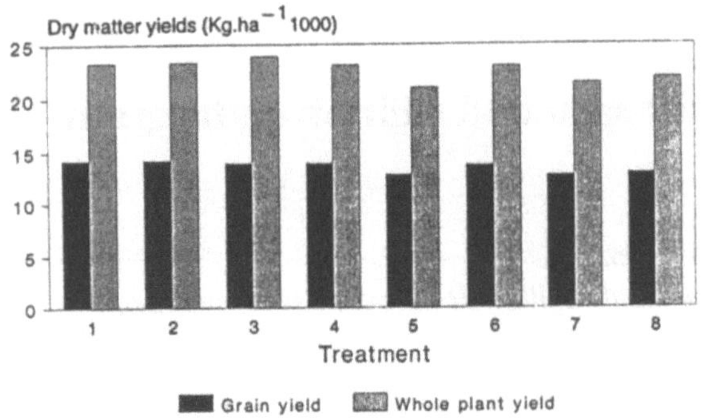

Fig. 1. Influence of irrigation water dose and type of N fertilizer on grain and above ground plant yield.

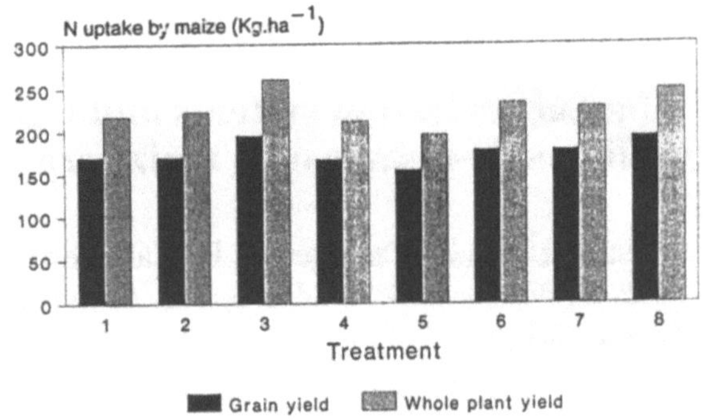

Fig. 2. Influence of irrigation water and type of N fertilizer on N uptake by grain and above ground plant in maize.

Top-dressing N was applied at the rate of 150 Kg ha^{-1} in accordance with soil analyses and was carried out with three different fertilizers as follows: urea (46%), Floranid 32 whose nitrogenous component is isobuthylidendiurea (a slow solubility compound) and a municipal waste compost. The latter was applied before maize planting at the rate or 27800 Kg ha^{-1} assuming 50% of N availability in the first year. Maize was also grown on unfertilized plots in order to make comparisons with those fertilized.

Two irrigation systems were used, conventional and optimized. Water was provided at the rate of 572 l m^{-2} (1.28 ET), distributed over 11 irrigation operations on the plots with conventional irrigation, whilst those under optimized irrigation were irrigated 20 times with 460 l m^{-2} (1.03 ET) total water volume. The rainfall during this period was 35 l m^{-2}. Moisture was monitored by a neutron probe and the water flow path through the soil profile by a set of tensiometers.

Aboveground biomass was measured by weighing the plants in 5 m of two adjacent rows in the middle of each single plot. Whole plant and plant parts were oven-dried at 60 °C for dry matter (DM) determination.

Total plant N content was analyzed using the Kjeldahl method. EUF NO_3^- was determined colorimetrically with N_1-Naphthylendiamine and EUF N (total N in extracts) by ultraviolet irradiations with potassium persulphate in an alkaline medium (Diez, 1988).

Results and discussion

Grain and plant yield was significantly higher ($p \leq 0.05$) in treatments with conventional irrigation than in optimized (Fig. 1).

The average grain yield on plants subjected to conventional irrigation exceeded 14000 kg ha^{-1} in each of the fertilizing treatments whilst with the optimized system, average productions between 12800 and 13800 kg ha^{-1} were obtained.

However, the effect of nitrogenous fertilization was slight in grain and above ground plant yield and there were no significant differences for either irrigation system between those treated with nitrogenous fertilization and those unfertilized at $p \leq 0.05$ (Table 1). This enables the deduction to be made that the level of N available in the top-soil was sufficient to adequately nourish the maize crop and it was unnecessary to apply 150 kg ha^{-1}.

The average values of N plant uptake and N grain uptake in each of the treatments are given in Fig. 2.

The type of irrigation system had no influence on N plant uptake and grain N (Table 1). Nevertheless, the use of nitrogenous fertilizers did have a clear influence on N plant uptake. In optimized irrigation, the average value for plots treated with urea, floranid 32 and compost was 233, 228 and 248 kg ha^{-1} respectively as against the 196 consumed on unfertilized plots. The fertilizer effect in the conventional system was only significantly higher at $p \leq 0.05$ on plots treated with the slow release fertilizer for which N plant uptake was 260 kg as against the 218 kg ha^{-1} of the unfertilized plots.

N lost in the top-soil (0–80 cm soil depth) was evaluated through the N balance in this zone (Cartagena *et al.*, 1995). The equation used is: $N_{lost} = N_s - N_{min} - N_{up}$, where N_{up} is the N taken up by the plant, N_{min} is the soil mineral N at harvest, and N_s is the N supplied, calculated as the sum of N fertilizer, N irrigation and N available in the top soil at the beginning of experiment (Huggins & Pan, 1993).

Table 1. Evaluation of the effects of different treatment using the F-ratio for some yield and N uptake in the experiment

	F-ratio			
	Grain yield	Plant yield	N uptake by grain	N uptake by plant
Fertilizers treatments	0.5	0.4	4.2*	6.2**
Irrigation system	6.2*	6.7*	0.0	0.1
Interactions	0.2	0.6	4.1*	6.4**

F-test significance at ** $p \leq 0.01$ and * $p \leq 0.05$.

Table 2. Δ N available and N lost in top-soil zone (0–80 soil depth) during crop season

Irrigation system	Fertilizer	N available $-N_{min}$	N lost
Conventional	Unfertilizer	338	120
	Urea	333	260
	Floranid 32	350	240
	Compost	313	251
Optimized	Unfertilizer	244	43
	Urea	248	165
	Floranid 32	245	162
	Compost	266	142

Table 2 shows available N values and N lost in the top-soil (0–80 cm soil depth). With conventional irrigation, total N losses were in the order of 120 kg N ha^{-1} on unfertilized plots and exceeded 240 kg N ha^{-1} on those fertilized. Highest losses occurred on urea treated plots. Losses with optimized irrigation were considerably less than with the conventional system, since there were no leaching losses in the former.

Conclusions

When endeavour is made to prevent nitrate losses through leaching by adjusting the irrigation dose to the maize crop's water requirements (optimized irrigation), grain and above ground plant yield slightly diminish in comparison with those obtained with conventional irrigation in which the irrigation dose exceeds the ET by 127 l m^{-2}. Providing nitrogenous fertilizer did not improve yields in comparison with plots without nitrogenous fertilization.

It was seen that N plant uptake was not modified by the type of irrigation system but, on the other hand, depends on the nitrogenous fertilizer provided.

N lost in each of the treatments, obtained via an N balance in the top soil area, was considerably higher on conventionally irrigated plots as a result of the excess water provided during irrigation.

References

Cartagena MC, Vallejo A, Diez JA, Bustos A, Caballero R and Roman R (1995) Effect of the type of fertilizer and source of irrigation water on N use in a maize crop. Field Crop Res (In press).

Diez JA (1988) Revisión del método de determinación automatizada del nitrógeno UV oxidable en extractos de suelo. Anal Edafol Agrobiol 47: 1029–1039.

Diez JA, Román R, Cartagena MC, Vallejo A, Bustos A and Caballero R (1994) Controlling nitrate pollution of aquifers by using diffemt nitrogenous controlled release fertilizers in maize crop. Agr Ecosyst Environ 48: 49–56.

Huggins Dr and Pan WL (1993) Nitrogen efficiency component analysis: An evaluation of cropping system differences in productivity. Agron J 85: 895–905.

Wicklicky L and Nemeth K (1981) Düngungsoptimierum mittels EUF-Bodernuntersuchung bei der zuckerrübe. Sonderdruck aus Band 106: 982–988.

C. Rodriguez-Barrueco (ed.), Fertilizers and Environment, 253–257.

Environmental advantages and disadvantages of different sources of nitrogen in agricultural systems

D.W. Widjajanto
Department of Animal Nutrition, Faculty of Animal Sciences, Diponegoro University, Semarang, Indonesia

Introduction

In agricultural systems, there are many sources of nitrogen such as mineralisation, mineral and organic fertilizer and legume-based agricultural systems. Concerning the environmental problem, all of them have advantages and disadvantages in providing soil nitrogen. For example, mineral fertilizers have many benefits compared with others like high nitrate flux in soil rooting, high vegetation uptake, high soil productivity and ease of calculation of nitrogen provision to plants. They, however, will influence ammonia emission, soil acidity and high nitrate leaching. Organic fertilizers, on the other hand, have advantages in term of releasing nitrogen to the soil and recycling; reducing supply of an expensive industrial product, although they influence air pollution by releasing ammonia and odour pollution, and it is difficult to predict the minimum rate, particularly in cool climates. Legume-based agricultural systems, furthermore, showed that they have many advantages like fixing atmospheric nitrogen, causing less nitrate leaching, no ammonia emission, but they result in a significant amount of emission of N_2O and NO to the atmosphere and leaching of surplus nitrate when legumes are ploughed.

Discussion

Regarding the advantages of using mineral fertilizers in agricultural systems, there is much evidence that application of mineral fertilizer influences the status of soil nitrogen. They have been illustrated in Table 1. All of them contain at least 21% nitrogen and urea itself contains 46% nitrogen. These circumstances, of course, influence the plant uptake and tend to affect the growth rate of plant. Estavillo *et al.* (1992), for example, reported that the production of natural grassland was

Table 1. The percentage of nitrogen in low and high-nutrient forms of inorganic fertilizers

Low-nutrient	% N	High-nutrient	% N
Ammonium sulphate	21	Ammonium nitrate	35
		Urea	46
		Liquid nitrogen	
		aqua ammonia	25–29
		Anhydrous ammonia	82

Tivy, (1990).

higher with ammonium nitrate (10.1–10.6 ton/hectare) compared with cattle slurry (9.0–9.4 ton/hectare), and the control (without fertilizer) reached 7.9 ton/hectare. However, soil nitrogen availability is often influenced by other factors such as irrigation and temperature. In addition, regarding the calculation of nitrogen provision to plants, urea is the easiest fertilizer compared with organic fertilizer and legumes as a source of nitrogen, because the percentage of N in urea is exactly known (46%). On the other hand, the nitrogen contents of organic fertilizer and legumes vary.

Organic fertilizer like FYM and slurry can be used to improve soil nitrogen status. As stated by Archer (1985) in Table 2, fresh slurry generally still contains nitrogen around 0.5–1.4% whereas that of FYM is between 0.6 and 4.2%. They, in addition, have benefits concerned with the releasing nitrogen to the soil. They take a long time in releasing nitrogen.

This has been proved by Glendining *et al.* (1990) who found that in long-term application of organic manure like FYM as much 35 ton/hectare for over 100 years, the soil currently contains nearly 3 times as much total nitrogen as that receiving from mineral fertilizer. Regarding the effect on the yield of cereals like winter wheat, Merzlaya *et al.* (1991), reported

Table 2. The nutrient composition of fresh slurry and FYM

Type	Percentage fresh weight			
	Approx. dry matter	N	P_2O_5	K_2O
Fresh Slurry:				
Cattle	10	0.5	0.2	0.05
Pigs				
dry meal fed	10	0.6	0.4	0.03
pipeline fed	6–10	0.5	0.2	0.03
whey fed	2–4	0.3	0.2	0.03
Poultry	25	1.4	1.4	0.12
FYM:				
Cattle	25	0.6	0.3	0.04
Pigs	25	0.6	0.6	0.04
Poultry				
deep litter	70	1.7	1.8	0.40
broiler litter	70	2.4	2.2	0.22
in-house air dried dropping	70	4.2	2.8	0.40

Archer, 1985.

that application of 15 ton/hectare FYM and 45 kg N mineral fertilizer during 3 years have the same effects. In comparison with the 3-year average yield without fertilizer, the yield increases because of nitrogen fertilizer were 0.96–1.12 ton/hectare, while those due to 15 and 75 ton/hectare FYM were 0.46 ton and 1.04 ton, respectively.

In comparison, using organic manure has advantages compared with mineral fertilizer and legumes, because recycling organic manure will reduce input nitrogen from mineral fertilizer. These tend to reduce the cost due to production of mineral fertilizer. In comparison with legumes, organic manure is more favourable because using legume needs extra cost for seed and management. Addiscott *et al.* (1982), showed (Table 3) the connection with animal waste in the UK and its value. The data implies that using the animal waste as a source of nitrogen in agricultural systems has a definite value.

Soil nitrogen status can be improved by legumes through nitrogen fixation and green manuring. There is much evidence that nitrogen fixation can improve soil nitrogen availability. This has been reported by White (1987) in Table 4. He indicated that temperate, tropical and subtropical crops can improve available soil nitrogen. He, furthermore, stated that temperate crops have higher value in nitrogen fixation than tropical and subtropical crops. It assumed that the conditions in temperate regions are more favourable than in tropical and subtropical regions for nitrogen fixation.

Nitrogen fixation is not only influenced by legumes themselves but is also affected by the strain of *Rhizobium*. Vasilas *et al.* (1993), proved that nitrogen fixation and total N in the shoot increased by 29% after inoculation by USDA 122 compared with indigenous bacteria. They showed that inoculation with USDA 122 also increased uptake of soil nitrogen and increased seed yield by 28 to 33% respectively. Hoque (1993), reported that the inoculants markedly increase nodule number, nodule mass, shoot weight and yield of crops compared with uninoculated control and urea-N-treatments. He indicated that inoculation increased yield 113% over the control and 49% over the urea-N-treatments for soybean; 36 and 11% for groundnut; 30 and 13% for lentil; and 47 and 7% for mung bean.

Legumes as a green manure is more efficient in the dry season than in the wet season. In the dry season the rates of mineralization increase because of higher temperature. Green manure was more efficient than urea particularly in wet season. Urea is more prone to leaching than green manure in wet season. Kundu *et al.* (1991) stated that the efficiency of urea was lower in the wet season than in the dry season. It happens because of increased nitrate leaching and denitrification in the wet season compared with in the dry season. Reddy *et al.* (1991), indicated that application of 75% *Leucaena leucocephala* mixed with 25% urea is better than application of 100% urea. They, furthermore, reported that the treatment increased the grain and green fodder yield of sorghum (*Sorghum bicolour*) by 74% and 59% respectively.

In comparison with organic manure such as FYM and slurry in term of releasing nitrogen, urea has advantages in short-term because it can release nitrate to the soil faster than organic manure, but for long-term use, organic manures are better than urea because they release nitrate into the soil very slowly.

Using mineral fertilizer, organic manure and legumes in agricultural systems have many advantages regard to soil nitrogen availability and crop production but all influence the environment through nitrogen losses. Losses can occur through denitrification, (N_2, N_2O), volatilization (NH_3) and leaching (NO_3). These determine the effectiveness of nitrogen sources in agricultural systems.

Regarding the environmental problem, there are two disadvantages of using mineral fertilizer like urea, such as volatilization and high leaching. These can be understood from urea's characteristics. Buresh *et al.*

Table 3. Numbers of animals in the UK and their waste

Animal	Numbers (10^6)	Dung and Urea (10^{12} g)	Water (%)	N (%)	Value[a] (£ 10^6)	Output(g N/day/animal)
Cattle	12	103	80	1.0	309	235
Sheep	43	42	65	1.7	214	19
Pigs	7.7	12.5	75	1.6	60	27
Poultry	131	4	78	4.8	192	1.2

Addiscott *et al.* (1991).
[a] Assuming all g N/day/animal available and price £ 0.30/kg.

Table 4. Estimated fixation of nitrogen by nodulated crops in temperate and tropical regions

Crops	Nitrogen fixation (kg/hectare/year)	
		median
Temperate		200
Clovers	55–600	
Lucerne	55–600	
Soy	90–200	
Pea	33–160	
Tropical and Subtropical		100
Grassed grass/legume pasture	10–129	
Beans	64	
Pigeon pea	97–152	

White, 1987.

(1993), reported that nitrogen loss through volatilization in young rice field with saturated or flooded soil was 35% of the fertilizer application. They indicated that in soil drying and flooding during the vegetative growth phase there were additional nitrogen losses from denitrification of around 14%. This study was supported by Watanabe *et al.* (1981). They found that the average loss of nitrogen through denitrification was 26 kg N/hectare annually; and this was increased to 123 kg N/hectare annually from both denitrification and volatilization. They, furthermore, reported that nitrogen loss through volatilization in alkaline rice soils annually was around 60% of the fertilizer application.

In comparison with mineral fertilizer the disadvantages of using organic manure like FYM and slurry are focused on volatilization, odour pollution, and the difficulties in prediction the minimum rate. In the United Kingdom, for example, the figure of ammonia emission coming from agriculture resulting from spreading livestock waste on land was around 35% (MAFF, 1992). Van den Abbeel *et al.* (1991) reported that nitrogen loss from slurry application through volatilization in winter and summer reached around 23% and 39% respectively. It can happen because in winter period the temperature is lower than in summer. Thompson *et al.* (1990) indicated that the mean total of ammonia loss from cattle, poultry and Dutch pig slurry was 35, 68 and 57% of the NH_4^+-N respectively. Van den Abbeel *et al.* (1991), on the other hand, found that loss through denitrification in winter was about 10% but in summer only 1.7%. It occurs because in winter the soil moisture is higher than in summer.

Legume-based agricultural systems have many advantages concerning the provision of available soil nitrogen compared with application of organic and mineral fertilizer. They, however, influence the environment through denitrification and leaching of simbiotically fixed nitrogen.

Using legumes in agricultural systems will increase soil nitrate concentration. It increases because legumes in association with *Rhizobium* bacteria can fix atmospheric nitrogen. These conditions are going to influence the process and result of denitrification. The

Table 5. Effect of nitrate concentration denitrification rate and the proportion of gas evolved as N_2 and N_2O

Concentration of added NO_3-N	Denitrification rate (μg N/mg/hour)	Percentage of total ^{15}N gas evolved	
		N_2O	N_2
0	–	95.2	4.8
0.5	0.54	93.9	6.1
2.0	0.73	89.8	10.2
20.0	1.15	85.4	14.6

Firestone *et al.*, 1980.

reduction of N_2O to N_2 will decrease when the soil nitrate concentration is high. Also, Firestone *et al.* (1980) reported that the reduction of N_2O to N_2 is usually inhibited when denitrification increases (Table 5).

Concerning the environmental problem, increasing the N_2O emission to the atmosphere tends to increase ozone layer destruction. Crutzen (1983), proved that a doubling of N_2O in the atmosphere causes a 12% decrease of the ozone total. This condition will increase UV radiation reaching the air surface. N_2O is also a "greenhouse gas", and so its emission will help to increase the earth surface temperature. It means that increasing N_2O emission to the atmosphere indirectly is very harmful for human beings.

Conclusion

Concerning the provision of available soil nitrogen, there is conflict between the benefits of using sources of nitrogen and the effects of those sources on the environment. Many sources can be used to provide soil nitrogen availability such as mineral fertilizers, organic manures and legumes. Unfortunately, all of them have advantages and disadvantages concerned with provision of nutrient nitrogen, and have associated environmental problems. Thus, it is very difficult to say that using mineral fertilizer, is for example, better than using the others. Mineral fertilizers have many advantages like high NO_3-N flux and ease calculation of application rates, but they influence the environment more dangerously than using organic fertilizer or legumes.

Regarding the problem of nitrogen, the important thing is how to maximize the soil nitrogen availability without creating problem in the environment. It can be done by using appropriate sources of nitrogen with proper timing and application.

References

Addiscott TM, Whitmore AP and Powlson DS (1991) Farming, Fertilisers and Nitrate Problem. CAB International, Wallingford, UK.

Archer J (1985) Crop Nutrition and Fertiliser Use. Farming Press, Ipswich, UK, pp 123–135.

Buresh RJ, Castillo EG and Datta SK de (1993) Nitrogen Losses in Puddled Soils as Affected by Timing of Water Deficit and Nitrogen Fertiliser. Plant Soil 157: 197–206.

Crutzen PJ (1983) Atmospheric Interactions-homogenous Gas Reactions of C, N and S Containing Compounds. In: Haynes RJ and Sherlock RR Gaseous Losses of Nitrogen; In: Kozlowski TT (ed.) Mineral Nitrogen in the Plants-Soil System Academic Press Inc., Harcourt Brace Jovanovich Publisher pp 242–302.

Estavillo JM, Gonzales-Murua C, and Rodriquez M (1992) Effect of Slurry on Yield and N Losses in Natural Pasture. Second Congress of European Society for Agronomy, Warwick University, UK. European for Agronomy pp 406–407.

Firestone MK, Firestone RB and Tiedje JM (1980) Nitrous Oxide from Soil Denitrification: Factors Controlling Its Biological Production. In: Haynes, RJ and Sherlock RR Gaseous Losses of Nitrogen; In: Kozlowski TT (ed) Mineral Nitrogen in the Plant-Soil System.

Glendining MJ, Powlson DS, and Poulton PR (1990) Some Agricultural and Environmental Consequences of Long-term Applications of Inorganic Fertiliser. In: Merckx R, Verrecken and Vlassak K. (eds) Fertilisation and Environment. Leuven University Press, Belgium pp 189–196.

Hoque MS (1993) Bradyrhizobium Technology: a Promosing Substitute for Chemical Nitrogen Fertiliser in Bangladesh Agriculture. Plant Soil 155–156: 337–340.

Kundu DK, Rao KV, and Pillai KG (1991) Agronomic Efficiency of Green Manure and Urea Nitrogen in Wetland Rice (*Oryza sativa*) as Influences by Seasonal Conditions. Indian Journal of Agricultural Sciences 61: 422–424.

Ministry of Agriculture, Fisheries and Food (1993) Solving the Nitrate Problem. MAFF Publications, London, United Kingdom.

Merzlaya GE, Gavrilova VA and Bulyga NL (1991) Effectiveness of Farmyard Manure and Mineral Fertilisers for Wheat Growing. Agrokhimiya 4: 35–39.

Reddy GS, Venkateswarlu B, and Vittal KPR (1991) Effect of Substitution of Fertiliser Nitrogen with Subabool (*Leucaena leucocephala*) Leaves on Growth and Yield of Rainfed Sorghum (*Sorghum bicolor*) in an Alfisol. Indian Journal of Agricultural Sciences 61: 316–319.

Thompson RB and Pain BF (1990) The Significance of Gaseous Losses of Nitrogen from Livestock Slurries Applied to Agricultural Land. In: Merckx R, Verrecken and Vlassak K. (eds) Fertilisation and Environment. Leuven University Press, Belgium pp 289–296.

Tivy J (1990) Agricultural Ecology. Longman Scientific and Technical Co, Wiley and Sons Inc, New York.

Van Den Abbeel R, Paulus D, De Ruysscher C, and Vlassak K (1990) Gaseous N Losses after the Application of Slurry: Important or Not? In: Merckx R, Verrecken and Vlassak K (eds) Fertilisation and Environment. Leuven University Press, Belgium pp 241–249.

Vasilas BL, Fuhrmann JJ (1993) Field Response of Soybean to Increase Dinitrogen Fixation. Crop Science 33 (4): 785–787.

Watanabe *et al.* (1981) Nitrogen Cycling in south-east Asia Wet Monsoonal Ecosystems. In: Scientific Committee on Problems of the Environment/United Nations Environment Programme. Australian Academy of Science, Canberra; In: Winteringham FPW (ed) Soil and Fertiliser Nitrogen, Technical Reports Series No. 244. International Atomic Energy Agency, Vienna.

White RE (1987) Introduction to The Principles and Practice Soil Science. 2nd Ed. Blackwell Scientific Publications, Oxford, London, Edinburgh, Boston, Palo, Alto, Melbourne pp 153–171.

C. Rodriguez-Barrueco (ed.), Fertilizers and Environment, 259–265.

Organic manure management and efficiency: Role of organic fertilizers and their management practices

E. Schnug[1], P. Oswald[2] & S. Haneklaus[2]
[1]*Institute of Plant Nutrition and Soil Science, Federal Agricultural Research Center (FAL), Bundesallee 50, D-38116 Braunschweig, Germany and* [2]*Ministry of Agriculture, Ref. 312, Rochusstrasse 1, D-53123 Bonn, Germany*

Key words: geographic information systems (GIS), global positioning systems (GPS), heavy metals, Local Resource-Management (LRM), nutrients, Organic fertilizers, sewage sludge, spatial variability, soil fertility, xenobiotics

Abstract

Increasing concerns about soil protection, human health and environmental quality in general will make the future use of slurries and sludges as organic fertilizers in agriculture very tough. Mechanisms and strategies need to be implemented in agricultural production which not only allow the control of the quality and quantity but also localize and manipulate the amounts of fertilizer used. More enhanced systems will also require features for variable application rates in order to the amount of organic fertilizers in line with the spatial variability of soil and crop parameters. Local Resource-Management (LRM) is an innovative concept in which traditional methods of field and laboratory soil science are combined with modern procedures and techniques of geostatistics, information handling (GIS) and interpretation (BOLIDES) togetlier with satellite aided positioning (GPS). This technology opens challenges for a control and spatial manipulation of organic fertilizers and provides the best chance that the problems arising from inherent variability of soils may be addressed successfully resulting in safe and efficient use of organic fertilizer materials.

Introduction

Organic fertilizers are quite a relatively heterogenous group of products comprising the wide range from single chemicals like urea via farmyard manure and slurry to highly complex and variable mixtures of organic and inorganic compounds like wastes from urban areas. Despite manufactured fertilizer products like urea, only wastes from animal production, raw sewage sludges and products processed from sewage and municipal wastes are of a greater interest for agricultural plant production. Therefore this contribution deals with slurries and sludges under the term 'organic fertilizers'. All organic fertilizers derived from the two sources mentioned above have in common that they are a mixture of various compounds, some with effect on physical soil properties, some with a nutritive value for higher plants but most time associated with possible pollutants of environmental concern. In many countries, especially in those with a higher degree of industrialization, the application of this kind of organic fertilizer is not primarily a matter of improving soil fertility, but with increasing economical welfare also quite an easy and welcome way to get rid of wastes from intensive animal production and urban areas. Increasing concerns relating to soil protection, human health and environmental quality in general, together with new legislation and restrictions on inputs in agriculture, will greatly restrict the future use or dispose of these sorts of organic fertilizers in agriculture (Campbell, 1995). But at the same time, the limited availability of natural resources encourage their more sustainable use in system based recycling. Mechanisms and strategies need to be introduced into agricultural production which will allow not only the control of quality and quantity but also the localization of the amounts of fertilizer used. More advanced systems require features for variable application rates in order to apply the amount of organic fertilizers in line with the spatial variability of soil and crop parameters. This contribution discusses selected problems concerning the efficient use of organic fertilizers in agriculture, proposes strategic and technical improvements, and introduces 'Local Resource-Management (LRM)' as an innovative concept for efficient management and control of organic fertilizers.

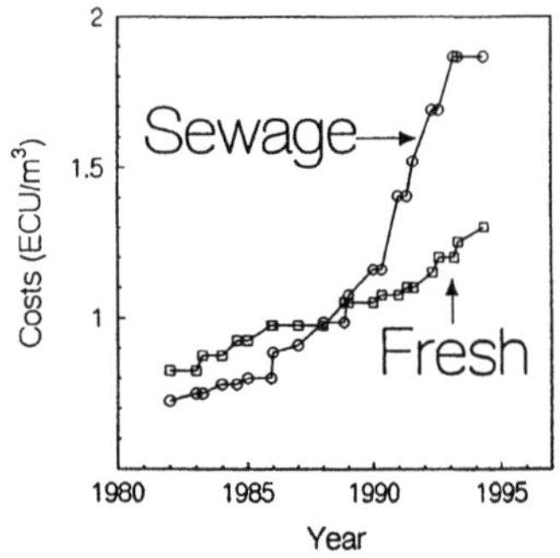

Fig. 1. Costs for freshwater and sewage in Braunschweig, Germany.

Sources of organic fertilizers with greater agricultural significance

There is no doubt about the positive effects of organic fertilization on soil fertility and the benefits of plant nutrients in organic fertilizers for plant growth. Historically the whole story of fertilization started with organic wastes, and in the time prior to manufactured mineral fertilizers, studies concentrated on simple techniques aimed at an efficient use of the nutrients in such sources. The rules for efficient nutrient management in organic manure are well proven can be found in older literature. Most of them are still valid today. One may call it the "forgotten art of fertilization". However the role of organic fertilizers is not what it was in fonner times. Formerly (but still also in most developing countries (Gasiunas *et al.*, 1995)) nutrients in organic fertilizers were of great benefit. However, today, and especially in developed countries, they are considered a burden. This is because of the surplus of nutrients, and the potential contamination of soils with undesirable compounds like heavy metals and xenobiotics.

Farmyard manure and slurries

The efficient integration of farmyard manure and slurries in the management of 'normal' farms, especially in terms of phosphorus, is a minor problem where the number of animals does not exceed 1-1.5 'livestock units' per hectare. At this stocking rate, there are no problems expected from incorporating nutrients from animal wastes into a balanced fertilization for whole farms. Major problems, however, arise in those areas where relatively cheap feedstuffs are imported and large animal production units are operating on areas of land too small to dispose their wastes in a way that nutrients can be efficiently recycled in crop production based on the concept of balanced fertilization (Steen, 1995). A simple rule is also that in almost all cases such nutrient surplus systems are operating on slurry. In contrast, farms operating on farmyard manure are normally smaller, with a much lower stocking rate and consequently have less impacts on the environment, as far as nutrients are concerned. The results of excessive use of slurries are well known: enrichment with phosphorous and, in case of certain pig slurries, copper in soils and losses of nitrogenous compounds to water resources or the atmosphere.

Sludges, composts and organic fertilizers derived from secondary resources

In 1991 in Germany, 3,002,000 T of sewage sludges were produced. In the same year 54% of the sludges ended up in landfills, 28% on agricultural land, 7% were burned and 11% were delivered to other forms of disposal or recycling (Anon, 1991). Thus, taking into account all restrictions for use on agricultural land, 9% of the potentially suitable land was treated with sludges. In the year 2000, the total amount of sludges is estimated to be 3,560,000 T per year. Some town councils in Germany are optimistic to deliver 86% of the sludges from sewage treatment to agriculture (Anon, 1994a), and therefore 31% of the potentially suitable agricultural land will be needed. Although the demand of agricultural land for sludge disposal is increasing, the acceptance of applying sludges to agricultural land is decreasing. Some food manufacturers and flourmills in Germany (Anon, 1994b) have banned products derived from land treated with sewage. Non-food manufacturers, for instance those making substrates for green roofing, also begin to object to even small amounts of sludges in their products. The result is that the costs for sewage disposal, and consequently the costs for sewage, are increasing much faster than the costs for fresh water (Fig. 1) reflecting the great efforts necessary to deal with increasing amounts of sludges with less space for landfills (in future dumping of materials with more than 5% organic matter will be banned in Germany) and a decreasing acceptance of burning and use on agricultural land. The problem with sludges is that, unlike in the case of animal wastes, not only nutrients and safe organic matter are applied to the soil but also potential harmful heavy metals and xenobiotics. A sewage plant collects wastewater from different sources. Therefore, compared to slurries from agriculture, sludges carry a lot more compounds than just the effluents of humans. One should keep in mind that in developed countries the potential number of "weekday chemicals" used is estimated at 50,000 - 100,000 (Mersch-Sundermann, 1994) all of which may end up in the sewage. There are two impressive examples of how the composition of sludges is influenced by socioeconomic factors: after the fall of the iron curtain and following the breakdown of the economy in the country, sludges in Vilnius (Lituania) showed reductions in heavy metal concentrations by 36% for Pb and Cr, and 62% for Cd (Gasiunas *et al.*, 1995). Also increasing human welfare has an impact on sludge composition: copper concentrations in sludges from Swedish councils seem to increase over the past 15 years, because during this period lots of older plumbing material, manufactured from Pb and Zn, were replaced by more expensive materials made from copper (Nilsson, 1993). A major problem with sludges is that the origin of constituents in sludges are diffuse. Thus there is virtually no chance to influence the composition of sludges. Contamination with heavy metals has been investigated now for over 20 years and we are still far away from a final evaluation of measures for acceptable or safe heavy metal loads in terms of soil protection (Witter, 1996). The problem with xenobiotics is even more complicated, as the number of relevant compounds is steadily increasing (Harms, 1995). To reduce waste volume, the separation of biodegradable organic material on the household level together with composting at a council level is a promising strategy. These sorts of biocomposts are easily, efficiently and safely managed in agriculture as long as no sewage sludge or trash is incorporated. Otherwise the problems of sludges, as mentioned above, are just diluted. Another new source of organic fertilizers will derive from "secondary" resources. In Germany, if residues are covered by the so called new "recycling and waste management law (KrW-/Abfg.)" and are found to be suitable for use on agricultural land (§3 KrW-/Abfg. (Abs. 1)) and to have a real market, they may be declared as "secondary-resources for agricultural use" and, if with a sufficient nutrient concentration, be accepted and traded as legal fertilizers. It is to be expected that, under this new law residues are treated legally, a great number of new organic fertilizers will appear in future providing new problems for fertilizer authorities, for example in identification (Feachem, 1983). However, these sorts of organic fertilizers are probably much easier to manage in agricultural practice than sludges, because their route from origin to land application can be tracked and quality parameters can be defined, monitored and maintained.

Key problems for a safe and effcient organic fertilizer management

Modern management practices for organic fertilizers must aim to maximize nutrient utilization with a minimal interference to ecosystems outside agriculture in order to keep agriculture sustainable and to maintain soil fertility. There are, however, several factors interfering with this overall aim. Some of the key problems involved in reaching this goal are the variability of nutrient concentrations in organic fertilizers, the spatial variability of soil fertility parameters, strategies for spatial management and spatial application techniques, and last but not least, reliable control mechanisms.

Variability of nutrient concentrations in organic fertilizers

Organic fertilizers are as inhomogenous as the sources from which they derive. But also within one sort of organic fertilizer, the variability of nutrient concentrations is considerably high. As an example, Stevens *et al.* (1994) reported coefficients of variation for nutrients in animal slurries from 29% for K to 68% for P. However, one important prerequisite for the efficient nutrient management in organic fertilizers is a detailed knowledge of the nutrient concentrations, mainly nitrogen and phosphorous. Efficient management of organic, fertilizers requires knowledge of how much nutrients are given to the soil and an assurance that what has been decided as an optimum dose is accurately delivered by the machinery employed. Therefore, besides an accurate application technique, some form of easy to run on-line measurement of major constituents in the fertilizer is required. It should be remarked that manipulating the dry matter content of slurries and sludges is one of the easiest ways to avoid any restrictions on dry matter loads to the land. Extended research, for instance conducted by Stevens *et al.* (1994), has shown that the electric conductivity explains up to 90% of the

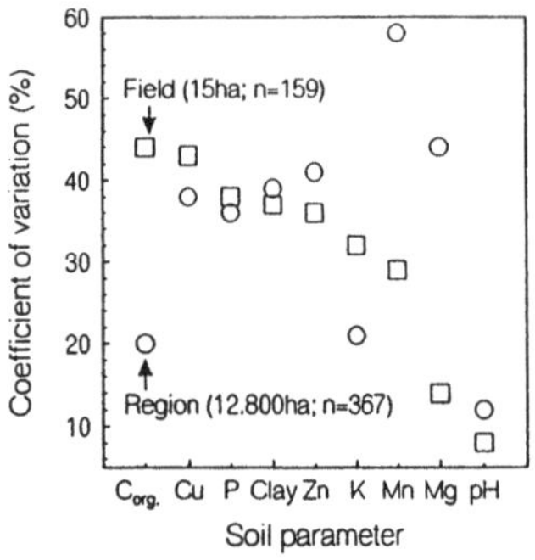

Fig. 2. Variability of soil fertility parameters on field and region level (Birkenmoor in Schleswig-Holstein, Northern Germany).

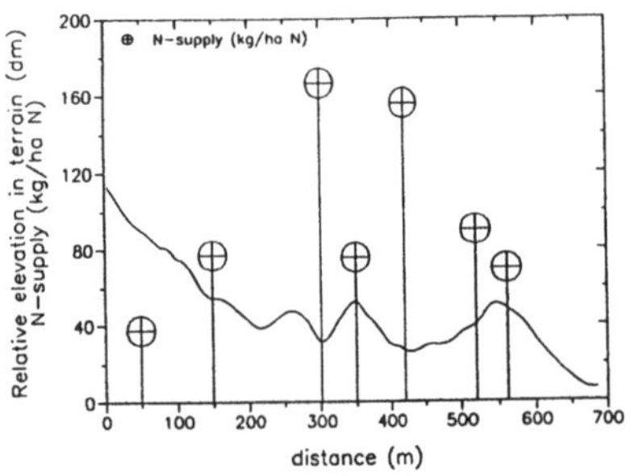

Fig. 3. Nitrogen supply depending upon elevation in a Brown Earth (sL, pH 6.4) (Rothenstein in Schleswig-Holstein, Northern Germany).

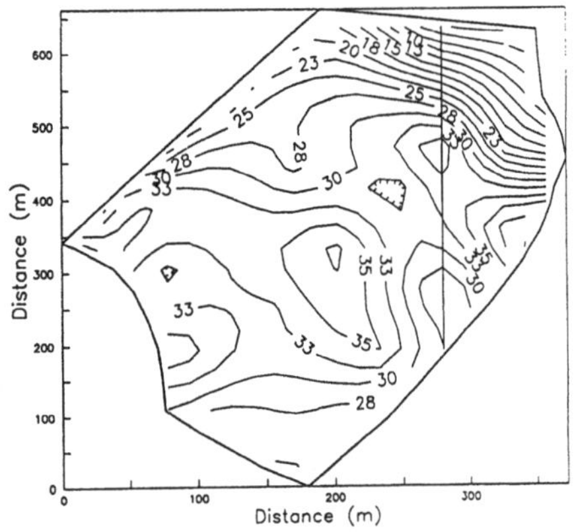

Fig. 4. Available (CAL) phosphate concentrations in die top layer of a Marsh soil (uT, pH 7.2 area weighted average: 25 mg $1OOg^{-1}$ P_2O_5) (Osterhof in Schleswig-Holstein, Northern Germany).

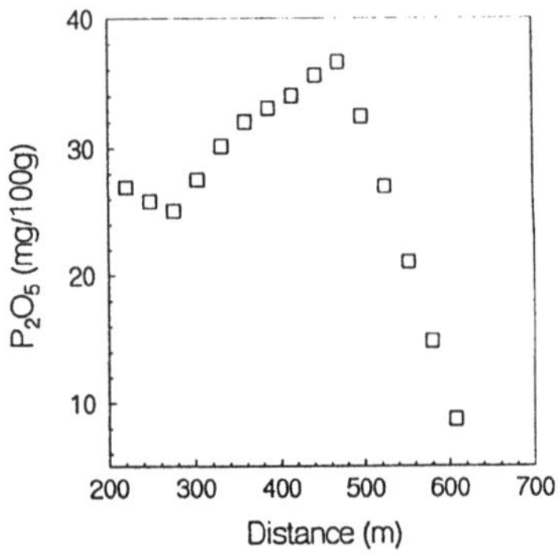

Fig. 5. Available (CAL) phosphate concentrations in the top layer of a Marsh soil (uT, pH 7.2, transect of Figure 4) (Osterhof in Schleswig-Holstein, Northern Germany).

variability of ammonia concentrations in animal slurries, and the dry matter content explains more than 83% of the phosphorus concentrations. Electric conductivity and dry matter content are parameters which are fairly easily be monitored by continuously working systems employed with the application machinery. Thus, by means of conductivity meters (Payne, 1984) and magnetic or gamma-ray flow meters (Thoustrup *et al.*, 1994) it is technically possible to maintain the amount of nutrients applied close to given values. With regard to sludges the content of heavy metals and xenobiotics need to be considered in order to meet the legal requirements for application on agricultural land (Anon, 1992). However, by law, sludges need to be analyzed for critical substances, before they can be distributed to farmers (Anon, 1992). If these analyses are expected to restrict a certain batch of sludge, the techniques mentioned above may also need to be used to estimate loads of heavy metals and xenobiotics.

Spatial variability of soil fertility parameters

One of the greatest barriers to efficient use of fertilizers is the spatial variability of soil fertility parameters. The variability of soil parameters within a single field can easily be higher than the variability of the same parameter within the surrounding landscape (Fig. 2). This is an old, but so far unsolved problem for the efficient use of almost any agricultural input. Figures 3 to 5 show some examples of what one may find anywhere else on agricultural land. Figure 3 shows the elevation profile of a transect through a field in Northern Germany plotted together with the amounts of available nitrogen found at the beginning of the vegetation period in the 0-90 cm layer of the soil. Over a distance of 600 meters, the variability of the natural nitrogen supply, ranging from less than 40 to more than 160 kg $Na^{-1}N$, is tremendously large. The area weighted average for the 15 ha field was 50 kg ha^{-1} N which, according

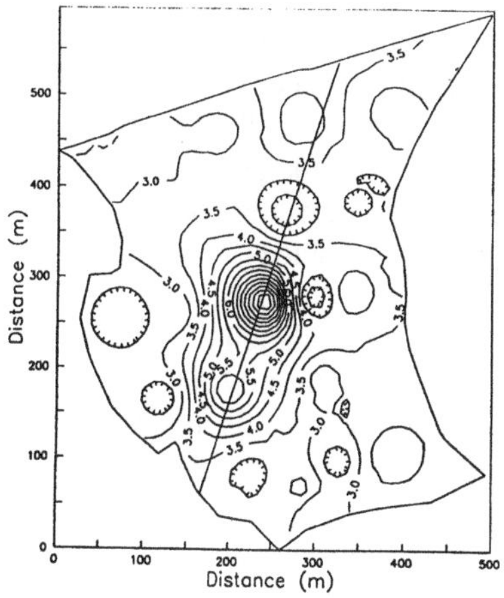

Fig. 6. Available (0,43m HNO_3) copper concentrations in the top layer of a Brown Earth (sL, pH 6.2, area weighted average 3.5 μg g^{-1} Cu) (Kuehren in Schleswig-Holstein, Northern Germany).

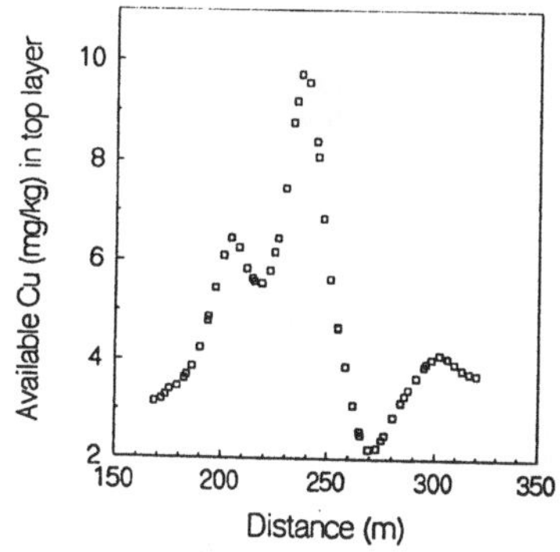

Fig. 7. Available (0,43m HNO_3) copper concentrations in the top layer of a Brown Earth (sL, pH 6.2, transect of Figure 6) (Kuehren in Schleswig-Holstein, Northern Germany).

to German recommendations, implies a nitrogen fertilization to wheat of 70 kg ha^{-1}. The effect is easy to imagine: overfertilization in those parts of the field where the natural N supply is already high, shortage of biomass production in those parts where the natural N supply is initially low. The result will be potentially increased losses of nitrogen compounds to the environment, side by side with underfertilization abolishing the potential productivity of the crop. Figures 4 to 7 give two more examples. In case of the field shown in Figures 4 and 5, the area weighted average for available phosphate of 250 mg kg^{-1} indicates no need for phosphate fertilization. This, however, would leave 20% of the crop in a suboptimal phosphate supply, because in the northeastern parts of the field the available phosphate concentrations are far below the field's average (Fig. 4). The last example shown in Figures 6 and 7 concerns the heavy metal copper, an element with nutritional but also environmental features. Much agricultural land shows yield limiting copper deficiency, but copper is also an element of concern as a potential pollutant with respect to the use of pig slurries and sludges. While large areas of the field used as an example in Figure 6 have only marginal concentrations of available copper, 5% of the mapped area have high copper concentrations, which need to be considered if any form of copper supply by fertilization is taken into account.

Spatial management and application techniques

The knowledge about limited fertilizer efficiency due to the spatial variability of soil fertility parameters is probably as old as fertilization itself. The chance to deal with this problem by technical means, however, is not only related to technical solutions on the application side, but much more to the problem of positioning, navigation and data processing. The great breakthrough came definitively with the availability of the 'Global Positioning System (GPS)' to civil users at the end of the eighties (Hurn, 1989) together with the very rapid development of personal computing, equipment. GPS is based on the a constellation of 24 satellites orbiting the globe at an altitude of 18,000-19,000 km. The basic principle of positioning by GPS relies on triangulation from satellites. To triangulate, GPS measures distances using the travel time of radio messages broadcast by the GPS satellites which carry very accurate clocks. The service is free and available to anybody anywhere on the globe. The systems employed in todays agricultural machinery mostly rely on the so called 'Differential GPS' which allows positioning with an accuracy between 3-5 m at a cost level below 10,000 ECU.

Control strategies

Control strategies are an important part of modern management practices for organic fertilizers. There are several reasons justifying control of organic fertilizers. The main one is probably that, as described above, the organic fertilizers mentioned here are wastes and there is more need to get rid of them rather than to use the nutrients in the fertilizers efficiently. These cir-

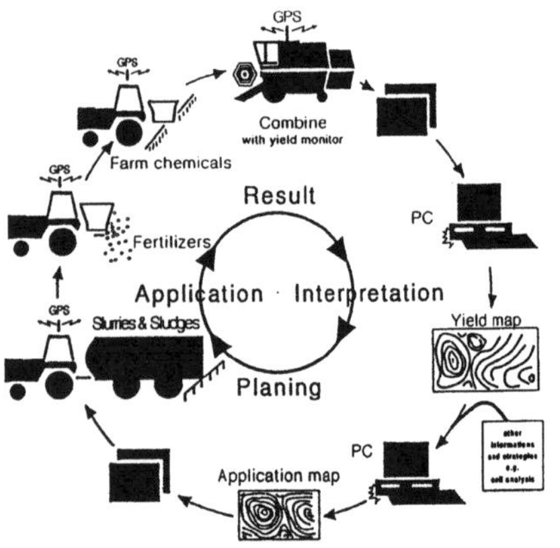

Fig. 8. Scheme of equipment and decision making strategy in 'Local Resource-Management (LRM)'.

cumstances always produce the chance of abuse and therefore consumers are concerned that food quality and sanitation may be affected by the use of wastes and residues on agricultural land. To protect soils and food and to increase the acceptance of waste recycling in agriculture it is necessary not only to analyze the quality of the fertilizer and of the land to which it will be applied but also to monitor the quantities coming from a particular source, the route by which they have arrived at a certain field and last but not least the quantity which has been applied to the land at a certain time. In the past, all this would have required tremendous administrative work, but with GPS and modern developments in so called 'Geographic Information Systems (GIS)' this can be achieved very easily and effectively.

The concept of Local Resource-Management (LRM)

Concepts for modern management practices in agricultural plant production must aim for a maximum of nutrient utilization and a minimum of interference with ecosystems outside agriculture. There must be full control over the type, source and destination of organic fertilizers. Local Resource-Management (LRM) is an innovative concept in which traditional methods of field and laboratory soil science are combined with modern procedures and techniques of geostatistics, information halidling (GIS) and interpretation (BOLIDES) together with satellite-aided positioning (GPS) (Schnug *et al.* 1993a,b, 1994a,b,c). The objective of LRM is simply to increase the efficiency of fertilizer inputs by identifying the spatial variability of soil fertility and addressing inputs according to the spatial variability of soil properties. A simplified scheme of the system is given in Figure 8 which explains the cyclic strategy how decisions in LRM are made, executed and controlled (Schnug *et al.*, 1993a,b, 1994a,b,c). This concept presents challenges for the control and spatial manipulation of organic fertilizers and provides the best opportunity that problems arising from inherent variability of soils may be successfully addressed resulting in save and efficient use of organic fertilizer materials. LRM and its techniques are no longer science fiction. In the past two years it has begun to be realized on more than 50 European farms and it is anticipated that its impact on agricultural practice will be as strong as for example the introduction of mineral fertilizers or farm mechanization in the past.

Acknowledgment

The authors thank Dr. K.C. Walker and Dr. P. Wightman for revising the language of this paper.

References

Anon (1991) Statistik der oeffentlichen Abwasserbeseitigung. Statistisches Bundesamt, Wiesbaden.

Anon (1992) Klaerschlammverordnung (AbfKlaerV). Bundesgesetzblatt, Teil 1.

Anon (1994a) Tonnenschwere Schlammkur für die Natur. Kölner Stadt-Anzeiger 192, 13.

Anon (1994b) Muehlen: Vorsicht beim Aufbringen von Klaerschlamm. Runscreiben des Verbandes Bayrischer Handelsmuehlen vom 24. Januar 1994.

Campbell P (1995) EC policy regarding pollution of agricultural origin; the nitates directive. In: C. Rodriguez-Barrueco (ed) Fertilizers and Environment, pp 539–544 (This volume).

Gasiunas V, Fleckenstein J und Schnug E (1995) Heavy metal problems in soil melioration by sludge application in Lituania. Landbauforsch Voelkenrode (In press).

Mersch-Sundermann V (1994) Zu streng oder nicht streng genug? - Kritische Beurteilung der Rueckstaende in Wasser. In: Macht unser taeglich Brot uns krank? Fortbildungsseminiar fuer Journalisten am 15 und 16 September 1994, Forum der Fal Braunschweig-Voelkenrode.

Feachem RG (1983) Sanitation and disease: Health aspects of excrete and wastewater management. World Bank Studies in Water Supply and Sanitation 3. John Wiley, New York.

Harms H (1994) Phytotoxizitaet, Metabolismus und Persistenz organischer Xenobotica aus kommunalen und industriellen Abfaellen in Pflanzen. In: Alef K, Fiedler H und Hutzinger O (eds) (Umweltmonitoring und Bioindikation, Band 5, pp 217–228.

Hurn J (1989) GPS A guide to the next utility. Trimble Navigation, Sunnyvale.

Nilsson P (1993) Data from "Foersoek medinarbetning av roetslam i Malmoe kommun" VA-Teknik and Vattenvard, Konglandsvaeg 3, S-27044 Broesarp (Pers commun).

Payne VWE (1984) Specific conductance of wastewater as an indicator of nutrient content. ASAE paper Nr. 84-4086, Michigan.

Schnug E, Murphy D, Evans E and Haneklaus S (1993a). Local Resource-Management in computer aided farming: A new approach for sustainable agriculture. In: Fragoso MAC and van Beusichem ML (eds) Optimization of Plant Nutrition, pp 657–663. Kluwer Acad. Publ. Dordrecht.

Schnug E, Murphy D, Evans E, Haneklaus S and Lamp J (1993b) Yield mapping and application of yield maps to computer aided local resource management. Proc. Workshop Soil Specific Crop Management, 87-93. ASA-CSSA-SSSA, Madison.

Schnug E, Haneklaus S and Murphy D (1994a) Equifertiles - an innovative concept for efficient sampling in the local resource management of agricultural soils. Asp Appl Biol 37 'Sampling to make decisions': 63–72.

Schnug E, Murphy D P and Haneklaus S (1994b) Innovative soil sampling and analysis procedures for the local resource management of agricultural soils. Proc. XV Int. Soil Sci. Conf., Vol. 6a, pp 613–630. Acapulco, Mexico.

Schnug E, Heym J and Murphy D P (1995) A boundary line determination technique (BOLIDES). Proc. Workshop Soil Specifc Crop Management. ASA-CSSA-SSSA, Madison (In press).

Steen I (1995) Putting the concept of environmental balanced fertilizers recommendations into practice on the farm. (This volume)

Stevens R J, O'Bric C J and Carton O T (1994) Estimating nutrient content of animal slurries using electrical conductivity. J. Agric. Sci. (Submitted).

Thoustrup A, Haneklaus S und Schnug E (1994) Moeglichkeiten der kontinuierlichen Erfassung, Speicherung und Auswertung von Ertragsdaten bei der Beerntung von Maehdruschfruechten. KTBL-Arbeitspapier 175: 170–174.

Witter E (1996) Towards zero accumulation of heavy metals in soils: justifiable or tenous? In: Rodriguez-Barrueco C (ed) Fertilizers and Environment, pp 397–404. Kluwer Acad. Publ. Dordrecht.

C. Rodriguez-Barrueco (ed.), Fertilizers and Environment, 267–274.

Soil effects due to sewage sludge application in agriculture

H. Häni, A. Siegenthaler & T. Candinas
Swiss Federal Research Station for Agricultural Chemistry and Hygiene of Environment, CH-3097 Liebefeld, Switzerland

Key words: sewage sludge, soil chemistry, soil biology, environmental risks, sludge management

Abstract

Field trials show that increased nutrients (mainly P) and heavy metals in soil due to agricultural use of sewage sludge have mainly to be expected when sludge is applied in too high amounts. The biological effects of heavy metals on plants are determined by their solubility which is increased by pH decrease and degradation of organic matter. Normally, it should be expected that the microbial activity in soil would increase with the application of organic matter in sewage sludge. But, also negative effects on the microflora by simultaneous enrichment of organic matter or inorganic and organic pollutants were found. The most important deleterious effects on soil microorganisms are the reduction in size of the total biomass, a reduced nitrogen fixing activity and changes in the composition of microbial populations of the soil. At what stage heavy metal toxicity to soil microorganisms or to microbial processes in soil is likely to become evident is unfortunately still uncertain. Again, the soluble fraction of heavy metals (also of organic pollutants) is a determing factor. To minimize the shown environmental risks an adequate sludge management is proposed.

Introduction

The use of sewage sludge in agriculture allows the exploitation of raw materials which otherwise would be incinerated/destroyed. At the same time an important contribution to reducing the urgent problem of waste disposal is achieved.

Because of water protection legislation, the amount of treated sewage produced increased dramatically at the beginning of the seventies. In important built up areas, this created problems due to too high application of sludges on agricultural soils where the surface area available for their disposal was too small. At the moment, the Swiss law allows a maximum application of 5 tons of dry matter per hectare of land within a time limit of 3 years, as long as the sludge content of phosphorus and nitrogen excludes an overfertilization with these elements (Ordinance Relating to Substances, Appendix 4.5, 1992). This amounts to less than one ton of organic matter per hectare and year. In the Liebefeld trial (Table 1), a little more than twice the legally accepted amount is applied in treatment SS2.

Besides the maximum allowed sludge application rate quality criteria are fixed in the above mentioned Ordinance as follows (limiting values in gramme per ton dry matter):

Lead	500
Cadmium	5
Chromium	500
Cobalt	60
Copper	600
Molybdenum	20
Nickel	80
Mercury	5
Zinc	2000
Adsorbed organohalogenes (AOx) (as guide level)	500

In the following paper effects from long-term sewage sludge field experiments on the chemistry and biology of soils are shown. Furthermore, measures to minimize environmental risks are proposed.

Table 1. Soil characteristics of the Liebefeld field trial after a duration of 15 years (1976–1990)

1990 Plot B (crop rotation)	Average value for each treatment				Total contents[c]		
	pH	C org %	P-Test[a]	K-Test[b]	Cu mg/kg	Zn mg/kg	Cd mg/kg
0	5.3	1.11	7.3	1.0	18.8	52.8	0.215
Min.	5.8	1.27	23.5	4.9	18.6	52.1	0.213
SS2[d]	6.5	1.60	16.0	4.5	26.4	89.6	0.413
SS5[e]	6.8	1.93	19.7	4.2	38.5	149.3	0.756

[a]P extracted with CO_2-saturated water (Test 1: 0.0356 mg P_2O_5/100 g).
[b]K extracted with CO_2-saturated water (Test 1: 1 mg K_2O/100g).
[c]Heavy metals extracted with 2 M HNO_3 (Ordinance Relating to Pollutants in Soil, 1986).
[d]2 t organic matter per ha and year.
[e]5 t organic matter per ha and year.

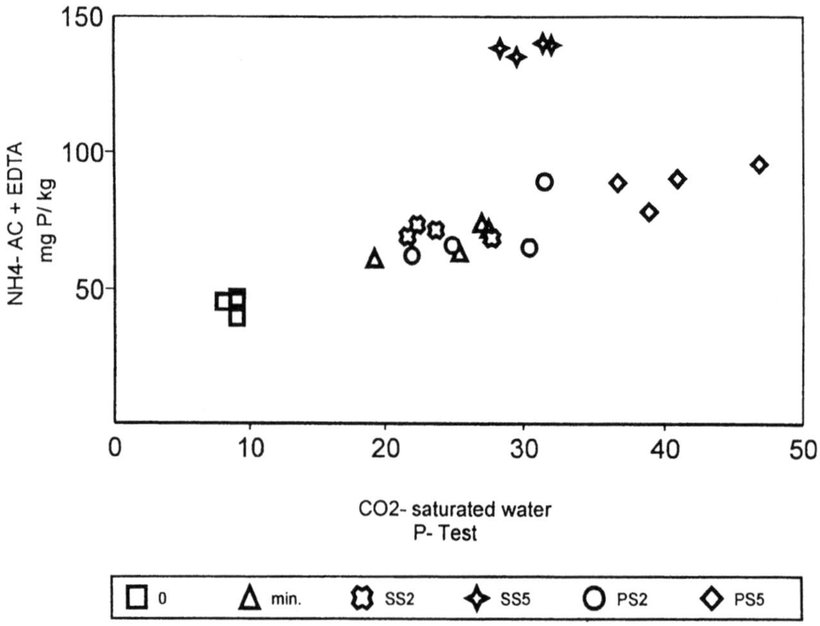

Fig. 1. Soil P-extraction with CO_2-saturated water and NH_4OA_c + EDTA (pH 4.65). (SS sewage sludge, PS pig slurry)

Results

Soil chemistry

The Liebefeld field trial is taken as an example for demonstrating the effects of sewage sludge application on chemical soil properties (for details of this trial see Siegenthaler *et al.*, 1994)

The comparison of sludge treatments with mineral fertilization shows an increase of pH, organic matter and heavy metal contents in the soil (Table 1). The pH increase is due to the high calciumcarbonate content of

Table 2. Influence of sewage sludge on soil microorganisms and their activities (results from field trials)

Field trial	Amounts of sludge (t organic matter/ha.a)	Sludge spiked with heavy metals	Period	Soil data	Influence on soil microorganisms and their activities	Reference
Liebefeld	2(SS2) 5 (SS5)	–	1977–1981	weakly acid sandy loam	Increase in the counts of heterotrophic soil microorganisms Stimulation of enzyme activities Reduction of the autotrophic soil algae and nitrogen fixing blue algae	Stadelmann (1982)
Upper Austria	2,5 (sludge I, III, IV) 11 (sludge II)	–	1980–1985	heavy silt	Soil biological activities are generally enhanced	Oehlinger *et al.* (1992)
Near Vienna	2,4		1977–1989	loamy silt	Increase of soil biomass was estimated to 0.6%	
	4,8				Increase of biological activities	
FAL Braunschweig	? 100 m^3		ten years	luvisol pH (KCl) 6.54	Positive effects on soil microbial parameters	Fliessbach und Reber (1992)
	? 100 m^3 + HM	Sludge contaminated with heavy metals twice its maximally admitted load		pH (KCl) 6.54	Positive effects less pronounced or even absent as compared to the conditions in non-sludged soil	
	? 300 m^3			pH (KCl) 6.30	Positive effects on soil microbial parameters	
	? 300 m^3 + HM	Sludge contaminated with heavy metals twice its maximally admitted load		pH (KCl) 6.08	Positive effects on soil microbial parameters less pronounced	
	? 100 m^3			cambisol pH (KCl) 5.71	Positive effects on soil microbial parameters	
	? 100 m^3 + HM	Sludge contaminated with heavy metals twice its maximally admitted load		pH (KCl) 5.54	Positive effects less pronounced	
	? 300 m^3			pH (KCl) 5.55	Positive effects on soil microbial parameters	
	? 300 m^3 + HM	Sludge contaminated with heavy metals twice its maximally admitted load		pH (KCl) 5.32	Positive effects less pronounced	

Table 2. continued

Uppsala	4 (added every other year)	–	more than 30 years	pH 5.3 % C 2.7 ; % N 0.28 Unfertilized soil	Reduced nitrogen fixing activity by freeliving diazotrophs and reduction in the occurence and activity of blue green algae	Martensson and Witter (1992)
				pH 6.2 % C 1.2	The size of the soil microbial biomass in the sludge treated soil had not increased above that found in the unfertilized control soil, which had received neither organic nor fertilized amendments	
Speyer	?	–	1987–1989	silty sand pH ($CaCl_2$) 6.3	Increase of microbial activity compared with the minerally fertilized control	Balzer and Ahrens (1992)
Braunschweig	2		1987–1989	silty loamy sand pH ($CaCl_2$) 6.4	Increase of microbial activity	
	2 + HM	addition of heavy metal salts to sludges at twice its permitted load		pH ($CaCl_2$) 6.4	Decreased microbial activity	
	6			pH ($CaCl_2$) 6.2		
	6 + HM			pH ($CaCl_2$) 6.0	Decreased microbial activity	
	1,25	–	1987–1989	Silty loam pH ($CaCl_2$) 6.5	Increase of microbial activity	
Ramisch-Holzhausen	2.5	–		pH ($CaCl_2$) 5.4	Less positive effects on soil microbial parameters than sludge application of 1.25 t/ha	

Table 2. continued

Braunschweig				C org. %	pH 0.1 M KCl	Cu mg/kg	Zn mg/kg	Ni mg/kg	Cd mg/kg		Fliessbach *et al.* (1994)
Arable soil	unsludged		1980–1990	0.79	6.6	15.2	48.2	6.0	0.23		
	2	–		0.98	6.4	23.0	91.9	8.1	0.43	Beneficial effects on biomass carbon and soil microbial activity	
	2	Sludge contaminated with Pb, Cd, Cr, Cu, Ni, Hg and Zn in order to reach the former German limits for heavy metals in soils		0.99	6.4	43.1	155	11.2	0.96	Beneficial effects on biomass carbon and soil microbial activity	
	6	–		1.37	6.2	40.6	204	12.1	0.83	Beneficial effects on biomass carbon and soil microbial activity	
	6	spiked		1.38	6.0	99.5	345	21.7	2.6	Decrease in biomass carbon; increase of qCO_2 (respiration per unit biomass)	
Former woodland soil	unsludged			1.62	5.7	8.8	41.0	6.0	0.17		
	2	–		1.75	5.7	16.3	88.3	8.3	0.43	Beneficial effects on biomass carbon and soil microbial activity	
	2	spiked		1.79	5.5	40.5	162.3	13.2	1.03	Decrease in biomass carbon; increase of qCO_2	
	6	–		2.19	5.56	32.5	190.9	13.0	0.83	Beneficial effects on biomass carbon and soil microbial activity	
	6	spiked		2.40	5.28	107.8	404.1	26.8	3.2	Decrease in biomass carbon; increase of qCO_2	

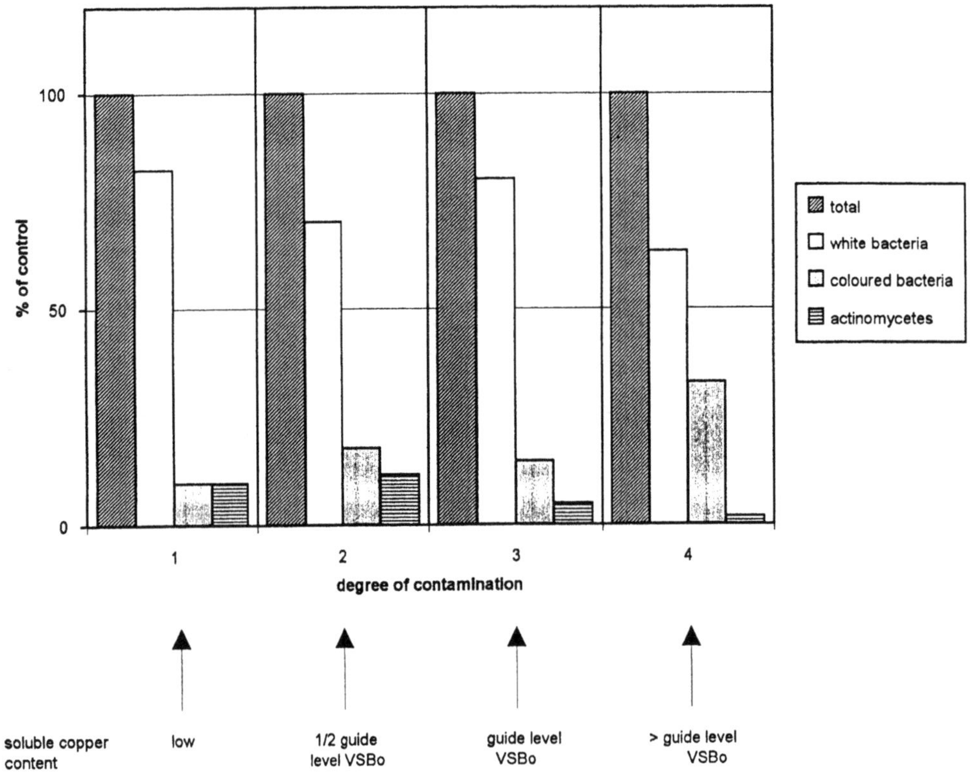

Fig. 2. Shift in the composition of the soil microbial population due to elevated soluble copper concentrations in a soil polluted by a heavy metal smelter

the used sludge. In other experiments, a pH decrease of the soil is often observed (Table 2).

For the pig slurry (also used in the Liebefeld trial) and the mineral fertilizer treatment, it is of interest to note that a rather good correlation is obtained when the P-contents of soil in NH_4OAc + EDTA (pH 4.65) are compared with those in CO_2-saturated water (see Figure 1). On the contrary, the P-amounts of the sludge-treated soils extracted by NH_4OAc + EDTA distinctly deviate upwards from this correlation. This behaviour could be explained by the assumption that a great part of phosphate in sludged soils is bound as stored phosphate.

After 15 years of sludge application (treatment SS5) the soil contents of copper, zinc and cadmium are near to the guide levels of the Swiss Ordinance Relating to Pollutants in Soil (1986):

Lead	50
Cadmium	0.8
Chromium	75
Cobalt	25
Copper	50
Molybdenum	5
Nickel	50
Mercury	0.8
Thallium	1
Zinc	200

Soil biology

From Table 2 it can be seen that microbial activity and organic matter levels are normally increased after long-term sewage sludge application. The most important deleterious effects on soil microorganisms, which have mainly to be expected when sludge is applied in too high amounts, are the following:

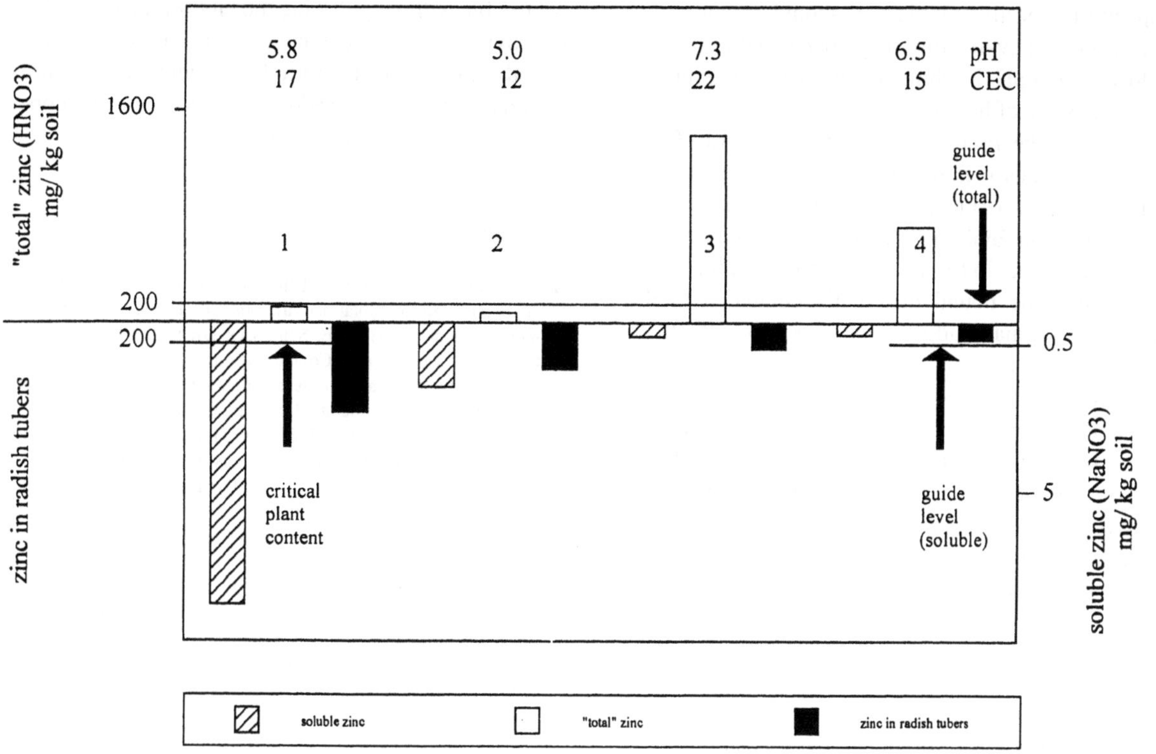

Fig. 3. Effect of total (2M HNO_3) zinc contents in polluted soils on the zinc content in radish tubers

- reduction in size of the total microbial biomass
- reduced nitrogen fixing activity
- changes in the composition of the soil microbial population

These negative effects are often related to a heavy metal toxicity. Martensson and Witter (1992) point out that such effects can already occur at soil metal concentrations close to the CEC guidelines.

Figure 2 gives an example of a shift in the composition of the soil microbial population due to elevated soluble copper concentrations in a soil polluted by a heavy metal smelter (Rudaz, 1987, unpublished). The actinomycetes have almost disappeared and the white colonies of bacteria were reduced at the expense of the coloured colonies at the highest soluble copper concentration.

Discussion

In the cited field trials, the amounts of applied sewage sludge are often twice or even several times higher than prescribed in the ordinances of different countries. Therefore, these results cannot easily be transfered to conditions under usual agricultural practice.

Moreover, in many of the experiments not enough is known about soil properties, heavy metal contents in soil and dynamics of the heavy metals. From some experiments, only the total contents of heavy metals are known. However, it is nowadays largely accepted that the easily soluble metal fraction represents a better means for estimating toxicity. Figure 3 shows an example of radish tubers grown in soils contaminated by zinc from a zinc coating plant (1), a motorway (2), compost (3) and sewage sludge (4). The plant content corresponds with the soluble (0.1 M $NaNO_3$) and not with the total content (Häni, 1990).

Thus, total values alone are never sufficient to define limits for soil pollution. These total contents are certainly a good method of laying down load restrictions, but in unfavourable cases, e.g. acid sandy soils, these values may be too high. In such a situation an additional control based on the easily soluble metal fraction is necessary. This fraction should never reach values which are toxic for sensitive plants or microbial soil processes. In soils with a higher metal-binding capacity the easily soluble part should be controlled because metal availability may change with time due to lowering of pH (leaching of lime, acid precipitation) or degradation of humus.

To finish the discussion about possible toxic effects of heavy metals it is important to emphasize that in many of the experiments shown in Table 2 negative effects on soil biology are only seen when spiked sludges are used. In these sludges, however, the heavy metals may be bound differently compared to the untreated original sludges what makes it difficult to interpret the results.

Similar to the heavy metals the toxicity of organic pollutants depends mainly on the proportion remaining in the soil solution. In a study about toxicity effects on soil microorganisms Welp and Brümmer (1992) showed that soil characteristics like the contents of organic matter, clay and iron-oxides and the pH value determine the absorbed amount of a given organic chemical in different soils. Considering all the possible interactions it has to be concluded that the bioavailability of organic pollutants cannot solely be predicted from solubility or adsorption parameters. The composition of the soil solution and the speciation of the chemicals and the other substances can also be of main importance for the effects on soil microflora.

Another environmental concern due to overfertilization in soils through organic wastes may have to be seen in the over-activities of soil microorganisms leading to N-losses through leaching or volatilization and N-accumulation in plants in form of nitrate (Stadelmann, 1982). A shift in the spectrum of the microflora due to too high amounts of organic matter could also occur.

To minimize the shown environmental risks the following requirements have to be fulfilled for a further use of sludge in agriculture:

- The applied sludge amounts have strictly to be limited to the need of nutrients for plants.
- Quality criteria with regard to the content of pollutants: Compared to the load of nutrients the load of heavy metals in waste fertilizers is still too high. The situation could immediatly be ameliorated if only the sludges of best quality were used.
- Surveillance of the sludge quality and advice for optimal use of sludge.

In Switzerland waste fertilizers are prefered to mineral fertilizers so far as they are of good quality. Waste products do exist. One of these is sewage sludge for which we need a solution. None of the disposal ways is possible without polluting the environment. Quality provided, it is of advantage to use a product which already exists instead of introducing new products into circulation which cause additional environmental contaminations (Frei *et al.* 1993).

References

Balzer W and Ahrens E (1992) Effects of long-term sewage sludge application on microbial activities in soils in relation to heavy metal content. In: Hall JE, Sauerbeck DR and P. L'Hermite P (eds) Effects of Organic Contaminations in Sewage Sludge on Soil Fertility, Plants and Animals. Office for Official Publications of the European Community, Luxembourg, 213–221.

Fliessbach A and Reber H (1992) Effects of long-term sewage sludge application on soil microbial parameters. In: Hall JE, Sauerbeck DR and L'Hermite P (eds) Effects of Organic Contaminations in Sewage Sludge on Soil Fertility, Plants and Animals. Office for Official Publications of the European Community, Luxembourg, 184–192.

Fliessbach A, Martens R and Reber H (1994) Soil microbial biomass and microbial activity in soils treated with heavy metal contaminated sewage sludge. Soil Biol Biochem 26: 1201–1205.

Frei U, Candinas T, Besson JM and Chardonnens M (1993) Stellung des Kompostes in der Düngung. FAC-Schriftenreihe Nr. 13, pp. 1–65. 3097 Liebefeld-Bern.

Häni H (1990) The analysis of inorganic and organic pollutants in soil with special regard to their bioavailability. Intern J Environ Anal Chem 39: 197–208.

Martensson A and Witter E (1992) Effects of long-term sewage sludge addition on nitrogen fixing microorganisms and on the size of the soil microbial biomass. In: Hall JE, Sauerbeck DR and L'Hermite P (eds) Effects of Organic Contaminations in Sewage Sludge on Soil Fertility, Plants and Animals. Office for Official Publications of the European Community, Luxembourg, 207–212.

Oehlinger R, Supersberger H and Cepuder P (1992) Effects of different contaminated sewage sludge on some soil microbial properties. In: Hall JE, Sauerbeck DR and L'Hermite P (eds) Effects of Organic Contaminations in Sewage Sludge on Soil Fertility, Plants and Animals. Office for Official Publications of the European Community, Luxembourg, 177–183.

Rudaz A (1987) Swiss Federal Research Station for Agricultural Chemistry and Hygiene of Environment (unpublished).

Siegenthaler A, Stauffer B, Stadelmann FX, Stauffer W and Häni H (1994) Excessive use of organic wastes in agriculture and field trials. Consultation of the European Cooperative Research Network on Animal Waste Utilization (in press).

Stadelmann FX (1982) Die Wirkung steigender Gaben von Klärschlamm und Schweinegülle in Feldversuchen. Schweiz Landw Fo 21 (3/4): 239–259.

Swiss Federal Council (1986) Ordinance Relating to Pollutants in Soil.

Swiss Federal Council (1992) Ordinance Relating to Substances, Appendix 4.5.

Welp G and Brümmer GW (1992) Toxicity of organic pollutants to soil microorganisms. In: Hall JE, Sauerbeck DR and L'Hermite P (eds) Effects of Organic Contaminations in Sewage Sludge on Soil Fertility, Plants and Animals. Office for Official Publications of the European Community, Luxembourg, 161–168.

C. Rodriguez-Barrueco (ed.), Fertilizers and Environment, 275–281.

Sub-surface injection of liquid waste

Paulo M P Águas[1] & Richard Godwin[2]
[1]*Escola Superior Agrária de Castelo Branco, 6000 Castelo Branco, Portugal;* [2]*Silsoe College, Cranfield University, Silsoe – Bedford, MK45 4DT, England*

Key words: growing crop, injection equipment, liquid waste

Abstract

The sub-surface injection of sewage sludge in grassland conditions has been investigated showing that injection is a viable alternative technique to surface applications. With the large amounts of existing liquid wastes, either municipal or agricultural, the land required for injection has risen over the available grassland. This study has considered the sub-surface injection of sewage sludge in a growing crop of spring wheat.

The agronomic response of spring wheat has been compared at three different levels of crop development. Five different techniques were used to inject the sludge, and compared with two control situations. Using a three leg injector, with the outside tines bent, and placing gaps between rows for the tines and wheels, the crop showed a better agronomic response at the final levels of development.

Analysing row by row the dry matter production in the injected plots, it was clear that the lower global production per square metre was due to the damage caused in the rows by the legs and traffic.

Introduction

Background

Sludge application to agricultural fields is accepted by farmers as a low-cost source of valuable nutrients (Department of the Environment/National Water Council, 1983), with the advantages of the required technology for application being well understood, and its nutrient value recognized (Hall, 1986). Unfortunately the surface spreading of slurry and sewage is every year responsible for reported causes of water pollution (MAFF, 1991), and complaints from the public about odour problems (MAFF, 1991).

Together with this situation, the Commission of the European Communities, since 1982, published E.C. Directives and Regulations on the disposal of sewage sludge (Department of the Environment/National Water Council, 1983). Largely as a result of these directives, sub-surface injection techniques have been a subject of increasing interest.

If slurry or sewage is injected into the ground, the loss of nitrogen in the form of ammonia will be less, reducing atmospheric pollution (Van Breeman, 1991). There will be more available nitrogen to the crop, improving the yields (Hall, 1986).

Due to the existing large amounts of municipal waste, the land area required for waste injection has risen dramatically, causing much attention to turn to injection into land growing cereal crops.

The injection procedure reduces the chance of odour problems, crop contamination and pathogen activity.

Soil Injection techniques prevent surface run-off into water courses, and the soil loosening effect breaks up plough pans, improving root development, water flow and aeration of the soil.

The closer the date of application of sludge is to the date of sowing, the higher the grain yield and the higher the nitrogen recovery (Hann *et al.*, 1992), hence a post-emergence injection would get better results. But injection into growing cereal crops causes great mechanical damage.

Considering the study of injection of sewage sludge in growing cereal crops, previous work on this subject has been restricted to row crops. Alternatively the treatment has been carried out in the period between harvest and crop establishment (Hall, 1986).

Objectives

In order to overcome land and timing restrictions for the use of sub-surface injection, when applied to cereal land, this project compares the feasibility of using available injecting techniques to apply sewage sludge to a growing cereal crop. This would be evaluated by assessing at three different stages of development:

(I) Crop response to different injection techniques, compared to surface applications of sewage sludge and inorganic granular fertilizer.

(II) Crop response to new crop arrangement, using a bent leg injector to minimize damage caused by traffic and tines.

(III) Spatial crop response for each injecting technique.

(IV) The effect on crop response of changing the injector press roll weight.

Injection equipment

Silsoe college, prototype injector

This injector consists of three tines, mounted on a toolbar, spacing at 0.77 m. Each tine is preceded by a disc coulter to minimize sward tear, and followed by an individual press wheel. A 45° forward raked leg is used, fabricated about a 0.04 m by 0.08 m box section steel tine, held in position within an outer sleeve, with which the depth of work can be controlled. The wings are bolted to the bottom of the leg, which extend 0.14 m from the leg. They have a lift height of 0.04 m, at a rake angle of 15°.

The leading tip extends 0.07 m ahead of the front lowest point of the leg, with the same rake angle as the wings.

BKW four tine umbilical injection system

Various manufacturers supply winged injection equipment with practically the same technical specifications as the Silsoe College prototype. Among them, BKW Pollution Control Engineering manufactures a four or six tine umbilical injector. The one used in this study has four tines, spaced at 0.67 m. Each tine is preceded by a disc coulter and followed by a heavy roll, with the full work width.

Bent legs injector

This injector uses the same base as the Silsoe College prototype, and the same central leg. The outside tines are bent, to try and deposit the manure under the crop root zone, causing less damage. These tines were designed, but never tried, to inject in growing cereal crops (Miller and Godwin, 1993), using a bed system. Each tine is preceded by a disc coulter, and followed by an individual press wheel.

The tine bent leg was designed with a shank of 40°, and a single wing, projecting further under the crop. The width of the single wing was equal to the sum of the two wings of the Silsoe College Prototype, i.e. 0.14 m.

Table 1. Treatments used

Treatment	Equipment	Rows of wheat per sample	Missing rows for tines
T 1	Bent leg injector + gap	8	Rows 3, 8, 13
T 2	Bent leg injector	9	Rows 3, 13
T 3	Control – Inorg. fertilizer	11	
T 4	BKW injector	11	
T 5	BKW injector (light)	11	
T 6	Control – Surface spread	11	
T 7	Silsoe injector	11	

Experiment design

Field layout

Five different treatments were selected to use with the given equipment and compared with two control situations (Table 1).

The total of seven treatments was compared at three different levels of crop development. The field used in this study had a total available area of 2100 m^2, (30.0 m by 70.0 m). With seven treatments, it was decided to use a randomised block experiment with three replicates, giving twenty degrees of freedom. The size of each plot was 2.5 m by 20.0 m. Each sample was 1.83 m wide, and 0.55 m long, providing an area of 1.0 m^2. 1.83 m represented the width of the tractor. Inside each plot, three samples were randomized, to be cut at three different levels of crop development.

Soil characteristics

A full test of the soil to be used was obtained. The texture of the soil was sandy loam, and none of the remaining information indicated that the selected field would not be suitable for injection. From each plot six samples of soil were collected; three at a depth of 0.15 m and three at a depth of 0.30 m. These samples were analysed by the Kjeldahl Method (Bremner and Mulvaney, 1982), using a Kjeltec System 1002 distilling unit, to determine the percentage of total nitrogen.

After analysis of variance, no significant difference was found between the levels of nitrogen in samples from the same depth.

Crop establishment

On the 10th of March 1993, the spring wheat *Triticum aestivum*, variety Baldus, with the label of seed HHI 806679, was sown at 99 kg ha^{-1}, to establish a target of 300 plants m^{-2}. A Massey Fergusson MF30 combine drill, 15 cultivators, with a work width of 2.5 m was used. Precautions were taken to avoid damage by pests and birds.

Treatments

In this study a single injection was planned. If the amount of nitrogen in the sludge was below the target value (140 kg ha^{-1}), an inorganic fertilizer application was made after injection. The injection of sewage sludge was carried out approximately eight weeks after the sowing operations, applying a rate of 80 m^3 ha^{-1} (Van Breemen, 1991). This application was selected after identifying that the experimental field was not a high risk area. The application rate of 50 m^3 ha^{-1} as advised by MAFF Environment Matters (1991) was, therefore, exceeded. All injections were done at a depth of work between 0.13 m and 0.15 m.

In order to achieve the target value of 140 kg ha^{-1} of nitrogen, an application of a 46% nitrogen fertilizer was made at an application rate of 90 kg of nitrogen per hectare. This application was carried out by hand to avoid supplementary crop damage. The surface application of inorganic nitrogen would not affect experimental results, as the aim of the project is to compare crop damage caused by different injection techniques.

Treatment 1 – bent leg injector + gap
The original objective of the bent leg design was to apply it to crops growing in bed systems, by establishing tramlines (Miller and Godwin, 1993) and placing the sludge under the bed without disturbing the crop. This process requires a more elaborate secondary tillage operation. The drill cultivators in front of the wheels were closed to provide tramlines. These gaps were used for tines entering the ground and were a practical alternative to the bed system. This allowed sludge deposition under the crop, with minimum damage to the crop. To avoid damage by the central tine, the drill cultivator in the centre (cultivator 8) was also closed, placing a gap in the middle. The drill was recalibrated for this treatment, to maintain the target population of 300 plants m^{-2}.

Treatment 2 – bent leg injector
This treatment used the same equipment as Treatment 1. The main difference was the absence of the gap at row 8, maintaining gaps at rows 3 and 13 for traffic and bent legs. The drill was also recalibrated for this treatment, to maintain the target population of 300 plants m^{-2}. The purpose of this treatment was to justify maintaining a gap at row 8, with a conventional tine operating at this row.

Treatment 3 – fertilizer control
In this treatment 140 kg of nitrogen was applied per hectare, using a 46% nitrogen fertilizer. This distribution was done by hand as fertilizer spreaders with a work width of 2.5 m could not be found. The tractor drove over rows 3 and 13 simulating a real situation.

Treatment 4 – BKW injector
This treatment used a 4 leg injector designed and widely tested for grassland applications in growing cereal.

Treatment 5 – light BKW injector
This treatment used the same equipment as Treatment 4 and was meant to overcome possible bulldozing problems, and damage caused by the press roll. The tractor hydraulic power was used to reduce the weight of the press roll to a minimum required weight to ensure efficient slot closure after injection.

Treatment 6 – surface spread
The surface application of the sewage sludge was done manually using buckets. This method gave a more even coverage than the alternative method of passing over the plot with the injector lifted out of the ground. The tractor drove over rows 3 and 13 to simulate a real situation.

Treatment 7 – Silsoe injector
This injector which was also originally designed to inject liquid waste in grassland, was tested for a cereal post emergence injection situation.

Results

Observations

The injection procedure with Treatment 1 (bent leg + gap) did not present major problems and the operator could easily find the gaps between rows of crop and match the tractor wheels with the gaps. The central tine and press roll did not cause any damage to rows 7 and 9.

With Treatment 2 (bent leg), the non-existence of the central gap gave guidance problems for the operator, leaving the tracks in a few situations. The damage caused by the central tine was relatively small. The crop population at Treatment 3 (fertilizer control) always looked very uniform, without any problem. With Treatment 4 (BKW), the tractor operator had great difficulty in following the rows due to a lack of guidance. Some of the rows coinciding with injection legs were highly damaged. Treatment 5 (light BKW) also accomplished as good a slot closure as Treatment 4, although with the same guidance problem.

With the surface application of sewage sludge, the crop always appeared in very good conditions. The Silsoe College injector caused much damage, partially caused by the difficulty of the operator keeping a straight line between rows of crop.

Cut 1 – plant dry weight
The data set for cut 1 was unclear concerning differences between treatments. The analysis of variance did not show a significant difference between the seven treatments. The task of gathering the samples all at the same height was very difficult, due to the early state of development of the crop. This is a possible explanation for an absence of significant differences between the seven treatments.

Cut 2 – plant dry weight
The analysis of variance for the second data collection showed significant differences between treatments. No significant difference was found between Treatment 1 and both control situations, but the controls showed a significantly higher productivity when compared with Treatments 4, 5 and 7. In this analysis, differences between the controls and bent leg treatments, compared with the grassland injectors, began to be evident.

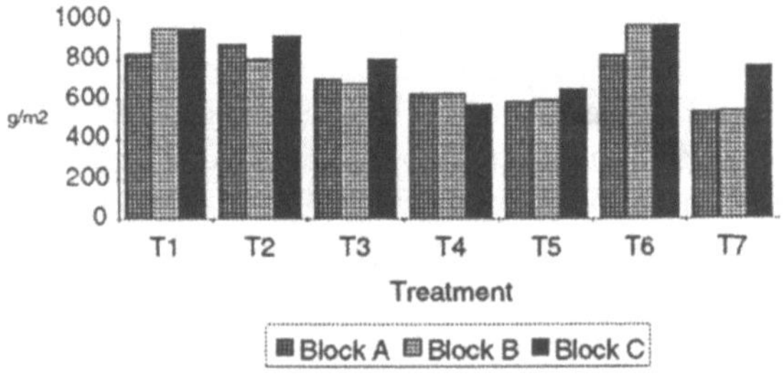

Fig. 1. Third data collection. Plant dry weight.

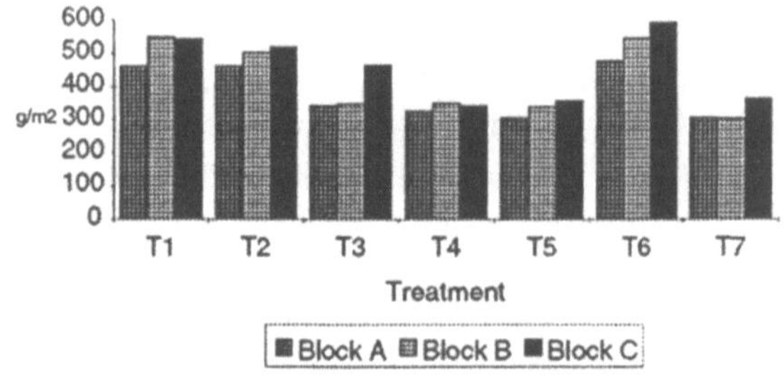

Fig. 2. Third data collection. Grain weight.

Cut 3 – plant dry weight
In the third data collection, differences between both BKW and Silsoe injector treatments, and the controls and both bent leg treatments, were evident. There was no significant difference between Treatments 4, 5 and 7. However these treatments showed a significantly lower productivity than controls and bent leg treatments.

No significant differences were found between both bent leg treatments and surface application of sewage. In this final stage of development, Treatment 3 showed a significantly lower productivity than Treatments 1, 2 and 6 (Fig. 1).

Grain production
The analysis of variance of the grain production confirmed the results from the plant dry weight production (Fig. 2).

Discussion

Observing the general state of each plot, it was noticed, mainly for Treatments 4, 5 and 7, that there was a dif-

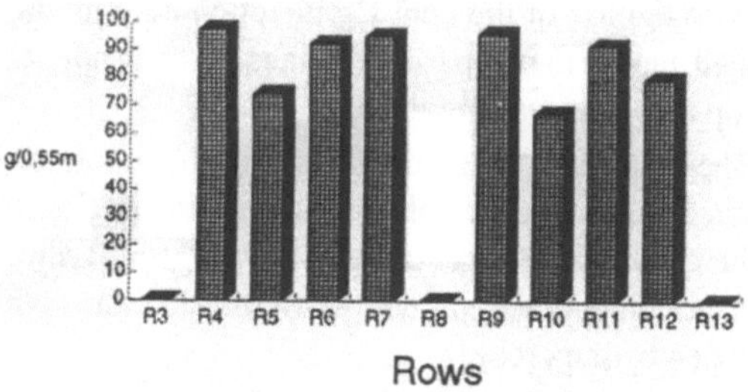

Fig. 3. Bent leg + gap. Plant dry weight, row by row.

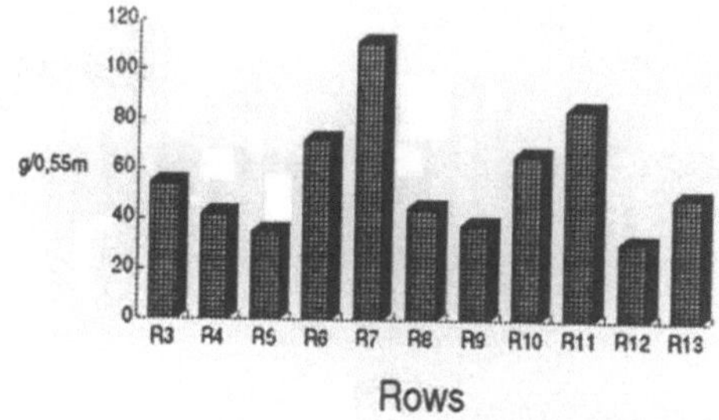

Fig. 5. BKW. Plant dry weight, row by row.

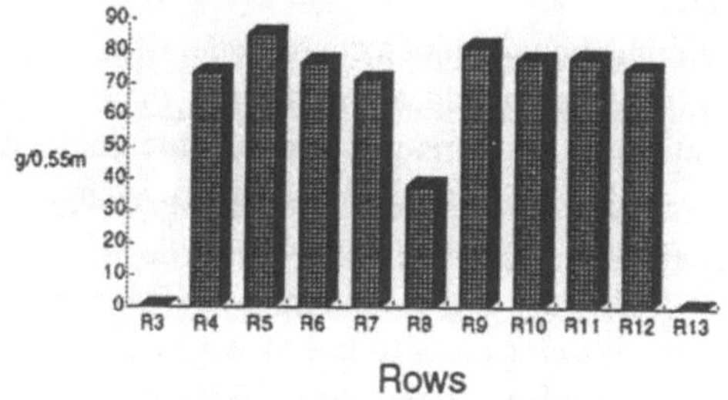

Fig. 4. Bent leg. Plant dry weight, row by row.

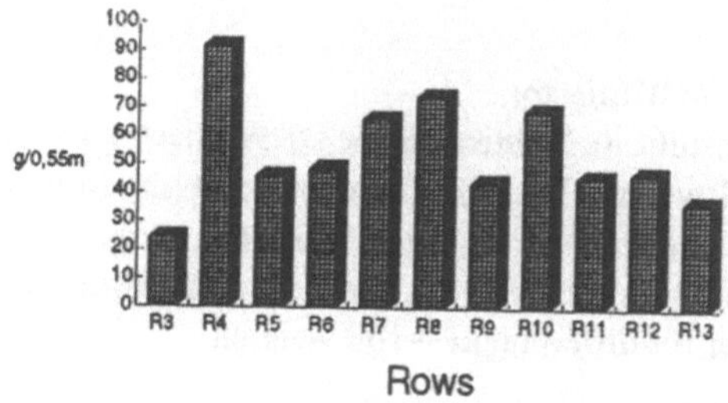

Fig. 6. Light BKW. Plant dry weight, row by row.

ference between rows of the same plot. The data collection of Cut 3 was done row by row in each sample. The objective of this procedure was to relate the lower productivity of traditionally injected plots with that due to tine damage.

Graphs showing productivity per rows for Block A of the experiment were chosen to represent the obtained results. Similar results were found for Blocks B and C.

Bent leg injector + gap

In this treatment, for the 3 blocks, the plant dry weights of each row were similar, and no major damage by tines and traffic was detected (Fig. 3).

The gaps in the crop, especially the central one, helped the tractor operator to keep the tractor in a perfect straight line, matching the tines and wheels with the gaps and avoiding major damage. Using this technique, the damage resulting from injection was overcome, obtaining the same productivity as with surface spreading of sewage.

Bent leg injector

As expected the 8th row, matching with the central tine of the injector, had for all blocks a lower productivity than the other rows (Fig. 4). In addition, its results are close to that of Treatment 1, and statistically there is no difference between Treatments 1 and 2, and surface application of sewage sludge.

For operational reasons Treatment 1 was preferred to Treatment 2, as the central gap is an excellent guide line for the operator, easily keeping the three tines matching with gaps. In Treatment 2 the operator, due to his position and crop development, had difficulty in seeing gaps at rows 3 and 13.

Fertilizer application

The final stage of development for Treatment 3 showed a significantly lower productivity than both bent leg treatments and the surface application of sewage. As no mechanical damage influenced Treatment 3, this was probably due to a lack of other nutrients, besides nitrogen, existing in the sewage sludge, such as phosphate.

BKW injector

The BKW injector obtained one of the lowest crop responses. The rows in this treatment, as observed in the field, did not have a similar productivity (Fig. 5).

The lowest productivity per row did not always match with the theoretical tine position. This may be due to the absence of guide lines, hence the operator probably had difficulty in keeping a straight line.

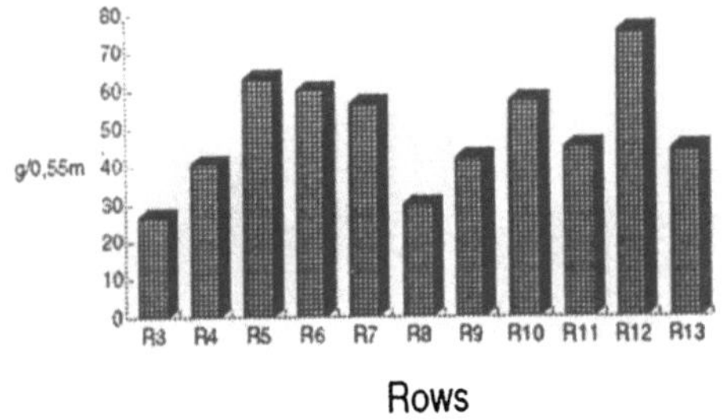

Fig. 7. Silsoe injector. Plant dry weight, row by row.

Light BKW injector
The results are identical to the BKW injector working at full weight (Fig. 6). The damage observed by the BKW injector for the soil conditions in this study is probably only caused by tine action, as results were similar for different press roll weights.

Surface spread application
Environmental reasons are certainly the main obstacle to the use of this technique in applying sewage in growing crops. No significant difference was found between this technique and the new ones, reaching the higher obtained productivities together with the bent leg equipment.

Silsoe College prototype
The rows where tines passed always had a lower productivity than the other rows, but each tine damaged more than one row (Fig. 7).

The difficulties in keeping tines between rows were as great as with both BKW treatments. A comparison of the crop response of using this injector, with 3 injecting tines with the BKW (4 tines), showed no significant differences.

Conclusions

(I) The use of angled tines for the injection of sewage sludge into a growing cereal crop reduces the mechanical damage to the crop.

(II) Sub-surface injection using the gaps in the crop to allow the passage of bent legs and wheels and, increasing plant population in the remaining rows, reduces the crop damage to a minimum. This technique produces the same yield as surface spreading without environmental hazard, and respects E.C. directives.

(III) With the use of the bent leg injection equipment, and 3 gaps in the crop for the 3 tines, an even spatial crop response was obtained.

(IV) The gaps act as guide lines for the operator, increasing the degree of accuracy in tine position. The existence of a central gap is very useful after further plant development, when the side gaps are hidden by the crop.

(V) Without guide lines, the degree of accuracy in tine positioning is very low. In this case the number of damaged rows is more than one per tine.

(VI) The implementation of a controlled traffic system using the gaps would allow the use of a tanker mounted injector. This would overcome the damage caused by the hose of an umbilical system used on grass land application on the growing crop.

(VII) For friable/dry soil conditions bulldozing effect from the injector press rolls was not noticed. The crop response for different weights of injector press rolls was the same.

(VIII) During sub-surface injection, most crop damage was due to tine action rather than to the press rolls.

(IX) The use of grass land sub-surface injection of sewage sludge techniques in growing cereal crops is not appropriate. The crop response for all stages of development, including final yield, was lower when compared with the bent leg system.

(X) A surface application of nitrogen fertilizer gave a lower crop response for the final development stages than a surface application of sewage sludge.

(XI) The use of adequate techniques permits sub-surface injection of sewage sludge without land or timing restrictions due to crop growth.

References

Bremner J M and Mulvaney C S (1982) Methods of Soil Analysis. Part II – Chemical and Microbiological Properties, second ed. Page A L, Miller R H and Keeney D R (eds.): 599–616

Department of the Environment/National Water Council (1983) Report of the Standing Committee on the Disposal of Sewage Sludge. Sewage Sludge Survey, 1980

Hall J E (1986) The Agriculture Value of Sewage Sludge. WRc Report N. ER 1220-M, WRc Environment, Henley Road, Medmenham, Bucks

Hann M J, Atckinson C J and Godwin R J (1992) Crop growth and nitrogen recovery resulting from the injection of sewage sludge. Department of Engineering for Agriculture, Silsoe College, unpublished Ph.D. Thesis

Miller J R and Godwin R J (1993) The design and evaluation of equipment for the injection of liquid wastes into post emergence cereal crops. Unpublished B. Eng. Thesis

MAFF (Ministry of Agriculture, Fisheries and Food) Code of good agricultural practice for the protection of water. Welsh Office Agriculture Department. MAFF Environmental Matters: 3–12.

Van Breemen N (1991) Ecological effects of ammonia deposition, in treatment and use of sewage sludge and liquid agricultural wastes. PL Hermit Elsevier Applied Science Publishers: 90–105.

C. Rodriguez-Barrueco (ed.), Fertilizers and Environment, 283–289.

Reducing N losses (NH_3, N_2O, N_2) and immobilization from slurry through optimized application techniques*

P. Dosch & R. Gutser
Institute of Plant Nutrition, Technical University Munich, 85350 Freising-Weihenstephan, Germany

Key words: denitrification, injection, NH_3 volatilization, N immobilization, slurry, surface application

Abstract

In model, pot and field trials the effect of C reduced slurries and different application techniques on N losses and N immobilization were investigated. The C reduced slurries were produced by mechanical separation. Ammonia losses from surface-applied and injected cattle slurry were measured under field conditions using a wind tunnel system. Injection of slurry was the most efficient way to reduce volatilization of ammonia. After 6 days the total loss from the injected slurry was only 9% of that from surface band application. Furthermore, additional losses of N may occur through denitrification, specially after injection of slurry which may create an anaerobic environment abundant in readily oxidizable C. Therefore denitrification measurements by the acetylene inhibition technique were conducted. Until 100 days after application the loss from the injected slurry was 7.3 kg compared to 4.5 kg N ha^{-1} from surface band applied slurry. After injection, denitrification was only 4.1 kg N ha^{-1} for C reduced compared to 6.5 kg N ha^{-1} for normal slurry. In pot trials the ammonium-^{15}N of normal slurry and C-reduced slurry was utilized by oats between 52 and 60%, the ammonium sulfate by 67%. The increased biomass C confirmed a greater immobilization of the NH_4-N of the normal slurry resulting in a lower initial efficiency.

Introduction

The utilization of potentially available nitrogen (NH_4-N) of slurries by growing plants is often unsatisfied, due to losses through N leaching and considerable gaseous losses through ammonia volatilization or biological denitrification. These losses are implicated with increasing costs for plant production and additionally, the emission of nitrogen gases (NH_3, N_2O) can be detrimental for natural ecosystems. The deposition of ammonia may lead beside the acidification of some forest soils in Europe (van Breemen *et al.*, 1987) also to undesired changes in oligotrophic ecosystems (Schulze *et al.*, 1989). Nitrous oxide has been shown to contribute to global warming and to the destruction of the stratospheric ozone when it is converted to nitric oxide (Crutzen, 1981).

In addition to the various losses of nitrogen an immobilization of the NH_4-N of slurry is responsible for a low efficiency (Gutser and Dosch, 1994). Therefore, the goal of best management practice for manuring is to obtain high N utilization of applied slurry in the first vegetation period. For this, strategies which minimize N losses and N immobilization are necessary.

Ammonia volatilization after application of slurry are reported to be 5–95% of applied ammonium N. The NH_3 loss rate is affected by climatic conditions, mode of application, chemical and physical soil properties (Rank *et al.*, 1987). It has also been shown, that the dry matter content of slurry affects NH_3 loss and in this way a lowered dry matter content of mechanical separated slurry lead to reduced NH_3 losses.

Immediate incorporation or the direct injection of slurry into the soil is known to minimize NH_3 losses to low values. Thompson *et al.* (1987) observed, that only 2% of the applied NH_4-N were lost by volatilization when slurry was injected to grassland. However, the injection of slurry increased the losses by denitrification up to 51 and 16% of the applied NH_4-N, after application in winter and spring, respectively.

In order to value the environmental compatibility of different strategies all losses of nitrogen must be considered, including N immobilization. The aim of the present study was to investigate the effect of C

* Dedicated to Prof. Dr. A. Amberger on his 75th birthday

Table 1. Composition of the cattle slurries (average)

Slurry	Dry matter	NH_4-N	Total N	Total C
		(% in fresh weight)		
Normal	7.6	0.21	0.37	3.3
Separated	5.3	0.22	0.36	2.3

reduced slurries in combination with different application techniques (injection, surface band application) on the extent of gaseous N losses and the N utilization by plants.

Methods

Site characteristics

Measurements were carried out on a brown earth (Dürnast soil) at the institute of Plant Nutrition in Weihenstephan near Munich. Some properties of the soil (0–25 cm): pH 6.4 ($CaCl_2$); clay 20%; silt 60%; total N 0.12%; organic C 1.11%; CEC 141 meq kg^{-1}.

Slurry

Throughout all experiments cattle slurries were used. Before each application, slurry was taken from the same tank to provide similar slurries and afterwards separated with a roller press separator. The separation resulted in a reduced dry matter and total C content but had no effect on the total N and NH_4-N content of slurry; the C/N ratio is decreased (Table 1). In each experiment slurries were applied on the basis of NH_4-N.

NH_4-N of slurries was determined with a NH_3-selective electrode (Orion model 95–12), total N by the Kjeldahl method.

Ammonia volatilization

Losses of N through ammonia volatilization were measured using a wind tunnel system like the system described by Lockyer (1984). The wind tunnel consists of an inverted U-shaped tunnel made from transparent polycarbonate, which covered an area of 1 m^2 (0.5 × 2.0 m) and a circular steel duct housing an electrically powered fan. Wind speed through the tunnels was controlled at 2.0 m s^{-1} in these experiments.

There were five wind tunnel units of which one was used for reference (measurement of the ambient ammonia level). All experiments were carried out in two treatments and two parallels.

An aliquant part of the airstream was sampled near the outlet of the steel duct using a diaphragm suction pump. The pump drew an airstream at a rate of 5 l min^{-1} through a 500 ml absorption flask fitted with a sintered gas distribution tube. The flask contained 300 ml 0.01 *M* H_2SO_4. Ammonia in the air was trapped in the sulfuric acid and determined in the laboratory by destillation and subsequent titration.

In the first experiment slurry treatments were (i) surface band application and (ii) injection of normal slurry. Injection depth was 100 to 150 mm. The normal slurry was applied at a rate equivalent to 70 kg NH_4-N ha^{-1} to the fallow experimental area. In the second experiment, slurry treatments were (i) surface application by broadly spreading the slurry (splash plates) and (ii) surface band application (trailing hoses) of separated and normal slurry. Both slurries were applied to growing winter wheat at shooting stage in a dose equivalent to 70 kg NH_4-N ha^{-1}.

Denitrification

Measurements of denitrification by the acetylene inhibition technique using a soil-cover method described by Hutchinson and Mosier (1981) were conducted in the field cropped with maize. 20-ml samples of the enclosed air were withdrawn 0, 60 and 120 min after placement and stored in vaccutainers until analysis of N_2O and CO_2 were made. Coated CaC_2 has been used as a source of C_2H_2 (Aulakh *et al.,* 1991).

Maize (*Zea mays*, L.) was planted 28 April 1993 in 66 cm rows. The first slurry application took place on 25 May 1993: surface band application with no incorporation spaced 200 mm apart and injection 100 mm deep, 200 mm apart, both as normal slurry. The second slurry application took place on 11 June 1993: injection of separated and normal slurry. In both experiments injected and surface-applied slurries were applied at a rate of 60 kg NH_4^-N ha^{-1}.

The vented chambers (diameter 120 mm) were placed directly over the surface or injection band. The measured flux can therefore be calculated for an area of 1818 m^2. Flux ha^{-1} was given additively by the measured flux of the control plot without slurry.

Soil samples (0–15 cm) were collected to each date as denitrification was measured. In order to obtain the whole slurry band (injection and surface band) a 200 mm wide soil block was sampled. The soil samples were extracted with 2 *M* KCl and 0.01 *M* $CaCl_2$ for NH_4 and NO_3, respectively. NH_4 was determined colorimetrically using the reaction with sodium hypochloride and sodium phanate and NO_3 by high pressure liquid chromatography as described by Vilsmeier (1984).

Pot trial

Field-moist soil from Dürnast was collected from the A horizon, passed through a 5 mm sieve and filled (6.5 kg) in pots (diameter 200 mm, 200 mm high). 450 mg NH_4-N as normal slurry, separate slurry or ammonium sulfate (atom% ^{15}N abundance of each treatment was 14.4) were thoroughly mixed with the soil. After one week of incubation oat (*Avena sativa*, L.) was seeded at 25 plants pot^{-1} and placed in the greenhouse. At flowering stage total N and ^{15}N analysis in soil or in plant samples were conducted with an automated mass spectrometer (Europa Scientific). Additionally, soil samples were removed 1, 4 and 8 weeks after manuring from pots without plants and analysed for microbial biomass (Anderson and Domsch, 1978).

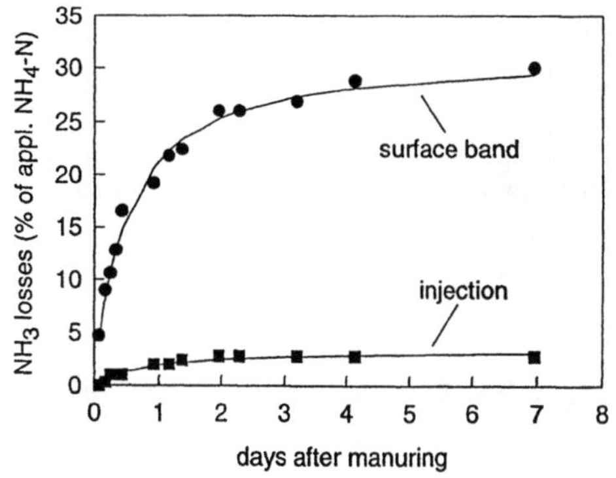

Fig. 1. Cumulative NH_3 losses after different application of slurry to soil without plants. N dose: 70 kg NH_4-N ha^{-1}.

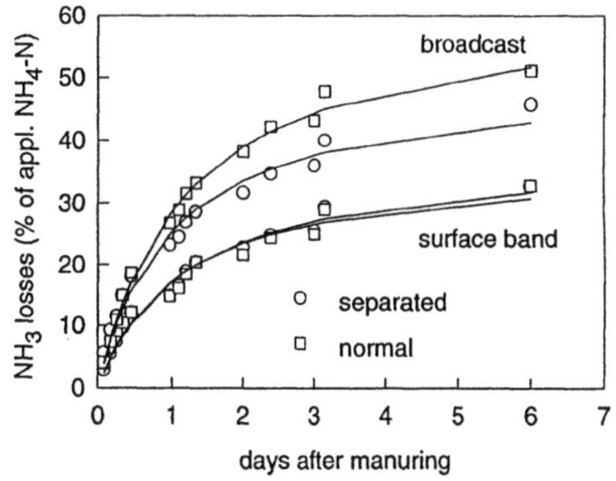

Fig. 2. Cumulative NH_3 losses after different application of normal and separated slurry to winter wheat at shooting stage. N dose: 70 kg NH_4-N ha^{-1}.

Results

Ammonia volatilization

The cumulative losses of N through volatilization after surface band application or injection of normal slurry are presented in Figure 1. The NH_3 losses of surface band applied slurry followed the typical pattern of lost ammonia after application of slurry. More than half of the total loss occuring during the 7-day period took place within the first day. The total NH_3 loss following surface application was 21.7 kg N ha^{-1} and was equivalent to 31% of the applied NH_4-N. Direct injection of slurry into the soil reduced NH_3 losses to low values. After 7 days the total loss from the injected slurry was only 9% of that from surface band application.

In growing winter wheat lower NH_3 losses were observed when slurry was applied in surface bands compared to broadcast application (Fig. 2). The total NH_3 loss following surface band application was 22.7 kg N ha^{-1}. The use of separated slurry showed in the broadcast treatments lower NH_3 emissions of 51 to 46% of the applied NH_4-N compared to normal slurry. This was equivalent to 35.8 and 32.0 kg N ha^{-1} for broadcast application of separated and normal slurry, respectively.

Denitrification

Denitrification measurements using the acetylene inhibition technique showed high N_2O-fluxes in the first 3 weeks after application of slurry (Fig. 3). The maximum rate observed was 625 g N_2O-N ha^{-1} d^{-1} from the injected slurry. After 21 days the rates of denitrification of both treatments remained < 35 g N_2O-N ha^{-1} d^{-1} in spite of high NO_3 concentrations in soil (Fig. 4a, b) and were within the range of the control plot. The total denitrification loss measured from injected slurry treatment was 7.3 kg N ha^{-1} compared to 4.5 kg N ha^{-1} from the surface banded slurry without incorporation.

The production of CO_2 followed a similar pattern as the rate of N losses by denitrification (Fig. 3). Therefore, the injected treatment led to a higher CO_2 emis-

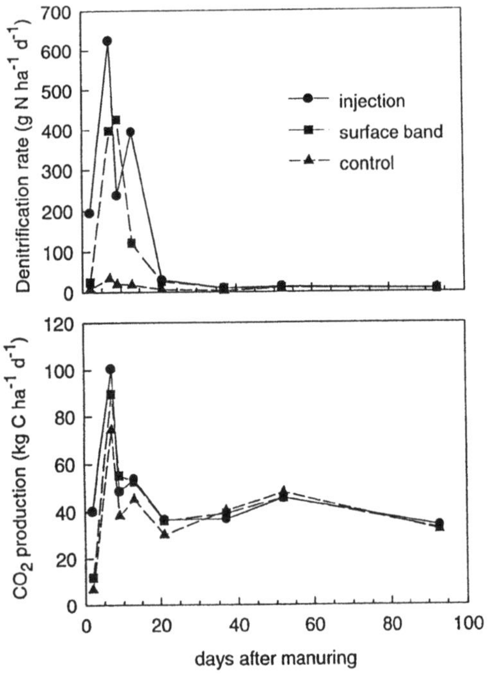

Fig. 3. Denitrification and CO_2 production rates from soil after different application of slurry to maize. N dose: 60 kg NH_4-N ha^{-1}.

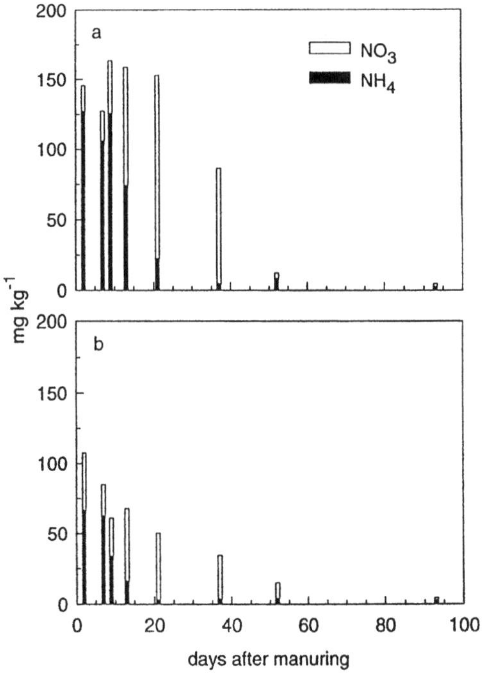

Fig. 4. Soil ammonium- and nitrate-N after injection (**a**) or surface band application (**b**) of slurry to maize.

sion than the surface band application. The similar N_2O-fluxes after 3 weeks of slurry treated and control plots may be a result of an exhaustion of added organic material, because the NO_3 content in soil was sufficiently high to allow denitrification.

Inorganic N concentrations in soil were lower after surface band application compared to injection (Fig. 4a, b). If we assume, that 2 days after manuring the N losses by denitrification were negligible, the lower N content after surface band application may probably due to higher N losses through volatilization of about 30% and confirmed the NH_3 losses measured in the wind tunnel. Decreasing inorganic N contents in the surface treated soil can be attributed to erosion of soil and nitrogen, caused by a heavy rainfall on the 4th day after manuring. After injection no decrease in anorganic N contents in the soil was observed. Therefore, surface band application demands incorporation of slurry, in order to minimize NH_3 losses as well as erosion of nitrogen.

With one exception higher denitrification rates up to 379 g N_2O-N ha^{-1} d^{-1} were observed for the injected normal slurry (Fig. 5). Afterwards the N losses by denitrification decreased strongly and reached the niveau of the control plot in day 28 after manuring. Until 90 days after injection of slurry, denitrification was only

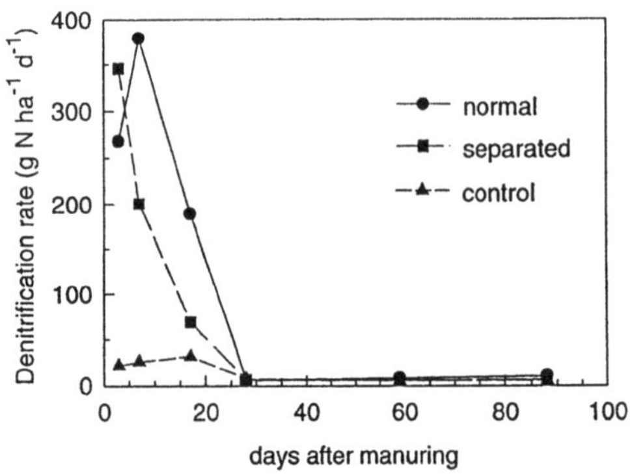

Fig. 5. Denitrification rate after injection of normal and separated slurry to maize. N dose: 60 kg NH_4-N ha^{-1}.

4.1 kg N ha^{-1} for the C reduced slurry compared to 6.5 kg N ha^{-1} for normal slurry. The rate of CO_2-C evolution at the first date was equal in both slurry treatments. But in the following dates a faster decrease was observed for the separated slurry, similar to the pattern of N_2O-fluxes. Substracting the CO_2 losses of the the control total CO_2 loss from the normal slurry was 342 kg C ha^{-1} compared to 209 kg C ha^{-1} from separated slurry.

Table 2. N utilization of slurry and mineral fertilizer (NH_4-^{15}N) in a pot trial with oat. N dose: 450 mg NH_4-N pot^{-1}

Treatment	N removal			^{15}N residue in soil
	Sum	Fertilizer	Soil*	
	(mg pot^{-1})			
Normal slurry	376^{b**}	235^{c} = 52%	141	27.2
Separated slurry	413^{ab}	272^{b} = 60%	145	19.6
$(NH_4)_2SO_4$	439^{a}	301^{a} = 67%	139	14.1

* includes N from organic N in slurry.
** Values with same letter within a column are not significantly different based on Tukey's multiple comparison test ($\alpha = 0.05$).

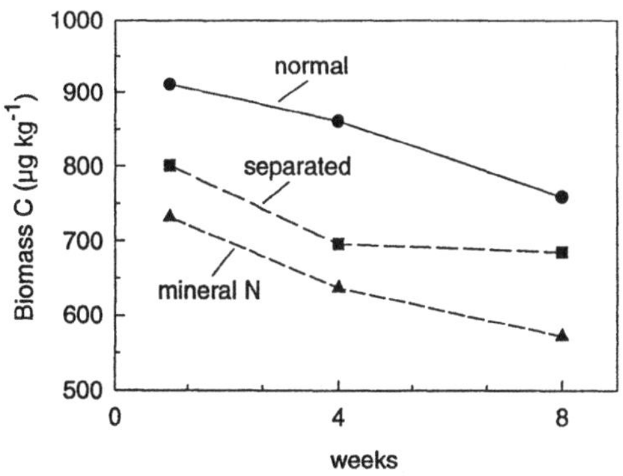

Fig. 6. Biomass C after application of slurry and $(NH_4)_2SO_4$.

Immobilization

A pot trial was conducted to examine the utilization of manured NH_4-N by plants as a function of different rates of organic C supplied either with separated or normal slurry and $(NH_4)_2SO_4$. The higher supply of organic C in the treatment 'normal slurry' resulted in a lower N removal compared to separated slurry (Table 2). The application of $(NH_4)_2SO_4$ led to the highest N removal and therefore, a significant greater amount of the added NH_4-N was available for plants.

After harvesting the green oat the normal slurry showed the highest fertilizer residue, probably due to a greater immobilization. The development of biomass C contents confirmed a biological fixing of nitrogen (Fig. 6). While addition of $(NH_4)_2SO_4$ led to the lowest biomass C contents, the supply of organic matter effected a strong increase of microbial activity. Table 2 showed, that in addition to NH_4-N of the normal slurry the mineralized N from soil or from organic N in slurry was partly immobilized, too. In another experiment on a more sandy soil this immobilization was markedly higher.

Discussion

Various experiments showed, that immediate incorporation of slurry considerably decreased the loss of N through volatilization (Rank *et al.,* 1987; van der Molen *et al.,* 1990). However, this precaution could not be realized in closed plant covers with exception of row cropping such as maize. But the utilization of applied NH_4-N of organic manures is highest during the growth period, probably due to a lower N leaching and N immobilization. Therefore, other modes of application must be used to minimize NH_3 losses.

Additionally to an immediate incorporation in row croppings, an injection of slurry direct into the soil reduced NH_3 losses to a negligible level. Sawyer and Hoeft (1990) reported that injection of slurry affected the maize rooting distribution due to changed soil properties, i.e. high pH, high NH_4-N and high NO_2-N. In our experiments a change of these soil properties were observed, but no alteration in rooting distribution could be demonstrated.

In growing winter wheat at shooting stage a surface band application of cattle slurry by trailing hoses caused lower NH_3 emissions of 30 and 37% compared to broadcast application of separated and normal cattle slurry, respectively. No smothering of leaves occured when slurry was applied in surface bands and therefore NH_4-N of the slurry can be retained greatest possible by cation exchange sites within the soil. The reduced NH_3 loss can also be ascribed to an absorption of NH_3 by plant leaves (Sommer *et al.,* 1992). Moreover, the closed plant cover may led to a lowered wind speed at the soil surface resulting in a smaller rate of NH_3 volatilization (Sommer *et al.*, 1991). The smaller difference in NH_3 emissions between broadcast and surface band application of separated slurry is due to a better run off from leaves onto respectively into the soil, caused by the lowered dry matter content by mechanical separation. Sommer and Olesen (1991) stated that a dry matter content higher than 12% or lower than 4% resulted in only small changes in NH_3 losses. In our experiments separation of slurry reduced the dry matter content of slurry even in the above mentioned range and therefore, NH_3 losses can be greatly influenced by a mechanical separation due to a better infiltration into the soil.

Compared to the losses through volatilization, denitrification losses are much lower. The rate of these losses were greatly influenced by the mode of application and the quantity of organic material present. Therefore, the higher rates were maintained by the injection of normal slurry into the soil. In this treatment high CO_2 emissions, stimulated by microbial respiration combined with a high O_2 consumption in a confined area, resulted in increased denitrification. In spite of a high CO_2 evolution from the injected treatment, denitrification rate was relatively low at the first date after manuring. Shortly after injection the extent of denitrification will be controlled by the NO_3-N in the soil (Comfort *et al.*, 1990). Later, when nitrification of the NH_4-N of slurry provides a large amount of NO_3-N in soil, denitrification is determined primarily by the oxidizable C in the injection zone. After 3 to 4 weeks the readily oxidizable C was metabolized and despite high NO_3-N concentrations in the soil only small denitrification losses occured. These data suggested that the potential for denitrification after injection decreases as soon as the readily oxidizable C has been removed. Immediately after injection of separated slurry emission of CO_2 was as high as for the normal slurry. But in the next dates a stronger decrease of CO_2 emissions was observed -the lower C input by separated slurry was faster metabolized- reduced N losses through denitrification were measured.

According to this results, the level of gaseous N losses can be largely minimized by plant specific application techniques. However, even an optimized application of slurry led to a lower availability of applied NH_4-N compared to ammonium sulfate, as shown in the pot trial. Slurry manuring is associated with the supply of readily oxidizable C resulting in an increased biological activity and a higher potential for immobilization of soil. The extent of this immobilization will

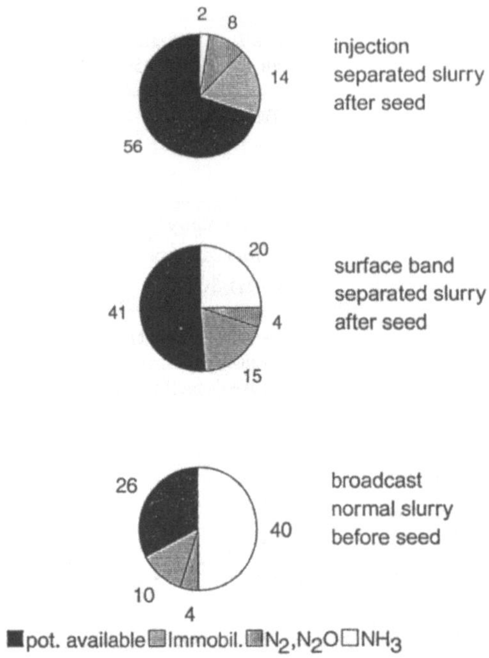

Fig. 7. Losses, immobilization and availability of slurry NH_4-N in different application systems for maize (kg NH_4-N ha^{-1}). N dose: 80 kg NH_4-N ha^{-1}.

be controlled by the quantity of supplied available C (Dosch and Gutser, 1993). Therefore, the C reduced separated slurry showed a lower N immobilization and consequently a better utilization of the applied NH_4-N already during the first vegetation period. The availability of immobilized N for plants and its effects to environment is discussed in detail by Gutser and Dosch (1994).

Figure 7 represents a calculation of the amount of N available for maize in different slurry managements, estimated by results of model and especially field trials. An important criterion is the prevailing reduction of N losses and N immobilization. As the diverse N losses may vary in the tested slurry managements, the sum of all N losses is important for an ecological and economical valuation. High N losses through volatilization of NH_3 were primarily responsible for the unsatisfied N utilization after broadcast application of slurry without incorporation before seed. A distinctly reduction of NH_3 losses is obtained by a surface band application of separated slurry, e.g. by trailing hoses in growing maize. Furthermore, the losses are markedly diminished after injection of separated slurry, while N losses due to denitrification increases only slightly. Considering a lower N immobilization of applied separated slurry the potentially available nitrogen in the treatment 'injection' come to 56 kg N ha^{-1} compared to 41 respectively 22 kg N ha^{-1} after surface band application and broadcast application. The supply of C reduced slurry (separation or anaerobic fermentation (Messner and Amberger, 1988)) in combination with special application techniques (injection in row croppings, trailing hoses in closed plant covers) increases the predictability of the manurial value of slurry and in consequence an optimized nutrition of plants by combination of mineral and organic fertilizers is possible. Moreover, these practices are favourable for the environment. Further investigations are needed to evaluate the production of N_2O, thereby taking into account both denitrification and nitrification.

References

Anderson JPE and Domsch KH (1978) A physiological method for the quantitative measurement of microbial biomass in soils. Soil Biol Biochem 10: 215–221

Aulakh MS, Doran JW and Mosier AR (1991) Field evaluation of four methods for measuring denitrification. Soil Sci Soc Am J 55: 1332–1338

Comfort SD, Kelling KA, Keeney DR and Converse JC (1990) Nitrous oxide production from injected liquid dairy manure. Soil Sci Soc Am J 54: 421–427

Crutzen PJ (1981) Atmospheric chemical processes of the oxides of nitrogen, including nitrous oxide. In: Delwiche C C (ed), Denitrification Nitrification and Atmospheric Nitrous Oxide, Wiley, New York pp 17–44.

Dosch P and Gutser R (1993) Strategien zur Optimierung der N-Wirkung der Gülle. Landw Forsch, Kongreßband 1993, VDLUFA-Schriftenreihe 37: 121–124

Gutser R and Dosch P (1994) Cattle-slurry-^{15}N turnover in a long-term lysimeter trial. Proceedings VIII. Int Symp CIEC on Fertilizers and Environment, Salamanca. In: Developments in Plant and Soil Sciences Series, Kluwer Acad. Publ., Dordrecht, Netherlands *(in press)*

Hutchinson GL and Mosier AR (1981) Improved soil cover method for field measurement of nitrous oxide fluxes. Soil Sci Soc Am J 45: 311–316

Lockyer DR (1984) A system for the measurement in the field of ammonia through volatilization. J Sci Food Agric 35: 837–848

Messner H and Amberger A (1988) Composition, nitrification and fertilizing effect of anaerobically fermented manure slurry. Proc. IV CIEC Symp. 1987, Braunschweig-Völkenrode 1: 125–130

Rank M, Huber J and Amberger A (1988) Model trials on the volatilization of ammonia following slurry application under controlled climate and field conditions. Proc. IV CIEC Symp. 1987, Braunschweig-Völkenrode 1: 315–320

Sawyer JE and Hoeft RG (1990) Effect of injected liquid beef manure on soil chemical properties and corn root distribution. J Prod Agric 3: 50–55

Schulze ED, de Vries W, Hanks M, Rosén K, Rasmussen L, Tamm SO and Nilsson J 1989 Critical loads for nitrogen deposition on forest ecosystems. Water Air Soil Pollut 48: 451–456

Sommer SG and Olesen JE (1991) Effects of dry matter content and temperature on ammonia loss from surface-applied cattle slurry. 20: 679–683

Sommer SG, Olesen JE and Christensen BT (1991) Effects of temperature, wind speed and air humidity on ammonia volatilization from surface-applied cattle slurry. J. Agric Sci 117: 91–100

Sommer SG, Jensen ES and Schørring JK (1992) Leaf absorption of gaseous ammonia after application of pig slurry on sand between rows of winter wheat. Air Pollution Research Report 39. Commission of the European Communities, Bruxelles, Belgium. pp 395–402

Thompson RB, Ryden JC and Lockyer DR (1987) Fate of nitrogen in cattle slurry following surface application or injection to grassland. J Soil Sci 38: 689–700

Van Breemen N, Mulder J and Van Grinsven JJM (1987) Impacts of acid atmospheric deposition on woodland soils in the Netherlands. II. Nitrogen transformations. Soil Sci Soc Am J 51: 1634–1640

Van der Molen J, Van Faassen HG, Leclerc MY, Vriesema R and Chardon WJ (1990) Ammonia volatilization from arable land after application of slurry. 1. Field estimates. Neth J Agric Sci 38: 145–158

Vilsmeier K (1984) Bestimmung von Dicyandiamid, Nitrat und Nitrit in Bodenextrakten mit Hochdruckflüssigkeitschromatographie. Z Pflanzenernähr Bodenkd 151: 459–473

C. Rodriguez-Barrueco (ed.), Fertilizers and Environment, 291–295.

Agricultural use of sewage sludge and its regulation

P. Soler-Rovira[1], J. Soler-Soler[2], J. Soler-Rovira[3] & A. Polo[1]
[1]*Centro de Ciencias Medioambientales, CSIC, Serrano 115 dpdo., Madrid, Spain;* [2]*Ministerio de Agricultura, Pesca y Alimentación, Pº Infanta Isabel 1, Madrid, Spain;* [3]*Escuela Universitaria Ingeniería Técnica Agrícola, Universidad Politécnica Madrid, Spain*

Key words: sewage sludge, organic matter, heavy metals, legislation

Abstract

Sewage sludge is waste for which the most positive outlet must be sought whilst, at the same time, making use of its qualities. In view of its organic matter and other nutrients content, it may be thought that the best system is to use it for agricultural purposes, although the load of heavy metals it may bear with it proves to be the main obstacle involved. Since it is a heterogeneous material whose composition depends on various factors, the content of such metals in sewage sludge will be highly variable. Thus it becomes necessary to always analyze sewage sludge before directing it towards agricultural use and also to know the heavy metal content of the soil where it is applied already so as to adjust this agronomical practice to the limits as set by current legislation in force. The provisions of law must also be complied with when certain substances deemed toxic or dangerous are present in sewage sludge. The purpose of all this is to prevent possible soil pollution and other kinds of risk involved in accumulation in the food chain.

Introduction

Water pollution has been increasing at the same rate as man's activity has developed on a domestic, industrial and agricultural level with increasing water consumption. This is why the need arose for regeneration to remove all its polluting load such as solid waste, microorganisms and a wide variety of soluble substances.

Thus, pollution is removed from water but a by-product is generated known as sewage sludge, which joins the long list of waste our society is producing. This is why an outlet for it was sought since otherwise it would involve an enormous loss of energy and general ecological imbalance. At the present time, this sewage sludge is disposed of in various ways but the end intended for it has to be perfectly controlled in order not to cause further pollution; an outlet has to be sought involving the least risks possible for the environment and where the best use can be made of its qualities.

In a country like Spain where, in general, soils are scant in organic matter and organic remains of an animal origin are becoming more and more scarce, it is natural to think that the best system for using this sewage sludge would be for agricultural purposes, with applications as a fertilizing material. Apart from providing organic matter (sewage sludge contains around 50% d.m.), it may also enrich soils with other nutrients like nitrogen (1–7% d.m.), phosphorus (1–5% P_2O_5 d.m.), potassium (0.1–3% K_2O d.m.) and other elements (most originating from sewage sludge stabilization treatments).

However, not all sewage sludge characteristics are positive in this sense since potentially toxic elements may be found within its composition. By definition, it is a heterogeneous material since it depends on wastewater, on the technology used and on the time of the year and thus industrial effluent will provide more polluted sewage sludge than that from treating wastewater of another type, mainly urban in nature.

Amongst negative aspects of applying sewage sludge as fertilizer its pathogen microorganism, heavy metal and certain organic compound content mainly stand out; whilst the former can be removed by treatment (composting) to obtain a well conditioned product, the latter have to be watched over both with respect to their content and to their possible transfer to the soil and living beings.

Sewage sludge displays a large amount and variety of microorganisms depending on its effluent of origin and how long they remain will depend on the conditions of the environment where they live and the treatments the sewage sludge has undergone. Those which have been stabilized by anaerobic digestion may have the number of their pathogens reduced by as much as 90%. These, then, are the best for agricultural use.

What are called heavy metals is a group of metal elements, metalloids and non-metals like As, B, Cd, Cr, Cu, Fe, Hg, Mn, Mo, Ni, Pb, Se and Zn, etc. which, though found in sewage sludge in relatively small amounts compared with its other components, may be excessive in many cases and represent a risk both in their toxicity in small concentrations and in their endurance and accumulative effect. The cycle such heavy metals undergo in the environment passes through any medium and may affect any of the beings populating it. Long term effects occur in the different links in the food chain. In addition, the antagonic and synergetic effects they have on other elements, which in most cases are essential to life, must not be forgotten.

Other substances which can be named as possible pollutants within sewer sewage sludge cover a wide group of organic compounds where most are characterized by being found in very low concentrations but they are very dangerous for their accumulative and enduring effects. Amongst such substances are PCBs (Polychlorobiphenyls), phytosanitary products, dioxins, organochlorated compounds and a long list of organic substances, all originating in industrial wastewater.

Legal regulations

The conclusion may be drawn, from all the foregoing, that the agricultural application of sewage sludge has to be made in a rational fashion, subject to prior experiments and always subjected to regulations which control this type of practice at any time and in all countries. This is why a series of regulations has been in force for some time now both at Community level and in each of the Member States dealing with regulating the agricultural use of this sewage sludge as best possible.

Sewage sludge used in agricultural operations was not included in Council Directive 75/422/EEC of 15 July, 1975 relating to waste. However, it was affected by the measures provided for in Council Directive 78/319/EEC of 20 March, 1978 relating to toxic and hazardous waste, since they are also applicable to sewage sludge inasmuch as it may contain or be polluted by matter or substances mentioned in Appendix A to said Directive, which represent a risk to human health or to the environment. For the same reason, in Spanish legislation, it is included in Basic Act 20/1986 of 14 May on toxic and hazardous waste. According to this Act's provisions, the management of sewage sludge or activities leading to its recovery must ensure that the final destination involves no danger at all whilst at the same time requiring such waste and its destination to be registered in order to facilitate information collection and monitoring by the authorities.

According to Directives 75/440/EEC and 80/68/EEC, sewage sludge must be used under such conditions that soil and surface and ground water protection is guaranteed.

This was why it was necessary to provide regulations ensuring they would be protected against the harmful effects of uncontrolled sewage sludge use. Whilst the latter displays useful agronomical properties, applying it must not harm soil quality and plant production, since certain heavy metals may be poisonous to plants and people. This has led to establishing limits with respect to their content in soil and in sewage sludge used.

Since it is necessary to control the quality of sewage sludge and the soil to which it will be applied, it is advisable to have a certain amount of related information which is to be periodically sent to the Commission. Taking this into account, the latter will make proposals if necessary to guarantee greater protection for the environment and soils.

With the purpose of standardizing all these circumstances, Council Directive 86/278/EEC of 12 June, 1986, was brought out. Its purpose was to regulate the use of sewage sludge in agriculture in order to prevent noxious effects in plants, soils, animals and human beings as well as promoting its correct use. According to this Directive, sewage sludge must be treated before agricultural use although under certain conditions, Member States may authorize the burial or injection of other types of sewage sludge into soil, provided no risk at all is involved. Its Appendices lay down certain limit values for heavy metals which it may contain:

a) Limit values for heavy metal concentration in soils: treated sewage sludge may not be applied to soils displaying a heavy metal concentration higher than that laid down.

b) Limit values for heavy metal concentration in sewage sludge earmarked for agricultural use : treated sewage sludge for application to soil shall not exceed the limit values as laid down in its heavy metal content.

c) Limit values for annual amounts of heavy metal which may be introduced into soils, based on a ten year average: the maximum amounts of sewage sludge which may be applied per $ha^{-1}year^{-1}$ will be those which do not exceed the limit values set in accordance with the heavy metal content of the soil and sewage sludge to be used.

In addition, sewage sludge and soil sampling and analysis reference methods are included and the frequency of such analysis and of the specific parameters to be determined in them is mentioned in each case.

Likewise, member States shall forbid the use of waste sewage sludge on grazing land or animal feed crops, on horticultural and fruit crops (except fruit trees) and on horticultural or fruit crops in contact with the soil which are eaten raw, within terms which shall be set as per the circumstances of application and the crop itself.

Likewise mention is made of the possibility of the pH of soil to which sewage sludge is applied being less than 6 and thus the mobility of the heavy metals increasing and the possibility of transfer to the crop being much greater, with which limit values as set could be reduced.

Article 16 of this Directive 86/278/EEC lays down that member States shall bring the legal, statutory and administrative provisions necessary to conform to the said Directive into force in a term of three years as from its notification and they shall inform the Commission of the text of the provisions of internal Law they may adopt on them.

Royal Decree 1310/1990 of 29 October (published in the B.O.E. of 1 November, 1990) whereby the use of sewage sludge in the agricultural sector is regulated, arose from the need to transpose the aforesaid Directive 86/278/EEC to internal Spanish Law with the purpose of harmonizing our legislation and applying the same within the terms provided for, whilst laying down a regulation framework enabling sewage sludge production and its agricultural use to be reconciled.

This Regulation proves to be a perfect transfer of the Community Directive as it almost literally reproduces the conditions as set thereby. In addition, it could not be expected that Spain, with conditions better than other countries for treating and using sewage sludge, was to be more demanding in its regulation. Spanish soils are generally scarcer in organic matter and have a serious desertification problem. Combined with high pH in most of them, using sewage sludge becomes a highly recommendable agricultural practice.

This Royal Decree's pertinent Appendices also lay down the limit values of heavy metals (Table 1) where soils with a pH above or below seven are differentiated. The latter are where the highest values included in Directive 86/278/EEC are authorized since under these conditions, the mobility of these metals noticeably diminishes. Only in the case of copper, nickel and zinc are the limit values of concentration in soils somewhat higher than the Community's and this is an optional responsibility of Member States contemplated in the Directive for soils with a pH constantly higher than seven.

This Royal Decree 1310/1990 empowers the Ministry of Agriculture, Fish and Food to decree the regulations necessary for developing and complying with the same. It establishes a series of controls which the Spanish Regional Governments must undertake to monitor the use of sewage sludge in agricultural operations and so the National Sewage Sludge Register was created, under the said Ministry.

Thus it became necessary to standardize information and lay down the channels for its final compilation in fulfilment of the Commission's mandate (Directive 86/278/EEC, article 17) for drawing up a synthesis report on the use of sewage sludge in agriculture, specifying amounts used, criteria used and difficulties found.

By article 5 of the Ministerial Order of 26 October, 1993 on the use of sewage sludge in the agricultural sector (B.O.E. of 5 November, 1993), also dealt with in article 8 of Royal Decree 1310/1990, the National Sewage Sludge Register is assigned to the Directorate General of Agricultural Production and Markets (Subdirectorate General of Agricultural Production Media). Likewise, it lays down that this register shall contain information relating to the national Census of treatment plants (annual amount of sewage sludge produced, its treatment and destination) and the Entities engaged in the agricultural use of sewage sludge (composition, treatment and application areas). The information which may be requested for drawing up this Register is detailed in Appendices I and II.

Favourable climate conditions in Spain enable waste treatment and final destination solutions to be easily found; another possibility is agricultural and

Table 1. Limit values of heavy metals

Limit values of heavy metal concentration in soils (mg kg^{-1} d.m.)		
Parameters	Limit values	
	Soils with pH less than 7	Soils with pH over 7
Cadmium	1	3.0
Copper	50	210.0
Nickel	30	112.0
Lead	50	300.0
Zinc	150	450.0
Mercury	1	1.5
Chrome	100	150.0
Limit values of heavy metal concentration in sewage sludge (mg kg^{-1} d.m.)		
Parameters	Limit values	
	Soils with pH less than 7	Soils with pH over 7
Cadmium	20	40
Copper	1000	1750
Nickel	300	400
Lead	750	1200
Zinc	2500	4000
Mercury	16	25
Chrome	1000	1500
Limit values for annual amounts of heavy metals which can be introduced into soils based on a ten year average (kg ha^{-1} $year^{-1}$)		
Parameters	Limit values	
Cadmium	0.15	
Copper	12.00	
Nickel	3.00	
Lead	15.00	
Zinc	30.00	
Mercury	0.10	
Chrome	3.00	

forestry use as an organic amendment of the compost produced from sewage sludge. Thus, if these are products marketed as such amendments, their use in agriculture is subject to Spanish legislation on fertilizers and the like appearing in Royal Decree 72/1988 of 5 February, 1988, subsequently modified by Royal Decree 877/1991 of 31 May, 1991, Ministerial Order of 14 June, 1991 (B.O.E. of 19 June, 1991), Resolution of the Secretariat General for Agricultural Productions and Markets of 31 July, 1991 (B.O.E. of 9 August, 1991) whereby the application for registration in the Fertilizers and Like Product Register is processed.

References

BOE (1975) Ley 42/1975, de 19 de noviembre, sobre desechos y RSU BOE 120

BOE (1986) Ley 20/1986, 14 de mayo, Básica de residuos tóxicos y peligrosos. BOE 120

BOE (1986) Real Decreto Legislativo 1163/1986, de 13 de junio, por el que se modifica la Ley 42/1975 sobre desechos y residuos sólidos urbanos. BOE 149

BOE (1990) Real Decreto 1310/1990, de 29 de octubre, por el que se regula la utilización de los lodos de depuración en el sector agrario. BOE 262

BOE (1993) Orden de 26 de octubre de 1993 sobre utilización de lodos de depuración en el sector agrario. BOE 265

Council of the European Communities (1975) Council directive relating to the management of sewage sludge under such conditions that soil and surface and ground water protection is guaranteed. Official Journal of the European Communities L194: 26

Council of the European Communities (1975) Council directive relating to waste. Official Journal of the European Communities L194: 39

Council of the European Communities (1978) Council directive relating to toxic and hazardous waste. Official Journal of the European Communities L84: 43

Council of the European Communities (1980) Council directive relating to the management of sewage sludge under such conditions that soil and surface and ground water protection is guaranteed. Official Journal of the European Communities L20: 43

Council of the European Communities (1986) Council directive on the protection of the environment, and in particular of the soil, when sewage sludge is used in agriculture. Official Journal of the European Communities L181: 6–12

C. Rodriguez-Barrueco (ed.), Fertilizers and Environment, 297–300.

Cotton fertilization with composts of (sugarbeet) vinasse and agricultural residues

E. Madejon[1], M.J Diaz[2], R. Lopez[2], C. Lozano[3] & F. Cabrera[2]
[1]*Ebro Agrícolas S.A. Apdo. 9, 41300 S. José de la Rinconada, Sevilla, Spain;* [2]*Instituto de Recursos Naturales y Agrobiología de Sevilla (CSIC). Apdo. 1052, 41080 Sevilla, Spain;* [3]*Eurosemillas S.A. Paseo de la Victoria, 31, 14004 Córdoba, Spain*

Key words: agricultural residues, cotton, fibre quality, vinasse

Abstract

A concentrated depotassified beet vinasse was mixed with each of ten solid agricultural residues. The ten mixtures were composted for 7 months. The composts obtained after this period were used to fertilize a cotton crop. A mineral treatment was used for comparison and a treatment without fertilization was used as control. The nitrate content of petiole determined before the first top dressing revealed significant differences between treatments. All treatments produced higher yields than the control. Analysis of fibre quality did not show significant differences between treatments.

Introduction

The recycling of certain organic wastes can jointly mitigate problems of both environmental pollution and soil degradation. Beet vinasse are beet molasses that are almost completely biochemically desugarized, distilled and sometimes concentrated. The high levels of N (3%), K (3%) and O.M. (35%) of the concentrated vinasse could be beneficial factors for the land application of this waste (López *et al.*, 1993; Murillo *et al.*, 1993). However, the vinasse has three major problems for direct application as fertilizer: (a) high salt content (up to 14% of total salt) (Murillo *et al.*, 1993), (b) low P content (0.012%) and (c) high density (1.23 g cm^{-3}).

One alternative to overcome these disadvantages is the co-composting of the vinasse with other agricultural organic solid wastes (Díaz *et al.*, 1993), obtaining composts easily handy with higher P content and lower salinity. Land composts application is recognized as an important alternative disposal method both in recycling nutrients essential for plant growth and in increasing soil fertility (Gallardo-Lara and Nogales, 1987). The present paper deals with the effect of ten composts of vinasse in cotton crop. Plant N status (nitrate concentration in petiole), nutritional status of leaves, yield (weight of fibre) and fibre quality were determined.

Materials and methods

Composts of vinasse

A concentrated vinasse was mixed with each of ten solid agricultural residues. The solids used for the mixtures and the initial proportions between the solid wastes and the vinasse are shown in Table 1. The mixtures were composted in controlled conditions of moisture and temperature during 4 months and left to mature for the following 3 months. The chemical characteristics of these composts obtained after this period are shown in Table 2.

Soil and treatments

The composts of vinasse were applied in deep fertilization to a soil whose characteristics are indicated in Table 3. The doses of composts used were 4000 kg ha^{-1} for mixtures C and T (treatments TC and TT respectively), 5000 kg ha^{-1} for mixtures O, C + L, S,

Table 1. Initial composition of the mixtures

Mix-ture	Vinasse (%)	Solid waste
O	50	Olive-pressed cake (50%)
O+L	50	Olive-pressed cake (25%) + Leonardite (25%)
C	75	Cotton gin trash (25%)
C+L	50	Cotton gin trash (25%) + Leonardite (25%)
M	35	Municipal solid wastes (65%)
T	60	Tobacco dust (40%)
E	30	Composted olive-oil-mill effluents (70%)
S	30	Olive-oil-mill sludge (70%)
G	50	Grape marc (50%)
R	50	Rice husk (50%)

Table 2. Chemical composition of the composts

	O.M (%)	Kjel-N (%)	C/N	NO_3^--N (%)	P_2O_5 (%)	K_2O (%)	Na (%)
O	53	3.45	11.5	0.10	0.14	3.46	4.11
O+L	40	1.50	13.5	0.10	0.13	2.50	3.50
C	24	3.57	10.2	0.20	0.23	5.94	4.10
C+L	34	2.32	12.1	0.16	0.21	5.11	3.14
M	12	2.19	5.6	0.02	0.87	2.49	2.01
T	25	3.67	9.6	0.21	0.28	6.27	3.50
E	15	2.06	8.5	0.01	0.72	3.23	1.72
S	32	3.55	10.5	0.02	0.37	3.78	2.30
G	36	3.38	12	0.09	0.26	5.24	3.10
R	35	2.39	13	0.11	0.11	5.16	2.50

G (treatments TO, T(C + L), TS and TG respectively) and 10000 kg ha^{-1} for mixtures O + L, M, E and R (treatments T(O + L), TM, TE and TR respectively). A mineral treatment (TF) consisting of 400 kg ha^{-1} of 15-15-15 was used for comparison and a treatment without fertilizaation (TB) was used as control. All treatments, except TB, received two top dressings of 150 kg ha^{-1} of urea (46% N).

The experiment was carried out in 10 m^2 plots in a complete randomized block with 3 replicates per treatment. Cotton, *Gossypium hirsutum* L, c.v. Alba was the test variety used in this experiment.

Table 3. Soil characteristics

Parameter		
pH		7.6
$CaCO_3$	%	22.5
O.M	%	1.30
Kjeldahl-N	mg kg^{-1}	980
Olsen-P	mg kg^{-1}	16.3
Available-K	mg kg^{-1}	340
C/N		7.0
Sand	%	59.8
Lime	%	20.0
Clay	%	20.2

Chemical analysis of plant material

Youngest fully mature leaves on main stem were collected prior to first bloom. The leaves and petioles were collected at 1025 degree days after sowing (2 days before the first top dressing).

Plant material was dried at 70 °C after decontamination by washing with deionized water and finely milled. Nitrogen was determined in leaves after Kjeldahl digestion. Mineral elements were determined according to Jones *et al.* (1990) following dry ashing and ash dissolution by treatment with conc. HCl on a hot plate. Sodium and K were determined by flame photometry, Ca, Mg, Fe, Cu and Zn were determined by atomic absorption spectrometry and P by colorimetric determination using the phosphovanadomolybd complex. Nitrate was extracted treating petioles (0.5 g of dry sample) for one hour with 50 ml of 0.1 *M* KCl solution. Filtered solution was analysed for NO_3-N using selective electrode (Davis *et al.*, 1972).

Yield and fibre quality

Yield (weight of fibre) was determined in the field. Analysis of fibre quality was performed according to the U.S. Department of Agriculture, (RAEA, 1993). The analysis was carried out in a H.V.I. (Height Volume Instrument), using International Standards. Several parameters were studied to evaluate the fibre quality: length (LEN), uniformity (UN), maturity (MIK), elongation (EL), resistance (STL), reflectance (RD) and yellowness grade (B).

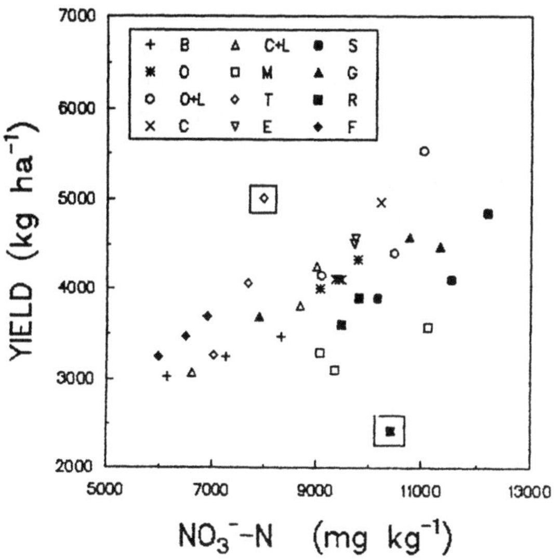

Fig. 1. Cotton fibre yield for the different treatments.

Table 4. NO_3^- N contents (mg kg^{-1}) in cotton petiole before the first top dressing

Treatments	NO_3-N (mg kg^{-1})	
TB	7243	ab
TO	9240	bcd
T(O+L)	10190	cd
TC	9677	bcd
T(C+L)	8107	abc
TM	9847	bcd
TT	7563	abc
TE	9613	bcd
TS	11300	d
TG	9990	bcd
TR	9897	bcd
TF	6470	a

*Values followed by the same letter in the same column do not differ significantly ($p < 0.05$).

Statistical analysis

The data of yield, nitrate content in petiole, parameters of fibre quality and nutrient content on leaves were analyzed by ANOVA and the differences among treatments were compared using Tukey's test. A significance level of $p < 0.05$ was considered throughout the study,

Results and discussion

Figure 1 shows yields (weight of fibre) for the different treatments. Weights of fibre for treatments TB, TM, TR and TF were lower than the mean value obtained for this variety of cotton in Andalusia (3700 kg ha^{-1}) (RAEA, 1993). However, for the remainder treatments, the weights of fibre were higher than that mean value. These high yields can be attributed to the moderate addition of Na through compost application, which appears to enhance the nitrogen uptake and yield of cotton crops (Perssarkli and Tucker, 1985)

Table 4 shows the nitrate level in petiole for all treatments. The nitrate level in petiole is suggested to be used as a criterion for plant N status (Jones *et al.*, 1990). All composts treatments, except TT, displayed the required NO_3^--N concentrations for mid-blossom stage (ca. 8000 mg kg^{-1}) reported by Jones *et al.*, (1990), while the mineral and control treatments (TF and TB) showed values below 8000 mg kg^{-1}.

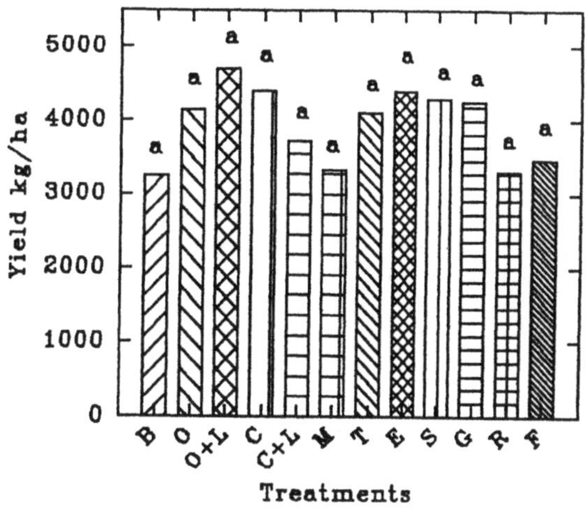

Fig. 2. Relationships between obtained yield (weight of fiber) and concentration of NO_3^--N in the petiole.

This indicates an adequate N supply through compost addition. A positive and highly significant relationship between the yield and the NO_3^--N content in petiole was found (Fig. 2.; $r = 0.71, p < 0.001$). Prediction of yield through this relationship has been suggested by Jones *et al.* (1990).

Macro and micronutrient analysis of leaves did not show any significant difference among treatments (Table 5). The nutrient concentrations of cotton top leaf were within the range reported by Jones *et al.* (1990).

Table 5. Nutritional content in cotton leaves collected prior to first bloom

Treatments	N (%)	P (%)	K (%)	Na (%)	Ca (%)	Mg (%)
TB	3.43 a	0.25 a	1.62 a	0.33 a	3.79 a	0.61 a
TO	3.75 a	0.27 a	1.74 a	0.31 a	3.78 a	0.55 a
T(O +L)	3.71 a	0.24 a	1.84 a	0.34 a	3.89 a	0.61 a
TC	3.77 a	0.24 a	1.96 a	0.32 a	4.13 a	0.65 a
T(C+L)	3.62 a	0.28 a	2.01 a	0.30 a	3.39 a	0.54 a
TM	3.75 a	0.25 a	1.58 a	0.38 a	4.15 a	0.64 a
TT	3.83 a	0.28 a	1.56 a	0.24 a	3.62 a	0.55 a
TE	3.75 a	0.25 a	1.42 a	0.20 a	3.69 a	0.58 a
TS	3.75 a	0.23 a	2.10 a	0.39 a	4.13 a	0.63 a
TG	3.60 a	0.26 a	1.96 a	0.33 a	3.65 a	0.57 a
TR	3.55 a	0.24 a	1.86 a	0.39 a	4.34 a	0.66 a
TF	3.69 a	0.23 a	1.91 a	0.37 a	3.78 a	0.59 a

Table 6. Parameters of fibre quality

Treatment	LEN	UN	STR	EL	MIK	B	RD
TB	1.12 a	84.1 a	29.1 a	6.53 a	3.97 a	7.90 a	69.3 a
TO	1.10 a	84.5 a	25.8 a	6.30 a	4.00 a	8.27 a	67.2 a
T(O + L)	1.14 a	84.1 a	25.6 a	6.43 a	4.10 a	8.17 a	68.7 a
TC	1.16 a	85.1 a	30.2 a	6.66a	4.10 a	7.97 a	70.5 a
T(C + L)	1.11 a	84.5 a	29.3 a	6.50 a	4.20 a	8.20 a	65.4 a
TM	1.30 a	84.4 a	26.3 a	6.37 a	3.80 a	8.17 a	65.6 a
TT	1.30 a	84.5 a	24.7 a	6.40 a	3.83 a	8.07	70.0 a
TE	1.19 a	83.9 a	27.4 a	6.45 a	3.95 a	8.05 a	71.13 a
TS	1.13 a	84.1 a	28.3 a	6.40 a	3.83 a	7.77 a	66.9 a
TG	1.12 a	84.0 a	27.7 a	6.27 a	3.93 a	8.33 a	67.1 a
TR	1.13 a	35.0 a	30.4 a	6.60 a	4.03 a	7.93 a	66.4 a
TF	1.09 a	84.3 a	25.7 a	6.60 a	3.90 a	8.20 a	56.6 a

* Values followed by the same letter in the same column do not differ significantly ($p < 0.05$).

Values of the parameters of fibre quality (Table 6) were within the normal values for this variety of cotton under Andalusian conditions (RAEA, 1993). No effect of fertilization on fibre quality was observed since no significant differences ($P<0.05$) among treatments were found for these parameters. These results are in agreement with those of RAEA (1993), which reported that generally, the fibre quality is more affected by crop variety and irrigation than by fertilization.

Acknowledgement

The authors thank D. Teodoro Nogueras for supervising this project, D. Manuel Fernández for sample collection and Dr. D. Ron for her helpful comments on the manuscript.

References

Cabrera F, Moreno F, Nacci S and De Arambarri P (1987) Utilization of wastes from olive and sugar beet processing industries in fertirrigation. Book of Proceedings, 4th International CIEC Symposium, Braunshweig, pp 75–483

Davies J E, Moody W G and Thomas J D (1972) Analyst 97, 87–94

Díaz M, Madejón E, López R, Cabrera F and Fernández M (1993) Nitrificación en el proceso de co-compostaje de la vinaza de remolacha. IX Nac. Cong. de Química de la ANQUE, Sevilla. Vol. II, pp 167–172

Gallardo-Lara, F. and Nogales, R. 1987. Effect on the application of town refuse compost on the soil-plant system: A review. Biol. Waste 19, 35–62

Jones B J Jr., Eck H V and Voss R (1990) Plant analysis as an aid in fertilizing cotton. In: Westermen RL (ed) Soil Testing and Plant Analysis, Chap. 18. Madison, SSSA

López R, Cabrera F, Murilo JM (1993) Effect of beet vinasse on radish seeding emergence and fresh weigth production. Acta Hortic 335: 115–119

Murillo J M, Cabrera F and López R (1993) Effect of beet vinasse on Germination and seedling performance of Ryegrass (*Lolium multiflorum*). J Sci Food Agric 61: 155–160

Perssarkli M and T C Tucker (1985). Uptake of N-15 by cotton under salt stress. Soil Sci Soc Am J 49: 149–152

RAEA (1993). Variedades de Algodón. Campaña 91/92 Consejería de Agricultura y Pesca de La Junta de Andulucía. Dirección General de Investigación y Extensión y Extensión Agrarias.

C. Rodriguez-Barrueco (ed.), Fertilizers and Environment, 301–303.

The effect of a special organic amendment on the development of pepper plants cultivated in a soil infested with *Verticillium dahliae*

Preliminary results

J.M. Garcia-Mina[2], R. Jordana[1], J. Aguirreolea[3] & M.A. Hernandez[1]
[1]*Department of Zoology and Ecology, University of Navarra, PO Box 273, E-31080, Pamplona;* [2]*Department of Research and Development, Inabonos-Roullier Group, C/ Ferrocarril s/no, E-31012 Pamplona;* [3]*Department of Plant Physiology, University of Navarra, PO Box 273, E-31080, Pamplona, (Navarra) Spain*

Key words: Organic amendment, pest control, pepper, *Verticillium dahliae*

Abstract

The effect of a new organic amendment of a natural origin (COA-N) with biostimulating and phytoprotecting properties, on pepper plants cultivated in a soil infested with *Verticillium dahliae* Kleb., has been studied. The different COA-N treatments led to increases in the average values of the number of fruit per plant, fruit size and fruit yield. Likewise, all COA-N treatments produced an important decrease in the number of plants affected by *Verticillium dahliae* under different experimental conditions. These preliminary results suggest that the special organic amendment could be an ecological and efficient means for the control of this plant disease.

Abbreviations: [COA-N(I)] – COA-N(I) treatment-experiment 1, [COA-N(II)] – COA-N(II) treatment-experiment 1, [COA-N] – COA-N(I) treatment-experiment 2, *V. dahliae* – *Verticillium dahliae* Kleb.

Introduction

The last three decades have seen a very important development throughout the chemical-organic industry dedicated to the search for new synthesized molecules of agricultural interest, as much for the control of plagues and diseases as in the field of fertilization. However, even when the use of these products is reported to be significantly beneficial in the short term, their generalized and continued use has resulted in producing significant negative effects on natural ecosystems and human health. For this reason, it is particularly interesting and necessary to develop research aimed towards the search for alternative products of a natural origin which have the ability to control these types of problems without affecting ecological systems.

Different studies have shown the capacity of a new organic amendment of a natural origin – commercially known as COA-Nematcil or COActyl-N – to effectively control disease caused by different phytoparasite nematodes and to stimulate plant development (Jordana *et al.*, 1994a, b).

The aim of this present research work is to study whether this new product can be useful in protecting plants from attack by various pathogenic fungi. For this the effect of different compositions of this special organic amendment on pepper plants cultivated in soil infested with *Verticillium dahliae* Kleb., in the region of Lodosa (Spain) was studied.

Materials and methods

Two different experiments were carried out:

Experiment 1

The effect of two compositions of the special organic amendments – COA-N(I) and COA-N(II) – on the production, quality and number of pepper plants affected by *V. dahliae* were studied.

To carry out the experiment one single plot homogeneously infested with *V. dahliae* was chosen. The degree of potential infection of the plot, estimated according to the results observed in previous years, was approximately 30% of infected plants.

Three treatments were carried out, each one with an average number of 550 plants: a control [C], treatment with COA-N(I) [COA-N(I)] and treatment with COA-N(II) [COA-N(II)]. All the treatments received the same mineral fertilizer: 450 kg ha^{-1} of 9–18–27–0.3B as a base, 250 kg ha^{-1} of urea and 300 kg ha^{-1} of potassium nitrate as cover. The treatments with COA-N also received 1500 kg ha^{-1} of the corresponding organic amendment (I or II), applied 2 days prior to transplanting without localizing.

The following parameters were evaluated: three months after transplanting, the number of fruit per plant on eighteen plants was counted and the average size (length and maximum diameter) of the fruits per plant on twenty plants was measured in each treatment, all plants measured were taken at random. The fruit yield (expressed as kg ha^{-1}) and % of infected plants were measured at the end of plant cycle. Statistical analysis was carried out using the LSD test ($p \leq 0.05$).

Experiment 2

With the aim of studying more closely the effect of COA-N on the control of the plants affected by *V. dahliae*, two blocks of 300 plants (three replicate of 100 plants) cultivated on a homogeneous soil with a degree of potential infection greater than 50% were chosen: a control block (C′) and a block treated with COA-N(I) (COA-N). Both blocks received the same mineral fertilizer (experiment 1), adding to COA-N treatment 400 kg ha^{-1} of COA-N(I) after transplanting, localized in the ridge.

The total number of plants affected by *V. dahliae* in each replicate of the different treatments was estimated, over a three week period, after the appearance of the visual symptoms of the disease (1st–3rd week of September). The rate of infection defined as the quotient between the number of infected plants per week and the number of days of the week was calculated. Statistical analysis was carried out using the LSD test ($p \leq 0.05$).

The soil of different plots (experiments 1 and 2) had a pH of 7,6 and $CaCO_3$ and organic matter contents of 6.2% and 1.45% respectively. Soil analyses were carried out by using standard methods (Page *et al.*, 1982).

Results and discussion

Experiment 1

Table 1 shows that both treatments with COA-N produced an increase in the average number of fruits per plant, although this was only significant in those corresponding to COA-N (II) ($p \leq 0.05$). Similarly, both compositions of COA-N produced a significant increase in the average size of the fruit ($p \leq 0.05$), in addition to a very important increase in the final production (fruit yield), estimated according to the real harvest obtained.

These results may be due to the direct biostimulating effect of some active substances contained in the special amendment (natural cytokinines – humic substances) on plant development, in addition to the ability to mobilize nutrients present in the soil in non-assimilable forms (Vaughan & Malcolm, 1985). With regard to the effects on the size of the fruits this could be the result of the correction of certain nutritional deficiencies associated with mono-culture management.

With regard to the disease caused by *V. dahliae*, treatment with COA-N produced an important decrease in the number of plants affected (Table 1).

Experiment 2

As can be seen in Table 2, treatment with COA-N produced a very significant reduction both in the number of plants affected by *V. dahliae* ($p \leq 0.05$) and in the rate of infection. This result, which confirms the results obtained in experiment 1, can be linked to the stimulating action of certain active substances contained in the special organic amendment on the disease resistance system of the plant. Similarly, the data obtained in the different experiments suggest that the application 3–4 days after transplanting localized on the ridge may allow a significant reduction in the dose while maintaining, or increasing, efficacy.

In conclusion, these results, in spite of their preliminary character, indicate that this new special organic amendment may constitute an efficient means of control of this plant disease.

Acknowledgements

The authors are grateful to Mr J Munilla, Mr JM Peña, Mr F Gainza and Miss Lopez for their valuable advice,

Table 1. Effect of two compositions of COA-N (I,II) on pepper plants cultivated in a soil infested with *Verticillium dahliae*. Effect on the production, quality and number of infected plants (Experiment 1).

Treatment	Fruit plant $^{-1}$ ($\bar{z} \pm$ SE)	Fruit size			
		Fruit yield (kg ha^{-1})	% Infected plants		
		Length ($\bar{x} \pm$ SE)	Diameter ($\bar{x} \pm$ SE)		
Control	6.24 ± 0,65	7.5 ± 0.24	3.85 ± 0.15	2872	25.32
COA-N (I)	7.65 ± 0.70	10.31±0.29	4.75 ± 0.14	6720	15.65
COA-N(II)	9.50 ± 0.90	10.38±0.28	4.89±0.11	7896	12.15
LSD ($p \leq 0.05$)	2.17	0.82	0.39		

Table 2. Effect of COA-N (composition I) on the number of infected plants and the infection rate in pepper cultivated in a soil infested by *Verticillium dahliae* (Experiment 2).

Treatment	Number of infected plants			Infection rate	
Week:	1	2	3	Week 1–2	Week 2–3
Control				5.40	6.43
Replicate 1	36	48	63		
Replicate 2	42	50	59		
Replicate 3	28	46	67		
$\bar{x} \pm$ SE	35.3 ± 4.06	48 ± 1.15	63 ± 2.31		
COA-N				2.14	1.28
Replicate 1	6	10	12		
Replicate 2	4	10	14		
Replicate 3	8	13	16		
$\bar{x} \pm$ SE	6 ± 1.15	11 ± 1	14 ± 1.15		
LSD ($p \leq 0.05$) Inter-week and treatment		6.51			

and to Mr S Cenoz and Mr J Lerga for their technical assistance. This research was supported by The Roullier Group.

References

Jordana R, Garcia-Mina JM and Hernandez MA (1994a) Organic amendment of natural origin capable of protecting plants from the agression of pathogens and of stimulating plant growth. EP 0 609 168 A1.

Jordana R, Garcia-Mina JM, Hernandez MA, Coello de Portugal D and Ariño A (1994b) The effect of an organic amendment on the phytoparasitic action of *Meloidogyne sp.* in a tobacco plantation. Afro-Asian J Nematol (In press).

Page AL, Miller RH and Keeney DR (1982) Methods of Soil Analysis. Part 2. – Chemical and Microbiological Properties. Soil Sci Soc Am Inc, Madison, Wisconsin.

Vaughan D and Malcolm RE (1985) Soil Organic Matter and Biological Activity. Martinus Nijhof/Dr W. Junk Publishers, Dordrecht, The Netherlands.

C. Rodriguez-Barrueco (ed.), Fertilizers and Environment, 305–307.

Evaluation of easily oxidizable organic matter in greenhouse soils through a biological technique

R.O. Pedraza, C.H. Bellone & S.C. de Bellone
Cátedra de Microbiología Agrícola, Facultad de Agronomía y Zootecnia, Universidad Nacional de Tucumán, Av Roca 1900, 4000 S.M. de Tucumán, Argentina

Key words: Organic matter easily oxidizable, greenhouse, Azotobacter spp., soil, biological technique

Abstract

In the province of Tucumán, Argentina, the production of vegetables in greenhouse grows in importance day by day, and with the incorporation of technology sprouts the need to know the ecological conditions and management of the crops, the physico-chemical and biological properties of and processes in the soil, and their relation to cultural practices.

These processes support an accelerated degradation of organic matter, higher than in soils under natural field conditions, regardless of the importance of the degradation of the easily assimilable fraction.

Using Azotobacter spp., heterotrophic bacteria capable of fixing molecular nitrogen as a test organism, in comparison with a chemical method to determine organic matter easily oxidizable, it was observed that Azotobacter spp. express sensitively the oxidation of those carbon compounds.

It is pointed out that it is possible to use Azotobacter as a test microorganism in order to measure organic matter easily oxidizable, proposing it as an indicative method to replace the wasted organic matter.

Introduction

The plaque method, based on the growth of naturally occurring Azotobacter, is used as a bioassay to sensitively determine deficiency of nutrients, P mainly being the one with more response to this technique because of the nitrogen fixation which requires proper concentrations of this and other elements such as Ca, S and Mg. However, this heterotrophic bacteria (Tchan and New, 1984), uses easily oxidizable carbon sources. Complex substances such as cellulose, lignine, hemicelulose, are not suitable to the bacteria, however their products of decomposition are. Thus, the deficiency of those carbon sources restricts bacterial growth as well as the rest of the heterotrophic microorganisms, even if the other nutritional elements are chemically available.

The influence of availability of organic matter in soil affects not only the bacterial metabolism, but also the physical and chemical soil properties and the level of available nutrients, which can affect the above processes and the general biological activity of soils (Alexander, 1980). Therefore, Azotobacter can be taken as a valid microorganism to obtain an orientation about the stock of mineral nutritive compounds and easily oxidizable soil organic matter (EOSOM).

Green (1933) relates the response of Azotobacter and plants to nutritional soil conditions. Equally, Amor Asunción *et al.* (1976) used this method to determine mineral deficiency, introducing many technical modifications to the original method of Winogradsky (1928). This method has never been used to diagnose deficiencies of EOSOM, but in this case it has been applied to greenhouse soil.

Materials and methods

Composite soil samples from 4 greenhouses located in Lules, Tucumán, were taken at random to make the bioassay. They correspond to a medium-textured soil with loam to sandy loam texture, pH medium acid. The samples represent in each case soils with 1, 2, 3 and 4 years of production with different vegetable crops; a fifth sample was taken as a control from fallow soil.

In every case the soil came from about 20 year of sugar cane production, and the assays were made on soils with inorganic fertilization and without organic matter incorporation during the years of cultivation in greenhouse.

In laboratory the samples were sieved, homogenized and divided into fractions in order to be treated as follows:

- 100 mg of soil was sown in Petri dishes with solid growth culture media for free nitrogen fixers, according to the technique of Winogradsky, making 5 replications for each treatment. They were incubated at 30 °C during 24 h. The Petri dishes were ranged in blocks with the number of colonies and soil treatments randomized within the block. Tukeys Honestly Significant Difference Test was used to evaluate the statistical significance (at $p =$ 5%) of the difference between 5 average values.
- The soil plaque method was performed with another part of the samples, using mannite as carbon source. Each Petri dish was put into a wet chamber at 30 °C during 24 h.
- The measurement of EOSOM, according to the method of Walkly-Blanck (Jackson, 1970), was made by digestion of soil, using potassium dichromate and sulfuric acid, utilizing half the acid concentration (density 1.845), and distilled water, in Erlenmeyers under a continuous flow of cold water and adding the acid slowly, in order to avoid a violent reaction and so to produce the oxidation of the EOSOM only.

The amount of chromic acid reduced by the organic matter was determined by photocolorimeter, using glucose as oxidizable carbohydrate.

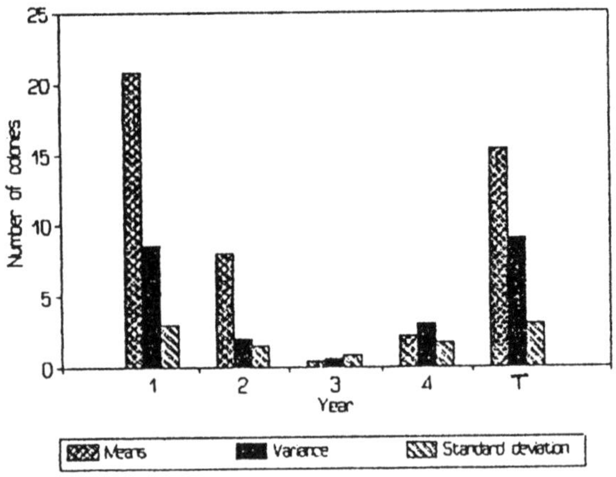

Fig. 1. Azotobacter in greenhouse soils characterized by statistics references.

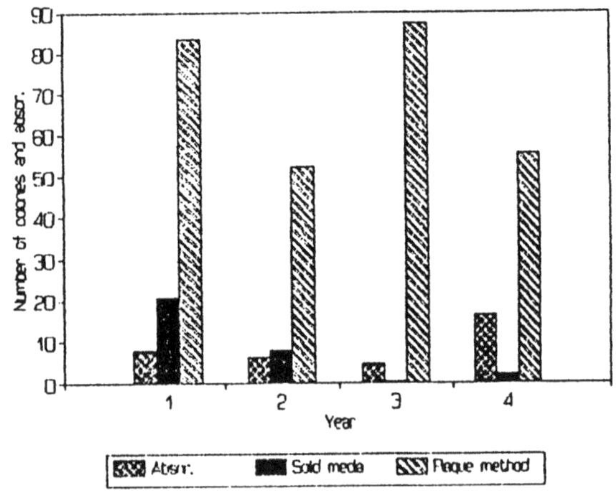

Fig. 2. Development of Azotobacter colonies compared to the amount of easily oxidizable organic matter (absorbance).

Results

The development of bacterial colonies in Winogradsky solid growth culture media shows a diminution in soils with 1, 2 and 3 year of cultivation; from the 3rd year there is an increment of colonies until the 4th year, while the values of development in the test are between the 1st and 2nd year.

The analysis of variance shows no differences between treatments, and very few between replications. Statistically, there are no significant differences between soils 3 and 4, while between soils 1, 2 and the control significant differences were found.

In the soil plaque method the development of Azotobacter colonies decreases from the 1st to the 2nd year of cultivation; then colonies development increases, exceeding the values of the 1st year, until the 3rd, after which it has a retraction and reaches the 4th year with values close to soils with 1 year of cultivation and the control.

The EOSOM diminishes from the 1st to the 3rd year. Then, it increases until the 4th year, surpassing the values of soils with 1 year of cultivation, while the control value exceeds 1st year soil a little.

Discussion

According to the results obtained, the diminution of EOSOM is correlative with the diminution in the number of colonies with Azotobacter in the first years, after which both increase.

This tendency could be the result of consumption of easily available carbon, as the microorganisms adapted to those compounds were unable to degrade cellulose, lignine and some other components of the organic matter, which would determine the death and gradual replacement by organisms that are able to use those components.

The desacceleration in the speed of carbon degradation with time is common in soil. In greenhouses it is more accelerated than in fallowed soil, and added to that, the natural replacement of organic matter makes carbon cycling possible, which does not happen in greenhouse where its oxidation is more accelerated because of the higher nutritional conditions and higher temperature, CO_2, humidity, etc. Therefore this situation can be due not only to a change in the chemical constitution of the fraction used by microorganisms in general and Azotobacter in particular, but also to metabolic products made by microorganisms, which are also resistant to decomposition.

The differences in the graphs showing the behavior of Azotobacter can be due to the methodology used in each case. When we spread soil on Winogradsky media, we used an artificial growth culture that, although valid for our experience, does not possess the advantages of using the soil plaque method, which is an ecological method. Thus, the development of Azotobacter rests exclusivly on the use of nutrients available in the soil, which are the same nutritive conditions for the crops. This makes it a method that permits an approach to a knowledge of real nutritional needs of plants expressed through the development of microorganisms capable of fixing molecular nitrogen. Nevertheless, both methods are complementary as in Winogradsky media the carbon source is included when using soil plaque. We added mannite to the soil in order to stimulate the growth of the bacteria that already existed in that soil.

Finally, we can suggest that Azotobacter counts can be used as a sensitive biological test to determine EOSOM in greenhouse soil.

References

Alexander M (1980) El ciclo del carbono. In Introducción a la Microbiología del Suelo. AGT Editor, S.A. Mexico. pp 163–203.

Amor Asunción MJ, Cusato M and Frontera G (1976/1977) Determinación de carencias minerales del suelo por la prueba del Azotobacter Rev de Investigaciones Agropecuarias. INTA XIII (2): 59–66.

Greene RA (1933) The relation of phosphorus to biological nitrogen fixation and the conformity to the law of decreasing increment Soil Science 36: 383–86.

Jackson ML (1970) Determinaciones de materia orgánica en los suelos. In Análisis Químico de los Suelos. 2ª edición, Ediciones OMEGA S.A., Barcelona. pp 294–300.

Tchan Y-T and New PB (1984) Genus Azotobacter. In N.R. Krieg et al. (ed), Bergey's Manual of Systematic Bacteriology. Vol 1 Williams and Wilkins, Baltimore. pp 220–229.

Winogradsky S (1928) Sur L application agronomique d une eprause microbiolgique. Comptes Rendues de l Academic de Sciences 14: 161–65.

C. Rodriguez-Barrueco (ed.), Fertilizers and Environment, 309–316.

Experimental study on agricultural recycling of LD slag

N. Balcázar[1], M. Pinto[2], G. Besga[2], M. Rodríguez[2] & F.A. López[1]
[1]*Dpt. of Materials Recycling, National Center for Metallurgical Research (CSIC), Avda. Gregorio del Amo, 8, 28040 Madrid, (Spain);* [2]*Servicio de Investigación y Mejora Agraria, Gobierno Vasco, Bo Arteaga, 26, 48016 Derio, Bizkaia (Spain)*

Key words: Linz-donawitz (LD) slag, recycling, pasture effects, pH, nutrients content, yield

Abstract

Linz-Donawitz (LD) slag, a by-product of the iron and steelmaking industry containing 40% CaO and 8% MgO, is produced in large quantities in Europe and poses a substantial disposal problem. In order to find a solution to this problem, a field trial was carried out to investigate the possibility of LD slag to be used as liming material for pastures. The effects of six LD slag rates (0, 1000, 1500, 3000, 5000 and 7500 kg LD slag/ha), both with and without fertilizer, on soil properties, botanical composition, and plant yield were investigated. The experimental field was set up at two locations in Bizkaia (northern Spain). These were a newly established pasture (*Dactylis glomerata, Trifolium repens* and *Lolium perenne*) at Derio and a resident pasture at Abadiano (dominated by *Holcus lanatus* and *Agrostis tenuis*).

The results after three years show that LD slag notably increases the pH of the soil. Application of LD slag increases linearly the soil pH at both sites. At Derio, the 7500 kg slag/ha rate increased the soil pH from 5.3 to 6.4, while at Abadiano the pH raised from 5.8 to 6.5 and from 5.8 to 6.7 with and without fertilization, respectively. This increase in pH was accompanied by a decrease in the percentage of aluminium saturation of cation exchange complex of the soil, with values under 10% when at least 3000 kg LD slag/ha was applied. At Derio (initial pH 5.3), soil exchangeable calcium increased in comparison with the control, by a factor of 1.8 and 2.3 with and without fertilization, respectively. Pasture yield increased with slag application, with the largest responses (41% increase with 3000 kg slag/ha) in second year. At Abadiano (initial pH 5.8), pasture yield did not respond to slag, but the botanical composition changed, decreasing the contribution of *Festuca rubra* L. and *Plantago lanceolata* L.

According to the results, LD appears to be a useful liming material for correcting soil acidity on pasture soils. As a consequence, the agronomic use of LD slag will contribute towards solving the environmental problem of their disposal.

Introduction

The iron and steel industry LD slag by-product comes from iron refining processes using LD converters. LD slag is basically made up of calcium silicates and ferrites, whose hydrolysis in water in the presence of H^+ ions in the soil, produces divalent ions (mainly Ca^{2+} and Mg^{2+}) which become part of the soil's exchange complex (López *et al.*, 1989). Table 1 shows the chemical composition of the LD slag used, with a grain size between 1 and 3 mm. The low P content of the LD slag avoids its use in agriculture as a phosphatic fertilizer, but the high total Ca (40%) and Mg (8%) contents make the LD slag a potential liming agent.

Experiments in several countries have demonstrated the ability of LD slag to raise the pH of acid soils, increasing at the same time Ca and Mg contents in the soil's exchange complex. Its use in agricultural soils with cereal crops and forestry has been studied in previous research (Formoso *et al.*, 1993; Piret, 1993). When slag is applied to soils with a initial pH of 4–5, it has been shown that it modifies its physical and chemical properties, with a 15% to 40% production increasing.

The objective of this work was to examine the influence of LD slag on soil pH, Ca, Mg, aluminium saturation of the cation exchange capacity (CEC), and

Table 1. Chemical composition of LD slag.

Sample	LD Slag
CaO_{total}	41.70
CaO_{free}	5.15
SiO_2	10.20
P_2O_5	1.84
MnO	3.25
MgO	7.74
Al_2O_3	15.43
Fe_{total}	22.05

Table 2. Initial characteristics of the working areas.

	Derio	Abadiano
pH (1:2.5 water)	4.8	5.6
Clay (%)	30.0 %	21.0 %
Organic matter (%)	3.1 %	4.3 %
total N (%)	0.2 %	0.32 %
CEC (meq/100g)	17.1	20.2
P (mg/kg)	n.i.*	6.8
K (mg/kg)	122	151
Al^{3+} (meq/100g)	4.3	0.6
Al % Saturation	36.0 %	5.6 %

* n.i. = non identifiable.

available P and K, as well as forage yield and pasture botanical composition.

Materials and methods

The trial was conducted at two locations, Derio (newly established pasture) and Abadiano (resident pasture). Table 2 shows the initial characteristics of the working areas.

At Derio a pinewood area was converted to meadow land using mechanical equipment and sowing a mixture of perennial ryegrass (*Lolium perenne*), cocksfoot (*Dactylis glomerata*) and white clover (*Trifolium repens*) in 1989. At Abadiano the resident pasture was dominated by yorkshire fog (*Holcus lanatus*), browntop (*Agrostis tenuis*), perennial ryegrass (*Lolium perenne*) and red fescue (*Festuca rubra*). Those botanical compositions is typical of Basque Country pastures with a poor fertilizer management.

The experimental design was a randomized complete block split-plot, with four replications. The two main plot treatments were plus or minus NPK fertilizer, and subplot treatments consisted of six rates of LD slag. Each subplot measured 2 × 5 m. The fertilizer treatment was an annual dressing of 120, 60 and 160 kg ha^{-1} of N, P, and K, respectively, applied as a mixture of calcium-ammonium nitrate, 0–14–14 NPK (normal superphosphate) and KCl. N was also applied after each forage cutting. The LD slag rates applied, only once at the time of trial establishment, were 0, 1000, 1500, 3000, 5000, and 7500 kg ha^{-1}. The slag was mixed with surface soil at Derio and broadcasted over the pasture at Abadiano.

Measurements were made between November 1989 and December 1992. Soil samples were analyzed for pH (1:2.5 water), extractable Ca, Mg and K, P, and Al. Yield was calculated by cutting and weighing a representative area (4.74 m^2) of each individual plot. At all cuttings herbage samples were taken for determination of Dry Matter (DM) content. Botanical composition was determined before each forage cutting using the step-point method (Evans & Love, 1957), only in 1992 at Derio and for the three years study period in Abadiano. All the results were treated using statistical criteria (SAS Institute, 1988).

Results and discussion

Results on soil

The effects of slag and NPK treatments on soil pH were similar during the three year study period. Soil pH increased linearly with slag application (Fig. 1). The pH increase was similar for plots with and without fertilization, but the pH values in fertilized plots were always lower than in unfertilized plots.

The increase in pH was associated with a marked decrease in the exchangeable Al. The response of Al saturation percentage to slag application fits a negative exponential curve, with a similar decrease in the amount of Al on the exchange complex over the three years (Fig. 2). The toxic effects of Al on plants include inhibition of lateral roots, disruption on P and Ca nutrition, and the inhibition of shoot growth. Bache & Ross (1991) considered a level not harmful to plant growth when Al percentage on the exchange complex is < 10%, at high soil P content, and < 2.5%, at low soil

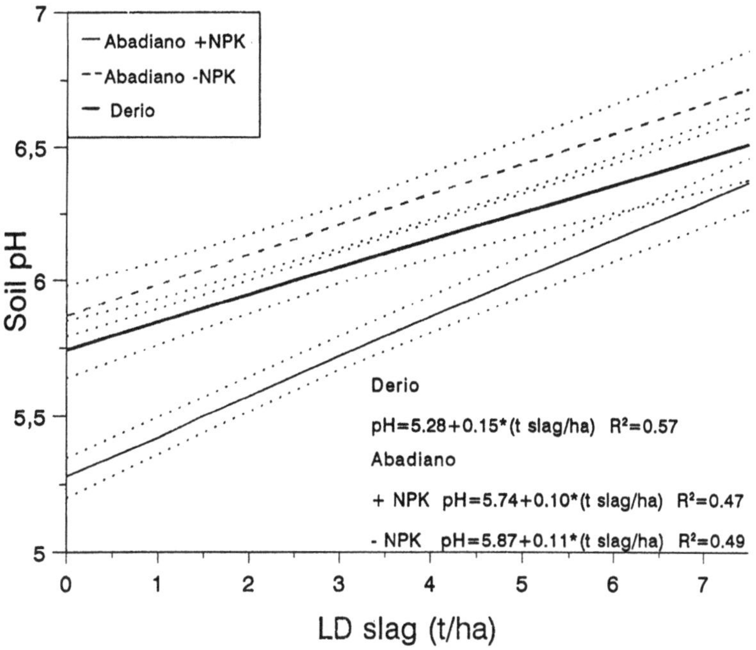

Fig. 1. Evolution of soil pH in the experimental sites. (Dotted lines represent confidence intervals at the 95% level.)

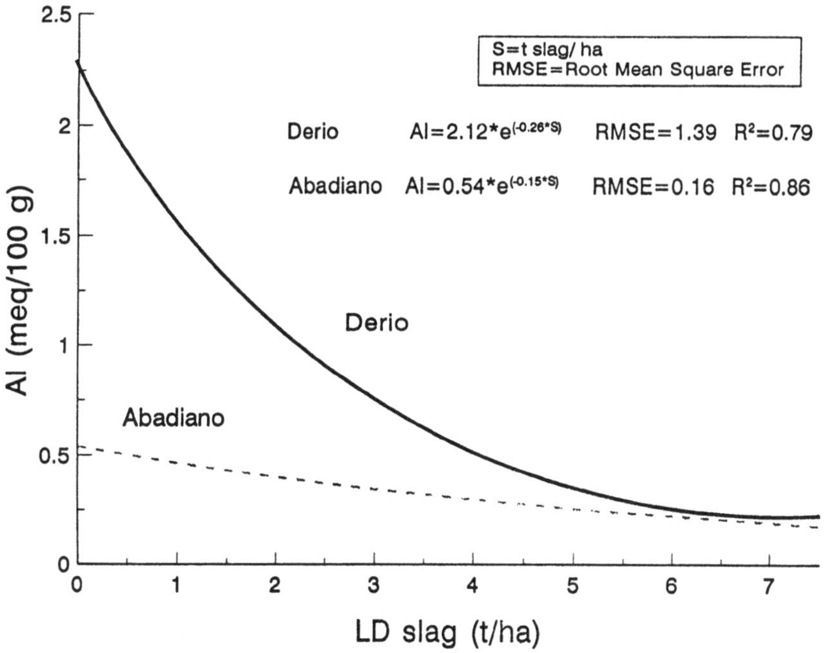

Fig. 2. Evolution of soil Al (meq/100 g) in Derio and Abadiano.

P content. At Derio, Al saturation of the CEC was reduced to 3% by 7500 kg slag ha^{-1}.

Exchangeable Ca increased notably as the amount of slag applied increased (Fig. 3). In both locations values for fertilized plots were slightly higher than for unfertilized plots at Derio, (Fig. 3a) and at Abadiano (Fig. 3b), respectively when 7500 kg slag ha^{-1} was applied. The smaller increases for fertilized plots could

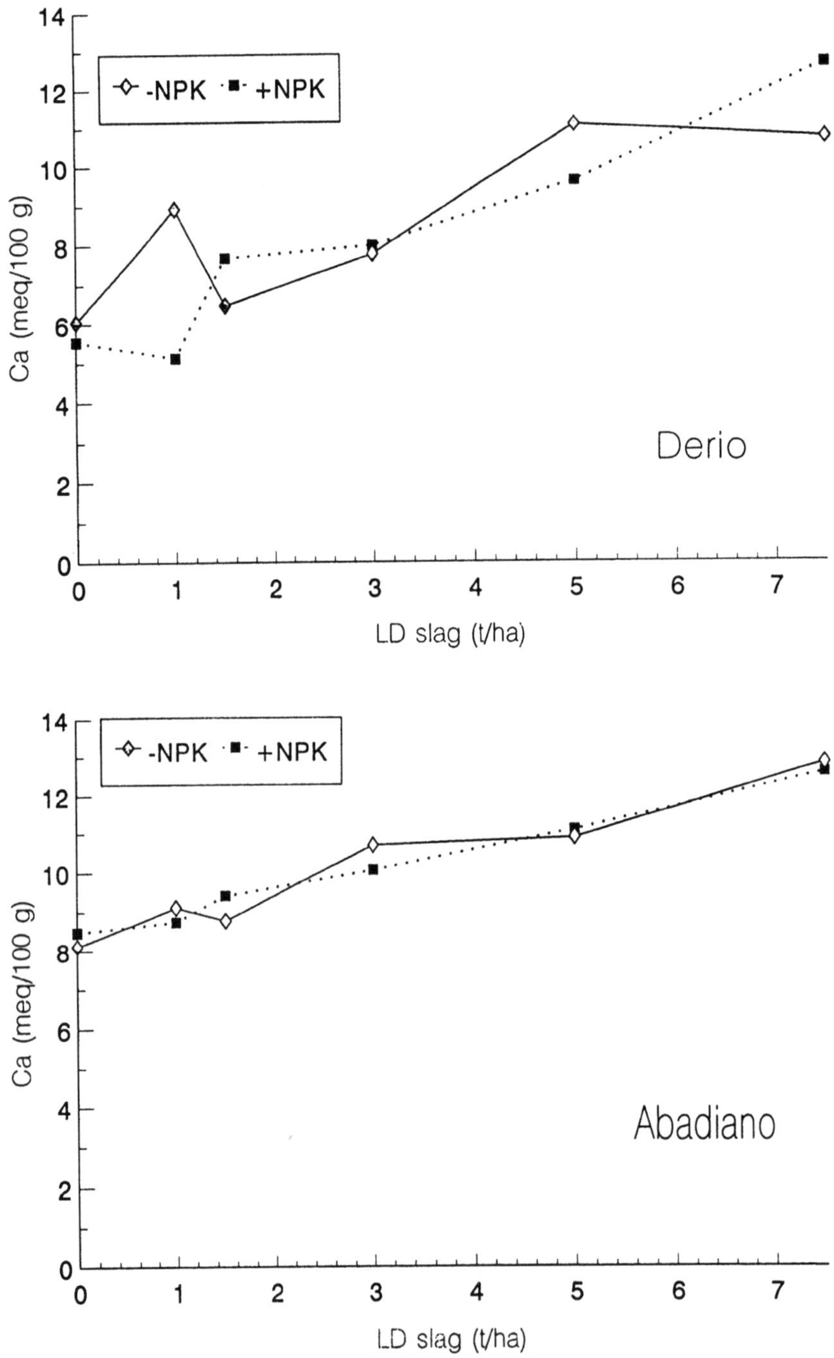

Fig. 3. Evolution of Ca contents of the soil with slag rates: a) Derio, b) Abadiano.

be explained by the different yield level for the plots with or without fertilization. Higher yields may be related to a greater Ca extraction from the soil.

Exchangeable Mg also increased with slag applications at both locations, but the values for exchangeable Mg were lower at Abadiano than at Derio (Fig. 4).

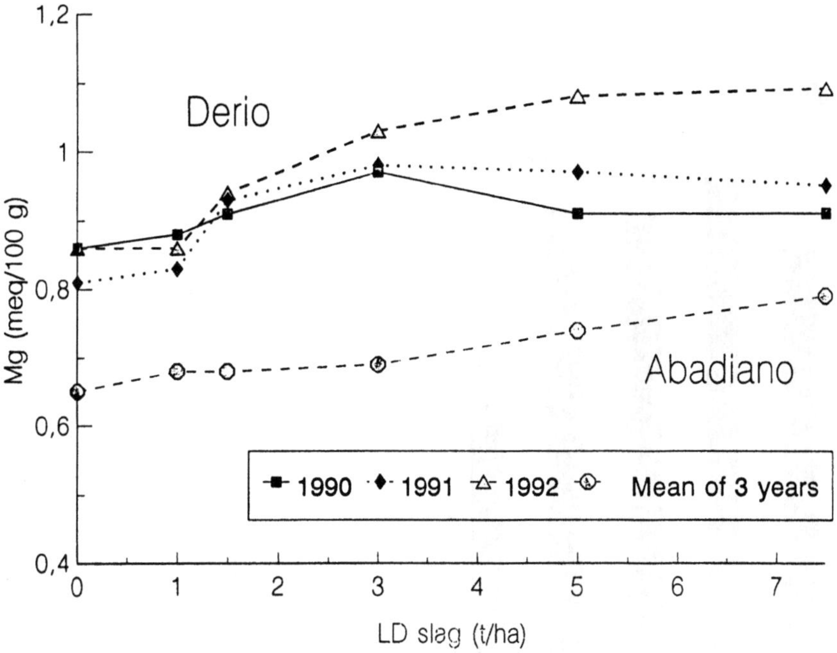

Fig. 4. Evolution of Mg contents of the soil with slag rates.

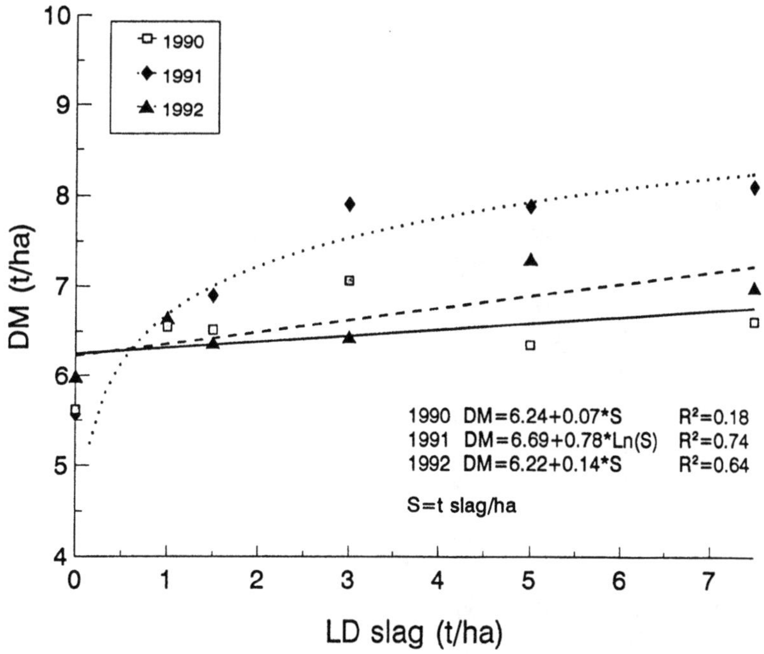

Fig. 5. Average yield in 1990, 1991 and 1992 in Derio.

Extractable P increased at both locations in the three-year study period, showing a different behaviour with the fertilization. An interaction with the slag in 2nd and 3rd years in the fertilized plots at Derio is shown. The application of slag significantly depressed extractable P. When there was enough available P in the soil and a high amount of Ca was applied, P may become not available by fixation (Adams, 1980).

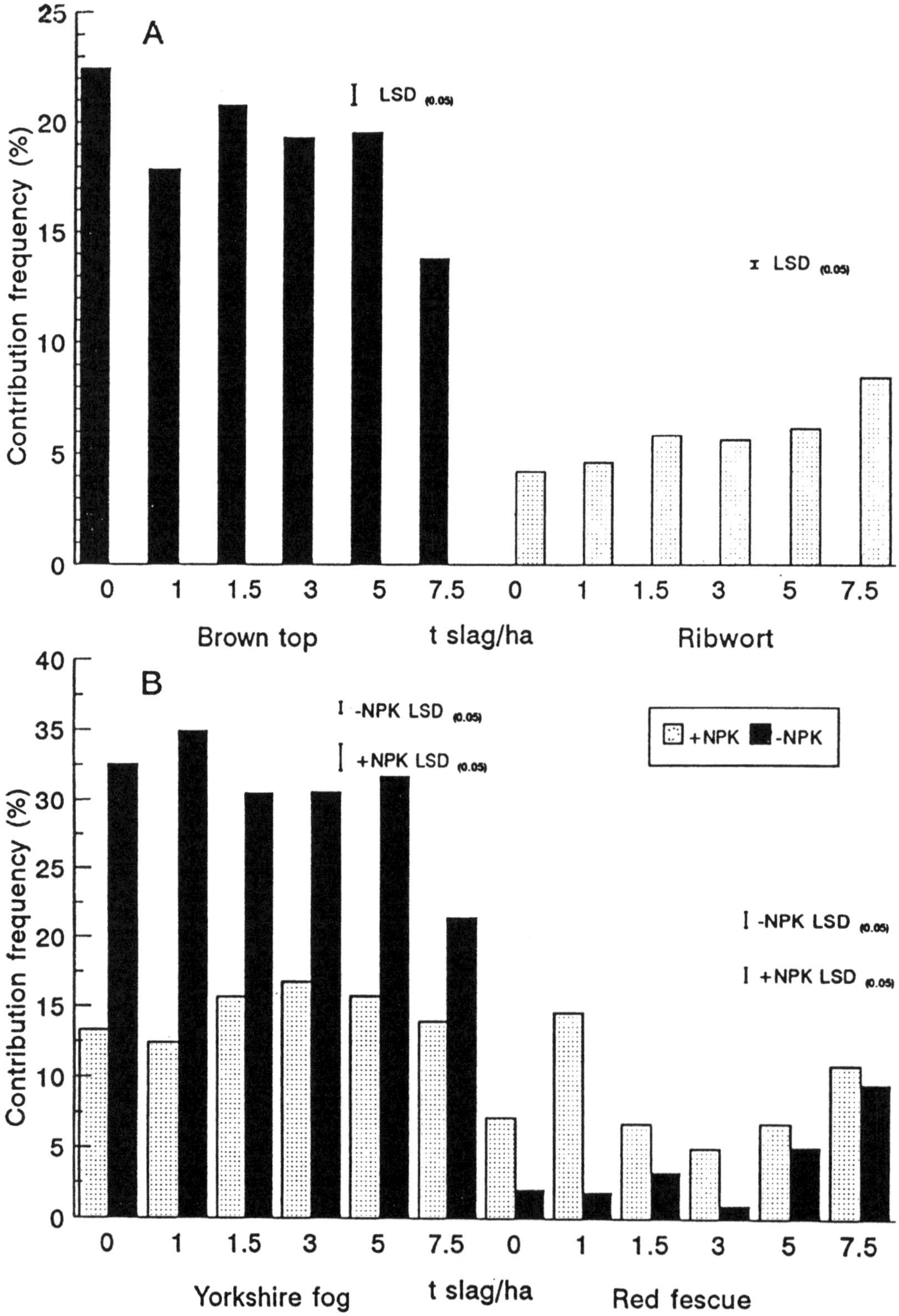

Fig. 6. Botanical composition in Abadiano (after three years).

However, Reeve & Sumner (1970) have reported that with high Ca levels there is an improved ability of plants to take up available P. At Abadiano, the increase in herbage production could render more P available through the decomposition of grass roots.

Application of slag decreased soil K content at Derio and was not affected at Abadiano. At Derio, this decrease may be related to the yield increase which would produce a higher K extraction from the soil, whereas at Abadiano neither the yield nor the CEC increased with slag application.

Pasture production

Dry matter responses to slag were different at the two locations. There was a non-significant interaction between NPK and slag, but application of fertilizer was the main factor influencing pasture yield in both sites.

At Derio, fertilization increased in 92%, 21% and 257% for 1990, 1991 and 1992, respectively. In 1990, and 1991 the pattern for yield increases fits very well in $R^2 = 0.74$ with the typical logarithmic response curve (Fig. 5), which shows small increases until flattening out between 3000 and 5000 kg slag ha^{-1}. In 1992 the response curve was linear ($R^2 = 0.64$) and gentle.

At Abadiano, the average yield for the three years was 79%, and in this case a significant effect of slag application on pasture dry matter yield was not observed.

The results recorded were related with the evolution of the changes in soil pH. When pH > 6.0, the pasture yield was slightly depressed. This coincides with the results of other authors (Edmeades *et al.*, 1984).

Botanical composition

In the newly established pasture, at Derio, all species responded clearly to slag application. In 1992 the individual contribution, in frequency to the sward, of the sown species (perennial ryegrass, cocksfoot and white clover) was 8%, 61% and 31%, respectively. White clover production responded to slag application significantly. The average white clover contribution in weight to pasture yield was lesser when NPK was applied (12.0 and 15.8 kg DM ha^{-1}, with and without fertilization). These changes in white clover content increase the feeding value of pastures that is an indirect beneficial consequence of liming (O'Connor & Edmeades, 1984).

In the resident pasture, at Abadiano, fertilization modified the botanical composition for most of the species, while slag only changed the contribution of browntop, yorkshire fog, red fescue and ribwort to the sward (Fig. 6). The increases in red fescue and ribwort contributions were counterbalanced by the decreases in browntop and yorkshire fog, but quality species like perennial ryegrass and clovers were not affected by slag application. It may be necessary to renew the pasture with quality species, that respond to improved soil pH and fertilization.

Conclusions

The application of LD slag in acid soils increases soil pH as other traditional liming materials (lime, dolomite, etc).The percentage of Al on the exchange complex decreased exponentially with slag rates. It also produces an increase in the Ca and Mg contents as raises the amount of slag applied. Available soil P and K contents show a general decrease, but the changes are not uniform with some interactions between soil P and other factors.

Yield responses in the new pasture are principally noticed when the initial soil pH is low, with the strong increases in pasture DM yields until the pH reaches 5.6. In the permanent pasture, there was not a noticeable response to slag application.

In respect to changes in botanical composition, at Derio it is observed that withe clover content increase with slag application. At Abadiano, with slag application a decrease in yorkshire fog and browntop is recorded, whereas the contribution to the botanical composition of red fescue and ribwort is increased.

The final results of these experiments, carried out on two types of pasture with acidic soils, show the ability of LD slag to be used as a liming material. This gives an opportunity for recycling in agronomic uses of a steel by-product such as LD slag.

References

Adams F (1980) Interaction of phosphorus with other elements in soils and in plants. In: Khasawneh, Sample and Kamprath (eds) The role of phosphorus in agriculture. ASA CSSA Soil Sci Soc Am.

Bache BW and Ross JAM (1991) Effect of phosphorus and aluminium in the response of spring barley to soil acidity. J Agric Sci 117: 299–305.

Edmeades DC, Pringle RM, Shannon PW and Mansell GP (1984) Effects of lime on pasture production on soils in the North island of New Zealand 4. Predicting lime responses. NZ J Agric Res 27: 371–382.

Evans RA and Love RM (1957) The step-point method of sampling – a practical tool in range research. J Range Man 10: 208–212.

Formoso A, López FA, Medina F and Balcázar N (1993) Agronomic Use of LD Slag. Final Report, ECSC Convention 7210-XA/931. CENIM, Madrid.

López FA, Formoso A and Medina F (1989) Escorias LD: Coproductos de las industrias siderúrgicas. I. Composición, tratamiento y aplicación. Rev Metal 25 (4): 247–254.

O'Connor MB and Edmeades DC (1984) A summary of MAF trials on the effects of lime on animal production. Jackson and Edmeades (eds) Workshop on Lime in New Zealand Agriculture. 17–19.

Piret J (1993) Utilization of LD slags. Proc 1st Europ Oxyg Steelm Cong Dusseldorf 296–297.

Reeve NG and Sumner ME (1970) Effects of aluminium toxicity and phosphorus fixation on crop growth on oxisols. Natal Soil Sci Soc Am Proc 34: 263–267.

SAS Institute (1988) The ANOVA procedure. In SAS User's Guide: Statistics. Version 6. SAS Inst North Carolina, USA.

C. Rodriguez-Barrueco (ed.), Fertilizers and Environment, 317–321.

Simultaneous use of pulp-mill sludge and poultry manure on rye-grass (*Lolium multiflorum* Lam.) fertilization

J.P. Carneiro[1] & J.Q. Dos Santos[2]
[1]*Escola Superior Agrária de Castelo Branco, Qta. Sra. Mércules, 6000 Castelo Branco, Portugal;* [2]*Secção Autónoma de Química Agrícola, Instituto Superior de Agronomia, 1399 Lisboa, Portugal*

Key words: Pollution, poultry manure, pulp-mill sludge, rye-grass, soil fertility

Abstract

With the objective of evaluating the influence of the application of different levels of pulp-mill sludge and/or poultry manure, on a soil which is acidic and poor in organic matter (cambisoil), a trial was carried out, in pots, using rye-grass (*Lolium multiflorum* Lam.).

The application of pulp, at all levels (30, 60, 90, and 120 t ha^{-1}), reduced the total production of rye-grass, while the application of manure (2, 4, and 6 t ha^{-1}) increased it. The maximum levels for each of these effects were not detected. It was also verified that, when applying both residues simultaneously, the manure did not totally eliminate the negative effect of the pulp on production especially at higher levels of sludge application.

The incorporation of these residues, especially the pulp, improved the nutritive value of the plants by increasing the amount of P, K, Ca, Mg and Na, and reducing the amount of Zn and Mn.

The evaluated soil characteristics after the trial, for both residues, mainly indicated the reduction of extractable Mn, and the increase in the value of pH and extractable amounts of Ca and Mg. The pulp was mainly responsible for the increment of the organic matter, P, extractable Na, exchangeable Ca and Na, total bases and base saturation.

Introduction

The low content of organic matter and high acidity of Portuguese soils are two major factors responsible for limitations in the production of most crops.

Therefore, we consider it worthwhile to use certain residues in agriculture, whose composition may prove to have a fertilizing effect in combination with possibly incorporating organic matter and correcting the acidity levels simultaneously. This is the case with primary pulp-mill sludge, an effluent produced in large quantities in Portugal – about 50,000 t of dry matter – which may constitute pollution foci in certain places if not used.

A characteristic of this by-product is a high ratio C/N, which may cause a decrease in the crop production. Therefore when considering using pulp-mill sludge, it is necessary to also increase the incorporation of nitrogen in the crop with the use of a nitrogen-rich residue such as poultry manure. This is a by-product produced in great amounts in Portugal, and which often gets accumulated in conditions which may cause severe environmental problems.

Materials and methods

In a greenhouse of the Agricultural Chemistry Department, a trial was carried out in pots, using rye-grass (*Lolium multiflorum* Lam.). The pots were filled with 11 kg taken from the superficial layer (fraction < 5 mm) of a Cambisoil.

Taking into consideration the characteristics of the soil, poultry manure, primary pulp-mill sludge (Table 1 and 2), and the results of previous experiments, twenty treatments were set-up. These were arranged in a completely randomized design and replicated three times, to include four levels of manure (E0, E1, E2 and E3, corresponding to 0, 2, 4 and 6 t ha^{-1}), and five levels of sludge (L0, L1, L2, L3 and L4, corresponding to 0, 30, 60, 90 and 120 t ha^{-1}). Both residues were applied after drying in the open air, and the rate of addition to

Table 1. Some soil characteristics

Parameters	Soil
pH (H_2O)	5.1
pH (KCl)	3.9
O.M. (%)	0.85
P_2O_5 (ppm)	37
K_2O (ppm)	98
Ca (ppm)	9.2
Mg (ppm)	23.2
Na (ppm)	42.0
Cu (ppm)	2.2
Fe (ppm)	54.8
Zn (ppm)	1.4
Mn (ppm)	51
H^+ (cmol(+). kg^{-1})	3.031
Ca^{2+} (cmol(+). Kg^{-1})	0.763
Mg^{2+} (cmol(+). kg^{-1})	0.194
Na^+ (cmol(+). kg^{-1})	0.107
K^+ (cmol(+). Kg^{-1})	0.140

Table 2. Some manure and sludge characteristics

Parameters	Manure	Sludge
H (% a 100° C)	40.24	7.74
pH (H_2O)	6.17	7.30
O.M. (%)	56.56	84.43
N (%)	2.25	0.31
P_2O_5 (%)	4.72	0.23
K_2O (%)	2.78	0.31
Ca (%)	11.86	2.49
Mg (%)	0.50	0.22
Na (%)	0.24	0.13
Cu (ppm)	57.50	46.30
Fe (ppm)	3107.50	4200.00
Zn (ppm)	402.50	74.00
Mn (ppm)	322.80	263.00
C/N	14.58	157.97
C/P	6.95	212.93

the pots was calculated taking into consideration the soil content of each pot.

In every treatment, a fertilizer was incorporated at the sowing stage, containing 0.5 g pot^{-1} of N, 1 g pot^{-1} of P_2O_5 and 1 g pot^{-1} of K_2O. After the second cut, 0.5 g pot^{-1} of N was also incorporated.

The manual sowing took place on 92/12/03. 1.5 g pot^{-1} of rye-grass seeds were sown in a four row disposition in each pot.

At the beginning of the experiment, pots were watered up to 60–70% of the soil field capacity, with demineralized water. After the initial wetting, the water content of the soil was controlled by weighing the pots three times per week.

The cuts of the forage were made four times: on the 21st January, 25th February, 18th March and 15th April, when the plants were approximately 30 cm high. After each cut, the rye-grass was weighed and put in a drying stove with artificial ventilation at 50 °C, until its weight remained constant. Afterwards the forage was weighed again and prepared for chemical analysis. Immediately after the fourth and last cut, a soil sample was taken from each pot, dried at a temperature below 40 °C, sieved through a 0.2 cm sieve, and analysed.

Organic matter in the soil was calculated by multiplication of percentage of organic carbon by the factor 1.724. Organic carbon was measured by dry combustion at 1200 °C, followed by measurement of the CO_2 evolved in a Ströhlein apparatus. Organic matter in sludge was determined by weight loss in a furnace at 400–500 °C, for 7–8 h, and the nitrogen in this residue was determined by the Kjeldhal method. Exchangeable cations in the soil were determined by atomic absorption spectrophotometry after extraction by the Mehlich method. Available phosphorus and potassium in soil were determined by colorimetric and flame emission photometric methods, respectively, after extraction by the Egner-Riehm method. This method was also used to determine the other mineral elements which were also considered in the analysis of the soil, by atomic absorption spectrophotometry. Phosphorus in the organic residues and in the plant tissues, was determined by the vanodomolybdophosphoric yellow colour method, while other mineral elements were determined by atomic absorption spectrophotometry in a Pye Unicam SP-9 apparatus, after hydrochloric acid (3 N) mineralization of the ashes.

The results of the study were analysed by ANOVA, followed by the Duncan method.

Results and discussion

From Fig. 1, it can be concluded that the treatment E0L4 (0 t ha^{-1} of manure and 120 t ha^{-1} of sludge) led to the lowest production (25.3 g pot^{-1}), while E3L0

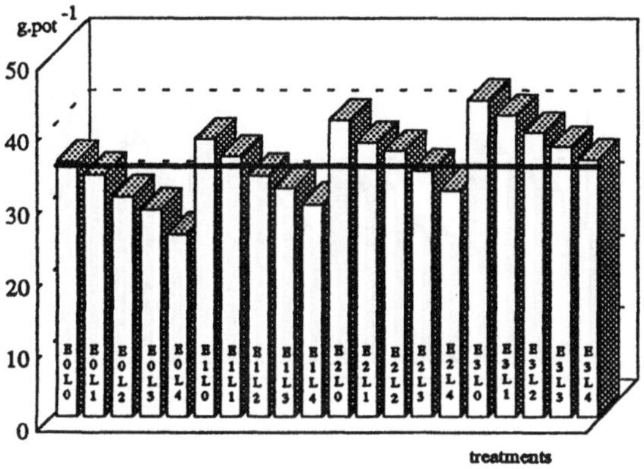

Fig. 1. Total production of rye-grass (D.M.)

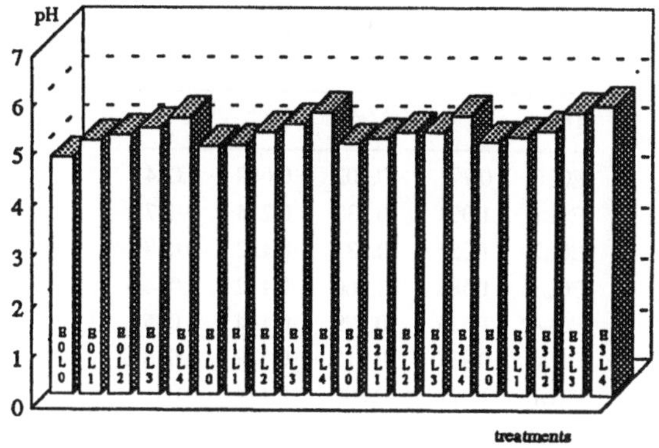

Fig. 2. pH (H_2O) at the end of the experiment

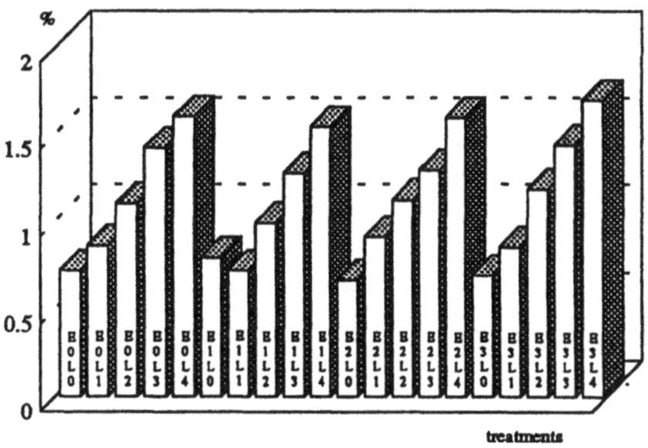

Fig. 3. Organic matter at the end of the experiment

(6 t ha^{-1} of manure and 0 t ha^{-1} of sludge) produced the best yield (43.77 g pot^{-1}).

It is also possible to verify that, the application of any level of sludge reduced the total production of rye-grass, whereas the incorporation of manure increased it.

The negative effect of the sludge may be due to the high C/N ratio of this residue, which is likely to affect the nitrogen uptake by the plants and thus decrease the forage production. However, sludge phenolic substances can also be responsible for this negative effect.

When applying both residues simultaneously, the addition of manure did not totally eliminate the negative effect of the sludge. However, it ensured that the negative effects only affected the production at higher levels of sludge application. Without manure, any level of sludge leads to a lower yield when compared with the control E0L0. When 6 t ha^{-1} of manure (E3) was incorporated, the yield was no lower than the one obtained in E0L0, in spite of an application of 120 t ha^{-1} of sludge.

Considering the composition of the plants, it was possible to conclude that the percentage of P and K in plants increased with the application of the residues. The improvement of some soil characteristics by the action of both residues (but initially the sludge), mainly the increment in pH values (Fig. 2), and P and K content, may explain such results. An increment of Ca and Mg levels could be found during the development of the rye-grass: Ca before second cut; Mg between the second and third cut. The lowest levels of these nutrients at the earliest stages of crop development, may be explained by the inclusion of K in the fertilization and because the plant used is monocotyledonous.

The results of the analyses of micronutrients (Fe, Mn, Zn and Cu), shows that their levels can be considered normal. However, if the sludge leads to a reduction of Mn level in the forage, which is beneficial, the increase of the Cu level, provided by manure, may be limited when larger amounts of this residue (over 6 t ha^{-1}) are applied.

From the results of the soil analysis, it is possible to conclude that the pH values (Fig. 2), as well as organic matter content (Fig. 3), increased with the application of sludge. The pH(H_2O) value was improved by 0.83 units and the percentage of organic matter by 40–50%. These improvements led to a reduction of the Mn level in the soil and the plants.

Levels of extractable Ca and Mg (Fig. 4 and 5), as well as the Ca^{2+} and Mg^{2+} values, improved with the application of increasing amounts of the residues. In fact, as Table 3 shows, the exchangeable cations levels increased, mainly, with the application of sludge.

The exchangeable Ca, which was one of the cations that proportionally underwent the greatest quantitative changes, increased by 30, 60 and 100%, when

Table 3. Exchangeable bases at end of the experiment

Treat.	ca^{2+}	Mg^{2+}	Na^{+}	K^{+}
E0L0	0.797	0.202	0.088	0.024
E0L1	1.063	0.232	0.234	0.037
E0L2	1.257	0.233	0.132	0.039
E0L3	1.585	0.260	0.172	0.044
E0L4	1.623	0.299	0.144	0.045
E1L0	0.933	0.215	0.179	0.038
E1L1	1.121	0.209	0.114	0.035
E1L2	1.319	0.244	0.151	0.045
E1L3	1.680	0.277	0.268	0.055
E1L4	1.8128	0.270	0.183	0.052
E2L0	0.800	0.197	0.176	0.018
E2L1	1.217	0.240	0.094	0.028
E2L2	1.228	0.222	0.205	0.028
E2L3	1.284	0.208	0.172	0.034
E2L4	1.721	0.294	0.260	0.052
E3L0	0.951	0.229	0.196	0.028
E3L1	1.123	0.251	0.157	0.026
E3L2	1.266	0.244	0.119	0.026
E3L3	1.411	0.249	0.276	0.047
E3L4	1.723	0.309	0.265	0.045

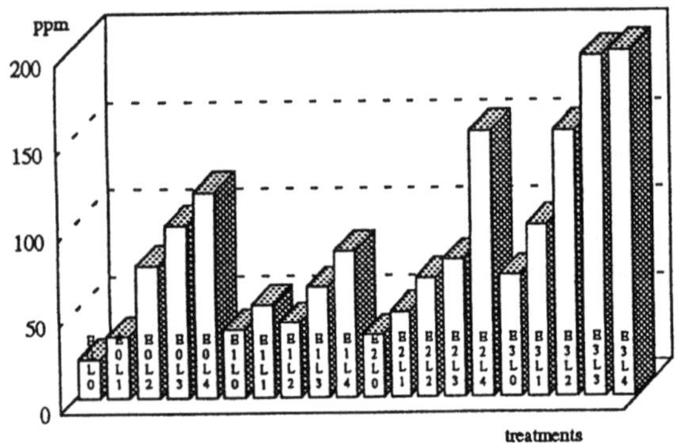

Fig. 4. Ca extractable levels at the end of the experiment

30, 60 and 90 t ha^{-1}, respectively, were incorporated. Although the value of the Mg^{2+} also increased in absolute terms, with the quantity of incorporated sludge, the value of the ratio of Ca^{2+}/Mg^{2+} increased by about 50%, when 90 t ha^{-1} of sludge (instead of 0 t ha^{-1}) were incorporated. Values higher than four were obtained, even with the application of 30 t ha^{-1}, which may point to a probable shortage of Mg in crops which have a greater need for this nutrient. Na^{+} increased also increased with the addition of sludge and/or manure.

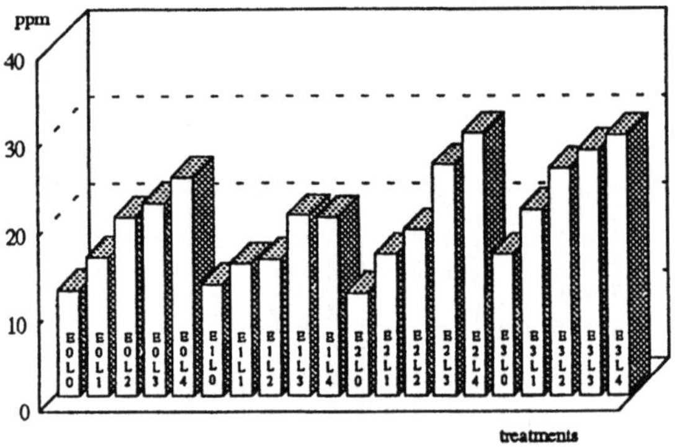

Fig. 5. Mg extractable levels at the end of the experiment

With the application of 2 t ha^{-1} of the latter residue, increases of 100% were obtained in the cation participation in the value of the total bases, and with the application of 6 t ha^{-1}, sodium occupied about 14% of the complex connections made by cations. This value can be considered high and limiting for the systematic application of manure, as was done in the experiment.

Conclusions

According to the results, the amounts of pulp-mill sludge applied to the soil, caused an important decrease in the yield of rye-grass. The main cause of the depressive effect, can be the high C/N ratio found in the sludge. However, sludge phenolic substances may also be responsible for these negative effects.

When applying sludge and manure simultaneously, the latter did not totally eliminate the negative effect of the sludge. However, it ensured that these negative effects can only affect the yield at higher levels of sludge application.

Considering the composition of the plants, it was possible to conclude that the nutritive value of the forage improved with the incorporation of the residues with the increase of the percentage of P, K, Ca and Mg, and reduction of the amounts of Mn and Zn. The increase of the Cu level in the plants when the amounts of manure are larger than 6 t ha^{-1}, suggests that the application of high amounts of manure may lead to very serious problems.

The improvement of the soil characteristics (the increment of pH value and the organic matter levels) due to the action of both residues (but mainly the sludge) bears special consideration for Portuguese soils.

References

Carneiro JP (1994) Interesse fertilizante da aplicação simultânea de lamas celulósicas e estrume de aviário. MSc (Master of Science) Thesis. Instituto Superior de Agronomia, Lisboa.

Martinho CFFG (1989) Estrmes de aviário – possibilidades da sua utilização. BSc Thesis. Universidade dos Açores.

Santos JQ, Vasconcelos E & Cabral F (1990) Utilização de lamas celulósicas como fertilizante. Proc. IV Encontro Nacional de Saneamento Básico, organised by Associação Portuguesa para Estudos de Saneamento Básico (APESB), Aveiro, Portugal 386–394.

Santos JQ & Falcão LA (1992) Interesse fertilizante das lamas celulósicas. Proc of Symposium "O controlo da poluição nas indústrias agrárias: Benefícios ambientais e económicos", Lisboa.

Vasconcelos E and Cabral F (1993) Use and environmental implications of pulp-mill sludge as an organic fertilizer. Environmental Pollution, 1993 80: 159–162.

C. Rodriguez-Barrueco (ed.), Fertilizers and Environment, 323–326.

Leachates from a vermicomposting process

A possible new liquid fertilizer?

E. Benitez[1], C. Elvira[2], M. Gomez[1], F. Gallardo-Lara[1] & R. Nogales[1]
[1]*Estación Experimental del Zaidín, CSIC, Box 419, 18080-Granada, Spain;* [2] *Departamento de Recursos Naturales y Medio Ambiente, Universidad de Vigo, 36200-Vigo, Spain*

Key words: Leachates, vermicomposting, germination, fertility

Abstract

The possible agricultural application of leachates collected during the vermicomposting process of sheep manure is discussed. Leachates do not have a negative effect on cress seed germination. Dry weight yields of spinach increased by up to 7% as compared to yields with a treatment containing mineral nitrogen, and leachates released N to plant efficiently, with a 55% effectiveness in comparison witn NH_4NO_3. The addition of leachates to the soil markedly raised levels of potassium in both plant and soil. However, phosphorus concentration in plants decreased appreciably, although in spinach no symptoms of deficiency were observed. Micronutrient levels in soils and plants were scarcely affected by the addition of these leachates.

Introduction

The use of earthworms to convert different organic wastes into more easily handled materials (vermicomposting) is on the increase (Lee, 1985). During the vermicomposting process, water must be added to the beds to prevent the mass from drying out (Fieldson, 1985). In some systems, the excess water added tends to drain through the mass, and should be collected in containers. Leachates may contain organic matter, nutrients, and varying concentrations of toxic compounds, depending on the origin of the waste. Until now, no information has been available on the composition of leachates, or whether they may be used as fertilizers. Research on other types of leachates such as those generated by percolation of water through solid wastes in municipal sanitary landfills (Winant *et al.*, 1981) and composts (Christensen, 1984) have aimed not only at determining their fertilizing capacity, but also at evaluating their possible contaminating effects on soil, plant and surface and ground waters (Hoeks, 1976; Menser *et al.*, 1983). This study reports initial results related to the possible use in agriculture of leachates collected during the vermicomposting of sheep manure.

Material and methods

The analytical composition of leachates were as follows: pH = 8.45, EC = 7.70 dS/m, total solids = 7.5 g l^{-1}, COT = 978 mg l^{-1}, TKN = 61 mg l^{-1}, P = 20 mg l^{-1}, K = 2.4 g l^{-1}, Fe = 163 mg l^{-1}, Mn = 14 mg l^{-1}, Cu = 9 mg l^{-1} and Zn = 13 mg l^{-1} (EPA, 1983).

Germination bioassay

Leachates whole or diluted to 30% and 10% were incubated (27 °C) in the dark for 24 and 48 h with cress seeds (*Lepidium sativum*, L.). Distilled water (0%) was used as a control. Fifteen seeds were placed in 7 cm diameter Petri dishes lined with filter paper, containing 1.6 ml of each dilution. Responses were evaluated as the germination index obtained by multiplying germination by root growth (Zucconi *et al.*, 1981a).

Fertility experiment

A greenhouse experiment was done using 2.5 l pots containing a calcareous Fluvisol. The treatments were: (C): nonfertilized soil, (N): Soil + nitrogen (130 kg N ha^{-1} as NH_4NO_3), and (L): Soil + leachates. Leachates were added to the soil in doses that provided the same quantity of N as the N supplied in mineral form. Pots of treatments (N) and (L) were treated with 200 kg P_2O_5 ha^{-1} and 110 kg K_2O ha^{-1}. Five pots per treatment were sown with spinach (*Spinacia oleracea* cv Polka). All pots were maintained at field capacity throughout the experimental period, and were harvested after 2 months. Total leaf area was measured (Areameter LICOR LI-3100), and the plant material was dried, weighed, ground and digested with H_2SO_4-H_2O_2 (Wolf, 1982). All samples were analyzed for N (Baethgen & Alley, 1989), P, K, Fe, Mn, Cu and Zn (CII, 1973). Soil samples collected from each pot after harvesting were air-dried and homogenized. P, K, Fe, Mn, Cu and Zn were extracted from the soil samples with 1 M NH_4HCO_3-0.005 M DTPA, pH 7.6 (AB-DTPA) (Soltampour & Schwab, 1977). Exchangeable NH_4 and NO_3 were extracted with 2 M KCl (Bremner, 1965).

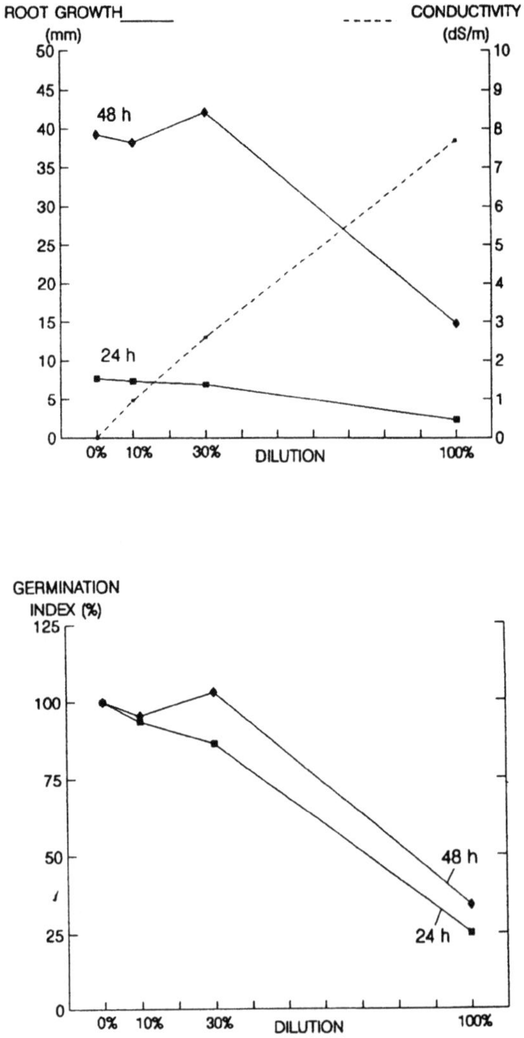

Fig. 1. Response of cress seed germination to different dilutions of the leachates. Above: root growth, below: germination index

Results and discussion

Germination index of cress seed was reduced when undiluted (100%) leachates were used (Fig. 1). However, when dilutions of 10% and 30% were used, germination index was unaffected. These results suggest that leachates from the vermicomposting of sheep manure had no toxic effect on seed germination, in accordance with the finding (Zucconi *et al.*, 1981b) that dilution to 30% led to germination indexes greater than 60%. Reduced germination with the undiluted leachate may be due to the high conductivity of this liquid, which may inhibit initial root growth (Fig. 1).

The application of leachates to the soil significantly increased leaf area and dry weight of spinach in comparison with other treatments (Fig. 2). The extent of the increase in dry weight was 39% as compared to control and 7% with respect to the treatment containing mineral nitrogen (N). This increase can be attributed directly to an increase in soil fertility resulting from the application of leachate.

In all treatments, low N concentrations in plant were detected, although no visual symptoms of deficiency were observed (Table 1). Spinach has high N requirements, and level of 35 g N Kg^{-1} has been reported as the minimun internal N requirement for normal growth in spinach (Benton Jones *et al.*, 1991). The addition of leachate increased N concentration and uptake by the plant and exchangeable ammonium in the soil, although these values were lower than recorded in treatments that contained mineral nitrogen. The estimated recovery of applied N by spinach from the leachates was 38%, as compared to 68% from the mineral nitrogen fertilizer.

The addition of leachates markedly increased potassium concentration and uptake by the plant (Table 1). Although the high level of potassium (126 g K kg^{-1}) detected in spinach indicates a luxury comsumption of this element, no detrimental effects on plant growth was observed. After harvesting, the high levels observed in the plant were correlated with high amounts of AB-DTPA extractable potassium in the soil.

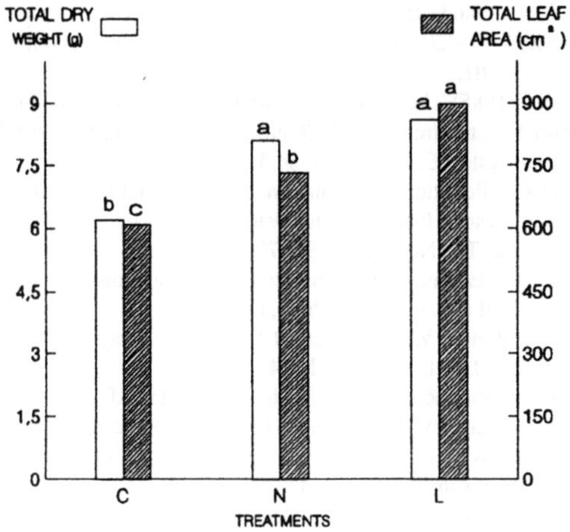

Fig. 2. Leaf area and dry weight of spinach in the different treatments assayed. Columns over the same letter are not significantly different at the 0.05 level.

Table 1. Nitrogen, phosphorus and potassium in spinach and in soil after harvesting.

Plant concentration g kg^{-1}	(C)	(N)	(L)
N	18c	27a	20b
P	6.7a	6.2a	5.1b
K	40b	44b	127a
Uptake mg/pot			
N	112c	219a	171b
P	42a	50a	44a
K	244c	351b	1093a
Soil exchangeable mg kg^{-1}			
NH_4-N	2.6 c	9.3a	4.8 b
NO_3-N	nd	nd	nd
Soil AB-DTPA extractable mg kg^{-1}			
P	46a	30c	40b
K	108b	84b	1507a

nd: not detected. Means within the same row followed by the same letter are not significantly different at the 0.05 level.

Treatments (N) and (L) led to lower levels of P concentration in spinach, and AB-DTPA extractable P from the soil, than the control treatment. These decreases were more pronounced when leachates were added to the soil (Table 1). These unexpected effects suggest that most of the P_2O_5 added to the soil in treatments (N) and (L) was not taken up by the plant, but was rapidly converted to insoluble forms in the soil, as a consequence of the calcareous characteristics of the soil used in this experiment. The same phenomenon may account for the P content in the leachates. Despite

Table 2. Iron, manganese, copper and zinc in spinach and in soil after harvesting

Plant concentration mg kg^{-1}	(C)	(N)	(L)
Fe	253b	301a	210c
Mn	45a	51a	48a
Cu	10a	10a	9b
Zn	167a	171a	187a
Uptake mg/pot			
Fe	1559b	2436a	1816b
Mn	278b	409a	416a
Cu	62b	83a	78a
Zn	1038c	1380b	1614a
Soil AB-DTPA extractable mg kg^{-1}			
Fe	50a	48ab	46b
Mn	11b	10b	14a
Cu	5a	5a	5a
Zn	7b	6c	8a

Means within the same row followed by the same letter are not significantly different at the 0.05 level.

the decreases in AB-DTPA extractable P from the soil in treatments (N) and (L), levels in the soil were very high (Olsen & Sommers, 1982).

The concentration of micronutrients in spinach (Table 2) were within adequate ranges in all treatments assayed (Reuter & Robinson, 1986). Leachate treatment significantly increased Zn uptake by the plant. In contrast, Fe concentration tended to decrease, probably because of dilution caused by greater vegetative growth in leachate-containing treatments. After the plants were harvested, AB-DTPA extractable micronutrients from the soil were nearly the same for (C), (N) and (L) treatments.

The data obtained in the present study support the following conclusions: Addition to the soil of leachates from the vermicomposting process of sheep manure does not impair seed germination, and increases the dry weight yield of plants. Leachates had a moderate capacity to immediately release N to the plant, with an effectiveness in this study of 55% in comparison with NH_4NO_3. Leachates can be considered potential sources of potassium to the soil-plant system. However, their agricultural use may induce P deficiency in the plant, especially if leachates are added to soils with low available P. Finally, the addition of leachates had little effect on the levels of micronutrients in soils and plants. Research is continuing with other soils and plants to verify the fertilizing capacity of this type of leachate.

Acknowledgements

The authors are grateful to the CICYT for financing this study through project no. AMB93–0423. We would also like to express our appreciation to Ms K. Shashok for improving the English style of the manuscript.

References

Baethgen WE and Alley MM (1989) A manual colorimetric procedure for measuring ammonium nitrogen in soil and plant kjeldahl digest. Commun Soil Sci Plant Anal 20: 961–969.

Benton Jones J, Wolf B and Mills HA (1991) Plant analysis handbook. Micro-Macro Publishing Inc, Athens, Georgia, 186 pp.

Bremner JM (1965) Inorganic forms of nitrogen. In: CA Black, DD Evans, JL White, LE Ensminger, FE Clark, RC Dinauer (eds) Methods of soil analysis, Part 2. ASA, Soil Sci Soc Am, Madison, Wis, Agron 9: 1179–1237.

Christensen TH (1984) Leaching from land disposed municipal compost: 3.Inorganic ions. Waste Manag and Res 2: 63–74.

CII (1973) Métodes de reférence pour le détermination de élements minéraux dans les vegetaux. Détermination des élements Ca, Fe, Mg, Cu, Mn et Zn par absorption atomique. Oleagineaux 28: 2663–2675.

EPA (1983) Methods for chemical analysis of water and wastes. EPA-600/4–79–020. US Environmental Protection Agency, Cincinnati, OH.

Fieldson RS (1985) The economic feasibility of earthworm culture on animal wastes. In: Gasser JKR (ed) Composting of agricultural and other wastes. Elsevier, London, UK, 243–255.

Hoeks J (1976) Pollution of soil and groundwater from land disposal of soil wastes. Inst Land and Water Manage Res Tech Bull Wageningen, The Netherlands 96: 70–86.

Lee KE (1985) Earthworms. Their ecology and relationships with soils and land use. Academic Press, London.

Menser HA, Winant WM and Bennet OL (1983) Spray Irrigation with Landfill Leachate. Biocycle 24: 463–467.

Olsen SR and Sommers LE (1982) Phosphorus. In: AL Page, RH Miller, DR Keeney, DE Baker, JD Rhoades, Roscoe Ellis, Jr. (eds) Methods of soil analysis Part 2. Agronomy ASA, Soil Sci Soc Am, Madison, Wis 9: 403–430.

Reuter DJ and Robinson JB (1986) Plant analysis. An interpretation manual. Intaka Press, Melbourne.

Soltampour PN and Shwab P (1977) A new soil test for simultaneous extraction of macro and micronutrients in alkaline soils. Commun Soil Sci Plant Anal 8: 195–207.

Winant WM, Menser HA and Bennett OL (1981) Effects of sanitary landfill leachate on some chemical properties. J Environ Qual 10: 318–322.

Wolf B (1982) A comprehensive system of leaf analysis and its we for diagnosing crop nutrition status. Comm Soil Sci Plant Anal 13: 1035–1059.

Zucconi F, Pera A, Forte M and De Bertoldi M (1981) Evaluating toxicity of inmature compost. Biocycle 22: 54–57.

Zucconi F, Forte M, Monaco A and De Bertoldi M (1981) Biological evaluation of compost maturity. Biocycle 22: 27–29.

C. Rodriguez-Barrueco (ed.), *Fertilizers and Environment*, 327–330.

Tobacco-waste obtained from cigarette factories to be used as organic fertilizer

A. Reşit Brohi[1], A. Aydeniz[1] & M.R. Karaman[1]
[1]*GOP. Üniv., Ziraat Fakültesi, Toprak Bölümü, Tokat, Turkey;* [2]*Adnan Menderes Üniv., Aydın, Turkey*

Key words: Tobacco-waste, ^{15}N trace technique, Ndff %, NdfT %, nitrogen recovery %, wheat crop

Abstract

Tobacco-waste obtained from Tokat cigarette factory was applied at different rates to see the effect of tobacco waste on the growth and N content of wheat crop. % Ndff (N-fertilizer), % NdfT (Tobacco-waste) and Nitrogen Recovery percent of tobacco-waste were determined by ^{15}N trace technique. Increasing rates of tobacco-waste together with nitrogen fertilization increased the dry matter yield of wheat crop from 8.78 to 14.01 grams/pot (40 tons/ha). The increase in N content of wheat crop at these rates increased from 1.47 to 2.09%, whereas, total N uptake increased from 131.43 mg to 293.68 mg/pot. Increasing rates of tobacco-waste increased the percent nitrogen derived from tobacco-waste (NdfT) from 25.19 (at 10 tons/ha) to 59.34 (at 40 tons/ha), whereas the % Ndff increased from 8.93 (20 kg N/ha) to 23.95 (at 80 kg N/ha). Maximum nitrogen recovery percentage obtained from tobacco-waste was about 11.28.

Introduction

In order to assess the available amount of nitrogen in tobacco-waste obtained from Tokat Cigarette Factory, it is necessary to quantify the entire N pool in the presence and absence of a source material. The A-value can be used to measure the amount of nutrients available to plants.

This type of study requires two treatments. The first has a fertilizer source which is labelled with ^{15}N as the standard treatment. The only other nutrient source in this case is unlabelled native soil N. The A-Value obtained from this treatment therefore represents only that of the soil. A second treatment consists of the addition of the same labelled N fertilizer source, plus the tobacco-waste. There are thus three nutrient sources, labelled fertilizer, unlabelled soil and unlabelled tobacco-waste from the second treatment therefore, the A-value obtained represents the sum of available soil N, and available tobacco-waste N measured in terms of labelled fertilizer equivalents. The difference between the A-value of the second and first treatments represents the A-value of tobacco-waste or the amount of N available for plant uptake in the tobacco-waste fertilizer (Danso *et al.*, 1983). The main object of this study is to evaluate the TW as a organic fertilizer in order to protect the environmental pollution and also to determine the % Ndf TW and also the N recovery % of TW in wheat crop by ^{15}N tracer technique.

Materials and methods

The experiment was laid out with completely randomized design with 4 replications. Pots used in this experiment contained 4 kgs of soil. The soil used was alkaline in reaction (pH 7.62) and clay-loam in texture with a CEC of 19.63 m.e./100 grams of soil. Tobacco waste at the rate of 0, 10, 20, 30 and 40 tons/ha was applied in the month of August 1991 and pots were kept at field capacity up to 20 March 1992 and on this date N-fertilization (NF) at the rate of 0, 20, 40, 60 and 80 kg/ha was applied to pots by sowing Bezostiya wheat variety. N-fertilization rates of 20 and 80 kg N/ha were labelled with ^{15}N at 10% atom excess as a $(NH_4)_2SO_4$. The above ground parts of wheat plants were harvested after 79 days of sowing. The plants were analyzed for total N by Kjeldhal method and ^{15}N by mass spectrometry by Turkish Atomic Energy Centre, Ankara.

Table 1. Dry matter yield of wheat crop affected by tobacco-waste and N fertilization

Tobacco Waste (tons/ha)	Nitrogen (kg/ha)					
	0	20[a]	40	60	80[a]	Av.
0	6.78	7.59	8.73	9.86	10.93	8.78 d
10	7.95	10.66	11.44	12.71	12.70	11.09 c
20	10.77	12.45	12.58	12.84	12.85	12.30 b
30	12.17	13.03	12.77	13.50	12.55	12.80 b
40	12.37	14.83	13.94	14.70	14.20	14.01 a
Av.	10.01 b	11.71 a	11.89 a	12.72 a	12.65 a	

[a] ^{15}N labelled N rates

Table 2. N-content of wheat crop affected by tobacco-waste and N fertilization

Tobacco Waste (tons/ha)	Nitrogen (kg/ha)					
	0	20[a]	40	60	80[a]	Av.
0	1.20	1.30	1.54	1.61	1.70	1.47 d
10	1.50	1.47	1.77	1.75	1.79	1.66 c
20	1.58	1.67	1.95	1.86	2.03	1.82 b
30	2.09	1.93	2.13	2.19	1.98	2.06 a
40	2.06	2.00	2.13	2.31	1.96	2.09 a
Av.	1.69 b	1.67 b	1.90 a	1.94 a	1.89 a	

[a] ^{15}N labelled N rates

Table 3. A-values calculated from tobacco-waste

N-fertilization (kg/ha)	Tobacco-waste rates (tons/ha)					
	0	10	20	30	40	Av.
20	-	5.00	10.40	16.07	23.81	13.82
80	-	7.10	12.97	22.83	28.16	17.77
Av.		6.05	11.69	19.45	25.99	

Table 4. Atom percent ^{15}N excess in upper parts of wheat

N-fert. (kg/ha)	Tobacco-waste (tons/ha)					
	0	10	20	30	40	Av.
20	1.414	1.045	0.815	0.662	0.527	0.893
80	3.605	2.731	2.275	1.777	1.589	2.395
Av.	2.510	1.888	1.545	1.220	1.058	

Results and discussion

Dry matter yield of wheat crop

The effect of tobacco-waste together with N fertilization on the dry matter yield of wheat crop is given in Table 1.

Tobacco-waste applied at different rates significantly (at 1% level) increased the dry matter yield of the wheat crop. It increased from 8.78 grams/pot (control) to a maximum of 14.01 grams/pot (40 tons

Table 5. Percent Nitrogen Derived from Tobacco-waste

N-fert. (kg/ha)	Tobacco-waste (tons/ha)					
	0	10	20	30	40	Av.
20	–	26.13	42.38	53.19	62.74	46.11
80	–	24.24	36.88	50.71	55.93	41.94
Av.	–	25.19	39.63	51.95	59.34	

T W/ha). Similar results have been obtained by Brohi and Durak (1986), Brohi (1988), Brohi (1991).

N-content and N-uptake of wheat crop

Effect of different rates of Tobacco-Waste together with N fertilization on the N-content of wheat is given in Table 2.

As seen from Table 2 increasing rates of tobacco-waste together with N fertilization increased the N-content of wheat crop. N-content of wheat crop was 1.47, 1.66, 1.82, 2.06 and 2.09% at 0, 10, 20, 30 and 40 tons/ha rates of tobacco-waste respectively. N rates of 40, 60 and 80 kg N/ha increased the N-content of wheat crop but according to Duncan test all three rates fall under the same group. Data is in confirmation with the findings of Aydeniz *et al.* (1977), Saxena (1984) and Özbek *et al.* (1984).

Increasing rates of tobacco-waste together with N fertilization increased total N uptake of wheat crop. N-uptake was 131.43, 185.12, 224.47, 264.06 and 293.68 mgs/pot at 0, 10, 20, 30 and 40 tons/ha rates of tobacco-waste respectively. N rates also increased N uptake of wheat crop. The highest N-uptake was at 60 kg N/ha, decreasing at 80 kg N/ha. Our data support previous findings (Raikova and Rankov, 1987; Caramete *et al.*, 1990).

A-values

A-values from fertilizer without tobacco-waste are 13.82 and 17.77 at 20 and 80 kg N/ha rates respectively. A-value calculated from tobacco-waste is given in Table 3.

Increasing rates of tobacco-waste increased the A-values from 6.05 (10 tons T W/ha) to 25.99 (40 tons T W/ha). This data is in consistence with Fried *et al.* (1983).

Atom % ^{15}N excess, % Ndff (N fertilizer) and % NdfT (tobacco-waste)

The atom % ^{15}N excess values of wheat crop are given in Table 4 and % NdfT are given in Table 5. The atom % ^{15}N excess of wheat crop increased with increasing rate of ^{15}N labelled $(NH_4)_2SO_4$ rates. It increased from 0.893 to 2.395, whereas tobacco-waste rates decreased it from 2.510 (control) to 1.058 (40 ton T W/ha). Our results are in consistence with the findings of Fried *et al.*, (1983) for barley.

Percent Ndff was 8.93 and 23.95 at 20 and 80 kg N/ha rates of N fertilization respectively. Percent of N in above ground parts of the wheat plants derived from TW applied at 10, 20, 30 and 40 tons/ha was 25.19, 39.63, 51.95 and 59.34 respectively (Table 5).

Nitrogen recovery percent

The recovery % of N by wheat of tobacco-waste applied at the rates of 10, 20, 30 and 40 tons/ha was found to be 10.14, 9.82, 11.28 and 9.14 respectively, whereas the recovery by wheat of N-fertilization applied at 20 and 80 kg/ha was found to be 38.0 and 35.63% respectively.

This means that much more of the N released from tobacco-waste might be retained in the soil. Müller (1988) found about 39–44% N recovery from red clover N when labelled with ^{15}N. However N recovery from tobacco-waste is low when compared with leguminous crops.

Acknowledgements

This research work is sponsored by Cumhuriyet and Gazios-manpaşa Univ. Research Funds and State Planning Commission of Turkey.

References

Aydeniz A, Dündar M and Brohi A (1977) Cycocel'li ve Cycocel'siz olarak değişik N-P düzeylerinde gübrelenen Bugday ve Arpa bitkisinin su tüketimi ve gelişmesine plastikle kaplama, perlitle örtmenin etkisi, I. Ulusal Perlit Kongresi Bildirileri MTA Türkiye Jeoloji Kurumu, Ankara, 1977.

Brohi A and Durak A (1986) Tütün tozunun organik gübre olarak değerlendirilmesi, Türkiye tütüncülüğü ve gelecegi sempozyumu: Tokat 261–278.

Brohi A (1988) Çimlerde tütün tozunun gübre olarak değerlendirilmesi, Tokat ZFD: 3–15.

Brohi A (1991) Tütün artıklarından gübre olarak yararlanma olanaklarının araştırılması, C.Ünv. Tokat ZF Yay. 7 Bilimsel araştırma ve incelemeler Tokat 4: 19–21.

Caramete C, Şuteu GE, Vısarıon M, Coroıanu M and Chıņa M (1990) Fertilizer rates on fixation of atm. N in soyabean grown on Brown-Reddish soil at Muora Domneasce Lucrari Stintifice. IANB Seria A Vol. XXXIII.

Danso SKA, Bole JB and Zapata F (1983) A guide to the use of Nitrogen-15 and Radioisotopes in studies of plant nutrition, calculations and interpretation of data. IAEA-TECDOC-288, Vienna.

Fried M, Danso SKA and Zapata F (1983) The methodology of measurement of N_2 fixation by nonlegumes as inferred from field expt with legumes. Can J Microbiol 29: 1053–1062.

Müller M (1988) The rate of clover-derived N (^{15}N) during decomposition under field conditions, effect of liming and fertilization. Plant and soil 111 (1): 121–126.

Özbek N, Halitligil MB and Korkmaz A (1984) Biological dinitrogen fixation and its economical importance for Agr. Nuclear Research center, 311-DI-SR-96, Ankara, Turkey.

Raikova L and Rainkov V (1987) Use of the isotope ^{15}N studying the N uptake by radishes. ESNA-Varna, Bulgaria.

Saxena MC (1984) Dinitrogen fixation in some food legumes under rainfed conditions of northern Syria seminar on the use of isotopes in studies on biological N fixation, Ankara.

Suten G, Caramete C and Negrescu I (1980) Discrimination of the N derived from the N reserve in the annual organs of perennial plants. Bulletin of the Academy of Agr and Forestry Sciences 14: 4–5.

C. Rodriguez-Barrueco (ed.), Fertilizers and Environment, 331–336.

Response of barley to sewage sludge manuring and irrigation in calcareous soils

M.S.A. Dahdoh & S. El-Demerdashe
Desert Research Center, Mataryia, Cairo, Egypt

Key words: Barley, sewage sludge, irrigation, calcareous soil

Abstract

A pot experiment was carried out to estimate the effect of adding sewage sludge rates to calcareous soil irrigated with sludge effluents with differing salinity content, using barley plants and tap water as a control treatment. Sludge water and sewage sludge were obtained from the Suez Experimental Station at Suez City.

Results show a significant increase of barley yield when using 2% sewage sludge but this decreased significantly when 4% sewage sludge was used. This was observed for all the sludge effluents as well as the tap water. The effect of the sludge effluents on yield differs according to salinity levels. In general, sludge effluent no. 4 resulted in the best yield relative to other effluents and tap water treatments.

The concentrations of N, P, K, Na, Fe, Zn, Cu, Mn and Ni in barley grown in sludge-amended soil are relatively higher compared with control (0 level of sewage sludge and irrigated with tap water). The content of N, Na, Zn, Cu and Ni in barley shoots increased with increasing salinity of the effluents, while Fe, Mn, P and K displayed the opposite trend. The elements uptake follow, to some extent, the same trend.

Computation of a Zn equivalent for the treatments studied shows that the rate of 4% sewage sludge is critical for all the sludge effluents; this may be considered as one of the reasons for decrease in the yield of barley shoots.

Introduction

Land application of sewage sludge often introduces large amounts of macro and micronutrients into agriculture soils. The quantity of sewage sludge, and consequently the induced elements, are expected to influence the yield and composition of grown crops. This, in fact, is dependent on the application rates of sludge, Pietz *et al.* (1983). Recycling and land utilization of sewage sludge has been going on for approximately the last three decades and there have been significant efforts to develop the use of sewage sludge with methods which are safe and without having a hazardous effect on soil, plant, animal and human beings. Although land application of sewage sludge is beneficial to crops as a source of essential macro and micronutrients, it is also considered to be a source of the non-essential and polluting elements, Cd, Pb, Ni, Co and Cr. The levels of these elements are dependent on the source of sewage sludge, but they all have a potentially hazardous effect on the growing crops and environments. Many experiments have been initiated to study the effects of sewage sludge on the soils that receive them and on the plants grown in the sludge-treated soils (Logan & Chaney, 1983). However, some heavy metals are ubiquitous in soils, but the concentrations indigenous to cropland soil do not appear harmful to plant growth.

The aim of this investigation is to evaluate the influence of land application of sewage sludge to calcareous soil irrigated with different sludge effluents having varying salinity levels, on barley yield, its metal content and uptake. The sludge effluents were obtained from the Suez Experimental Station.

Materials and methods

An anaerobically digested sewage sludge produced from the waste water treatment station of Suez City and calcareous soil samples from Sinai (Ras Sudr) were used in the study. The main characteristics of the soil, sewage sludge and sludge effluents are given in Tables (1 & 2).

Table (1) Some physical and chemical properties of the studied soil.

Mechanical analysis								Chemical properties															
								Available elements (ppm)								Solution extract (me/L)							
																Cations				Anions			
C S.	F.S.	Silt	Clay	Tex.	$CaCO_3$ (%)	pH in paste	EC dS/m	Fe	Zn	Mn	Cu	Ni	N	P	K	Ca^{++}	Mg^{++}	Na^+	K^+	HCO_3^-	Cl^-	SO_4^{--}	CO_3^{--}
19	46	22	11	S.C.L.	52.5	7.3	5.8	1.4	0.48	0.68	0.12	0.12	12	8	30	15.3	11.6	30.6	0.5	1.8	52.1	4.1	-

Table (2) Some properties of sewage sludge and sludge effluents

(A) Sewage sludge

EC dS/m	pH	Moisture (%)	O.M. (%)	Total									Available (ppm)								
				%					ppm												
				N	P	K	Na	Fe	Zn	Mn	Cu	Ni	N	P	K	Na	Fe	Zn	Mn	Cu	Ni
4.7	6.8	6	37.3	1.3	0.09	0.17	0.38	5.0	1060	188	310	120	468	132	310	468	168	260	20	39	8

(B) Sludge effluents

	(me/L)							(me/L)										
Elements / S.Eff	N	P	Fe	Zn	Mn	Cu	Ni	Na^+	K^+	Ca^{++}	Mg^{++}	HCO_3^-	Cl^-	CO_3^-	SO_4^{--}	pH	EC dS/m
1	1.12	0.18	0.65	0.15	0.25	0.38	0.13	50.5	1.1	9.9	8.6	1.9	65.8	0.5	2.2	6.9	6.8
2	0.98	0.15	0.45	0.13	0.22	0.31	0.12	27.3	0.6	5.3	3.8	1.0	34.1	0.4	1.2	7.2	3.7
3	0.83	0.10	0.35	0.10	0.17	0.29	0.10	14.7	0.3	2.9	1.9	0.5	18.6	0.3	0.6	7.2	2.0
4	0.80	0.09	0.32	0.08	0.15	0.21	0.08	9.5	0.2	1.9	1.4	0.3	12.1	0.2	0.4	7.5	1.3

A pot experiment was conducted in a greenhouse using plastic pots of 1 kg capacity arranged in a split block design with 4 replications. Different rates of sewage sludge, i.e., 0, 2 and 4% (soil basis) were applied to the soil of each pot and thoroughly mixed.

Ten grains of barley (*Hordeum vulgare*, Giza 163) were germinated in each pot and irrigated with tap water for 7 days, then the plants were thinned to 5 per

Table (3) Effect of sewage sludge and sludge effluents on the dry matter yield of barley shoots (g/pot).

S.S.	S. Eff.				Tap water
	(1)	(2)	(3)	(4)	
0	1.26	2.88	3.09	3.74	3.63
2%	1.40	2.98	3.32	5.16	3.94
4%	0.99	2.37	2.51	3.55	2.88

L.S.D. 0.05 S.S. S.Eff. S.S×S.Eff.
0.10 0.30 0.30
S.Eff. = Sludge effluent S.S. = Sewage sludge

Table (4) Effect of sewage sludge and sludge effluents on the concentration of some elements and Zn equivalent in barley shoots.

	S.Eff.												Tap water		
	(1)			(2)			(3)			(4)					
S.S.(%) Element	0	2	4	0	2	4	0	2	4	0	2	4	0	2	4
N %	2.08	2.59	2.68	1.91	2.38	2.40	1.91	2.07	2.27	1.59	1.93	2.00	1.42	1.63	1.83
P %	0.14	0.19	0.29	0.18	0.23	0.32	0.19	0.28	0.36	0.23	0.33	0.45	0.17	0.32	0.38
K %	1.1	2.3	2.9	1.4	2.8	3.1	1.8	2.9	3.2	1.9	3.2	3.3	1.6	2.7	2.9
Na %	2.0	2.3	2.8	1.4	1.6	2.0	1.1	1.2	1.9	1.0	1.1	1.9	0.8	1.1	1.6
Zn ppm	40.0	60.0	70.0	45.0	68.0	75.0	40.0	80.0	86.0	50.0	88.0	103.0	38.0	80.0	88.0
Cu ppm	9.8	10.0	11.5	9.5	10.2	13.0	9.5	10.5	14.0	9.0	12.0	15.0	8.5	10.0	11.5
Fe ppm	360	420	500	380	450	520	450	460	560	505	555	650	480	500	62.0
Mn ppm	30.0	40.0	60	38	43	58	58	44	55	40	48	50	42	45	48.0
Ni ppm	9.4	20.6	22.9	7.3	20.0	20.6	6.5	14.3	19.4	5.3	12.3	17.1	5.0	12.3	15.7
Zn equivalent	135	244	276	122	248	266	111	215	269	110	210	270	95	206	237

pot. After 15 days of cultivation, the plants were irrigated with four different sludge effluents having variable salinity levels (6.8, 3.7, 2.0 and 1.7 mmhos/cm) and tap water (control) as 75% of field capacity.

After 45 days, the barley shoots (1 cm above the soil surface) were harvested, washed with distilled water, then dried at 70 °C for 48 hrs in an oven and the dry yield recorded. The dried material was ground, and wet digested (Thomas *et al.*, 1967). P was determined colorimetrically using the ascorbic acid method (Watanabe & Olsen, 1965); N by the Kjeldahl method; K and Na by flamephotometer (Black, 1983). Fe, Mn, Zn, Cu and Ni were measured by atomic absorption spectrophotometer, Perkin Elmer, 2380. Statistical analysis of the analytical data was carried out according to Snedecor and Cochran (1973).

Results and discussion

The dry matter yield of barley shoots grown on sludge-amended soil and irrigated with different sludge effluents is significantly increased by 2% sludge application and decreased at 4% sludge application (Table 3). Moreover, the increasing salinity of sludge effluent resulted in a depressive effect on the barley yield. This

Table (5) Effect of sewage sludge and sludge effluents on some elements uptaken in barley shoots.

	S.Eff. (1)			(2)			(3)			(4)			Tap water		
S.S.(%) / Element	0	2	4	0	2	4	0	2	4	0	2	4	0	2	4
N mg/pot	2.62	3.62	2.65	5.50	7.09	5.68	5.90	6.87	5.69	5.94	9.95	7.1	5.15	6.42	5.27
P mg/pot	0.17	0.26	0.28	0.51	0.68	0.75	0.55	0.92	0.90	0.86	1.7	1.59	0.61	1.26	1.09
K mg/pot	1.38	3.22	2.87	4.03	8.34	7.34	8.65	9.62	8.03	10.84	16.51	11.71	9.43	11.42	8.35
Na mg/pot	2.52	3.22	2.77	4.03	4.76	4.74	3.39	3.98	4.76	3.74	5.67	6.74	2.90	4.33	4.60
Zn µg/pot	50	84	69	130	203	178	124	266	216	187	454	366	138	315	253
Cu µg/pot	12.3	14	11.4	27.3	30.4	30.8	29.4	349	35.1	33.7	61.9	53.3	30.1	39.4	33.1
Fe µg/pot	454	588	495	1094	1341	1232	1391	1527	1406	1889	2864	2308	1742	1970	1785
Mn µg/pot	38	56	59	109	128	137	117	146	138	150	248	178	152	177	138
Ni µg/pot	11.8	288	22.7	21.02	59.6	48.8	20.1	47.5	48.7	19.8	63.5	60.7	18.2	48.5	45.2

L.S.D. (0.05)

	N	P	K	Na	Zn	Fe	Cu	Mn	Ni
S.Eff.	0.22	0.04	1.38	1.10	18	58	3.5	18	3.18
S.S.	0.18	0.05	1.12	0.18	16	31	1.8	13	2.65
S.Eff. x S.S.	0.39	0.07	1.62	0.98	23	63	4.18	28	4.23

Table (6) The "C"* values of the studied elements as affected by adding sewage sludge and sludge effluents to calcareous soil.

	S.Eff. (1)			(2)			(3)			(4)			Tap water		
S.S.(%) / Element	0	2	4	0	2	4	0	2	4	0	2	4	0	2	4
N	1.46	1.82	1.88	1.34	1.67	1.69	1.34	1.45	1.59	1.11	1.35	1.4	1.0	1.14	1.28
P	0.82	1.11	1.70	1.05	1.35	1.88	1.11	1.64	2.11	1.35	1.94	2.64	1.0	1.88	2.23
K	0.68	1.43	1.81	0.87	1.73	1.93	1.12	1.81	2.0	1.18	2.0	2.06	1.0	1.68	1.81
Na	2.50	2.87	3.5	1.75	2.0	2.5	1.37	1.5	2.37	1.25	1.37	2.37	1.0	1.37	2.0
Zn	1.05	1.57	1.84	1.18	1.78	1.97	1.0	2.1	2.26	1.31	2.31	2.71	1.0	2.1	2.31
Cu	1.15	1.17	1.35	1.11	1.2	1.52	1.11	1.23	1.64	1.05	1.41	1.76	1.0	1.17	1.35
Fe	0.75	0.87	1.04	0.79	0.93	1.06	0.93	0.95	1.16	1.05	1.15	1.35	1.0	1.04	1.29
Mn	0.71	0.95	1.42	0.90	1.02	1.38	0.9	1.04	1.30	0.95	1.14	1.19	1.0	1.07	1.14
Ni	1.88	4.12	4.58	1.46	4.0	4.12	1.3	2.86	3.88	1.06	2.46	3.42	1.0	2.46	3.14

$$*C = \frac{\text{Element concentration in sludge-treated plant}}{\text{Element concentration in control (zero level of S.S. and irrigated with tap water)}}$$

finding stands in agreement with Hassan *et al.* (1970), Mass *et al.* (1972); Dahyia & Singh (1976) and Fayed (1988) who found that increasing the salinity of irrigation water decreases the yield of crops. However, it is necessary to estimate the actual tolerance of plants to high application rates of sewage sludge and higher salinity levels of sludge effluents especially when they are added to the fine-textured and calcareous soils. For this purpose, several trials have been undertaken by Page *et al.* (1972), Bingham *et al.* (1975), Soon *et al.* (1980) and El-Sokkary & El-Keiy (1989).

Elemental content in barley shoots

Application of sewage sludge increased the concentration and significantly increased the uptake of the studied elements for all the sludge effluents. For instance, the concentrations of N, P, K, Na, Fe, Zn, Cu, Mn and Ni (Table 4) in barley shoots grown on sludge-amended soil are relatively higher than that of critical levels reported by Melsted *et al.* (1969). Adding sludge effluents also increased N, Na, Zn, Cu and Ni concentrations, while for Fe, Mn, P and K the opposite trend was observed when compared with tap water treatments for all the applied rates of sludge.

The discrepancy in behaviour of some elements which display the opposite trend is expected to be due to the multireactions between salinity and sludge components on the one hand and nutritive elements on the other. It must also be taken into account that competitive and antagonistic effects may also take place even within the nutritive elements themselves.

The uptake of elements is significantly affected by adding both sewage sludge and sludge effluents and follows nearly the same trends as observed for the concentration (Table 5).

The values of concentration indices (C) given in Table (6), show that the magnitude of the increase of element concentration in sludge-treated plants varies widely relative to the control under the sludge effluents irrigation. The "C" values of the elements corresponding to sludge effluents could be arranged in the descending order; S.Eff. (1), S.Eff. (2), S.Eff. (3), S.Eff. (4), Tap water. This order coincides very well with salinity concentrations of sludge effluents.

The toxicity of Zn, Cu and Ni differs in specific plants in specific soils. Generally, Cu is twice as toxic as Zn, and Ni is eight times as toxic as Zn. In this respect, Chumbley (1971) introduced the "Zn equivalent" factor where ppm Zn equivalent = ppm Zn + 2 $\times$ ppm Cu + 8 $\times$ ppm Ni, to take into account the differences among metals. He suggested that no more than 250 ppm Zn (equivalent) should be added to agricultural soils (with pH maintained > 6.5). With regard to the data obtained, it was observed that the Zn (equivalent) is more than the 250 ppm value when sewage sludge is added at a rate of 4% to the soil under all the sludge effluents. This means that this critical rate is not suitable to be added to soil under the experimental conditions. Undoubtedly, any rate higher than 4% should not be applied at all.

In conclusion, it could be deduced that the data obtained for barley shoots, as an indicator for its yield, are significantly affected by sludge amendments and sludge effluent irrigation. The sludge composition and salinity of the effluent affects the element content, uptake and yield. The critical levels for such an effect are indicated. However, the element uptake by plants is a character that can be improved if genetically controlled.

References

Bingham ET, Page AL, Mahler RJ and Canje TJ (1973) Growth and Cd accumulation in plants grown on a soil treated with cadmium enriched sewage sludge. J Environ Qual 4: 207–211.

Black CA (ed) (1983) Methods of Soil Analysis Part 1, Agron Series No 9. Am Soc Agron, Mad Wisc, USA.

Chumbley GG (1971) Permissible levels of toxic metals in sewage used in agricultural land. Agricultural Development and Advisor Service Report No 10. Ministry of Agriculture Fisheries and Food, London.

Dahyia SS and Singh M (1976) Effect of salinity alkalinity and iron application on the availability of Mn, P and Na in (*Pisum sativum* L.) crop. Plant and Soil 44: 697–702.

El-Sokkary IA and El-Keiy OM (1989) Response of alfalfa, wheat, faba bean, soybean and sordan to sewage sludge applied to sandy soil. Egypt J soil Sci Special Issue, 375–386.

Fayed RH (1988) Salinity effect on some nutrients status in some soils of Egypt. Ph.D. Thesis, Fac Agric, Al-Azhar Univ, Egypt.

Hassan NAK, Drew JV, Knudsen D and Olsen RA (1970) Influence of soil salinity on production of dry matter and uptake and distribution of nutrients in barley and corn. I Barley (*Hordeum vulgare* L.). Agron J 62: 43–45.

Logan TJ and Chaney RL (1983) Utilization of municipal wastewater and sludge on Land-Metals. In: Page AL, Gleason TL, Smith Jr JE, Iskandar IK and Sommers LE (eds) Utilization of municipal wastewater and sludge on land. Univ. of California, Riverside, CA 92521, U.S.A., 235–326.

Mass EV, Ogata G and Garber MJ (1972) Influence of salinity on Fe, Mn and Zn uptake by plants. Agron J 64: 793–795.

Melsted SW, Motto HL and Peck TR (1969) Critical plant nutrient composition values useful in interpreting plant analysis data. Agric J 61: 17–20.

Page AL, Bingham FT and Nelson C (1972) Cadmium absorption and growth of various plant species as influenced by solution cadmium concentration. J Environ Qual 1: 288–291.

Pietz RI, Pterson JR, Hinesly TD, Ziegler EL, Redborg KE and Lue-Hing C (1983) Sewage sludge application to calcareous strip mine spoil. II – Effect on soil and corn, cadmium, copper, nickel and zinc. J Environ Qual 12: 463–467.

Snedecor CW and Cochran G (1973) Statistical Methods. Iowa State Univ Press, USA.

Soon YK, Bates TE & Mayer JR (1980) Land application of chemically treated sewage sludge: III – Effect on soil and plant heavy metals content. J Environ Qual 9: 497–504.

Thomas RL, Shearel RW and Moyer ZR (1967) Comparison of conventional and automated procedures for nitrogen, phosphorus and potassium analysis of plant material using single digestion. Agron J 59: 240–243.

Watanabe FS and Olsen SR (1965) Test of an ascorbic acid method for determining phosphorus in water and $NaHCO_3$ extracts from soil. Soil Sci Soc Am Proc 29: 677–678.

C. Rodriguez-Barrueco (ed.), Fertilizers and Environment, 337–340.

Study on nitrogen mineralization in municipal solid waste compost using electroultrafiltration

I. Diaz-Marcote, J.A. Diez & A. Polo
Centro de Ciencias Medioambientales, CSIC, Serrano, 115 dpdo, 28006 Madrid, Spain

Key words: Compost, incubation, mineralization, nitrogen, EUF

Abstract

The nitrogen mineralization of municipal solid waste (MSW) compost (CI) and a mixture of MSW with sewage sludge compost (CL) applied to a typical soil of the Central region of Spain has been studied. Doses of 20 and 80 t ha^{-1} were applied under aerobic incubation conditions at 28 °C for six months. Nitrogen was studied by the electroultrafiltration method. Such a method enables mineral nitrogen (EUF-N_{NO3}) and organic nitrogen (EUF-N_{org}) corresponding to easily mineralized nitrogen (low molecular weight organic compounds) to be determined. The results are based on the evolution of the net mineralization rate and the ratio of potential for mobilizing nitrogen reserves.

The data obtained indicates that adding compost (CL) in the first stage of the incubation process promoted soil nitrogen mineralization. However, in treatments with CI, no effect on soil mineralization was found when compared with the control. This is explained because easily mineralized nitrogen compounds are more abundant in CL, whilst in CI, the nitrogen is found in more stable forms. After 90 days, a very intense mobilization phase commences in MSW compost treatments.

Introduction

Adding urban waste compost to agricultural soil is a practice which can be recommended mainly because of its organic nature, (a component scarce in Spanish soils) and its nutritional and trace element content (Bussinelli *et al.*, 1990; Gallardo-Lara *et al.*, 1979).

The macroelements in the compost and other organic wastes are mainly associated with organic matter and their availability is therefore not immediate. The transformations which occur on the organic matter are highly complex making it difficult to determine the availability. Of the existing chemical and biological methods, incubation tests are the ones most frequently used as they enable the evolution of available nitrogen fractions over a certain period to be studied. This paper deals with an incubation test controlled over six months using the electroultrafiltration technique (EUF) for measuring the mineralized nitrogen.

Materials and methods

A typic haploxeralf soil from the Central Area of Spain was used originating from the CSIC owned "La Higueruela" experimental farm located in Santa Olalla (Toledo). Its main features are: C: 0.8%; N: 0.07%; pH: 5.8; CIC: 10.1 $cmol_c\ kg^{-1}$. The composts used came from the Valdemingómez MSW treatment plant in Madrid (CI) and also a compost made by mixing MSW and sewage sludge in a 1:1 ratio (CL) which was carried out on a pilot plant scale at the CEBAS (Murcia) (Hernández *et al.* 1992). Some of the general characteristics of the compost used are:

CI: C: 18.4%; N: 1.8%; water-soluble N: 303 mg kg^{-1}; C/N: 10.0; pH: 8.2

CL: C: 11.7%; N: 1.2%; water-soluble N: 438 mg kg^{-1}; C/N: 9.9 C/N; pH: 7.5

Incubation conditions

The experiment was carried out in a chamber measuring 2 × 3 × 2.5 m^3 under controlled conditions (28 ± 2 °C, 80% relative humidity). Pots with 3 kg of

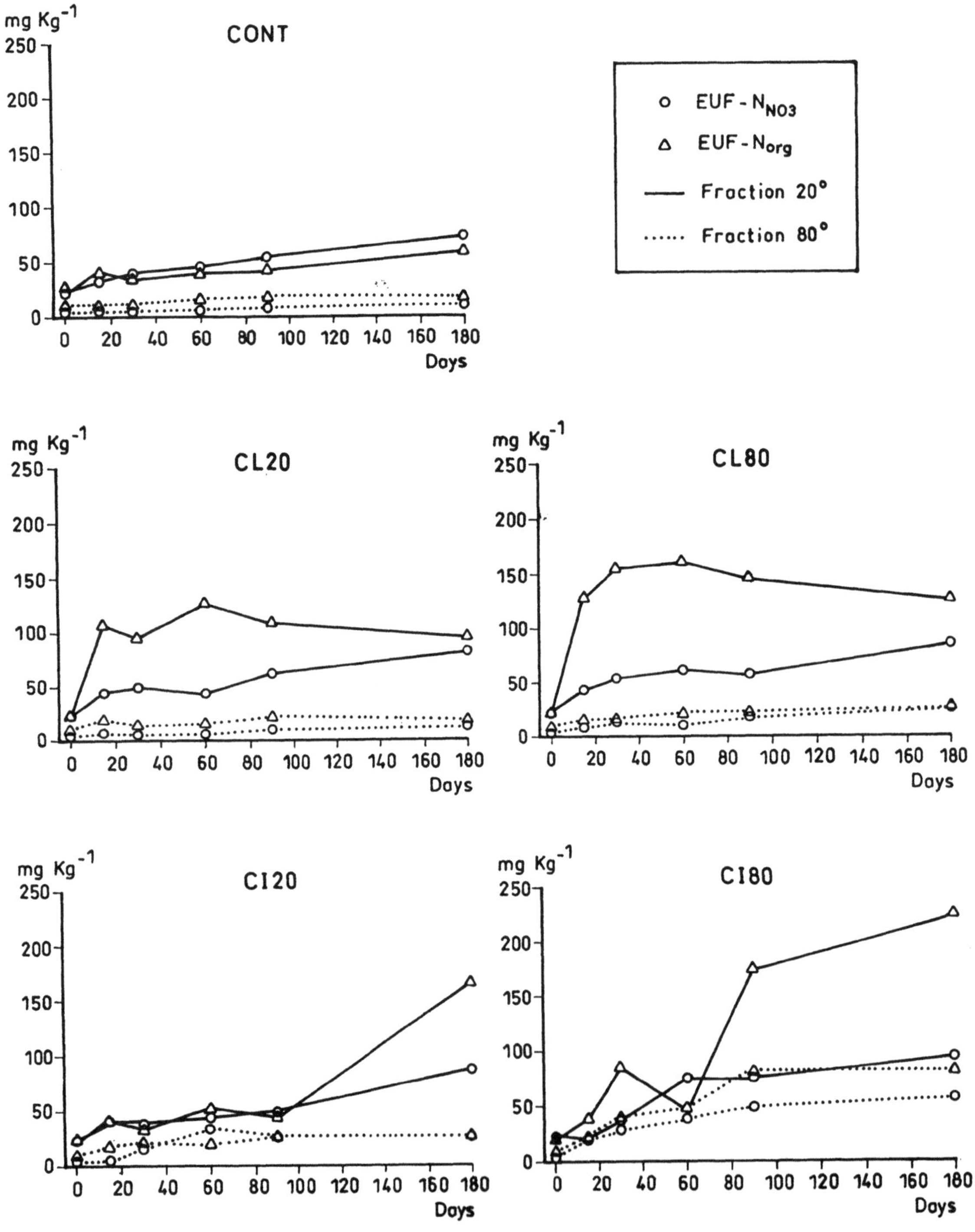

Fig. 1. Evolution of EUF-N_{NO3} and EUF-N_{ORG} in the fractions extracted at 20 and 80 °C.

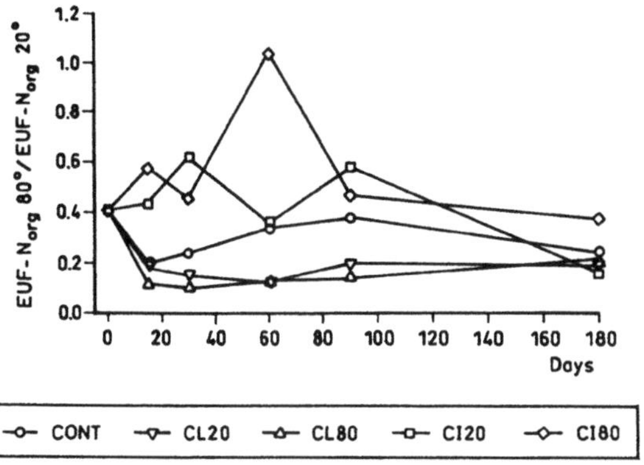

Fig. 2. Evolution of the EUF-N_{ORG} 80 °/EUF-N_{ORG} 20 ° index.

soil were used and waste was added in doses of 20 and 80 t ha^{-1}. The treatments were labelled: CI20, CI80, CL20, CL80 and control (CONT). The mixtures were watered up to 75% of water holding capacity and kept for 6 months under the same conditions by regular watering and turning over. Samples were taken after 15, 30, 60, 90 and 180 days.

Method of determining N by electroultrafiltration

A 5 g soil sample (< 1 mm) was taken and 50 ml of distilled water was added. An electric field was applied to this suspension which complied with a voltage and temperature programme (Nemeth, 1979). Two fractions were obtained:

Fraction I: 200 V, 20 °C, 30 minutes, $\leq$ 15mA

Fraction II: 400 V, 80 °C, 5 minutes, $\leq$ 150 mA

Total nitrogen (EUF-N) and nitrate contents (EUF-N_{NO3}) were determined in the extracts according to Díez (1985). Organic nitrogen (EUF-N_{org}) which corresponds to easily mineralized nitrogen was obtained by difference. The organic nitrogen was made up mainly of low molecular weight compounds, together with interlaminar ammonium nitrogen or nitrogen from the hydrolysis of organic compounds during extraction.

Results and discussion

An increase in the nitrogen mobilization occurred in all treatments (Fig. 1) due to favourable microbe development incubation conditions. By differentiating the fractions of EUF-N_{NO3} and EUF-N_{org} obtained at 20 and 80 °C, it can be seen that the CL treatment provides a large amount of low molecular weight organic nitrogen (amino acids, etc.) in the fraction extracted at 20 °C. This diminishes with incubation time at the same time as EUF-N_{NO3} increases. These low molecular weight nitrogenous compounds come from the decomposition of proteic material. Nevertheless, in CI treatments, low values of both EUF-N_{NO3} and EUF-N_{org} were maintained until 60 or 90 days when they begin to rise sharply. These differences demonstrate the different mineralization of the nitrogen in the two composts, which can be explained by nitrogen compounds of a lesser molecular weight abounding more in CL due to greater proteolytic activity in this kind of waste (García *et al.*, 1993; Lerch *et al.*, 1992).

The tendency of the nitrogen mobilization process in the compost was quantified by the net mobilization rate (NM) (Table 1). In CL compost treatments, the NM is very high in the first period and it diminishes at the end of incubation showing that part of the mobilized nitrogen has been incorporated into less EUF extractable forms. On the other hand, in CI compost treatments, the NM in the initial phase is approximately 20% and increases up to very high percentages at end of the experiment (63% in CI80 and 121% in CI20 respectively). A value over 100% implies that an amount of nitrogen greater than that applied has been mobilized. This is known as the "priming effect" and has been detected by several authors (Nemeth and Recke, 1985).

It is interesting to see that the value of EUF-N increases with the dose for each compost whilst the NM decreases. Other authors have also observed that the reduction of mineral nitrogen accumulation in soil, when higher doses are applied, occurs through the stimulation of gas losses via ammonification (Fine *et al.*, 1989) or by a process of denitrification caused by point anaerobic conditions in the incubated soil (Lindemann and Cardenas, 1984).

Relations between the different EUF-N forms are often used as indices (Nemeth, 1979) helping to simplify the study of the dynamics of nitrogen mineralization in soil. Figure 2 shows the values obtained with the EUF-N_{org}80 °C/EUF-N_{org}20 °C ratio which indicates the soil's capacity for mineralizing nitrogen reserves (Nemeth, 1988). Treatments with CI compost display the highest values demonstrating high mineralizing potential. In the first phase, over 50% of nitrogen is immobilized in organic forms of a more complex structure than those extracted at 20 °C. Only after 90 days does mineralization of organic nitrogen become

Table 1. Values of EUF-N (mg kg^{-1}) and net mobilization rates (NM) for each treatment after 30 and 180 incubation days

	30 days EUF-N	180 days EUF-N	$N_{APPLIED}$	30 days NM[a]	180 days NM
CONT	85	155	–	–	–
CL20	164	205	80	98.8	62.5
CL80	236	262	320	47.2	33.4
CI20	108	300	120	19.2	120.8
CI80	188	459	480	21.5	63.3

[a]NM = $(\text{EUF-N}_{TREATMENT} - \text{EUF-N}_{CONT}) / N_{APPLIED} \times 100$.

evident in these treatments (reduction of the index). However, the low values found in CL treatments lead to the assumption that a large part of the nitrogen which was incorporated with the said matter was very labile and mineralized in the first few days. The increase in the index in CL treatments implies that a nitrogen immobilization process occurred at the end of the incubation period.

Acknowledgement

Supported by CICYT (93-0429).

References

Bussinelli M, Gigliotti G and Giusquiani PL (1990) Applicazione del compost da RSU in agricoltura I. Effetto sulla produttivita del mais e destino dei nutrienti e dei metalli pesanti nel vegetale. Agrochimica 34: 454–466.

Díez JA (1985) Determinación de nitrógeno UV oxidable y de nitrato, en extractos de suelo mediante autoanalizador. An Edafol Agrobiol 44: 1453–1464.

Fine P, Mingelgrin U and Feigin A (1989) Incubation studies of the rate of organic nitrogen in soils amended with activated sludge. Soil Sci Soc Am J 53: 444–450.

Gallardo-Lara F, Azcon M, Gómez M and Esteban E (1979) Poder fertilizante de un compost de basura urbana I. Capacidad de suministro de macronutrientes. An Edafol Agrobiol 38: 1747–1764.

García C, Hernández T, Costa F, Ceccanti B, Masciandaro G and Ciardi C (1993) A study of biochemical parameters of composted and fresh municipal wastes. Biores Technol 44: 17–23.

Hernández T, García C, Costa F, Valero JA and Ayuso M (1992) Utilización de residuos urbanos como fertilizantes orgánicos. Suelo y Planta 2: 373–383.

Lerch RN, Barbarick KA, Sommers LE and Westfall DG (1992) Sewage sludge proteins as labile carbon and nitrogen sources. Soil Sci Soc Am J 56: 1470–1476.

Lindemann WC and Cardenas M (1984) Nitrogen mineralization potencial and nitrogen transformations of sludge-amended soil. Soil Sci Soc Am J 48: 1072–1077.

Nemeth K (1979) The availability of nutrients in the soil as determined by electro-ultrafiltration (EUF). Advances in Agronomy 31: 155–188.

Nemeth K and Recke H (1985) EUF-N fractions in different soils during a vegetation period in pot and field experiments. Plant Soil 83: 39–46.

Nemeth K (1988) Grundlagen der EUF-Stickstoffempfehlung zu Getreide und Hackfrüchten. In Neue Forschungsergebnisse der EUF-Arbeitsgemeinschaft zur Förderung der Bodenfruchtbarkeit und Bodengesundheit, Mannheim: 47–70.

C. Rodriguez-Barrueco (ed.), Fertilizers and Environment, 341–343.

Stabilization of sludges from a dairy processing plant using vermicomposting

P. Gratelly[1], E. Benitez[1], C. Elvira[2], A. Polo[3] & R. Nogales[1]
[1]*Estación Experimental del Zaidín, CSIC, 18080-Granada, Spain* [2]*Departamento de Medio Ambiente y Recursos Naturales, Universidad de Vigo, 36200, Vigo, Spain* [3]*Centro de Ciencias Medioambientales, CSIC, 28006-Madrid. Spain*

Key words: Vermicomposting, *Eisenia andrei*, dairy processing sludge, organic matter, elements

Abstract

In an attempt to obtain an organic amendment to soils, research was undertaken to assess the feasibility of using earthworms to stabilize sludge obtained from wastewater treatment at a dairy processing plant. Total biomass, growth and reproductivity of the earthworms that had fed on dairy sludge were found to be higher than in earthworms placed in sheep manure. After the vermicomposting period, about 39% of the total organic carbon contained in the dairy sludge was lost. The vermicomposted dairy sludge was characterized by a high humic acid content, high degree of humification, high HA/FA ratios and high concentration of macro- and micronutrients.

Introduction

Land application and incineration are the most widespread methods of disposal of sludges, including those from treated wastewater of dairy processing plants. Before the sludge can be applied to land, it must be stabilized to avoid negative effects on the soil-plant system. A possible management method applicable to this sludge is vermicomposting, an ecobiotechnological process used for waste stabilization, as already described for animal slurries (Hand and Hayes, 1983; Albanell et al., 1988), liquid dewatered municipal sludges (Neuhauser *et al.*, 1988), urban solid wastes (Graziano and Casalicchio, 1987), paper mill sludges (Elvira, 1994) and other industrial wastes (Edwards, 1988). This study assesses the feasibility of using earthworms to stabilize sludges from a dairy processing plant to obtain an organic amendment to soils; data on earthworm growth and reproductivity and the changes in the composition of these sludge are also presented.

Table 1. Analysis of the initial products used in the special system

	Dairy sludge	Inoculum	Sheep manure
Moisture %	86	49	66
pH	8.2	8.1	8.3
EC dS/m	0.42	0.62	3.9
TOC g kg^{-1}	339	207	360
TKN g kg^{-1}	75	17	15
C/N	4.5	20	23
P g kg^{-1}	8.1	7.2	5
K g kg^{-1}	3	9	22
Ca g kg^{-1}	87	90	24
Mg g kg^{-1}	4.3	28	6.1
Na g kg^{-1}	5.8	1.5	5.4
Fe g kg^{-1}	2.6	12	9
Mn mg kg^{-1}	25	428	121
Cu mg kg^{-1}	193	62	37
Zn mg kg^{-1}	558	167	82
Ni mg kg^{-1}	25	27	7

Material and methods

The sludges were obtained from a wastewater treatment plant of a dairy processing factory (Puleva, Granada). The main characteristics of these sludges are shown in Table 1.

In a preliminary experiment, earthworms were reared in dairy sludge and in four mixtures of this waste with different bulking materials (straw, wood shavings, cardboard and tobacco wastes) at a ratio 1:1 (v/v). Five non-clitellated earthworms (*Eisenia andrei*) were placed in a 2 l plastic container, with 1500 g fresh weight of each feed. After 24 h all the worms had died. This experience was repeated weekly for 50 days with these products, and the same negative results were always obtained.

This lethal effect was avoided with the development of a special system. A bed of sheep manure vermicompost (800 g f.w.) (Table 1) was placed in the container to act as a microbial inoculum and to provide a suitable habitat for earthworms. Twenty worms (weighing between 0.15 and 0.25 g) were placed in this material and the container was then filled with a layer of dairy sludge (1000 g f.w.). A treatment containing this inoculum and a layer of fresh sheep manure (30 days old) (Table 1) was used as control. During the vermicomposting period (70 days), survival, growth, clitellium development and cocoon production were evaluated weekly. At the end of this period, and after separation of earthworms, the content of each container was thoroughly mixed, dried and analyzed. Total organic matter, organic matter fractions (Sequi et al., 1986), humification parameters (Ciavatta et al. 1988) and elemental composition of this vermicomposted material were determined.

Results and discussion

The negative effect observed in the preliminary experiment led to the conclusion that this type of sludge was lethal to the worms, probably because of the high ammonia concentration. In the mixture including tobacco waste, the earthworms died especially quickly, probably because the tobacco waste contains high concentrations of salts and a certain amount of nicotine, which is toxic for worms (Verdonck et al., 1985).

In the special system, the earthworms achieved maximum growth in 52 to 60 days, after which size and weight declined during the remaining ten days (Figure 1). Total biomass was 13.5 g for earthworms that had fed on dairy sludge, that is, 1.6 times heavier in comparison with those which had fed on fresh sheep manure. All the earthworms fed with dairy sludge had reached sexual maturity by 25 days (100% clitellated, worms), whereas in the control treatment only 60% of the worms showed clitellum development. Cocoon production was also higher when the worms were kept in the medium containing dairy sludge (Figure 1). In this treatment, the production of cocoons per clitellated

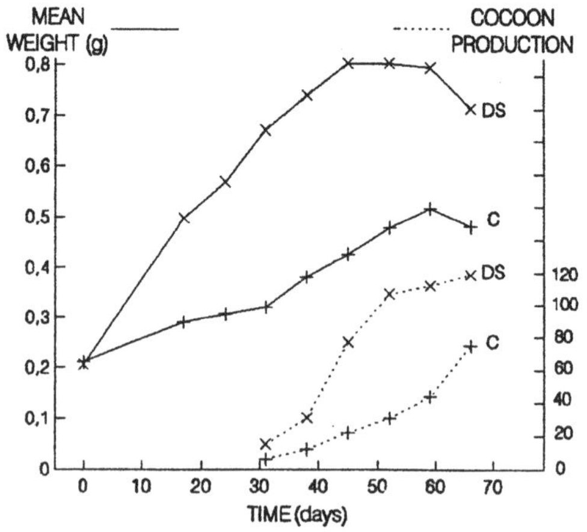

Fig. 1. Earthworm weights and cocoon production in the control (C) and dairy sludge (DS) during the vermicomposting period.

worm and week was 1.06 in comparison with c. 0.80 in the control treatment.

At the end of the vermicomposting process, about 39% of the total organic carbon (TOC) contained initially in the dairy sludge was lost, due to organic matter mineralization, and, to a lesser extent, to dilution with the organic matter content in the inoculum. Likewise, total extractable carbon, nonhumified fractions (NH) and fulvic acids (FA) decreased appreciably, whereas amounts of humic acids (HA) were scarcely affected (Table 2). For these reasons, the vermicomposted dairy sludge showed a higher degree of humification (DH), higher AH/AF ratio, and lower humification index (HI) than the initial dairy sludge, representing levels similar to those recorded in the control treatment and to other stabilized and humified organic materials (Casalicchio and Graziano, 1987; De Nobili et al. 1989).

Elemental composition of the dairy sludge was appreciably modified after the vermicomposting period (Table 3). The changes observed were most likely due to the following effects: i) Concentration of these elements caused by the mineralization of organic matter in the sludge, ii) Volatilization of some N contained in the sludge as NH_4, iii) Dilution or concentration of some elements induced by the mixture of this sludge with the initial vermicompost used as inoculum. As a result of these effects, the levels of N, P and Cu decreased, whereas the levels of K, Ca, Mg, Fe, Mn and Zn increased in the vermicomposted sludge as compared with the initial dairy sludge. Cadmium and lead were not detected in vermicomposted dairy sludge.

From data obtained we may conclude that sludges obtained from treated wastewater at a dairy processing

Table 2. Fractionation of organic matter and parameters of humification in the different products. Humifica ion parameters: HI=HA+FA/NH, DH=HA+FA/TEC × 100

	Dairy sludge	Vermicomposted dairy sludge	Control
OM%	75	43	45
TOC%	34	20	19
TEC%	15	4.3	5.5
NH mg/100 g	6552	922	862
FA mg/100 g	6186	691	2748
HA mg/100 g	2697	2582	1867
% TEC/TOC	45	20	28
HI	0.74	0.28	0.19
DH	57	78	84
AH/AF	0.43	3.73	0.68

Table 3. Analysis of the products after vermicomposting

	Vermicomposted dairy sludge	Control
Moisture % pH	7.95	8.40
EC dS/m	1.22	0.97
TKN %	2.9	2.5
C/N	7.1	7.8
P%	0.75	0.77
C/P	27	25
K%	0.73	1.3
Ca%	10	9.3
Mg%	3.1	3.1
Na%	0.25	0.20
Fe%	1.2	1.2
Mn mg kg^{-1}	403	405
Cu mg kg^{-1}	93	63
Zn mg kg^{-1}	237	191
Ni mg kg^{-1}	29	22

plant can be used as a suitable material for vermicomposting, if the sludge is layered over a previous vermicomposted material. In this system, growth and reproductivity of earthworms were optimum. The vermicomposted dairy sludges produced with this system contained large amounts of humic acids, macro-and micronutrients, and showed a high degree of humification. They were therefore suitable for use as organic soil conditioners.

Acknowledgements

The authors are grateful to the CICYT for financing this study through project no. AMB93-0423. P. Gratelly thanks the ICI-CSIC for funding his stay at the EEZ, CSIC, Granada. We would also like to express our appreciation to Ms. K Shashok for improving the English style of the manuscript.

References

Albanell E Plaxats J and Cabrero T (1988) Chemical changes during vermicomposting (*Eisenia fetida*) of sheep manure mixed with cotton industrial wastes. Biol. Fertil Soils, 6, 266–269.

Casalicchio G and Graziano PL (1987) A comparison of the chemical properties of compost and wormcasting from solid municipal wastee and sewage sludge. In AMB Pagliai and P Omodeo (eds). On Earthworms. Selected Symposia and Monographs U.Z.I., pp 437–457 Muchi, Modena, Italy.

Ciavatta C, Antisari L and Sequi P (1988) A first approach to the characterization of the presence of humified materials in organic fertilizers. Agrochimica, 32, 510–517.

De Nobili M, Ciavatta C and Sequi P (1989) La valutazione del grado di maturazione della sostanza organica del compost mediante la determinazione di parametri dell'humificazione e per electtrofocalizzazione. In Proceedings of the Int. Symp. on Compost: Production and Use. pp 328–342. San Michele all'Adige, Italy.

Edwards CA (1988) Breakdown of animal, vegetable and industrial organic wastes by earthworms. In: CA Edwards and EF Neuhauser (eds). Earthworms in waste and environmental management. pp 21–31. Academic Publishing, The Hague.

Elvira C (1994) Vermicompostaje de residuos organicos. Su aplicación a lodos generados por la industria papelera. Thesis, Univ. Santiago de Compostela, España, 292 p.

Graziano PL and Casalicchio G (1987). Use of worm-casting techniques on sludges and municipal wastes: development and application. In: AMB Pagliai and P Omodeo (eds) On Earthworms. Selected Symposia and Monographs U.Z.I., pp 459–464. Muchi, Modena Italy.

Hand P and Hayes WA (1983). The composting of slurries by earthworms. In: International Conference on composting of solid waste and slurries. pp 246–260. University of Leeds, England.

Neuhauser EF, Loehr EC and Malek MR (1988). The potential of earthworms for managing sewage sludge. In: CA Edwards and EF Neuhauser (eds). Earthworms in waste and environmental management. pp 9–20. Academic Publishing, The Hague.

Sequi P, de Nobili M, Leita L and Cercignani G (1986). A new index of humification. Agrochimica 30, 175–179.

Verdonck O, De Bott M, Stradiot P and Penninck R (1985). The use of tree bark and tobacco waste in agriculture and horticulture. In: JKR Gasser (ed) Composting of Agricultural and other wastes. pp 203–215. Elsevier Applied Science Publishers, New York.

C. Rodriguez-Barrueco (ed.), Fertilizers and Environment, 345–350.

Cattle-slurry – ^{15}N turnover in a long-term lysimeter trial*

R. Gutser & P. Dosch
Institute of Plant Nutrition, Technical University Munich, D-85350 Freising-Weihenstephan, Germany

Key words: Slurry, lysimeter, N utilization, N leaching, N immobilization, N accumulation

Abstract

The usual application systems for slurry leads to unsatisfying N utilization by plants. Causes are not only the low availability of the organic N and losses of NH_3, but also the low efficiency of potentially plant available NH_4-N of slurry due to immobilization. In an 11 year lysimeter trial (^{15}N) the utilization of the NH_4-N of optimally applied slurry by plants in the first year was only 32% as compared to 59% from mineral fertilizers (CAN). In the next years the rates of remobilization of immobilized fertilizer N of both treatments were nearly the same with 9 to 11% in the 2nd year and 4 to 10% thereafter, depending on the crop. The soil analysis confirmed the strong accumulation of nitrogen in soil by slurry (N_t = 0.165%; CAN 0.130%). According to a simplified calculation the fertilizer specific N accumulation in soil would approach its equilibrium after a continuous manuring for 40 to 60 years. The N accumulation in soil not only increases its fertility, but also the danger of uncontrolled N losses; long-term application of slurry (on average 120 kg N ha^{-1}, slurry only NH_4-N) enhanced the N leaching by on average 10 kg N ha^{-1} yr^{-1} compared to mineral fertilization. Modifying the composition of the slurry by reducing its carbon content decreased the immobilization of NH_4-N in soil with consequence of a reduced N accumulation in soil by long-term supply.

Introduction

The actual cultivation methods of conventional agriculture lead to an increasing load of our environment; negative effects on hydrosphere, atmosphere, pedosphere and biosphere are known (Isermann, 1993). Farms with high density of livestock contribute to a N eutrophication of environment. The N surplus between fertilizer input and removal by harvest products frequently exceeds 100 kg N ha^{-1}. The calculation of the effect of organic and inorganic nitrogen in organic manure is very difficult (Gutser and Dosch, 1992). The knowledge of the N availability of organic and inorganic fertilizer in the year of application and later is necessary, not only for regulating the specific N nutrition of plants but also for a coarse regulating of the fertilizer management for several years; thus the decrease of N surplus is possible without considerably yield losses. Reduced N losses during storage and in the field directly after application are suppositions for an optimized N effect of organic manures.

In this paper we studied the short- and long-term turnover of slurry N in lysimeter and field trials as well as in model trials and deduced recognitions for optimized strategies of slurry with special regard to agricultural – environmental relationships. Additional results are found in Dosch and Gutser (1995) in the same proceedings.

Methods

Lysimeter trial: In a lysimeter trial (1980–1993) in Weihenstephan (794 mm mean precipitation, 7.4 °C air temperature) 40 km northern to Munich, the N turnover of mineral fertilizer (CAN = calcium ammonium nitrate) and cattle slurry were tested (Gutser *et al.*, 1988; Vilsmeier and Gutser, 1989; Gutser and Claassen, 1993).

Soil: silty loam (20% clay, 70% silt), pH($CaCl_2$) = 6.5, 1978 filled in lysimeter plots conform to the natural stratification (horizons of 20 cm) of a brown earth.

* Dedicated to Prof. Dr. A. Amberger on his 75th birthday.

Crop rotation: sugar beet – winter wheat – (winter barley); leaves and straw remained on plots. Catch crops were only cultivated in the treatment "slurry to catch crop".

Fertilizing: CAN: on average 120 resp. 165 kg N ha^{-1} yr^{-1}; slurry: on average 70 (NH_4) = 134 (total) kg N ha^{-1} yr^{-1}, adequate to a density of livestock of 1.8 livestock units (LU) ha^{-1}; the slurry treatments were supplied with on average 50 kg N as CAN ha^{-1} yr^{-1}.

Application time of slurry: (i) to main crop in early spring or (ii) to cash crop in autumn in combination with straw manuring – immediate incorporation or surface band application in growing cereals reduced NH_3 losses.

Since 1988 the utilization of NH_4-N of slurry in comparison with CAN by use of ^{15}N labelled fertilizers (for slurry an addition of $(NH_4)_2SO_4$) was tested – labelling: 12–15 atom% ^{15}N. The effect in the first year and later (5 years = 1989–1993) were studied and the yearly available N (= sum of N uptake by plants + N leaching) were determined and converted in % of fertilizer residue in soil at harvest time of the last crop (including N in leaves or straw).

Field trial: similar treatments as in lysimeter trial; begin: 1979; soil: brown earth (silty loam, pH ($CaCl_2$) = 6.2); crop rotation: silage maize-cereals (1–2 ×). In the first period (1979–1987) slurry was given exclusively to maize, later to each main crop by special application techniques to reduce NH_3 losses.

Slurry dose: on average 83 NH_4 = 148 (total) kg N ha^{-1} yr^{-1}, adequate to a density of livestock of 2 LU ha^{-1}. Till 1988 cereals (winter wheat) received an unit supply of on average 90 kg N as CAN ha^{-1}. Straw remained on field.

Model trial: incubation trial to study the immobilization of ^{15}N labelled slurries (NH_4-N) and CAN; 3 soils differing in clay contents (20, 9 and 6%) – soil moisture 60% of maximum water capacity; fertilizing: 0.2 mg NH_4-N g^{-1} soil – incubation period: 4, 11, 21 weeks.

N immobilization = ^{15}N fertilization minus $^{15}N_{min}$(2 M KCl extractable NO_3- and NH_4-N) or ^{15}N residue in soil (see Vilsmeier and Gutser, 1989).

Methods for ^{15}N in plants, soils, drainage water: $^{15}N/^{14}N$ ratios were determined with ^{15}N analysator NO1 Fa. Jasco, Japan resp. mass spectrometer Europa Scientific.

Special methods for soil analysis: microbial biomass (Anderson and Domsch, 1978)

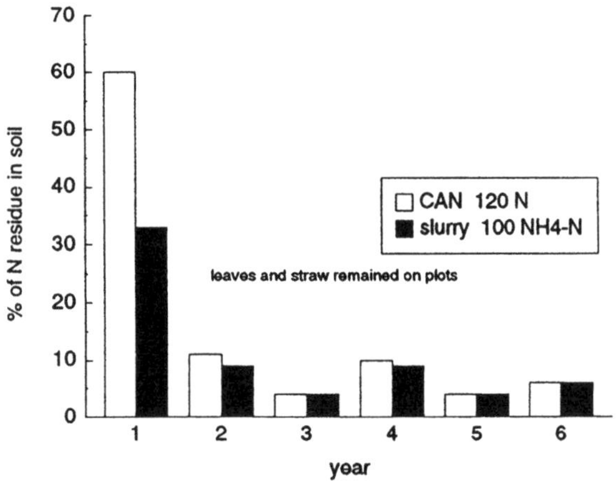

Fig. 1. Available fertilizer N (^{15}N) in the first year (1988) and later (1989–1993)
available N = N uptake (beets, leaves, grain) + N leaching.

N_{org} = soluble organic N, extractable by 0.01 M $CaCl_2$ (Gutser *et al.* 1990) or EUF (electro-ultra-filtration, Nemeth, 1982).

Results

In a 15 years old field trial the known unsatisfying N utilization of slurry by plants was confirmed (Table 1). The average N utilization of slurry by a crop rotation maize/wheat was 16% in the beginning (1979–1987) due to high N losses after broadcast application without incorporation and predominant supply to maize. Afterwards, a slightly but not sufficient increase of N utilization (1987–1993: 20%) was reached by the improvement of application techniques and supply to both crops (equivalence of slurry-N to mineral fertilizer N was near to 50%).

Slurry attained a better result in an 11 years old lysimeter trial with an equivalence to mineral fertilizer of 72% (base: NH_4-N) – (Table 2). This increased utilization should be contributed to minimized NH_3 losses due to better application techniques (immediate incorporation and surface band application).

On base of NH_4-N of slurry the two variants slurry/CAN and CAN alone reached similar yields and N removals; however, N leaching and nitrate content of drainage water were increased by slurry and the increase will extent in time (Fig. 2). On base of total N slurry obtained a 8% lower yield as CAN; the differences in N leaching were unimportant. Corresponding to Fig. 2 the deficit in yields will decrease in time.

Table 1. Utilization of slurry N by a crop rotation maize-wheat (1979–1993)

average N-dose: 83 (NH_4) resp. 148 (total) kg N ha^{-1} yr^{-1}

calculation base	utilization %*		
	1979–1987	1979–1993	1989–1993
NH_4-N	28.9	32.7	35.9
total N	16.2	18.3	20.2
CAN		mean 50–60	

utilization = N uptake (maize shoots + grain) minus control in % of N input by slurry

Table 2. N effect of mineral fertilizer (CAN) and slurry/mineral fertilizer in a lysimeter trial (1983–1993)

measured parameters	control	Base: NH_4-N of slurry		Base: total N of slurry	
	N_o	CAN	slurry	CAN	slurry
fertilizing (kg N ha^{-1} yr^{-1})	–	120	120	165	184
			(70 + 50)		(134 + 50)
yield (relative)	54	= 100	102	111	102
				= 100	92
removal (grain, beets) (kg N ha^{-1} yr^{-1})	42	94	91	119	91
utilization of fert. (%)		43	31	46	26
equivalence to min. fert. (%)	–	= 100	72	= 100	57
leaching (kg N ha^{-1} yr^{-1})	37	35	45	42	45
avg. mg NO_3 l^{-1} drain. water	46	58	75	72	75

Causes for the low utilization of slurry are not only the slow mineralization of organic N (Amberger *et al.*, 1982) or N losses (NH_3, NO_3), but also the low efficiency of NH_4-N in the first year.

In 7th year of trial slurry and CAN reached nearly the same N uptake in sugar beets (Table 3), because of the high N mineralization in soil long-term treated with slurry. The utilization of slurry N (only NH_4-N ^{15}N labelled) applied to beet was 32% compared to 59% of CAN; the slurry to catch crops attained only an efficiency of 18% in the first year. It is remarkable that the leached N always comes from the N pool of soil and only to an unimportant extent from the fertilizer, directly applied to the main crop.

The availability of the immobilized NH_4-N of slurry in following years was not different from that of immobilized mineral fertilizer (Fig. 1); depending on crop the mineralization rate reached 4 to 10% of the actual residue of fertilizer N in soil.

In 6 years (2 crop rotations) only 24 to 30% of 1988 applied NH_4-N of slurry were utilized in harvest products in relation to 51% in case of CAN (Table 4). Accordingly the N losses by leaching in slurry plots were higher than that of CAN plots. Without considering gaseous losses the calculated residues of slurry N in soil after 6 years came to 60 resp. 68% of applied NH_4-N (100 kg N ha^{-1}); the relative value for CAN was 42%.

The low utilization of NH_4-N and especially that of organic N (on average 50% of total N) led to an evident N accumulation in soil (Table 5); the required carbon was delivered by slurry in addition to roots and harvest residues (see control and CAN treatment) – the contents increased like N.

The differences in N contents of soil are well explained by calculated N balances; of course a correct calculation of N amount in soil is not possible in consequence of mistakes in soil analysis (sampling, plant

Table 3. Utilization of NH_4-N (^{15}N) of slurry by sugar beets in relation to mineral fertilizer (CAN) 7th year of lysimeter trial (1988) – kg N ha^{-1}

treatment	N-uptake (beets + leaves)			N leaching	
	fertilizer	soil	total	fertilizer	soil
slurry (march)*	32 = 32%	156	188	1	82
slurry (catch crop)*	18 = 18%	182	200	3	34
CAN	71 = 59%	111	182	1	61
control	–	97	97	–	57

*+40N as CAN.

Table 4. Utilization of ^{15}N-fertilizer in a lysimeter trial (1988–1993) fertilizing 1988 (kg N ha^{-1}): slurry: 100 NH_4-N, CAN 120 N

treatment	% of fertilized N		
	leaching	removal (beets, grain)	residue in soil calculated*
slurry march	10	30	60
slurry catch crop	8	24	68
CAN	7	51	42

*without considering gaseous N losses.

Table 5. N balance (1983–1993) resp. N and C contents of soil (1991) in a lysimeter trial
N balance = fertilizer input – removal (beets, grain) – leaching

treatment	balance kg N ha^{-1}	soil (0–25 cm)%	
		N_t	C_{org}
control	– 855	0.130	1.17
slurry spring	+ 459	0.165	1.60
slurry catch crop	+ 639	0.170	1.62
CAN	0	0.138	1.45
at begin (1982)	–	0.127	1.12

Table 6. N leaching after application of slurry in relation to mineral fertilizer – ^{15}N lysimeter trial 1988 (= 7th year): sugar beets – (kg N ha^{-1})

treatment		N leaching		
		fertilizer	soil	total
control	N_0	–	57	57
slurry march	N_{100}	1	82	83
slurry catch crop	N_{100}	3	34	37
CAN	N_{120}	1	61	62

residues in soil) resp. the limitation only for surface soil.

The results from lysimeter trial with ^{15}N showed, that the yearly N leaching is less dependent from the fertilizer given to the growing crop than really more from the N pool of soil specific accumulated by long-term fertilizing management (Table 6). Supposition for this establishment is a correctly terminated application of fertilizers.

Therefore the reason of an increased N leaching in slurry plot (application in march) was not the direct effect of the last manuring, but the mobilization from N pool of soil enriched by a long-term application of slurry. The low N leaching on plots with "slurry to catch crop" emphasized the evident N conserving effect of catch crops (other treatments without catch crops).

The N mobilization of N pool in soil, accumulated by manuring as a function of its time, is not only important for the N losses by leaching, but very substantial for the N nutrition of plants (see Table 3). To realize

Table 7. Suitability of microbiological and chemical methods to recognize the increased N potential of soils by long-term application of slurry 11th year of field trials, surface soil (0–25 cm) – October 1988 – mg N resp. C kg^{-1} dry soil

treatment	N_t	aerobic incubation (12 weeks)	biomass-C	N_{org}	
				EUF	$CaCl_2$
Control	1180	33	370	13	6
slurry to catch crop	1300	48	480	16	6
slurry to main crop	1360	51	430	12	6

Table 8. Immobilization of slurry NH_4-N as a function of C content model trial (^{15}N) – avg. 4 to 12 weeks incubation time

treatment	C % fresh w.	N immobilization (% of fertil. dose)	
		sand	loess
slurry	4.4	24	44
slurry	2.0	16	34
slurry	0.6	13	30
CAN	–	7	13

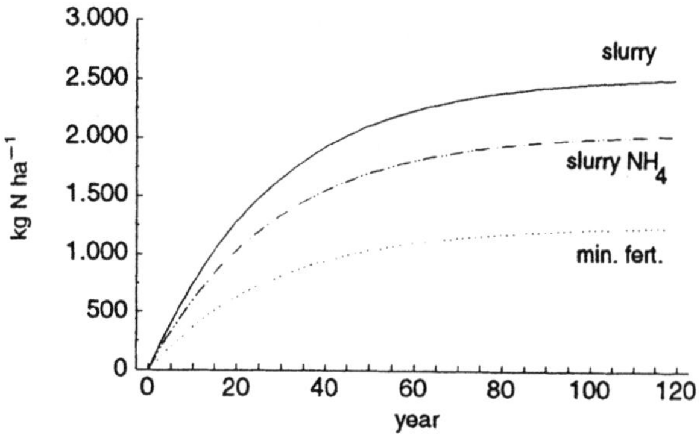

Fig. 2. N accumulation of soil by long-term application of slurry and mineral fertilizer (120 kg ha^{-1}) $N_{acc} = z \cdot r^{-1} \cdot (1 - e^{-r \cdot t})$ Z = input: fertilizer residue in soil after 2nd year r = yearly rate of mineralization (constant 0.035) t = years mineralization rate ($1^{st}/2^{nd}$ yr): slurry 0.22 / 0.06 slurry NH_4: 0.35 / 0.08 min. fert.: 0.60 / 0.08.

a favorable fertilizing practice, that means to optimize the year – and plant-specific N fertilization, it would be helpful to know the N mineralization potential of soils. This N potential is hardly determinable with chemical methods, e.g.the determination of soluble organic N (N_{org}) after extraction with $CaCl_2$ or EUF. While microbiological methods should be more successful, their high analytical expense make them unsuitable for routine analysis (Table 7).

Table 3 shows the difference between the utilization of NH_4-N of slurry and mineral fertilizer. The most important cause for the low efficiency of the applied NH_4-N of slurry may be due to an immobilization in soil (Vilsmeier and Gutser, 1989).

In an incubation trial with cattle slurry (Table 8) of different carbon contents (0.6–4.4% in fresh weight) the positive correlation between C content of manure and extent of N immobilization could be demonstrated. These results point to a soil-specific immobilization potential.

Discussion

Due to the strong dependence from N losses (NH_3, NO_3, N_2, N_2O) and immobilization of the applied NH_4-N the N efficiency in the first year is hardly to calculate. Moreover the organic N of slurry shows a low availability. Therefore, slurry impedes an optimized N fertilization of agricultural crops. In long-term trials the utilization of mineral N of slurry reaches 18 to 26%, but mineral fertilizers attain 50% in average.

Therefore, fertilizer managements will produce on base of the same total N supply in case of slurry or slurry/mineral fertilizer combinations smaller yields than through mineral fertilizer alone. On base of NH_4-N of slurry the problems may be in higher N losses by leaching, increasing with long-term manuring.

Not only the level of fertilization but also the extent of immobilization of slurry N in the first year determine the intensity of N accumulation in soil. In a simplified calculation the different effects of slurry and mineral fertilizer on N accumulation of soil is clearly demonstrated (Fig. 2). Details of this calculation are found in Amberger *et al.* (1982) and particularly Gutser and Claassen (1994).

As shown in the lysimeter trial the mineralization rate of fertilizer residues in soil (immobilized N of slurry) after two years and later is nearly the same between slurry and CAN; in the calculation of Fig. 2 a constant rate of 0.035 was fixed (the problems for different conditions of soil, climate and cultivation systems are known). The main reason for the differences in specific accumulation curves are the extent of fertilizer residues in soil after the first year and partly second year; in case of fertilizer dose of 120 kg N ha^{-1} the residues after 2nd year reach 44 (mineral fertilizer) or 89 (cattle slurry) kg N ha^{-1}. The fertilizer specific N accumulation in soil would approach its equilibrium after a continuous manuring for 40 to 60 years. Under practical conditions after about 40 years there should be nearly the same amounts of available N in soil (mineral fertilizer 109, slurry 98 kg N ha^{-1}). The predominant part of the available N comes in case of mineral fertilizer directly from fertilizer (66%), in case of slurry from N pool in soil (73%).

The N accumulation in soil not only improves its fertility, but also the dangers of uncontrolled N losses. The extent of N accumulation is limitable by a reduced N fertilization (densities of livestock should not exceed 1.5 LU ha^{-1}) (Enquete Kommission, 1994) as well as an increased utilization of slurry N in the first year. The reduction of C contents of slurry by separation techniques or anaerobic fermentation (Messner and Amberger, 1988) should be a possibility to decrease the immobilization in soil and increase N utilization by plants. For an optimized strategy of slurry with special regards to agricultural-environmental relationship a high efficiency of slurry N in the first year is necessary, which is attainable by reducing N losses, but also N immobilization (see Dosch and Gutser (1995) in the same proceedings).

References

Amberger A, Vilsmeier K and Gutser R (1982) Stickstofffraktionen verschiedener Güllen und deren Wirkung im Pflanzenversuch. Z Pflanzenern u Bodenkde 145: 325–336.

Anderson JPE and Domsch KH (1978) A physiological method of the quantitative measurement of microbial biomass. Soil Biol Biochem 10: 215–221.

Dosch P and Gutser R (1996) Reducing N losses (NH_3, N_2O, N_2) and immobilization from slurry through optimized application techniques. In: Rodriguez-Barrueco C. (ed) Fertilizers and Environment. pp 269–276. Kluwer Acad. Publ., Dordrecht.

Enquete Kommission des Deutschen Bundestages (1994) "Schutz der Erdatmosphäre". Agra-Europe, August 1994, 35.

Gutser R, Amberger A, Vilsmeier K (1989) N-Bilanzen nach Mineral- und Gülledüngung mit Dicyandiamid-Zusatz in langjährigen Lysimeter- und Feldversuchen. Landw Forsch, Kongreßband 1988, VDLUFA-Schriftenreihe 28: 503–516.

Gutser R, Teicher K and Maier S (1991) Gefäßversuche zur Aussagekraft des N_{org}-Stickstoffs für das N-Potential von Böden. Landw Forsch, Kongreßband 1990, VDLUFA-Schriftenreihe 32: 243–248.

Gutser R and Dosch P (1992) Strategien zur Optimierung der Stickstoffwirkung von Flüssigmist. Tagung "Neue Techniken zum umweltgerechten und wirtschaftlichen Einsatz von mineralischen und organischen Düngern", 11 Nov 1992, Neumarkt, Landtechnik Weihenstephan, Schrift 2, 38–48.

Gutser R and Claassen N (1994) Langzeitversuche zum N-Umsatz von Wirtschaftsdüngern und kommunalen Komposten. Mitt Dtsch Bodenkundl Gesellsch 73: 47–50.

Isermann K (1993) Territorial, continental and global aspects of C, N, P and S emissions from agricultural ecosystems. In: Wollast R, Mackenzie FT and Chou L (eds) Interactions of C, N, P and S biogeochemical cycles and global change. Springer-Verlag, Heidelberg: NATO AS, Series 14, 79–121.

Messner H and Amberger A (1988) Composition, nitrification and fertilizing effect of anaerobically fermented manure slurry. Proc IV CIEC Symp 1987, Braunschweig-Völkenrode 1: 125–130.

Nemeth K (1982) Electro-ultrafiltration of aqueous soil suspension with simultaneously varying temperature and voltage. Plant Soil 64: 7–23.

Vilsmeier K, Gutser R and Amberger A (1988) Zur Wirkung des NH_4-^{15}N aus Gülle und Ammoniumsulfat in Modell- und Gefäßversuchen. Landw Forsch, Kongreßband 1987, VDLUFA-Schriftenreihe 23: 297–312.

Vilsmeier K and Gutser R (1990) N-Ausnutzung nach Mineral- und Gülledüngung (^{15}N) im Weihenstephaner Lysimeter. Landw Forsch, Kongreßband 1989, VDLUFA – Schriftenreihe 30: 175–180.

C. Rodriguez-Barrueco (ed.), Fertilizers and Environment, 351–353.

An evaluation of three municipal solid wastes composts

H.M.F. Ribeiro[1], E.F. d'Almeida Duarte[1], M. Baião[2], E. Rola[2] & M.C. Vaz[2]
[1]*Instituto Superior de Agronomia, Tapada da Ajuda, 1399 Lisboa Codex, Portugal;* [2] *Laboratório de Análises, Instituto Superior Técnico, Lisboa, Portugal*

Key words: Municipal solid waste compost, nutrients, heavy metals

Abstract

Treatment of municipal solid wastes by the composting technology is increasing in Portugal, as it seems to be a good alternative for the valorization of this wastes. In Portugal this technology started in the 1970s and at present three "industrial composting units" are implemented.

The main purpose of this paper is to evaluate the level of nutrients and heavy metals in compost samples produced in these three units. The correct determination of heavy metals content implies the optimization of samples pre-treatment conditions (microwave and classical wet digestion systems) and the application of different instrumental analytical methods (ICP and AA) in order to get reliable data.

The results obtained in this study allow us to advise about the correct utilization of this type of compost as a soil amendment, avoiding soil pollution.

Introduction

Soil compaction, soil structure deterioration and an accompanying decrease of soil fertility are world-wide problems (Barnes *et al.*, 1971; Wood, 1985). These are related to the decrease of the organic matter in cultivated soils (Mann, 1986). There are many reasons for turning the organic fraction of waste into a compost product, refined to a greater or lesser degree depending on its use. First, composting can have a favorable impact on other disposal options being employed (incineration, for example). Second, with today's escalating landfill costs and the diminishing space that can be licensed for landfill, composting can save landfill space. Advantages of composting organic wastes include (Thurgood, 1994):

- Volume reduction by 30 to 40 percent;
- Creation of a fertilizer/soil conditioner product;
- Sanitation and stabilization of wastes.

In Portugal the use of compost as soil conditioners seems very interesting to improve the quality of Portuguese soils once they are, in most of the cases, very poor in organic matter.

Materials and methods

The samples were collected in three industrial composting units implemented in different regions of Portugal.

The samples were dried to constant weight and subsamples were ground using a stainless steel centrifugal mill.

For the determination of metals each sample was digested by two different processes:

- Classical (nitric/hydrochloric acids);
- Microwave (Milestone 1200 mega unit).

Atomic absorption and inductively coupled plasma (ICP) were used to quantify calcium, magnesium and heavy metals (cadmium, chromium, copper, iron, lead, manganese, nickel and zinc).

All the other parameters – dry matter, humidity, total carbon, conductivity, pH, nitrogen, phosphorus, sodium, potassium – were quantified by the usual techniques described in Official Methods of Analysis (AOAC, 1990).

Results

The results obtained in this work are presented in the following tables:

Table 1. Physical and chemical characteristics of composts

	Compost 1	Compost 2	Compost 3
Humidity (%)	29.1	48.5	49.9
Organic matter (% D.M.)	44.7	50.7	48.3
Ashes (% D.M.)	55.3	49.3	51.7
Ratio C/N	13	21	20
Conductivity (mS/cm) (1:5)	4.6	4.3	6.3
pH	8.4	8.0	6.4

Table 2. Chemical characteristics of humic substances from compost

	Compost 1	Compost 2	Compost 1
Total organic carbon (Corg) (%)	25.8	29.3	27.9
Total humic carbon (Cext) (%)	9.8	7.7	11.5
Humic acid carbon (CHA) (%)	5.7	3.8	4.0
Fulvic acid carbon (CFA) (%)	4.0	4.0	7.5
CHA/CFA	1.43	0.95	0.53
Humification index (CHA × 100/Corg) (%)	22.1	13.0	14.3

Discussion

All the results are going to be compared with the acceptable and quality limits presented in Table 6 (Regione Piemonte, 1992).

- Water content; compost 2 and 3 present a high level of humidity.

Table 3. Nutrients expressed as % of D.M.

	Compost 1	Compost 2	Compost 3
Total nitrogen – N kjh.	1.97	1.40	1.40
Total phosphorus P_2O_5	1.44	1.09	0.63
Total potassium – K	0.59	0.34	0.29
Total calcium – Ca	6.09	2.92	3.16
Total magnesium – Mg	0.24	0.22	0.24
Total sodium – Na	0.49	0.41	0.58

Table 4. Total content of aluminium and heavy-metals (mg/kg D.M.)

	Compost 1	Compost 2	Compost 3
Aluminium	11760	5508	2958
Cadmium	4	2	2
Chromium	70	39	19
Copper	987	615	986
Iron	21530	13409	8505
Lead	882	497	290
Manganese	348	189	184
Nickel	210	35	19
Zinc	1470	1028	1003

- pH; as the Portuguese soils are in general acid the application of alkaline compost as amendment is possible.
- Organic Matter; the organic matter present in the three samples is, in average, 50%, so in accordance with the recommended levels.

Table 5. Available metals – extraction with EDTA and water (mg/kg)

	Water	EDTA	Water	EDTA	Water	EDTA
Aluminium	0.5	1.3	0.5	2.7	0.7	4.9
Cadmium	–	30	–	44	–	48
Chromium	0.8	1.2	1.0	2.8	2.8	6.5
Copper	4.2	28.5	2.4	24.2	1.9	11.1
Iron	0.4	6.6	0.3	22	0.5	24
Lead	0.8	77	0.3	87	–	72
Manganese	1.1	48	1.0	67	1.1	72
Nickel	1.2	3.4	3.1	19	15	27
Zinc	1.2	43.4	1.4	64	0.9	40

Table 6. Analytical Characteristics of Quality Compost (Regione Piemonte, 1992)

Parameters	Acceptable limits	Quality limits
pH	5.5–8	5.5–8
Chloride (mg/kg)	5000	2000
Sulphate (mg/kg)	10000	5000
Humidity (%)	40	40
Organic matter (%)	> 40	> 40
Organic carbon (%)	> 20	> 20
Nitrogen tot. min. (%)	0.8	1.7
Nitrogen tot. max. (%)	2	3
Nitrogen amm. (%)	0.06	0.06
Nitrogen nitric (%)	> 0.04	> 0,04
C/N ratio	30	20
Humidity degree (%)	> 30	> 50
Humic/Fulvic acids (%)	> 1	> 1.5
Phosphorus tot. (%)	> 0.5	> 1
Potassium tot. (%)	> 0.4	> 0.7
TOTAL METALS		
Arsenic (mg/kg)	10	2.5
Cadmium (mg/kg)	20	5
Chromium (mg/kg)	750	500
Copper (mg/Kg)	1000	500
Mercury (mg/kg)	10	2.5
Nickel (mg/kg)	300	150
Lead (mg/kg)	750	350
Selenium (mg/kg)	2.5	2.5
Zinc (mg/kg)	3000	1500
Boron (mg/kg)	100	40

- C/N ratio; the value of this ratio is lower than the quality limit in one of the samples (compost 1).
- Nitrogen; nitrogen is one of the essential nutrients in plant production; all the values are acceptable.
- Phosphorus; the P_2O_5 values found range around 1%, so in accordance with recommended values.
- Potassium; the content of potassium of these samples is low.
- Calcium and Magnesium; these values (4% CaO and 0.8% MgO) are in accordance with established values.
- Heavy metals; the mean values are in the range of acceptable limits, but in general are very high according to quality limits.

Once the available fraction of heavy metals is lower than the total content, we have simulated extraction with water and EDTA (Table 5). The values obtained show that only a low fraction is available.

Conclusions

According to the agronomic value we conclude that compost 1 shows a greater degree of maturation once the index of humification and the ratio CHA/CFA are higher combined with a low C/N ratio (Tables 1 and 3). Compost 2 and 3 shows a low degree of maturation.

Although the results obtained are interesting we need to complete this work doing the determination of other parameters such as arsenic, boron, mercury, selenium and organic chemicals (PAH, PCB's and Pesticides) to evaluate the quality of the compost.

On the other hand we have to improve the sample preparation (digestion process) to optimize the analytical results. In this work we obtained similar results with classical and microwave digestion processes so, in future work, we will choose the last one that is quicker.

The application of compost as a soil amendment needs a correct characterization of soils and compost in order that the right quantities are applied.

References

Barnes KK, Carlton WM, Taylor HM, Throckmorton RI and Vanden Berg GE (1971) Compaction of Agricultural Soils. Amer Soc Agric Eng, St. Joseph, MI, USA.

Mann LK (1986) Changes in soil carbon stooge after cultivation. Soil Science 142: 279–288.

AOAC (1980), Official Methods of Analysis of the Association of Official Analytical Chemists 13th ed. AOAC, Washington, D.C., U.S.A.

Regione Piemonte (1992) Metodi di Analisi dei Compost. Collana Ambienti. Regione Piemont, Torino, Italy.

Thurgood M (1994) Technology Creates New Era For Composting. World Wastes, International Edition 2 (2): 20–22.

Wood AW (1985) Soil degradation and management under intensive sugarcane cultivation in north Queensland. Soil Use and Management 1: 120–124.

C. Rodriguez-Barrueco (ed.), Fertilizers and Environment, 355–361.

Effect of agricultural practices on the nitrogen losses to the environment

C. Ramos
Instituto Valenciano de Investigaciones Agrarias, Moncada, Spain

Key words: agriculture, environment, nitrogen cycle

The environmental significance of nitrogen emissions from agriculture

Nitrogen losses from agricultural fields affect the quality of both water and air. In relation to water quality the main concerns are the nitrate concentration in groundwater and total N concentration in surface waters. Nitrate concentrations above 50 mg l^{-1} make water not appropriate for drinking because it can produce methaemoglobinaemia in infants (Fraser and Chilves, 1981); another health problem associated to high nitrate concentration in drinking water is stomach cancer (Joossens and Geboers, 1981) although the evidence is not conclusive (Addiscott *et al*, 1991).

Eutrophication or nutrient enrichment of surface waters, mainly in nitrogen and phosphorus, is an important environmental problem in many areas because it produces excessive algal growth that can cause death of fish by creating anoxia conditions when the algae die and bacteria decompose them using the dissolved oxygen; also, algal growth is a nuisance for the recreational uses of these waters. In addition, eutrophication changes the proportion of aquatic organisms: macrophytes decrease and algae proliferate; some of the latter are toxic to fish and mammals and can cause taints in drinking water (O'Riordan and Bentham, 1993).

The main gaseous nitrogen compounds produced in agricultural soils that have atmospheric relevance are the nitrogen oxides and ammonia. Nitrous oxide (N_2O) and nitric oxide (NO) are produced in the denitrification reaction, mostly a microbial mediated process, and also in the nitrification reactions.

Nitrous oxide has, indirectly, a negative effect on the stratospheric ozone layer since there it can be converted to nitric oxide that, in turn, reacts with ozone.

The concentration of N_2O in the atmosphere was 310 ppbv in 1990 and its rate of increase has been estimated in 0.25% yr^{-1}. Although N_2O absorbs infrared radiation, its contribution to global warming has been estimated to be only 6% (Houghton *et al.*, 1990). NO and NO_2 (nitrogen dioxide) are also involved in acid rain (Fowler *et al.*, 1982).

Atmospheric ammonia plays a key role in atmospheric acidity and therefore in the 'acid rain' problem; it has been estimated that NH_3 emissions to the atmosphere are capable of neutralizing about 70% of the acid originally present in acid rain in large parts of Europe (EMEP/CEC, 1984). More complex effects of NH_3 on atmospheric chemistry are discussed in Derwent *et al.*, 1988).

Direct toxic effects of ammonia in plants, mainly conifers, have been observed especially when high NH_3 concentrations are combined with low temperatures (Roelofs and Houdijk, 1990).

Ammonia in the atmosphere is rapidly dissolved in the water microdroplets and converted to NH_4^+ with SO_4^{2-}, Cl^- and NO_3^- as the main counterions. Land deposition of NH_3/NH_4^+ from the atmosphere may provide about 10–20 kg N ha^{-1} yr^{-1} (Derwent *et al*, 1988), although values of 35–40 kg N ha^{-1} yr^{-1} have been reported (Goulding, 1990).

The contribution of fertilizers to total NH_3 emissions from agricultural systems (including livestock operations) is small; in a study in U.K. this contribution was estimated to be about 5%, in comparison to about 33% from land spreading of livestock wastes (Jarvis and Pain, 1990).

The nitrogen cycle in agricultural soils

To understand the effect of agricultural practices on nitrogen losses to the environment it is necessary to know the various processes that interact in the transport and transformation of N in soils, also known as

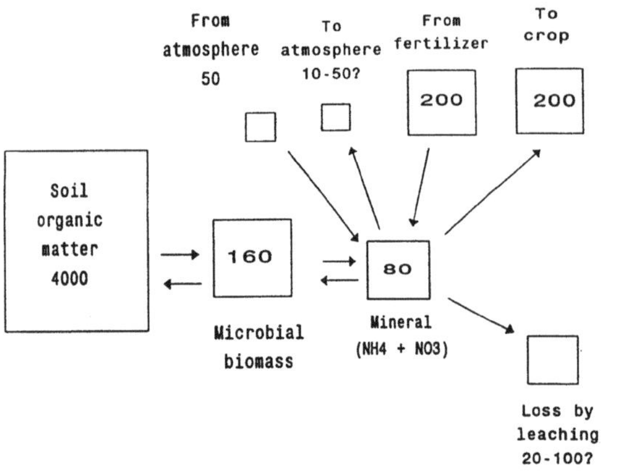

Fig. 1. The nitrogen cycle in soil for an arable field. The quantity of N in each pool (kg ha^{-1}) or in each process (kg ha^{-1} yr^{-1}) is proportional to the size of the square (after Powlson, 1993)

the 'nitrogen cycle'. Two recent reviews of this cycle are in Powlson (1993) and Vinten and Smith (1993). On a global scale the N cycle has been discussed by Jenkinson (1990).

Some important points to consider in the N cycle in agricultural soils are:

1) In most soils, organic matter contains about 95% or more of the total N. This implies that the amount of organic N in the first 50 cm of soil ranges from 6000 to 8000 kg ha^{-1}.

2) Small changes in the organic N pool by mineralization or immobilization result in significant amounts of mineral N in comparison to N fertilizer inputs.

3) The main processes of N loss from soil are nitrate leaching, denitrification and ammonia volatilization.

4) Plant growth has an important role in the N cycle since it can be the main sink of soil N and also affects the soil water balance that, in turn, influences directly nitrate leaching.

Fig. 1 shows a diagram of the N cycle for an arable field in U.K. that could be representative of many arable fields from northern-west Europe (Powlson, 1993). In this example, the main unknown values correspond to nitrate leaching and gaseous losses, that are difficult to measure (Powlson, 1993).

Next, the effects of agricultural practices on the main processes in the N cycle will be discussed.

Nitrate leaching

This is one of the main leaks of the N cycle in soils. Its importance depends on type of crop, cultivation practices, soil and climate. Recent reviews on nitrate leaching are those by Smith *et al.* (1990), Addiscott *et al.* (1991), Vinten and Smith (1993), Parkinson (1993), Powlson (1993) and Shepherd *et al.* (1993).

Fertilizer effects

Three main aspects will be considered: fertilizer rate, chemical form, and timing.

Fertilizer rate

The general relationship between fertilizer rate and nitrate leaching potential is presented in Fig. 2. In this figure, residual nitrogen is that nitrate remaining in soil at the end of the cropping period (harvest time); this nitrate can be leached by the winter rains. Fig. 2 shows that residual nitrate increases rapidly when fertilizer rates are above those that give maximum yield and N uptake by the plant. Similar evidence to that in Fig. 2 has been obtained by others (Bock and Hergert, 1991; Chaney, 1990; Pratt, 1984).

The assumption that the residual nitrate in soil after harvest is related to nitrate leaching induced by rain during the following months is supported by the results of Bergström and Brink (1986). In some cases, the observed increment in the residual mineral N at harvest with N fertilizer application rate is not as high as that in Fig. 2 (Neeteson, 1995).

Chemical form

There are not many studies on the effect of the chemical form of N in fertilizer on nitrate leaching, probably because the more reduced forms of N (ureic, ammonium) are converted to nitrate in the soil quite rapidly, depending on climate and soil factors. In temperate climates, transformation of urea or ammnonium fertilizers to nitrate is fast enough to prevent observing any difference in nitrate leaching, except when some important rain or irrigation occur shortly after fertilizer application.

Dilz (1988) reviewed the effect of timing and N source type on the N-use efficiency for several important crops in the Netherlands and concluded that calcium ammonium urea was, in general, superior to urea, and that splitting of applications tended to increase N-use efficiency and/or reduce nitrate leaching. Baud-

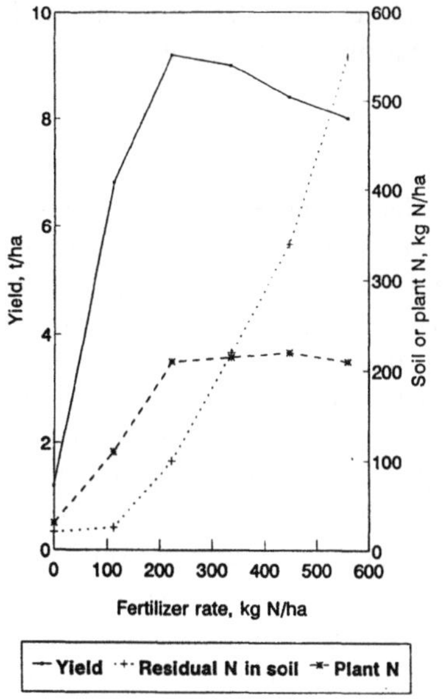

Fig. 2. Yield and N uptake by a maize crop, and residual nitrate in soil (0–240 cm) at harvest in relation to fertilizer application (after Broadbent and Carlton, 1978).

er and Montgomery (1979) reported greater leaching losses when fertilizing with nitrate than with NH_4^+ or urea.

The use of nitrification inhibitors such as nitrapyrin and dicyandiamide (DCD) can delay nitrification of urea and ammonium fertilizers and reduce nitrate leaching (Owens, 1987; Serna *et al.*, 1994).

Timing

This is an important factor for two main reasons: 1) it can increase nitrate leaching if the N fertilizer application is before the rain season and, 2) it can greatly influence N utilization by the crop and, indirectly, nitrate leaching. Juergens-Schwind (1989) reviewed the evidence on the advantages of split N applications to match crop needs in cereals, in relation to nitrate accumulation in soil and, therefore, nitrate leaching potential.

Varshney *et al.* (1993) found that a single application of 175 kg N ha^{-1} to maize resulted in a higher amount of residual nitrate than three applications totalling 125 kg N ha^{-1}, and no difference in yield between these two treatments was observed.

Organic manures

The production of organic manures has increased in the last decades considerably (Wadman *et al.*, 1987) and its inappropriate use on agricultural land can cause important nitrate leaching problems due to (Prins and Wadman, 1990):

- the large amounts of manure applied to land on farms with a high manure/land ratio (i.e. pig and poultry farms).
- the variable N content of manures (average values in U.K. are given by Archer (1991)).
- the uncertainties in mineral N content and mineralization rate of the organic N.
- the uneven distribution at land spreading.
- the time of application.

The effectiveness of N in manure in comparison to that in fertilizer is variable but ranges about 30 to 60% (Smith and Chambers, 1992; Wadman and Neeteson, 1992). Typical values for the fraction of total N in manures that is available for the plant in the season of application are 25% for farmyard manure and 60% for pig slurry (Davies and Archer, 1990).

Nitrate leaching can be one of the main losses from the soil when large amounts of N are applied with slurry or poultry manure, mostly if applied to good drainage soils in autumn (Smith and Chambers, 1993).

Factors that affect the efficiency of manure application and, indirectly, nitrate leaching are: type of application (surface or injected, and delay in incorporation to soil when surface applied), time of application in relation to the rain season, and dry matter content of slurry (Smith and Chambers, 1992).

Irrigation

Irrigation influences nitrate leaching because it determines the amount of water flowing down below the root zone. Experimental evidence relating nitrate leaching to the amount of drainage or deep percolation is given by Pratt (1984) and shows that nitrate leaching increases with drainage. The effects of water management on nitrate leaching have been reviewed by Ferguson *et al.* (1990). In a study on irrigated maize, Ritter *et al.* (1991) found that nitrate leaching was related to drainage volume and that it occurred mainly during the fall and winter months, when most drainage took place; leaching losses during the growing season varied from 34 kg N ha^{-1} for the less irrigated treatment (about 50% of maximum crop evapotranspiration) to 139 kg N ha^{-1} for the well irrigated treatment.

The importance of fall and winter rains for nitrate leaching prompted Smith *et al.* (1990) to propose that irrigation should be managed to obtain a soil water deficit at harvest as high as possible (without reducing yield), since this would reduce drainage in the following rainy months.

Cover crops

Cover crops are legumes, cereals or other crops grown to protect the soil from erosion, to improve soil physical properties, to reduce insects and pathogens through increasing biodiversity, and to reduce groundwater pollution by nutrients or pesticides. These crops are not grown for harvest but to fill periods of time when cash crops would leave the ground bare (Lal *et al.*, 1991).

In a review on the effects of cover crops on nitrate leaching, Meisinger *et al.* (1991) concluded that these crops can reduce the mass of N leached and the nitrate concentration of the leachate by 20–80% in comparison with no cover crop, and that grasses (mainly rye) and brassicas (i.e. mustard, rape, radish) are two to three times more efficient than legumes. However, there are reports showing no effects of cover crops on nitrate leaching (Vinten *et al.*, 1991).

Green manure cover crops contain adequate N to be mineralized shortly after incorporation to soil; the amount of available N that these crops can provide to the following crops ranges about 20–100 kg ha^{-1} (Smith *et al.*, 1990).

Crop residues management

Crop residues contain variable amounts of N that should be taken into account when planning the fertilization rates. A review of the N content of plant residues was made by Meisinger and Randall (1991). Residues of non-leguminous crops do not usually contribute significantly to N supply for the following crops because of their low N content and high C/N ratio. Straw incorporation to soil reduced leaching up to 30–40% (Jarvis *et al.*, 1989) but the long term use of this practice to reduce nitrate leaching has been questioned (Powlson *et al.*, 1987; Catt *et al.*, 1992). Bremer and van Kessel (1992) studied the lentil and wheat residues as a source of N for subsequent crops and found that only the lentil green manure provided a significant amount of N (40% of its N content). Similarly, Smith and Sharpley (1993) found that mineralization of sorghum and wheat residues provided less than 18 kg N ha^{-1} whereas alfalfa residues gave 55 kg N ha^{-1}.

Residues with less than 1.0–1.2% N usually immobilize mineral N (Smith *et al.*, 1990; Vigil and Kissel, 1991); this reduces the potential for nitrate leaching but can create temporary N deficiency.

Tillage

Tillage can affect nitrate leaching because it can modify the soil water balance and soil mineralization rate. Vinten and Smith (1993) suggested that the main effect of tillage on nitrate leaching is that it promotes soil aeration and, therefore, increases mineralization and decreases denitrification. Vinten *et al.* (1991) measured higher nitrate leaching losses from plots that had been cultivated (chisel ploughed and subsoiled) than from plots left in stubble over the winter. Similarly, Goss (1990) found that shallow cultivation or direct drilling reduced leaching losses compared to ploughing. Other researchers have observed small differences in nitrate leaching between conventional tillage and no-till treatments (Ritter *et al.*, 1993).

Denitrification

This is the major biological process by which the nitrogen cycle is completed and fixed N_2 is returned to the atmosphere as NO, N_2O and N_2. Reviews on N losses from soil by denitrification have been published (Aulakh *et al.*, 1992; Smith and Arah, 1990). Measured denitrification losses range from 5 to 50% of the applied N (Nieder *et al.*, 1989).

Denitrification requires the presence of nitrate metabolizable carbon compounds, and anaerobic conditions. In some cases, anaerobic conditions can occur inside soil aggregates even when the bulk soil oxygen content is adequate. Denitrification rates increase with temperature up to 40 °C or higher, and the optimum pH range is 6 to 8. From the environmental point of view, the relevant value is not the absolute denitrification rate but the N_2O production. The ratio of N_2O to N_2 in the gaseous products of denitrification depends on soil water content, amount of available C, and nitrate concentration (Weier *et al.*, 1993). Nitrous oxide can also be produced in the nitrification process, and there are different opinions about the relative importance of these two processes in the production of N_2O: some scientists think that denitrification is a more important source (Wild, 1993), whereas others argue that the relative importance of these pathways may depend on soil conditions, so that, in wet soils, denitrification would be more important but, in well aireated soils, nitrification would be the main source (Smith and Arah, 1990). Recent evidence for this latter view was obtained by Hutchinson *et al.* (1993), who found that, in soils with water potentials ranging from -10 to -1000 kPa (implying good aeration), denitrification made no significant contribution to nitrogen oxide emissions that were attributed to a chemoautotrophic NH_4^+ oxidation by nitrifying microorganisms.

N_2O emissions from soils are weakly related to nitrogen fertilizer application rates (Eichner, 1990). Denitrification losses of 20–40 kg N ha^{-1} yr^{-1} were measured in vegetable fields receiving large amounts of fertilizer N and frequently irrigated (Ryden and Lund, 1980). In another field experiment, farmyard manure application resulted in N losses by denitrification about six times those of the untreated control (Webster and Goulding, 1989).

Tillage also can influence denitrification. Aulakh *et al.* (1992) concluded that denitrification losses from conventional fallow fields are two to seven times higher than from conventional cropped fields, and that, generally, conventional cultivated cropped soils have lower denitrification losses than reduced or no-till soil management.

Straw incorporation to soil stimulates denitrification by supplying high energy material to the denitrifying microorganisms and by increasing soil moisture (Ball, 1990).

Ammonia volatilization

Ammonia volatilization can be an important loss of N to the atmosphere in calcareous soils where high pH conditions are frequent. N losses by NH_3 volatilization when using urea can be very high; losses up to 80% of the N applied as urea have been reported (Fenn and Hossner, 1985; Gould *et al.*, 1986). The more effective way to reduce NH_3 emissions after urea application is by injecting it into the soil (Stevens and Laughlin, 1989). Surface application of an urea solution did not reduce volatilization losses in comparison to prilled urea (Watson *et al.*, 1992). NH_3 volatilization losses of up to 35% of the applied N have been observed after urea application to rice in flooded soils (Buresh *et al*, 1993).

Ammonia losses can be very important when adding organic wastes and manures to the soil. The kind of machinery used in the field influences the volatilization losses during the application of slurry to land but, the more important losses occur after the application (Phillips *et al.*, 1990). Up to 70% of the total loss can occur within the first 24 hr (Jarvis and Pain, 1990). The main factors involved in NH_3 losses after land application of slurry are soil moisture and total solids content of the slurry (Smith and Chambers, 1992). Some of the proposed methods to decrease NH_3 volatilization include: slurry dilution (Sommer and Olsen, 1991), soil incorporation immediately after application (Bless *et al.*, 1991), removal of slurry solids by filtration, acidification, and dilution with water (Stevens *et al.*, 1992).

Nitrogen losses in runoff water

Nitrogen in runoff can be in organic form, associated with the eroded soil, or in soluble form, mostly nitrate. Therefore, all practices that affect soil erosion will have an impact on N losses in runoff. These losses are small, in general, and can be lower than the N input in precipitation (Keeney, 1983). Important losses, however, may occur when a major runoff event takes place shortly after surface application of N fertilizer (Smith *et al.*, 1990).

Tillage is one of the agricultural practices that has a major influence on N losses in runoff. In a six year study to compare no-till and conventional chisel-till soil management in relation to runoff, Chichester and Richardson (1992) found that N losses in runoff were, on average, 4 and 8 kg ha^{-1} yr^{-1}, for no-till and conventional till, respectively. Sharpley *et al.* (1991) measured N in runoff in sorghum watersheds during five years for conventional tillage, reduced tillage, and no-till, and found losses of 7.3, 1.0 and 0.8 kg N ha^{-1} yr^{-1}, respectively.

Conclusions

It is difficult to predict in a given situation the appropriate agricultural practices to reduce N losses, without decreasing yield, because of the complex interactions of the processes involved in the N cycle in soil. To integrate all these phenomena many models have been developed, (i.e. Addiscott *et al.,* 1991; Groot *et al.,* 1991; Hutson and Wagenet, 1991; Shaffer *et al,* 1991) but their practical application is still rare due to different reasons including the high number of data they require, and their limited testing. However, this is a field of active research and good advances in the robustness of these models and in their user 'friendlines' are to be expected. In the mean time, the judicious application of the more general knowledge of the different aspects of the nitrogen cycle in soils should help in making possible a profitable agriculture with a low degradation of the environment.

Acknowledgement

Thanks are due to A.L. Lidón and R. Gazzola for their help in preparing the typescript.

References

Addiscott TM, Whitmore AP and Powlson DS (1991) Farming, Fertilizers and the Nitrate problem. CAB International, Wallingford, UK

Archer JR (1991) Management of livestock wastes for nitrate control. In: Pollution - the New Standard. RASE/ADAS Conference, 6 Feb 1991 (Cited in Parkinson, 1993)

Aulakh MS, Doran JW and Mosier AR (1992) Soil denitrification - Significance, measurement, and effects of management. Adv Soil Sci 18: 1–57

Ball BC (1990) Straw incorporation and tillage methods: straw decomposition, denitrification and growth and yield of winter barley. J Agric Eng Res 46: 223–243

Bauder JW and Montgomery BR (1979) Overwinter redistribution and leaching of fall applied nitrogen. Soil Sci Soc Am J 43: 744–747

Bergström L and Brink N (1986) Effects of differentiated applications of fertilizer N on leaching losses and distribution of inorganic N in the soil. Plant and Soil 93: 333–345

Bless HG, Beinhauer R and Sattelmacher B (1991) Ammonia emission from slurry applied to wheat stubble and rape in North Germany. J Agric Sci 117: 225–231

Bock BR and Hergert GW (1991) Fertilizer nitrogen management. In: Follet RF, Keeney DR and Cruse RM (eds) Managing Nitrogen for Groundwater Quality and Farm Profitability, pp 139–164. Soil Sci Soc Am, Madison, Wisconsin

Bremer E and Van Kessel C (1992) Plant-available nitrogen from lentil and wheat residues during a subsequent growing season. Soil Sci Soc Am J 56: 1155–1160

Broadbent FE and Carlton AB (1978) Field trials with isotopically labeled nitrogen fertilizer. In: Nielsen DR and MacDonald JG (eds) Nitrogen and the Environment, pp 1–41. Academic Press

Buresh RJ, Castillo EG and De Katta SK (1993) Nitrogen losses in puddled soils as affected by timing of water deficit and nitrogen fertilization. Plant and Soil 157: 197–206

Catt JA, Christian DG, Goss MJ, Harris GL and Howse KR (1992) Strategies to reduce nitrate leaching by crop rotation, minimal cultivation and straw incorporation in the Brimstone Farm experiment, Oxfordshire. Aspects of Applied Biology. Nitrate and Farming Systems, pp 255–262

Chaney K (1990) Effect of nitrogen fertilizer rate on soil nitrogen content after harvesting winter wheat. J Agric Sci 114: 171–176

Chichester FW and Richardson CW (1992) Sediment and nutrient loss from clay soils as affected by tillage. J Environ Qual 21: 587–590

Davies DB and Archer JR (1990) Nitrate management in the United Kingdom. In: Calvet R (ed) Nitrates, Agriculture, Eau, pp 511–525. INRA, Paris

Derwent RG, Dollard GJ and Metcalfe SE (1988) On the nitrogen budgets for the United Kingdom and North West Europe. Quart J Royal Met Soc 114: 1127–1152

Dilz K (1988) Efficiency of uptake and utilization of fertilizer nitrogen by plants. In: Jenkinson DS and Smith KA (eds) Nitrogen Efficiency in Agricultural Soils, pp 1–26. Elsevier Applied Sci, London

Eichner MJ (1990) Nitrous oxide emissions from fertilized soils: summary of available data. J Environ Qual 19: 272–280

EMEP/CEC (1984) Summary report from the chemical coordinating centre for the second phase of EMEP rpt 2/84 (cited in Smith and Arah 1990)

Fenn LB and Hossner LB (1985) Ammonia volatilization from ammonium or ammonium-forming nitrogen fertilizers. Adv Soil Sci 1: 123–169

Ferguson RB, Eisenhauer DE, Bockstadter TL, Krull DH and Buttermore G (1990) Water and nitrogen management in Central Platte Valley of Nebraska. J Irrig Drain Eng ASCE 116: 557–565

Fowler D, Cape JN, Leith ID, Paterson IS, Kinnaird JW and Nicholson IA (1982) Rainfall acidity in northern Britain. Nature 297: 383–386

Fraser P and Chilvers C (1981) Health aspects of nitrate in drinking water. Sci Total Environ 18: 103–116

Goss MJ (1990) The effects of soil and crop management on the leaching of nitrates. In: Calvet R (ed) Nitrates, Agriculture, Eau, pp 389–394. INRA, Paris

Gould WD, Hagedorn C and McCready RGL (1986) Urea transformations and fertilizer efficiency in soil. Adv Agron 40: 209–238

Goulding KWT (1990) Nitrogen deposition to land from the atmosphere. Soil Use Manage 6: 61–63

Groot JJR, de Willigen P and Verberne ELJ (eds) (1991) Nitrogen turnover in the soil-crop system. Kluwer Acad Publ, The Netherlands

Houghton JT, Jenkins GJ and Ephraums JJ (eds) (1990) Intergovernmental Panel on Climate Change. Cambridge University Press

Hutchinson GL, Guenzi WD and Livingston GP (1993) Soil water controls on aerobic soil emission of gaseous nitrogen oxides. Soil Biol Biochem 25: 1–9

Hutson JL and Wagenet RJ (1991) Simulating nitrogen dynamics in soils using a deterministic model. Soil Use Manage 7: 74–78

Jarvis S, Barraclough D, Unwin RJ and Royle SM (1989) Nitrate leaching from grazed grassland and after straw incorporation in arable soils. In: Germon JC (ed) Management Systems to Reduce Impact of Nitrates, pp 110–125. Elsevier, Amsterdam

Jarvis SC and Pain BF (1990) Ammonia volatilisation from agricultural land. The Fertiliser Soc, Proc N° 298, London

Jenkinson DS (1990) An introduction to the global nitrogen cycle. Soil Use Manage 6: 56–61

Joossens JV and Geboers J (1981) Nutrition and gastric cancer. Proc Nutr Soc 40: 37–46

Juergens-Gschwind S (1989) Ground water nitrates in other developed countries (Europe) -Relationships to land use patterns. In: Follet RF (ed) Nitrogen Management and Ground Water Protection, pp 75–138. Elsevier, Amsterdam

Keeney DR (1983) Transformations and transport of nitrogen. In: Schaller FW and Bailey GW (eds) Agricultural Management and Water Quality. Iowa State University Press, Ames

Lal R, Regnier E, Eckert DJ, Edwards WM and Hammond R (1991) Expectations of cover crops for sustainable agriculture. In: Hargrove WL (ed) Cover Crops for Clean Water, pp 1–11. Soil and Water Conserv Soc, Ankeny, Iowa

Meisinger JJ, Hargrove WL, Mikkelsen RL, Williams JR and Benson VW (1991) Effects of cover crops on groundwater quality. In: Hargrove WL (ed) Cover Crops for Clean Water, pp 57–68. Soil and Water Conserv Soc, Ankeny, Iowa

Meisinger JJ and Randall GW (1991) Estimating nitrogen budgets for soil-crop systems. In: Follet RF, Keeney DR and Cruse RM (eds) Managing Nitrogen for Groundwater Quality and Farm Profitability, pp 85–124. Soil Sci Soc Am, Madison

Neeteson JJ (1995) Residual soil nitrate after application of nitrogen fertilizers to crops. In: Adriano DC., Iskandar AK and Murarka IP (eds) Contamination of Groundwaters, pp 347–365. Science Reviews, Northwood, UK

Nieder R, Schollmayer G and Richter J (1989) Denitrification in the rooting zone of cropped soils with regard to methodology and climate: a review. Biol Fertil Soils 8: 219–226

O'Riordan T and Bentham G (1993) The politics of nitrate in the UK. In: Burt TP, Heathwaite AL and Trudgill ST (eds) Nitrate: Processes, Patterns and Management, pp 403–416. John Wiley, Chichester, UK

Owens LB (1987) Nitrate leaching losses from monolith lysimeters as influenced by nitrapyrin. J Environ Qual 16: 34–38

Parkinson RJ (1993) Changes in agricultural practice. In: Burt TP, Heathwaite AL and Trudgill ST (eds) Nitrate: Processes, Patterns and Management, pp 321–339. John Wiley, Chichester, UK

Phillips VR, Pain BF and Klarenbeek JV (1990) Factors influencing the odour and ammonia emissions during and after the land spreading of animal slurries. In: Nielsen VC, Pain BF and Hartung J (eds) Odour and Ammonia Emissions from Livestock Production, pp 98–106. Elsevier Appl Sci, London

Powlson DS (1993) Understanding the soil nitrogen cycle. Soil Use Manage 9: 86–94

Powlson DS, Brookes PC and Christensen BT (1987) Measurement of soil microbial biomass provides an early indication of changes in total soil organic matter due to straw incorporation. Soil Biol Biochem 19: 154–164

Pratt PF (1984) Nitrogen use and nitrate leaching in irrigated agriculture. In: Hauck RD (ed) Nitrogen in Crop Production, pp 319–333. Am Soc Agron, Madison, Wisconsin

Prins WH and Wadman WP (1990) Management of animal manures to reduce nitrate pollution of groundwater. In: Calvet R (ed) Nitrates, Agriculture, Eau, pp 313–322. INRA, Paris

Ritter WF, Scarborough RW and Chirnside AEM (1991) Nitrate leaching under irrigation on coastal plain soils. J Irrig Drain Eng ASCE 117: 490–502

Ritter WF, Scarborough RW and Chirnside AEM (1993) Nitrate leaching under irrigated corn. J Irrig Drain Eng ASCE 119: 544–553

Roelofs JGM and Houdijk ALFM (1991) Effects of ammonia. In: Nielsen VC, Voorburg JH and L'Hermite P (eds) Ammonia and Odour Emission from Livestock Production, pp 10–16. Elsevier Appl Sci, London

Ryden JC and Lund LJ (1980) Nature and extent of directly measured denitrification losses from some irrigated vegetable crop production units. Soil Sci Soc Am J 44: 505–511

Serna MD, Legaz F and Primo-Millo E (1994) Efficacy of dicyandiamide as a soil nitrification inhibitor in citrus production. Soil Sci Soc Am J 58: 1817–1824

Shaffer MJ, Halvorson AD and Pierce FJ (1991) Nitrate leaching and economic analysis package (NLEAP): model description and application. In: Follet RF, Keeney DR and Cruse RM (eds) Managing Nitrogen for Groundwater Quality and Farm Profitability, pp 85–124. Soil Sci Soc Am, Madison

Sharpley AN, Smith SJ, Williams JR, Jones OR and Coleman GA (1991) Water quality impacts associated with sorghum culture in the Southern Plains. J Environ Qual 20: 239–244

Shepherd MA, Davies DB and Jonhson PA (1993) Minimizing nitrate losses from arable soils. Soil Use Manage 9: 94–99

Smith KA and Arah JRM (1990) Losses of nitrogen by denitrification and emissions of nitrogen oxides from soils. Fertilizer Soc Proc No. 299, 34 p

Smith KA and Chambers BJ (1992) Improved utilization of slurry nitrogen for arable cropping. In: Archer *et al.* (eds) Aspects of Applied Biology No. 30. Nitrate and Farming Systems, pp 127–134

Smith KA and Chambers BJ (1993) Utilizing the nitrogen content of organic manures on farms - problems and practical solutions. Soil Use Manage 9: 105–112

Smith SJ, Schepers JS and Porter LK (1990) Assessing and managing agricultural nitrogen losses to the environment. Adv Soil Sci 14: 1–43

Smith SJ and Sharpley AN (1993) Nitrogen availability from surface-applied and soil-incorporated crop residues. Agron J 85: 776–778

Sommer SG and Olsen JE (1991) Effects of dry matter content and temperature on ammonia loss from surface-applied cattle slurry. J Environ Qual 20: 679–683

Stevens RJ and Laughlin RJ (1989) A microplot study of the fate of ^{15}N-labelled ammonium nitrate and urea applied at two rates to ryegrass in spring. Fert Res 20: 33–39

Stevens RJ, Laughlin RJ and Frost JP (1992) Effects of separation, dilution, washing and acidification on ammonia volatilization from surface-applied cattle slurry. J Agric Sci 119: 383–389

Varshney P, Kanwar RS, Baker JL and Anderson CE (1993) Tillage and nitrogen management effects on nitrate-nitrogen in the soil profile. Trans ASAE 36: 783–789

Vigil MF and Kissel DE (1991) Equations for estimating the amount of nitrogen mineralized from crop residues. Soil Sci Soc Am J 55: 757–761

Vinten AJA, Howard RS and Redman MH (1991) Measurement of nitrate losses from arable plots under different nitrogen input regimes. Soil Use Manage 7: 3–13

Vinten AJA and Smith KA (1993) Nitrogen cycling in agricultural soils. In: Burt TP, Heathwaite AL and Trudgill ST (eds) Nitrate: Processes, Patterns and Management, pp 39–73. John Wile Chichester UK

Wadman P and Neeteson JJ (1992) Leaching losses from organic manures - the Dutch experience. In: Archer *et al.* (eds) Aspects of Applied Biology No. 30. Nitrate and Farming Systems, pp 117–126

Wadman WP, Sluijsmans CMJ and de la Lande Cremer LCN (1987) Value of animal manures: changes in perception. In: Van der Meer HG, Unwin RJ, Van Dijk TA and Ennik GC (eds) Animal Manures on Grassland and Fodder Crops. Fertilizer or Waste pp 1–16. Dordrecht, Martinus Nijhoff Publ.

Watson CJ, Stevens RJ, Laughlin RJ and Poland P (1992) Volatilization of ammonia from solid and liquid urea surface-applied to perennial ryegrass. J Agric Sci 119: 223–226

Webster CP and Goulding KWT 1989 Influence of soil carbon content on denitrification from fallow land during autumn. J Sci Food Agric 49: 131–142

Weier KL, Doran JW, Power JF and Walters DT (1993) Denitrification and the dinitrogen/nitrous oxide ratio as affected by soil water, available carbon, and nitrate. Soil Sci Soc Am J 57: 66–72

Wild A (1993) Soils and the Environment. An Introduction. Cambridge University Press, Cambridge, UK

C. Rodriguez-Barrueco (ed.), Fertilizers and Environment, 363–367.

Control of nitrate pollution by application of controlled release fertilizer (CRF), compost and an optimized irrigation system

J.A. Diez[1], R. Caballero[1], A. Bustos[1], R. Roman[1], M.C. Cartagena[2] & A. Vallejo[2]
[1]*Centro de Ciencias Medioambientales CSIC, C/ Serrano 115 Madrid, Spain;* [2]*E.T.S.I., Agronomos U.P.M., Madrid, Spain*

Key words: nitrate leaching, pollution, fertilizer, irrigation

Abstract

A nitrogenous controlled release fertilizer (Floranid 32) and a treatment of municipal organic waste compost were tested under two irrigation managements (conventional and ET-adjusted irrigation rates) with the aim of assessing risk of nitrate leaching to the aquifer. A check without N fertilizer was introduced. The experiment was carried out at La Poveda Field Station (30 km SE Madrid, Spain) in alluvial soils with water table depth at 4 m and under maize cropping. The experiment was laid out in a randomized complete block design with three replications, allocating 12 plots to each irrigation management. Although N fertilizer rate (150 kg ha^{-1}) was reduced at half as related to a previous experiment, no difference in grain yields was observed. This result relates to a high content of soil-N. Floranid showed promising results in controlling N-leaching in comparison with urea that exhibited an accelerated rate of N release which finally determines low use of N by the plant and marked NO_3^- leaching. Treatment of municipal waste compost showed NO_3^- concentrations in the soil water solution of similar values as those of urea at 140 cm. ET-adjusted irrigation showed no drainage during the corn growing season and lower NO_3^- concentrations in the soil water solution which could indicate a general lower rate of N solubilization.

Introduction

Widespread and conventional cultivation of irrigated maize in the mid-basin of the Jarama river (Central Spain), includes N application rates above crop requirements and irrigation rates unmatched to crop consumption. NO_3^--N concentration in the main aquifer is near the EEC tolerance limit for human consumption (European Economic Community, 1980).

The effects of both conventional practices have been evaluated in a previous work (Díez *et al.*, 1994). The amount of nitrate leaching to groundwater depends on both the seasonal volume of drainage and on the correspondent concentration of NO_3^--N in the soil-water solution at the deep-drainage layer. Percolation loss of 20% of irrigation water and NO_3^--N concentration in the drainage water up to 20 times the EEC tolerance limit for human consumption were found.

Lower N application rates and alternative sources of fertilizer N have been tested in this work in combination with conventional and ET-matched rates of irrigation. The main objective was to test if these alternative cropping practices would alter percolation loss of NO_3^--N as related to conventional cultivation practices.

Material and methods

An experimental field of 4000 m^2 was laid out at La Poveda Field Station (30 km SE Madrid). Twenty-four 100 m^2 plots were distributed to irrigation management (two levels) and sources of fertilizer N (four levels) with three replications.

The experimental field showed a sandy-loam texture in the first 50 cm of soil-depth and a sand-increasing rate as soil-depth progresses below surface. A dominant gravel-layer appears at variable soil-depth (1.3 to 2.2 m) within the experimental field being the water table at the site located at 4 to 4.5 m below the soil surface.

Top-dressing N was applied at the rate of 150 kg ha^{-1} in agreement with electro ultrafiltration (EUF) soil-N reserve (Nemeth, 1979) and available soil-N

(Wiklicky *et al.*, 1981). Sources of fertilizer N include: urea (46% N), Floranid 32 (a low solubility compound whose nitrogenous component is isobuthylidendiurea) and a municipal waste compost. The latter was applied before maize planting at the rate of 27800 kg ha^{-1} assuming 50% of N availability in the first year. A check without fertilizer N was randomized within the treatment structure in both irrigation systems.

Maize cv. Juanita (Pioneer) was sown in April 1993 in rows 75 cm spacing and density of 70000 plants ha^{-1} and harvested on early December. Prior to seedbed preparation, a compound fertilizer (0N,14P_2O_5,7K_2O) and K_2SO_4 (50%) were applied to the experimental field at the rates of 714 and 100 kg ha^{-1}, respectively.

Conventional irrigation plots received 11 irrigations at mean rates of 52 l m^{-2} and total irrigation water of 572 l m^{-2} (1.28 ET). Optimized irrigation plots received 20 irrigations to a total of 1.03 ET. Total rainfall was 35 l m^{-2} during the ET controlled period.

Volume of soil-water and hydraulic charge profiles were seasonally monitored by a neutron probe and a set of tensiometers, respectively (Roman *et al.*, 1994). The NO_3^--N concentrations in the soil-water solution were measured in samples extracted with a ceramic candel system with tips at soil-depths of 50, 90 and 140 cm. Two ceramic candels were sampled at 140 cm soil-depth.

Water flow and soil moisture profiles were monitored 36 and 55 times in conventional and ET-matched irrigation systems, respectively. These measurements allowed for determination of periods and amounts of water percolation. When downwards flow of water was detected from the deepest soil layer, the volume of drainage water was multiplied by correspondent NO_3^--N concentration at 140 cm. Seasonal integration of drainage periods allowed for an estimation of amounts of NO_3^--N leached to the groundwater (Letey *et al.*, 1987).

Total N in EUF extracts (EUF-N) from soil samples, was determined by digestion with UV radiation and subsequent oxidation with potassium persulphate in alkaline medium (Díez, 1988). The NO_3^- determinations both of the soil-water solution from the ceramic cups, and the EUF extracts, were performed colorimetrically using a Technicon AAII Autoanalyzer with N1 naphtylethylenediamine. P was determined using an ammonium molybdate reagent and K by flame emission photometry.

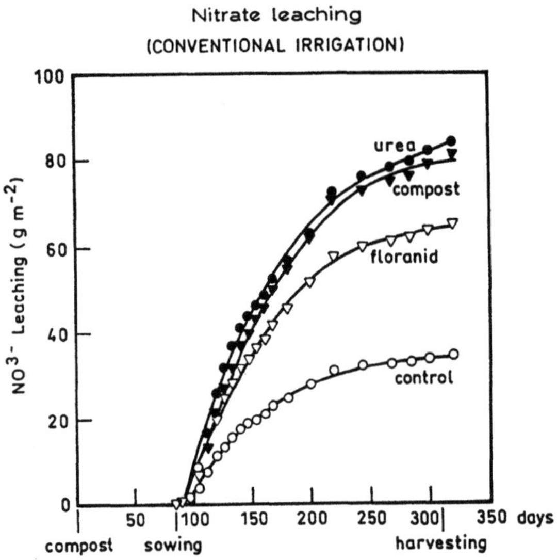

Fig. 1. Cumulative curves of nitrate leaching for each of the treatment during the experiment

Results and discussion

Source of fertilizer N treatments had the greatest impact on NO_3^--N concentration in the soil solution. As expected, the unfertilized check exhibited the lowest NO_3^--N concentration across any soil-depths sampling or irrigation managements.

At 50 cm of soil-depth, the Floranid 32 and compost treatments induced lower NO_3^--N concentrations than the urea treatment when conventional irrigation was applied. However, at this soil-depth, samples of the soil-solution showed similar NO_3^--N concentrations amongst the three fertilizer treatments where optimized irrigation (ET-matched) was applied (Table 1).

At 140 cm of soil-depth, the compost treatment exhibited lower NO_3^--N concentration than both the urea or floranid when the optimized irrigation system was applied. The differences amongst the three sources of fertilizers N were not significant with the conventional irrigation system (Table 1) although Floranid 32 showed lower cumulative leaching trend than the other two sources of N fertilizer (Fig. 1).

A significant irrigation management by source of fertilizer N was found at both 50 and 140 cm of soil-depth sampling. The analysis of variance at the 140 cm of soil-depth (Table 2) also indicated a significant spatial variation within the experimental field which is consistent with the wide range of variation exhibited in this type of experiments when conducted under field conditions (Lund, 1982).

Table 1. Mean values of NO_3^--N (mg l^{-1}) in the soil-water solution at two soil-depths under different irrigation systems throughout the maize growing season

Treatments	Conventional irrigation		Optimized irrigation	
	50 cm	140 cm	50 cm	140 cm
Unfertilized check	92.7 ± 86	193 ± 33	59.3 ± 43	–
Urea	447.2 ± 90	425 ± 53	215.2 ± 71	413 ± 55
Floranid	188.9 ± 54	341 ± 31	170.4 ± 46	375 ± 84
Compost	131.1 ± 61	404 ± 48	187.7 ± 65	219 ± 53

Table 2. Analysis of variance for nitrate concentation (mg l^{-1}) at 140 cm

Source of variation	d.f	Mean square	F. ratio
Irrigation management	1	525.8	0.023
Source of fertilizer N	3	1141000.9	49.515[c]
Replications	5	248028.8	10.764[b]
Interaction	3	463049.1	20.095[c]
Residual	625	23043.3	
Total	637		

[a], [b], [c] signification levels at 5, 1 and 0.1 respectively.

Lower soil moisture with the optimized irrigation system did not lead to higher concentrations of NO_3^--N in the soil solution. By the opposite, it seems that the solubilization process was hampered by low level of soil moisture. This effect was more apparent at the 50 cm than at the 140 cm of soil-depth sampling. At this last level, only the compost treatment exhibited lower NO_3^--N concentrations than the other two sources of fertilizer N. Additional research would be needed to assess the cause of this interesting effect.

Nitrate leaching was evaluated periodically for each of the treatment throughout the experiment. The results corresponding to conventional irrigation management are presented as cumulative curves (Fig. 1). The urea treatments showed more NO_3^--N leaching than the Floranid treatment. It seems that higher values of soil moisture with this irrigation management favours high-solubility fertilizers such as urea, leading to higher levels of nitrate leaching.

Seasonal volume of drainage water when conventional irrigation was applied is shown in Fig. 2. Total percolation to the groundwater was 120 l m^{-2} during the irrigation season. If an extended period is considered (from February to January of the following year) this volume reached 202 l m^{-2}. These last dates corresponded to the application of the compost treatment and to the sowing of the following crop (wheat), respectively. Losses of water to the aquifer were in the order of 20 % of irrigation water applied which confirmed results of a previous experiment (Díez *et al.*, 1994) with conventional irrigation management.

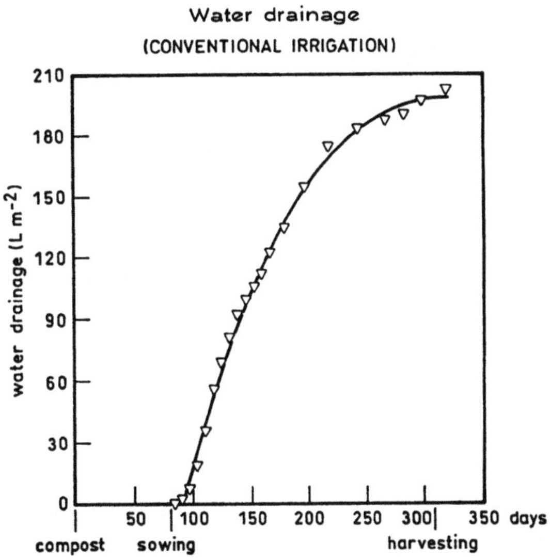

Fig. 2. Seasonal volume of water drainage with conventional irrigation

When the irrigation depth was adjusted to seasonal crop consumption as measured by an erlier experiment (Roman *et al.*, 1994, unpublished) little water volumes reached the aquifer during the crop growing season and thus, nitrate leaching was avoided when this optimized irrigation system was applied.

The unfertilized check showed similar crop production which indicate a high level of soil N reserve before the experiment. Mean values of EUF-N, EUF-NO_3^- and available N in the root zone before sowing

Table 3. Average figures of EUF-N (20 °C + 80 °C), EUF-NO_3^- (20 °C + 80 °C) and N available before sowing (B) and after maize crop harvesting (A)

Treatments	EUF-N		EUF-NO_3		N available	
	mg N 100 g^{-1} of soil				kg N ha^{-1}	
	B	A	B	A	B	A
Conventional irrigation						
Unfertilized	4.70	3.39	1.88	1.46	197	140
Urea	5.18	3.80	2.38	1.55	211	159
Floranid 32	5.42	3.55	2.25	1.49	226	147
Compost	4.72	4.26	1.88	2.09	198	171
Optimized irrigation						
Unfertilized	4.36	3.38	1.67	1.42	184	140
Urea	4.28	3.51	1.51	1.65	183	142
Floranid	5.41	3.45	2.12	1.71	228	138
Compost	4.81	3.37	1.90	1.61	202	136

and after harvesting, indicated a decrease in the soil N reserve across all fertilizer or irrigation treatments (Table 3).

Conclusions

Leaching losses of NO_3^--N must be evaluated by both the interception of water below the root zone and concentration of NO_3^--N on the drainage water. NO_3^--N concentrations at different soil-depth must be of interest to trace the fate of NO_3^--N in the mid or long-run.

Floranid showed better results than the urea in controlling nitrate leaching when conventional irrigation practices were used. The compost treatment showed the lowest NO_3^--N concentration at 50 cm of soil-depth which must be an indication of low potential for nitrate leaching in the mid or long-run.

The optimized irrigation system avoided losses of water through drainage during the maize growing season. Interestingly, somewhat lower concentrations of NO_3^--N in the soil-water solution were found with this treatment than with conventional irrigation system across most sources of fertilizer N.

Acknowledgements

The cooperation by the Regional Government of Madrid for funding Proyect No 191/92 is gratefully acknowledged. Technical assistance and most field works were performed by A. Caballero and P. J. Hernaíz attached to La Poveda Field Station (CCMA, CSIC).

References

Diez JA (1988) Revisión del método de determinación automatizada de Nitrógeno U.V. oxidable en extractos de suelo. Anal Edafol Agrobiol 47: 1029–1039

Diez JA, Roman R, Cartagena MC, Vallejo A, Bustos A and Caballero R (1994) Controlling nitrate pollution of aquifers by using different controlled release fertilizers in maize crop. Agr Ecos Env 48: 49–56

European Economic Community 1980. Council directive on the quality of water for human consumption. Official J 23 No 80/778 EEC L229: 11–29

Letey J, Bigar JW, Stolzy LH and Ayers RS (1978) Effect of water management on nitrate leaching. In Prat PF (ed) Nat. Conf. of Management of Nitrogen in Irrigated Agriculture. Dept Soil and Environ. Sci. Univ. of California Riverside CA 231–249

Lund LJ (1982) Variations in nitrate and chloride concentrations below selected agricultural fields. Soil Sci Soc Amer J 46: 1062–1066

Nemeth K (1979) The availability of nutrients in the soil as determined by electro ultrafiltration(EUF). Adv Agron 31: 155–187

Nielsen DR, Simmons DS and Bigar JW (1979) Flux of nitrate from an spatially variable field soil. In P.F. Pratt (Ed) Nitrate in effluents from irrigate lands. Rep. to Natl. Sci. Fonndation. Nat. Tech. Information Service, Springfield VA. 487–501

Roman R, Caballero R, Bustos A, Diez JA, Cartagena MC and Vallejo A (1994) Field scale fluxes of water and nutrients under conventional corn cropping practices in Central Spain. I Water Balance Soil Sci Soc Amer J (in review)

Wiklicky L and Nemeth K (1981) Düngungsoptimierum mittels EUF-Bodernuntersuchung bei der Zükerrube. Sonderdruck aus Band 106: 982–998

C. Rodriguez-Barrueco (ed.), Fertilizers and Environment, 369–373.

Nitrogen losses by denitrification and leaching in grassland

The effect of cow slurry application

J. M. Estavillo[1], M. Rodriguez[2] & C. Gonzalez-Murua[1]
[1]*Dpto. de Biología Vegetal y Ecología, Universidad del País Vasco, Apdo. 644, 48080 Bilbao, Spain;* [2]*Servicio de Investigación y Mejora Agraria, Dpto. de Agricultura y Pesca, Gobierno Vasco, 48016 Derio (Bizkaia), Spain*

Key words: denitrification, grassland, nitrate leaching, slurry

Abstract

In this study nitrogen losses by denitrification and leaching have been quantified in a natural grassland, comparing the losses produced by a mineral fertilization (calcium ammonium nitrate) and an organic fertilization (cow slurry). It has been stated that in the edafoclimatic conditions of the Basque Country denitrification losses are more important than those produced by leaching, spring and autumn being the seasons with a greater risk of losses. These losses are great and immediately occurring after mineral fertilization and small but distributed along the time when organic fertilization is applied.

Introduction

The inefficient use of nitrogen fertilizers can lead to great N losses that, apart from the economic loss, can lead to non desirable ecological problems. Nitrate leaching can be responsible for the eutrofication of rivers, lakes and seas and the pollution of drinking waters. On the other hand, the nitrous oxide (N_2O) produced by denitrification can play an important role in enhancing the global warming and the ozone layer depletion. Cow slurry is traditionally applied to grasslands and this utilization means a recycling of nutrients in the farm as well as avoiding the utilization of other kind of more expensive fertilizers. In this way, it is important to study and compare which is the fate of the N supplied by slurry and that of the N supplied by a mineral fertilization in order to be able to do future fertilization recommendations.

In the Basque Country there has been no previous study on N losses from grasslands. In this work we present the results from a two-year study in which N losses by denitrification (Estavillo *et al.*, 1994) and leaching were compared between a mineral fertilization as ammonium nitrate and an organic one as cow-slurry.

Table 1. Nitrogen fertilizer application rates (kg N ha^{-1}) as calcium ammonium nitrate (N1, N2) or cow slurry (P1, P2)

Treatment	1991	1992
Control	0	0
N1, P1	145	120
N2, P2	290	240

Methods

The trial was established at Derio (Bizkaia) on a natural grassland and a poorly drained clay loam soil representative of the coastal region of the Basque Country. Rainfall was 1234 mm in 1991 and 1686 mm in 1992.

A cutting system was used in a randomised block design with six replicates and a plot size of 2 × 6 (12 m^2). Two kinds of fertilization were applied: an inorganic one as calcium ammonium nitrate (treatments N1 and N2) and an organic one as cow slurry (treatments P1 and P2) at two different rates. A control plot received no nitrogen. The rates of application are shown in Table 1. These rates were achieved in four different applications of 30 and 60 kg N ha^{-1} along the year. The slurry was analyzed for N, P and K content before each application. Phosphorus (P_2O_5), potassium (K_2O) and water were applied to all except P2 plots in order to make up the P, K and water to the same level on all plots and study the differences due only to nitrogen.

Denitrification was determined weekly or fortnightly in three of the blocks using the acetylene inhibition technique (Ryden *et al.*, 1987). Six soil cores (2.3 cm diameter × 10 depth) per plot were incubated in a 1 l glass jar. A 5% acetylene atmosphere was created and the jars were incubated at ambient soil temperature in holes adjacent to the experimental plots. A sample from the jars' atmosphere was taken after a 24 h incubation period and analysed for N_2O by gas chromatography.

Nitrate leaching was estimated by the utilization of porous ceramic cups. One cup per plot was inserted to 70 cm depth in four of the blocks. Vacuum was made and weekly samples were taken from soil solution and stored at 4 °C to be analyzed for nitrate content (Cawse, 1967). Drainage through the soil profile was estimated by balance taking into account rainfall, potential evapotranspiration and soil water status.

Table 2. Nitrogen losses by denitrification (kg N ha^{-1}) and percentage of the N applied lost by this process (%)

Treatment	1991		1992	
	kg N ha^{-1}	%	kg N ha^{-1}	%
Control	3 c	–	5 c	–
N1	25 b	17	18 a	15
N2	37 a	13	22 a	9
P1	6 c	4	13 b	10
P2	8 c	3	22 a	9

Figures followed by the same letter are not significantly different at $p < 0.05$

Table 3. Percentage (%) of the N applied at each fertilization lost by denitrification

Treatment	1991	1992	
	spring	spring	autumn
N1	37	23	33
N2	26	15	18
P1	5	5	29
P2	3	8	21

Results

Figure 1 shows denitrification daily rates for treatments N2 and P2 along the study. 85 % of the denitrification daily rates were very small (<0.1 kg N ha^{-1} day^{-1}). The greatest losses occurred in spring and autumn, accounting for most of the losses in the year. Although N2 treatment was the one producing the greatest denitrification daily rates, in the second year it can be observed that in the periods of time when fertilizations are not applied it is the P2 treatment showing the greatest denitrification rates when soil is saturated with water (July, October, November and December 1992).

Table 2 shows nitrogen losses by denitrification as an integration along the time. In 1991 losses for the slurry treatments were not different from the control and accounted for less than a 5 % of the N applied. This year losses were 4 times greater for the mineral treatments than for the slurry ones. In 1992, however, losses for the slurry treatments were greater than the previous year and similar to their respective mineral treatments.

Losses in spring and autumn accounted for up to 20–40 % of the nitrogen applied at each fertilization in the mineral treatments as well as in the slurry ones in the second year (Table 3).

Soil water content was the main factor controlling this kind of losses. When soil water content was smaller than 30 % no rate greater than 0.04 kg N ha^{-1} day^{-1} was found (Figure 1).

Figure 2 shows nitrate concentrations in drainage water along the study. These concentrations were greater for N2 treatment. Nevertheless, they were nev-

Table 4. Nitrogen losses by leaching (kg N ha^{-1}) and percentage of the N applied lost by this process (%)

Treatment	1991		1992	
	kg N ha^{-1}	%	kg N ha^{-1}	%
Control	2.0	–	4.5	–
N1	3.8	2.6	4.0	3.3
N2	9.7	3.3	6.9	2.9
P1	3.8	2.6	6.2	5.1
P2	4.5	1.6	5.1	2.1

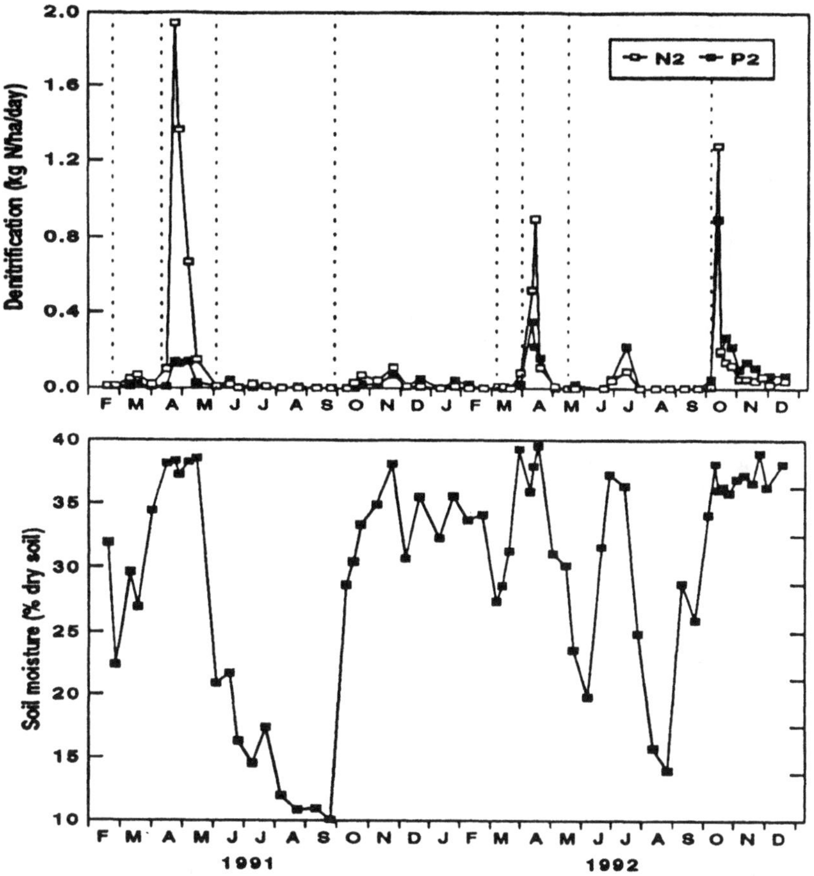

Fig. 1. Daily denitrification rates for N2 and P2 treatments and soil moisture at 10 cm depth along the years 1991 and 1992. Dotted lines indicate the dates of fertilizer applications (Estavillo *et al.*, 1994).

er greater than the 11.3 mg N l^{-1} imposed by the European Community for drinking water. In our case, the highest concentration detected was 9 mg l^{-1} and the mean nitrate concentration for the two-year study was not greater than 4 mg N l^{-1}. When losses were calculated for the two years (Table 4), it was observed that, although there was no significant difference between treatments, losses were greater for N2 treatment, particularly in 1991. However, these losses did never account for more than 5 % of the N applied.

When comparing nitrogen losses by denitrification and leaching (Table 5), they resulted to be similar for the control treatment while denitrification losses were two times greater than leaching ones for P1 treatment, three times greater for P2 and N2 and five times greater for N1 treatment.

Table 5. Number of times that N-losses due to denitrification exceed those due to leaching.

Treatment	1991	1992	Media
Control	1.6	1.1	1.3
N1	6.5	4.6	5.5
N2	3.8	3.2	3.5
P1	1.7	2.0	1.9
P2	1.8	4.3	3.1

Discussion

The small N leaching losses observed in this study are similar to those reported by other authors (Barraclough *et al.*, 1983; Barraclough *et al.*, 1984) who concluded that N application rates under the optimum for grass

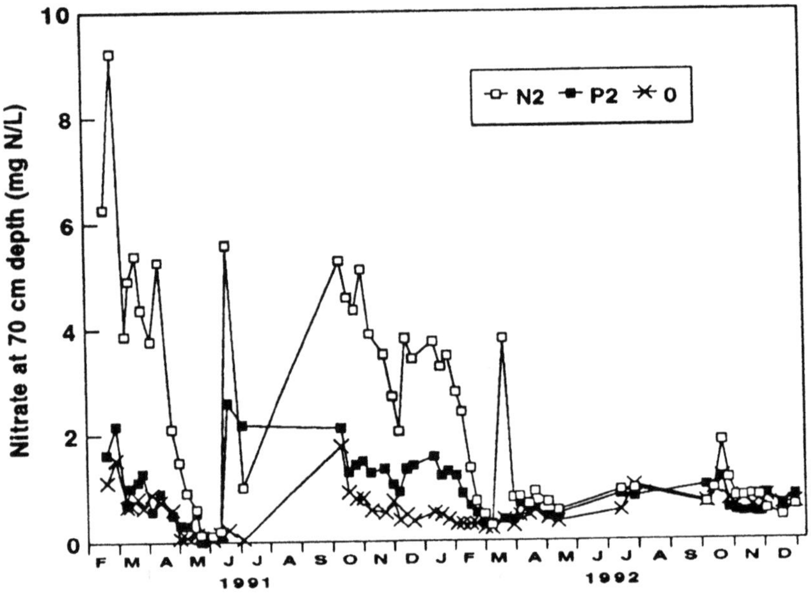

Fig. 2. Nitrate concentration in soil solution at 70 cm depth along the years 1991 and 1992.

production (400 kg N ha^{-1}) were not likely to produce serious ecological problems by nitrate leaching. In our case, the optimum N application rate is well below the 400 kg N ha^{-1} rate reported for ryegrass swards. 240 kg N ha^{-1} may be the optimum for a natural grassland (Gómez-Ibarlucea *et al.*, 1981) being 90–120 kg N ha^{-1}, the recommended rate in order to maintain clover content (Rodríguez and Domingo, 1987).

In this study it is clearly concluded that in our edafoclimatic conditions denitrification is the main way of N losses and so the one to be avoided in order to improve N efficiency. This conclusion concurs with the idea that in a heavy soil greater losses by denitrification than by leaching should be expected because drainage is restricted while flooding is favoured.

The occurrence of similar amount of losses by denitrification for both kind of treatments in the second year but not in the first one was due to soil saturation with water in the periods between fertilizations when the slurry was showing its residual effect because of the mineralization of the organic N supplied. Thus, the basic difference between the slurry and the mineral fertilization is that slurry acts as a slow release fertilizer which supplies nitrate continuously at a small rate while nitrate supply is immediate after mineral fertilization. This is translated into great immediate denitrification losses by the mineral fertilization in the moments of application if soil is saturated with water. On the contrary, slurry applications will produce small losses but extended along the time.

The temporal pattern followed by denitrification in this study fits in with the generalized seasonal pattern of denitrification rates for northern temperate soils commented by Tiedje *et al.* (1989). In this way, although the determination of the kind of fertilization that will produce greater total losses by denitrification will depend on the atmospheric conditions occurring each year, it seems that the greatest risk for losses corresponds to the mineral fertilization, as spring and autumn are seasons in which heavy rainfalls occur coinciding with fertilizer applications. Furthermore, the N_2O / N_2 ratio in denitrification losses has been reported to be greater when high nitrate concentrations occur in the soil (Blackmer and Bremner, 1978; Jenkinson, 1990), which is the case of a mineral fertilization. Thus, although the ratio N_2O / N_2 has not been measured in this study, the potential atmospheric pollution risk of slurry seems to be similar or smaller than that of the mineral fertilization. This, together with the need that exists to recycle animal manures and the reduction of costs when slurries are applied, makes the utilization of slurries an interesting management system from both an economic and an ecological point of view.

Conclusions

In our edafoclimatic conditions N-losses due to denitrification are more important than those due to leaching.

Spring and autumn are the seasons with a highest risk for denitrification, particularly for mineral fertilizations, as applications are made in periods of soil saturation with water.

There is a high risk of immediate losses by denitrification after the application of mineral fertilizations. For slurry applications there is a lower risk of immediate losses but the period of risk is extended due to the mineralization of the organic nitrogen applied.

Nitrate leaching losses are very small (less than 5 % of the N applied) in both kinds of treatment for application rates of up to 240–290 kg N ha^{-1}, being N concentrations in drainage water below the 11.3 mg N l^{-1} limit imposed by the European Commission for drinking water.

References

Barraclough D, Hyden MJ and Davies GP (1983) Fate of fertilizer nitrogen applied to grassland. I. Field leaching results. J Soil Sci 34: 483–497.

Barraclough D, Geens EL and Maggs JM (1984) Fate of fertilizer nitrogen applied to grassland. II. Nitrogen-15 leaching results. J Soil Sci 35: 191–199.

Blackmer AM and Bremner JM (1978) Inhibitory effect of nitrate on reduction of nitrous oxide to N_2 by soil micro-organisms. Soil Biol Biochem 10: 187–191.

Cawse PA (1976) The determination of nitrate in soil solutions by ultraviolet spectrophotometry. Analyst 92: 311–315.

Estavillo JM, Rodriguez M, Domingo M, Muñoz-Rueda A and Gonzalez-Murua C (1994) Denitrification losses from a natural grassland in the Basque Country under organic and inorganic fertilization. Plant and Soil 162: 19–29.

Gómez-Ibarlucea C, García-Gómez A, Prieto V and González-Arranz E (1981) Fertilización fosfo-potásica en praderas naturales de Galicia y Asturias. Anales del INIA. Serie Agrícola 16: 45–55.

Jenkinson DS (1990) An introduction to the global nitrogen cycle. Soil Use and Management 6: 56–60.

Rodríguez M and Domingo M (1987) Fertilización nitro-fosfo-potásica en praderas naturales del País Vasco. Pastos 17: 203–218.

Ryden JC, Skinner JH and Nixon KJ (1987) Soil core incubation system for the field measurement of denitrification using acetylene-inhibition. Soil Biol Biochem 19: 753–757.

Tiedje JM, Simkins S and Groffman PM (1989) Perspectives on measurement of denitrification in the field including recommended protocols for acetylene based methods. Plant and Soil 115: 261–284.

C. Rodriguez-Barrueco (ed.), Fertilizers and Environment, 375–379.

Considerations on main factors which take part in nitrate contamination of ground water in Spain with relationship to other EU countries

S. Ruano Criado
Fertiberia, S.L. Juan Hurtado de Mendoza, 4 28036 Madrid, Spain

Key words: fertilization rates, manure, fertirrigation

Abstract

Analysis of different factors affecting nitrate pollution of waters in Spain and other E.U. countries is presented. Average pluviometry in Spain is ca. 560 l/m^2, whereas in the agricultural countries is over 750 l/m^2. Fertirrigation in Spain, in drip irrigation, is over 200 M ha. Average rates of mineral fertilization are 36–19–15, in kg/ha N-P_2O_5-K_2O, whereas in the E.U. they are 69–28–30. Animal liquid sludges contribution is in relation to the livestock population and, in Spain, density of bovine and swine heads per ha is 0.8, whereas in the E.U. it is 1.3. Presently situation of underground water pollution in Spain is lower than in the E.U. Northern countries but efforts must be enhanced to improve water quality.

Development

To understand correctly the problems of nitrate leaching, nitrogen balance must be taken into consideration, with very clear available data on inputs and outputs in soils.

Humus content in soils is in the range of 1–2%, N content is 5% approx. and the annual mineralization rate is about 1.5%, depending mainly on soil texture. Thus, N supply from O.M. can be around 50 kg/ha per year.

O.M. in the soil is the primary source of N (since it contains 90% of total N) but, in addition, possible supply from solid and liquid manures, crop haulms, straw and green manure (legumes in general) must be taken into account. If we add the supply of N coming from irrigation and rain waters, we get all possible inputs.

Major outputs are also clear. First of all, crop removal, denitrification or gaseous N conversion to nitrate, volatilization or gaseous ammonia emissions to the atmosphere and erosion by washout of the soil upper layer, are becoming a major problem in Spain.

With this scope, direct losses by nitrate leaching can be due partly to the applied fertilizer and partly to the N coming from other sources. Establishing for each 'situation' the balance in the 'time' is essential in order to work towards reducing pollution.

Table 1. Pluviometric data Spain /E.U

Country	Annual rain		
	Mean	Max.	Min.
Germany	742	1739	513
Belgium/Lux	968	1509	852
Denmark	713	1433	553
France	763	1475	546
Netherlands	735	767	689
Italy	769	1529	290
North Spain	654	1189	337
South Spain	464	664	233
Global Spain	556	1189	233

(Mean 30 years – liters/m^2)

From all factors to be considered when we analyse nitrate leaching, rainfall, soil management, chemical fertilization rates, livestock population/organic manuring and agricultural practices/fertirrigation are fundamental. We will analyse these factors comparing Spain to other E.U. countries.

Pluviometry in Spain is different to that in Northern E.U. countries, as it is located in the Mediterranean area (Table 1). Average annual rains in Spain are close to

Table 2. Land distributions 1993

Country	Annual crops	Perennial crops	Grassland	Fallow	Total land
Germany	10400	200	5300		15900
Belgium/Lux	740	20	680		1440
Denmark	2030		480		2510
France	16360	1180	10820	430	28790
Greece	2400	1050	5190	60	8700
Netherlands	850	30	1050		1930
Ireland	360		5180		5540
Italy	8370	2730	4500	5100	20700
Portugal	1190	440	2070	240	3940
U.K.	4440	50	12600	40	17130
Spain	10320	4750	6980	4000	26050
Europe-12	57460	10450	54850	9870	132630

Thousand Has

Table 3. Fertilization rates Spain/E.U. by crop groups (kg/ha nutrients)

Country	Annual crops			Perennial crops			Total ag. area		
	N	P_2O_5	K_2O	N	P_2O_5	K_2O	N	P_2O_5	K_2O
Germany	113	39	56	40	28	60	106	31	43
Belgium/Lux	105	39	88	50	40	60	118	36	70
Denmark	112	26	36				129	25	42
France	121	54	65	30	15	56	76	36	47
Greece	128	52	16	80	50	18	45	21	7
Netherlands	120	53	85	100	40	80	199	36	44
Ireland	116	101	114				64	24	32
Italy	88	58	32	60	45	44	58	39	25
Portugal	84	46	27	48	25	25	33	18	12
U.K.	140	53	62	62	20	28	71	20	25
Spain	62	33	24	32	15	17	36	19	15
Total E.U.	104	47	48	45	27	30	69	28	30

550 mm, while in most European countries they are over 750 mm. In addition, in the last years, rains have decreased sharply due to prolonged drought, wherefore differences are bigger. These climatic conditions in Spain, so much negative for agriculture, are on the contrary positive for possible nitrate leaching.

Land utilization in Spain is also different. Table 2 shows land distribution in several countries. Share of perennial crops and fallow, which recently have increased more than 1 M ha due to 'set aside' policy for small grains and wines, is very high in relation to the total land, and this influences leaching.

Tree crops are a positive factor to prevent leaching, while the extensive acreage without vegetal layer, mainly from fallow, is negative.

One of the points emphasized, we think with very good judgement, by the "Code of Best Agricultural Practices", ready to be published in Spain, is land management referred to crop rotation systems and to land acreage for perennial crops share in relation to annual crops share, as well as keeping during rainy seasons a minimum vegetal layer to absorb N from soils.

The influence of mineral fertilization depends on the application rates as much as on the chemical form of the N applied, timing and method of application. Although fertilization rate is the first condition and perhaps the most important. We must take into consideration the chemical form in which N is applied; it will depend on the time the crop needs the fertilizer and the higher or lower speed of nitrification in the area.

Also, application time is very important to fit the needs of plants, although we must admit that N split applications are always positive. Finally, application method has to be considered since there will be big differences between fertilizer incorporation and homogeneous or not homogeneous broadcasting with standard applicators.

Mineral fertilizer rates in Spain for main crops are much below those used in E.U. countries, basically because yields are lower, but, in some cases, because in Spain some crops are poorly fertilized, such as olive trees, sunflowers, fruit trees, etc. However, other crops are fertilized in excess.

In Table 3, we present details of fertilization rates N-P_2O_5-K_2O per ha for annual and perennial crops and for the total agricultural area in every E.U. country. For the total agricultural area, rates in Spain are 36–19–15, whereas for the total E.U. area they are 69–28–30. Only in Portugal rates are slightly lower.

If we consider fertilization rates in several Autonomic Communities of Spain, referred to cultivated areas (Table 4), the situation is different in Valencia, Murcia, La Rioja and Cataluña, where rates are higher.

If we look at N fertilization rates for annual crops (Table 5), we can see, as it is logical, that rates for wheat and barley are lower in Spain, in accordance to yields. Neither sunflower nor forage crops are properly fertilized. However, irrigated crops are fertilized more in line with other E.U. countries.

We must carry out fertilization tests in our country to find out, for each fertilization rate, the marginal cost and the marginal productivity, clearing up from these tests the technical optimum and the economic optimum, which depend on:

- production potential of the selected variety.
- climatic conditions of the area.
- selling price of the agricultural product.
- purchasing price of the applied fertilizers.

Table 4. N.P.K. Fertilization split by different autonomic communities kg/ha nutrients

Auton. comm.	N	P_2O_5	K_2O
1. Andalúcìa	49	20	16
2. Extremadura	21	10	11
3. Murcia	80	41	48
4. Castilla Mancha	20	11	7
5. Valenciana	124	57	49
7. Castilla León	58	33	20
12. Navarra	73	75	27
13. La Rioja	98	79	51
14. Aragón	49	26	27
15. Cataluña	80	26	50

Cultivated areas in 1990 are considered, not including grassland and pastures.

On the other hand, we have to think over that CAP, GATT, OCM, etc., influence in different ways on the revenue of our crops. Also, fertilizer price increase has been cut, looking at its evolution along the past 5 years. It means that some curves previously calculated do not settle the present situation, its revision being needed.

Manure and liquid animal sludges, supplied to the crops, are directly related to the livestock density in each country. Table 6 shows, for the agricultural season 92/93, that the number of bovine heads in Spain was 5.100 M and swine 18.000 M, lower than in Germany and other countries. As for sheep and poultry, number of heads in Spain was higher than in other E.U. countries.

These data are more meaningful if they are related to cultivated land. Figure 1 shows the number of bovine + swine + sheep heads per ha of cultivated land and it is evident that in 3 European countries (Netherlands, Belgium and Denmark), livestock density is much higher that in the rest of the E.U.

I would like to stop for a moment to look at the liquid animal sludges supply and its importance. The tremendous amount of manure produced in the E.U., whose 20% N is lost by volatilization, requires a tight control, even more if its efficiency is very limited, about 30–50%, depending on the type of cattle (bovine, swine and poultry in descending order), time of application (spring-fall) and its use in grassland and other crops (lower in grassland where 40% of the total is incorporated).

The level of organic residues put in the agriculture today is much higher than in the past years, and

Table 5. Nitrogen fertilization rates (kg N/ha) Spain/E.U. in the major annual crops

Country	Wheat	Bar.	Corn	Pot.	S.B.	Sun.	Leg.	Veg.	For.	Other	Total a.crops
Germany	131	125	127	115	130	47	30	160	76	70	113
Belgium/Lux	130	90	80	150	120		15	100	90	20	105
Denmark	155	90		120	110			150	110	110	112
France	160	130	160	170	145	60	6	100	95	50	121
Greece	110	95	280	190	160	50	60	230	50	140	128
Netherlands	175	90	60	180	130		15	150	50	100	120
Ireland	124	104		140	185				120	60	116
Italy	85	70	185	90	75	33	7	84	74	15	88
Portugal	80	60	160	100			5	125	80		84
U.K.	183	115		181	110	54	4	150	66	61	140
Spain	78	71	170	111	190	5	1	190	15	18	62
Europe-12	130	100	169	138	130	24	7	146	80	61	104

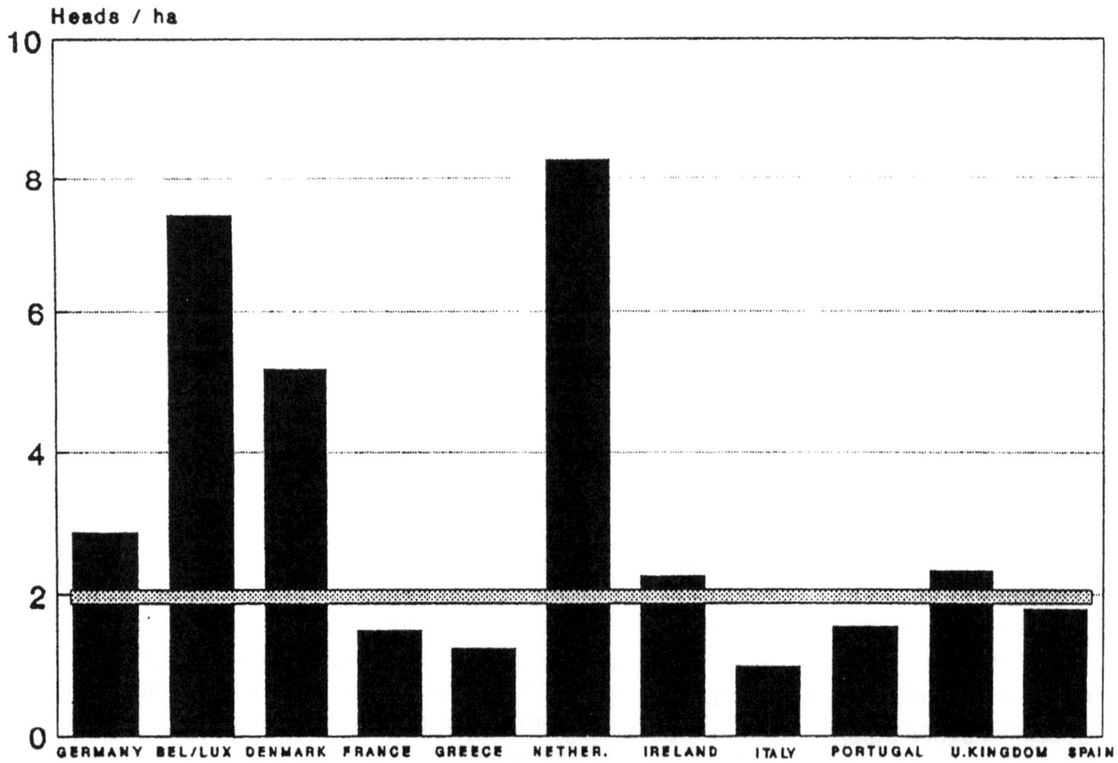

Fig. 1. Weight incidence of livestock (bovine + swine + sheep) by cultivated land (annual + perennial + grassland + fallow)

Table 6. Livestock heads Spain/E.U.

Country	Bovine	Swine	Sheep	Poultry
Germany	16200	26800	3000	104000
Belgium/Lux	3300	7200	200	27000
Denmark	2200	10600	200	13800
France	20300	12900	10300	298100
Greece	500	800	9600	43000
Netherlands	4800	9300	1900	98400
Ireland	6300	1400	4800	11500
Italy	7800	5700	7200	130000
Portugal	1400	2500	2200	156100
U.K.	11800	7800	20600	132700
Spain	5100	18000	24000	585200
Europe-12	79700	103000	84000	159980

1992/93 (.000 heads)

we must pay more and more attention to the nutrient inputs in this way, using the 'method of balances'. In Spanish cereal farms, this method, allowing any farmer to estimate quickly and easily possible N inputs and outputs and thus determine N needs in top-dressing applications, is not well known yet.

One of the agricultural practices which more positively can work to face nitrate pollution is fertirrigation, more and more important in our country, with no resemblance to any European country.

Fertirrigation, mainly drip irrigation, is used in Spain in agricultural areas where there are likely pollution problems and also it is used in crops with higher nutrient needs, thus receiving higher nutrient supply. At present, there are over 200,000 ha, mainly in Eastern coast area and Andalucia; citrus, fruit trees, vegetables and olive trees (more recently) are the most fertirrigated crops.

Fertirrigation allows to apply important nutrient amounts, practically without losses; it is suitable for crops with high frequency and low volume waterings.

Nitrate pollution levels in underground waters in Spain (30% of the drinking water total, whereas the percentage in nearly all E.U. countries is much higher) is generally lower than in northern European countries. More affected areas are: Júcar basin (Castellón plains, littoral plains, Eastern Mancha), Eastern Pyrenees (Maresme), Guadalquivir (Seville, Carmona), Guadiana (Western Mancha, Tomelloso, Daimiel), South (El Ejido, Motril, Salobreña) and Segura (Segura valley and Guadalentín, Cartagena fields).

We must finish our presentation supporting the role that our society plays as environment defender, promoting all actions leading to reduce the pollution level of waters. Implementing in Spain the EEC Directive for protection of waters against pollution, with the prompt publication of the "Code of Best Agricultural Practices" and the definition of 'vulnerable areas', will be surely important steps.

Among other actions leading to reduce pollution, we have already mentioned the method of the balances, better application of liquid animal sludges, fertirrigation promotion and development of sustainable agriculture, a large concept in which many meanings can be introduced, and we support those related to minimum tillage and direct sowing in suitable areas.

References

ABREGO: Fertilizer use magazine edited by ENFERSA. La materia orgánica, el humus y los fertilizantes minerales.

ANFFE: National Fertilizer Manufacturers Association. Desarrollo del mercado de fertilizantes en España.

MAPA: Ministry of Agriculture, Fisheries and Food. Agricultural Statistics Year Book. La agricultura y la contaminación de las aguas por nitratos.

EFMA: European Fertilizer Manufacturers Association. Fertilizer consumption for agricultural use.

UERT: Unión Española de Explosivos, S.A. La fertilización de los cereales.

C. Rodriguez-Barrueco (ed.), Fertilizers and Environment, 381–387.

Phosphate enrichment in the sandy loam soils of West-Flanders, Belgium

J. De Smet[1], G. Hofman[1], J. Vanderdeelen[2], M. Van Meirvenne[1] & L. Baert[2]
[1] *University of Ghent, Dept. Soil Management and Soil Care, Coupure Links 653, B-9000 Gent, Belgium;*
[2] *University of Ghent, Dept. Applied Analytical and Physical Chemistry, Coupure Links 653, B-9000 Gent, Belgium*

Key words: phosphorus, saturation, inventory, leaching, eutrophication

Abstract

The last three decades, pig breeding has evolved towards a specialised, large scaled, land independent bio-industry in the province of West-Flanders. Subsequently, in certain regions, very high amounts of liquid pig manure are produced each year. This pig slurry is used as a fertilizer at a rate which very often exceeds normal agricultural practices. Because of the nonequilibrium between the phosphorus crop requirements and the P-inputs, phosphates accumulate in the soil. However, the phosphate sorption capacity of a soil is limited. Once the sorption capacity is exceeded, phosphates will start leaching through the soil profile. Since, during winter, in these areas, the groundwater table is situated at a depth of less than 1.0 m, phosphate breakthrough might take place. In the sandy loam soil region ($\pm$ 1000 km^2) of the province, an inventory of the P status of the soil was made. The region was sampled according to a regular grid with 2 km intervals. At random, some sample points were only 500 m apart. This resulted in a total of 296 samplings. In view of fertilizer recommendations, lactate extractable P of the plough layer (0–30 cm) was determined. A maximum value of 101 mg P 100 g^{-1} of air dry soil, a minimum value of 6 mg P 100 g^{-1} and a median value of 31 mg P 100 g^{-1} were found, indicating that for half of the spots monitored, the P status of the soil is high to very high. An oxalate extraction was done to investigate the phosphate saturation of the soil profile (0–90 cm). Based on a critical phosphate saturation degree of 30%, more than half of the soil profiles are phosphate saturated. Phosphate leaching at a rate higher than 0.1 mg ortho-P l^{-1} at a depth of 90 cm can be expected. Therefore, a restriction of the P fertilization should be highly recommended. The geostatistical processing of the data using block kriging resulted in a spatial continuous estimate of the phosphate saturation degree. A good agreement was found between the pig density and the phosphate saturation degree of the soil profile.

Introduction

Since the seventies, pig breeding has become an important agricultural activity in West-Flanders. In the province, approximately 3.32 million pigs are present. This intensive animal husbandry is practised in specialised, large scaled, land independent farms. As such, huge amounts of liquid pig manure are produced. Each year, the pig population produces 5.33 million tonnes of slurry, representing a P amount of 6.7 10^6 kg (VLM, 1993). Inasmuch as the slurry is used as a fertilizer on the fields, phosphorus is applied at an average rate of 32 kg P ha^{-1} $year^{-1}$. If the total production of animal manure, including cattle and chicken manure, is considered, a rate of 63 kg P ha^{-1} $year^{-1}$ is reached. Supplemented with some mineral P fertilizers, this amount is far beyond the phosphorus requirements of the crops.

As the soil possesses a capacity to sorb orthophosphates, the excess of phosphate is accumulated in the top soil (Beek and van Riemsdijk, 1982). However, this sorption capacity is limited. Subsequently, repeated excessive applications result in a displacement of phosphates towards the deeper soil layers and towards the groundwater table.

In West-Flanders, the intensive animal husbandry is mainly found in regions with sandy soils. As the P mobility in these soils is relatively high, due to the low sorption capacity, a eutrophication problem may arise in the near future. Already, at some spots, high concentrations of orthophosphates in the drainagewater were measured during winter (De Smet *et al.*, 1990).

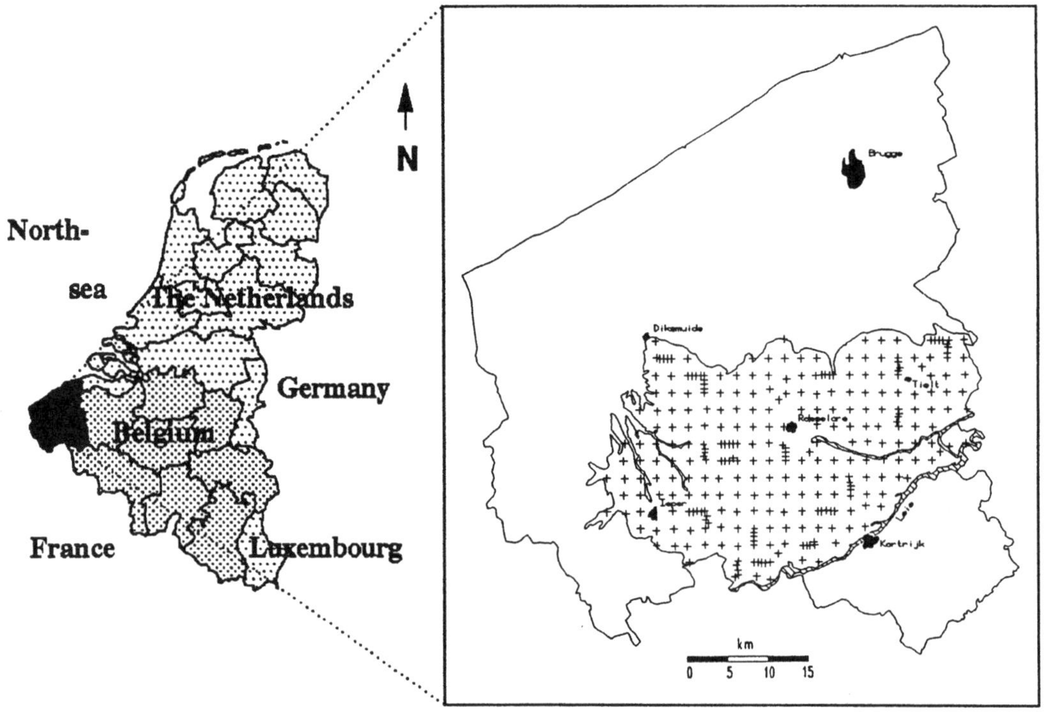

Fig. 1. The light sandy loam soil region of the province of West-Flanders with indication of the sampling locations.

Sorption and desorption of orthophosphate has been studied intensively in order to predict the displacement of orthophosphate in the soil profile (Barrow, 1983; Enfield *et al.*, 1981; Van der Zee and van Riemsdijk, 1986). Van der Zee *et al.* (1990a) describe a model to quantify phosphate leaching in acid sandy soils.

The scope of this study is to make a regional inventory of the phosphorus status of the arable land, both from an agricultural and an environmental point of view.

Materials and methods

The sandy loam soil region of West-Flanders was chosen as study area. The soils occurring in this region are mainly classified as Typic Eutrochrepts or Typic Haplumbrepts according to Soil Taxonomy. The area, 956 km^2, covers 30% of the total surface of the province. Within the region, a regular sampling grid with 2 km intervals was designed. At random, some sampling locations were provided at a short distance (500 m) (Figure 1).

In total, 296 coordinates were sampled. The soil sampling consisted of 5 augerings inside a circular area of approximately 30 m^2. The soil profile was sampled to a depth of 90 cm in steps of 30 cm. Each 30 cm layer, a mixed sample of the five subsamples was made. At the laboratory, the soil samples were dried to the air, ground and sieved to retain only soil particles < 2 mm.

With respect to the plant available phosphorus content of the soil, the samples were extracted with a mixture of ammonia, acetic acid and lactic acid (Egner *et al.*, 1960). The phosphorus extracted is referred to as lactate extractable P.

With respect to phosphorus displacement, the phosphate saturation degree (PSD) of the soil was determined according to Van der Zee *et al.* (1990a):

$$PSD = \left(\frac{P_{ox}}{\frac{Al_{ox} + Fe_{ox}}{2}} \right) \times 100(\%) \qquad (1)$$

P_{ox}, Fe_{ox} and Al_{ox} indicate the number of moles of P, Fe and Al per kg of soil extracted with ammoniumoxalate-oxalic acid (Schwertmann, 1964). In this formula, the ratio $(Al_{ox} + Fe_{ox})/2$ is an estimate of the total phosphate sorption capacity of the soil. The phosphorus concentration in both extracts

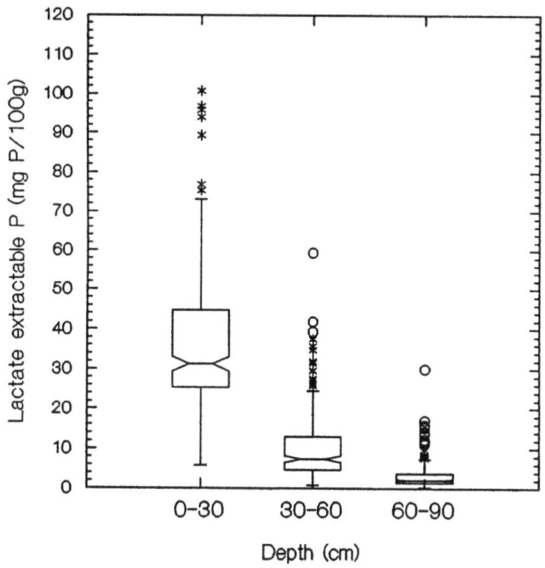

Fig. 2. Box plot presentation of the lactate extractable phosphorus (mg P 100 g^{-1}) in the three soil layers.

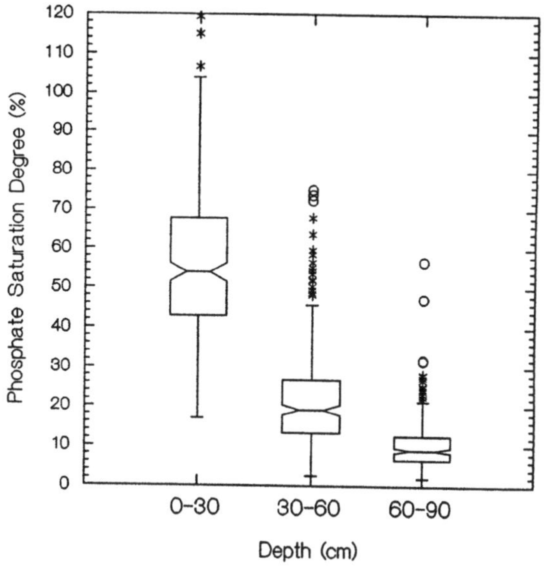

Fig. 3. Box plot presentation of the phosphate saturation degree (%) of the three soil layers.

was measured according to the colorimetric procedure of Scheel (1936). Oxalate extractable P was measured after destruction of the aliquot. The iron and aluminum concentration in the oxalic acid extract was determined by atomic absorption spectrophotometry.

The point data of the phosphate saturation degree were statistically and geostatistically processed. After characterization of the variogram (Burgess and Webster, 1980a), a spatial interpolation was carried out using block kriging (Burgess and Webster, 1980b). This resulted in a spatial continuous display of the phosphate saturation degree of the soil profile.

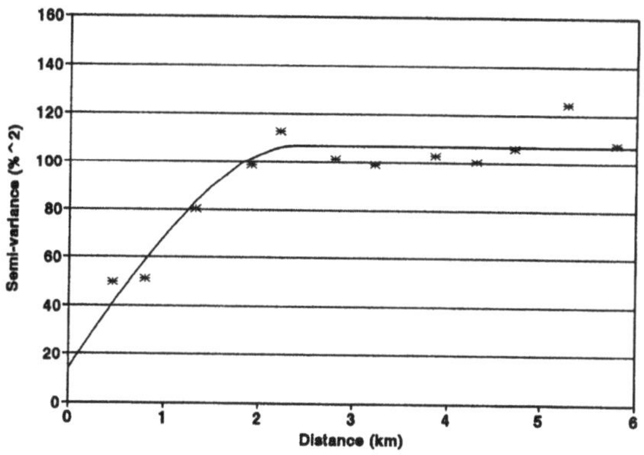

Fig. 4. Semi-variance (%^{2}) of the PSD in function of the distance (km).

Results and discussion

For the three soil layers separately, the distribution of the lactate extractable P is shown in Figure 2. The distribution of the data is presented by means of a box plot. The 25th and 75th percentile comprise the edges of the central box. Inside the box, the median and its 95% confidence interval are indicated. The asterisks and the empty circles represent the so called outside and far outside values (Systat, 1990).

In the layer 0–30 cm, the minimum and maximum amount of lactate extractable P was respectively 6 and 101 mg P 100 g^{-1}. 50% of the measurements were situated between 25 mg P 100 g^{-1} (25th percentile) and 45 mg P 100 g^{-1} (75th percentile). The median, 31 mg P 100 g^{-1}, is lower than the mean, 36 mg P 100 g^{-1}, as the batch of data is not normally distributed (tail to the right).

Regarding the distribution of phosphorus in the soil profile, an accumulation in the top layer is noticed. In the layers underneath, the mean P content drops to 10 resp. 3 mg P 100 g^{-1}.

The agricultural interpretation of the data is done using the evaluation limits proposed by Hendrickx *et al.* (1992) (Table 1).

These limits are based on 50,000 soil analyses performed in the Flemish part of Belgium during the period 1989–1991. In practice, when the plough layer con-

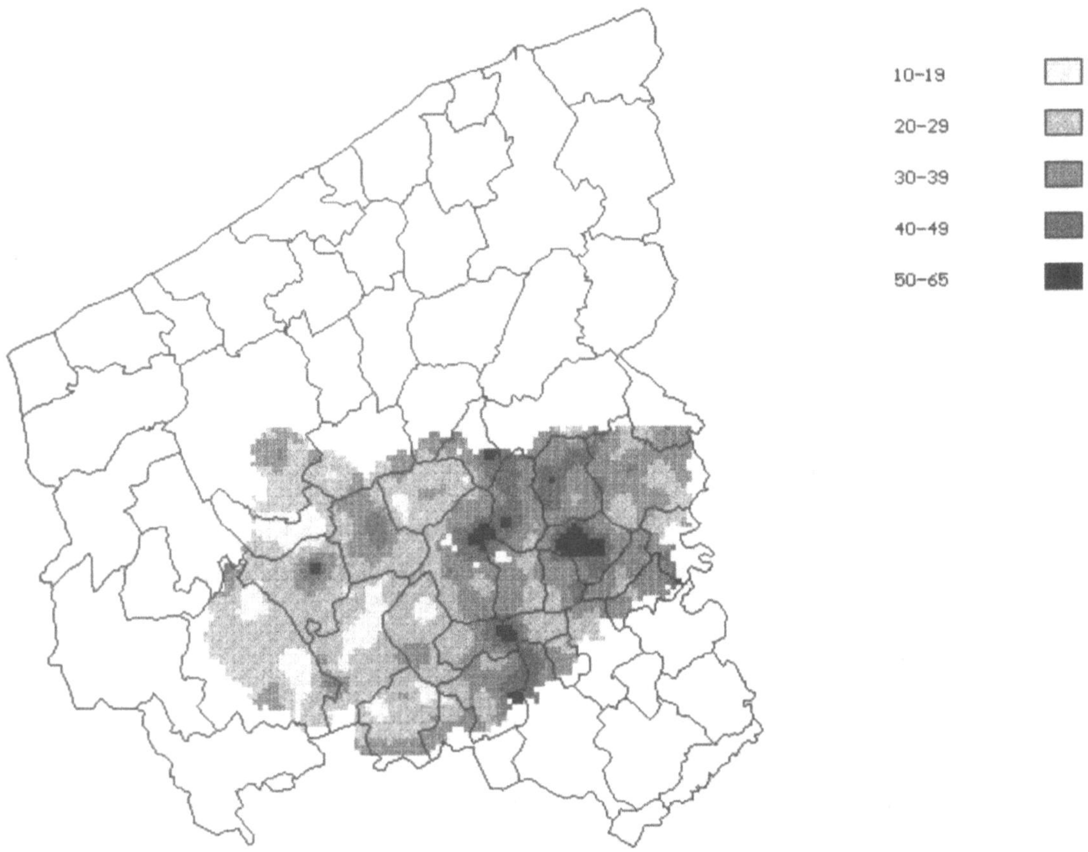

Fig. 5. **The phosphate saturation degree (%) of the profile 0–90 cm for units 500 × 500 m, estimated on basis of 296 measurements.**

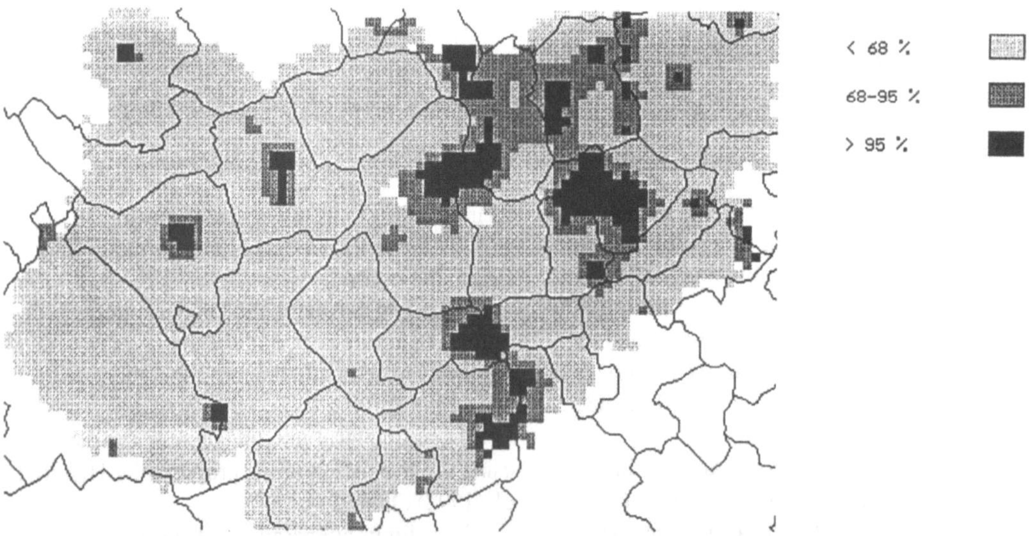

Fig. 6. Probability map (68 % and 95 % probability) of the critical PSD of 30 %.

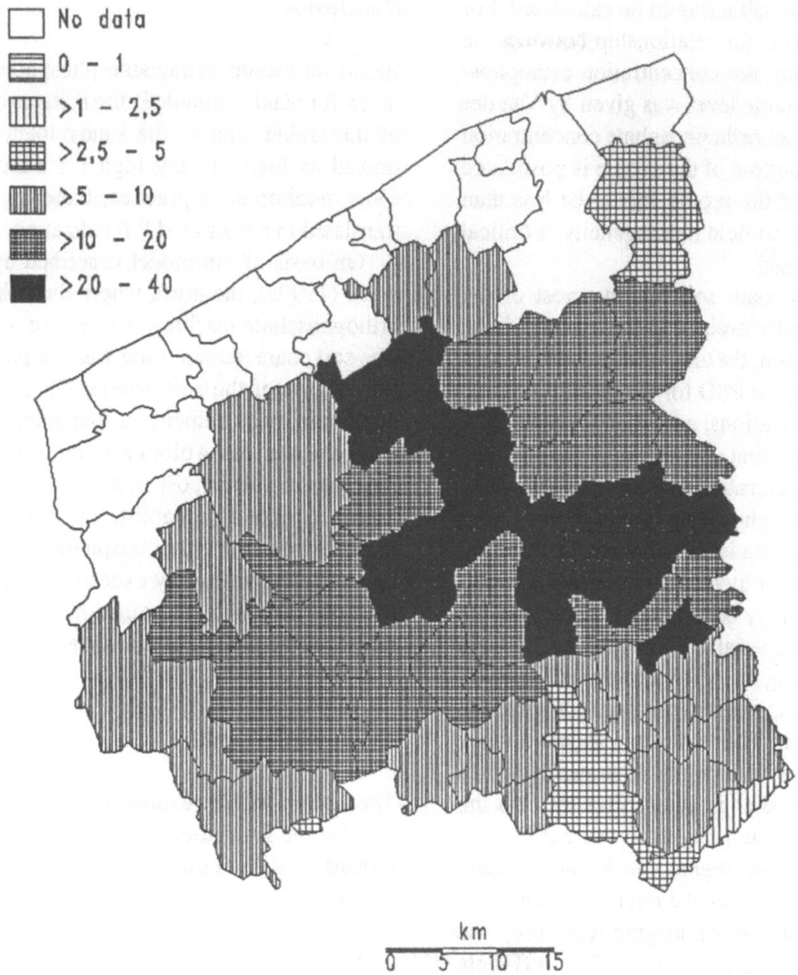

Fig. 7. Spatial distribution of the pig population in West-Flanders (number of pigs per ha cultivated land).

Table 1. Evaluation limits of the lactate extractable phosphorus (mg P 100 g^{-1}) in the top soil of arable land used by the Belgian Pedological Service

Limits (mg P 100 g^{-1})	Evaluation
< 5	very low
5–8	low
9–11	rather low
12–18	optimal
19–30	rather high
31–50	high
> 50	very high

tains more than 30 mg lactate extractable P per 100 g of soil, the phosphorus status is ranked as high and phosphorus fertilization is restricted. As Figure 2 shows, at almost half of the visited spots, P fertilization should be restricted.

In order to evaluate the environmental impact of those high phosphorus contents, the phosphate saturation degree (PSD) of the soil was determined. The descriptive statistical analyses of the PSD of the samples are presented in Figure 3.

In agreement with the quantities of lactate extractable P, in the top soil, very high phosphate saturations were measured. A mean PSD of 57% was found. In the layers 30–60 cm and 60–90 cm, this value drops to 22% and 11% respectively.

As phosphates desorbing from the top layer will be adsorbed in the soil layers below before reaching the groundwater table, the PSD of the total soil profile above the groundwater table should be calculated. For acid light textured soils the relationship between the PSD of the profile and the concentration orthophosphate at groundwatertable level was given by Van der Zee *et al.* (1990b). If an orthophosphate concentration of 0.1 mg l^{-1} at the bottom of the profile is postulated as a limit, the PSD of the profile should be less than 24%. In practice, due to field heterogeneity, a critical PSD of 30% is proposed.

Within the sandy loam soil region, most of the arable land is artificially drained at the average depth of 90 cm. For that reason, the total profile depth was set to 90 cm. Calculating the PSD for the profile 0–90 cm at the 296 different locations, a median value of 28% was found. This means that almost half of the sampled profiles is phosphate saturated. At these spots, elevated concentrations of orthophosphates are expected.

By means of a geostatistical analysis of the data, the areas where a risk for increased phosphate leaching during winter exists, may be indicated. Therefore, the global structure of the spatial variability of the PSD was analyzed by calculating the semivariogram (Figure 4). A spherical model was fitted according to a weighted least squares criterion. The variogram shows a range of 2.4 km: the PSD at two points separated less than 2.4 km one from another is auto-correlated. As the unit dimension of the sampling grid was 2.0 km, the PSD at any point in the region can be interpolated with a precision smaller than the overall variance. As interpolation technique, block kriging was used. The PSD was estimated for units of 500 × 500 m (Figure 5).

For each unit, the PSD is calculated with an accompanying block kriging standard deviation. This standard deviation ranged from 5 to 10%. Assuming a normal distribution of the error around the estimated value, the 68% and 95% confidence intervals around the critical level of 30% can be stipulated (Figure 6). That way, it is calculated that 19% of the region is phosphate saturated with a probability of more than 68%. If a probability level of 95% is postulated, 6% of the area is phosphate saturated. Especially within these critical areas, in which increased phosphate leaching may occur during winter, phosphate fertilization should be reduced.

The spatial distribution of the PSD as presented in Figure 5 reveals a good agreement with the spatial distribution of the pig population in the province (Figure 7). This correlation indicates that the intensive pig breeding is one of the major causes of the phosphate enrichment of the soil in West-Flanders.

Conclusion

Based on lactate extractable phosphorus, used as an index for plant available P, the P status of the top layer of the arable land in the sandy loam soil region is ranked as high to very high for almost half of the spots monitored. In practice, these high amounts are translated in a restricted P fertilization.

On basis of the model described by Van der Zee *et al.* (1990a), the areas where the risk for increased orthophosphate leaching is effective, can be isolated. The soil characteristic used was the phosphate saturation degree of the soil profile. On basis of a limited number of measurements, a spatial continuous display was generated using block kriging. Taking into account the estimation error, 6% of the study area is phosphate saturated with a probability of 95%. In these areas, during winter, the orthophosphate concentration in the upper groundwater may exceed 0.1 mg P l^{-1}. At that moment, the artificial drained fields are an immediate threat for the quality of the surface waters.

Acknowledgements

The authors wish to express their gratitude to the Institute for the Encouragement of Scientific Research in Industry and Agriculture (IWONL) for subsidizing this research.

References

Barrow NJ (1983) A mechanistic model for describing the sorption and desorption of phosphate by soil. J Soil Sci 34: 733–750

Beek J and van Riemsdijk WH (1982) Interactions of orthophosphate ions with soil. In: Bolt GH (ed) Soil Chemistry B, Physico-Chemical Models. Elsevier, Amsterdam, pp. 259–284

Burgess TM and Webster R (1980a) Optimal interpolation and isarithmic mapping of soil properties. I: The semi-variogram and punctual kriging. Journal of Soil Science 31: 315–331

Burgess TM and Webster R (1980b) Optimal interpolation and isarithmic mapping of soil properties. II: Block kriging. Journal of Soil Science 31: 333–341

De Smet J, Hartmann R and De Boodt M (1990) De fosfortoestand van de bodem en het grondwater binnen het arrondissement Tielt. Rijksuniversiteit Gent. Med Fac Landbouwwetenschappen 55: 17–23

Egner H, Riehm H and Domingo WR (1960) Untersuchungen über die chemische Bodenanalyse als Grundlage für die Beürteilung des Nährstoffzustandes der Böden. II. Chemische Extraktionsmethoden zur Phosphor- und Kaliumbestimmung. K. Lantbrukshögsk. Ann. 26: 199–346

Enfield CG, Phan T, Walters DM and Ellis R Jr. (1981) Kinetic model for phosphate transport and transformation in calcareous soils. I: Kinetics of transformation. Soil Sci Soc Am J 45: 1059–1064.

Hendrickx G, Boon W, Bries J, Kempeneers L, Vandendriessche H, Deckers S and Geypens M (1992) De chemische bodemvruchtbarrheid van het Vlaamse akkerbouw- en weilandareaal 1989–1991. Bodemkundige Dienst van België, Heverlee

Scheel KC (1936) Colorimetric determination of phosphoric acid in fertilizers with the Aulfrich photometer. Zeitschrift für Analytische Chemie 105: 256–259

Schwertmann U (1964) Differenzierung der Eisenoxide des Bodens durch Extraktion mit Ammoniumoxalat-Lösung. Z. Pflanzenernährung, Düngung und Bodenkunde 105: 194–202

Systat (1990) Systat: The system for statistics. Evanston, IL: SYSTAT, Inc.

Van der Zee SEATM and van Riemsdijk WH (1986) Sorption kinetics and transport of phosphate in sandy soil. Geoderma 38: 293–309

Van der Zee SEATM, van Riemsdijk WH and de Haan FAM (1990a) Het protokol fosfaatverzadigde gronden. Deel I: Toelichting. Landbouwuniversiteit Wageningen, Vakgroep Bodemkunde en Plantevoeding

Van der Zee SEATM, van Riemsdijk WH and de Haan FAM (1990b) Het protokol fosfaatverzadigde gronden. Deel II: Technische uitwerking. Landbouwuniversiteit Wageningen, Vakgroep Bodemkunde en Plantevoeding

VLM (1993) Mestactieplan, Brussels

C. Rodriguez-Barrueco (ed.), Fertilizers and Environment, 389–391.

Nitrogen and phosphorus losses from soils treated with suspension fertilizers by effect of water draining

C. Benítez, E. Bellido, M.M. Dobao, M. Tejada, J.L. Ruiz & J.L. González
Departamento de Química Agrícola y Edafología, Universidad de Córdoba, Avda San Alberto Magno s/n, E-14004 Córdoba, Spain

Introduction

The vast amounts of fertilizers, particularly those containing nitrogen, that are currently applied to cultivated land call for improved fertilization practices, which essentially entails increasing nutrient availability to plants and hence decrease nutrient losses.

While nitrogen losses from soils are by far the highest among nutrients (soils rarely use more than 60% of all nitrogen in conventional fertilizers), those of other elements typically contained in traditional fertilizers (e.g. phosphorus and potassium) can occasionally be quite significant (e.g. in slightly acidic soils (P) and coarse soils (K), respectively).

Leaching accounts for the greatest nutrient losses from irrigated land supplied with conventional fertilizers. This decreases fertilization efficiency and introduces waterways and aquifers pollution that may eventually lead to eutrophication of continental waters, which has fostered the search for such alternatives as slow-release fertilizers, whose high costs still make them unaffordable for extensive farming.

Also, the ceaseless impoverishment of agricultural soils in organic matter has propitiated the use of a variety of organic fertilizers; while these substantially improve soil properties and fertility, they take some time to mineralize and make their nutrients available to plants, which are therefore subject to nutritional stress at the early vegetative stages. In addition, some of these products (e.g. liquid manure) loose abundant gaseous nitrogen.

All these problems are currently being addressed by testing stable organic-mineral fertilizers in order to provide soils with the amounts of organic matter and inorganic nutrients they require to preserve their fertility and productivity (Rozsypal and Baier, 1989; Espejo and González, 1993) while decreasing the environmental impact of fertilizer element losses.

In this work, we extended previous research (Barón *et al.*, 1992) by studying nitrogen and phosphorus leaching losses from a typical soil of the Guadalquivir Valley that was treated with three types of fertilizer, viz. a granulated mineral fertilizer, a suspension mineral fertilizer and a suspension organic-mineral fertilizer, in order to determine which resulted in the smallest leaching losses and hence was the most effective.

Material and methods

Experiments were run in triplicate by using percolation columns packed with soil treated with one of the three fertilizers studied (conventional fertilizer C-5-5-0, suspension fertilizer 5-5-50 and suspension organo-mineral fertilizer 5-5-0+EH), which were added in a proportion of 75 kg/ha for both nitrogen and P_2O_5. The organic-mineral fertilizer also provided 37.5 kg/ha humic extract. A blank run was also performed on unfertilized soil. The soils were supplied with an overall 2000 mm of water that was evenly distributed between 10 irrigations. The fertilizers used were all commercially available from Agrifluide SA.

Nitrogen fractions were quantified by using the methods of Bremner (1965) and Kempers (1974), while available phosphorus was measured by that of Williams and Stewart as implemented by Guitian and Carballas (1976). All the results are relative to dry soil matter.

Results and discussion

The results obtained in the soil percolation tests revealed that most nitrogen was lost as nitrate. Also, leaching losses (at least 50%) were highest on the first

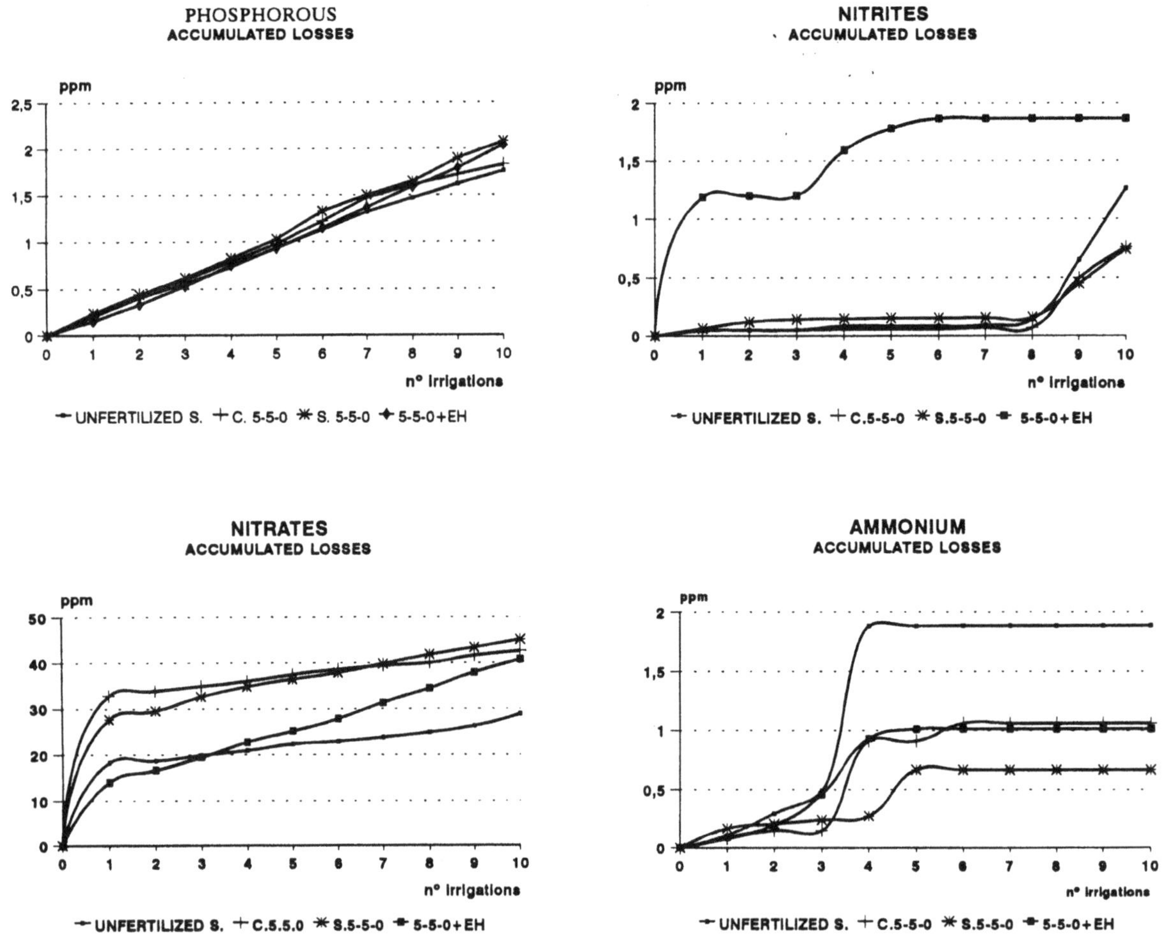

Fig. 1.

irrigation. While the final losses ranged from 30 to 50 ppm N-NO_3^- for the fertilized soils, they were more gradual for the suspension fertilizers (both the mineral and the organic-mineral variety) than they were for the granulated fertilizer, so the former two must be more efficiently used by the soil thanks to their greater resistance to leaching.

The N-NO_2^- levels in the leachates were lower than 2 ppm in every case and were only appreciable for the organic-mineral fertilizer after the first irrigation and a water supply of 1600 mm of water was exceeded. Such low levels for this nitrogen fraction rule out a phytotoxic effect of nitrite and suggest that the experimental procedure avoided anoxia in the percolation columns.

The leaching losses of N-NH_4^+ were always smaller than 2 ppm and became appreciable above a water supply of 800–1000 mm. Also, they were somewhat more marked for the unfertilized soil.

As regards phosphorus leaching losses, they never exceeded 2.5 ppm. Also, they were more gradual than those of the nitrogen fractions; in fact, they increased virtually linearly with the amount of water supplied, which suggests a higher solubility of phosphorus in suspension fertilizers. This is highly consequential for calcareous soils, where retrogradation of phosphorus fertilizers can be quite significant.

Finally, irrespective of the fertilizer, the N-NO_3^-/P ratio was never lower than 90 after the first leaching. This suggests that the leachates do not make an appropriate source for biological production and hence that they will hinder eutrophication, which typically calls for an N/P ratio of ca. 25 but never exceeding 60 (Catalán, 1981). This in turn suggests that the main source of the phosphorus that gives rise to eutrophication in basins and waterways next to calcareous, alkaline agricultural soils is phosphoric fertilizers.

References

Baron R, Benitez IC and Gonzalez JL (1992) Estudio lisimétrico de un suelo fertilizado con abonos orgánicos y organominerales. 24 Reunion Bienal Real Sociedad Española de Química, Torremolinos, España.

Bremner JM (1965) Inorganic forms of nitrogen. In CA Black, DD Evans, JL White, LE Ensminger and FE Clark (eds). Am Soc Agron. Madison, Wisconsin (USA).

Catalan JG (1981) Química del Agua" 2nd Ed. Catalán, Madrid, Spain.

Espejo JA and Gonzalez JL (1993) Influencia del abonado organomineral en algunos aspectos productivos y de calidad del espárrago verde". II Congreso Ibérico de Ciencias Hortícolas. Zaragoza, España.

Guitian Ojea F and Carballas T (1976) Técnicas de análisis de suelos. Editorial Picro Sacro, Santiago de Compostela, España.

Kempers AJ (1974) Determination of submicro-quantities of ammonium and nitrates in soil with phenol, sodium nitroprusside and hypoclorite. Geoderma: 12–20.

Rozsypal R and Baier J (1989) Effect of long-term dfferentiated fertilitation on potassium content in soil. Rostlinná Vyroba 35: 629–637.

C. Rodriguez-Barrueco (ed.), Fertilizers and Environment, 393–395.

Preliminary study of the effect of suspended particulate matter in the atmosphere on solar radiation in Salamanca (Spain)

M.R. Fidalgo Martínez[1], A.M. García González[2] & A. Blanco De Pablos[3]
[1]*Dpto Física General y de la Atmósfera;* [2]*Dpto Física Aplicada, Facultad de Ciencias, Universidad de Salamanca;* [3]*Instituto de Recursos Naturales y Agrobiología, CSIC, Salamanca*

Abstract

This paper reports on the preliminary results for Salamanca (Spain) of the relationship between the different components of solar radiation and suspended particulate matter in the atmosphere, characterized as black smoke, during the period from January to September, 1990. Both the total number of days period and clear days only were considered with a view to eliminating the effect of cloud, above all, on the diffuse component of solar radiation. The relationships with diffuse/global and diffuse/direct indices were also studied.

Introduction

Among other parameters, the Earth's climate depends on the chemical composition of the atmosphere as this regulates the radiation balance of the Earth-Atmosphere system. Atmospheric pollution affects this balance of radiation. In addition, atmospheric aerosols absorb and disperse solar and terrestrial radiation.

Several authors have shown that the different components of radiation are disturbed by the emissions from aerosols and by particles arising from industrial activity and traffic, primarily in urban areas (Galindo, 1984).

The aim of this paper was therefore to study the possible effects of atmospheric aerosols on solar radiation in Salamanca.

Experimental

Daily samples were taken in 1990 (from January to September). Suspended particulate matter – characterized as black smoke – was determined by reflectometry and its concentrations expressed in μg m^{-3}. The site where the samples for the meteorological data were taken was the Institute of Natural Resources and Agrobiology of the CSIC in Salamanca, where pollution measurements were also made.

The atmospheric variables in Salamanca (longitude 5° 40′ W; latitude 40° 58′ N) were measured with a UNIDATA model 6003 automatic data collection system provided with model 6504 C sensors for measuring wind velocity, mean temperature and global solar radiation and using model 6505 for measuring diffuse solar radiation following the addition of a model BS-100 shade band. The radiation was measured in units of W m^{-2}.

Results

The Solar radiation – atmospheric pollution relationship

Initially, a qualitative study was made of the relationship between the daily mean values of global, direct and diffuse radiation and the concentration of black smoke, thereafter, the results which were obtained were checked with regression analysis.

Two series of data were considered:

1) The total number of days studied (series A)
2) Only clear days (series B).

Clear days were considered to be those on which the sum of the three daily observations of cloud ranged from 0 to 4 octas; if this sum lay within 5–19 octas the day was considered to be cloudy and at more than 20 octas it was deemed covered. The aim in this series was to eliminate the effect of cloud on diffuse radiation.

The number of clear days was less than or equal to 5 for each of the months studied with the exception of July and August (17 and 11 days, respectively). According to the climatological study of Roldán

(1987) for Salamanca, the number of clear days in the year studied (1990) was below the annual mean, although the number of hours of sun during the period was greater than the mean.

In order to observe the evolution of direct, diffuse and global solar radiation as the suspended particulate matter in the atmosphere increased, five intervals were set up within series A and series B for the concentrations of black smoke: < 15; 16–25; 26–35; 36–45 and $> 45\ \mu g\ m^{-3}$. For each interval the mean value of the three variables were calculated.

It may be seen that from a smoke concentration of $> 25\ \mu g\ m^{-3}$ the global and direct components of radiation decrease in proportion to the increase in the concentration of smoke while, by contrast, diffuse radiation increases as the values of such concentrations increase. This is more apparent in the case of the series of clear days (B). These findings are consistent with those of work carried out by other authors at different sites (Karras, 1990).

Solar radiation – atmospheric pollution correlations

In order to quantify the relationships between suspended particulate matter and the radiation variables, simple linear regression analysis was performed with the equation:

$$C_h = a + b\,V_r \qquad (1)$$

where C_h is the concentration of black smoke and V_r is the radiation variable considered.

With respect to the A and B series, for series B, the clear days were separated into polluted and unpolluted days. The distinction between these two types of clear days was established at $25\ \mu g\ m^{-3}$ according to the concentration reached by this pollutant in Salamanca during the sampling time. It should be noted that compared with other more polluted areas, these values are not high, although at the same time, it should be mentioned that the period studied in 1990 was less polluted than the corresponding period in previous years.

Table 1 shows the correlations obtained between black smoke and the different radiation variables, including the D/G and D/I ratios. Many authors consider these to be indices of the amount of aerosols or suspended particulate matter linked to atmospheric turbidity (Aguado, 1990).

For the A series, the radiation variables do not show a significant correlation with black smoke, at least with regard to the mean daily values. However, for the B series, these correlations increase when the effect of cloud is eliminated, above all in diffuse radiation, reaching a correlation coefficient with this variable of 0.65. In the case of the D/G and D/I ratios, the correlations obtained were 0.63 and 0.59, respectively. The correlations that proved to be significant at 99.5%, according to the Snedecor F test, are shown in bold type together with their correlation coefficient. There were no significant correlations to the series of unpolluted clear days.

When considering the series of clear days, the increase in the correlation coefficients of black smoke for all the variables is patent. This is especially the case with diffuse radiation (0.81) and with the D/G ratio (0.82), which in the particular case of Salamanca seems to point to an index of suspended particulate matter in the atmosphere better than the D/I (0.75). The simple regression equations for diffuse radiation (D) and D/G are, respectively:

$$C_{smoke} = 0.53D + 12.94; \quad r = 0.81 \qquad (2)$$

$$C_{smoke} = 98.63D/G + 17.90; \quad r = 0.82 \qquad (3)$$

These equations describe the percentages of variance of the dependent variable of 66 and 68%, respectively. For this same series, the correlation coefficient with direct radiation reached a value of −0.60, pointing to the attenuation of this component with the increase in the concentration of suspended particulate matter in the atmosphere.

Conclusions

From this preliminary study of the relationship between the components of solar radiation and suspended particulate matter in the atmosphere in Salamanca, a clear inverse relationship can be seen between global and direct radiation. In the case of diffuse radiation, the relationship is direct and increases with the increase in the concentration of particles. This is confirmed by the signs of the regression coefficients on quantifying these relationships by regression analysis.

For the study of the diffuse radiation, it is clear that only the clear days should be considered in order to avoid, among other factors, the strong effect of clouds. Global and direct radiation do not show any correlation

Table 1. Coefficients of correlation: black smoke – radiation variables

	Complete	Clear days		
		Total	Polluted	Unpolluted
Global radiation (G)	−0.06	−0.15	−0.38	0.39
Diffuse radiation (D)	0.06	**0.65**	**0.81**	0.45
Direct radiation (I)	−0.07	−0.34	**−0.60**	0.25
D/G	0.03	**0.63**	**0.82**	0.18
D/I	−0.06	**0.59**	**0.75**	0.15

worthy of mention, at least when considering the daily mean values, with the exception of direct radiation on polluted clear days.

It may be seen that, for clear days, apart from diffuse radiation itself, the D/G ratio is a good index (better than D/I) of the concentration of suspended particulate matter in the atmosphere of Salamanca. Although the present study covers the period from January to September 1990, it should be noted that the number of clear days useful for the study decreased, in part, owing to the meteorological peculiarities of that period. This, together with the low concentrations measured with respect to other years, and the fact that we were also working with daily mean values, meant that the correlations were not even higher.

Acknowledgements

The authors thank D. Miguel Tapia and D. Agustin Rincón from the Institute of Natural Resources and Agrobiology of the CSIC in Salamanca, without whose help this work would not have been possible.

References

Aguado E (1990) Effect of advected pollutants on solar radiation attenuation: Mojave desert, California. Atmos Envir 24B (1): 153–157.

Galindo I (1984) Antropogenic aerosols and their regional scale climatic effects. In: HE Gerber and A Deepak (eds) Aerosols and their climatic effects. ISBN 0-937194-06-9.

Karras GS, Pissimanis DK and Notaridou VA (1990) On the trend of the transmittance of direct solar irradiance in Athens during the summer. Atmos Envir 24B: 221–225.

Roldán A (1987) Notas para una Climatología de Salamanca. INM Publicación, Madrid, serie K, no. 20.

C. Rodriguez-Barrueco (ed.), Fertilizers and Environment, 397–400.

Dynamics of nutrient transport in a small river of an agricultural water catchment in Northern Germany

M. Kücke & K. Hasenpusch
Institute of Plant Nutrition and Soil Science, Federal Agricultural Research Centre Braunschweig, Bundesallee 50, 38116 Braunschweig, Germany

Key words: river water, nutrient fluxes, dissolved nutrients, solid bound nutrients, rainstorm

Abstract

Nutrient and solid fluxes were recorded in a small river of a water catchment with intensive agriculture from February to June 1993 using sampling intervals of 3 hours to quantify the importance of rainstorm events for the fluxes. Solid concentrations were 0.01 to 0.05 g/l in period with low stream flows but reached 5.8 g/l during 3 hours of one high stream flow event (150 l/s compared to 5 l/s in average). During this event, the concentrations of sediment bound P and K increased in line with stream flow while the peak of dissolve P and K concentrations were recorded with a 3 hours delay after the peak of stream flow. The nitrate concentration was not increased during this event. The majority of the solid, P and K load of the monitoring period was transported during this single rainstorm event (93% of solids, 91% of P and 84% of the K load). Compared to this, only 3% of the total NO_3-N losses occurred during this one day.

Introduction

Nutrient input from arable soils into brooks occurs by drainage water and ground water discharge as well as by surface water runoff. For aquatic life, sediments are argued to be the major pollutant (Granillo *et al.* 1985). The intensity of the nutrient input and turnover in surface water systems either as dissolved nutrients or as sediment bound nutrients depends on the climatic situation (esp. season and intensity of rainstorm events), geology (e.g. soils and their water storage capacity), hydrology, relief of the catchment, and on agronomic practice such as dominant crops, fertilisation intensity and catch stripes around the border of surface waters.

One problem in such investigations is the importance of short-term rainfall events which significantly influence the water discharge, the concentration of suspended sediments and the nutrient flow. The aim of this paper is to point out the change in nutrient and solid concentrations in a brook system during short term rainfall events with special regard to the solid bound nutrients and their importance for nutrient losses which leave the catchments via the brook systems.

Methods

The catchment of the small river Ohebach covers about 0.9 km^2 and is located in the northern foothills of the Harz mountains in Lower Saxony, Germany about 50 km in south of the city of Braunschweig. The surrounding area is exclusively arable land on loam soils derived from loess, predominantly gleyic luvisols (FAO). Most of these soils are drained. The Ohebach is about 1 km long and usually not deeper or wider than 1 m. It gets most of its water from the drainage systems of the surrounding fields. Details of the location, the dominant soil types, geology and the agronomic situation are described by Othmer and Bork (1989), by Kücke *et al.* (1995) and by Kücke and Kleeberg (this volume).

From February 16 to June 21, 1993, brook water was sampled at 3 hour intervals by use of 4 automatic water samplers and analysed after filtration (0.45 μm) for the concentrations of dissolved nutrients (NO_3 spectrophotometrically after enzymatic reduction to nitrite (Kücke and Przemeck 1985) by use of a continuous flow autoanalyser, P spectrophotometrically according to Murphy and Riley (1962), and K using emission spectrometry). Solid concentrations were calculated from the weighted amount of solids after filtration. Nutrient content of the solids was analysed separately after digestion with concentrated nitric acid and perchloric acid (4:1) and expressed as nutrient content in the solids ($P_{in\ solids}$, $K_{in\ solids}$) or as solid bound nutrient content in the water (solid $P_{in\ water}$, solid $K_{in\ water}$). The stream flow (l/s) in the brook was recorded by use of venturi dams.

Results

Stream flow and nutrient concentrations

During the investigation period, the water discharge in the brook Ohebach decreased steadily from 28 l/s at February 21 to around 3 l/s in mid May. This general pattern was interrupted by slight increases after rainfall events from March to mid May (Fig. 1). The rainfall events during this period usually had an intensity of less than 10 mm/day. Between May 21 and June 2, 4 short term increases in the water stream occurred giving stream flow intensities of 7 l/s (May 21), 13 l/s (May 27), 10 l/s (May 28) and 150 l/s (June 2). These events corresponded to heavy rainfall with intensities > 10 mm/day.

Precipitation did not increase the stream flow intensity in every case but effected the solid concentration in most cases (Fig. 1). Solid concentrations were 0.01 to 0.05 g/l in periods with low water stream but reached values up to 5.8 g/l during high stream flow events. The highest solid concentration was recorded during the heavy rainfall event on June 2 (Table 1).

The average nitrate concentration in the water of the Ohebach was 10 mg NO_3-N/l (Table 1). During the monitoring period a slight decrease was detected. No clear relationship was detected between precipitation and nitrate concentrations in the water over the monitoring period.

The dissolved K concentration (K_{diss}) was rather constant (1 ppm K), but some peaks occurred at some precipitation events. However, not every rainfall event was accompanied by an increase of the potassium concentration. During periods with low stream flow the concentrations of dissolved K (K_{diss}) were clearly higher than those of the solid bound K in the water (solid $K_{in\ water}$). However, during the rainstorm on June 2 the concentrations of the solid bound K (solid $K_{in\ water}$) in water was about 15 times higher than that of the dissolved K (K_{diss}).

Dissolved P was detected in 18% of the water samples only, and the highest recorded concentration was 0.25 mg P/l. The concentration of solid bound P (solid $P_{in\ water}$) was about 2 times higher than that of the dissolved P (P_{diss}) and reached its maximum (6.2 mg/l) during the rainstorm on June 2.

The highest values for water discharge, solid concentration and solid bound nutrient concentrations (solid $P_{in\ water}$, solid $K_{in\ water}$) recorded during the monitoring period (Table 1) were detected between 6 and 9 o'clock p.m. on June 2, while the dissolved P and K concentrations (P_{diss}, K_{diss}) reached their maximum 3 hours later (9 and 12 o'clock p.m.). P and K concentrations in the solids ($P_{in\ solids}$, $K_{in\ solids}$) only increased slightly during this event and had therefore only a low effect on the increasing solid nutrient contents in the water.

Nutrient charges

10 t solids, 565 kg of NO_3-N, 11 kg P and 44 kg K were transported by the river water during the monitoring period (Fig. 2). However, the majority of the solid, P and K load was transported during one single rainstorm event at the 2. June (93% of solids, 91% of P and 84% of the K charge). Compared to this, only 3% of the total NO_3-N losses occurred during this one day.

The solid bound P charge exceeded that of the dissolved P load in general over the monitoring period. However, the effect of the rainstorm event on the potassium transport was different: During periods with low water discharge only 16% of the potassium transported was carried as solid bound K while this kind of transport significantly dominated during the rainstorm event (91% on June 2, 6 to 9 o'clock p.m.).

Discussion

The results indicate that sampling and analytical requirements are extremely high if reliable data on the nutrient turnover of a river system are to be gained. NO_3 charges can be quantified quite easy by analysing the river water about once per week (Lammel 1990) and calculating the fluxes from continuously measured water flows. However, P is predominantly transported solid bound, and the concentrations of suspended solids in such a small river system can change significantly within minutes, especially during heavy rainfall events. Reliable data on the solid concentrations can only be obtained with short term sampling intervals and after complete digestion of the nutrient compounds

Table 1. Precipitation, solid and nutrient content in the Ohebach brook during the monitoring period and during the rainstorm at 2. June

	Precipitation (mm/day)	Stream (l/s)	solid (g/l)	P (mg/kg)	K (mg/kg)	solid P (mg/l)	solid K (mg/l)	P_{diss} (mg/l)	K_{diss} (mg/l)	NO_3-N (mg/l)
Min	0	2.15	0	45	177	0	0	0	0.64	5.82
Max	30.5	150	5.85	5903	73000	6.18	26.28	0.25	5.96	17.15
Average	1.4	4.60	0.02	977	2144	0.02	0.06	0.01	1.05	10.32
2 June 1994										
6–9 p.m.	22.0	150	5.85	1057	4495	6.18	26.28	0.055	1.79	9.41
9–12 p.m.		13	0.35	1021	5351	0.36	1.87	0.25	5.96	11.95

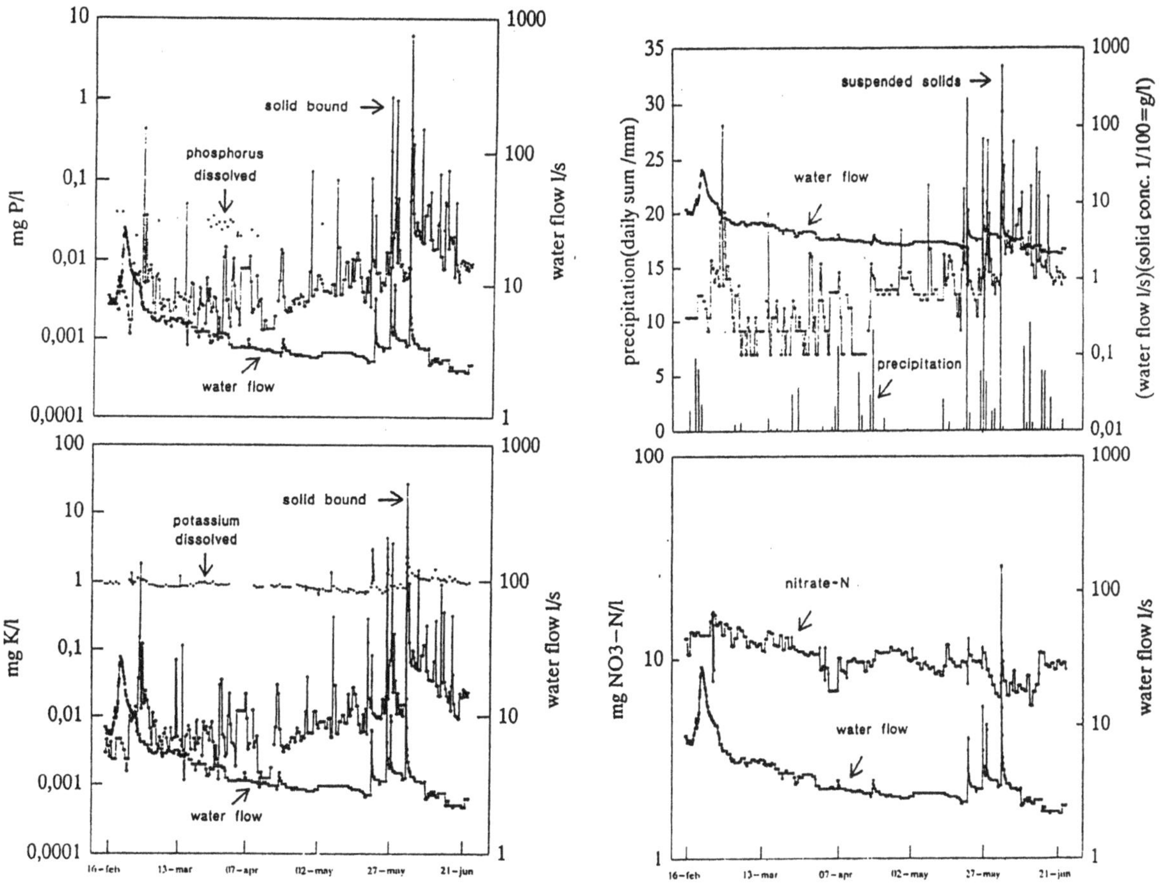

Fig. 1. Concentrations of solids and nutrients (NO_3-N, dissolved P and K, solid bound P and K) in the small river Ohebach in relation to precipitation and stream flow intensity.

which makes these determinations extremely time and labour consuming. If the short term sampling had not been carried out in this study, the short term changes of the concentrations during the most important 3 hours interval at the 2 of June would not have been detected and only 10% of the total P charges would have been determined.

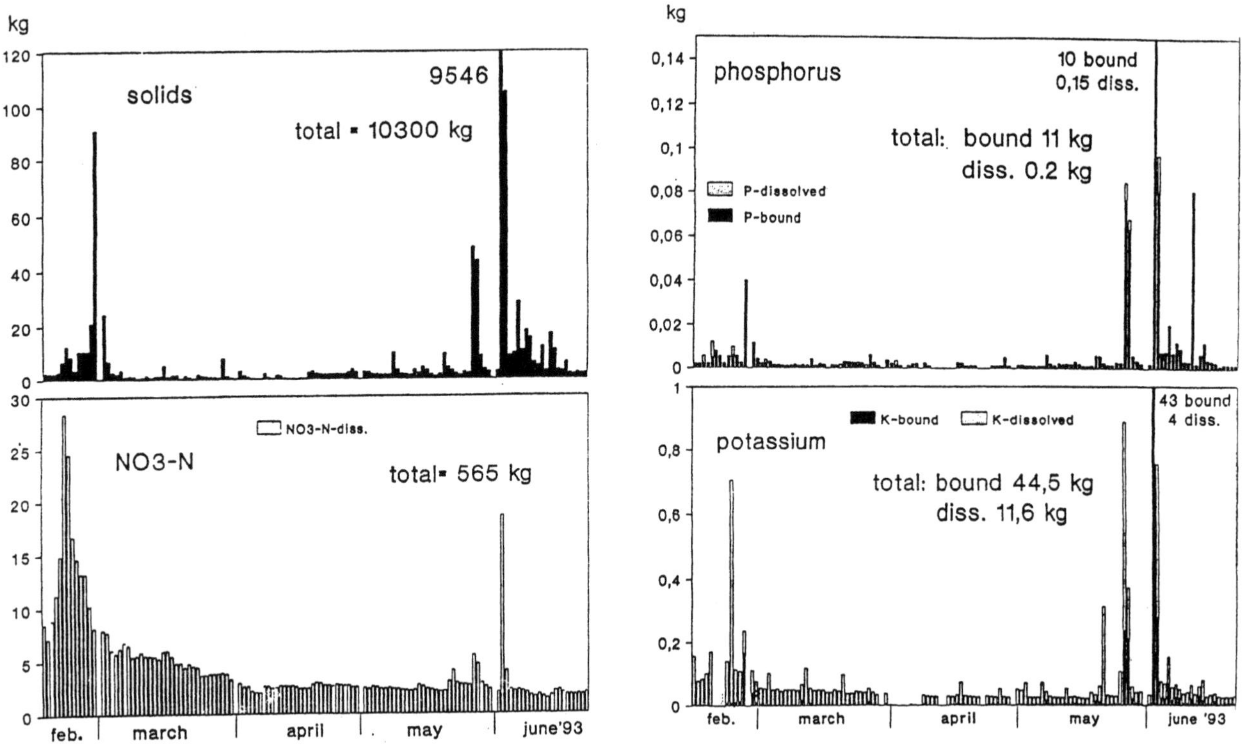

Fig. 2.

From other studies it is concluded that potassium is predominantly transported as dissolved K in river systems because the concentrations of dissolved K are in most cases higher than those of the solid bound K. This study clearly indicates that this is only true for low stream flow situations. During rainstorm event solid bound K becomes the dominant transport form (> 80%).

Acknowledgement

This investigations have been funded by the German Research Organisation in the interdisciplinary research group "Water and Matter Dynamics in Agroecosystems" running since 1986. The authors thank Mr. B. Bosse for assisting the samplings, Miss S. Herz and Miss U. Krüger and Miss R. Rietz for analytical and technical assistance.

References

Granillo AB, Chang M, Rashin EB (1985) Correlation between suspended sediment and other water quality parameters in small streams of forested east Texas. The Texas Journal of Science 27: 227–234.

Kücke M and Przemeck E (1985) A Microbiological Nitrate Test for Serial Analysis of Plant Extracts. Plant Research and Development 21: 100–112.

Kücke M Schmid H and Spieß A (1995) Applicability and Comparability of Methods for Root Growth Measurements in different Soils. Plant and Soil 172: 63–71.

Kücke M and Kleeberg P (this volume) N Fertilisation Balances in two agriculturally used water catchment areas.

Lammel J (1990) Nährstoffausträge durch Dräne und Vorfluter. Angewandte Wissenschaft 386 (J of the German Ministry of Alimentation, Agriculture and Forestry), Landwirtschaftsverlag Münster-Hiltrup.

Murphy J and Riley JP (1962) A modified single solution method for the determination of phosphate in natural waters. Anal Chim Acta 27: 31–36.

Othmer H and Bork HR (1989) Characterisation of the investigation sites. Landschaftsgenese und Landschaftsökologie 16: 73–86.

C. Rodriguez-Barrueco (ed.), Fertilizers and Environment, 401–404.

Nitrogen leaching in a maize field trial of Lower Mondego Valley (Coimbra)-lysimeters studies

Maria Isabel Ferreira Magalhaes-Martins
Direccao Regional de Agricultura da Beira Litoral, Avenida Fernao de Magalhaes, 465, 3000 Coimbra, Portugal

Key words: Field trials, leaching, lysimeters Ebermayer, maize, nitrates, nitrogen

Abstract

With the main purpose of increasing the knowledge about nitrogen leaching, lixiviates were collected in lysimeters, during two years (1992 and 1993), in a nitrogen fertilization trial of maize on the Lower Mondego Valley (Magalhaes Martins, 1994).

The field was divided into two big plots, with and without nitrogen fertilization (NO = 0 kg N ha^{-1}; N3 = 300 kg N ha^{-l}), and each was subdivided into two plots, with and without maize plants. In these four plots, the lysimeters were installed at depths of 0.30 and 0.60 m.

The maize yields and N-NO_3 contents of the lixiviates colleted in the lysimeters gave some indications about the losses of N-$N0_3$ by leaching:

- the greatest nitrogen losses occurred from the plots without maize, the bare ground being the main responsible for the high content of the subterranean waters
- these lixiviates, in some plots without maize, reached contents of 76 mg N-$N0_3l^{-1}$ at level N3, and 63 mg N-$N0_3l^{-1}$ with no nitrogen = NO application
- on the cultivated plots the nitrates content in the lixiviates were very low, even in those plots with 300 kg N ha^{-1} applied.

Introduction

The research with lysimeters, to investigate nitrogen leaching, has got an inceasing importance in the elucidation of the percolation of nitrates through soil and the forecast of losses of nitrogen that may occur, which can give rise to water pollution and present economic and environmental losses.

With the objective of clarifying some aspects of nitrogen dynamics in an Eutric Fluvisols, typical of the majority of the soils of the Lower Mondego Valley, namely its contribution to the nitrogen nutrition of maize and the nitrogen losses by leaching, 12 lysimeters Ebermayer type were installed, in a nitrogen fertilization trial of maize, for the collection of the waters draining from the maize root zone, comparing their nitrate (NO_3-N) contents with the waters which drain from a bare soil, without any cultivation.

Materials and methods

The tests were done in a representative Eutric Fluvisols of alluvial type, in a field with the following geographical coordinates - latitude 40º 13'35" N, longitude 8º 29' 32 W and altitude 10 m, which for many years was cultivated with rice.

On the profile of this soil can be seen only one horizon, with three layers: the 1st layer, with a depth of 0.32 m has a silt loam texture and a granular structure; the 2nd layer, can go down to 0.65 or 1.05 m and has a typical aspect of reduction between depths of 0.32 to 0.42 m, the main texture being silt loam without structure; the 3rd layer shows lumps of sand and coarse particles, the texture being a mixture of sandy loam, loamy sand and sandy. The water table is located at depths of 0.90 to 1.20 m.

Before starting the tests, a composite soil sample was taken at several depths.The layer 0 to 0.30m shows a pH (H20) value of 5.5, a medium content of organic matter (2.5-3.0%), medium content of available phosphorus (45–60mg kg^{-1} P_2O_5, Egner-Rhiem method) and high content of available potassium (90–130mg kg^{-1} K_2O, Egner-Rhiem method).

The tests were done for two consecutive years, running on the same plot. Each plot, with maize, 4.5 m wide and 12 m long, was sown with six rows of maize, of the cultivar BLACK (a simple hybrid, FAO 600), in a spacing of 0.75 ×0.165m (80 800 seeds ha^{-1}) The yields were evaluated at the two central rows, on the plots with maize. The plots without maize had the same dimensions. The necessary phytosanitary treatments were done to control the main pests and weeds existing in this field. The irrigation was done by independent furrows in each treatment. Each year three irrigations were done with variable amounts of water from 540 m^3 ha^{-1} to 930 m^3 ha^{-1}. The harvest, threshing and evaluation of the yields was done during the first fortnight of November. The grain yields took into account a humidity of 14%.

In the experimental layout to collect the lixiviates, horizontal collectors were used, type Ebermayer lysimeters (Addiscott *et al.*, 1991), without suction. They were made of rigid PVC, cilindric, 0.34 m diameter at the upper surface, between 0.14 m high at the back and 0.18 m at the front, and capacity of 14.5 litres. The upper surface had some holes and was covered with a protective cloth.

The field was divided into two big plots, with and without nitrogen fertilizition, and each one was divided again into two plots, with and without maize plants. Therefore there were four plots:

- Plot *without maize* (no maize = bare soil) and *nitrogen fertilization* N3 = 300 kg N ha^{-1}
- Plot *without maize* (no maize = bare soil) and *no nitrogen fertilization* NO = 0 kg N ha^{-1}
- Plot *with maize* and *nitrogen fertilization* N3 = 300 kg N ha^{-1}
- Plot *with maize* and *no nitrogen fertilization* NO = 0 kg N ha^{-1}

At these four plots 12 lysimeters were installed at depths of 0.30 m and 0.60 m. They were installed under the ridges and under the furrows. Two lysimeters at each plot with bare soil and four lysimeters at each plot with maize. The soil remained undisturbed. The sampling from the lysimeters was done after irrigation and after strong rainfall.

The rate of fertilization, before sowing, was 135 kg ha^{-1} of P_2O_5, 50 kg ha^{-1} K_2O in all plots and one third of nitrogen (100 kg N ha^{-1}), in each plot with N3 level of fertilization. The remaining two thirds of nitrogen (200 kg N ha^{-1}), were applied on top-dressing fertilization, before raising. The NO_3-N contents in water were determined by the potentiometric method of the ion-specific electrode (Byrne, 1979; Carranca, 1988).

Results and discussion

During 1992 and 1993 several samples of the lixiviates from the lysimeters were taken and analysed for N-NO_3.

The results of these analyses show a good correlation with the treatments and therefore this method for collecting water and the nitrate that it contans seemed to give results that can be interpreted. Other positive aspects were its ease of execution and rather low costs of the equipment, its set up and use. The 12 lysimeters were put in place as follows:

N. 1 - Under the plot *without* maize - level N3 - at 0.30 m - below *the ridge*
N. 2 - Under the plot *without* maize - level N3 - at 0.30 m - below *furrow*
N. 3 - Under the plot *without* maize - level N3 - at 0.60m - below *the ridge*
N. 4 - Under the plot *without* maize - level N3 - at 0.60m - below *furrow*
N. 5 - Under the plot *without* maize - level NO - at 0.30m - below the *ridge*
N. 6 - Under the plot *without* maize - level NO - at 0.60m - below *the ridge*
N. 7 - Under the plot *with* maize - level N3 - at 0.30m - below *the ridge*
N. 8 - Under the plot *with* maize - level N3 - at 0.30m - below it furrow
N. 9 - Under the plot *with* maize - level N3 - at 0.60m - below *the ridge*
N.10 - Under the plot *with* maize - level N3 - at 0.60m - below *furrow*
N.11 - Under the plot *with* maize - level NO - at 0.30m - below *the ridge*
N.12 - Under the plot *with* maize - level NO - at 0.60m - below *the ridge*

In 1992 eight samples of water were taken from the lysimeters and six samples in 1993. The sampling dates and the concentrations in N-NO3 are indicated on Tables 1 and 2, and four samples can be seen in Figures 1 to 4. The two first samples of 1993 were spoiled by the malfunction of the suction pump, and its replacement made possible the catchment of the remaining samples. The nitrate contents of the irrigation water,

Table 1. Concentrations in N-N0$_3$ (mg l^{-1}) in the lixiviates collected from lysimeters in 1992

Lysimeter number	1st 29/7	2nd 30/7	3rd 03/08	4th 20/08	5th 24/08	6th 26/08	7th 15/09	8th 21/10
1 - No maize-N3-0.30m	67.0	43.7	13.7	5.7	–	29.6	65.8	–
2 - No maize-N3-0.30m	54.7	64.7	21.6	6.9	–	23.9	76.3	–
3 - No maize-N3-0.60m	21.3	18.2	6.0	17.5	19.8	29.7	62.0	34.1
4 - No maize-N3-0.60m	22.4	13.7	4.7	17.0	15.2	76.0	55.0	34.3
5 - No maize-NO-0.30m	28.5	27.8	7.5	4.0	5.4	–	44.0	21.3
6 - No maize-NO-0.60m	47.2*	37.2	9.8	7.7	19.2	47.7	62.8	14.0
7 - With maize N3-0.30m	18.2	14.5	3.3	–	–	11.4	22.7	–
8 - With maize N3-0.30m	13.0	9.0	1.0	1.5	–	7.6	10.8	5.9
9 - With maize N3-0.60m	14.0	3.9	1.1	–	2.0	11.0	18.7	10.7
10 - With maize N3-0.60m	10.0	5.4	1.1	–	1.7	13.0	17.6	13.0
11 - With maize NO-0.30m	5.7	5.9	5.1	1.5	–	–	27.8	3.8
12 - With maize NO-0.60m	12.6*	15.6*	2.6	6.9*	4.6	28.4	41.8	13.5

* The amount of water collected was higher than the capacity of the lysimeters

Table 2. Concentrations in N-N0$_3$ (mgl^{-1}) in the lixiviates collected from lysimeters in 1993

Lysimeters number	1st 30/06	2nd 27/07	3rd 20/08	4th 02/09	5th 08/09	6th 10/09
1 - No maize-N3-0.30m	40.4	75.5	43.5	54.5	39.0	44.2
2 - No maize-N3-0.30m	10.7	–	25.0	54.1	39.0	46.0
3 - No maize-N3-0.60m	10.7	–	26.4	30.2	22.1	17.3
4 - No maize-N3-0.60m	–	23.8	28.0	38.5	23.2	27.3
5 - No maize-NO-0.30m	1.3	14.9	21.0	6.4	23.7	19.3
6 - No maize-NO-0.60m	–	–	–	12.2	15.0	13.4
7 - With maize-N3-0.30m	–	14.8	2.5	3.2	5.8	3.3
8 - With maize-N3-0.30m	0.7	–	–	2.5	0.7	0.4
9 - With maize-N3-0.60m	–	–	2.5	3.8	2.0	1.4
10 - With maize-NO-0.60m	–	10.7	–	1.7	2.0	1.2
11 - With maize-NO-0.30m	–	2.0	0.4	1.8	0.4	0.4
12 - With maize-NO-0.60m	–	–	0.5	2.0	0.5*	0.5

* The amount of water collected was higher than the capacity of the lysimeters

taken directly from Mondego river, varied between 1.3 and 2.4 mg N-N0$_3$ l^{-l}.

In 1992, the water sampling were done in dates as follows:

- 1st sampling (29/07) - just after the first irrigation;
- 2nd sampling (30/07) - in the second day after the first irrigation, owing the field inundation;
- 3rd sampling (03/08) - after strong rainfall;
- 4th sampling (20/08) - just after the second irrigation;
- 5th sampling (24/08) - four days after the second irrigation;
- 6th sampling (26/08) - after one small additional irrigation, to collect water in all lysimeters;
- 7th sampling (15/09) - just after the third irrigation;
- 8th sampling (21/10) - after strong rainfall.

In 1993, the water sampling were done in dates as follows:

Waters samples from lysimeters in 1992
N-NO_3^- (mg L^{-1}) contents

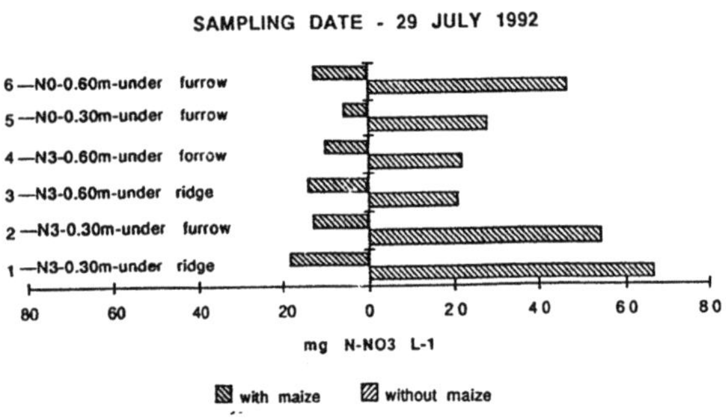

Fig. 1.

Waters samples from lysimeters in 1993
N-NO_3^- (mg L^{-1}) contents

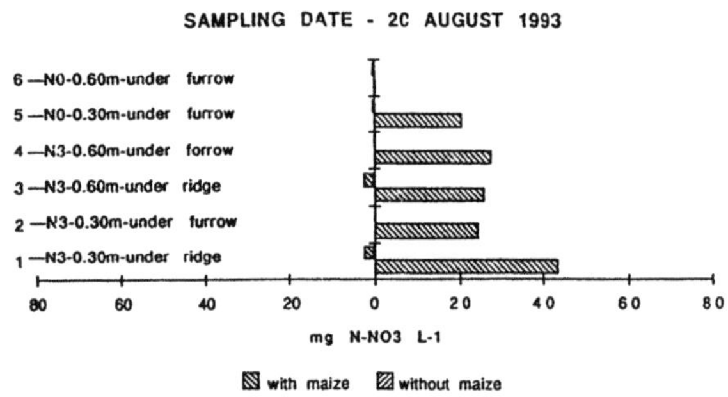

Fig. 3.

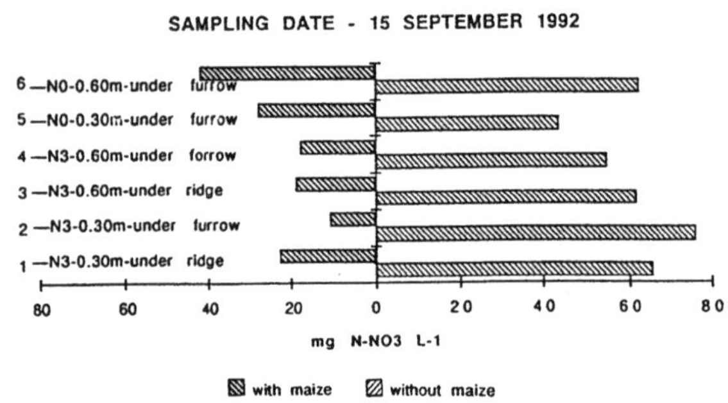

Fig. 2.

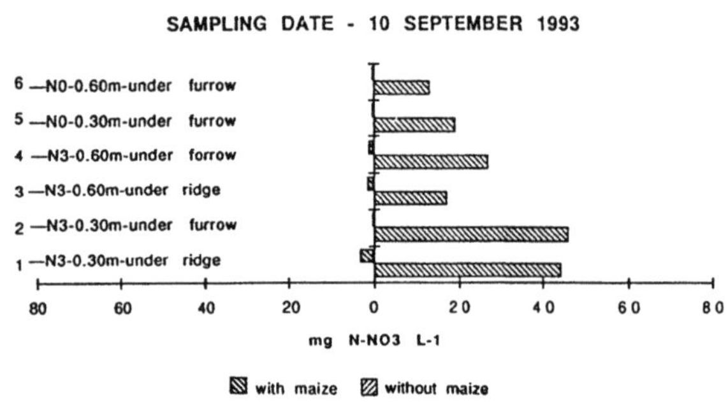

Fig. 4.

- 1st sampling (30/06) - to take the water accumulated, owing the strong rainfall of June;
- 2nd sampling (27/07) - just after the first irrigation;
- 3rd sampling (20/08) - just after the second irrigation;
- 4th sampling (02/09) - just after the third irrigation;
- 5th sampling (08/09) - after strong rainfall;
- 6th sampling (10/09) - after strong rainfall.

The contents in nitrates in the lixiviates seem to show the effect of maize in absorption of nitrates avoiding its carrying in the lixiviates, as the greatest differences were between the lixiviates from the plots without maize and the lixiviates from the plots with maize. Nitrates losses by leaching at the plots with maize, even with high application of nitrogen, are relatively unimportant, specialy at the fertilization with 300 kg N ha^{-1}. The lixiviates at the non fertilized plots, without maize, showed nitrate contents rather higher than the fertilized plots, with maize.

At the plots with bare soil the nitrate content in the lixiviates, in September and October, increased to very high values, but at the plots cultivated with maize, in that period, the nitrate content in the lixiviates were much lower.

References

Addiscott TM, Whitmore AP and Powlson DS (1991) Farming, Fertilizers and the Nitrate problem. 2nd ed. CAB International, Wallingford, Oxon, UK.170 p.

Byrne E (1979) Chemical analysis of agricultural materials. An Foras Taluntais, Johnstown Castle Research Centre, Wexford, Ireland 320 p.

Carranca CF (1988) Determinação do teor em nitratos pelo método do eléctrodo selectivo, em solos, plantas e águas. Pedologia 23(2) 135–153.

Magalhães Martins MIF (1994) A fertilização azotada do milho em solos do Baixo Mondego. Thesis for M.Sc. in "Plant Nutrition, Soil Fertilily and Fertilization". Instituto Superior de Agrononia, Universidade Técnica de Lisboa, Lisboa, Portugal. 203 p.

C. Rodriguez-Barrueco (ed.), Fertilizers and Environment, 405–411.

Aluminium phytotoxicity

A challenge for plant scientists

J. Barceló, Ch. Poschenrieder, M.D. Vázquez & B. Gunsé
Laboratorio de Fisiología Vegetal, Facultad de Ciencias, Universidad Autónoma de Barcelona, E-08193 Bellaterra, Spain

Key words: aluminium, phytotoxicity, tolerance

Abstract

Breeding for Al tolerance is considered a useful way to improve crop performance on acid soils. However, a successful selection of Al-tolerant genotypes, which will render high yields on acid soils, requires the development of screening methods based on the knowledge of Al-toxicity and Al-tolerance mechanisms. In this paper the present stage of knowledge on Al phytotoxicity and tolerance is comprehensively reviewed. Special emphasis will be made on primary mechanisms of Al toxicity, the interactions of Al with membranes, and the role of Ca in the Al-toxicity syndrome.

Introduction

Aluminium toxicity is considered the most important metal toxicity problem on acid soils (pH $\leq$5.0) (Foy *et al.*, 1978). As nearly half of the non irrigated arable lands in the world are acidic soils, Al toxicity is a severe problem for crop production (Clark 1982).

The Al species, present in the solutions of acid soils, that are responsible for phytotoxicity, are not clearly established. Among the mononuclear Al-species, complexes of Al with SO_4^{2-} (Kinraide and Parker, 1987), F^- (Wright *et al.*, 1989), and organic ligands (Hue *et al.*, 1986) are considered to be nontoxic. The rhizotoxicity of mononuclear hydroxy-Al species remains unclear, while Al^{3+} is toxic, as are other polycations (Kinraide, 1991). Although the soluble polymeric Al species, Al_{13}, has been shown to cause severe rhizotoxicity in solution culture (Wagatsuma and Ezoe, 1985; Parker *et al.*, 1989), its importance in natural soil solutions has been questioned, because suspended solids would readily sorb polymeric Al (Driscoll and Schecher, 1988).

As Al toxicity only affects crop productivity on acid soils, it seems obvious, that poor crop productivity, due to Al toxicity, can be avoided by liming. If low cost sources of lime are available, increase of pH in the plow layer is readily achieved by conventional procedures. Unfortunately, not only in tropical, but also in many subtropical and temperate latitudes, subsoil acidity occurs.

The problem of subsoil acidity in relation to amelioration by gypsum has been reviewed by Shainberg *et al.*, 1989. Under conditions of subsoils acidity, conventional surface liming is of little use, in most cases, because of the low mobility of the alkaline component of lime.

To improve crop performance under these circumstances, principally three strategies have been proposed (Shainberg *et al.*, 1989): 1, use of equipment to incorporate lime into subsoil, 2, surface incorporation of gypsum and 3, use of Al-tolerant genotypes.

The incorporation of lime into subsoils has a high power requirement and, from the technical point of view, such an incorporation is not possible under all circumstances. Surface incorporation of gypsum seems to be a better solution, but only if this material is available locally and soils are not susceptible to leaching of Mg.

The development of Al-tolerant genotypes is time consuming and expensive. Moreover, other factors than Al toxicity, such as toxicity of Mn, Fe and H^+ and deficiency of Ca, Mg, P and K may also limit crop performance on acidic soils (Foy, 1984). Nevertheless, on a long-term scale the development of Al-tolerant

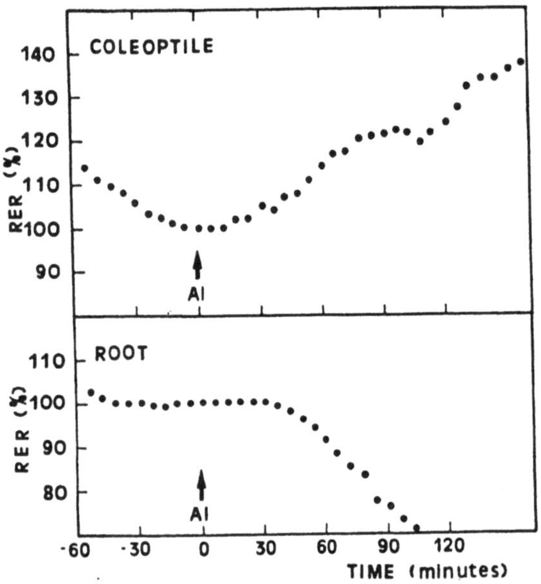

Fig. 1. Relative elongation rates (RER) of coleoptiles and roots of maize seedlings (n=4) from 60 min prior to 150 min after the supply (t=0) of 20 μM Al. (Redrawn from Llugany, 1994 and Llugany *et al.* 1995)

genotypes seems to be the most feasible solution, from the economical point of view.

Screening programs for Al tolerant germplasm would be substantially improved if the primary mechanisms of Al-toxicity in plants were better understood. Such a fundamental knowledge would facilitate the choice of early selection criterions and so speed up the selection process.

Here we will make a comprehensive review of recent developments in the knowledge on Al toxicity and tolerance mechanisms and give a short outlook on future research necessities in this field.

Aluminium toxicity

Inhibition of root elongation

Many studies have demonstrated that the inhibition of root elongation is the most rapid, visible symptom of Al toxicity in plants, while decrease of shoot growth is a later occurring effect, that is thought to be mainly a consequence of the toxic Al effects in roots. Figure 1 shows the inhibitory effect of Al on root and coleoptile elongation of maize seedlings. While 20 μM Al required less than 60 minutes to induce a significant decrease of the root elongation rate, the coleoptile continued with a high elongation rate after more than 2.5 h exposure to Al (Llugany, 1994; Llugany *et al.*, 1995).

The mechanisms of this Al-induced inhibition of root elongation are not fully understood. It is well known that Al can interact with nucleic acids and Al has been localized in nuclei of root cells (Matsumoto *et al.*, 1976). Inhibition by Al of mitosis in apical meristems of roots has been demonstrated (Clarkson, 1965; Morimura *et al.*, 1978). Less clear is the role of decreased cell division as the primary mechanism of the Al-induced inhibition of root elongation. A convincing argument against this hypothesis is the fact that Al, as a metallic ligand with preference for oxygen donors and high affinity for carboxyl and phosphate groups, can bind to many other cell components, before entering the nucleus. In the apoplast, cell walls and the plasma membrane surface are targets for Al binding and Al is likely to be toxic, even before entering root cells (Horst *et al.*, 1991). Moreover, due to the antagonism between Al and Ca and the potential of Al^{3+} for interacting with any system that uses Mg^{2+} or any form of phosphate (Martin, 1988), Al can inhibit a large array of key metabolic processes regulated by Ca (Rengel, 1992), calmodulin (Haug and Caldwell, 1985), GTP-binding regulatory proteins and hexokinases (Martin, 1988). Although, cell division finally would be affected by such metabolic alterations, the celerity of the inhibitory effect of Al on root elongation (e.g. average response time in maize cultivar BR 201 F exposed to 20 μM Al was only 33 min; Llugany *et al.*, 1995), suggests that the inhibition of root cell expansion may be initially responsible for Al-induced decrease of root elongation rates.

Under steady state conditions, cell expansion growth can be described by the modified Lockhart's equation (Dale and Sutcliffe 1986), according to which the increase of cell volume depends on the apparent hydraulic conductance, the extensibility of the cell wall, the difference of water potential between the cell and its surroundings, the threshold turgor pressure and the actual turgor pressure. An influence of Al on any of these parameters would alter the rate of cell expansion growth (Barceló and Poschenrieder, 1990).

It has been suggested that Al may inhibit root elongation by cross-linking of the carboxyl groups of the pectin fraction of cell walls, causing a decrease of cell wall extensibility and cell wall elasticity (Klimashevski and Dedov, 1975). The cross-linking hypothesis for Al-induced inhibition of root growth is contradicted by the observation, that Al was not directly associated with the pectin fraction of root cell walls in peas (Mat-

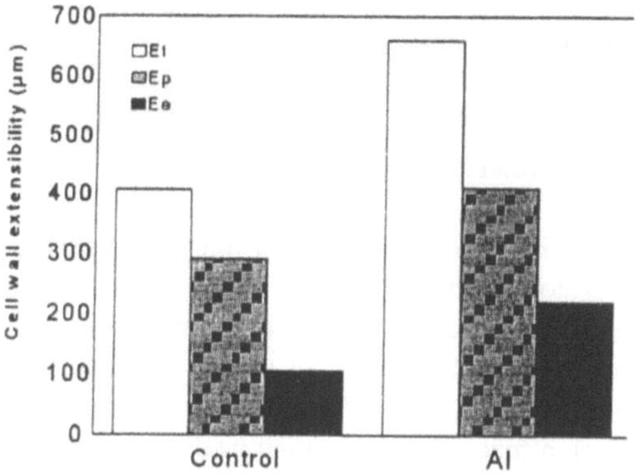

Fig. 2. Total (Et), plastic (Ep) and elastic (Ee) extensibilities of maize coleoptiles floated for 4 h on solution containing 0.4 mM Ca and 0 (Control) or 0.1 mM Al.

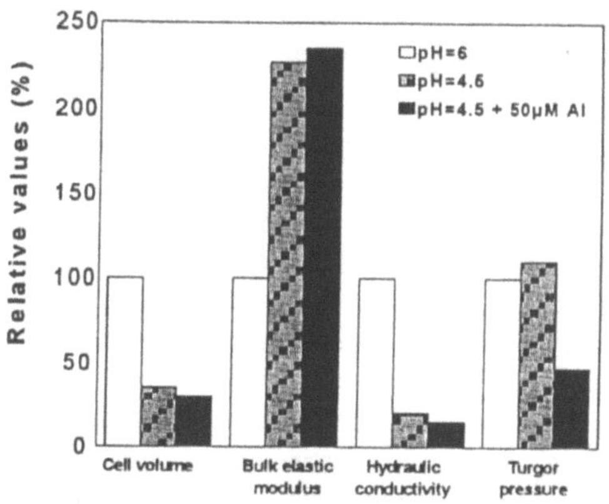

Fig. 3. Relative values (treatment pH 6, 100%) of cell volume, bulk elastic modulus, hydraulic conductivity and turgor pressure of maize roots in nutrient solutions with different pH and Al treatments.

sumoto *et al.*, 1977). Direct measurements of cell wall extensibility by the Instron technique failed to detect a decrease of wall extensibility in maize coleoptiles floated on Al-containing solutions (Figure 2) (Llugany *et al.*, 1992).

Several parameters related to expansion growth in roots can be determined by pressure probe measurements. The first results we got in our laboratory with the root pressure probe, using maize plants exposed for 24 h to different pH or 50 μM Al, are shown in Figure 3. Both acid pH and Al decreased the volume of root cells and these treatments caused a significant increase of the bulk elastic modulus of the system. This result suggests that both low pH and Al decrease the elasticity of maize roots. Aluminium but not acid pH decreased the pressure potential of maize roots, but the most important effect of both acid pH and Al was the severe reduction of the apparent hydraulic conductivity of the roots (Figure 3). These first results indicate that Al severely affects water relation parameters of roots. Further investigations, using the cell pressure probe technique, are in progress, in order to see the possible relation between Al-induced decrease of cell volume, the elastic properties of cell walls and the cell turgor pressure.

The negative influence of Al on the hydraulic conductivity of roots, which has also been observed by other authors (Zhao *et al.*, 1987), supports the hypothesis that the plasma membrane is a primary site for Al-induced injury. The surface of the plasma membrane is negatively charged due to the presence of phosphate groups and carboxyl chains. The plasma membrane is stabilized by Ca ions. The chemical properties of Ca^{2+} enable this cation to form electrostatic bridges between individual neighboring anionic phospholipids and between phospholipids and the carboxyl chains of membrane proteins. This Ca-binding determines the rigidity of the membrane. Under normal growth conditions, it is assumed that not all the anionic sites in a membrane are bound with Ca^{2+}, yet an excess of Ca-binding would make the membrane exceedingly rigid (Leshem, 1992). In fact, high Ca concentrations are known to stiffen plasma membranes and to decrease cell elongation.

Aluminium is thought to affect membrane fluidity by altering the chemical environment of membrane lipids probably by cross-linking the polar regions of phospholipids (Zhao *et al.*, 1987). Such a rigidification of the membrane would directly or indirectly lead to a large array of alterations concerning the function of membrane-bound enzymes and ion transport systems.

If the binding of Al^{3+} with the negative surface charges of the plasmalemma is the primary cause of Al-phytotoxicity, the Al-induced damage would depend on both the activity of Al^{3+} at the membrane surface and the surface electrical potential. This view is supported by the observation, that amelioration of Al^{3+}-toxicity by Ca^{2+} is not an ion specific phenomenon and can be achieved also by other cations, including H^+, which reduce surface negativity (Kinraide *et al.*, 1992). Moreover, it has been found that surface negativity and the binding of basic dyes to protoplasts correspond to the Al sensitivity of parent roots (Wagatsuma & Akiba, 1989).

Further circumstancial evidence for the hypothesis that the interaction of Al^{3+} with the plasma membrane is a primary mechanism of Al-toxicity is provided by the fact that one of the most rapid effects of toxic

Al concentrations is the induction of callose synthesis (Wissemeier *et al.*, 1992). Polycations such as chitosans, but also polyamines and Al^{3+}, are potent elicitors of callose synthesis. These polycations by binding to the negative charges of the membrane surface cause a decrease of fluidity. The change in membrane fluidity may open stretch-activated Ca^{2+} channels, causing a small increase of the cytoplasmic Ca^{2+} concentration, which would activate callose synthesis (Kauss, 1990). Although, at present, this mode of callose induction by Al is only a working hypothesis, the observation that both Al concentration in root tips and Al-induced inhibition of root elongation were correlated to callose concentration in maize root tips, indicates that callose may be used as an early indicator of Al-sensitivity in screening programs (Llugany *et al.*, 1994).

Aluminium - calcium interaction

It is well known, that Al decreases Ca concentrations in plants. In long-term studies, Al-toxicity is often expressed as Ca deficiency and species adapted to acid soils frequently present high Ca efficiency (Foy, 1984). The effect of Al^{3+} on Ca^{2+} transport is rapid, reversible and consistent with competitive inhibition. Moreover, it has been shown that Al-induced inhibition of Ca^{2+} uptake was more affected in Al-sensitive than in Al-tolerant wheat cultivars (Huang *et al.*, 1992; Ryan & Kochian, 1993). Those observations lead to the hypothesis that Al-induced inhibition of root elongation may be caused by inhibition of Ca^{2+} transport. However, recent investigations using ion-selective microelectrodes or ^{45}Ca for measuring Ca^{2+} fluxes in root apices of Al-sensitive wheat plants, grown in solutions with constant activities of Al^{3+} and Ca^{2+}, clearly show that root elongation is not directly coupled with Ca^{2+} uptake into the root and translocation to the shoot . This result contradicts the hypothesis that decreased Ca^{2+} transport is directly responsible for Al-induced inhibition of root elongation (Ryan *et al.*, 1994). Investigations from the same research groups also contradict the Ca^{2+} displacement hypothesis (Kinraide *et al.*, 1994), according to which Al causes inhibition of root elongation by displacing Ca from binding sites in the membrane.

However, these results neither exclude the possibility, that binding of Al to the surface of plasma membrane may be a primary mechanism of toxicity nor that alternative interactions between Al^{3+} and Ca^{2+} or an Al-induced disruption of Ca homeostasis play an important role in the initial events leading to Al-induced inhibition of root elongation.

Aluminium tolerance

The uncertainties concerning the primary mechanisms of Al rhizotoxicity are intimately related to the fact that the present knowledge on Al-tolerance mechanisms is also inconclusive. Several types of mechanisms have been proposed (Taylor, 1988), which can be classified in exclusion mechanisms (immobilization at the cell wall, selective permeability at the plasma membrane, plant induced pH barrier in the rhizosphere, exudation of chelate ligands) and internal tolerance mechanisms (chelation in the cytosol, compartmentation in the vacuole, aluminium binding proteins, evolution of Al-tolerant enzymes). Within the frame of this review it is impossible to discuss in detail these possible mechanisms and only some recent developments in the fields of both exclusion mechanisms and compartmentation of Al in cereals will be considered.

Recent studies with wheat and maize cultivars differing in Al tolerance have clearly shown that Al-tolerant cultivars contain lower Al concentrations in root tips than Al-sensitive cultivars (Tice *et al.*,1992; Delhaize *et al.*, 1993a; Llugany *et al.*, 1994). Although, at present, no techniques are available which inequivocously allow an exact distinction between apoplastic and symplastic Al in roots, studies in wheat using morin fluorescence staining or fractionation techniques have shown that tolerant cultivars accumulate less Al in the apoplasm than sensitive cultivars, while the concentration of Al in the symplasm associated with a 50% inhibition of root elongation was similar in Al-tolerant and Al-sensitive cultivars (Tice *et al.*, 1992). These results support the hypothesis that exclusion of Al from root tips is an important mechanism for avoiding Al toxicity, while detoxification in the cytoplasm does not seem to play a role. Exclusion of Al from root tips in tolerant wheat may be achieved by increased malate extrusion from the tip region, leading to Al chelation and thus to a lower activity of Al^{3+} in the rooting medium (Delhaize *et al.*, 1992b).

It is less clear if, in addition to exclusion of Al from the apoplast, compartmentation of Al in root vacuoles may play a role in Al-tolerance. Aluminium was not detectable by morin staining in root tip vacuoles of either tolerant or sensitive wheat exposed for 48 h to Al (Tice *et al.*, 1992). In our laboratory, Al was detected by X-ray microanalysis in electron-dense deposits

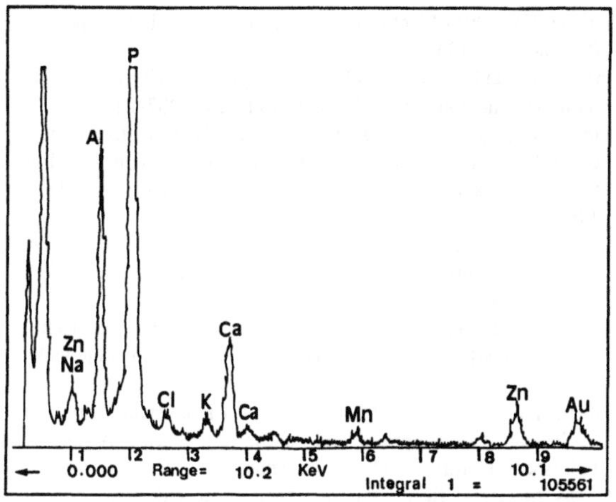

Fig. 4. EDX spectrum from electron-dense precipitates found in small root tip vacuoles of maize seedlings exposed to 20 μM Al for 96 h.

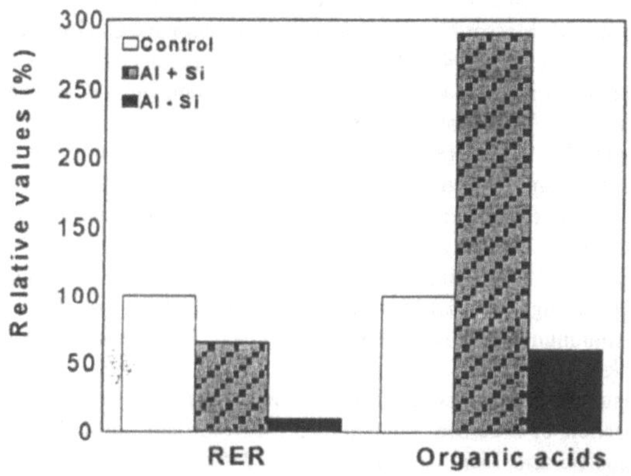

Fig. 5. Relative root elongation rates and relative organic acid concentration in roots of teosinte plants grown in solution with (4 μM) or without (control) silicon. Al concentration in + Si solution: nominal value, 120 μM; Al_{mono}, 35 $\pm$ 1.5 μM. Al in $-$Si solution: nominal value, 60 μM; Al_{mono}, 33 $\pm$5.0 μM. (Modified from Barceló *et al.*, 1993).

found in the small root tip vacuoles of maize plants treated with Al for 24 to 120 h, (Figure 4). Aluminium was associated with high P concentrations. Differences between cultivars were found but, based on preliminary results, no relation between Al concentration in the vacuoles and tolerance index based on root elongation could be established.

Conclusions

Our short revision shows that investigations into Al phytotoxicity and tolerance mechanisms are extremely complex. In general, progress in the field of root stress is severely hampered by our restricted knowledge on the mechanisms underlying root growth, membrane functioning and signal perception and transduction in plants. In the special case of Al, the lack of an adequate radioactive tracer severely hampers the investigation of uptake and compartmentation mechanisms. Even so, important progress has been made in the recent years and it has got clear that the activity of Al^{3+} at the plasma membrane is essential for Al-toxicity to occur. Further investigations have to address, at the molecular level, the question if differences in Al-tolerance between genotypes are due to differences in either or both plasma membrane composition and metabolic pathways leading to enhanced capacity for Al quelation. For these investigations to be useful, from the practical point of view, the plant mechanisms leading to a decrease of Al^{3+} activity at the membrane level have to be studied under growth conditions which really reflect the ionic environment of the rhizosphere soil under field conditions. Several soil factors, that generally are not reproduced in solution culture studies, have been shown to have a significant influence on the response of plants to Al-toxicity. As an example, seedlings of teosinte, cultivated in solutions with similar concentrations of monomeric Al species, but differing in silicon supply, showed substantially improved Al-tolerance in the presence of Si. Lower Al concentrations and increased levels of organic acids were found in the Si treated plants (Figure 5) (Barceló *et al.*, 1993). Other soil factors which can influence Al-tolerance responses in plants and which, in general, are not taken into account in conventional nutrient solution studies, are soil compactation (Horst *et al.*, 1990), mycorrhizal infection and soil organic matter. The disregard of these factors may lead to severe discrepancies between results got in solution screening and plant performance under field conditions. To avoid this severe problem, Al tolerance, in addition to a challenge for plant molecular biology, has to be more and more a target for cooperation between plant and soil scientists.

Acknowledgements

The experimental results from the authors shown in the review are from projects CE TS*CT92–0071 and DGICYT PB 91–664.

References

Alva AK, Blamey FPC, Edwards DG and Asher CJ (1986) An evaluation of aluminum indices to predict aluminum toxicity to plants grown in nutrient solutions. Commun Soil Sci Plant Anal 17: 1271–1280.

Barceló J, Guevara P and Poschenrieder C (1993) Silicon amelioration of aluminium toxicity in teosinte (*Zea mays* L. ssp mexicana). Plant and Soil 154: 249–255.

Barceló J and Poschenrieder C (1990) Plant water relations as affected by heavy metal stress: a review. J Plant Nutr 13: 1–37.

Clark RB (1982) Plant response to mineral element toxicity and deficiency. In: Christiansen MN and Lewis CF (eds) Breeding Plants for Less Favorable Environments, pp 71–142. John Wiley & Sons, New York.

Clarkson DT (1965) The effect of aluminium and some other trivalent metal cations on cell division in the root apices of *Allium cepa*. Ann Bot 29: 309–315.

Dale JE and Sutcliffe JF (1986) Water relations of plant cells. In: Steward FC, Sutcliffe JF and Dale JE (eds) Plant Physiology, A Treatise. Vol. 9. Water and Solutes in Plants, pp 1–48. Academic Press, Inc., Orlando.

Delhaize E, Craig S, Beaton CD, Bennet RJ, Jagadish VC and Randall PJ (1993a) Aluminum tolerance in wheat (*Triticum aestivum* L.). I. Uptake and distribution of aluminum in root apices. Plant Physiol 103: 685–693.

Delhaize E, Ryan PR and Randall PJ (1993b) Aluminum tolerance in wheat (*Triticum aestivum* L.). II. Aluminum-stimulated excretion of malic acid from root apices. Plant Physiol 103: 695–702.

Driscoll CT and Schecher WD (1988) Aluminum in the environment. In: Sigel H and Sigel A (eds) Metal Ions in Biological Systems. Vol 24. Aluminum and its Role in Biology, pp 59–122. Marcel Dekker, Inc., New York.

Foy CD (1984) Physiological effects of hydrogen, aluminum, and manganese toxicities in acid soil. In: Adams F (ed) Soil Acidity and Liming. Agronomy Monograph no. 12. 2nd edition, pp 57–97. ASA-CSSA-SSSA Publisher, Madison, Wisconsin.

Foy CD, Chaney RL and White MC (1978) The physiology of metal toxicity in plants. Annu Rev Plant Physiol 29: 511–566.

Haug AR and Caldwell CR (1985) Aluminum toxicity in plants: the role of the root plasma membrane and calmodulin. In: St. John JB, Berlin E and Jackson PJ (eds) Frontiers of Membrane Research in Agriculture, pp 359–381. Rowan & Allanheld, Totowa.

Horst WJ, Asher CJ, Cakmak J, Szulkiewicz P and Wissemeier AH (1991) Short-term responses of soybean roots to aluminium. In: Wright RJ, Baligar VC and Murrmann RP (eds) Plant-Soil Interactions at Low pH, pp 733–739. Kluwer Academic Publishers, Dordrecht, The Netherlands.

Horst WJ, Klotz F and Szulkiewicz P (1990) Mechanical impedance increases aluminium tolerance of soybean (*Glycine max*) roots. In: van Beusichem, ML (ed) Plant Nutrition – Physiology and Applications, pp 351–355. Kluwer Academic Publishers, Dordrecht, The Netherlands.

Huang JW, Shaff JE, Grunes DL and Kochian LV (1992) Aluminum effects on calcium fluxes at the root apex of aluminum-tolerant and aluminum-sensitive wheat cultivars. Plant Physiol 98: 230–237.

Hue NV, Craddock GR and Adams F (1986) Effect of organic acids on aluminum toxicity in subsoils. Soil Sci Am J 50: 28–34.

Kauss H (1990) Role of the plasma membrane in host-patogen interactions. In: Larsson C and Moller IM (eds) The Plant Plasma Membrane, Structure, Function and Molecular Biology, pp 320–350. Springer-Verlag, Berlin.

Kinraide TB (1991) Identity of the rhizotoxic aluminium species. Plant and Soil 134: 167–178.

Kinraide TB and Parker DR (1987) Non-phytotoxicity of the aluminum sulfate ion, $AlSO_4^+$. Physiol Plant 71: 207–212.

Kinraide TB, Ryan PR and Kochian LV (1992) Interactive effects of Al^{3+}, H^+ and other cations on root elongation considered in terms of cell-surface electrical potential. Plant Physiol 99: 1461–1468.

Kinraide TB, Ryan PR and Kochian LV (1994) Al^{3+}-Ca^{2+} interactions in aluminum rhizotoxicity. II. Evaluating the Ca^{2+}-desplacement hypothesis. Planta 192: 104–109.

Klimashevski EL and Dedov VM (1975) Localization of the mechanism of growth inhibiting action of Al^{3+} in elongating cell walls. Fiziol Rast 22: 1040–1046.

Leshem YY (1992) Plant Membranes: A Biophysical Approach to Structure, Development and Senescence. Kluwer Academic Publishers, Dordrecht, The Netherlands.

Llugany M (1994) Respuestas diferenciales de cultivares de *Zea mays* L. a la toxicidad por aluminio. Ph. D. Thesis, Universidad Autónoma de Barcelona, Barcelona, Spain.

Llugany M, Gunsé B, Poschenrieder C and Barceló J (1992) Total, plastic and elastic extensibility of *Zea mays* coleoptiles exposed to aluminum *in vitro*. Physiol Plant 85 (part II), A 76.

Llugany M, Massot N, Wissemeier A, Poschenrieder C, Horst WJ and Barceló J (1994) Differences in aluminium tolerance between maize varieties as assessed by callose formation and root elongation. Z. Pflanzenernähr Bodenk 157: 447–451.

Llugany M, Poschenrieder C and Barceló J (1995) Monitoring of aluminium-induced inhibition of root elongation in four maize cultivars differing in tolerance to Al and proton toxicity. Physiol Plant 93: 265–271.

Martin RB (1988) Bioinorganic chemistry of aluminum. In: Sigel H and Sigel A (eds) Metal Ions in Biological Systems. Vol. 24. Aluminum and Its Role in Biology, pp 1–57. Marcel Dekker, Inc., New York.

Matsumoto H, Hirasawa F, Torikai H and Takahashi E (1976) Localization of absorbed aluminium in pea root and its binding to nucleic acid. Plant Cell Physiol 17: 627–631.

Matsumoto H, Morimura S and Takahashi E (1977) Less involvement of pectin in the precipitation of aluminium in pea roots. Plant Cell Physiol 18: 325–335.

Morimura S, Takahashi E and Matsumoto H (1978) Association of aluminium with nuclei and inhibition of cell division in onion (*Allium cepa*) roots. Z Pflanzenphysiol 88: 395–401.

Parker DR, Kinraide TB and Zelazny LW (1989) On the phytotoxicity of polynuclear hydroxy-aluminum complexes. Soil Sci Soc Am J 53: 789–796.

Parker DR, Zelazny LW and Kinraide TB (1989) Chemical speciation and plant toxicity of aqueous aluminum. In: Lewis TE (ed) Environmental Chemistry and Toxicology of Aluminum, pp 117–145. Lewis Publishers, Inc., Michigan.

Rengel Z (1992) Disturbance in cell Ca^{2+} homeostasis as a primary trigger of aluminium toxicity syndrome. Plant Cell Environ 15: 931–938.

Ryan PR, Kinraide TB and Kochian LV (1994) Al^{3+}-Ca^{2+} interactions in aluminum rhizotoxicity. I. Inhibition of root growth is not caused by reduction of calcium uptake. Planta 192: 98–103.

Ryan PP and Kochian LV (1993) Interaction between aluminum toxicity and calcium uptake at the root apex in near-isogenic lines of wheat (*Triticum aestivum*) differing in aluminum tolerance. Plant Physiol 102: 975–982.

Shainberg I, Sumner ME, Miller WP, Farina MPW, Pavan MA and Fey MV (1989) Use of gypsum on soils: a review. In: Stewart

BA (ed) Advances in Soil Science. Vol. 9, pp 1–111. Springer Verlag, New York.

Taylor GJ (1988) The physiology of Al tolerance. In: Sigel H and Sigel A (eds) Metal Ions in Biological Systems. Vol. 24. Aluminum and its Role in Biology, pp 165–188. Marcel Dekker Inc., New York.

Tice KR, Parker DR and DeMason DA (1992) Operationally defined apoplastic and symplastic aluminum fractions in root tips of aluminum-intoxicated wheat. Plant Physiol 100: 309–318.

Wagatsuma T and Ezoe Y (1985) Effects of pH on ionic species of aluminum in medium and on aluminum toxicity under solution culture. Soil Sci Plant Nutr 31: 457–561.

Wagatsuma T and Akiba R (1989) Low surface negativity of root protoplasts from aluminum-tolerant species. Soil Sci Plant Nutr 35: 443–452.

Wissemeier AH, Diening A, Hergenröder A, Horst WJ and Mix-Wagner G (1992) Callose formation as parameter for assessing genotypical plant tolerance of aluminium and manganese. Plant and Soil 146: 67–75.

Wright RJ, Baligar VC and Wright SF (1987) Estimation of phytotoxic aluminum in soil solution using three spectrophotometric methods. Soil Sci 144: 224–232.

Zhao XJ, Sucoff E and Stadelmann EJ (1987) Al^{3+} and Ca^{2+} alteration of membrane permeability of *Quercus rubra* root cortex cells. Plant Physiol 83: 159–162.

C. Rodriguez-Barrueco (ed.), Fertilizers and Environment, 413–421.

Towards zero accumulation of heavy metals in soils

An imperative or a fad?

E. Witter

Department of Soil Sciences, Swedish University of Agricultural Sciences, Box 7014, S-75007 Uppsala, Sweden

Key words: sewage sludge, heavy metals, limits, agriculture, soils

Abstract

Sweden has recently introduced new guidelines for the agricultural use of sewage sludge which contain successively more restrictive metal loading rates for soils. At the same time further efforts are made to reduce the input of heavy metals through atmospheric deposition and fertiliser use. The proposed metal loading limits are among the most restrictive in the world, although other scandinavian countries and the Netherlands have proposed similarly low limits. It is the aim of the Swedish Environmental Protection Board to successively reduce the metal load of soils until a situation of zero accumulation is reached. With the possible exception for Cd, there is apparently no scientific evidence at the moment to suggest that zero accumulation of metals in soil is required to adequately protect soil productivity, the environment, and human and animal health. A policy which steers towards zero accumulation may therefore seem excessively cautious. It is, however, also a policy which recognises the practically irreversible nature of elevated heavy metal concentrations and their effects in soil, the deficiencies in the evidence currently available with which to establish safe metal loadings for soils, as well as the need to preserve the agronomic value of soils for many years to come. It is argued that the use of restrictive annual metal loading rates can be used to effectively ensure that maximum soil concentrations or cumulative pollutant loadings, considered to be safe are not reached in the foreseeable future.

Introduction

With industrialisation has come the widespread use of heavy metals in society. This has resulted in the release of heavy metals into the environment from both point and diffuse sources. Amongst others, this has resulted in the accumulation of heavy metals in soils and the contamination of sewage sludge, a traditional source of plant nutrients and organic matter in agriculture. The development of an extensive sewerage system and sewage treatment works alleviated waterways from serious pollution problems, but created a new disposal problem in the form of the polluted sewage sludge. Given the historic use of human manure, agriculture was a natural outlet for the sludge produced at the treatment works, although in Sweden dumping at sea was practiced in Malmö and Stockholm until 1967 when it was prohibited. No statistics were held in Sweden over the disposal routes for sewage sludge until 1970. Results from a survey in 1969 showed that more than 60% of the sludge produced was used in agriculture (Socialstyrelsen, 1970), a figure which has slowly declined to 40% in the 1980's and to about 30% in 1993.

Because sewage sludge is an important source of heavy metals for those soils to which it is applied, most countries have introduced ceilings to the heavy metal load when sewage sludge is used in agriculture. The stated aim of such limits in virtually all countries is to protect the productivity of soils, the environment, as well as human and animal health. Despite this similarity of aim there are large differences in heavy metal limits for soils between countries. Differences in the philosophy behind environmental protection and in the choices of which organisms to protect have been put forward as an explanation for such diverse metal limits (McGrath *et al.*, 1994a). Some of the Scandinavian countries and the Netherlands have metal limits which go beyond the immediate protection of target organisms, but aim at minimising metal accumulation in soils as far as is practicably possible. In this article I will look at the history of guidelines and regulations affecting metal accumulation in agricultural soils in Sweden, and discuss in how far the recently formulated goal of zero accumulation of heavy metals in soils (Notter, 1993) can be justified on basis of the scientific evidence available.

The persistence of sludge derived heavy metals and their effects in soils

In most countries neither monitoring of the metal content of sewage sludge, nor of its disposal was introduced until the 1970s. Earlier evidence of the accumulation of heavy metals in soils regularly receiving sewage sludge must therefore be found in experimental trials. Only a few of the earliest of such experiments which are currently still maintained date back to before 1960. Evidence from these experiments on the persistence of heavy metals in soils after sewage sludge application may at first seem contradictory. Juste & Mench (1992) compiled metal balance data for several long-term sludge experiments and found that mass balances of metals added to the soil ranged from 30 to 90%. It is unlikely that such low recoveries less than 30 years after metal addition is due to removal of metals through crop uptake or leaching losses. These authors concluded that the hypotheses brought forward by McGrath and Lane (1989) that low recoveries are generally due to lateral movement of soil in the experimental plots was the most plausible explanation for low recoveries. Crop uptake of metals applied with sewage sludge between 1942 and 1961 at the Woburn experiments in the U.K. ranged from 0.6 to 0.03% for Zn and Cr over a 20 year period after the cessation of sludge application, whereas movement of metals down the soil profile was negligible (McGrath, 1987). Also the application of sludge with a metal content lower than that of the sludge used in the Woburn experiment, and more typical of sewage sludge derived currently from domestic sources, results in crop uptake which is only a small percentage of the amounts of metals applied. Over an eight year period with annual applications of 7 tonne sand-free sewage sludge (dm) per hectare, crop removal ranged from about 5% for Cu, Cd and Zn to less than 1% for Cr and Pb (Larsen and Petersen, 1993). In a parallel lysimeter experiment it was found that the metals applied with the sewage sludge persisted in the topsoil over an 11 year period, with non-detectable metal concentrations in the leachate (Larsen and Petersen, 1993).

The low rates of metal loss mean that their residence time in soil is extremely long, and has been estimated to lie in the order of hundreds to thousands of years (Bowen, 1977, 1984; McGrath, 1987). A simple calculation shows that it would take between 20 (Zn) and 55 (Ni) years (Table 1) to increase soil metal concentrations from current Swedish background levels to the maximum EC concentration for soils, if sewage sludge were to be applied at the maximum EC annual metal loading rate (Commission of the European Communities, 1986). In contrast, it would take between 2,400 (Zn) and 375,000 (Pb) years before soil concentrations would be even halved from the EC metal limits for soil (Table 1). In this calculation a number of simplifying assumptions is made (see Table 1), but it illustrates in terms of order of magnitude how easily metals can accumulate in soils, and how long it takes to reduce soil metal concentrations.

It is likely that the effects of elevated soil metal concentrations on crops and soil organisms will persist concomitantly. Results from long-term field experiments show that crops grown on previously sludged soils still have higher metal concentrations than plants grown on the control plots many years after the last application of sludge, even though the metal content of the crop may both decrease, increase or remain the same immediately after cessation of sludge application (Sauerbeck and Styperek, 1986). There is so far only limited evidence available on the persistence of heavy metal toxicity effects on soil microbes. Most evidence from the field comes from sites where sewage sludge was still being applied, or had only been stopped a few years before measurements were carried out. The best evidence comes from the site at Woburn where effects have persisted more than 30 years after the last sludge applications were made. The evidence of metal toxicity effects on soil microorganisms from this and other long-term field experiments with sewage sludge application was recently reviewed by McGrath *et al.* (1994b).

Even though for the small percentage of agricultural soils receiving sewage sludge it is usually by far the most significant source of heavy metal input, the per acreage smaller input of several heavy metals through fertiliser use and atmospheric deposition are important sources nationwide and are in themselves large enough to result in gradually increasing soil metal concentrations. In this context cadmium is of particular concern. Jones *et al.* (1988) found that at Rothamsted Experimental Station (U.K.) soil Cd had increased by approximately 50% since the middle of the last cen-

Table 1. Years required to increase soil metal concentrations to EC soil limits when sewage sludge is applied to agricultural land according to maximum annual loading rate in the CEC directive (Commission of the European Communities, 1986), and number of years required to subsequently decrease soil metal concentrations to half the EC limit

	Cd	Cr	Cu	Hg	Ni	Pb	Zn
Years to increase[a]	47	–	26	35	55	47	20
Years to decrease[b]	12500	187500	6500	94000	13000	375000	2400

[a] In the calculations it is assumed that soil background concentrations are average for Swedish agricultural soils (Andersson, 1992). Metal losses are insignificant relative to the rates of metal input, and ignored in the calculations.
[b] Rates of metal loss due to leaching and crop removal are for Swedish arable soils (Andersson, 1992). It is further assumed that there is no further metal input during this time, and that rates of metal loss are independent of soil concentrations.

tury, with a very marked increase from the middle of this century. Andersson (1992) calculated approximate metal balances for Swedish agricultural soils since the beginning of this century on basis of inputs from atmospheric deposition, fertilisers and lime, feed additives, and fungicides, and losses through crop uptake and leaching. As an average for Swedish agricultural soils Cd and Hg concentrations were calculated as having increased between 1900 and 1990 by 25 and 31%, respectively, Pb by 12%, whereas for Cu, Cr, Mn, Ni and Zn the increases were relatively small at less than 10% (Andersson, 1992). In the study at Rothamsted the authors were unable to detect an increase in wheat grain Cd concentrations parallel to the increasing soil concentrations due to effects of variable soil, crop, and seasonal factors on Cd concentrations in the grain. It is, however, likely that Cd concentrations in grain increase with increasing soil Cd concentrations. After application of sewage sludge linear relationships between soil Cd and crop Cd have been found at both high soil Cd concentrations of up to 10–20 mg kg^{-1} (Davis *et al.,* 1983) and low concentrations of up to 1 mg kg^{-1} (Hansen and Tjell, 1983). But also the very gradual, and smaller, increase in soil Cd concentrations due to atmospheric deposition and fertiliser use is likely to increase Cd concentrations in crops. An analysis of trends and annual variations in Cd concentration of wheat in Sweden indicated that concentrations have approximately doubled since the beginning of this century (Andersson and Bingefors, 1985). Due to the large number of soil, climatic and plant factors affecting Cd concentrations in the crop it is difficult to relate soil Cd concentrations to those found in cereal grains. Nevertheless, soil Cd is positively related to Cd concentrations in grain of oat and winter wheat in Sweden (Eriksson *et al.,* 1990; Eriksson, 1990).

Swedish guidelines for agricultural practices affecting metal accumulation in soils

Concern in Sweden about the accumulation of heavy metals in agricultural soils, and in the environment in general, dates back to the latter half of this century. Elevated concentrations of mercury in fresh water fish as well as symptoms of mercury poisoning in wild birds were very noticeable sings of the increased release of heavy metals in the environment during the 1950s and 1960s. This resulted in a ban on the use of methyl- and ethylmercuric compounds in agriculture, and restrictions on the use of mercuric compounds in the paper manufacturing industries. Nationwide surveys of metal concentrations in moss were started in 1968 and revealed elevated concentrations of in particular Cd and Pb in southwestern Sweden (Bernes, 1987). Even though much of the deposited Cd and Pb originated from abroad, steps were undertaken in 1970 to reduce the Pb content of petrol, and in 1982 restrictions on the use of Cd in a range of products were introduced (Bernes, 1987).

Despite this early awareness of, and response to, the problems caused by the release of heavy metals in the environment, the accumulation of heavy metals in agricultural soils did not receive attention until several years later. In 1977 results were published of the first survey of heavy metal concentrations in Swedish agricultural soils (Andersson, 1977), and in 1992 an assessment of the extent of metal accumulation in agricultural soils was carried out (Andersson, 1992). Over the period 1900–1990 the most significant sources of heavy metal input to agricultural soils in Sweden were commercial fertilisers (Cd), feed additives (Cu and Zn), fungicides (Cu and Hg), and atmospheric deposition (Cd, Hg and Pb), whilst sewage sludge was an important source for those soils which had received regular applications (Andersson, 1992). Sweden has recently introduced maximum permitted concentrations for heavy metals in soils when sewage sludge is used in agriculture (SNFS, 1994), but makes use of other regulations and guidelines to control the metal load of agricultural soils, a number of which relate directly to agricultural practices. The first of such regulations, mentioned earlier, prohibited the use of fungicides containing methyl- and ethylmercuric compounds, which was followed by a ban on the use of all mercuric compounds in agriculture in 1988. The use of Cu containing fungicides is currently still permitted, although their use has drastically declined in the last twenty years. Although introduced for reasons of animal health, restrictions on the use of Cu and Zn as additives to animal feeds has contributed to the relatively low concentrations of these elements in animal manures. Considerable efforts have been made to reduce the release of Cd into the environment (Notter, 1993), and the introduction of limits on the Cd content of fertilisers in combination with a progressive environmental taxation related to their Cd content has recently been introduced following an official inquiry (Jörnstedt, 1992). Earlier, through voluntary action on the side of the fertiliser industry and importers the Cd content was, however, on average already halved since the 1970's to its current level of around 60 mg Cd kg^{-1} P.

Because the application of sewage sludge results in very high local metal loadings, guidelines for the maximum permitted metal concentration of sewage sludge for agricultural use has received particular attention. Such guidelines were first formulated in 1973 (Socialstyrelsen, 1973) and contained maximum concentrations for Ag, Cd, Co, Cr, Cu, Hg, Mn, Ni, Pb and Zn (Table 2). In the absence of scientific evidence of which metal concentrations were safe, these first limits were entirely pragmatic and represented the then 'normal' concentrations in sewage sludge derived from mainly domestic sources. A loading limit of 1 tonne dm sludge per ha was recommended in order to limit the heavy metal load. In the discussions leading to these guidelines it was already recognised that "*with continued and extensive agricultural use of sewage sludge one can expect that sooner or later toxic effects will become evident. These may occur in the growing crop, soil microorganisms, or in humans or animals consuming the crops*" (Socialstyrelsen, 1970). This cautionary principle and realisation that high metal loads will result in the accumulation of heavy metals in soils with possible negative consequences in the long-term has been a guiding principle in the formulation of guidelines for the agriculture of sewage sludge ever since. Of the disposal routes available for sewage sludge, agriculture was generally deemed to be the most desirable from an environmental point of view. In order that this disposal route would remain viable in the long-term efforts were made, mainly through initiative by the Swedish Environmental Protection Board (SNV), to reduce the metal load of sewage treatment works. This has led to successively reduced concentrations of heavy metals in sewage sludge, with particularly Cd, Pb and Hg showing a sharp decline (Figure 1).

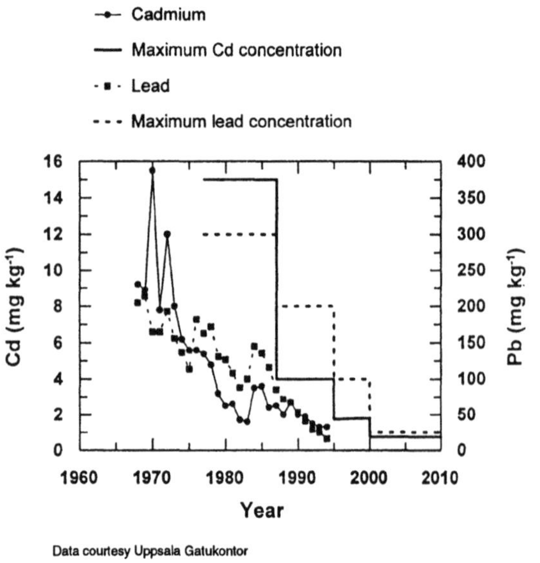

Fig. 1. Concentrations of cadmium and lead in sewage sludge produced at the Uppsala sewage treatment works serving a population of ca. 220,000 between 1968 and 1994, in relation to the maximum permitted concentrations for sludge to be used in agriculture.

The guidelines for the agricultural use of sewage sludge were reviewed in 1987 (Berglund, 1987), 1989 (Palm, 1989; Witter, 1989) and 1993 (Naturvårdsverket, 1993). On each occasion it was concluded that there was no scientific evidence to suggest that with current guidelines for the agricultural use of sewage sludge there was any immediate risk to either human health or the environment, but that work should continue to reduce the metal contents of sewage sludge. The metal limits for sewage sludge to

Table 2. Maximum concentrations of heavy metals in sewage sludge for use in agriculture (mg kg^{-1}). Until 1995 a maximum of 1 tonne dm sludge can be applied $ha^{-1}.yr^{-1}$

Reference	Year	Cd	Co	Cr	Cu	Hg	Mn	Ni	Pb	Zn
Socialstyrelsen 1973	1973	5–15	8–20	50–200	500–1500	4–8	200–500	25–100	100–300	1000–3000
Naturvårdsverket 1987	1987	4	–[b]	150	600	5	–[b]	100	200	1500
Naturvårdsverket 1990	1990	2[a]	–[b]	150	600[a]	2.5[a]	–[b]	100	100[a]	1500
SNFS 1994[c]	1995	1.75	–[b]	100	600[d]	2.5	–[b]	50	100	800
	2000	0.75	–[b]	40	300[d]	1.5	–[b]	25	25	600

[a]Concentrations may be exceeded by up to twice this value until 1 Januari 1995.
[b]Maximum permitted concentration for this metal was dropped.
[c]The maximum permitted metal concentration for sewage sludge is dropped. Instead there are limits on the yearly metal, P and ammonium-N load, which in practice means that generally less than 1 tonne ds sludge will be applied per hectare and year. The figures in the table are based on an application rate of 1 tonne ds sludge $ha^{-1}yr^{-1}$.
[d]The metal load for copper may be exceeded if there is a known copper deficiency in the soil.

Table 3. Heavy metal loads of agricultural soils resulting from the application of a hypothetical sewage sludge which heavy metal content is derived from foodstuffs and drinking water only, sewage sludge with the maximum metal content proposed in Sweden for the year 2000, and the application of farmyard manure and commercial fertilisers, and the heavy metal load through atmospheric deposition. The annual application rates for sewage sludge is 1 ton (dm) ha^{-1}, for farmyard manure 2 ton (dm) ha^{-1}, and for the commercial fertiliser 20 kg P ha^{-1}. Atmospheric deposition is for southwestern Sweden

Metal	Via foodstuffs and drinking water	Sewage sludge Swedish limits year 2000	Farmyard manure[a]	Commercial fertilisers	Atmospheric deposition
	g $ha^{-1}yr^{-1}$				
Cd	0.5	0.75	0.64	0.8[d]	0.5
Cr	3.7	40	10	36	1.2
Cu	90[b]	300	82	3.3(12)[e]	12
Hg	1[c]	1.5	0.18	0.04	0.20
Ni	15	25	16	4.3	3.0
Pn	7	25	6	0.6	35
Zn	250	600	430	25 (30)[e]	110

[a]Includes metals from feed additives (mainly Cu and Zn).
[b]This does not include Cu derived through corrosion of Cu water pipes, as this contribution is highly variable.
[c]The main contributor to Hg in the sludge is losses from dental amalgam.
[d]Average for 1992.
[e]Figures in brackets are for Cu- and Zn- enriched fertilisers.

Adapted from Naturvårdsverket (1993).

be used in agriculture have therefore been successively reduced (Table 2). At the maximum permitted metal loading rates for soils for the year 2000 there will still be a net accumulation of metals in soil, but at a much reduced rate. For Cd, input from sludge will be close to average losses through crop removal and leaching (Andersson, 1992). To achieve a balance was deemed to be especially critical for Cd as in certain regions in Sweden the limit for Cd in cereals and pulses of 0.1 mg Cd kg^{-1} discussed by WHO/FAO's CODEX-Alimentarious Commission is occasionally exceeded. The proposed limits for the year 2000 are as low as was deemed practically possible. Of the non-essential trace elements, further reductions in the concentration of Pb in sewage sludge are likely to occur as the use of lead-free petrol will eventually replace all leaded petrol. Much of the mercury in sludge is thought to originate from dental amalgam, and as can be seen in

Table 3, the proposed limit of 1.5 mg kg^{-1} is close to the estimated concentration of Hg in sludge when all Hg originates from dental amalgam, foodstuffs and drinking water. At the metal limits for the year 2000, an average application of sewage sludge will result in a metal load similar to that from other sources (Table 3).

Towards zero accumulation of metals in soil: Is it an imperative?

With the possible exception for Cd, there is apparently no scientific evidence at the moment to suggest that zero accumulation is required to adequately protect soil productivity, the environment, and human and animal health. On the basis of the scientific evidence available, SNV concluded (Naturvårdsverket, 1993) that at the 1987 and 1995 metal limits for sewage sludge in Sweden the metal loads are such that the cumulative soil metal load which might negatively affect human health or the environment would not be reached for several hundred years for all metals, except Cd. Similarly, in a recent review of the rules for sewage sludge application to agricultural land in the U.K., which has among the highest limits for metal concentrations in soil in Europe, it was recommended that, with the exception for Zn, there was no scientific evidence to warrant a revision of existing soil limits (MAFF, 1993a; MAFF, 1993b). Except for Cd, the Swedish metal limits for sewage sludge do therefore not represent the maximum loading limit considered safe for both human health and the environment, as do the U.K. and U.S. soil and cumulative loading limits, but rather a desire to keep metal accumulation in soils at an as low as possible level.

For Cd it is considered that even the current average dietary intake of 10–35 μg Cd person^{-1} day^{-1} (about 15–50% of the WHO/FAO recommended maximum intake) may put the most vulnerable part of the population at risk because of variations in sensitivity and exposure to Cd in the population, and should therefore not be further increased (Jörnstedt, 1992; Boström *et al.*, 1993). A similar concern has been expressed earlier by others (Hansen and Tjell, 1983). Moreover, Cd concentrations in wheat grain in Sweden occasionally exceed 0.1 mg kg^{-1} which virtually has become a *de facto* limit in Sweden and has been formulated as an environmental aim for Cd by SNV (Notter, 1993). These arguments, rather than the long-term aim of zero accumulation in itself, are the motivation for the proposed limits for Cd in sludge. The U.K. Department of Environment and the U.S. Environmental Protection Agency use a different approach than Sweden in the setting of a limit for Cd. Both work on basis of the WHO/FAO provisional tolerable intake of 60 μg Cd person^{-1} day^{-1} (JECFA, 1972) as the maximum acceptable intake, rather than a limit on the concentration of Cd in grain. Also their assessment of the health risks associated with Cd intake and their assessment of Cd exposure of the population differs from the Swedish approach, which may to a large extent explain the widely differing maximum acceptable loading rates for Cd between Sweden, the U.K. and the U.S.A.

Towards zero accumulation of metals in soil: Or is it merely a fad?

The maximum annual metal loading rates when sewage sludge is used in agriculture recently introduced in Denmark, the Netherlands and Sweden are nearly 2–3 orders of magnitude lower than the maximum loading rates in the CEC directive, the U.K. and the U.S.A. (Table 4). When comparing these figures it must, however, be born in mind that whereas in the former countries metal accumulation in soils as a result of sewage sludge application is regulated through loading limits, in many other countries this is through maximum concentrations of metals in soils. The Swedish metal limits for sewage sludge aim to minimise metal accumulation in soil, eventually leading to zero accumulation (Naturvårdsverket, 1993). This approach to the setting of metal limits differs markedly from those used by several other European countries as well as the United States who have set metal limits in accordance with scientific evidence of metal toxicity to target organisms, an assessment of risk and exposure, and their aims of protecting the environment and human health (McGrath *et al.*, 1994a). By setting limits at the maximum level consistent with the aims of protection of the environment and human health it is implicitly accepted that there is sufficient evidence to define such safe levels, or that the use of precautionary principles when deriving limits can sufficiently compensate for the deficiencies in the evidence available. The scientific evidence available for setting metal limits is incomplete, and will probably remain so for some time to come. Moreover, there are differences in opinion as to the selection of target organisms, the selection of the type of evidence that is relevant, and how this should be interpreted (McGrath *et al.*, 1994a). For example, despite the fact that the metal limits proposed by the U.S. EPA are based on one of the most

Table 4. Annual metal loading limits when sewage sludge is applied to agricultural land in some European countries and the USA

Country	Year	Cd	Cr	Cu	Hg	Ni	Pb	Zn
EU member states	1986	0.15	–	12	0.1	3	15	30
Denmark	1995	0.008	1	10	0.008	0.3	1.2	40
Sweden	1995	0.00175	0.1	0.6	0.0025	0.05	0.1	0.8
The Netherlands (arable land)	1995	0.0025	0.15	0.15	0.0015	0.06	0.2	0.6
United Kingdom	1989	0.15	–	7.5	0.1	3	15	15
USA	1993	1.9	150	75	0.85	21	15	140

Sources: Commission of the European Communities, 1986; Miljøministeriet, 1989; VROM, 1993; SNFS, 1994; U.K. DoE, 1989; U.S. Environmental Protection Agency, 1993.

comprehensive reviews of the scientific evidence available (U.S.Environmental Protection Agency, 1989), no account was taken of the harmful effects of heavy metals on soil microorganisms, which may have significantly affected the limits proposed for some of the metals (McGrath *et al.,* 1994a). Also the soil limits in the CEC directive (Commission of the European Communities, 1986) do not take into account metal toxicity to soil microorganisms, which they may not adequately protect (McGrath *et al.,* 1994b). As discussed in the previous section there are also differences in opinion as to the health risks associated with current dietary Cd intake.

There is a large number of uncertainties involved when deriving safe metal limits from the incomplete, and sometimes apparently contradictory evidence available. New evidence may show that limits previously thought to be safe, are in fact not. This should not deter attempts to establish safe limits, indeed this is imperative as long as metal accumulation continues to take place in our soils. But given the persistence of heavy metals in soils great caution should be exercised in how such limits are applied. It must always be remembered that such limits are only safe in the context of our current knowledge and our current objectives for protection of the environment and human health. As the simple calculation in Table 1 shows, the EC, U.K. and U.S.A. annual metal loading limits allow a rapid accumulation of metals in soils until maximum soil concentrations or cumulative metal loadings are reached. At the same time, it would take hundreds to thousands of years before such soil concentrations become significantly reduced through leaching losses and crop removal. Both our knowledge and objectives for soil protections may change long before the metals, which were allowed to accumulate in the knowledge that it was safe to do so, have been lost from the soil.

The need to reduce the metal load of soils as far as possible was also recognised by the Steering Group and the Independent Scientific Committee reviewing the U.K. guidelines for sewage sludge application to agricultural land. Both recommended that efforts should be made to further reduce the metal load of soils, despite the fact that they could conclude that current U.K. metal limits for soils, except for Zn, need not be revised in the light of new evidence (MAFF, 1993a; MAFF, 1993b). In the words of the Steering Group: " [....] *it must be borne in mind that agricultural land is an asset held in trust for the future. Contamination of such land by metals should be regarded as irreversible and must be kept to the lowest practicable level. There is the possibility that further information on these metals may lead to the lowering of acceptable concentrations in soil and tolerable dietary intakes*" (MAFF, 1993b).

The use of restrictive annual metal loading rates is an effective way to minimise the rate of metal accumulation, which many countries do not appear to make effective use of (see Table 4). It ensures that the currently considered safe limits are not reached in the foreseeable future. Moreover, it provides an effective incentive to reduce the metal loads of sewage treatment works, and to reduce the release of metals in the environment in general. In essence, this is the policy which has dictated the setting of metal limits for sewage sludge in Sweden. Experience in Sweden has shown that such low metal limits need not be a major hindrance to the disposal of sewage sludge on agricultural land (see Figure 1). In contrast, it is concern on the side of the farming community and the consumers about the presence of pollutants in sewage sludge, rather than the metal limits for sewage sludge, which explains the declining use of sewage sludge in agriculture in Sweden (personal communication O. Palm, SNV).

Conclusions

In most countries metal loading limits are set to ensure that cumulative loading rates or soil metal concentrations do not exceed maximum levels considered to be safe. There is often less emphasis on limits which regulate the annual metal loading rates for soil. The evidence on which safe metal limits are based is controversial and incomplete. Because metal accumulation in soils is essentially irreversible, it reduces the margin of safety left for future generations before potentially harmful concentrations in soil are reached. A policy which steers towards zero accumulation of metals in soils may seem prudent on basis of current knowledge of the potentially toxic effects of elevated metal concentrations in soil. It is, however, also a policy which recognises both the persistence of heavy metals and their effects in soil, as well as the need to preserve the agronomic value of soils for many years to come.

Acknowledgements

The author wishes to thank the Commission of the European Community for partial financial support from the Science and Technology for Environmental Protection Programme.

References

Andersson A (1977) Heavy metals in Swedish soils: On their retention, distribution and amounts. Swedish J Agric Sci 7: 7–20.

Andersson A (1992) Trace elements in agricultural soils – fluxes, balances and background values. Naturvårdsverket Rapport 4077. Statens Naturvårdsverk, Solna.

Andersson A and Bihgeforj S (1985) Trends and annual variations in Cd concentrations in grain of winter wheat. Acta Agric Scand 35: 339–344.

Berglund S (1987) Användning av slam på åkermark. *Naturvårdsverket Rapport* 3391. Statens Naturvårdsverk, Solna.

Bernes C (1987) Monitor 1987. Heavy metals – occurrence and turnover in the environment. Naturvårdsverket, Solna.

Boström CE, Heikensten ML, Mark-Berglund T, Sandberg E and Slorach S (1993) Miljöstörningar och hälsa. *Naturvårdsverket Rapport* 4139. Statens Naturvårdsverk, Solna.

Bowen HJM (1977) Residence times of heavy metals in the environment. International Conference on heavy metals in the environment 1: 1–19.

Bowen HJM (1984) Environmental chemistry of the elements. London, Academic Press.

Commission of the European Communities (1986) Council Directive of 12 June 1986 on the protection of the environment, and in particular of the soil, when sewage sludge is used in agriculture. Official Journal of the European Communities L181 (86/278/EEC): 6–12.

Davis RD, Stark JH and Carlton-Smith CH (1983) Cadmium in sludge-treated soil in relation to potential human dietary intake of cadmium. In: Davis RD, Hucker G and L'Hermite P (eds) Environmental effects of organic and inorganic contaminants in sewage sludge, pp 137–146. D. Reidel, Dordrecht.

U.K. Department of Environment (1989) Code of practice for the agricultural use of sewage sludge. Her Majesty's Stationary Office, London.

Eriksson JE (1990) A field study on factors influencing Cd levels in soils and in grain of oats and winter wheat. Water, Air, and Soil Pollut 53: 69–81.

Eriksson JE, Andersson A and Wenblad A (1990) Cd, Ni and Zn contents of oat grain as related to soil factors and precipitation. Swedish J Agric Res 20: 81–87.

Hansen JA and Tjell JC (1983) Sludge application to land – overview of the cadmium problem. In: Davis RD, Hucker G and L'Hermite P (eds) Environmental effects of organic and inorganic contaminants in sewage sludge, pp 91–112. D. Reidel, Dordrecht.

JECFA (1972) Evaluation of certain food additives and the contaminants mercury, lead and cadmium 1972. WHO Technical Report Series 505.

Jones KC, Symon CJ and Johnston AE (1988) Atmospheric inputs of cadmium to an arable agricultural system. In: Mathy P (ed) Air Pollution and Ecosystems, pp 548–553. D.Reidell Publishing Company, Dordrecht.

Jörnstedt O (1992) Mindre kadmium i handelsgödsel. Statens Offentliga Utredningar 1992:14. Jordbruksdepartementet, Stockholm

Juste C and Mench M (1992) Long-term application of sewage sludge and its effect on metal uptake by crops. In: Adriano DC (ed) Biogeochemistry of trace metals, pp 159–194. CRC Press, Inc, Boca Raton.

Larsen KE and Petersen J (1993) Langvarige mark- og lysimeterforsog med årlig tilforsel af store mængder tungmetalholdigt spildevandsslam. Statens Planteavlforsøg, Rapport 3.

MAFF (1993a) Review of the rules for sewage sludge application to agricultural land. Soil fertility aspects of potentially toxic elements. Report of the independent scientific committee. MAFF Publications, London.

MAFF (1993b) Review of the rules for sewage sludge application to agricultural land. Food safety and relevant animal health aspects of potentially toxic elements. MAFF Publications, London.

McGrath SP (1987) Long term studies of metal transfers following application of sewage sludge. In: Coughtrey PJ, Martin MH and Unsworth MH (eds) Pollutant transport and fate in ecosystems. Vol 6, pp 301–317. British Ecological Society, London.

McGrath SP, Chang AC, Page AL and Witter E (1994a) Land application of sewage sludge – Scientific perspectives of heavy metal loading limits in Europe and the United States. Env Reviews 2: 1–11.

McGrath SP, Chaudri AM and Giller KE (1994b) Long-term effect of land application of sewage sludge: Soils, microorganisms, and plants. In: Proceedings of the 15th International Cogress of Soil Science, Mexico (in press).

McGrath SP and Lane PW (1989) An explanation for the apparent losses of metals in a long-term field experiment with sewage sludge. Environ Poll 60: 235–256.

Miljøministeriet (1989) Bekendtgørelse om anvendelse af slam, spildevand og kompost m.v. til jordbrugsformål. Miljøministeriets bekendtgorelse nr. 736 af 26 oktober 1989.

Naturvårdsverket (1987) Hantering av slam från kommunala reningsverk. Naturvårdsverket Allmänna Råd 87:9. Statens Naturvårdsverk, Solna.

Naturvårdsverket (1990) Slam från kommunala avloppsreningsverk. Naturvårdsverket Allmänna Råd 90:13. Statens Naturvårdsverk, Solna.

Naturvårdsverket (1993) Renare slam – Åtgärder för kommunala avloppsreningsverk. Naturvårdsverket Rapport 4251. Statens Naturvårdsverk, Solna.

Notter M (1993) Metallerna och miljön. Naturvårdsverket Rapport 4135. Statens Naturvårdsverk, Solna.

Palm O (1989) Slam – Hantering och miljöproblem. Naturvårdsverket Rapport 3632. Statens Naturvårdsverk, Solna.

Sauerbeck D. R. and Styperek P. (1986) Long-term effects of contaminants. In: Processing and use of organic sludge and liquid agricultural wastes. Proceedings of the 4th International Symposium, Rome (P. L'Hermite, Ed), pp. 318–325. Reidel, Dordrecht.

SNFS (1994) Kungörelse med föreskrifter för miljön, särskild marken, när avloppsslam används i jordbruket. SNFS 1994:2 MS:72. Statens naturvårdsverks författningssamling. Miljöskydd, Stockholm.

Socialstyrelsen (1970) Slam som jordförbättringsmedel. Socialstyrelsen, Stockholm.

Socialstyrelsen (1973) Användning av rötslam som jordförbättringsmedel. Råd och Anvisningar 30. Socialstyrelsen, Stockholm.

U.S. Environmental Protection Agency (1989) Land application and distribution and marketing of sewage sludge. Technical support document. Office of Water and Regulations and Standards. Washington, DC.

U.S. Environmental Protection Agency (1993) Standards for the use and disposal of sewage sludge. Federal Register 58, 9248–9415. Washington, DC

VROM Nederland. Ministerie van Volkshuisvesting Ruimtelijke Ordening en Milieubeheer (1993) Het besluit kwaliteit en gebruik van zuiveringsslib, compost en zwarte grond. Ministerie VROM, Den Haag.

Witter E (1989) Agricultural use of sewage sludge: Controlling metal contamination of soils. Naturvårdsverket Rapport 3620. Statens Naturvårdsverk, Solna.

C. Rodriguez-Barrueco (ed.), Fertilizers and Environment, 423–430.

Phytotoxicity of heavy metals

W.H.O. Ernst
Department of Ecology and Ecotoxicology, Faculty of Biology, Vrije Universiteit, De Boelelaan 1087, 1081 HV Amsterdam, The Netherlands

Key words: Cadmium, copper, zinc, membrane interaction, compartmentation, cytosolic free metals, metal competition

Abstract

During the evolution of plants only a few heavy metals were incorporated in metabolic processes. Phytotoxicity of plants to various heavy metals occurs by surpassing critical levels. It depends on the capability of species, cultivars and genotypes to handle appropriately the uptake, translocation, incorporation into organic compounds and cellular compartmentation of these metals. These capabilities are not distributed in a random manner. Sensitivity to specific heavy metals is determined only by one to a few genes. Several principles are elaborated: at the cellular level it is the importance of the plasmamembrane integrity at the exposition to elevated concentrations of Ag, Cu, and Hg, the regulation of the concentrations of free metals in the cytosol, and the cytosolic decontamination by an enhanced transport across the tonoplast into the vacuole (compartmentation); at the whole plant level it is the translocation velocity from root to shoot and the ability of leaves to accumulate a high amount of metals.

Introduction

A surplus of heavy metals in agricultural soils originates very seldomly from geogenetic processes (weathering of ore outcrops; Ernst, 1974; Lanaras *et al.*, 1993), but it is frequently caused by anthropogenic activities, either by metal smelting and processing, even by the cement industry, or by an excessive application of fertilizers or impurified fertilizers (He and Singh, 1994; Scholten and Zuur, 1994), sewage sludges (Smith, 1994) and metal-based pesticides (Cox, 1986). For non-adapted plants these metal-enriched soils may be hazardous or even toxic. In Dutch agricultural and forest soils the Cd concentration in the soil solution varies between 0.02 and 0.05 μM (Salomons, 1993); these concentrations are already in the range of the EC10 and EC25 values affecting the glutathione content of maize (Tukendorf and Rauser, 1990). The toxicity of heavy metals in the soil, however, does not depend on its total concentration, but on the bioavailable fraction which may be modified by rhizosphere processes or artificially by the addition of phosphate, lime, organic matter and other soil additives (Mench *et al.*, 1994).

Phytotoxicity of heavy metals is the result of the imbalance between the uptake of an element and the incapability of the metabolism to cope with its cellular, especially cytosolic concentration. As soon as a critical concentration is achieved or surpassed the heavy metal can hamper directly metabolic processes or indirectly by a substitution of another metal in metalloproteins. The external concentration of a metal in the soil solution and its chemical speciation (ionic or complexed) determines the uptake process. Genotypic and species-specific patterns, however, decide on the impact of the external concentration on the physiological processes at the cellular and whole plant level. For a better understanding of phytotoxic levels of a metal it looks reasonable to consider some evolutionary processes in angiosperms which may determine the differences in uptake pattern by either a low affinity (often called incorrectly "excluder"; Baker, 1981) or a very high affinity (called "hyperaccumulators" resulting in metal concentrations above 1000 mg per kg dry mass, Brooks *et al.*, 1979).

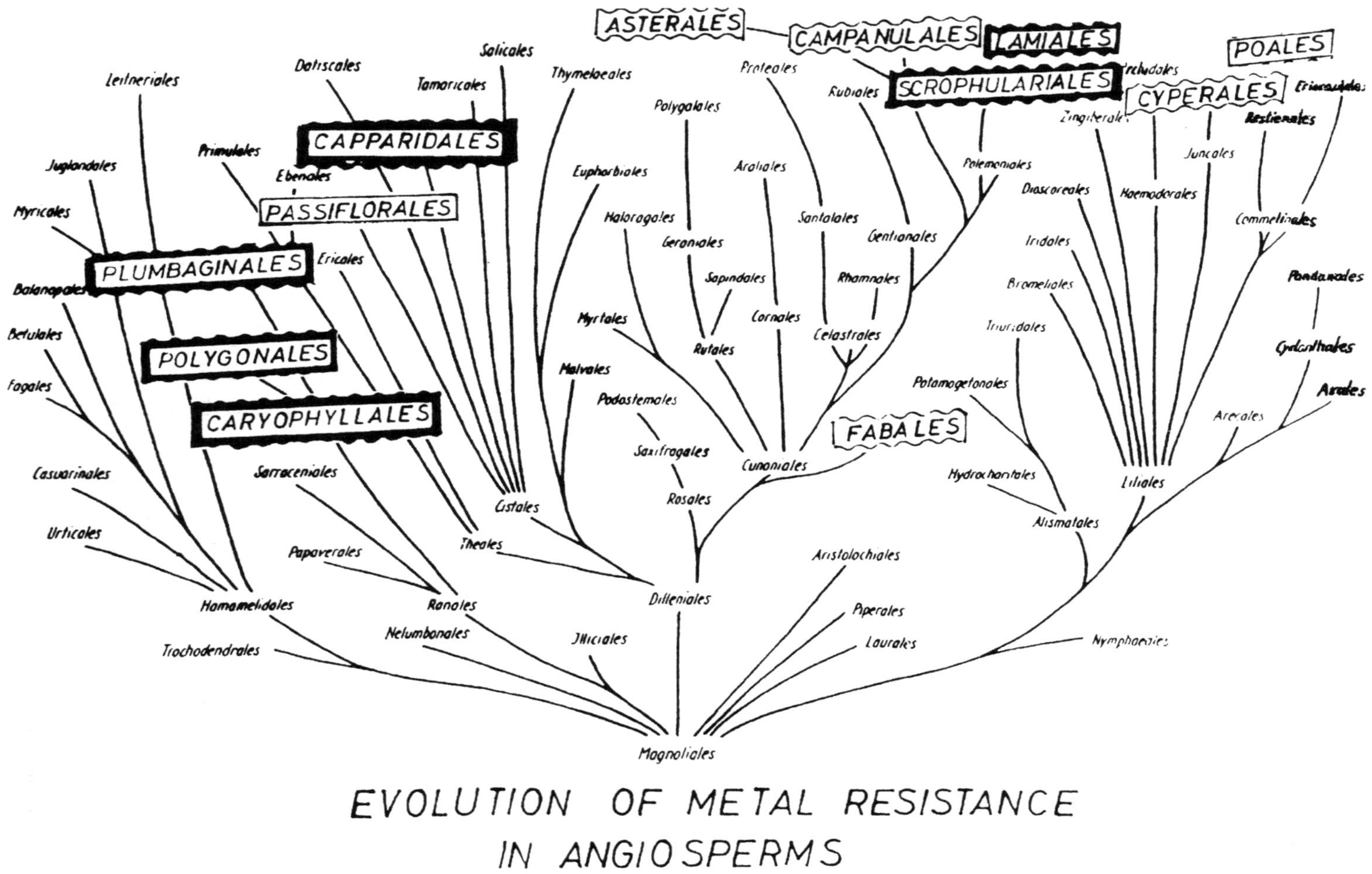

Fig. 1 Evolutionary branches of metal-hyperaccumulators (marked by thick lines) and metal resistance without hyperaccumulation. The evolutionary tree is taken from Takhtajan (1959).

Table 1. Heavy metal concentrations in leaves (umol g^{-1} dry mass) of heavy metal-resistant plants in healthy and metal-stressed (chlorotic) state on a zinc-mine at Blankenrode (FRG)

Plant species	State	Fe	Mn	Zn	Cu	Cd	Pb
Agrostis capillaris	healthy	1.4	0.90	6.6	0.15	0.01	0.01
	chlorotic	1.4	0.34	17.3	0.12	0.03	0.01
Festuca ovina	healthy	1.9	1.52	2.1	0.05	0.01	0.01
	chlorotic	1.3	0.24	17.1	0.11	0.02	0.02
Cardaminopsis halleri	healthy	3.8	2.28	97.9	0.12	0.08	0.83
	chlorotic	2.2	0.49	382.4	0.12	0.50	0.16

Evolutionary pattern in metal accumulation

During the evolution only a few heavy metals have achieved an essential role in metabolic processes as micronutrients; they are Co, Cu, Fe, Mn, Mo, Ni, and Zn. The demand for these metals is very metal-specific; a high demand exists for iron due to its central role in heme and Fe-S complexes, whereas a low demand is present for nickel, only supporting the activity of one enzyme in Fabaceae. In addition to the metal-specific demand plant species and even genotypes differ strongly (Ernst, 1974), finally resulting in a high diversity of metal concentrations in plant species and their various tissues. The accumulation of heavy metals in plant tissues without adverse metabolic effects is the final result of a cellular compartmentation and an appropriate reaction pattern of a tissue and a whole plant as soon as the cellular compartment is metal saturated.

First of all I will consider the evolution of hyperaccumulation of heavy metals (Fig. 1). Only a few evolutionary branches of the angiospermous tree sensu Takhtajan (1959) have realized this accumulation behaviour. To these branches belong the branch with the Caryophyllales, Polygonales, and Plumbaginales, the branch evolved from the Cistales with Capparidales, Passiflorales, and Salicales, and the branch with the Rubiales, Scrophulariales, Lamiales, Campanulales and Asterales. All monocotyledones are excluded from the evolution of hyperaccumulation. But within an order, and even within a family the evolution of hyperaccumulation is not randomly distributed, thus indicating that taxonomic generalisations should not be taken too far. Within the Capparidales and the family Brassicaceae as an example the genera Alyssum, Cardaminopsis, and Thlaspi contain extraordinarily many hyperaccumulators, all without an association with vesicular-arbuscular mycorrhizal fungi. The functional reason for this high accumulation, however, is still unclear: one option is the defense against herbivores and pathogens, as shown for the leaves of several hyperaccumulators (Ernst *et al.*, 1990; Boyd *et al.*, 1994). The absence of a strong impact of pierid caterpillars on hyperaccumulating Brassicaceae in contrast to non-accumulating ones supports this hypothesis. In the context of agricultural crops cereals belong to a family which has never developed metal-hyperaccumulation, thus protecting man and animals from a high metal intake. But at the same time it is quite obvious that hyperaccumulation is related to a high metal resistance and accompanied by a low metal-based phytotoxicity (cf. Table 1).

In contrast to the evolution of hyperaccumulators is the development of plants with a very low accumulation behaviour, often only realized by a low translocation efficiency (Clark, 1983) of metals from roots to shoots. This diminished metal translocation can be accompanied by metal resistance. The evolution of metal resistance in non-metal-hyperaccumulating plants follows a similar non-random taxonomic pattern. In the case of grasses only a few genera, i.e. Agrostis, Deschampsia and Festuca, contain a lot of species which have evolved resistance for a number of heavy metals, often in coevolution with their mycorrhizal partners (Griffioen *et al.*, 1994) and with a restriction of the translocation of the heavy metals from root to shoot (Ernst, 1974). The metal sensitivity of many cereals and grass cultivars may be the result of their selection from genera with a lack of metal resistance and of restricted translocation from the root to the shoot. Changes in translocation efficiency can also enhance phytotoxicity as demonstrated for Cu-sensitive genotypes of Silene vulgaris (Ernst, 1972; Lolkema *et al.*, 1984).

The genetic control of metal sensitivity and tolerance is based only on a few genes, mostly one or two major genes (Schat and Ten Bookum, 1992). Comparably single major genes, distributed over different chromosomes, control the Cu and Mn efficiency in rye (Graham, 1984). In the case of copper it has been demonstrated that the gene for copper tolerance is present in all Cu-tolerant plants (Macnair, 1983; Schat *et al.*, 1993), but that the degree of tolerance can be enhanced by a second gene. It is conceivable that the gene for Cu-tolerance controls the integrity of the plasmamembrane (Strange and Macnair, 1991). In that case it is understandable that Cu-tolerant plants have a co-tolerance for Ag and Hg. The involvement of only one gene in the case of Zn-tolerance is known for a long time (Broeker, 1963), and has been confirmed recently. This gene has the control over a rapid compartmentation of Zn from the cytoplasm into the vacuole.

Phytotoxic impacts of heavy metals at the cellular level

For an understanding of metal sensitivity it is necessary to consider first the cellular level, because the root cell is the first target for nearly all heavy metals accumulated in higher plants.

Biomembrane. The first interaction of a heavy metal with the root cell (and after translocation to the shoot) with the shoot cell is the plasmamembrane. It will receive signals of the metal concentration. Then it may react either directly by transducing the signal to the biochemical machinery of the cell (Cunning and Tomsett, 1992) or it may be affected in its integrity. An increased external concentration of sulfhydryl active metals, i.e. Ag, Cu, and Hg, will damage the stability of the plasmamembrane, as it can be easily measured as K^+-efflux and later as lipid peroxidation (De Vos *et al.*, 1991, 1993; Meharg, 1993). Another option explaining the disturbance of the integrity of the plasmamembrane is its depolarization by affecting the operation of voltage-gated ion channels. Due to the insensitivity of recent ion uptake techniques, i.e. patchclamp, for the study of heavy metals their impact on this process is not known. It has been postulated that structural changes in the plasmamembrane are protected by an enhanced synthesis of asparagine and glutamine, as a result of an impaired water potential (Costa and Morel, 1994). Non-sulfhydryl reactive heavy metals such as Co, Ni and Zn will not affect the plasmamembrane, but their uptake by plant roots may be modified by the presence of other cations. The best investigated example of competition for the same carrier is that of arsenate and phosphate by As-resistant and As-sensitive genotypes of the grass species Agrostis capillaris, A. delicatula, A. castellana, Deschampsia cespitosa, and Holcus lanatus (Meharg and Macnair, 1991a, b; De Koe, 1994). The toxicity of arsenate strongly depends on the phosphate concentration present in the soil solution or supplied by vesicular-arbuscular mycorrhizal fungi and on differences in phosphate and arsenate affinity of the genotype.

Still insufficiently known is the transport of heavy metals across the plasmamembrane. It has been proposed that they are transported by similar carriers (Clarkson and Luettge, 1989) due to competitive inhibition of Cu and Zn (Bowen, 1987), Ni and Cu as well as Ni and Zn (Cataldo *et al.*, 1978). For an explanation of different uptake kinetics between metal-resistant and non-resistant genotypes, however, the hypothesis of similar carriers demands a change of the affinity of the carrier for a specific metal or – in the case of hyperaccumulators – the presence of more than the conventional number of ion channels.

Another essential biomembrane in the regulation of the metal concentration in the cytoplasm is the tonoplast. It is very obvious that plants with an increased metal tolerance are capable to remove the metal rapidly from the cytoplasm into the vacuole. Various mechanisms have been proposed for these processes, ranging from a specific Cd/H^+ antiport activity of the tonoplast to metal sequestration models for Zn, Ni, Cu and Cd with organic acids or phytochelatins as metal shuttles (Ernst *et al.*,1992; Salt and Wagner, 1993; Cumming and Tomsett, 1992). Independent of the models there is general agreement that either the activities or the number of ion channels in the tonoplast has to be increased in metal-tolerant genotypes. In addition a strong complexation of the metal in the vacuole is necessary to protect the integrity of the tonoplast and to hamper the possibility of the vacuole-stored metals to reenter the cytoplasm.

Cytoplasm. Even if the plasmamembrane has altered its biochemistry and thus sensitivity to certain heavy metals the toxicity of the metals will remain as soon as they enter the cytoplasm. One huge gap in the understanding of metals in the cytoplasm is the knowledge on the preferent ion channels in the cytoplasm to sites of specific metal demands such as to cytosolic metallo-enzymes or cell organelles, for instance Cu, Mn, Fe and Zn into the chloroplast. It has been demonstrated that metal transport to the chloroplast is independent in genotypes sensitive or tolerant to Cu and Zn, whereas the Zn supply to the enzyme carboanhydrase has to compete with Zn compartmentation to the vacuole in Zn-tolerant genotypes (Ernst *et al.*, 1992).

Free heavy metals in the cytosol can immediately react with cytosolic enzymes and decrease their activity (Mathys, 1975). One of the most sensitive enzymes is the nitrate reductase; it may explain the rapid breakdown of the protein metabolism in metal-affected cells (Weber *et al.*, 1991). The stimulation of peroxidases by heavy metals is a general response to the presence of peroxides (Mathys, 1975; van Assche and Clijsters, 1990). At the exposure to elevated metal concentrations the production of metal-binding compounds is stimulated in the cell. As soon as the cytosolic metal concentration surpasses a critical level, phytochelatins are synthesized by activating the phytochelatin synthase (Grill *et al.*, 1989). Unfortunately the stimulation of this enzyme by heavy metals demands such a high surplus of free metals in the cytosol, that phytochelatins can not contribute to a meaningful detoxification of these metals (Schultz and Hutchinson, 1988; Davies *et al.*, 1991; Schat and Kalff, 1992; De Knecht *et al.*, 1992, 1994; Harmens *et al.*, 1993). Therefore phytochelatins may be only a very short remediation of a metal surplus in the cytosol, especially also due to their impact on the glutathione metabolism and its consequences for an appropriate management of oxidative stress (De Vos *et al.*, 1992).

A surplus of free metals in the cytoplasm will have a lot of other spin-offs. By conflicting with normal cellular activity they hamper the cell division, as shown for the impact of zinc exposure to Zn-tolerant and non-tolerant genotypes of Festuca rubra (Powell *et al.*, 1986a, b). Such a delay of cell division and cell elongation is the reason for the stunted growth of metal-affected roots and primarily the principle behind the root growth elongation test for the analysis of metal resistance in higher plants (Gregory and Bradshaw, 1965). A diminished cell growth affects the demands for proteins, which may explain the accumulation of proline at high Zn and Cu exposure (Bassi and Sharma, 1993).

Another impact on the metabolism is demonstrated by the induction of the mRNAs for heat-shock proteins by heavy metals, as demonstrated for Cd in seedlings of Glycine max (Czarnecka *et al.*, 1984) and in cell cultures of Lycopersicon esculentum (Neumann *et al.*, 1994). Their function may be an involvement in the repair of the metal stress in the plasmamembrane.

Cell organelles. A lot of experiments have carried out to investigate the impact of heavy metals on the chloroplasts (for a review: Van Assche and Clijsters, 1990). Unfortunately, in vitro-experiments are often done without a proper control of the metal availability in the reaction vessel(cf. Hampp *et al.*, 1973; Renganathan and Bose, 1990) and with irrealistic metal concentrations (Angelov *et al.*, 1993), obviously to overcome the metal precipitation and/or complexing in the reaction mixture (Hsu and Lee, 1988). In all proper experiments there is a good agreement that an excess of copper has the strongest effect of all heavy metals to chloroplasts.

After an exposure of whole plants to an excess of copper, all measurements indicate an inhibitory effect on the acceptorside of PS II by a down-regulation of the electron transport (Ouzounidou *et al.*, 1993; Maksymiece *et al.*, 1994) and by structural changes of the thylakoid membranes (Droppa and Horvath, 1990). The latter effect is also caused by Cd (Maksymiec and Baszynski, 1988). In contrast to Cd and Cu, Zn is expected to affect the carboxylase activity and the hydrolase activity by substituting Mg and Mn, respectively (van Assche and Clijsters, 1990).

Impact on the whole plant

The various changes in metabolic activity of roots and shoots under metal stress finally result in a reduction of biomass production under laboratory (Ernst, 1974; Verkleij and Prast, 1989) and field conditions (Lanaras *et al.*, 1993; Smith, 1994). Such a growth reduction may be accompanied by visible symptoms such as chlorosis and necrosis (Ernst, 1974), but at a low level of metal toxicity they may be absent. In annual plants such as cereals and pulses a diminished biomass production will affect the quality and/or the quantity of the harvest up to a complete failure (Ernst, 1974). Generally, an increase in metal concentration in leaf tissue is a first indication of an increase in metal availability. A negative linear relation between yield and metal concentration in the leaves, as shown recently for ryegrass (Smith, 1994) will only occur if the metal concentration has reached a toxic level. The reaction pattern at the whole plant-level is an integration of disturbances of the water economy (Barcelo and Poschenrieder, 1990), the mineral nutrient balance (Ernst, 1972), and the pho-

Table 2. Sequence of physiological processes due to an environmental stress by Cd

Physiological parameter	Reaction time	Tissue	Reference
Cd-concentration	minutes	roots	Cataldo *et al.* 1983
ATP-level	hours	roots	Keck 1978
Ethylene increase	hours	leaves	Fuhrer 1982
Glutathion decrease	hours	roots	Rauser 1987
Phytochelatin increase	days	roots	De Knecht *et al.* 1993
Peroxidase increase	days	leaves	Fuhrer 1982
Insoluble phenolics	days	leaves	Fuhrer 1982
Water deficit	days	leaves	Lamoreaux & Chaney 1978
RuBP carboxylase	days	leaves	Weigel 1985
Cd-concentration	days	all tissues	Ernst 1972
Phytochelatin increase	weeks	leaves	Vögeli-Lange & Wagner 1990
Abscisic acid	weeks	leaves	Barcelo *et al.* 1986
Photosynthesis decrease	weeks	leaves	Clijsters & Van Assche 1985
Senescence	weeks	leaves	Vázquez *et al.* 1989
Plant growth	weeks	all tissues	Ernst *et al.* 1974

tosynthetic source-sink relationship (Greger and Lindberg, 1986).

A surplus of heavy metals is known to disrupt plant water relations by diminishing transpiration rates and stomatal conductance (Barcelo and Poschenrieder, 1990; Kastori *et al.*, 1992; Costa and Morel, 1994). Stomatal behaviour is strongly controlled by Ca-mediated signals in guard cells (Mansfield *et al.*, 1990). The well-known impact of increased metal concentrations in the soil solution on the uptake of calcium (Ernst, 1992) may be the reason for the disruption of the water balance of plants under metal stress.

A further impact of a surplus of heavy metals on the metabolism of a whole plant is due to changes in metal translocation and compartmentation. It is quite often observed that plants under metal stress have chlorotic leaves. Such a chlorosis may be caused by an impaired iron uptake and translocation and/or by a substitution of metals in reactioncenters, as the already mentioned substitution of Mn by Zn in the hydrolase. Several of these processes may coincide (Table 1). As a consequence photosynthesis and growth are diminished, often associated with the production of small cells and finally of small leaves (Schwanitz and Hahn, 1954; Maksymiec *et al.*, 1994).

Conclusion

There is a sequence of phytotoxic effects of a surplus of heavy metals concerning the plant tissue and the metabolic processes (Table 2). Several steps in this sequence depend on metal-specific reaction patterns. The root and within the root the plasmamembrane may be the first target of metal toxicity. The plasmamembrane of the root cells looses its integrity already a few minutes after exposure to a surplus of sulfhydryl-active heavy metals, followed up by possible changes of the electric potential and the functioning of ion channels; the latter two aspects can be caused by all heavy metals. Due to the assumed genetic control of plasmamembrane integrity by a gene coding for Cu-tolerance, selection of agricultural crops for growth on slightly metal contaminated soils gives perspectives, as long as these crops do not translocate the metal excess to consumable plant parts. As soon as a surplus of free metals is accumulated in the cytoplasm they can deregulate the cell metabolism by interacting with the reaction centre of proteins, and stimulation of the production of stress peptides like phytochelatins and stress proteins, i.e. polyamines and heat shock proteins.

Due to the delay of metal translocation from root to shoot, shoots seem to be still healthy when roots are already heavily affected by a surplus of heavy metals.

Therefore the impact of metal stress on photosynthesis and water economy by deregulating ion balances, and finally the reduction of yield are a very late expression of metal toxicity, occurring sometimes several weeks after root damage.

With regard to long-term application of fertilizers, partially impurified by non-essential heavy metals, and sewage sludge the availability of Cd, Ni, Zn, and Cu, the latter especially on fen soils and in crops with a high demand for pest control, will enhance the problem of phytotoxicity. As soon as the acidification of soils will be further enhanced by acid rain (Salomons, 1993) the bioavailable fraction of heavy metals in soils will further increase and thus the potential for plant injury by a surplus of heavy metals.

References

Angelov M, Tsonev T, Uzunova A and Gaidardjieva K (1993) Cu^{2+} effect upon photosynthesis, chloroplast structure, RNA and proteinsynthesis of pea plants. Photosynthetica 28: 341–350.

Baker AJM (1981) Accumulators and excluders – strategies in the response of plants to heavy metals. J Plant Nutr 3: 643.

Barcelo J and Poschenrieder C (1990) Plant water relations as affected by heavy metal stress: a review. J Plant Nutr 13: 1–37.

Bassi R and Sharma SS (1993) Changes in proline content accompanying the uptake of zinc and copper by Lemna minor. Ann Bot 72: 151–154.

Bowen JE (1987) Physiology of genetic differences in zinc and copper uptake in rice and tomato. Plant Soil 99: 15–125.

Boyd RS, Shaw JJ and Martens SN (1994) Nickel hyperaccumulation defends Streptanthus polygaloides (Brassicaceae) against pathogens. Am J Bot 81: 294–300.

Broeker W (1963) Genetisch-physiologische Untersuchungen ueber die Zinkvertraeglichkeit von Silene inflata Sm. Flora (Jena) 153: 122–156.

Brooks RR, Morrison RS, Reeves RD, Dudley TR and Akman Y (1979) Hyperaccumulation of nickel by Alyssum Linnaeus (Cruciferae). Proc Roy Soc Lond B 203: 387–403.

Cataldo DA, Garland TR and Wildung RE (1978) Nickel in plants: 1. Uptake kinetics using intact soybean seedlings. Plant Physiol 62: 563–565.

Clark RB (1983) Plant genotype differences in the uptake, translocation, accumulation, and use of mineral elements required for plant growth. Plant Soil 72: 175–196.

Clarkson DT and Luettge U (1989) Divalent cations, transport and compartmentation. Progr Bot 51: 93–112.

Costa G and Morel JL (1994) Water relations, gas exchange and aminoacid content in Cd-treated lettuce. Plant Physiol Biochem 32: 561–570.

Cox RM (1986) Contamination and effects of cadmium in native plants. In: Mislin H and Ravera O (eds) Cadmium in the Environment, pp 101–109. Birkhaeuser Verlag, Basel.

Cumming JR and Tomsett AB (1992) Metal tolerance in plants: signal transduction and acclimation mechanism. In: Adriano C (ed) Biogeochemistry of Trace Metals, pp 329–364. Lewis Publishers, Boca Raton.

Davies KL, Davies MS and Fracis D (1991) The influence of an inhibitor of phytochelatin synthesis on root growth and root meristimatic activity in Festuca rubra L. in response to zinc. New Phytol 118: 565–570.

De Knecht JA, Koevoets PLM, Verkleij JAC and Ernst WHO (1992) Evidence against a role for phytochelatins in naturally selected increased cadmium tolerance in Silene vulgaris (Moench)Garcke. New Phytol 122: 681–688.

De Knecht JA, Van Dillen M, Koevoets PJM, Schat H, Verkleij JAC and Ernst WHO (1994) Phytochelatins in cadmium-sensitive and cadmium-tolerant Silene vulgaris. Plant Physiol 104: 255–261.

De Koe Tj (1994) Arsenic resistance in submediterranean Agrostis species. Doctorate Thesis, Vrije Universiteit Amsterdam.

De Vos CHR, Schat H, De Waal MAM, Vooijs R and Ernst WHO (1991) Increased resistance to copper-induced damage of the root cell plasmalemma in copper tolerant Silene cucubalus. Physiol Plant 82: 523–528.

De Vos CHR, Ten Bookum WM, Vooijs R, Schat H and De Kok LJ (1993) Effect of copper on fatty acid composition and peroxidation of lipids in the roots of copper tolerant and sensitive Silene cucubalus. Plant Physiol Biochem 31: 151–158.

De Vos CHR, Vonk MJ, Vooijs R and Schat H (1992) Glutathione depletion due to copper-induced phytochelatin synthesis causes oxidative stress in Silene cucubalus. Plant Physiol 98: 853–858.

Droppa M and Horvath G (1990) The role of copper in photosynthesis. CRC Crit Rev Plant Sci 9: 111–123.

Ernst WHO (1974) Schwermetallvegetation der Erde. G.Fischer Verlag Stuttgart.

Ernst WHO, Schat H and Verkleij JAC (1990) Evolutionary biology of metal resistance in Silene vulgaris. Evol Trends Plants 4: 45–51.

Ernst WHO, Verkleij JAC and Schat H (1992) Metal tolerance in plants. Acta Bot Neerl 41: 229–248.

Ernst WHO (1972) Schwermetallresistenz und Mineralstoffhaushalt. Westdeutscher Verlag, Opladen.

Graham RD (1984) Breeding for nutritional characteristics in cereals. Adv Plant Nutr 1: 57–102.

Greger M and Lindberg S (1986) Effects of Cd^{2+} and EDTA on young sugar beets Beta vulgaris. I. Cd^{2+} uptake and sugar accumulation. Physiol Plant 66: 69–74.

Gregory RPG and Bradshaw AD (1965) Heavy metal tolerance in populations of Agrostis tenuis Sibth. and other grasses. New Phytol 64: 131–143.

Griffioen WAJ, Ietswaart JH and Ernst WHO (1994) Mycorrhizal infection of an Agrostis capillaris population on a copper contaminated soil. Plant Soil 158: 83–89.

Grill E, Loeffler S, Winnacker EL and Zenk MH (1989) Phytochelatins, the heavy-metal binding peptides of plants, are synthesized from glutathione by a specific y-glutamylcysteine dipeptidyl transpeptidase (phytochelatin synthase). Proc Natl Acad Sci USA 86: 6838–6843.

Harmens H, Den Hartog PR, Ten Bookum WM and Verkleij JAC (1993) Increased zinc tolerance in Silene vulgaris (Moench)Garcke is not due to increased production of phytochelatins. Plant Physiol 103: 1305–1309.

He QB and Singh BR (1994) Crop uptake of cadmium from phosphorus fertilizers. I. Yield and cadmium content. Water, Air, Soil Pollut 74: 251–265.

Hsu BD and Lee JY (1988) Toxic effects of copper on photosystem II of spinach chloroplasts. Plant Physiol 87: 116–119.

Kastori R, Petrovic M and Petrovic N (1992) Effects of excess lead, cadmium, copper and zinc on water relations in sunflower. J Plant Nutr 15: 2427–2439.

Lanaras T, Moustakas M, Symenoidis L, Diamantoglou S and Karataglis S (1993) Plant metal content, growth responses and some photosynthetic measurements on field-cultivated wheat growing on ore bodies enriched in Cu. Physiol Plant 88: 307–314.

Lolkema PC, Donker MH, Schouten AJ and Ernst WHO (1984) The possible role of metallothionein in copper tolerance of Silene cucubalus. Planta 162: 174–179.

Macnair MR (1983) The genetic control of copper tolerance in the yellow monkey flower, Mimulus guttatus. Heredity 50: 283–293.

Maksymiec W and Basynski T (1988) The effect of Cd^{2+} on the release of proteins from thylakoid membranes of tomato leaves. Acta Soc Bot Pol 57: 465–474.

Maksymiec W, Russa R, Urbanyk-Sypniewska T and Basynski T (1994) Effect of excess Cu on the photosynthetic apparatus of runner bean leaves treated at two diferent stages. Physiol Plant 91: 715–721.

Mansfield TA, Etherington AM and Atkinson CJ (1990) Some current aspects of stomatal physiology. Annu Rev Plant Physiol Plant Mol Biol 41: 55–75.

Mathys W (1975) Enzymes of heavy metal-resistant and non-resistant populations of Silene cucubalus and their interaction with some heavy metals in vitro and in vivo. Physiol Plant 33: 161–165.

Meharg AA (1993) The role of the plasmalemma in metal tolerance in angiosperms. Physiol Plant 88: 191–198.

Meharg AA and Macnair MR (1991a) Uptake, accumulation and translocation of arsenate in arsenate-tolerant and non-tolerant Holcus lanatus. New Phytol 117: 225–231.

Meharg AA and Macnair MR (1991b) The mechanisms of arsenate tolerance in Deschampsia cespitosa (l.) Beauv. and Agrostis capilllaris L.Adaptation of the arsenate uptake system. New Phytol 119: 291–297.

Mench MJ, Didier VL, Loeffler M, Gomez A and Masson P (1994) A mimicked in-situ remediation study of metal-contaminated soils with emphasis on cadmium and lead. J Environ Qual 23: 58–63.

Ouzounidou G, Lannoye R and Karataglis S (1993) Photoacoustic measurements of photosynthetic activities in intact leaves under copper stress. Plant Sci 89: 221–226.

Powell MJ, Davies MS and Francis D (1986a) The influence of zinc on the cell cycle in the root meristem of a zinc-tolerant and a non-tolerant cultivar of Festuca rubra L. New Phytol 102: 419–428.

Powell MJ, Davies MS and Francis D (1986b) Effects of zinc on cell, nuclear and nucleolar size, and on RNA and protein content in the root meristem of a zinc-tolerant and a non-tolerant cultivar of Festuca rubra L. New Phytol 104: 671–679.

Rengonathan M and Bose S (1990) Inhibition of photosystem II activity by Cu^{2+} ion. Choice of buffer and reagent is critical. Photosynth Res 23: 95–99.

Salomons W (1993) Lange-termijngedrag van stoffen in de bodem. Bodem 4: 196–202.

Salt DE and Wagner GJ (1993) Cadmium transport acros tonoplast of vesicles from oat roots. J Biol Chem 268: 12297–12302.

Schat H and Kalff MMA (1992) Are phytochelatins involved in differential metal tolerance or do they merely reflect metal-imposed strain. Plant Physiol 99: 1475–1480.

Schat H and Ten Bookum WM (1992) Genetic control of copper in Silene vulgaris. Heredity 68: 219–229.

Schat H, Kuiper E, Ten Bookum WM and Voijs R (1993) A general model for the genetic control of copper tolerance in Silene vulgaris: evidence from crosses between plants from different tolerant populations. Heredity 70: 142–147.

Scholten LC and Roelofs LMM (1994) Radiological problems at ore processing industries in the Netherlands. Portmouths 94 Proceedings, pp 423–426. Nuclear Technology Publishing.

Schultz CL and Hutchinson TC (1988) Evidence against a key role for metallothionein-like protein in the copper tolerance mechanism of Deschampsia cespitosa (L.) Beauv. New Phytol 110: 163–171.

Schwanitz F and Hahn H (1954) Genetisch-entwicklungsphysiologische Untersuchungen an Galmeipflanzen. I. Pflanzengroesse und Resistenz gegen Zinksulfat bei Viola lutea Hudson, Alsine verna L. und Silene inflata Sm. Z Bot 42: 179–190.

Smith SR (1994) Effect of soil pH on availability to crops of metals in sewage sludge-treated soils: Nickel, copper and zinc and toxicity to ryegrass. Environ Pollut 85: 321–327.

Strange J and Macnair MR (1991) Evidence for a role for the cell membrane in copper tolerance of Mimulus guttatus Fisher ex DC. New Phytol 119: 383–388.

Takhtajan (1959) Die Evolution der Angiospermen.Fischer Verlag.

Tukendorf A and Rauser WE (1990) Changes in glutathione and phytochelatins in roots of maize seedlings exposed to cadmium. Plant Science 70: 155–166.

Van Assche F and Clijsters H (1990) Effects of metals on enzyme activity in plants. Plant Cell Environ 13: 195–206.

Verkleij JAC and Prast JE (1989) Cadmium tolerance and co-tolerance in Silene vulgaris (Moench)Garcke [=S.cucubalus (L.) Wib.}. New Phytol 111: 637–645.

Weber MB, Schat H and Ten Bookum-vander Maarel WM (1991) The effect of copper toxicity on the contents of nitrogen compounds in Silene vulgaris (Moench)Garcke. Plant Soil 133: 101–109.

C. Rodriguez-Barrueco (ed.), Fertilizers and Environment, 431–436.

Heavy metal transfer in the food chain to humans

H.-J. Hapke
School of Veterinary Medicine Hannover Bünteweg 17, D-30559 Hannover Germany

Key words: Heavy metals, food chain, food toxicology

Abstract

Heavy metals are ubiquitous and chemically stable, so they can be expected to be present in all parts of the biotic and abiotic matter. The pathway within the food chain starts in case of cadmium in the soil via roots to plants, in case of lead by the air dust and in case of arsenic and mercury in the polluted water. During the transfer from one link in the chain to another some heavy metals may be accumulated up to the final link, man. To avoid an increase of unwanted toxic heavy metals in food for human consumption it is necessary to limit the concentrations in the lower links of the food chain, i.e. in water and soil, by establishing tolerable levels or by a limitation of industrial emissions.

Introduction

Food for human nutrition as well as animal feed contains many chemical compounds without any nutritive value. Most of them are unavoidable as they are the result of a general environmental contamination from air, soil, and water as well as from plants and animals which are used for food production.

The actual occurrence of most environmental pollutants in food like heavy metals depends on factors regarding the physico-chemical properties like *bioavailability* in soil and water, in plants and animals, *absorption* rate through biological membranes, metabolic *stability* of the compound in the biotic and abiotic environment, *distribution* in different tissues of plant and animal organisms, and last but not least on the abilitiy to form *deposits*, for example in bones, liver, kidneys or in fat tissue. The final factor in this respect is the rate of *excretion* via bile, feces or urine. These factors are totally independant on pharmacodynamic and toxicological properties, but only on the toxicokinetic behaviour, that is the movement of chemicals through biological systems, whether unchanged as mother substance or after metabolisation in the biological system.

Food chain

Out of the large group of chemical pollutants in the environment only persistent chemicals accumulate within the food chain from step to step. That means that concentrations in a higher link of the chain are comparatively higher than those of the lower links. Those chemicals which are quickly excreted decumulate and therefore are of low toxicological significance.

Food chain is the way of transfer of persistent pollutants from soil and water via plants and animals to men as the final link, followed by mother milk and the suckling infant. In case of heavy metals the food chain stops in human tissues without subsequent excretion in mother milk, in contrast to organochlorines which are present in the fat compartment of mother milk in remarkable amounts. Heavy metals are not excreted preferentially by milk in an important quantity (less than 0.01 mg lead per one liter milk) in all mammals including man, even not after a high exposition of the lactating animal or the nursing mother.

The pharmacokinetic or toxicokinetic behaviour of most compounds from one link to another within the food chain is not fully investigated. For many substances only one single transfer step from one chain member to another of the higher level has been analysed, for instance the passage from animal feed to animal tissues, which are used as food for human nutrition, including milk, eggs, or even honey.

We are only generally informed that fixed relations of concentrations in animal feed and animal tissues which are used for food production, can be recognised easily by experimental investigations in food producing animals. We know that a lead concentration of 30 mg/kg dry matter in animal feed results in a concentration of not more than 0.8 mg/kg in the animal's kidney (fresh matter), as a calculated mean value. On the other hand only 2 mg cadmium/kg dry matter in feed leads to the same concentration of 0.8 mg/kg kidney. The consequences of these investigations in cows and pigs, the most important meat producing animals in most countries, are the tolerances for lead of 30 and for cadmium of 2 mg/kg dry matter of feed. Therefore it is necessary to observe the heavy metal content of animal feed to avoid a high concentration in edible tissues.

The results after reaching a steady state between intake and elimination allow the basis to calculate so called carry over factors. A carry over factor of less than 1 means that no accumulation occurs, carry over factors of accumulating pollutans are between 1 and 10 or even more, if the expressions of both concentrations in feed and animal tissues are identical. In practice this is not the case, as concentrations in animal feed are expressed as content in dry matter (88%), but those in food as content in fresh matter, according to the official regulations of tolerances in feed and in food respectively.

Within the food chain, only two groups of pollutants are of actual relevance, as far as we know today, regarding the above mentioned reasons:

- Many heavy metals like lead, cadmium, arsenic, mercury, chromium, thallium (but not iron, copper, nickel and zinc)
- Most aromatic organochlorine compounds like DDT, dioxines and PCBs (but not aliphatic organochlorine compounds like chloroform).

Metal load of feed and food depends on the conditions under which they are produced. Some heavy metals are without any nutritional value and their presence may have only toxicological significance: lead, cadmium, mercury, arsenic, in this decreasing significance.

Accumulation takes place only in some target organs which play a role as deposits of these elements. Increase of concentrations of heavy metals in feed or food depends:

1. on their concentration in the environment,
2. on the duration of exposure from some days in plants to many years in animals and
3. on the bioavailibility, i.e. on the physico-chemical properties of the individual compound.

In many cases either the dose is very low or the exposure time is short or the bioavailability is insufficient for an accumulation process.

Since individual heavy metals and their organic and inorganic compounds exhibit distinguished behaviour, a specified description of the process of accumulation in the food chain is necessary.

Differing reactivities of heavy metals promote deposit formation by binding of ionized metals to storage structures in the organic matter of the environment or in living beings. Both absorption and distribution depend on the water- and/or lipid-solubility of the individual ccimpound which is remarkably influenced by many conditions, present in the biosphere, particularly in soil, water and organisms. Heavy metals are not present in a pure metallic form but as different compounds like oxides, sulfides or carbonates. As the physico-chemical properties vary from one compound species type to another, and this situation is further influenced by pH-value of the surrounding site, it is very difficult or even impossible to quantify the pathway of a certain metal and to describe the transer from water or soil into plants or animals, if not the real situation is recognized. And that changes from time to time and from area to area.

That is the reason that we can not describe a single pathway of a heavy metal in the food chain. In the following part only some examples can be mentioned, so far as we have knowledge about the complicated material. The description will be restricted to arsencic, cadmium, lead and mercury to mention the most important ones. Nickel and zinc have no accumulating properties, in so far they are not mentioned here in detail. Arsenic and mercury enter the food chain preferentially from the water into marine organisms, cadmium from the soil into roots of plants, lead from the air dust into plants and leaves.

Arsenic

With respect to foodstuffs, the transfer of arsenic from soil to plants is of little importance. Health problems only may occur, if any, after an industrial or agricultural pollution of the soil. In contrast, arsenic is continously absorbed by marine organisms, where it accumulates to values of 0.1 to 5 or more mg/kg (mean 2.6) in fish. Concentrations of 10 mg/kg fresh matter have been found in estuaries and offshore areas. By way of the accumulation in plankton, small and medium-sized fish and therefore fishmeal, animal feed from this origin may contain arsenic concentrations of 0.1 to 0.5 mg/kg. Thus arsenic is to be expected in those tissues of farm animals, that are fed with fishmeal, namely poultry and pigs:

Water	0.05	mg/l
Fish	2.60	mg/kg
Fishmeal	0.10	mg/kg
Poultry	0.20	mg/kg
Pork	0.10	mg/kg
Kidney	0.50	mg/kg
Liver	0.50	mg/kg
Beef	<0.01	mg/kg
Milk	0.03	mg/kg

To decrease the arsenic content of food for humans only those procedures are sufficiently effective which are working in water protection measures. Highly contaminated fish like tuna (> 1 mg/kg) should not be used for human consumption.

Cadmium

Only about 1% of the daily intake in animals as well as in man is absorbed from the intestine and 99% leaves the organism again with feces. But most of this small part is stored in the organism. Cadmium concentrations in the storage tissues (kidneys, liver) will increase continously, because the excretion of the absorbed part is lower than the absorption of the ingested part. After a time the concentration will come up to an equilibrium. Cadmium is comparatively immobile and enters the food chain at the level of the polluted soil, depending additionally on the geological formation, composition and physical properties of the soil and the use of some fertilizers in the agriculture (phosphates from special geological formations). The uptake of this heavy metal by the plant roots depends on the pH value of the soil at that site. Due to a short vegetation period of food plants, however, only small scale accumulation is observed in plants. Thus the cadmium contents of vegetable feeds vary from 0.001 to 0.1 mg/kg and sometimes in few cases much more than that.

As a result of contaminated animal feed, cadmium is a normal constituent of food from animal origin. Since the rate of even a small oral uptake in farm animals as well as in man exceeds excretion, accumulation results in some tissues, depending on the time of exposure, that is the age of food producing animals or of man. Pigs which are going to be slaughtered after a fattening period of about 6 months are lower contaminated with cadmium (<0.5 mg/kg kidney cortex; <0.08 mg/kg liver) than old animals with an age of many years (more than 5 mg/kg kidney; 0.1 to 1.0 mg/kg liver respectively). An accumulation of cadmium in meat and milk is unlikely, except in horses.

If the animal feed contains no more than 0.5 mg/kg dry matter, a critical concentration of 0.8 mg/kg in kidney of aninials within 3 years can be avoided. After this period the concentration may increase to more than 1 mg/kg kidney, which is discussed to be the tolerance level for human nutrition. Thus the cadmium content of animal feed should be limited to 0.5 mg/kg dry matter. For a period of less than 3 years the limited concentration may be up to 2 mg/kg.

0,8 mg/kg	0,5 mg/kg
fresh matter	dry matter
kidney	animal feed
(0,5 - 1,0)	(0,2 - 0,8)

An impressive example is given by the time- and dose-dependent accumulation of cadmium in broilers (Fig. 1). After the results of feeding experiments in these animals a threshold level in their feed of 0.4 mg/kg dry matter is obvious, using liver, a target tissue for cadmium. Hares living in a polluted area have remarkably higher concentrations of cadmium in musculature, liver and kidneys than those from unpolluted areas. Additionally the increasing age of these animals (up to 3 years) increases the cadmium contents in these tissues.

If the cadmium concentration in the feed for pigs and cattle is less than 0.5 mg/kg dry matter, the following concentrations in animal tissues will be the result:

	Liver	Kidney	Meat	Fat	Milk
for pigs:	0.05	0.5	0.01	0.001	-
for cattle:	0.1	0.5	0.01	0.001	0.0005

Lead

Transfer of lead from the soil to plants takes place only if lead concentrations of the soil are extremely high, like in special geological formations or after an industrial pollution, i.e. more than 3000 mg/kg dry matter. Obviously there is no appreciable pathway of lead from the soil through roots to plants.

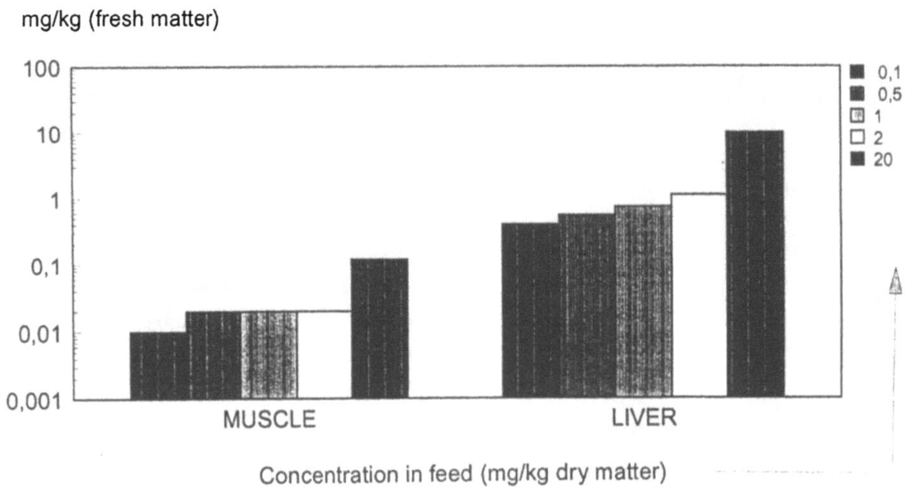

Fig. 1. Cadmium accumulation in broilers.

Plants which are used for animal feed and human food show lead concentrations ranging from less than 0.1 up to 5 or even 10 mg/kg dry matter. The reason is not an uptake from the soil but preferentially by dust from the air, that means that only an external pollution is observed, in contrast to cadmium. It can be removed by cleaning the plants, which is the case usual in food processing.

Lead contents of animal tissues, in particular of liver and kidneys, not of other organs like meat, are caused mainly by the uptake of contaminated feed, only a small part added by inhalation.

Experiments in broilers show a dose-dependent increase of lead in liver, kidneys and bones (Fig. 2). Only after a high dose of 40 mg/kg feed there is a slight increase of lead content in the muscle. Bones are representative for a lead burden quantification (Fig. 3). If the fodder for farm animals does not contain lead concentrations of more than 30 mg/kg dry matter, the content of lead in the liver of these animals will not exceed 1 mg/kg fresh tissue, which is tolerated for human consumption.

1 mg fresh matter liver, kidney	30 mg/kg dry matter animal feed

As the feed of food producing animals shows a concentration of much less than 30 mg/kg, the tissue contents in animals, which are used for human nutrition are really very low.

Mercury

Mercury is a toxic heavy metal of which the food chain mainly starts in the water. The pathway from polluted soil, like after an industrial or agricultural contamination (i.e. some fungicides) in the past, is very small and is the exception today. Food and animal feed derived from plants usually show mercury concentrations between 0.001 and 0.03 mg/kg and are insofar without significance for human health.

By industrial usage, mercury sometimes is included into bodies of water where this metal exhibits a behaviour different to that in soil, which may lead to accumulation in the food chain.

As is the case for arsenic, various inorganic and organic species of mercury must be distinguished. Marine organisms are especially able to transform inorganic mercury cornpounds into organic ones with quite other physico-chemical properties including higher lipid solubility. As an organic compound, mercury is more easily transferred through the aquatic food chain than as an inorganic compound or as metal itself, which in fact does not play a role in food toxicology.

As a result, marine organisms can show mercury levels of up to 5 mg/kg, mostly bound as methylmercury, depending on age and species of the organism. Accumulation factors in the marine food chain are 100 to 1000, compared to only 2 to 5 in the terrestrial food chain. Sea food is thus a particular source of mercury burden in man and animals. Fresh water and estuarine fish such as mussels and eels show high-

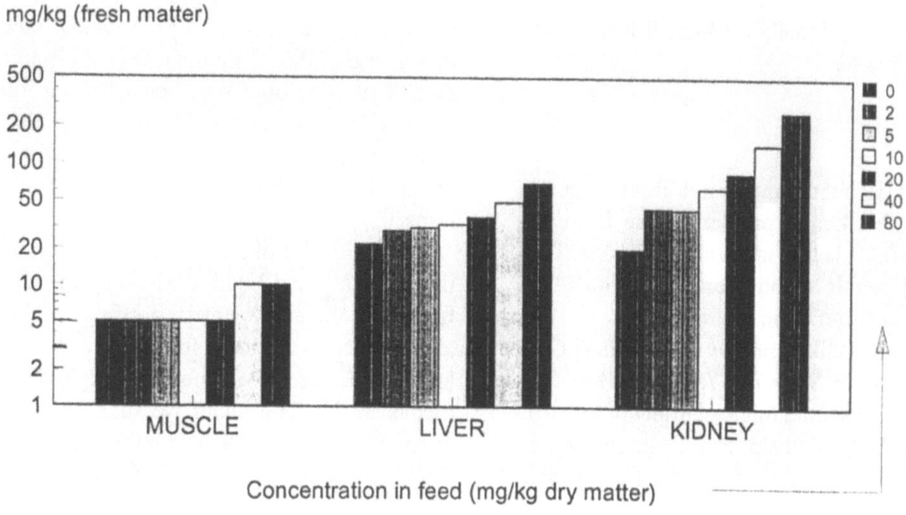

Fig. 2. Lead accumulation in broilers.

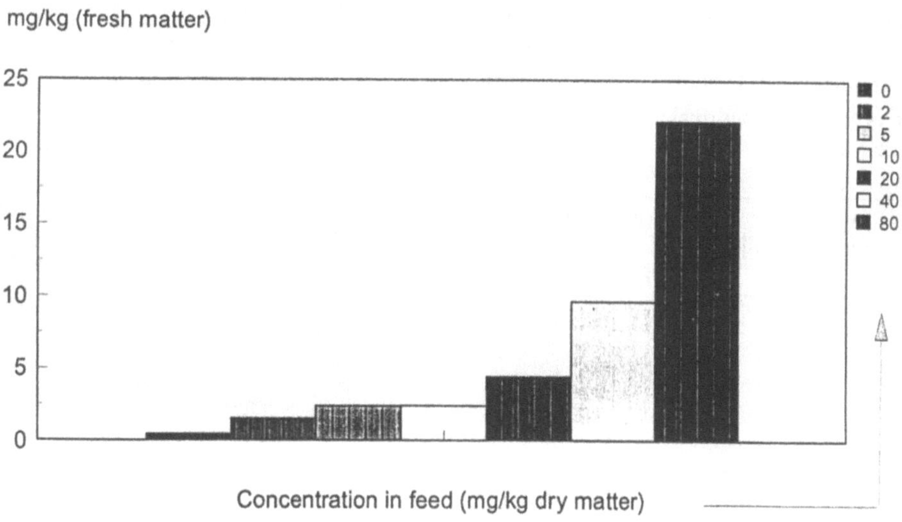

Fig. 3. Lead accumulation in bones of broilers.

er concentrations (up to 3 mg/kg), depending on the higher pollution of their residential district, than fish caught in the open sea (0.005 mg/kg).

Mercury passes to domestic animals and animal derived food products via fish meal, which is used as feed additive in animal production. Swine liver may contain up to 0.1 mg/kg and pork meat 0.05 mg/kg. Beef products on the other hand contain only very small amounts due to a demethylation of organic mercury compounds by microorganisms in the rumen of cattle. The resulting inorganic compounds are rnuch less lipid-soluble and the enteric absorption is very poor. Beef contains 0.001 to 0.02 mg/kg, milk no more than 0.01 mg/l. Poultry and eggs show higher concentrations as a result of the widespread use of fish meal as feed in hens and broilers.

So the nutritional mercury burden ofman derives mainly from the consumption of fish and other aquatic animals. To avoid a mercury burden of man, the content of mercury in fish and fish-derived food should be limited to 0.1 or exceptionally for some fish species to 1.0 mg/kg fresh matter.

Zinc, nickel

Zinc is an essential metal and is not accumulated in the food chain. No health problems arise from this compound. The same is true in the case of nickel. Thus these two heavy metals are not worth mentioning here as a problem of accumulation.

Many other metals like *Antimony, Chromium, Copper Iron, Molybdenum, Selenium, Thallium, Vanadium* may play a role as essential micro-nutrient or as a toxic compound in the food chain. But the knowledge of these elements is very poor, so that no further information is available.

Conclusion

A comparison of the actual occurence of heavy metals in animal feed with those concentrations which are propably the limited levels in feed show that today the real situation is not dangerous or alarming, as is shown in the following table.

	Heavy metals in feed (mg/kg dry matter)	
	actual	limit
Lead	0,1 - 5,0	30-40
Cadmium	0,0001-0,05	1-2
Mercury	0,0001-0,01	102
Arsenic	0,1-1,0	1-2
Chromium	0,02-10	1000?
Thallium	0,02-0,1	1-2
Nickel	0,1-10	150-250
Zinc	1-50	150-250
Copper	5-20	10-50
Iron	1-100	500-1000

If in special cases heavy metal concentrations in feed are too high some procedures are necessary to reduce the bioavailibility of the special metal for the further use of this contaminated feed for animal nutrition. This procedure may consist in an introduction of chemical substances which change the physicochemical properties of the heavy metal within the food chain. We call this method the use of ecochemical traps.

There is a lot of further research work to be done in Veterinary and Food Toxicology in the very next future, for the aim to ensure a safe animal production and production of food from animal and plant origin in an industrial age. If we know the single steps of heavy metals in the food chain we shall be able to avoid transfers. This is necessazy for the health protection of the consumer, to whom we all belong.

References

Deutsche Gesellschaft für Ernährung (1992) Ernährungsbericht 1992 (*Nutrition Report*), Frankfurt/M.

Hapke HJ (1991) Metal accumulation in the food chain and load of feed and food. In: Merian, E (ed) Metals and their compounds in the environment. VCH Verl ges Weinheim.

Hapke HJ (in press) Belastung der Umweltmedien - Lebensmittel. In: Wichmann, Schlipköter and Fülgraff (ed) Handbuch der Umweltmedizin. Ecomed Verl ges Landsberg.

Hecht H (1986) Zur Cadmiumbelastung und des Carry overs beim freilebenden Tier. In: Bundesministeirium für Ernährung: Zum Carry over von Cadmium. Angew Wiss Heft 335: 117–136.

Vemmer R and Petersen H (1979) Untersuchungen über den Einfluß steigender Cadmiumzulagen auf die Entwicklung von Mastschweinen und auf die Rückstandsbildung. Landwirtsch Forsch **32**: 303–315.

Vogt H and Nezel K (1981) Der Einfluß von Bleizusätzen zum Broiler- und Legehennenfutter auf die Leistung der Tiere. In: Bundesminister für Ernährung: Zum Carry-over von Blei. Angew Wiss Heft 254: 25–32.

C. Rodriguez-Barrueco (ed.), Fertilizers and Environment, 437–442.

Dilution effect of plant biomass on plant cadmium concentration as induced by application of phosphate fertilizers

S.H. Chien & R.G. Menon
Research and Development Division, International Fertilizer Development Center (IFDC), P.O. Box 2040, Muscle Shoals, AL 35662, U.S.A.

Key words: plant Cd concentration, P fertilizers, dilution effect, P and Cd uptake by maize, P solubility

Abstract

Two greenhouse experiments were conducted to investigate the dilution effect of plant biomass on Cd concentration in plant tissue as induced by application of P fertilizers. In the first experiment, 18 soils varying widely in soil properties were treated with 200 g kg^{-1} of Cd in $CdCl_2$ solution. The soils then received either no P or 200 mg P kg^{-1} as TSP. Maize was grown for 6 weeks after planting. Significant P response was observed with all the soils. Plant Cd concentration was significantly lower in the TSP-treated pots than in the control pots. A decrease of plant Cd concentration with an increase in dry-matter yield followed a curvilinear semi-log function. Soil DTPA-extractable Cd was about the same with and without TSP added. In the second experiment, three sources of P fertilizers were used: North Carolina phosphate rock (NCPR), Togo phosphate rock partially acidulated with H_2SO_4 at 50% level (PAPR), and a commercial-grade TSP. The Cd contents in NCPR, Togo PAPR, and TSP were 47, 40 and 7 mg Cd kg^{-1}, respectively. The P fertilizers were added to Hartsells silt loam (pH 4.7) at rates of 0, 100, 500, 1000, and 2000 mg P kg^{-1}. Maize was grown for 6 weeks after planting. The results show that plant Cd uptake depended on the P solubility and Cd content of the P fertilizers used. Plant P uptake increased with increasing dry-matter yield, whereas plant Cd concentration decreased with increase in dry-matter yield.

Introduction

One of the environmental problems associated with the use of P fertilizers is the potential accumulation of Cd in soil and plant. Uptake of Cd by plants that may be consumed directly or indirectly by man is one avenue by which Cd enters the human food chain from the applied P fertilizers. Numerous studies have investigated the Cd uptake by crops from the applied Cd-containing P fertilizers (Mortvedt *et al.*, 1981; Mortvedt, 1987; Mortvedt and Osborn, 1982).

In a study on the effect of long-term application of P fertilizers on Cd bioavailability, Mulla *et al.* (1980) observed that Cd concentrations in barley leaves from plants grown on unfertilized plots were greater than those observed for leaves from plants grown on the P-treated plots. They assumed that the very high levels of P applied rendered Cd more plant unavailable in treated than in control plots. The DTPA-extractable Cd from the P-treated plots, however, was significantly greater than that from the control plots. Thus, this observation did not support their explanation that added P rendered Cd less available by the formation of insoluble Cd-P compounds, as reported by Street *et al.* (1978) and Pezzarossa *et al.* (1990).

Alternatively, a decreased plant Cd concentration associated with P application may be due to the dilution effect (Mortvedt *et al.*, 1981; Mortvedt, 1987). If a soil is deficient in available P, application of P to the soil will increase the plant's growth and produce higher yield, which frequently decreases plant Cd concentration. The implication is that high available P in soil not only increases plant yield but also reduces plant Cd concentration. However, little information is available on direct evidence to separate the dilution effect in plants from the chemical effects in soils in terms of explaining the relationship between P application and Cd availability.

Table 1. Selected soil properties of the soils used

Soil	pH	Organi.matter	Clay	ECEC[a]	P_iP^b
		%		cmol kg^{-1}	mg P kg^{-1}
Hartsells	4.7	4.4	21.4	4.3	4.0
Crowley	4.7	3.3	23.9	6.9	9.8
Bladen	4.8	2.2	6.8	2.6	3.3
Mountview	4.8	2.5	20.2	3.2	6.3
Venango	5.1	4.4	13.2	6.8	9.9
Guthrie	5.5	2.2	29.6	8.2	27.4
Guin	5.8	0.9	20.0	3.6	22.6
Sharky	6.0	2.6	64.0	29.5	33.2
Lakeland	6.3	1.1	5.6	1.6	4.3
Josephine	6.3	6.0	23.0	11.7	10.1
Troup	6.4	0.8	4.6	0.5	28.8
Windthorst (sub)	6.5	0.8	42.7	19.1	2.9
Savannah	6.7	1.2	18.7	5.6	18.7
Greenville	6.7	1.7	16.5	6.2	18.2
Windthorst (top)	7.7	1.0	20.2	11.9	2.8
Bromfield	8.1	0.2	10.9	6.6	5.1
Vernon (sub)	8.4	1.3	42.2	48.8	13.7
Vernon (top)	8.5	1.1	46.4	37.2	13.1

[a] Effective cation exchange capacity (ECEC).
[b] Extraction with iron oxide–impregnated paper (Pi paper) (Menon *et al.*, 1990).

The objective of the present study was to investigate the possibility that a decreased plant Cd concentration associated with P application could be due to the dilution effect in plant tissue rather than due to the formation of insoluble Cd-P compounds in soil.

Materials and methods

Experiment 1

Eighteen soils varying widely in soil properties were used in this experiment. A summary of these properties is shown in Table 1. The soils were air-dried and screened to less than 2 mm before potting.

Cadmium was added to 4 kg of each soil at a rate of 200 μg Cd kg^{-1} in the form of $CdCl_2$ solution. Phosphorus was mixed with one set of soil samples at a rate of 200 mg P kg^{-1} as TSP, whereas the other set of soil samples received no P. The ratio of concentrations of Cd and P added to the soils was equivalent to 1000 μg Cd g^{-1} P. This is comparable to the high Cd concentrations in the two commercial DAP and TSP fertilizers that were used by Mortvedt and Osborn (1982). Adequate levels of fertilizers containing N, K, and micronutrients were added to each pot at constant rates.

Six seeds of maize (*Zea mays L.*, Variety Funks G5757) were planted, and water was added to the soils on a daily basis to bring the soil moisture level to approximately 80% of field capacity. Treatments were replicated three times in a randomized complete block design. After germination, the plants were thinned to three plants per pot. Six weeks after planting, the aboveground portions of the plants were harvested, dried in an oven at 60 °C, and weighed. Soil samples were taken from each pot for chemical analyses.

The plant materials were ground using a Wiley mill. Plant samples were digested with H_2SO_4-H_2O_2 mixture. Cadmium was extracted from the soil samples with DTPA solution. To determine Cd uptake by plants, the plant samples were ashed in a furnace at 400 °C. The ash was then dissolved with 5N HNO_3 at 50–60 °C.

Phosphorus in the plant digests was determined using the ammonium molybdate-ascorbic acid method.

Table 2. Chemical properties of three P sources used

P source	Total P	Water-soluble P[a]	Citrate-soluble P[a]	Total Cd
	%			mg Cd kg^{-1}
North Carolina PR	13.3	–	20.2	47
Togo PAPR	12.3	40.1	5.3	40
TSP	20.8	83.0	17.0	7

[a] As % of total P.

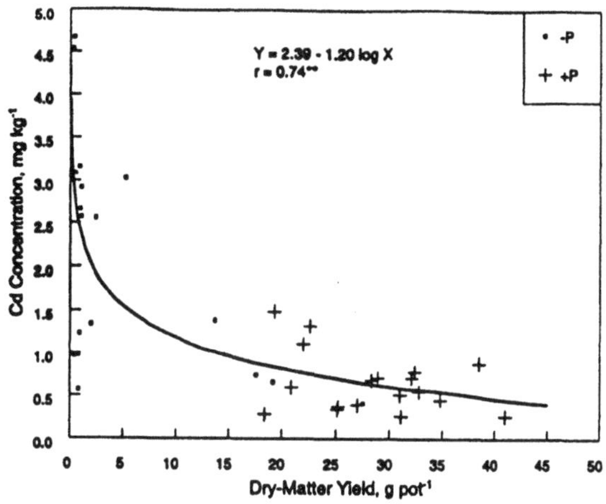

Fig. 1. Relationship between plant Cd concentration and dry matter yield obtained with or without P application.

Cadmium in the ashed plant samples and soil DTPA extracts was determined by inductively coupled plasma (ICP).

Experiment 2

The soil used was a Hartsells silt loam classified as Typic Hapludults. Its pertinent chemical properties are shown in Table 1 (soil No. 1). Three P sources varying in P solubility and Cd content were used. These were North Carolina phosphate rock (NCPR), Togo PR partially acidulated with H_2SO_4 at 50% level (PAPR), and a commercial-grade TSP. Their chemical properties are shown in Table 2.

Phosphorus fertilizers in the amounts required to give total P rates of 0, 100, 500, 1000, and 2000 mg P kg^{-1} were mixed with 4 kg of soil. The very high P rates were used on the assumption that the time span required for evaluating the cumulative effect of applying small doses of P fertilizers for several seasons can be shortened by applying a very large amount of fertilizers in one dose. The remaining experimental procedures were the same as those described in Experiment 1.

Results and discussion

Experiment 1

A significant increase in dry-matter yield of maize due to the application of P fertilizer was observed with all the soils, even though some soils produced high dry-matter yield without added P (Table 3). Concentration of Cd was consistently higher in plants grown in soils not fertilized with P than in those grown in soils treated with P fertilizers (Table 3). The relation between plant Cd concentration and plant dry-matter yield shown in Fig. 1 clearly shows that plant Cd concentration decreases with an increase in dry-matter yield a dilution effect of plant biomass on Cd concentration in a semi-log function. The data also show that all the soils treated with TSP produced low plant Cd concentration ≤ 1.50 mg kg^{-1}. In other words, in a soil with low available P and highly polluted with Cd, application of P fertilizer not only increases plant yield but also reduces plant Cd concentration due to the dilution effect.

There was no significant difference in DTPA-extractable Cd content from soils treated with and without P fertilizer (Table 3). This indicates that there was no formation of insoluble Cd-P compounds in soils treated with Cd and P. This observation further supports the contention that the decrease in plant Cd concentration as induced by P fertilizer is not due to the Cd-P precipitation in soils but is mainly due to the dilution effect as a result of increasing plant biomass obtained with P treatment. This explanation contradicts Mulla *et al.* (1980) who assumed that, although DTPA-extractable Cd was greater in the P-treated than

Table 3. Dry-matter yield of maize, plant Cd concentration and P uptake, and DTPA-extractable Cd obtained with or without TSP treatments

Soil	Dry matter yield g pot^{-1}		Cd concentration mg Cd kg^{-1}		P uptake mg P pot^{-1}		DTPA-Cd mg Cd kg^{-1}	
	−P	+P	−P	+P	−P	+P	−P	+P
Hartsells	0.33	22.0	4.54	1.11	0.3	42.6	0.14	0.13
Crowley	27.5	41.0	0.41	0.26	48.8	105.7	0.18	0.20
Bladen	0.57	22.6	3.09	1.32	0.6	71.7	0.13	0.11
Mountview	1.07	32.8	2.57	0.55	1.1	70.9	0.16	0.15
Venango	1.08	32.4	2.92	0.78	1.2	74.8	0.44	0.41
Guthrie	17.6	27.0	0.74	0.39	37.5	63.7	0.24	0.23
Guin	5.23	32.1	3.03	0.71	5.8	118.0	0.17	o.15
Sharky	19.2	25.1	0.66	0.34	34.9	84.8	0.24	0.25
Lakeland	0.97	28.3	2.66	0.68	0.9	114.6	0.18	0.15
Josephine	1.94	31.1	1.34	0.26	2.1	65.6	0.18	0.17
Troup	0.40	19.3	4.68	1.49	0.5	75.0	0.16	0.11
Windthorst (sub)	0.90	38.5	3.16	0.88	0.9	101.6	0.12	0.12
Savannah	13.7	28.9	1.38	0.71	20.8	94.7	0.18	0.14
Greenville	2.40	31.0	2.56	0.51	2.7	87.7	0.16	0.17
Windthorst (top)	0.89	34.8	1.23	0.45	0.8	86.4	0.17	0.14
Bromfield	0.80	20.9	0.98	0.60	0.7	55.6	0.14	0.12
Vernon (sub)	0.40	18.4	0.97	0.28	0.4	36.7	0.12	0.10
Vernon (top)	0.81	25.2	0.57	0.37	0.7	37.9	0.16	0.12

in the control plots, applied P rendered Cd more plant unavailable in P-treated than in control plots so that plant Cd concentrations were decreased due to P treatments.

Unlike plant Cd concentration, plant P concentration tended to increase with increasing plant dry-matter yield ($\gamma = 0.78^{**}$). The plant P concentration was consistently higher in the TSP-treated soils than in the soils without added P. However, correlation between plant P concentration and dry-matter yield was rather poor (r = 0.05) among the P-treated soils. The correlation was greatly improved (r = 0.93**) when plant P uptake, rather than plant P concentration, was plotted against dry-matter yield for all the treatments. However, the correlation was not good when plant Cd uptake was plotted against dry-matter yield, although the trend shows that Cd uptake tended to increase with increasing dry-matter yield for the treatments without P. It should be noted that plant P uptake depends mainly on the soil available P level whereas plant Cd uptake is influenced by soil pH and exchangeable cations (McBride *et al.*, 1981). This probably explains why the pattern of plant P uptake differs from that of plant Cd uptake as related to plant biomass.

Experiment 2

Dry-matter yield of maize increased with increased rate of P applied for all three P sources. Maximum dry-matter yield (38 g pot^{-1}) was obtained with TSP at P rate of 1000 mg kg^{-1}. The agronomic effectiveness of the P sources used follows TSP > Togo PAPR > NCPR, an order similar to that for P water solubility (Table 2). Uptake of P by maize followed the same order of TSP > Togo PAPR > NCPR. It is interesting to note that P uptake with TSP and Togo PAPR increased to the 2000 mg P kg^{-1} rate of application whereas P uptake with NCPR reached the maximum at the 500 mg P kg^{-1} rate. Because the solubility of a given PR is governed by a constant solubility-product (Chien and Black, 1976), a maximum solubility would be fixed at a P rate beyond which the solubility of PR would not be increased. This may explain why the maximum P uptake with NCPR was fixed at the 500 mg P kg^{-1} rate.

Uptake of Cd by maize also increased with increase in rate of P applied and followed the order of Togo PAPR > NCPR > TSP. For TSP, maximum Cd uptake was observed with a P rate at 500 mg P kg^{-1}. The Cd

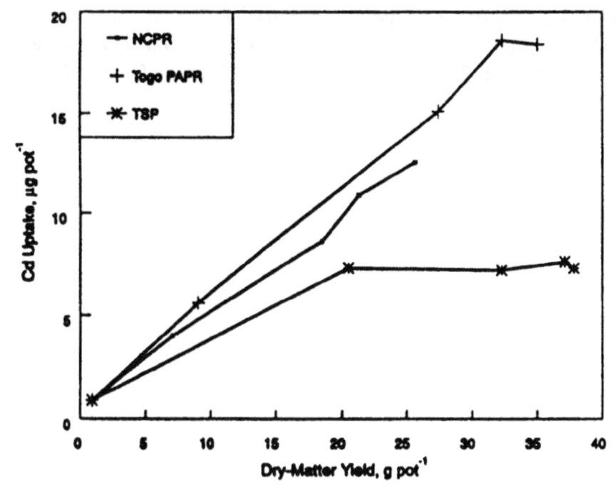

Fig. 2. Cd uptake by maize obtained with three sources of P fertilizers.

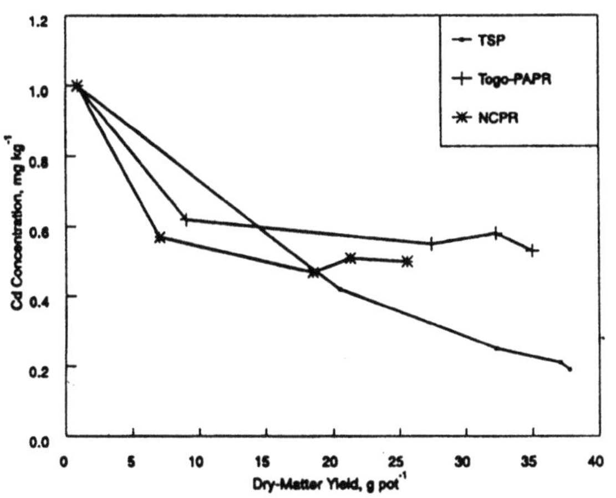

Fig. 3. Plant Cd concentration as related to dry matter yield of maize obtained with three sources of P fertilizer.

content in Togo PAPR and NCPR was higher than that in TSP (Table 2) and was responsible for the higher Cd uptake from Togo PAPR and NCPR than from TSP. However, for P fertilizers containing about the same Cd content, it was P availability that determined Cd uptake. The higher P solubility associated with Togo PAPR than with NCPR explains why the Cd uptake from Togo PAPR was higher, even though more Cd was applied to the soil from NCPR (3.5–70.0 mg kg^{-1}) than from Togo PAPR (3.3–65.0 mg kg^{-1}).

Because both dry-matter yield and Cd uptake increased with rate of P application, Cd uptake also increased with increase in dry-matter yield (Fig. 2). However, when plant Cd concentration was plotted against dry-matter yield, Cd concentration decreased with increase in dry-matter yield (Fig. 3). Again, the dilution effect of plant biomass was responsible for the decrease in plant Cd concentration. Because there was interaction between P solubility and Cd content with respect to plant Cd uptake from P fertilizers, the relationship between plant Cd concentration and dry-matter yield differed for the three sources, as shown by the three distinct curves.

Summary and conclusion

Data from these two greenhouse experiments showed that plant Cd concentration varied with P solubility and Cd content of the P fertilizers used and with plant yield produced after P application. In general, plant Cd uptake increased with increase in dry- matter yield, whereas plant Cd concentration decreased with increasing dry-matter yield due to the dilution effect of plant biomass. When a soil is low in available P and is polluted with Cd, application of P fertilizer not only increases plant yield but also reduces plant Cd concentration.

Acknowledgements

The financial support given by the United Nations Development Programme (UNDP) for the research is hereby gratefully acknowledged.

References

Chien SH and Black CA (1976) Free energy of formation of carbonate apatites in some phosphate rocks. Soil Sci Soc Am J 40: 234–239.

McBride MB, Tyler LD and Hovde DA (1981) Cadmium adsorption by soils and uptake by plants as affected by soil chemical properties. Soil Sci Soc Am J 45: 739–744.

Menon RG, Chien SH and Hammond LL (1990) Development and evaluation of the Pi soil test for plant available phosphorus. Commun in Soil Sci Plant Anal 21: 1131–1150.

Mortvedt JJ, Mays DA and Osborn G (1981) Uptake by wheat of cadmium and other heavy metal contaminants in phosphate fertilizers. J Environ Qual 10: 193–197.

Mortvedt JJ and Osborn G (1982) Studies on the chemical form of cadmium contaminants in phosphate fertilizers. Soil Sci 134: 185–192.

Mortvedt JJ (1987) Cadmium levels in soils and plants from some long-term soil fertility experiments in the United States of America. J Environ Qual 16: 137–142.

Mulla DJ, Page AL and Ganje T (1980) Cadmium accumulations and availability in soils from long-term phosphorus fertilization. J Environ Qual 9: 408–412.

Pezzarossa B, Malorgio F, Lubrano L, Tognoni F and Petruzzelli G (1990) Phosphate fertilizers as a source of heavy metals in protected cultivation. Commun Soil Sci Plant Anal 21: 737–751.

Street JJ, Sabey BR and Lindsay WL (1978) Influence of pH, phosphorus, cadmium, sewage sludge, and incubation time on the solubility and plant uptake of cadmium. J Environ Qual 7: 286–293.

C. Rodriguez-Barrueco (ed.), Fertilizers and Environment, 443–448.

Thallium concentration in soils and crops and critical values with respect to food chain

Ch. Makridis[1] & A. Amberger[2]
[1]*Agricultural Research Centre – N.AG.RE.F., Thefrastou 1, 41335 Larissa, Greece;* [2]*Institute Plant Nutrition, Technical University Munich, Weihenstefan, Germany.*

Abstract

1. Thallium is known as an environment chemical, associated with dust from cement industry, sometimes used as K-fertilizer (30% K_2O).

2. In hydroponic experiments Tl was taken up by crops very easily, competing with potassium, and was accumulated in shoots. Toxicity levels (10% and 25% yield reduction) were 130 and 340 mg Tl for beans, however 800 and 1900 mg Tl/kg dry matter for rape.

3. Toxicity symptoms appeared as chlorosis along the leaf nervature.

4. In pot experiments cement dust was incorporated into soils (6.5 kg/pot) at pH 5.4, 6.5, 7.0 and an annual deposition of 125 g/m^2 The critical concentrations in plants (25% yield reduction) were 20 mg Tl for rye grass, 15 mg Tl for beans and 575 mg Tl/kg for rape on dry matter basis. Toxicity limits (25% yield reduction) in soils were: 8 mg Tl for rye grass, 11 mg Tl for beans and 10 mg Tl/kg soil for green rape.

5. Between thallium concentration in soils and plants there were positive correlations (rye grass r^2 =0.63, beans r^2 =0.78, rape r^2 =0.58).

6. CONCLUSIONS: Considering the tendency of green rape (as a fodder crop) to accumulating Tl in high concentrations the generally accepted limit of 2.5 mg Tl/kg dry matter seems too high with respect to great risks for food chain. Cement dust should be eliminated as K-fertilizer.

Introduction

The critical concentration of heavy metals in soil and in plant tissues has long been a subject of great interest for crop production and for public health. This regards the toxicity effects on plants and health hazards to human as well as to animal life and other forms of life (Allison and Dzialo, 1981; Page, 1981; Hall, 1972; Sauerbeck, 1982; Zitko *et al.*, 1975; Beckett and Davis, 1977).

Thallium is considered as toxic element for the environment and it is known that its presence is associated with dust from cement industry (Achenbach *et al.*, 1979; Heckt, 1982; Smith and Carson, 1977; Prinz *et al.*, 1979; Bambauer and Schaefer, 1981).

The use of dust from cement industry, rich in potassium, as fertilizer involves the hazardous effects of thallium on animals and human beings.

For this study the 2.5 mg Tl/kg dry matter in plant tissues, as a safe and general accepted limit with respect to food chain, was the reference point to determine the phytotoxic level of thallium concentration in soils and plants. At the same time the specific toxicity symptoms produced on different crops, were investigated, after the application of $TlCl_3 \cdot H_2O$ in nutrient solution.

Material and methods

Nutrient solution experiments

In plastic pots (4 L) which contain Hoagland solution, the plants of each crop that were previously sprouted in Blaehton were transplanted, four plants of bush beans (var. marona), six plants of green rape (var. akela), respectively. After eleven days for bush beans and sixteen days for the green rape the solution was changed with one that contained thallium in different concentrations of 0, 1, 2, 5, 10 and 20 mg Tl L^{-1} as $TlCl_3 \cdot H_2O$.

The plants were left to grow for 21 days for bush beans and 34 days for the green rape and then harvested for analysis. The nutrient solution was changed a few

Table 1. Characteristics of experimental soils

		Soil		
		I	II	III
Clay	%	18	17	13
Silt		55	55	60
Sand		27	28	27
pH		5.6	6.2	7.0
Organic matter	%	3.2	3.7	3.7
Thallium	mg/kg (d.m. soil)	0.42	0.37	0.67
CEC		12.9	14.9	12.8
Ex-Na		0.02	0.1	0.02
Ex-K	meq/100 g	0.5	0.9	0.3
Ex-Ca	soil	8.4	12.0	10.8
Ex-Mg	1.8	1.9	1.7	
Sum		10.7	14.9	12.8
Base saturation	%	82.9	100	100

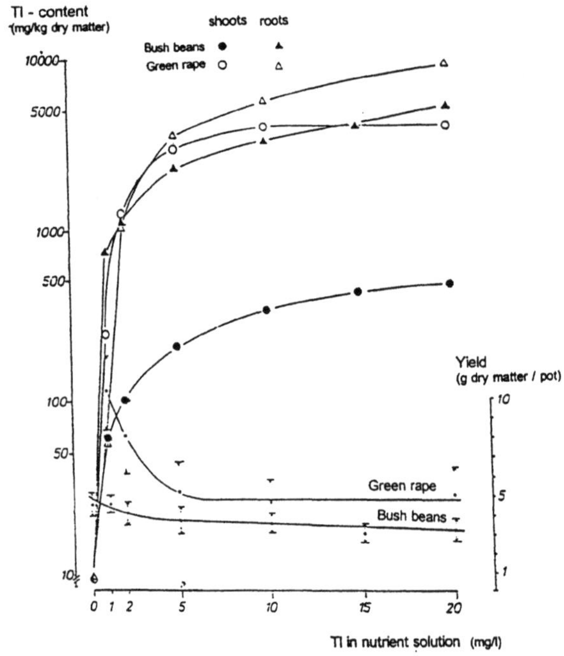

Fig. 1. Thallium contents and yield of bush beans and green rape as they are affected of thallium supply in nutrient solution.

times during the growing period in order to keep the concentration constant.

Soil experiments

The soil experiments involves the use of three different soils with pH 5.6, 6.2 and 7.0 (sandy-silty loam), as Table 1 shows. The dusts were mixed with soil 6.5 kg per pot corresponding to an annual deposition of 125 g/m^2 and the latter one was increased up to ninefold amount. Because the dust contains also calcium and potassium the corresponding amount of these elements was added to the control pots. Each crop was cultivated in the pots, in a greenhouse, and the next crop was planted in the same pots in a series, bush beans(var. marona) – green rape (var akela) – rye grass (var. perma). The plants were harvested at the beginning of flowering (bush bean), at the end of vegetative period (green rape) and at the 25 cm height (rye grass). The total thallium concentration was ranged between 0.4 to 13.1 mg Tl per kg of soil (Tl-soil+Tl-dust). The amount of Tl added as Tl-dust was ranged between 0.05 to 12.1 mg Tl per kg of soil. The crops received the common fertilization practice but concerning the estimated amount of nutrient added, the amount of nutrient the dust contained was taken into account.

Thallium was extracted from plant tissues by concentrated HNO_3 and from the soil by HNO_3 and H_2SO4 and determined photometrically with brilliant green (Scholl, 1981).

Results

Experiments with nutrient solution

Table 2 shows the results of nutrient culture. Yield reduction, statistically significant, was observed at 2 mg Tl/l. This reduction of bush beans was about 11% for shoots and 20% for roots. The corresponding reductions for the green rape were 22% and 23%.

Thallium was accumulated in high amounts in the plant tissues, Table 3, even at a concentration of 1 mg Tl/l and reached 62 ppm in the shoots and 742 in the roots of bush beans. The corresponding amounts for the green rape were 244 and 57 ppm.

After the manipulation of the available data, the concentration of Tl in the plant tissues of bush beans was 130 and 340 mg/l Tl when the corresponding yield reduction was 10% and 25%. The latter reduction was estimated for a growth period of 4 weeks from which the last ten days the plants were transplanted into the Tl

Table 2. Yield production of bush beans and green rape (After 10 days and 18 days influence of $TlCl_3 \cdot H_2O$ correspondingly)

Tl-conc. (mg/l in N.S.)	Yield (g dry matter per pot)			
	Bush beans		Green rape	
	shoots	roots	shoots	roots
Control	4.5	1.0	10.3 a*	1.3
1	4.5	0.9	11.0 a	1.2
2	4.0	0.8	8.0 b	1.0
5	3.7	0.7	5.2 c	0.7
10	3.5	0.6	4.7 c	0.7
20	3.2	0.6	5.1 c	0.7
p 5%	0.6	–	*Variance a > b > c	–

Table 3. Thallium concentration and uptake of bush beans and green rape from nutrient solution

Tl-conc. (mg/l in N.S.)	Thallium concentration in plant (mg Tl/kg dry matter)			
	Bush bean		Green rape	
	shoots	roots	shoots	roots
Control	<1	3	1	8
1	62	742	244	57
2	104	1119	1134	991
5	212	2311	2983	3517
10	337	3336	4057	5822
20	487	5207	4163	6701

rich solution. With regard to green rape the concentration of Tl in the plant tissues was 800 and 1900 mg/l Tl when the corresponding yield reduction was 10% and 25%. The latter reduction was estimated for a growth period of 5 weeks from which the last eighteen days the plants were transplanted into the Tl rich solution.

If the Tl uptake (mg Tl per kg dry matter) is plotted against Tl concentration (mg Tl/l) in the nutrient solution (semilogarithmic scale), the concentration effect can be seen, Fig. 1.

From these plots one can see that the Tl uptake is a rapid one at the very low concentration of Tl in the nutrient solution and reaches already maximum concentration at about 1 mg Tl/l, whilst the plants seem not to be harmed by the presence of Tl, but at the same time is taking up large amounts of available Tl in solution.

From 2 mg Tl/l on, the plants continue to take up Tl, so that the Tl accumulation in plant tissues reaches the toxicity level and toxicity symptoms start appearing. The reduction of plant growth is very obvious and can be explained as "concentration effect".

In this figure one can see that in case of green rape at the concentration of 2 mg Tl/l the concentration of Tl in the shoot is higher than that in roots. The situation is reversed at the concentration of 5 mg Tl/l. If we look to the bush beans crop, even at the very low concentration of Tl in the nutrient solution, the concentration of Tl in the roots is always very much ahead to that of the shoots. Looking at the curves that show Tl concentration in shoots and roots for the two crops, the corresponding curves for the green rape go very close together while the curves for the bush beans are far away from each other.

(a)

(b)

Fig. 2. (a) Symptoms of thallium toxicity in bush beans at 10 mg Tl/l nutrient solution. (b) Symptoms of thallium toxicity in green rape at 20 mg Tl/l nutrient solution.

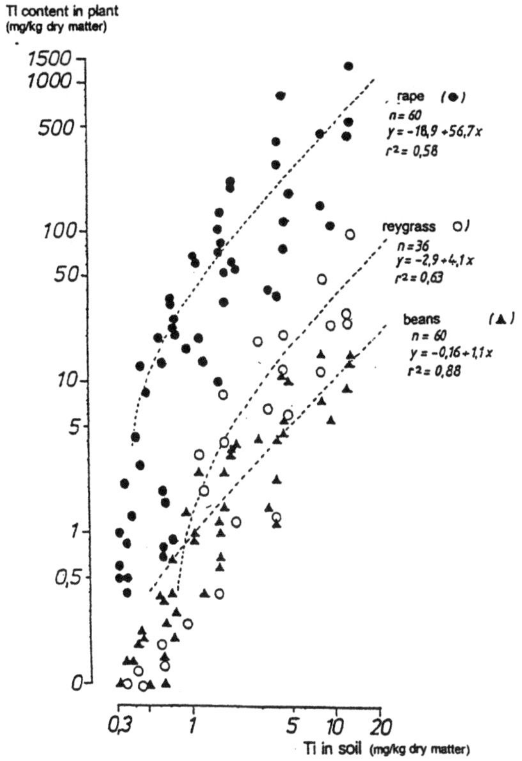

Fig. 3. Relation between thallium content in soils and crops.

This reveals as well a difficulty of Tl transport to shoots of bush beans as that there is a big difference with regard to Tl uptake between bush beans and green rape.

Figures 2a-b show the characteristic symptoms of Tl toxicity. For the bush beans the symptoms appear as chlorosis along the leaf nervature and for the green rape the symptoms are necrotic spots along the leaf nervature as well. These spots in severe situations are extended to the whole area between the leaf nervature.

With respect to phytotoxicity, beans proved to be very sensitive to Tl, compared with green rape which is very tolerant in spite of much high thallium uptake and concentration in leaves.

Experiments with soils

The pot experimental data (Fig. 3) revealed again a positive linear relationship between thallium supply to soil and thallium concentration in plants.

However a negative correlation exists between the thallium concentration in plant tissues and crop yield regardless the pH values. The differences in Tl uptake between the crops are distinguishable in the order of green rape > rye grass > bush beans with the regression coefficients $r^2 = 0.58$, $r^2 = 0.63$ and $r^2 = 0.78$.

From pot experiment it has been found that the addition even of 0.05 mg Tl/kg soil (total Tl content 0.47 mg Tl/kg soil) has significantly increased the Tl concentration in green rape. In contrast, 0.15 mg Tl/kg soil (total Tl content 0.56 mg Tl/kg soil) has to be added for the bush beans crop to produce a such significant increase. Tl availability is higher in acid soils and 18% of the added amount in dust was taken up by the plants.

A reduction of 25% in yield occurred at a concentration of 20 mg Tl/kg dry matter of plant tissues for rye grass, 15 mg/kg for beans and 575 mg/kg for rape.

The critical concentration of Tl in the soil at which a 25% reduction in yield occurred is: 8 mg Tl/kg dry soil for rye grass, 11 mg/kg for bush beans and 10 mg/kg for green rape.

Discussion

The experiments of the present study show a specific behaviour of the crops with regard to Tl. Green rape appeared to possess an ability to take up greater amounts of thallium than the rye grass and bush beans. These differences in uptake have been reported by Hoffmann *et al.*, 1982; Schweiger and Hoffmann, 1983.

Our experimental data from soil and nutrient solution show that Tl in green rape is more mobile than in bush beans. Differences in Tl mobility have also been mentioned by Hoffmann *et al.*, 1982; Kick *et al.*, 1981; Cataldo and Wildung, 1978.

The minimum concentration of an element that produces 10% yield reduction, is considered (Davis *et al.*, 1978) from the phytophysiological point of view, as "critical value" of plant toxicity. Such critical values, at which 25% of yield reduction is observed for the tested crops in nutrient solution culture, are 340, 1900 mg Tl/kg dry matter of plant tissues for bush beans and green rape respectively. However in soil culture experiments such reduction (25%) occurred when the Tl concentration in plant tissues is 15, 575 and 20 mg/kg dry matter for bush beans, green rape and rye grass correspondingly. The differences in the size of the "critical values" for toxicity, reported elsewhere, have their source from the level of yield reduction that has been adopted (10% or 25%).

Davis *et al.*, 1978 reported that the critical concentration for 10% yield reduction is 20 mg Tl/kg dry matter plant tissues for burley crop in sand culture experiments. Hoffmann *et al.*,1982 working with green rape and cabbage in soil culture, where the Tl was given as $TlNO_3$, mentioned 1000 and 30 mg Tl/kg dry matter as critical values.

In our experiments with bush beans and green rape, when the toxicity symptoms were observed, the corresponding concentration of Tl was 200 and 3000 mg Tl/kg of shoots.

The critical concentrations of Tl in our soil are 11, 10 and 8 mg Tl/kg for bush beans, green rape and rye grass. The latter ones are close to the ones reported for other crops (Hoffmann *et al.*,1982).

The concentration of Tl in soil at which the crops are severely contaminated must be investigated for each individual crop. The general accepted limit concerning the Tl content in soil is 1 mg Tl/kg. The limit for fodder crops and eatable crops is 0.25 mg Tl/kg fresh plant tissues and 2.5 mg Tl/kg dry matter.

It is an important result of our experiments with green rape that even a concentration of 1 mg Tl/kg soil was high enough to exceed the accepted limit of Tl content in plants. This limit is much more easier to be exceeded when the soil-thallium is brought high by using the dust from dust industry.

Conclusions

The use of dust from cement industry as K-fertilizer has to be considered very seriously. The contamination of soil and plants with high amounts of thallium can exceed often the general accepted limit of 1 mg Tl/kg soil and 2.5 mg Tl/kg plant dry matter for human or animal nutrition. Considering the great tendency of green rape (a fodder crop) to accumulate very high amounts of thallium without visible toxicity symptoms it is a great risk to food chain when this crop is fertilized with byproducts. Therefore cement dust should be eliminated as K-fertilizer.

References

Achenbach CO, Hauswurt C, Heidrichs R, Ziskoven F, Koehler J, Smend J and Kowalewski S (1979) Toxizitaet und Teratogenitaet von Thallium, Dt Aerztebl 48: 3189–3192.

Allinson DW and Dzialo C (1981) The influence of lead, cadmium and nickel on the growth of ryegras and oats, Plant and Soil 62: 81–89.

Cataldo DA and Wildung RE (1978) Soil and plant factors influencing the accumulation of heavy metals by plants, Environ Health Persp 27: 149–159.

Bambawer HV and Schaefer (1981) Der minerabestand eines Thallium-haltigen Reingasstaubes aus der Zementherstellung, Naturwiss 68: 571–572.

Beckett PHT and Davis RD (1977) Upper critical Levels of toxic elements in plants, New Phytol 79: 95–106.

Davis RD, Beckett PHT and Wollan E (1978) Critical levels of twenty potentially toxic elements in young spring barley, Plant Soil 49: 395–408.

Hall SK (1972) Pollution and Poisoning, Environ Sci Technol 6: 31–34.

Hoffmann G, Schweiger P and Scholl W (1982) Aufnahme von thallium durch landwirtschaftliche und gartnerische Nutzpflanzen, Landw Forsch 35: 45–54.

Heckt H (1982) Die verschiedenen Wege und Moeglichkeiten des Eintrags von Schwermetallen in die Futtermittel und die dabei ru beachtenden Belastungsgrenzen, Landw Forsch 39: 94–107.

Kick H, Buerger H and Sommer K (1981) Vegetationsversuche zur Aufnahme von Berillium und Thallium durch Sommergerste und Raps, Landw Forsch 34: 186–190.

Page AL (1981) Heavy metals in the Environment, Proceedings at an international conference in Amsterdam, CEP Consultant Ltd., Edinburgh, p 206.

Prinz B, Krause CjHM and Stratmann H (1979) Thallium schaeden in der Umgebung der Dyckerhoff-Zementwerke AG in Lengerich, Westfalen, Staub-Reinhalt Luft 39: 457–462.

Sauerbeek D (1982) Welche Schwermetallgehalte in Pflanzen duerfen nicht ueberschritten werden,um Wachstumsbeein traechtigungen zu vemeiden?, Landw Forsch 39: 108–129.

Scholl W (1981) Bestimmung von Thallium in verschiedenen anorganischen und organischen Matrizes–ein einfaches photometrisches Routineverfahren mit Brillant gruen, Landw Forsch 35: 216–223.

Schweiger P, und Hoffmann G (1983) Thallium-Aufnahme verschiedener Pflanzen in Hydrokultur, Kali Briefe 16: 391.

Smith IC and Carson BL (1977) Trace metals in the environment, Vol. 1, Thallium, Ann Arbor science Publ Inc, Michigan.

Zitko V, Carson MV and Carson WG (1975) Thallium occurence in the enviroment and toxicity to fish, Bull Environ Contam Toxical 13: 23.

C. Rodriguez-Barrueco (ed.), Fertilizers and Environment, 449–452.

Soil research on heavy metal pollution in a Belgian risk-bearing region with intensive agriculture

L. Vanongeval, M. Geypens & H. Vandendriessche
Soil Service of Belgium, W. de Croylaan 48, B-3001 Leuven, Belgium

Key words: fertilization, heavy metals, pollution, soil fertility

Abstract

To quantify the influence of the intensive agricultural practices in the western part of Belgium on soil fertility and heavy metal content of the topsoil, 1341 soil samples from the province of West-Vlaanderen are analysed. Soil fertility, represented by pH, % carbon and available P, K, Mg and Ca content, is evaluated using the expert system BEMEX (developed by the Soil Service of Belgium). Particularly in the districts with intensive agriculture, more than 70 % of the soil samples has a high (31–50 mg P/100 g soil) to very high (>50 mg P/100 g soil) available P content. Most of the soil samples have a 'normal' Cu, Pb, and Zn content. An important amount of the heavy textured soil samples has a Cd content exceeding 1 ppm. The heavy metal content of the top soil varies much more between the soil types than between the districts : no extremely high Cu, Pb, Zn or Cd contents are measured in the districts with high available P content.

Introduction

Agriculture in the western part of Belgium (province of West-Vlaanderen) has undergone an important intensification due to the explosive growth of non soil-linked fattening of cattle, the conversion to more intensively cultivated vegetables and the increased application of organic and inorganic fertilizers. The intensive pig farming is concentrated in the districts of Tielt and Roeselare, with densities exceeding 80 animals/ha in some municipalities (NIS, 1992). The growing field vegetables expanded mainly in the district of Roeselare.

This evolution has consequences on the chemical soil fertility in this area. The results of soil analysis on macronutrients in the topsoil, carried out by the Soil Service of Belgium during many years, show the evolution of the chemical soil fertility of the Belgian arable land and grassland. For the period 1989–1991, 90308 soil samples were analysed for pH, C, P, K, Mg, Ca and Na. The samples with high and very high available P content are concentrated in the regions with intensive pig and cattle farming and growing field vegetables (Vandendriessche *et al.*, 1993).

To quantify the influence of the intensive agricultural practices on heavy metal content of the topsoil, soil samples from the province of West-Vlaanderen were analysed for Cu, Pb, Zn and Cd.

Materials and methods

Regarding to this research, a set of soil samples from the province of West-Vlaanderen are restrained during one year (1988). These are soil samples from practice, analysed for pH, carbon and the macronutrients P, K, Mg and Ca, in order to calculate fertilizer recommendations for a specific crop rotation. In total, 1341 soil samples (1086 arable land and 255 grassland samples) are also analysed for heavy metals Cu, Pb, Zn and Cd. The soil samples were not selected on the base of their location or the previous history of the corresponding parcel.

The soil samples are taken down to a depth of 23 cm for arable land and 6 cm for grassland. The samples are air dried and passed through a 2 mm sieve. Carbon is determined using the modified Walkley and Black method and pH is measured in a KCl-solution. The extraction of the elements P, K, Mg and Ca is done in ammonium lactate (pH=3.75). K, Mg and Ca are measured with Atomic Absorption Spectrophotometry (A.A.S.). Determination of P is done by colorime-

Table 1. Frequency distribution (%) of the arable land and grassland samples, in 7 fertility classes for pH, C and macronutrients

Classification	pH KCl	Carbon	P	K	Mg	Ca
Arable land						
Very low	2.0[a]	4.1	0.0	0.5	0.8	0.6
Low	19.4	6.1	0.9	5.7	6.8	10.8
Rather low	30.4	16.0	1.5	9.9	20.2	28.6
Target zone	31.0	49.9	9.9	37.3	43.6	53.6
Rather high	11.8	21.8	32.8	40.0	16.9	4.5
High	3.7	1.9	44.4	6.0	9.0	1.3
Very high	1.7	0.2[b]	10.5	0.6	2.7	0.6
Grassland						
Very low	2.0[a]	3.1	0.0	0.0	0.8	2.4
Low	4.3	11.8	3.1	0.8	6.3	8.2
Rather low	18.0	15.3	7.8	10.2	24.7	15.7
Target zone	29.4	40.8	21.2	35.3	43.2	54.5
Rather high	23.5	19.6	34.2	22.0	12.9	9.0
High	17.3	6.7	23.1	23.5	7.8	8.6
Very high	5.5	2.7[b]	10.6	8.2	4.3	1.6

[a] Strongly acid.
[b] Peaty.

try. Soil type is determined by palpation. Cu, Pb, Zn and Cd are measured with Inductively Coupled Plasma (I.C.P.), after 'aqua regia' destruction.

For the evaluation of the chemical soil fertility, the Soil Service of Belgium uses the expert system BEMEX (Geypens *et al.*, 1989). Soil fertility, represented by pH, % carbon, P, K, Mg and Ca content, is evaluated in seven fertility classes, taking into account the soil texture and the carbon content of the soil.

Results and discussion

The frequency distribution of the soil fertility for the arable land and grassland samples is illustrated in Table 1. Less than one third of the analysed soil samples can be classified in the target zone for pH. For exchangeable K, the evaluation classes with the highest frequency are the target zone and the rather high class. More than 40 % of the samples is evaluated within the target zone for exchangeable Mg, more than half of the analysed samples has a Ca content within the target zone. The frequency distribution regarding the evolution of the available P content shows that more than 90 % of the analysed samples is classified in the classes from target zone to very high.

This evaluation of soil fertility corresponds to the soil fertility inventory study of the period 1989–1991 (Hendrickx *et al.*, 1992; Vandendriessche *et al.*, 1993). As a result, the studied dataset can be considered as a good sample.

Table 2 illustrates the frequency distribution of the arable land samples in the fertility classes for P, by district. In all districts, most of the soil samples have an available P content higher than the target zone. Particularly in the districts of Roeselare and Tielt, more than 70% of the analysed samples have a high (31–50 mg P/100 g soil) to very high (>51 mg P/100 g soil) available P content.

Table 3 shows the 95% confidence intervals of the mean Cu, Pb, Zn and Cd content of the analysed arable land and grassland samples as a function of soil texture. The median value is added to illustrate right-skewness of the distribution. The Cu content of the arable land samples is higher on the loamy soils, compared to the sandy and sandy loam samples.

Clay and polder samples, on the other hand, show a rather low Cu content. Soil texture influences Pb content very little: polder and loamy soils have on

Table 2. Frequency distribution (%) of arable land samples, in 7 fertility classes for P, by district

District	Very low	Low	Rather low	Target zone	Rather high	High	Very high
Brugge	0.0	0.0	0.0	5.3	28.1	53.4	13.2
Diksmuide	0.0	0.0	0.7	8.9	27.4	52.0	11.0
Ieper	0.0	2.9	2.2	14.7	48.7	29.7	1.8
Kortrijk	0.0	0.0	0.0	7.4	39.3	40.2	13.1
Oostende	0.0	0.0	7.9	18.4	21.1	50.0	2.6
Roeselare	0.0	0.0	1.2	4.1	22.2	57.4	15.1
Tielt	0.0	0.0	0.0	3.5	19.0	55.7	21.8
Veurne	0.0	2.7	5.5	26.0	37.0	23.2	5.5

Table 3. 95 % confidence interval of the mean Cu, Pb, Zn and Cd content (ppm) and corresponding median value (in brackets) of the arable land and grassland samples as a function of soil texture

Soil texture	Cu	Pb	Zn	Cd
Arable land				
Sand (260)[a]	12.0–13.8 (11.0)	23.3–26.7 (22.1)	28.8–33.8 (27.0)	0.46–0.50 (0.45)
Sandy loam (304)	13.5–15.5 (12.9)	25.2–30.0 (24.1)	35.4–38.2 (34.6)	0.59–0.63 (0.61)
Loam (419)	16.0–19.2 (13.8)	30.5–32.9 (28.8)	42.2–45.0 (40.6)	0.77–0.81 (1.20)
Clay (13)	5.5–11.1 (7.8)	20.0–34.6 (25.3)	32.4–46.6 (36.8)	0.59–0.81 (0.91)
Polder (90)	6.6–7.8 (7.2)	29.2–31.8 (30.5)	42.5–48.7 (46.6)	0.98–1.12 (1.42)
Grassland				
Sand (103)[a]	11.2–14.8 (10.6)	25.5–31.3 (25.5)	31.5–38.1 (31.2)	0.51–0.67 (0.49)
Sandy loam (64)	12.3–14.9 (12.0)	31.6–37.8 (33.1)	44.6–54.0 (42.7)	0.78–0.94 (0.80)
Loam (60)	12.9–16.1 (13.3)	31.6–36.6 (32.1)	45.2–53.4 (45.5)	0.86–1.02 (0.91)
Clay (10)	8.7–28–7 (13.9)	30.8–76.2 (43.6	36.3–137.5 (67.6)	0.60–1.76 (0.87)
Polder (18)	7.5–13.3 (9.0)	32.3–42.1 (35.3)	55.1–83.9 (65.5)	0.98–1.22 (1.13)

[a] number of soil samples.

Table 4. Percentage of the arable land samples with heavy metal concentrations below the 'normal' background concentration[a,b]

Soil texture	Cu		Pb		Zn		Cd	
	(1)	(2)	(1)	(2)	(1)	(2)	(1)	(2)
Sand	71.7	99.2	95.4	99.6	96.9	98.8	98.1	98.1
Sandy loam	94.7	98.4	95.1	99.0	99.7	100.0	98.7	98.7
Loam	89.5	96.7	91.9	100.0	99.5	99.5	81.6	81.6
Polder	98.9	100.0	98.9	98.9	100.0	100.0	42.2	42.2

(1) Percentage of the soil samples with a Cu, Pb, Zn, Cd concentration below the upper limit of the 'normal' background concentration (for soil layer 0–10 cm), defined by De Temmerman *et al.*, 1982. Cu : 15 ppm for sand, 25 ppm for sandy loam and loam, 30 ppm for clay; Pb : 50 ppm for all soil types; Zn : 100 ppm for sand and sandy loam, 150 ppm for loam and clay; Cd : 1 ppm for all soil types.
(2) Percentage of the soil samples with a Cu, Pb, Zn, Cd concentration below the 'normal' background concentration, defined by Flemish legislation (Anonymus, 1992). Cu: 50 ppm; Pb: 100 ppm; Zn: 150 ppm; Cd: 1 ppm for all soil types.

Table 5. Median value of Cu, Pb, Zn and Cd content and frequency distribution of soil texture of the arable land soil samples, by district

District	Heavy metal content (ppm)				Frequency distribution (%) of soil texture			
	Cu	Pb	Zn	Cd	Sand	Sandy loam	Loam	Clay+Polder
Brugge	9.0	24.7	26.2	0.50	67.5	1.5	0.0	21.0
Diksmuide	11.0	22.6	31.3	0.54	45.2	31.5	14.4	8.9
Ieper	15.6	31.1	43.8	0.81	0.4	13.6	898.0	0.0
Kortrijk	11.1	25.1	35.2	0.64	23.0	32.8	44.2	0.0
Oostende	6.7	30.2	42.2	0.93	15.8	0.0	0.0	84.2
Roeselare	13.3	23.9	35.7	0.61	18.0	57.0	25.0	0.0
Tielt	14.4	22.3	31.7	0.54	33.8	46.5	19.0	0.7
Veurne	7.6	29.6	37.6	0.76	4.1	5.5	45.2	45.2

average only a slightly higher Pb content than sandy soils. Heavy textured soils have a clearly higher Zn and Cd content than light textured soils. The average Cd content is twice as high on polder soil samples as on sandy soil samples.

Most of the soil samples have a Cu, Pb and Zn content below the upper limit of the 'normal' background concentration (Table 4), determined on different soil types of Belgium by De Temmerman *et al.*, 1982. Compared to the 'normal' background concentration, defined by the Flemish legislation (Anonymus, 1992), only a few samples exceed the limit of 50 ppm for Cu, 100 ppm for Pb and 150 ppm for Zn. On the other hand, an important amount of the loam and clay samples have a Cd content exceeding 1 ppm. This indicates that the fixed limit for Cd at least needs a differentiation for soil type. The median value of the Cu content (Table 5) is low in the districts of Brugge, with mainly sandy soils, and Oostende and Veurne, with an important amount of heavy textured soils. On the other hand, the district of Ieper, with mostly sandy loam and loamy soil samples, is characterized by the highest Cu content. The Pb, Zn and Cd content is clearly higher in the districts of Ieper, Oostende and Veurne, with a large percentage of loam, clay or polder samples, compared to the other districts. As a result, the variation in heavy metal content between the districts is largely due to the difference in soil textures between the corresponding districts. The soil samples from districts with high available P levels, related to intensive agricultural practices, can not be distinguished by their heavy metal content of the topsoil.

References

Anonymus (1992) Vademecum Milieurecht II – Vlarem II – Besluit van de Vlaamse Executieve van 7 januari 1992 houdende vaststelling van het Vlaams Reglement inzake milieuvoorwaarden voor hinderlijke inrichtingen, gewijzigd bij besluit van de Vlaamse Executieve van 31 juli 1992, Ministerie van de Vlaamse Gemeenschap, Leefmilieu en Infrastructuur, AMINAL, Brussel

De Temmerman LO, Istas JR, Hoenig M, Dupire S, Ledent G, Van Elsen Y, Baeten H, De Meyer A (1982) Onderzoek naar de 'normale' gehalten aan spoorelementen in een aantal Belgische bodems als basis voor de detectie en het onderzoek naar bodemvervuiling. Landbouwtijdschrift 35(2): 1883–1911

Geypens M, Boon W and De Wijngaert K (1989) BEMEX : programma en handleiding. Interne publicatie Bodemkundige Dienst van België, Leuven, p 53

Hendrickx G, Boon W, Bries J, Kempeneers L, Vandendriessche H, Deckers S and Geypens M (1992) De chemische bodemvruchtbaarheid van het Vlaamse akkerbouw- en weide-areaal (1989–1991). Bodemkundige Dienst van België, Leuven, p 143

NIS (1992) Landbouwstatistieken. Nationaal Instituut voor de Statistiek, Brussel

Vandendriessche H, Hendrickx G, Bries J and Geypens M (1993) Soil fertility and adjusted fertilizer recommendations for arable land and grassland in Belgium : a review for the period 1989–1991. Bull. Rech. Agron. Gembloux 28 (2–3), 377–391.

C. Rodriguez-Barrueco (ed.), Fertilizers and Environment, 453–455.

Effect of pyrite residue amendment on sulphur availability in a calcareous soil cropped with sown pasture

M.E. Balsa[1], M.G. Serrão[1], M.I.M. Martins[2], M.A. Castelo-Branco[1], M.R. Gusmão[1] & M.L. Fernandes[1]

[1] *Estação Agronómica Nacional, Quinta do Marquês, 2780 Oeiras, Portugal;* [2]*Direcção Regional de Agricultura da Beira Litoral, Av. Fernão de Magalhães, 465, 3007 Coimbra, Portugal*

Key words: Critical limits, extractable SO_4-S, pyrite, total plant S

Abstract

To evaluate an iron pyrite as a soil amendment in calcareous soils of low fertility, a field experiment was carried out during three years in a Calcaric Cambisol. The experiment was established in a split plot design with four pyrite rates (0, 3, 6 and 9 t ha^{-1}), three P rates (0, 76.5 and 153 kg P ha^{-1}) applied as ordinary superphosphate and three replicates. The soil was cropped with a mixed grass-clover pasture. The changes of soil extractable SO_4-S that occurred over time at the different rates of pyrite and superphosphate were evaluated and the effect of treatments on plant S concentration and S uptake in the 1st cut was estimated. The use of pyrite increased soil extractable SO_4-S to levels higher than the critical limit of 10 mg S kg^{-1}. Also, plant total S in the 1st cut was affected positively by pyrite application. The highest mean plant S concentration was observed when pyrite was applied at the rate of 6 t ha^{-1}.

Introduction

The mineral iron pyrite (iron sulphide) has been used as an amendment to ameliorate both chemical and physical properties of alkali soils (Somani, 1986). As this material contains appreciable amounts of sulphur, it can also be a source of this element for most of the crops, therefore preventing sulphur deficiency, namely in soils presenting low extractable SO_4-S levels (Tiwari *et al.*, 1985). As pyrite is a waste mining product, its expanded use may contribute to decrease the expenses with sulphur fertilizers and amendment materials.

Under favorable conditions for oxidation of pyrites in soil, iron sulphide is oxidized into sulphuric acid and ferrous sulphate according to the schematic equation (Starkey, 1966):

$$4FeS_2 + 2H_2O + 15O_2 \rightarrow 2Fe_2(SO_4)_3 + 2H_2SO_4$$

A number of factors may affect the oxidation of pyrites, namely particle size and type of placement, soil moisture, aeration conditions, soil temperature, organic matter content, presence of sulphur oxidizing bacteria (Somani, 1986; Tiwari *et al.*, 1984), calcium carbonate content (Tiwari *et al.*, 1984), soil pH (Somani, 1986), and concentration of phosphate ions (Quispel *et al.*, 1952; Wallace and Wallace, 1992). Reduced particle size, surface application of pyrite and keeping the soil moisture around field capacity is in general advisable for faster oxidation (Somani, 1986).

A field experiment has been carried out to establish the appropriate rates of both pyrite and ordinary superphosphate to be applied in a calcareous soil cropped with a mixed grass-clover pasture. At the end of three years, a high statistically significant effect of superphosphate and no significant effects of pyrite amendment on the yield pasture were observed (Serrão *et al.*, 1994). However, the combination of the rates of 6 t ha^{-1} of pyrite and 76.5 kg P ha^{-1} as superphosphate led to the highest yield. This suggested that sulphate mainly derived from oxidation of pyrite might have influenced positively crop growth. The objective of this work was to evaluate the changes of soil extractable SO_4-S that occurred with time at the different rates of pyrite and superphosphate. Also, the effect of treatments on plant S concentration and S uptake in the 1st cut was evaluated.

Methods

The field experiment has been carried out, since the Autumn of 1990, in a high calcareous soil located in the middle western coast of Portugal with a low organic matter content (15 g kg^{-1}), total carbonates content ranging from 50 to 70%, about 20% of "active" calcium carbonate content, and with an appreciably high pH value (8.2 in a 1:2.5 soil:water suspension). The experiment was established in a split plot design with four rates of pyrite (0, 3, 6 and 9 t ha^{-1}), three rates of P (0, 76.5 and 153 kg ha^{-1}) applied as ordinary superphosphate, and three replicates. Pyrite and superphosphate were applied only once, at the beginning of the experiment. Some characteristics of the pyrite used are given in Table 1.

Surface soil samples (0–20 cm) were taken up from all plots in October of 1990 (before the application of pyrite and superphosphate), in November of 1992, and also in June of 1993 (after harvest). Soil extractable SO_4-S was determined by extraction with a $Ca(H_2PO_4)_2$ solution containing 500 mg P dm^{-3} (Fox *et al.*, 1964). The extracts were analyzed for SO_4 using a turbidimetric method (Chesnin & Yien, 1950).

The cutting of the mixed grass-clover pasture was carried out in March of 1993. Plant total S was determined by wet digestion with nitric and perchloric acids and subsequent turbidimetric analysis (Blanchar *et al.*, 1965).

The amounts of soil extractable SO_4-S, and plant S concentrations and S uptake were compared using analysis of variance and the LSD test.

Table 1. Physical and chemical composition of pyrite*

Mean diameter (μm)	36.87
Copper (mg kg^{-1})	1200
Zinc (mg kg^{-1})	2800
Sulph ur (g kg^{-1})	387
Iron (g kg^{-1})	308
MgO (g kg^{-1})	0.8
CaO (g kg^{-1})	5.0
Al_2O_3 (g kg^{-1})	3.8
SiO_2 (g kg^{-1})	21.4

*Obtained from high-grade copper mine of Neves-Corvo, SOMINCOR.

Results

Before applying pyrite and superphosphate (1990), the amounts of soil extractable SO_4-S wer lower than the critical limit of 10 mg S kg^{-1} (Bansal *et al.*, 1983), ranging from 3 to 5 mg S kg^{-1}. In the plots with 0 t ha^{-1} of pyrite, soil extractable SO_4-S remained close to that critical value, although a slight increase was observed over time, probably due to superphosphate application (Fig. 1).

When pyrite was applied, at both soil samplings of 1992 and 1993, the amounts of soil extractable SO_4-S were much higher than the critical value, with mean values ranging from 12 to 51 mg S kg^{-1} (Fig. 1).

In the treatments with pyrite, mean values of extractable SO_4-S in both soil samplings of 1992 and 1993 were similar, but were significantly different from the values of the initial sampling ($P < 0.001$) indicating that after S removal by the crop a similar and appreciable amount of extractable SO_4-S remained in the soil surface layer for the subsequent cutting of pasture. There was a significant effect of pyrite on soil extractable SO_4-S ($P < 0.001$), as mean values at the rate of 0 t ha^{-1} were statistically different from the mean values obtained at the rates 6 and 9 t ha^{-1} (considering the mean values of all replicates, dates of sampling, and rates of superphosphate). The amounts of soil extractable SO_4-S at the rate of 6 t ha^{-1} did not differ from the values observed at the rates of 3 and 9 t ha^{-1}. There was no significant effect of superphosphate on extractable SO_4-S. Also, the interactions rates of superphosphate × rates of pyrite × time of sampling were not significant ($P > 0.05$).

Mean value of S concentration in plant at the rate of 0 t ha^{-1} of pyrite was lower than that critical value 0.35% referred by Syers *et al.* (1987) for mixed grass-clover swards. On the contrary, pyrite increased significantly S concentration in plant ($P < 0.001$) to higher values than that critical value. However, plant S concentrations at the rates of 3, 6 and 9 t ha^{-1} were statistically similar.

The highest plant S concentration was found at the rate of 6 t ha^{-1} (0.42% S). Plant total S at the rates of 3 and 9 t ha^{-1} was 0.35% S and 0.39% S, respectively.

Superphosphate did not influence plant total S. Also, the interaction between rates of superphosphate and rates of pyrite was not significant.

The use of different rates of pyrite did not influence S uptake ($P = 0.073$). Only superphosphate had a highly significant effect on S uptake ($P < 0.001$).

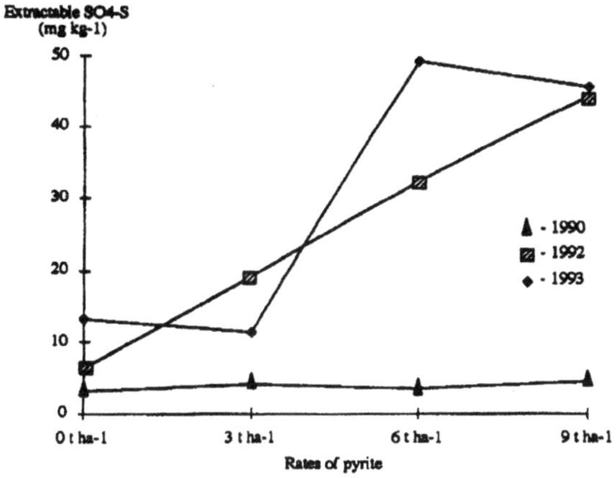

Fig. 1. Mean values of soil extractable SO_4-S over time as affected by rates of pyrite.

The interaction between rates of pyrite and rates of superphosphate and S uptake was not significant ($P = 0.069$).

Discussion and conclusions

From the results, it seems clear that the significant increases in soil extractable SO_4-S and plant S concentrations, observed at the rates of 6 and 9 t ha^{-1} of pyrite, were only due to this material. In fact, there was no significant effect of superphosphate on soil extractable SO_4-S and plant total S, indicating that sulphur in superphosphate did not influence these parameters.

The positive and significant effect of superphosphate on the yields earlier reported (Serrão *et al.*, 1994) may explain the significant effect of superphosphate on S uptake.

At the end of three years, the pyrite did not influence significantly the yields of pasture, although its application increased plant S concentrations to satisfactory levels. Climatic conditions have not been favorable for fast oxidation of pyrite, as too much dry weather occurred during the winter of the first two years. Perhaps the observed response to the pyrite was from oxidation that has occurred before application (Wallace and Wallace, 1992). This can explain the absence of significant responses of S uptake by application of different rates of pyrite.

The rate of 6 t ha^{-1} of pyrite and 76.5 kg P ha^{-1} showed to be the treatment at which it was obtained both highest yield and plant S concentration. Although plant S concentration at the rate of 9 t ha^{-1} of pyrite also indicated a normal nutritional sulphur status, this rate seems to be excessive as the yield was less than that obtained with 6 t ha^{-1}.

Acknowledgements

The authors thank Mr J.M. Santos and Mr J.D. Gama for field assistance and to Mrs M. de Lurdes Cravo, M. Celeste Campos and M. Augusta Pinto for help in the chemical analysis.

References

Bansal KN, Motiramani DP and Pal AR (1983) Studies on sulphur in vertisols. I. Soil and plant tests for diagnosing sulphur deficiency in soybean (*Glycine max* L. Merr.). Plant Soil 70: 133–140.

Blanchar RW, Rehm G and Caldwell AC (1965) Sulfur in plant materials by digestion with nitric and perchloric acid. Soil Sci Soc Am Proc 29: 71–72.

Chesnin L and Yien CH (1950) Turbidimetric determination of available sulfates. Soil Sci Soc Am Proc 15: 149–151.

Fox RL, Olson RA and Rhoades HF (1964) Evaluating the sulphur status of soils by plant and soil tests. Soil Sci Soc Am Proc 28: 243–246.

Quispel A, Harmsen GW and Otzen D (1952) Contribution to the chemical and bacteriological oxidation of pyrite in soil. Plant Soil 4: 43–55.

Serrão MG, Castelo Branco MA and Fernandes ML (1994) Residual extractable phosphorus as affected by phosphorus fertilizer and pyrite residue amendment in a calcareous soil. Proc 3rd ESA Congress 526–527.

Somani LL (1986) Use of pyrites as a soil amendment. Sulphur in Agriculture 10: 16–20.

Starkey RL (1966) Oxidation and reduction of sulphur compounds in soils. Soil Sci 101: 297–304.

Syers JK, Skinner RJ and Curtin D (1987) Soil and fertiliser sulphur in UK agriculture. The Fertiliser Society, Proceedings No 264.

Tiwari KN, Dwivedi BS and Pathak AN (1985) Iron pyrites as sulphur fertilizer for legumes. Plant Soil 86: 295–298.

Tiwari KN, Dwivedi BS, Upadhyay GP and Pathak AN (1984) Sedimentary iron pyrites as amendment for sodic soils and carrier of fertiliser sulphur and iron – A Review. Fertiliser News: 31–41.

Wallace A and Wallace GA (1992) Factors influencing oxidation of iron pyrite in soil. J Plant Nutr 15: 1579–1587.

[illegible] response to the swards was [illegible] cation that has occurred before application (Wallace and Wallace, 1992). This can explain the absence of significant responses of S uptake by application of different rates of pyrite.

Estimating the flux of nitrate in a sandy loam soil under corn

E. Fernández-Boy, F. Cabrera, J.M. Murillo, F. Moreno, J.A. Cayuela & J.E. Fernández
Instituto de Recursos Naturales y Agrobiología de Sevilla, CSIC, Apartado 1052, 41080 Sevilla, Spain

Key words: Nitrate leaching, monolith lysimeter, water flux, hydraulic conductivity, soil water content

Abstract

Drainage and nitrate leaching were determined in an undisturbed sandy loam soil contained in two monolith lysimeters (L1 and L2; 1.2 m deep) cropped with corn, irrigated, and fertilized with 510 and 175 kg N ha^{-1} in L1 and L2 respectively. Drainage was estimated at 0.9 m depth using Darcy's law and the hydraulic conductivity-soil water content relationship. Nitrate leaching at the same depth was obtained multiplying the estimated drainage by the nitrate concentration of the soil solution extracted with suction cups. A comparison between results from measurements at the bottom of lysimeters and that estimated at 0.9 m depth was made for the whole experimental period (March 1992–February 1993). The NO_3-N concentrations in the drainage water ranged between 2.8 and 151 mg l^{-1} in L1 and between 0.1 and 34.4 mg l^{-1} in L2. In the soil solution, they ranged between 5.9 and 240.3 mg NO_3-N l^{-1} in L1 and between 0.1 and 42.3 mg NO_3-N l^{-1} in L2. The total drainage at the bottom of lysimeters amounted to 139 and 203 mm and that estimated at 0.9 m depth amounted to 151 and 245 mm, in L1 and L2 respectively. Total NO_3-N leached in L1 and L2 was 80 and 30 kg ha^{-1} respectively. The estimated amounts of NO_3-N leached below 0.9 depth were 64 and 42 Kg ha^{-1} in L1 and L2 respectively. Total drainage and nitrate leached measured at the bottom were of the same order as the estimated values.

Introduction

In southern Spain, nitrogen management to increase crop yields usually involves high rates of N-fertilizer inputs. The use of N-fertilization rates that exceed the N requirements of crops increases the amounts of potentially leachable nitrate in the soil, and consequently a considerable amount of N-fertilizer can be lost from the soil (Roth and Fox, 1990). The NO_3-N leached below the root zone is one of the major pollutants of the groundwater. In large areas leaching of nitrate from agricultural land is the main no-point source for nitrate in groundwater (Duynisveld *et al.*, 1988; Cabrera *et al.*, 1994).

Jolley & Pierre (1977) sampled long-term corn fertility plots and found that 25 to 46% of the applied N fertilizer was lost from the soil after the third year of application. Losses of NO_3-N increase as N fertilizer rate or amount of subsurface drainage increase (Baker and Timmons, 1994; Fernández-Boy *et al.*, 1994).

Measurement of nitrate losses under field conditions is difficult. For both direct measurement and estimation of nitrate losses, at a depth below the root zone, the water flux draining from the upper layers and the nitrate concentration of this water must be known. Good approaches are (i) the use of monolith lysimeter containing undisturbed soil (Dowdell *et al.*, 1984; Owens, 1990; Bergtröm and Johansson, 1991) in which direct measurement of drainage water and its nitrate concentration is possible; (ii) the water flow, at a determinate depth in the soil, can be estimated using Darcy's law and the relationship between the soil hydraulic conductivity (K) and the soil water content (θ) determined *in situ*, and the concentration of nitrate can be determined in the soil solution extracted with suction cups. In both cases, the amount of nitrate leached can be calculated multiplying the amount of drainage water by the concentration of nitrate in this water.

The objective of this study was to determine the nitrate leached under a cropped soil using undisturbed monolith lysimeters. A comparison was also made between the nitrate leached in the drainage water at the bottom of the lysimeter and that estimated below 0.9 m depth.

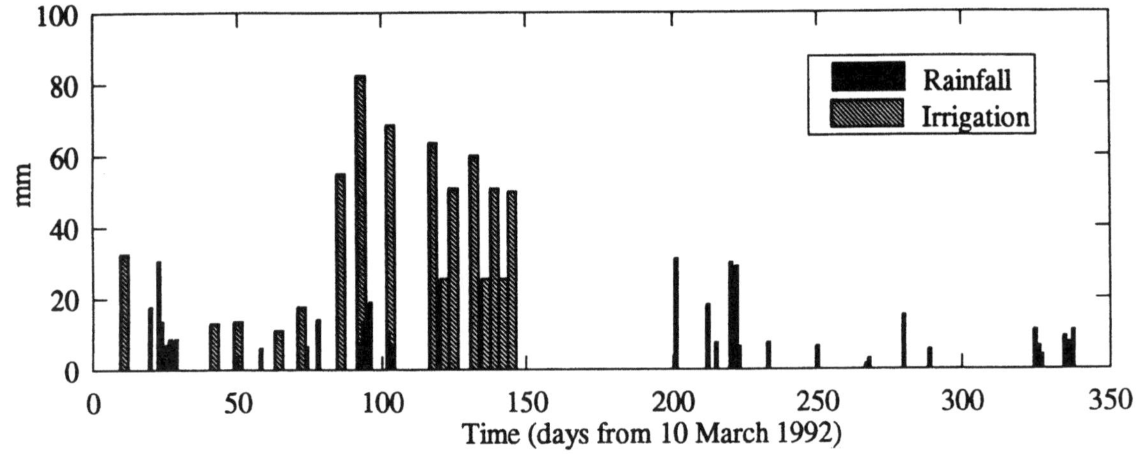

Fig. 1. Rainfall and irrigation during the experiment period

Materials and methods

The experiment on NO_3-N leaching was conducted from March 1992 until February 1993 in two lysimeters in the Experimental Farm "La Hampa" of the Instituto de Recursos Naturales y Agrobiología de Sevilla (CSIC) in Coria del Rio (province of Seville, SW Spain). The soil of the lysimeters is a sandy loam (Xerochrept; pH(H_2O) 7.2 and 7.1; $CaCO_3$ 5.2 and 3.1%; OM 0.88 and 0.69%; Kjeldahl-N 599 and 454 kg ha^{-1}; NO_3-N 44 and 61 kg ha^{-1}; at 0–0.5 and 0.5–1.0 m depth respectively).

Two monolith lysimeters (L1 and L2, 1 m in diameter and 1.5 m deep) were installed *in situ* without disturbing the soil profile. The lysimeters were provided with a system to collect the drainage water at 1.2 m depth, an access tube for the neutron probe, two tensiometers at 0.7 and 0.9 m depth, and a suction cup at 0.9 m depth. They were sown with corn (*Zea mays* L., cv. PRISMA (G-4730)) on 24th March 1992 with a density of 75,000 plant ha^{-1}. The corn was irrigated following the same schedule as the surrounding plots – also cropped with corn. Dates and quantity of irrigation and rainfall are shown in Fig. 1. The total amount of water applied by irrigation was 661 mm. Rainfall was monitored throughout the experimental period (total rainfall 394.5 mm). The characteristics of the irrigation water were: E.C. 2.2 dS m^{-1}; NO_3-N 9.8 mg l^{-1}; Cl 446.6 mg l^{-1}; SO_4-S 44.7 mg l^{-1}.

Two different fertilizations were used. L1 received one of the highest N-fertilization rates used by farmers of SW Andalucia: a deep fertilization with 1000 kg ha^{-1} of 15-15-15 complex NPK fertilizer and two top dressings with 400 kg ha^{-1} of urea (46% N), equivalent to ca. 510 kg N ha^{-1}. The fertilization applied to L2 was one third of that applied to L1: a deep fertilization with 340 kg ha^{-1} of complex NPK fertilizer and two top dressing with 130 kg ha^{-1} of urea (ca. 170 kg N ha^{-1}).

Soil water content was measured periodically with a neutron probe and soil solution was extracted by suction cups and analysed for nitrate content by ionic chromatography using a solution of 0.0013 M borate–0.0013 M gluconate in acetonitrile (12% v/v) at pH 8.5 as eluent (Waters Ion Chromatography Cookbook, 1989). In addition, leachates were collected and volumes measured following each rain or irrigation event when they were sufficient to create a leaching. Nitrate was also analysed in drainage water.

Water flow, at 0.9 m depth in the lysimeter, was estimated using Darcy's law, and the hydraulic conductivity(K)-soil water content(θ) relationships determined in situ (Cayuela *et al.*, 1994).

Results and discussion

The concentrations of nitrate in the drainage water collected at the bottom of the lysimeters (1.2 m depth) ranged between 2.8 and 151 mg l^{-1} in L1, and between 0.1 and 34.4 mg l^{-1} in L2. These concentrations were sometimes much higher than the maximum admissible concentration allowed by the EC for drinking water – 11.3 mg NO_3-N l^{-1} (Council of the European Community, 1980). The cumulative drainage water and the cumulative NO_3-N leached during the experimental

Table 1. Cumulative drainage water and nitrate leaching measured at the bottom of the lysimeter (at 1.2 m depth) and estimated at 0.9 m depth.

Period	drained water mm				NO_3-N losses Kg ha^{-1}			
	measured 1.2 m		estimated 0.9 m		measured 1.2 m		estimated 0.9 m	
	L1	L2	L1	L2	L1	L2	L1	L2
Crop season	131	137	125	175	77	26	47	35
Dry season	–	11	–	–	–	0.1	–	–
Rainy season	8	55	26	70	3	4	17	7
TOTAL	139	203	151	245	80	30	64	42

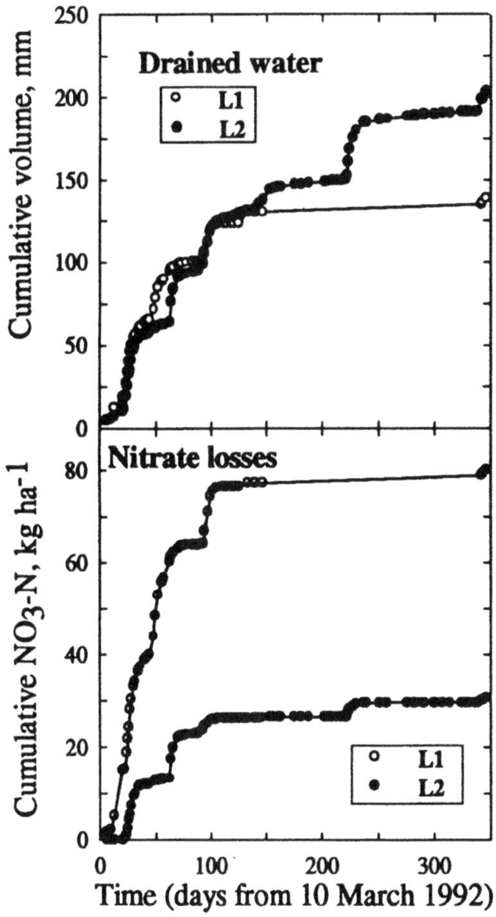

Fig. 2. Cumulative drained water and nitrate losses measured at the bottom of the lysimeters

period, in both lysimeters, are shown in Fig. 2. Three seasons can be distinguished during the experimental period, in the same way as has been reported by Cabrera *et al.* (1994) in previous experiments in the same lysimeters: i) 'Crop season' (10 March 1992–11 August 1992); ii) 'Dry season' (11 August 1992–15 September 1992); iii) 'Rainy season' (15 September 1992–21 February 1993) in which the soil was bare. During the 'crop season' the lysimeters received a total of 661 mm of water by irrigation and 181 mm by rainfall. In this season the total water drainage was 131 mm and 137 mm in L1 and L2 respectively (Table 1), and the total NO_3-N leached was 77 and 26 kg ha^{-1} in L1 and L2 respectively. The considerable nitrate leaching observed during the early growth period in the crop season (Fig. 2) was due to the rainfall (90 mm) when the soil was wet from previous irrigations. During the 'dry season' there was practically no drainage, and consequently no leaching of NO_3-N was observed in L1, and only a very small quantity (0.1 kg ha^{-1}) in L2 (Table 1).

During the 'rainy season', in which the lysimeters received 213 mm of rainfall, the water drainage was 8 and 55 mm in L1 and L2 respectively (Table 1). The NO_3-N leached was 3 and 4 kg ha^{-1} in L1 and L2 respectively. The drainage observed in L1 was lower than in L2 due to the fact that at the end of the dry season the soil water content in the deepest layer (0.9–1.2 m depth) of L1 was much lower than in L2; thus the hydraulic conductivity was much lower in this layer of L1 than in the same one in L2, as has been shown by Cayuela *et al.* (1994) for the $K(\theta)$ function for this soil. Even though the drainage was different in the two lysimeters, the NO_3-N leached was practically the same, due to the higher NO_3-N concentrations in the drainage water in L1 than in L2.

The total drainage during the whole period was 139 and 203 mm in L1 and L2 respectively (Table 1), and

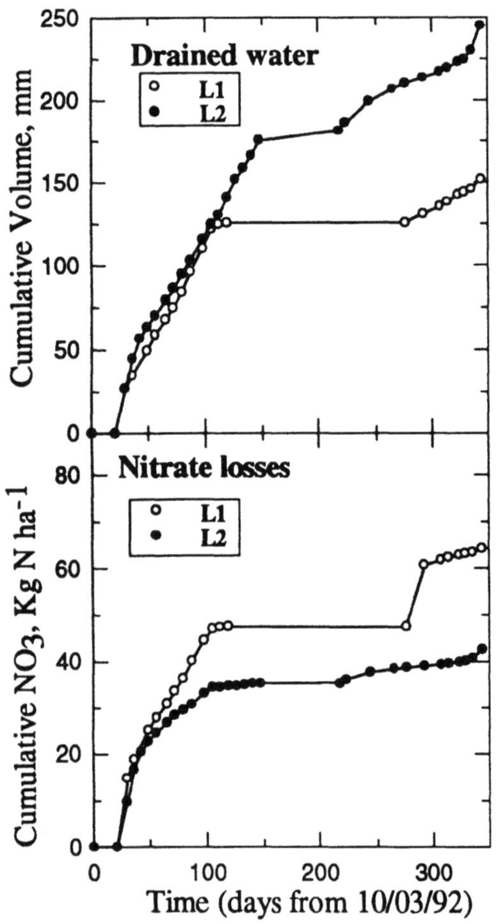

Fig. 3. Cumulative drained water and nitrate losses estimated at 0.9 m depth

the total NO_3-N leached was 80 and 30 kg ha^{-1} in L1 and L2 respectively.

Nitrate concentrations in the soil solution at 0.9 m depth were always higher in L1 than in L2. These concentrations ranged between 5.9 and 240.3 mg NO_3-N l^{-1} in L1, and between 0.1 and 42.3 mg NO_3-N l^{-1} in L2. The cumulative drainage and NO_3-N leached below 0.9 m depth are shown in Fig. 3. These results show similar patterns to those obtained at the bottom of the lysimeters (Fig. 2).

Considering the same three seasons mentioned above, the drainage estimated at 0.9 m depth during the 'crop season' amounted to 137 mm and 175 mm in L1 and L2 respectively (Table 1). In the case of L1 the estimated drainage is practically the same as that measured at the bottom of the lysimeter. In contrast, for L2 the estimated drainage was higher than that measured at the bottom. The NO_3-N leached estimated from the water flow and the concentration of the soil solution amounted to 47 and 35 kg ha^{-1} in L1 and L2 respectively. In the case of L1 the estimated value is much lower than that measured at the bottom. This may be due to the leaching of the residual NO_3-N from the fertilization of the previous year (Cabrera *et al.*, 1993) in the deepest layer (0.9–1.2 m depth). During the autumn-winter period of 1991–1992 the rainfall was very low and the observed drainage was also very low. This did not occur in L2 because no fertilization was applied to this lysimeter in the previous year.

During the 'dry season', drainage and NO_3-N leaching were not observed at 0.9 m depth, which agrees with the results at the bottom of the lysimeters (Table 1). In the case of the 'rainy season' the estimated drainage (Table 1) followed the same pattern as at the bottom of the lysimeters, but with higher values in both lysimeters.

Conclusions

The results obtained in this study show that in an irrigated corn crop, in SW Spain, a strong reduction in the N-fertilization rate used by farmers produces a considerable decrease in nitrate leaching. The volumes of drainage water measured at the bottom of the lysimeters (1.2 m depth) are of the same order as those estimated at 0.9 m depth using Darcy's law and the hydraulic conductivity-soil water content relationship. In both lysimeters the amount of nitrate leached at the bottom of the lysimeters agrees with the estimated amount of nitrate leached below 0.9 m depth.

Acknowledgement

Research carried out in the framework of the contract No. STEP-CT90–0032 of the C.E.

References

Baker JL and Timmons DR (1994) Fertilizer management effects on leaching of labeled nitrogen for no-tillage corn in field lysimeters. J Environ Qual 23: 305–310.

Bergtröm L. and Johansson R (1991) Leaching of nitrate from monolith lysimeters of different types of agricultural soils. J Environ Qual 20: 801–807.

Cabrera F, Fernández-Boy E, Aparicio M.G, Murillo JM and Moreno F (1995) Leaching of nitrate from a sandy loam soil under corn and two N-fertilizations. Fresenius Environ Bull 4: 250–255.

Cabrera F, Reyes A, Fernández-Boy E, Cayuela JA, Murillo JM and Moreno F (1993) Losses of nitrate on a sandy loam soil under corn: lysimeter experiment. Acta Horticulturae 335: 59–64.

Cayuela JA, Fernández JE, Moreno F, Moreno JM and Cabrera F (1994) Estimación de las pérdidas de nitrato en un suelo con cultivo de maíz y riego. Riego y Drenajes XXI 75: 30–34.

Council of the European Community. Council Directive of 15 July 1980 80/778/EEC. Official Journal of the European Community 1980, L229 11–29.

Duynisveld WHM, Strebel O and Böttcher J (1988) Are nitrate leaching from arable land and nitrate pollution of ground water avoidable? Ecol Bull 39: 116–125.

Dowdell RJ, Webster CP, Hill D, and Mercer ER (1984) A lysimeter study of the fate of nitrogen in spring barley crops grown on a shallow soil overlying Chalk: crop uptake and leaching losses. J Soil Sci 35: 183–190.

Fernández Boy E, Aparicio MG, Cabrera F, Murillo JM and Moreno F (1994) Lixiviado de nitrato en un suelo francoarenoso cultivado de maíz bajo riego y dos fertilizaciones. Riegos y Drenajes XXI 74: 8–13.

Jolley VD and Pierre WH (1977) Profile accumulation of fertilizer-derived nitrate and total nitrogen recovery in two long-term nitrogen rate experiment with corn. Soil Sci Soc Am J 41: 373–378.

Owens LB (1990) Nitrate-nitrogen concentration in percolate from lysimeters planted to a legume-grass mixture. J Environ Qual 19: 131–135.

Roth LW and Fox RH (1990) Soil nitrate accumulations following nitrogen-fertilized corn in Pennsylvania. J Environ Qual 19: 131–135.

Smith SJ, Schepers JS and Porter KL (1990) Assessing and managing agricultural nitrogen losses to the environment. Adv Soil Sci 14: 1–43.

Waters Ion Chromatography Cookbook (1989) Waters, Millipore, Milford, Massachusetts.

C. Rodriguez-Barrueco (ed.), Fertilizers and Environment, 463–465.

Long term effects of wheat straw incorporation compared with burning on wheat yield and soil properties

I. Walter, R. Miralles de Imperial and M. Bigeriego
Area de Conservación del Medio Natural, CIT-INIA, Apdo. de correos 8111, 28080 Madrid, Spain

Key words: Winter wheat, grain yields, straw burning, straw incorporation, soil properties

Abstract

Two manipulations of wheat straw, burning and soil incorporation, were studied to determine the long-term effects on winter wheat (*Triticum aestivum* L.) yields and soil properties. The experiment was carried out on a basic calcareous soil of the semiarid Central Region of Spain, and initiated in 1979. Burning versus not burning wheat stubble was examined at three N levels on sowing time and top fertilization; 0, 28 and 56 kg ha^{-1} and 0, 50 and 100 kg ha^{-1} respectively, over 12 years. A basic inorganic fertilizer of P and K was also applied in all treatments. Stubble burning had no effect on grain yield and on grain N uptake in any of the 12 years. Significant difference between both treatments were not found. Nitrogen fertilization increased grain yield and N grain uptake.Long term burning reduced the oxidable organic matter, but this decrease was not significative. Total and inorganic N in soil did not significantly changed in the plot with burning straw as compared to the plots with straw incorporated. Soil samples from straw-incorporated treatments showed a light decrease in soil pH, and plots where straw was burnt the cation exchangeable capacity (CEC), decreased significantly ($p < 0.05$), in all treatments.

Introduction

The maintenance of soil fertility by the return of cereal residues to the soil should be the main objective in the dry areas of countries with a Mediterranean type climate. Burning post-harvest crop residues is still a common practice used by farmers in these countries, in spite of this being not recommendable because of vegetative cover losses leading to a reduction in soil protection against erosion (Biederbeck *et al.*, 1980). Dry land cereal regions are generally deficient in N and OM, and straw burning may produce large losses increasing fertilizer needs (Raison and McGarity, 1979; Kalburtzi *et al.*, 1990).

Many studies on burning and no burning crop straws, have reported that burning did not affect average wheat yields. Where burning practices are used repeatedly on the same land the OM and NO_3-N contents of the soil are reduced although available P is increased. On the other hand, inorganic N is temporarily immobilized by microorganisms during the decomposition of straw cereals when they are incorporated into the soil, and decreased yields due to N deficiencies may be occur (Rasmussen and Rohde, 1988).

Table 1. Characteristics of the top soil used

Characteristics	
pH (H_2O)	8.4
Oxi. C %	0.46
total N %	0.04
NO_3-N mg kg^{-1}	5.40
NH_4-N mg kg^{-1}	2.10
P (Olsen) mg kg^{-1}	10.3
K (NH_4-Acetate) mg kg^{-1}	185
$CaCO_3$ %	16.5
CEC meq per 100 g	9.75
Texture (ISSS)	clay-sandy loam

The purpose of this study, started in 1979, was to determine the effect of burnt or unburnt stubble on wheat yield and on the soil characteristics under semiarid condition.

Table 2. Effect of stubble burning vs. stubble soil incorporation and N fertilization on grain yield of winter wheat and on grain N uptake. Average 12 years. Kg ha^{-1}

Treatment N fertilizer		straw burnt			straw unburnt		
		0	50	100	0	50	100
0	grain yield	1000	1140	1340	850	1060	1100
	N uptake	20.50	25.08	28.80	17.85	22.60	24.00
28	grain yield	1330	1300	1300	1100	1250	1315
	N uptake	26.60	27.95	32.50	24.20	27.50	29.20
56	grain yield	1350	1400	1600	1350	1390	1650
	N uptake	31.05	32.20	36.80	29.26	30.3	34.70

Materials and methods

The field study, initiated in 1979, consisted of three N rates, 0, 28 and 56 kg ha^{-1} applied prior to wheat seeding, three N rates 0, 50 and 100 kg ha^{-1} applied in spring, two burnt conditions (burnt and straw incorporation), and three replicates. The soil at the experimental site is a *calcisol* (Table 1) with an average annual precipitation of 500 mm, 70% of which usually falls between October and May. Wheat stubble was burnt and incorporated to the soil on designated plots in September. All plots were then moldboard plowed 0.22 m deep. Nitrogen as NH_4NO_3 and a basic P and K inorganic fertilization was also applied in all treatments. N-urea was broadcast between 1 and 15 April.

Winter wheat was seeded in 0.25 m rows at 70 kg ha^{-1} about 10th November. Wheat was harvested in July. Grain yield for 12 successive crops was recorded annually. Nitrogen in grain subsample was determined by Kjeldahl. Soil samples were take after 12 years from each plot, air-dried and ground to pass a 2 mm sieve. Soil pH in water, oxidizable C by Walkley, total N by Kjeldahl, NH_4-N and NO_3-N by Bremmer, CEC by Na-Acetate solution 1.0 N and soluble P by Olsen solution were determined for each plot.

All data were subjected to analysis of variance by BMDP statistical software Inc.

Table 3. Effect of stubble burning vs. stubble soil incorporation on soil physico-chemical properties. Average of N fertilizer. After 12 years.

Treatment	straw burnt	straw unburnt
pH (H_2O)	8.4	8.2
oxi. C %	0.44	0.51
total N %	0.035	0.040
NO_3-N mg kg^{-1}	5.77	6.37
NH_4-N mg kg^{-1}	2.21	2.40
P available mg kg^{-1}	12.90	11.75
CEC* meq per 100 g	9.83	10.70

* Significant difference $p = 0.027$.

Results and discussion

Grain yield

The 12 years average wheat yields showed increases in grain yield as a result of burning, although differences were not significative. Grain yield average of 1.44 and 1.40 Mg ha^{-1} for straw burnt and straw unburnt respectively were found. Burning tended to increase yields in the early years but it tended to depress yields thenceforth.

Grain N uptake tended to be slightly higher with burnt treatment also, but this increase was not significant. That beneficial effect suggests greater immobilization of N during spring, with eventual release later in the growing season when straw was incorporated.

In both tillage treatments grain yield responded directly to both N fertilizer doses (Table 2).

Effect on physico-chemical soil properties

The values of the different physico-chemical soil properties found are in the Table 3. Small increases of soil organic matter occurred after 12 years when straw was returned to the soil annually. However changes were not significantly different.

Soil pH was lower under unburnt plots although no significant difference was found. Many years of repeated burning did not affect total and inorganic N. However NO_3-N and NH_4-N levels showed a slight increase in unburnt plots. No significant changes in $NaHCO_3$-extractable P of soil in either of the treatments were found. Cation exchangeable capacity of the soil was affected by the different treatments only. CEC was significantly greater in the plots with straw incorporation than that in the straw burning plots, ($p = 0.027$).

The results show that after 12 years of annual straw return to the soil no clear effect on grain wheat yields and on the physico-chemical properties of the soil was observed. Both beneficial as well as detrimental effects due to burning were shown. Beneficial effects such as minor grass weed risk and seedbed preparation for the next season's crop are the only reasons why straw is burnt after harvesting.

References

Bierderbeck VO, Campbell CA, Bowren KE, Schnitzer M and McIver RN (1980) Effect of burning cereal straw on soil properties and grain yields in Saskatchewan. Soil Sci Am J 44: 103–111.

Kalburtzi DS, Veresoglou DS and Gerakis PA (1990) Effects of burnt or unburnt straw on wheat and fababens as influenced by N fertilization. Agriculture, Ecosystems and Environment 31: 173–185.

Raison R and McGarity JW (1979) Changes in C and N fractions of graminaceous straws after burning. Plant Soil 51: 311–318.

Rasmussen PE and Rohde CR (1988) Stubble burning effects on winter wheat yield and N utilization under semiarid conditions. Agron J 80: 940–942.

C. Rodriguez-Barrueco (ed.), Fertilizers and Environment, 467–469.

Loads on soil due to agricultural activities in Hungary

O. Pálmai
Plant Health and Soil Conservation Station of County Fejér, 2481 Velence, Hungary

Key words: Fertilizers, pesticides, sludge, rates, environmental risks, trends in last decade

Abstract

Cultivated lands are continuously loaded by different chemicals during agricultural production in Hungary. Nowadays the countries with developed agriculture obviously and necessarily use various amounts of pesticides and fertilizers to reach the desirable level of production. Disposal of waste waters and sewage sludges on agricultural lands also has to be considered in several cases. However, this can be made only under controlled and supervised conditions in order to maintain and protect the quality of the soils. Situations of fertilizer and pesticide use, and of application of sewage sludges to agricultural lands in Hungary during the last decade are summarized. Impact of socio-economic changes on environmental issues are also considered.

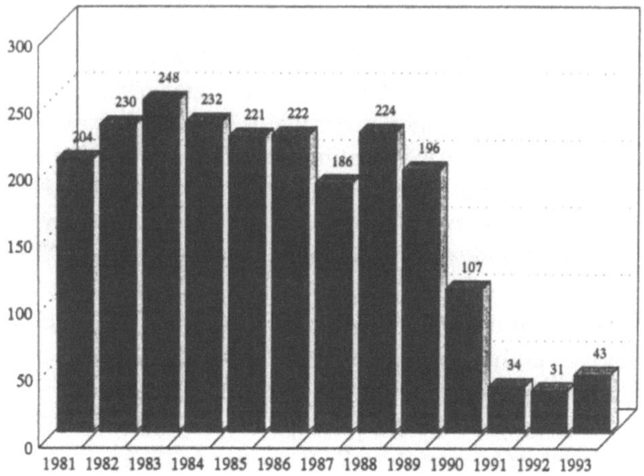

Fig. 1. Fertilizer use in Hungary expressed as active ingredient in kg/ha

Application of fertilizers

Agricultural production in Hungary reached its intensive phase by the mid-1970s, i.e. the application of fertilizers tripled within 10 years resulting in significant yield increase. Between 1975 and 1985, a national average of 250 kg/ha N:P:K (active ingredient) was used.

It exceeded 400 kg/ha for the major field crops. Then, after a slow decrease in consumption, a a dramatic drop was observed from 1990 (Fig.1). This was primarily caused by the significant increase in fertilizer prices, but also by the change of ownership in the Hungarian agriculture.

The effect of the losses by storage on the environment is negligible because transportation of fertilizers is made directly from almost 100 modern storage facilities. Due to fertilizer application, low level N losses might occur, but the increase in nitrate levels of the living and groundwaters originates basically from municipal and industrial sources. Ninety per cent of the Hungarian population consumes drinking waters through conductive pipes, however slightly more than 50% of the communities have their own canalization system. In addition, we have 3000 waste deposits at various communities.

Lake Velence is the second largest resort place in Hungary with a water surface of 27 km^2. Sources of the contaminations reaching the lake have been continuously tested for 6 years in its 60,000 ha catchment area. There are 14 villages and 35,000 ha of agriculturally cultivated areas. The 9 animal keeping sites of large populations work with conventional technologies, i.e. a minimum volume of liquid manures is produced. Studies of the N and P forms of the water streams feeding the lake as well as that of the anionactive detergents obviously points to the municipal origin of the contaminations. This is due to the poor canalization system of the inhabited areas, where there are villages with no sewage conductions at all. Waters flowing through the

Table 1. Acceptable maximal volume of toxic elements in fertilizers (mg/kg dry matter)

Arsenic (Ar)	10
Cadmium (Cd)	4
Cobalt (Co)	50
Chromium (Cr)	100
Mercury (Hg)	1
Nickel (Ni)	50
Lead (Pb)	100
Selenium (Se)	10

Table 2. Detected pesticide residues (1976–1991)

active ingredient	No. of positive samples	conc. ranges μg/L
Aktinit PK	159	0.1–80
Aktinit DT	2	10–25
2,4–D	10	0.5–4.0
2,4,5-T	5	0.6–2.0
MCPA	2	0.5–7.5
Metidation	1	0.2–0.3
Triazophos	1	0.1–2.0

agricultural areas "purify themselves", and concentration values of contaminants decrease.

The soil acidification effect of fertilizer application should also be noted because natural soil heavy metals become soluble and thus they may reach the food chain (chemical time bomb). Of the 6.6 million hectares of agricultural land of Hungary 2.3 million ha are acid areas needing some treatments. The necessary liming may provide effective control of the acidification due to fertilizer application. The power stations produce 800,000 t/year atmospheric S contamination in Hungary, therefore pH values of 3.5–4.5 have already been measured several times in precipitation waters. Thus, it is rather difficult to evaluate the soil acidification rate caused by fertilizers.

The fertilizers contain high volumes of carriers and other contaminants, causing a special load on the soil. Nitrogen-fertilizers are the cleanest as their origin is the atmospheric nitrogen. The contamination is produced during manufacturing and depends on the technology employed. In 1982 we had calnitro (Pétisó) on the market that contained 0.24% Sr accompanied by high content of Ca and Mg. The P fertilizers contain % volumes of Ca, Mg, S and decimal % of Al, Fe, K, Na, Si and Sr. In Hungarian superphosphate, about 1% of Sr has already been measured, and also similar values of Al in thermophosphate. The Cd content of some P basic materials may also be high. The Hg, Cr, Cd, V, Ti and Mn concentrations were extremely high in the Thomas phosphate manufactured in ex-East-Germany. As far as the trace elements and heavy metals are concerned, the K-fertilizers are clean, only low levels of Cu, Cd, Ba, Sr, Zn and Mn were found and others can even not be detected. The heavy metal contents of the fertilizers put on the market from last year have been regulated, i.e. the maximum acceptable concentrations have been determined (Table 1).

Chemical pesticides

The volumes of applied chemical pesticides have shown significant changes. In 1985 the 5 kg of pesticide active ingredients used on each hectare, have been lowered to only 2.5–3 kg/ha, which is still being high. The share of herbicides has increased the total pesticide consumption.

At present 550 pesticides are registered in Hungary, which contain 300 different active ingredients. The Hungarian registration procedure is very strict, based on wide technical information and fixed in legislation. Distribution, purchase and application of the registered chemicals is linked to special education/training. More than 3000 university graduated plant protection specialists are responsible for this activity. During application and distribution, the quality of the chemicals as well as the agricultural produces are regularly controlled.

Hungary was the first to prohibit application of chlorinated hydrocarbons 26 years ago. In 1972 the use of triazins was drastically reduced (max. 1.5 kg/ha), and 2 years ago we banned the application of Gramoxone.

Pesticide residues are annually tested in 7000–8000 food samples taken from the production places and the markets. In addition, the nitrate and heavy metal contents have also been tested for the last years. In over 95% of the samples no residues could be detected and the concentration of the pesticide residues was near or reaching the accepted limits only in less than 2% of the samples.

From 1976, the pesticide residue content of the living waters in the catchment area of lake Balaton has also been regularly tested. Pesticide contents could be detected only in 180 of 3000 water samples, i.e. in

6% of the samples under study (Table 2). The measured values did not, however, reach, in any case, the concentration range toxic to fish; they were generally lower by several degrees of magnitude.

Because there is no practical alternative to the use of chemical pesticides, the key factor of the usage should be the strict legislation equal to the requirements of the EU member countries.

Pesticide wastes mean a permanent risk together with the problem of disposal of packing materials. It is actually a very important and urgent task, born from the actual transformation of agricultural ownership, to improve the new farmers' technical knowledge and "plant protection expertise".

Table 3. Maximum volume of microelements accepted for arable land (kg/ha/year)

Zinc (Zn)	30
Copper (Cu)	10
Cadmium (Cd)	15
Nickel (Ni)	2
Lead (Pb)	10
Chromium (Cr)	15
Mercury (Hg)	15
Manganese (Mn)	30
Arsenic (As)	3
Selenium (Se)	1

Sewage sludge

In Hungary, 1 billion cubic meters of waste waters in produced annually, 60% of which gets without treatment to the sites or to other deposits. In the late 1980s, 380 sewage treatment plants of different size were built at various communities and 3 million m3 of sewage sludges were expected. Considering Hungary's geographic properties, it is obvious to place the sewages on agricultural areas. On the other hand, the quality control of the land requires that only substances not decreasing production safety should be applied to agriculture. Therefore these substances can be applied for agricultural purposes only if they meet the technical requirements and limits specified by the Measures and Technical Guidelines for the Deposit of Waste Waters and Sewage Sludges (Table 3). Prior to any application a pedological study should be made.

Application of fertilizers and pesticides as well as deposition of sewage sludges on agricultural fields are officialy supervised and technically coordinated by the country plant protection and soil conservation stations of the Ministry of Agriculture. These county stations have trained specialists and well-equipped laboratory facilities to perform regulatory tasks and to prepare the necessary study reports. International accreditation of the laboratories has already started.

Uniform principles were worked out to perform regularly, on a 3-year basis, from 1978, the soil tests of the whole agricultural area in order to provide reasonable plant nutrition. From the 1980's, tests were carried out every 5 year and heavy metal contents in the soil have also been measured. This activity has been slowed down because of the transformation of the Hungarian agriculture, however, the computerised data-base of all information is available for all the cultivated areas. The Soil Conservation Information Monitoring System (TIM) has been operated for 3 years. On 1200 designated sampling sites the soil status and the changes in their physical and chemical properties have been continuously controlled and examined.

Summary

In spite of the fact that soils in Hungary are adequately used, many different loads reach them. However, they can be well maintained. The actual conditions of the agricultural areas in Hungary satisfy the majority of criteria. Our soils are not contaminated and they can meet the ever stricter requirements of environmental protection. Our long-term objective is to maintain this favourable condition. This aim is supported by a Law on the Land effective from summer of 1994.

C. Rodriguez-Barrueco (ed.), Fertilizers and Environment, 471–475.

The effects of pulp-mill sludge on leaching of mineral nitrogen

M.J. Cabrita[1], E. Vasconcelos[2] & F. Cabral[2]
[1] *Universidade de Évora, Largo dos Colegiais, 2. 7000 Évora Portugal;* [2] *Agricultural Chemistry Department. Instituto Superior de Agronomia, Tapada da Ajuda, 1399 Lisboa Codex. Portugal*

Key words: Leaching of mineral nitrogen; pig slurry; pulp-mill sludge; *Triticum aestivum* L.

Abstract

An experiment was carried out with wheat plants using Mitscherlich pots filled with a Cambic Arenosol and a Cromic Vertisol, in order to study the effect of pulp-mill sludge application on leaching of mineral nitrogen provided by pig slurry and/or mineral nitrogen fertilizers. From the results obtained it is possible to conclude that the application of pulp-mill sludge led to a decrease of $N\text{-}NO_3^-$ and $N\text{-}NH_4^+$ leached. On the other hand, the use of pig slurry in order to substitute nitrogen mineral fertilization at planting, led to an increase of $N\text{-}NO_3^-$ leached. The amount of mineral nitrogen leached was greater in Cambic Arenosol that in Cromic Vertisol essentially due to the retention of $N\text{-}NH_4^+$ in the latter soil.

Introduction

In Portugal the output of pig slurry has been steadily increasing and the environmental hazards caused by this byproduct is a subject of great concern, namely in what respects leaching of nitrates and the consequent superficial and groundwater contamination, after pig slurry spreading in the soils.

To overcome or at least to minimise those risks some experiments have been carried out by combining the use of organic manure of a high C/N ratio (straw) with slurry, in order to block biologically nitrogen protecting it from leaching. (Amberger, 1987; Rauhe, 1987). In Portugal, the paper processing industry has undergone a considerable growth recently and the production of pulp-mill sludge is one of the main byproducts of this industry, that shows a high biodegradable organic matter content as well as a variable C/N ratio according to the treatments of the effluents (Cabral and Vasconcelos, 1993; Vasconcelos and Cabral, 1993).

The main purpose of this work is to study in two soils with very different physical and chemical characteristics the effect of pulp-mill sludge application on leaching of mineral nitrogen provided by pig slurry and/or mineral nitrogen fertilizers.

Table 1. Soils physical and chemical characteristics

Determinations	soil A	soil B
Organic matter (%C × 1.724)	0.86	3.14
pH (H_2O)	5.60	6.70
pH (KCl)	5.10	5.70
$N_{NO_3^-}$ ($\mu g\ g^{-1}$)	1.05	52.50
$N_{NH_4^+}$ ($\mu g\ g^{-1}$)	11.98	4.39
K (Egner-Riehm) ($\mu g\ g^{-1}$)	65.70	138.60
P (Egner-Riehm) ($\mu g\ g^{-1}$)	7.90	42.30
Exc. cations (cmol (+) kg^{-1})		
Ca	1.54	15.25
Mg	0.35	11.96
K	0.12	0.10
Na	0.44	1.70

Material and methods

Experimental details

An experiment out doors with wheat plants (*Triticum aestivum* L. cv. Ansa) was carried out using Mitscherlich pots (0.03 m^2 of available area and 0.2 m in depth), filled with 7.0 kg and 5.5 kg of the superficial layer

Table 2. The physical and chemical composition of pig slurry

Organic matter	g l^{-1}	3.10
pH		8.10
Conductivity	mS cm^{-1}	15.80
P	mg l^{-1}	36.30
K	mg l^{-1}	1320.00
$N_{NH_4^+}$	mg l^{-1}	2082.20
$N_{NO_3^-}$	mg l^{-1}	235.80
N Kjeldhal	mg l^{-1}	2400.00
Mg	mg l^{-1}	33.80
Ca	mg l^{-1}	367.50
Na	mg l^{-1}	492.50
Cu	mg l^{-1}	4.20
Fe	mg l^{-1}	10.90
Mn	mg l^{-1}	1.00
Zn	mg l^{-1}	3.80

Table 3. The physical and chemical composition of pulp-mill sludge (dry matter)

Moisture	%	69.50
pH (dil 1:10)		7.23
C/N ratio		80.30
Organic matter	%	54.00
P	%	0.18
N_{Kj}	%	0.39
K	%	0.03
Na	%	0.19
Mg	%	0.40
Ca	%	5.50
Cu	$\mu g\ g^{-1}$	87.60
Fe	$\mu g\ g^{-1}$	1221.30
Zn	$\mu g\ g^{-1}$	52.20
Mn	$\mu g\ g^{-1}$	212.30

(fraction < 5 mm) of a Cambic Arenosol (soil A) and a Cromic Vertisol (soil B) (Table 1) with a completely randomized design.

Treatments replicated four times consisted of:

A-NPK; B-NPK + 50 t ha^{-1} of pulpmill sludge; C-nitrogen applied at planting fertilization provided by pig slurry + N'P'K'; D-nitrogen applied at planting fertilization provided by pig slurry + 50 t ha^{-1} of pulp-mill sludge + N'P'K'.

The rate of addition of nutrients consisted of:

Table 4. Watering calendar

Month	Days	Volume (ml)
November	14,16,21	100
December	28	250
January	27,30	100
February	1,4,7,8,11,17,19, 22,23,25,26,27,28	100
March	1,4,5	100
	6,8,10,12,14,17,19, 22,24,26,28,31	200
April	10,13,15,18,20,22,23	200

Table 5. Yield obtained to wheat plants (g/pot)

Treatments	soil A		soil B	
	grain	straw	grain	straw
A	23.30 a*	20.25 b	32.50 a	40.60 a
B	23.40 a	25.30 a	33.10 a	38.40 ab
C	24.40 a	26.10 a	32.20 a	37.40 ab
D	23.80 a	24.90 a	33.20 a	35.40 b

N - 1.25 g of nitrogen per pot being 0.625 g applied at planting (ammonium sulphate) and 2 × 0.312 g at top-dressing fertilization (ammonium nitrate).

N' - 2 × 0.312 g of nitrogen per pot applied at top-dressing fertilization (ammonium nitrate)

P - 0.437 g of phosphorus per pot (super phosphate 18%)

P' - 0.437-*x* g of phosphorus, *x* being the phosphorus provided by pig slurry application

K - 0.830 g of potassium per pot (potassium chloride)

K' - 0.830 - *y* g of potassium, *y* being the potassium provided by pig slurry application.

Tables 2 and 3 show the composition of pig slurry and pulp-mill sludge.

Fifty wheat seeds were sown per pot (11 November 1992). After emergence a selection was made in order to leave the twenty five most uniform plants in the pots (22 November 1992).

At the beginning of the experiment pots were watered with an amount of demineralized water enough to allow seeds germination. During the experiment some waterings were carried out (Table 4), in

Table 6. Amounts of leached nitrogen (mg/pot).

Treat.	soil A			soil B		
	$N_{NH_4^+}$	$N_{NO_3^-}$	N_{total}	$N_{NH_4^+}$	$N_{NO_3^-}$	N_{total}
A	476.3 a	48.8 b*	525.1 a	5.8 a	176.3 bc	182.7bc
B	335.2 b	57.2 b	392.4 b	3.4 b	142.7 b	146.1 b
C	298.5 b	115.5 b	414.0 b	4.4 ab	256.8 a	261.2 a
D	168.3 c	82.9 c	251.2 c	3.7 b	198.2 c	201.9 c

Table 7. Nitrogen content and nitrogen uptake by wheat plans

Treat.	N(%)				N uptake (mg/pot)	
	soil A		soil B		soil A	soil B
	straw	grain	straw	grain		
A	1.26 a*	1.88 b	1.14 a	2.13 b	693 a	1156 a
B	1.17 a	1.89 b	1.19 a	2.27 ab	739 a	1208 a
C	1.11 a	1.88 b	1.07 a	2.29 ab	749 a	1137 a
D	1.17 a	2.14 a	1.05 a	2.33 a	801 a	1045 a

order to allow a regular development of wheat plants, while precipitation values were abnormally low.

Leachates from each pot were collected, rigorously measured and analysed.

Analytical methods

Organic matter in the soil was calculated by multiplication of the percentage of organic carbon by the factor 1.724, based on the assumption that soil organic matter is 58% carbon. Organic carbon was measured by dry combustion at 1200 °C, followed by measurement of the CO_2 evolved by Ströhlein equipment.

Organic matter in the sludge was determined by weight loss in a furnace at 350–400 °C after 7–8 h.

Exchangeable cations in the soil were determined by atomic absorption spectrophotometry, after extraction by the Mehlich method, using a barium chloride-triethanolamine solution of pH 8.1 (Mehlich, 1953). Available phosphorus and potassium in the soil were determined by colorimetric and flame emission photometric methods respectively, after extraction using an ammonium lactateacetic acid solution of pH 3.75 (Egner *et al.*, 1960). Nitrogen in sludge and plant tissues, was determined by the Kjeldhal method (Jackson, 1958), using a Tecator equipment.

All the other mineral elements, were determined by atomic absorption spectrophotometry (Pye-Unicam SP-9) after hydrochloric mineralization of the ash (Marti and Muñoz, 1957), except for phosphorus which was, determined by the vanadomolybdophosphoric yellow colour method (Koening and Johnson, 1942) in a Hitachi U-2000 spectrophotometer.

N-NH_4^+ on leachates was determined by distillation on a Kjeltec apparatus.

N-NO_3^- on leachates was determined using an electrode technique (Byrne, 1979)

Statistical analysis

Results from the study were analysed by one-way ANOVA, followed by Scheffe F-test at P=0.05.

Results and discussion

Table 5 shows the yield obtained to wheat plants at the end of the experiment.

From the results obtained it is possible to conclude that there was no significant differences between treatments concerning grain yield. On the other hand grain

Table 8. Balance of mineral nitrogen (mg/pot) in both soils

Treat.	N applied	N leached				N uptake		N uptake / N applied (%)	
		soil A		soil B		soil A	soil B	soil A	soil B
		NH_4^+	NO_3^-	NH_4^+	NO_3^-				
A	1250	476.3	48.8	5.8	186.3	693	1156	55.4	92.5
B	1250	335.2	57.2	3.4	142.7	739	1208	59.1	96.6
C	1250	298.5	115.5	4.4	256.8	749	1137	59.9	90.9
D	1250	168.3	82.9	3.7	198.2	801	1145	64.1	91.6

yield was higher in soil B, that is in perfect accordance with its higher productive capacity.

Table 6 shows the amount of leached nitrogen in both soils.

From the observation of this table it is possible to conclude that the application of pulp-mill sludge led to a decrease of total nitrogen leached in both soils. As a matter of fact, treatments where pulp-mill sludge was applied (B and D) show significantly lower amounts of total nitrogen leached compaired with other treatments (A and C). On the other hand, the application of pig slurry led to an increase of nitrate nitrogen leached.

The decrease of nitrate nitrogen leached when pulp-mill sludge is applied shows significant importance only in the treatments where pig slurry was applied as well. Comparing both soils it is possible to observe a very different behaviour concerning the origin of nitrogen leached. In fact in soil A (sandy soil with a low capacity to keep nutrients) the main form of leached nitrogen is the ammoniacal form, while in soil B (clayish soil with a high capacity to keep nutrients) nitrate form takes advantage. Otherwise, the greatest leaching of total nitrogen shown in treatment A, is essentially due to nitrogen ammoniacal form.

The decrease of leached mineral nitrogen observed in both soils when pulp-mill sludge is applied, may be explained taking into consideration the temporary immobilisation of available nitrogen in soil. The results obtained are in accordance with those obtained by Zibilske (1987).

Table 7 shows nitrogen content in wheat plants as well as nitrogen uptake by this crop.

Concerning grain, the highest content of nitrogen in both soils is observed in the treatment where pig slurry and pulp-mill sludge were together applied (treat. D). Other values did not show significant differences.

Table 8 shows the mineral nitrogen balance in both soils. Concerning the efficiency of nitrogen fertilization it was significantly higher in soil B (≈90–95%) that in soil A (≈50–60%). This difference is essentially due to high retention of the ammoniacal nitrogen in soil B. Treatments where pulp-mill sludge was applied (B and D), show a greatest efficiency of nitrogen in both soils. The higher leaching of nitrate nitrogen in the treatments where pig slurry was applied it was expected because a fraction of nitrogen in pig slurry is in the nitrate form while nitrogen fertilization at planting was achieved using an ammoniacal fertilizer (ammonium sulphate).

From the results obtained it is possible to draw the following conclusions:

Pig slurry application in order to substitute nitrogen fertilization at planting led to an increase of leached nitrogen;

Pulp-mill sludge application decreases nitrate nitrogen leaching specially when pig slurry is applied as well;

Leaching of ammoniacal nitrogen decreases with the simultaneously application of sludge and pig slurry. However both soils show a different behaviour: in soil B because of its high capacity to keep nutrients it is evident the ammoniacal nitrogen retention, while in soil A ammoniacal nitrogen is the main responsable for the great amounts of total nitrogen leached.

When soils show a low cation exchangeable capacity (CEC) such as soil A, the use of an organic fertilizer, like pulp-mill sludge, may lead to a significant leaching decrease of mineral nitrogen, essentially due to the retention of ammoniacal nitrogen.

References

Amberger A (1987) Utilization of organic wastes and its environmental implications. In: E Welte and I Szabolcs (eds) Agricultural Waste Management and Environmental Protection. Proc. 4th Int. Symp. of the International Scientific Center of Fertilizers, vol. 1, p 37–54.

Byrne E (1979) Ed. Chemical analysis of Agriculture material. An Foras Taluntais.

Cabral F and Vasconcelos E (1993) Agriculture use of combined primary/secondary pulp-mill sludge. Agrochimica XXXVII: 409–417

Egner H Riehm H Domingo WR (1960) Untersuchungen über die chemische Bodenanalyse als Grundlage für die Beurteilung des Nährstoff-zustandes der Böden. II Chemische Extrakionsmethoden zur Phosphor und Kaliumbestimmung. Kungl Lantbr Höngsk Ann 26: 199.

Jackson M L (1958) Soil Chemical Analysis. Prentice-Hall, Inc. Englewood Clifs, N.Y.

Koening R A and Johnson C R (1942) Colorimetric determination of phosphorus in biological materials. Ind Eng Chem Anal 14: 155.

Marti F B and Muñoz J R (1957) Flame photometry. Elsevier Amsterdam.

Mehlich A (1953) Rapid determination of cation and anion exchange properties and pH of soils. J AOAC 36: 445–447.

Rauhe K (1987) Effects of organic manuring and cropping on soil humus and fertility. In: E Welte and I Szabolcs (eds) Agricultural Waste Management and Environmental Protection. Proc 4th Int. Symp. of the International Scientific Center of Fertilizers, vol. 1, 55–75.

Vasconcelos E Cabral F (1993) Use and environmental implications of pulp-mill sludge as an organic fertilizer. Environ Pollut 80: 159–162.

Zibilske L M (1987) Dinamics of nitrogen and carbon in soil during papermill sludge decomposition. Soil Science 1: 26–33.

C. Rodriguez-Barrueco (ed.), Fertilizers and Environment, 477–480.

Alpechín as fertilizer

A new way to eliminate its phytotoxicity

I. Franco, F. Gallardo-Lara, M. Hurtarte & M.T. Baca
Estación Experimental del zaidín, CSIC, PO Box 419, E-18080 Granada, Spain

Key words: Alpechín, organic mater, phytotoxicity, organic fertilizer

Abstract

Composted organic matter is well known as soil improver. Its high cation exchange and chelating capacities aid to eliminate the harmful effects of toxic organic substances. Alpechín is the wastewater produced during olive oil extraction and it is characterized by its high biotoxicity. Alpechín was treated to eliminate the phytotoxicity by absorbing it with organic matter and posterior incubation. When this wastewater was absorbed on compost (SCA), a great part of its phytotoxicity was chemically neutralized. At the end of the incubation period, plants grew better than the control and soil biomass increased in a mixture of soil-compost-alpechín (SCA). When the incubations were conducted with a parallel plant cultivation the evolution of the parameters studied became increased.

Introduction

In the world there is superproduction of solid residues a great part of which are organic materials that can be transformed and applied to the soil as amendments. But they cannot be directly added to the soil because these residues contain pathogens and parasites to dangerous man, animals and plants. Besides, the initial process of this decomposition causes rapid metabolic changes that may damage soil and crops because of newlyformed highly reactive biotoxic substances. For this reason it is necessary to submit the organic residues to an stabilization process before incorporating them to the soil (Baca *et al.*, 1990, 1992). Alpechín is a waste-water produced during olive oil extraction. It is rich in mineral nutrients but also has a high biotoxicity due to its high content of organic matter (40–80 g/l) rich in polyphenolic substances and with a high BOD_5. The disposal of this type of waste-water is a particularly pressing problem in Southern Spain and other mediterranean countries because of its phytotoxicity and difficulty of handling. The addition of this waste-water to soil has harmfull effects on seed germination and plants development (Perez *et al.*, 1986). On the other hand, composted organic matter is well known as soil improvement since its high cation exchange capacity and chelating capacity aid to eliminate the harmful effects of heavy metals and toxic organic substances of the media. Soil microbial biomass plays an important role in the degradation of soil organic matter which is consumed as a carbon source in their metabolic processes. This process has beneficial effects, contributing nutrients and avoiding in an effective manner the formation of pathogens which cause the 'sick soils'. The aim of this study was a) to eliminate the phytotoxicity of alpechín by absorbing it with solid organic matter in two different degrees of transformation, (raw and composted) and b) to accelerate the above cited degradation processes introducing a root system in order to increase the soil biomass and activity.

Material and methods

The soil used in our experiments was a silty loamy plough layer with an organic matter content of 3.5%. The soil was incubated being admixtured with differents organic residues. The organic materials utilized were: alpechín (A) with 9.7% of dry matter; compost of cotton waste (C) with 72% of volatil solids and raw cotton waste (R) with 91% of volatil solids.

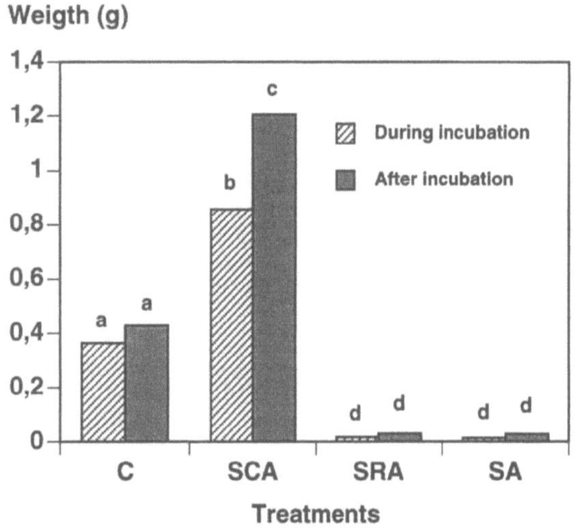

Fig. 1. Comparison between dry weight of shoots of plants grown during the incubation and that of plants cultivated after the incubation period of 50 days in the same treatments without plants

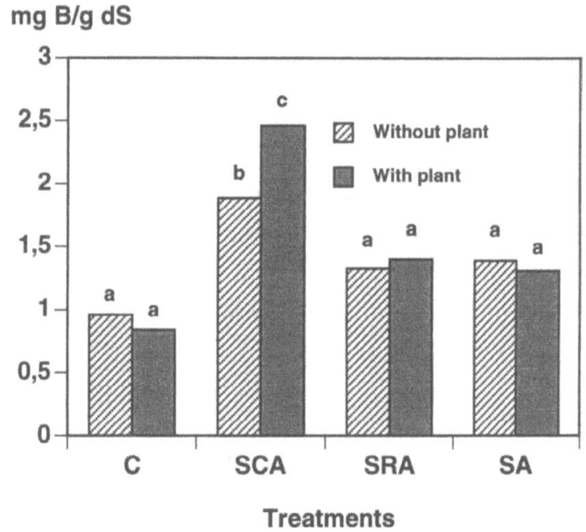

Fig. 2. Biomass (B) content expressed as mg/g dry substrates (dS) in the treatments without and with parallel plant growth determined at the end of the incubation period.

The treatments were: soil (C), soil-compost-alpechín (SCA), soil-raw cotton waste-alpechín (SRA) and soil-alpechín (SA). Soil-organic matter proportions used were; 80 ml soil:15 ml alpechín and 5 g of compost or raw organic matter depending on the treatment.

Data were obtained after 50 days of soil-organic matter incubation with or without parallel plant growth; the incubation with plant was performed in 100 ml plastic pots in which seeds of *Lactuca sativa* L. were sown. The experimental design was: 1 plant/pot and 5 repetitions per treatment. The assay was performed in a greenhouse under controlled conditions of light, temperature and humidity.

The organic carbon was extracted with $Na_4P_2O_5$ + NaOH 0.1 M with a ratio 1:10 (w:v) after shaking during 30 min. under N_2 atm. and posterior centrifugation. Humified and non humified fractions were separated by column chromatograpy of PVP (Sequi *et al.*, 1986). The carbon content was determined by wet oxidation. Biomass content was determined by the 'fumigation-extraction' method proposed by Ocio and Brookes (1990).

At the end of the incubation period a new experiment was carried out in the treatment in which the soil was incubated without plant growth, in the same conditions of the assay described above, and the dry weight of shoots was again determined.

Results and discussion

Alpechín showed a strong inhibitory effect on plant growth in treatments SRA and SA where the phytotoxicity of the wastewater was not neutralized when mixing it with soil or soil-cotton waste. At the end of the incubation period, only a little part of the phytotoxicity was eliminated as shown by the plant growth (Fig. 1).

Controls did not present any significant change in plant growth with the incubation, but only for a slight (not significant) increase probably due to the previous reactivation of the biomass caused by the soil conditioning.

When alpechín was absorbed on the composted organic matter it seemed that most of the phytotoxicity disappeared. Plants duplicated their shoot dry weight as compared with the control and triplicated this weight when planted at the end of the incubation period (Fig. 1).

Quantitatively, under treatments C, SRA and SA, the biomass content did not present any significant difference when measured with and without plant growth (Fig. 2). SRA and SA had a slightly superior biomass content than the control (C), although qualitatively this biomass was different from the control because of the phytotoxic effect of the waste water add. Toxicity affected plants and probably soil microorganisms,

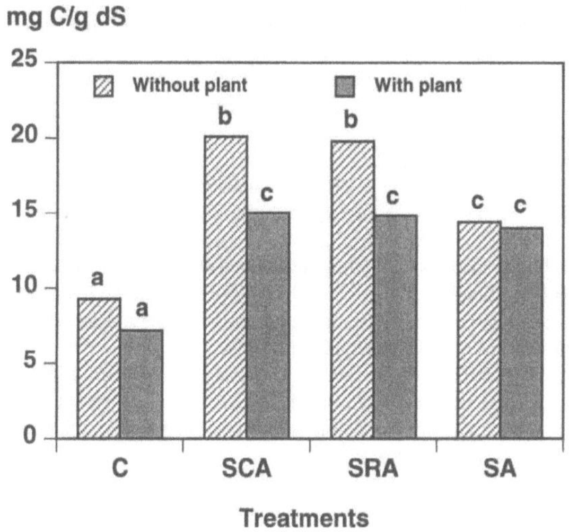

Fig. 3. Carbon extracted in alkaline solution (mgC/g), determined in the treatments without and with parallel plant growth at the end of the incubation period.

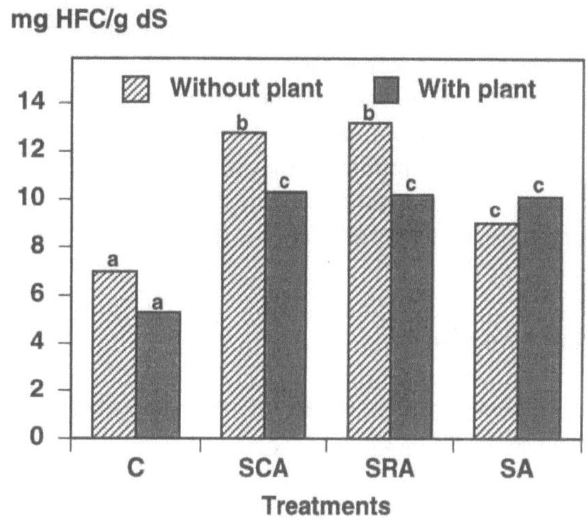

Fig. 4. Carbon of the humified (HF) fraction extracted in alkaline solution and separated by chromatography in PVP resin, determined at the end of the incubation period in the treatments without cultivation and with cultivation of plant.

surviving in the treated soils only those microbes able to adapt to the adverse environment generated.

After incubation, treatment SCA duplicated the biomass content of the control when incubated without plant growth, and triplicated in the treatment with plant growth. This biomass did not interfere negatively on plant growth otherwise shown when adding raw compost to cultivated soils (Baca *et al.*, 1990).

Total organic carbon (TOC) extracted in alkaline solution decreased during the incubation due to the biological activity (respiration). TOC became mineralized between 24–25% more in C, SCA and SRA when incubated in the presence of plants. Treatment (SA) did not present any variation in TOC after incubation (Fig. 3).

Soil incubated alone (control) decreased its total organic carbon content (Fig. 3). Microorganisms utilized as carbon source not only the less humified fraction (NH) but even the humified one (HF). A similar situation takes place in cultivated soils under the modern tillage practices where the soils become impoverished in organic matter as well as in many of their physicochemical characteristics.

The carbon content of the humified fraction (HF) diminished after the incubation in all cases but the presence of roots produced a decrease in C, SCA and SRA between 19 and 24% more than the treatments incubated without plant growth (Fig. 4).

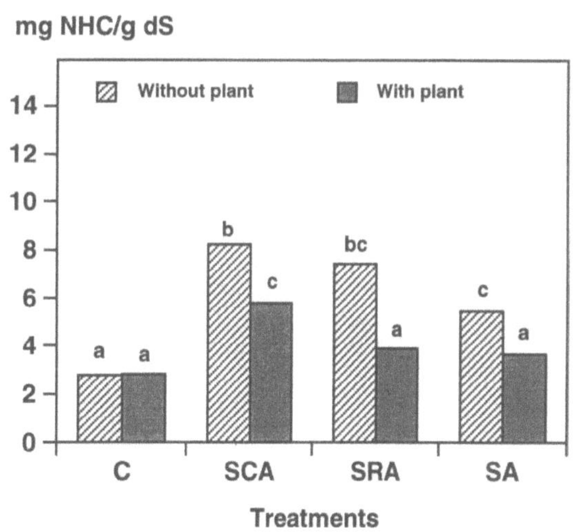

Fig. 5. Carbon of the non humified (NH) fraction extracted in alkaline solution and separated by chromatography in PVP resin, determined at the end of the incubation period in the treatments without cultivation and with cultivation of plant.

The second fraction, non humified (NH) was strongly consumed during the incubation in all the treatments decreasing their contents to 30% in the case of SCA; 33% in SA and even 48% in SRA treatment.

This parameter did not change in the control (Fig. 5). More significant changes happened in treatment SRA probably because of the high lignocellulose content presented of the cotton waste, which was easily consumed by the microorganisms.

These and above results point to a new way of disposal for at least a part of this waste-water transforming an ecological problem in an ecological solution, but many studies must still be made to obtain more information about the feasibility to this proposal.

Acknowledgement

This work was made with the financial support of the project: NAT 90-0823

References

Baca MT, Fornasier F and de Nobili M (1992) Mineralization and humification pathways in two composting processes applied to cotton wastes. J of Fermentation and Bioeng. 74: 179–184.

Baca MT, Delgado IC, Sanchez-Raya J and Gallardo-Lara, F (1990) Comparative use of cress seeds germination and physiological parameters of Helianthus annuus L. to assess compost maturation. Biological Wastes 33: 251–261.

De Nobili M and Petrussi F (1988) Humification index (HI) as evaluation of the stabilization degree during composting. J Fermentation Technol. 66: 577–583.

Ocio JA and Brookes P (1990) An evaluation of methods for measuring the microbial biomass in soils following recent additions of wheat straw and the characterization of the biomass thats develops. Soil Biol and Biochem 22: 685–694.

Perez JD, Esteban E, Gomez M and Gallardo-Lara F (1986). Effects of wastewater from olive processing on seed germination and early plant growth of different vegetable species J of Environ Sci and Health. b21: 349–357.

Sequi P, de Nobili M, Cercignani G and Leita L (1986) A new index of humification. Agrochimica 30: 175–186.

C. Rodriguez-Barrueco (ed.), Fertilizers and Environment, 481–483.

Effect of cadmium concentration in the nutrient solution on lettuce growth

M.E. García López De Sa
Centro de Ciencias Medioambientales, CSIC, Serrano 115-bis. E-28003 Madrid, Spain

Key words: Cd, nutrient solution, lettuce

Abstract

Lettuce plants grown in nutrient solution which has Cd added to it are contaminated by this element. However, at low levels of Cd there is an increase in plant development whilst at 10 ppm Cd and above, there is a depressive effect on the lettuce development. The Cd contamination in lettuce plants increases with increasing levels in the nutrient solution.

The presence of Cd in the nutrient solution produces an increase in absorption of other ions such as Mn, Fe and Zn but when the Cd concentration is increased from 5 to 10 ppm, the Mn content stays the same whilst the levels of Fe and Zn in the lettuce leaves diminish, which is when a depression in yield occurs. Moreover, an increase from 10 to 15 ppm of Cd produces a slight increase in development, although it is still less than in plants grown with uncontaminated solution.

Introduction

Cd is a zootoxic element, i.e., more toxic to animals than plants (Davis & Coker, 1979). Plant uptake of metals is the first step in the food chain contamination.

The degree of Cd toxicity depends on the absorption capacity of the plants as well as on synergic and antagonistic effects taking place with other nutrient elements (Barceló & Poschenrieder, 1992).

The lettuce plant is a Cd accumulator (Alloway *et al.*, 1990) and as this heavy metal is one of the most dangerous elements for human health (FAO, 1978), it is interesting to see the effects that Cd contamination has both on the lettuce plant yield and the Fe, Mn, Zn and Cd content.

Materials and method

A glasshouse experiment was performed with lettuce plants. The lettuce were grown in an unsupported hydroponic culture. The nutrient solution was kept circulating in 50 litre capacity tanks with the purpose of minimizing variations occurring in the nutrient solution composition. The circulating system works by raising the nutrient solution by the "absorption through the air" method to the level of the highest pot from where it is distributed by gravity (Van Dried, 1964). The nutrient solution then returns to the tank, thus minimizing the variations occurring in element concentration.

Germination was carried out on acid washed quartz sand, initially wetted by capillarity with distilled water and then with a ten times diluted nutrient solution.

The nutrient solution used was Hoagland and Snyder's for macronutrients and Arnon's for micronutrients (Hewitt, 1966).

When the plants were approximately 5 cm tall, they were transplanted to pots with a normal nutrient solution. The treatments performed provided different degrees of Cd contamination to the nutrient solution with Cd levels of 0 ppm, 5 ppm, 10 ppm and 15 ppm.

Fresh and dry matter yield was determined in the plants and then a dry attack was carried out followed by a wet attack with HNO_3-$HClO_4$. The Fe, Mn and Zn content was determined by atomic absorption and Cd by ICP.

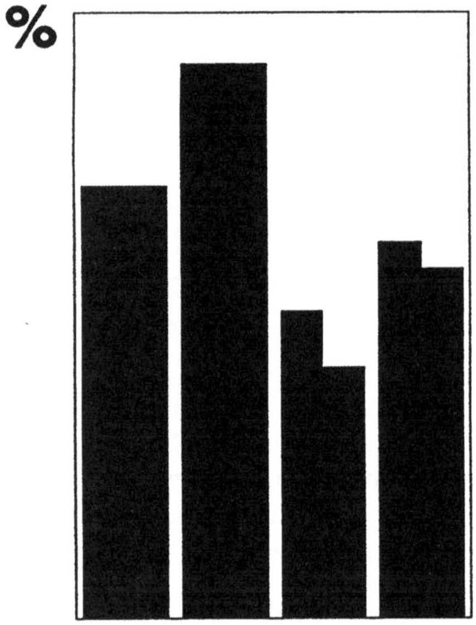

Fig. 1. Effect of Cd contamination in lettuce yield.

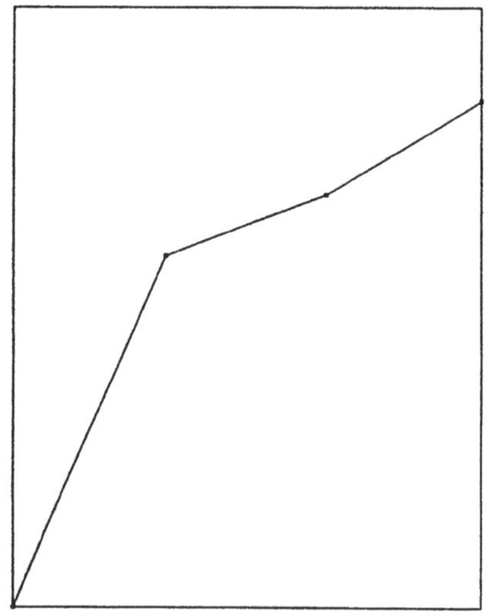

Fig. 2. Cd concentration in lettuce grown under different levels of Cd in the nutrient in the nutrient solution.

Results and discussion

Fig. 1 displays the dry and fresh weight percentages of the lettuce plant developed in a nutrient solution with different degrees of Cd contamination.

The lettuce plants display a greater development at low contamination levels, whilst lesser plant development is seen at the 10 ppm and higher Cd levels.

It can be seen from the results obtained, that 5 ppm Cd contamination is present in some lettuce plants with no sign of a possible risk for their consumption. More highly developed plants are also seen.

When Cd contamination is 10 ppm, there is a depressive effect on the development of the plants with a 29% drop in yield. This leads us to think that the Cd present begins to cause alterations in the metabolic processes due to exchanges with other metals in the different cofactors of the enzymes.

The effect of a yield increase with the low Cd contamination level is reported by other researchers both for lettuce and other crops (Moral *et al.*, 1994).

Fig. 2 demonstrates the Cd concentration in lettuce developed in the nutrient solution for all Cd concentrations.

Lettuce grown in a nutrient solution in a 5 ppm Cd concentration gives a value of 40 ppm of Cd in the parts above the ground. This reveals that although plants with that level of contamination do not display symptoms of phytotoxicity, they have a high concentration of Cd in their tissue.

From the results obtained, lettuce plants are shown to hold a great capacity for absorbing Cd from the nutrient solution. Most of this heavy metal accumulates in the leaves, the edible part of the plant, and therefore, contamination of this element in the culture medium may cause unwanted consequences for man (as Cd is a zootoxic metal). Moreover, in the concentration range studied, a higher concentration of Cd in the nutrient solution leads to a higher concentration of Cd in the parts of the plant above ground. Although, at levels of 10 and 15 ppm of Cd, no plants with phytotoxicity symptoms were observed.

Fig. 3 shows the Fe content in the above ground part of lettuce developed with different Cd concentrations in the nutrient solution.

The effect of Cd contamination on Fe absorption depends on the degree of the existing contamination. Thus, at the 5 ppm Cd level, the plant shows a higher Fe content, whilst at the 10 ppm level, the Fe content diminishes to lower values than those existing in contaminated plants. Moreover, an increase from 10 ppm to 15 ppm of Cd in the nutrient solution has no influence on the lettuce plant Fe content.

The reduction in the Fe content in the plant tissue coincides with a depression in the plant development.

Zn content in lettuce plants grown under different Cd levels is shown in Fig. 4. Like Fe, the Zn content in lettuce shoots increases at the 5 ppm Cd level in the nutrient solution, whilst at the 10 and 15 ppm of Cd, the Zn concentration in the plant tissue is less than in

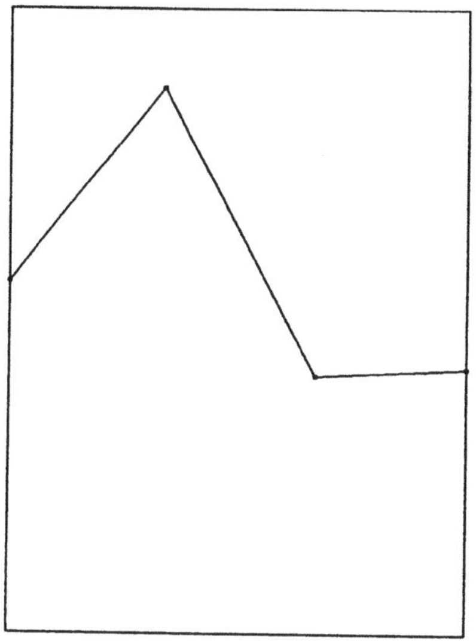

Fig. 3. Fe content in lettuce grown at different levels of Cd.

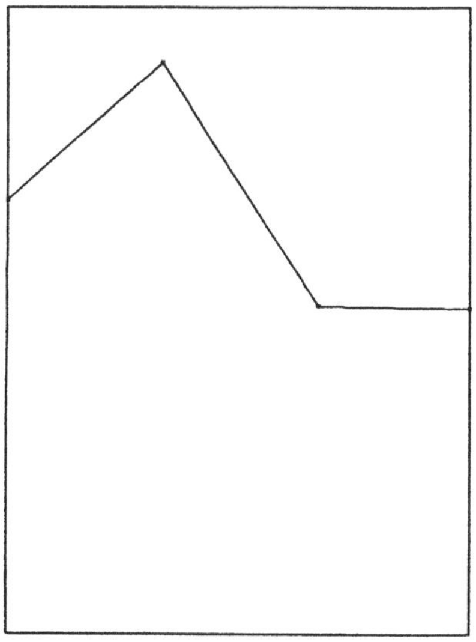

Fig. 4. Zn concentration in lettuce as affected by Cd in the nutrient solution.

the control plants. Thus the presence of Cd may replace Zn and affect the plant physiological functions (Kim *et al.*, 1988).

Fig. 5 shows the Mn concentration in Cd contaminated plants. A 5 ppm Cd concentration produces an increase in the Mn content, whereas a 5 ppm to 10 ppm increase of Cd in the nutrient solution produces no variations in the Mn concentration in the lettuce plants. On the contrary, a 15 ppm increase in Cd produces Mn figures lower than in the control plants.

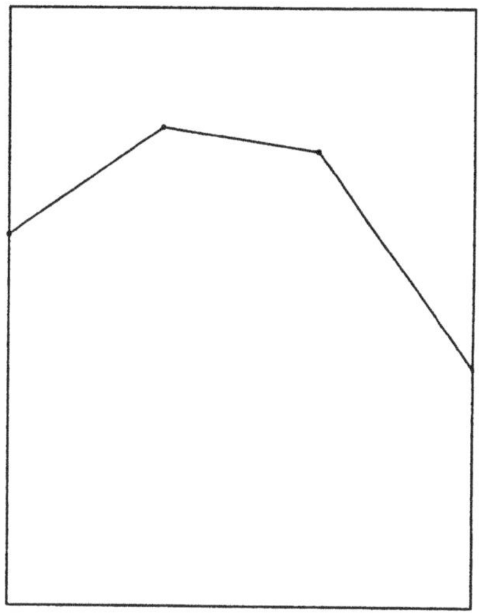

Fig. 5. Mn concentration in lettuce as affected by Cd in the nutrient solution.

Therefore, at the 5 ppm Cd level there is an increase in the Fe, Mn and Zn content in lettuce. At the 10 ppm level, the concentrations of Fe and Zn fall whilst the concentration of Mn is maintained. Finally, Fe and Zn are maintained at the 15 ppm Cd level, and the Mn concentration in lettuce shoots is reduced.

References

Alloway BJ, Tiles AR and Morgan H (1990). The accumulation of Cd by vegetable grown on soil contaminated from a variety of sources. Sci Total Environ 91: 223–236.

Barceló, J and CH Poschenrieder 1992 Respuestas de las plantas a la contaminación por metales pesados. Proc IV Simp Nac Nutr Min Plantas (SEFV Ed.). Vol I, 46 Alicante. Spain.

Davis RD and Coker EG (1979) Proc Int Conf Management and Control of Heavy Metals in the Environment, 553. CEP Consultants Ltd, Edinburgh, UK.

FAO/WHO (1978) List of Maximum Levels Recommendations for Contaminants by Joint FAO/WHO Codex Alimentarius Commission. CAC/Fal Third Series, Rome, Italy 4–178.

Hewitt EJ (1966) Sand and water culture methods used in the study of plant nutrition. Tech Comm no 22, Commonwealth Agric Bureau, London.

Kim SJ, Chang AC Page AL and Warpake JE (1988) Relative concentrations of cadmium and zinc in tissue of selected food plants grown on sludge-treated soils. J Environ Qual 17: 568–573.

Moral R, Gómez I, Navarro Pedreno J and Mataix J (1994) Effects of cadmium on nutrient distribution, yield, and growth of tomato grown in soilless culture. J Plant Nutr 17: 953–962.

Van Dried W (1964) The effect of iron ethylenediaminotetracetic acid on the growth and metabolism of tomato plant in water culture. Plant Soil 20: 85.

C. Rodriguez-Barrueco (ed.), Fertilizers and Environment, 485–489.

As and Se in soils and plants from abandoned mining areas of the Salamanca province, Spain

A. García Sánchez[1], I. Santa Regina[1], N. Rodríguez[2], O. Jiménez[2] & J.F. Antona[1]
[1]*Instituto de Recursos Naturales y Agrobiología, CSIC, Salamanca, Apto. 257, Spain;* [2]*Dpt. Analytical Chemistry, Universidad de Salamanca, Spain*

Key words: Arsenic, selenium, soil, plants

Abstract

Four mining zones were chosen for this study. The Morille, Montejo and Saucelle mining zones which produce tin and tungsten ores with sulphides hosted in shales. The Cabaco mining zone which produces tungsten ore with pyrite and arsenopyrite hosted in granites.

In the Salamanca province, the arsenic background is less than 10 ppm in soils formed from granite and shales. Anomalous soils contain more than 10 ppm and range from 10 to > 1000 ppm, due to arsenopyrite alteration in mining areas.

High and very high arsenic content in plants (99–40669 ppb, some of them probably phytotoxic) were collected in these mining zones.

The arsenic content in *Vulpia bromoides* (L) SF Gray, show a correlation with the arsenic content in the soils; nevertheless for *Agrostis castellana* Bois & Rent, other pedological factors (such as pH) must have a greater effect than the arsenic content in the soil.

Dactylis glomerata L. has a higher degree of arsenic accumulation than *Ag. castellana* and *Cynosurus echinatus* L. in the same As-anomalous soil.

On the other hand, the selenium contents in plants are low, with most of the data showing values lower than 50 ppb (which is considered to be deficient for animal nutrition) and ranging from 2 to 260. These values are in agreement with other data obtained previously in the Salamanca province. *V. bromoides* contain higher selenium levels than *Ag. castellana* in most of the soils studied.

Introduction

There are about 200 occurrences of mineralized zones in the Salamanca province, Spain, most of which are mined and smelted until very recently. Sn and W in arsenical ores, mainly arsenopyrite, are the most occurring minerals. The mining activities in these zones have led to contamination of the agricultural land, with total soil As values as high as 1000–2000 ppm due to weathering of Arsenopyrite to Scorodite. In contrast, the average soil concentration of As worldwide, is 6 ppm, with a typical range of 0.1 to 40 ppm (Bowen, 1979).

Once Arsenopyrite oxidation ceases, scorodite dissolves incongruently to iron hydroxide and arsenate ion solutions (Dove & Rimstidt, 1985). Pasture herbage growing in contaminated soils in Cornwall (UK) contained up to five times more As than herbage from control sites elsewhere (Davies, 1980). In general, the amounts of As in grass increase with increasing As content in the soil, but the As uptake by plants is greatly dependent on other pedological factors (pH, OM, clay content, Fe and Mn oxides content, etc.).

Plant toxicity due to As is often reached prior to the accumulation of toxic levels for the wildlife which ingest the plants (Tamaki & Frankenberger, 1992). However, some individual plant species on high As soil have shown to accumulate As to extreme levels which could cause physiological disfunctions or poisoning in grazing livestock. Also, the poisoning could be caused by ingestion of As-rich soils or mine waste.

Selenium behaves in on of two ways depending on its concentration and chemical form. Selenium can be an essential element or a toxin to plants, livestock and

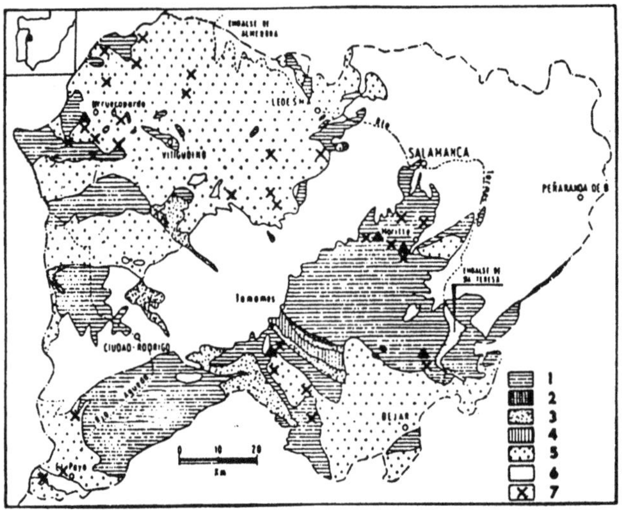

Fig. 1. Study areas (triangles) in the Salamanca province and geological sketch (Carcía Sánchez *et al.*, 1980). 1 – Schists grau wake complex (Precambric-cambric. 2 – Limestones (Cambric). 3 – Quarzites (Arenig). 4 – Shales. 5 – Granites. 6 – Terciary Cover. 7 – As anomalies.

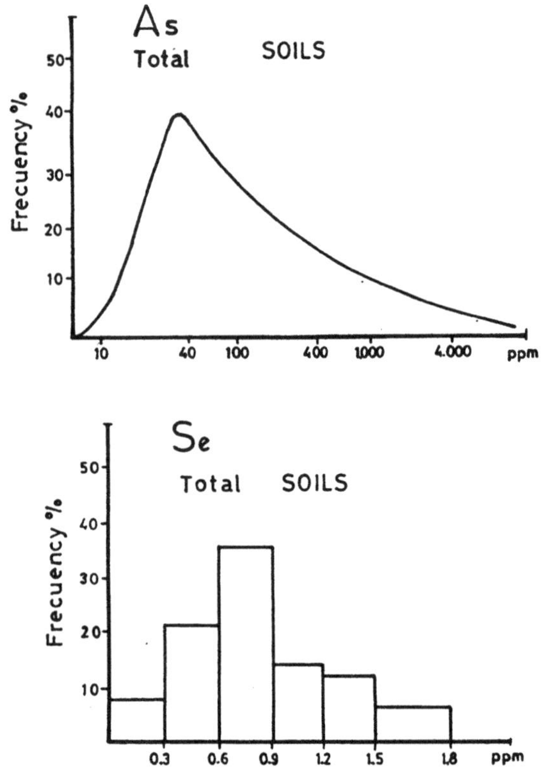

Fig. 2. Frequency distribution of As and Se contents in soils.

humans. Selenium toxicity in livestock is known in a few areas around the world, but Se deficiency has turned out to be much more common than Se toxicity.

The desirable Se level in forages and cereals is > 50 ppb (Gissel-Nielsen *et al.*, 1984) and to avoid symptoms of deficiency, Se levels in grain and ley vegetation should be > 100 ppb, (Frank *et al.*, 1986). More than 450 ppb Se is the minimum concentration required in soils if feed crops are to provide animals with adequate amounts of Se (Whatkinson, 1963). Although the Se concentration in most soils is not specially low, on neutral to acid mineral-soils with relatively high contents of Fe and O.M., it is fixed strongly in the soil making its availability to plants very low.

Material and methods

Soils were sampled from 0–15 cm topsoil, from the following topologies: *district cambisol, humic cambisol, district leptosol, humic leptosol* with the selected characteristics as shown in Table 1. The situation of samples are given in Fig. 1

The plants were sampled in June 1994 and the following species were selected: *Aira cariophyllea, Agrostis castellana, Bromus mollis, Cynosurus echinatus, Dactylis glomerata, Taeniatherum caput-medusae, Hordeum murinum, Senecio jacobaea, Vulpia bromoides.*

Table 1.

Sample N°	Coarse Sand %	Fine Sand %	Silt %	Clay %	pH	C %	O.M. %
28598	21.5	31.5	28.5	15.9	5.54	2.03	3.49
28599	25.0	34.5	21.3	15.5	5.75	1.81	3.12
28600	19.0	38.0	21.5	13.1	5.57	2.96	5.10
28601	24.5	26.0	21.0	20.3	5.14	2.81	4.85
28602	16.5	42.0	23.9	11.8	4.97	3.15	5.43
28603	14.0	50.5	13.5	19.4	4.94	1.76	3.03
28604	17.0	58.0	9.5	13.8	5.80	0.64	1.10
28605	22.5	40.5	12.5	18.5	4.43	2.51	4.83
28606	32.0	51.5	6.5	9.2	5.02	0.75	1.29
28607	22.5	49.5	12.2	10.8	5.13	0.51	0.87
28608	33.5	48.0	4.6	15.0	5.87	0.45	0.78
28609	12.0	63.0	10.7	13.4	5.20	2.93	5.04
28610	7.5	64.5	12.3	11.0	4.11	1.97	3.40
28611	26.5	56.5	6.6	11.4	6.67	0.32	0.55
28612	63.0	22.5	6.8	9.2	5.39	1.09	1.87
28613	19.5	60.0	8.7	8.2	5.15	0.47	0.81
28614	35.5	32.5	14.3	13.4	6.08	1.59	2.74

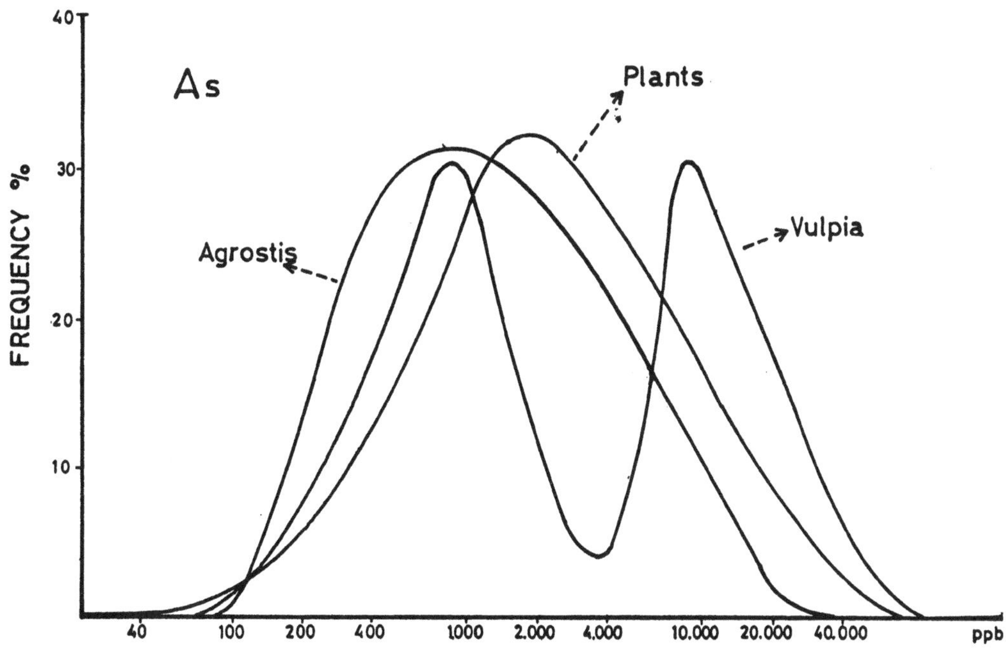

Fig. 3. Frequency distributions of As in plants (all species pooled), Vulpia and Agrostis.

Chemical analysis

Arsenic and Se in plants, and Se in soils were determined by hidride generation and AA spectrophotometry in a flow system (FIA, HG-AAS) and a standard addition method. The detection limits were: 0.8 ppb for As and 1 ppb for Se. As in soil was determined by

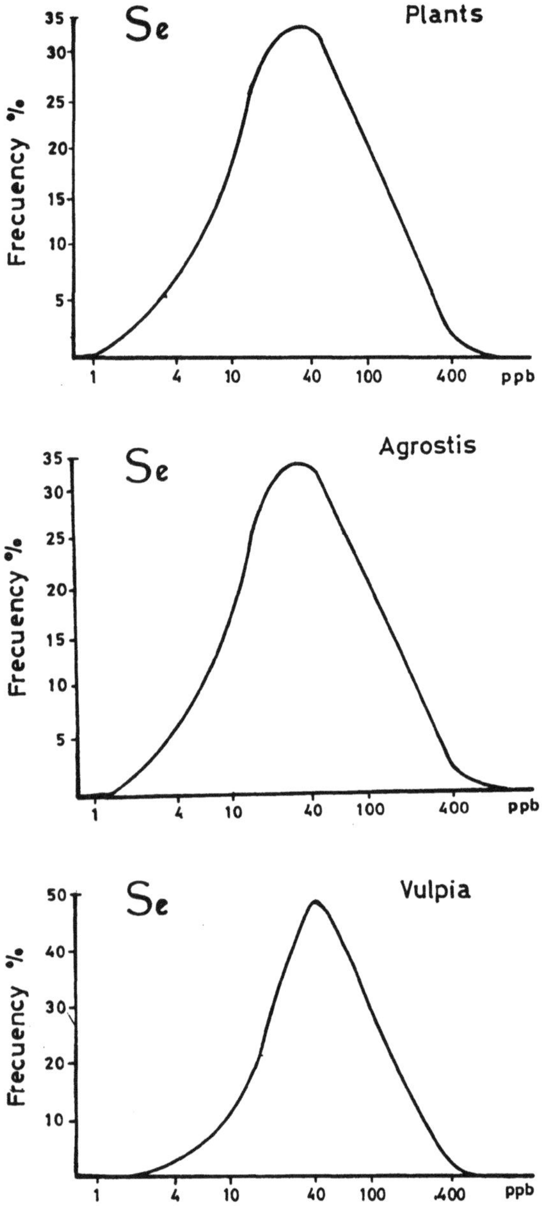

Fig. 4. Frequency distributions of Se in plants.

X-ray fluorescence spectrometry. The detection limit was 5 ppm.

Results

The chemical analysis of soils and plants are presented as histograms in Figs. 2, 3 and 4.

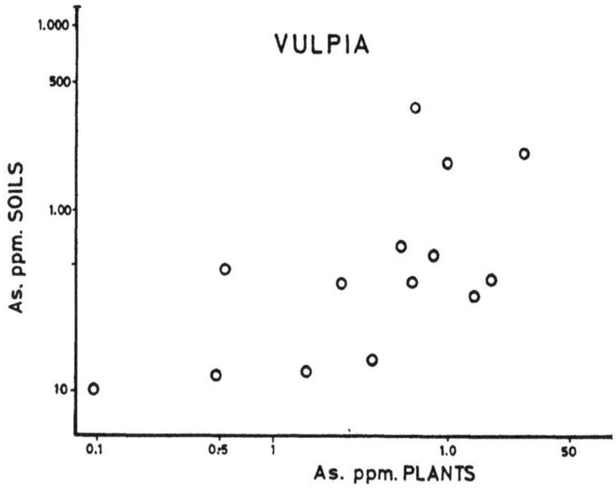

Fig. 5. Correlation between total As in soils with plant As.

Discussion

The total As ranges in herbage are very wide (99–40699 ppb). *Agrostis* shows a distribution lognormal with mode around 1000 ppb, and in *Vulpia* is bimodal with modes of 1000 and 10000 ppb. For some plants the As contents are probably phytotoxic (Sheppard, 1992) and thus act in protecting livestock. The As contents in herbage are similar to those in other mining zones, e.g. Cornwall (Davies, 1980).

Only *Vulpia* shows some correlation between its As contents and total As concentrations in soils (Fig. 5).

Soil pH seems to act as a regulator of As accumulation in *Agrostis* (Fig. 6), although the correlation is rather poor.

The degree of As accumulation in plants occurs in the following sequence: *Aira* > *Cynosurus* > *Elymus* > *Dactylis* > *Vulpia* > *Senecio* > *Agrostis*.

Physiological disfunctions in livestock may be due to ingestion of As-rich soil or mine waste more than to the amounts of As in the herbage.

The variation in the range of Se contents in plants is narrow (2–260 ppb) with a lognormal distribution and modes around 30–40 ppb. These values are too low to be considered deficient for animal nutrition (Frank *et al.*, 1986). The findings for the Se contents are similar to other previous data collected for the Salamanca province (Montalvo *et al.*, 1983) and corresponds to the data obtained for wheat and maize in thirty countries as reported by Sillanpää & Jansson, (1992).

The degree of Se uptake in the studied plants occurs in the following sequence: *Vulpia* > *Elymus* = *Agrostis* > *Senecio* > *Cynosurus*

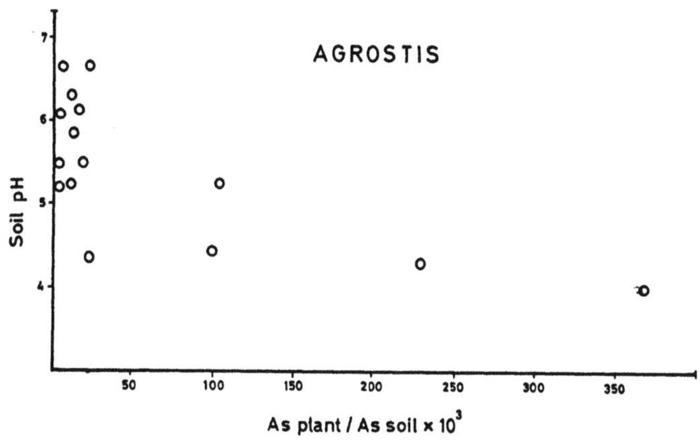

Fig. 6. Correlation between soil pH with As accumulation degree in Agrostis.

Senecio *jacobaea* shows the following distribution in the different organs:
Se: leaves > flowers > roots > branches
As: leaves > roots > flowers > branches

References

Bowen MJU (1979) Environmental Chemistry of the Elements. Academic Press, New York.

Davies BE (1980) Applied soil trace elements. John Wiley, Chichester.

Dove PM and Rimstidt JD (1985) The solubility and stability of scorodite, $FeAsO_4 \cdot 2H_2O$. Amer Miner 70: 838–844.

Frank A, Pehrson B and Petersson LR (1986) Concentrations of some important elements in liver of young cattle supplemented with selenate enriched feed J Vet Med A 33: 422–425.

García Sánchez A and Saavedra J (1980) Sobre los recursos minerales de la provincia de Salamanca, IOATO- Diputación de Salamanca

Gissel-Nielsen G, Gupta UC, Lamand M and Westesmarck T (1984) Selenium in soils and plants and its importance in livestock and human nutrition. Adv in Agron 37: 397–461.

Montalvo MI, García B and Lamand M (1983) Contenido de selenio en pastizales de zonas semiáridas. Revista Ayma 53–59.

Sheppard SC (1992) Sumary of phytotoxic levels of soil Arsenic. Water, Air and Soil Pollution 64: 539–550.

Sillanpää M and Jansson M (1992) Status of Cd, Pb, Co and Se in soils and plants of thirty countries. FAO Soils Bull 65: 195 pp.

Tamaki S and Frankenbenger WT (1992) Environmental Biochemistry of Arsenic. Rev Environ Contam and Toxic 124: 79–110.

Whatkinson JM (1963) Soil Selenium and animal health. Intern Soil Conf Lower Mult NZ, 1963: 149–154.

C. Rodriguez-Barrueco (ed.), Fertilizers and Environment, 491–493.

Incidence of heavy metals in the application of inorganic fertilizers to rice farming soils (Valencia, Spain)

E. Gimeno-García[1], V. Andreu[1] & R. Boluda[2]
[1]Dpt. de Desertificación IATA (CSIC), Jaime Roig 11, 46010 Valencia, Spain; [2] Dpt. Biología Vegetal, Universidad de Valencia, Avda. Vicente Andrés Estellés s/n, 46100 Burjassot, Valencia, Spain

Key words: Heavy metals, fertilizer, soil, contamination indexes

Abstract

The presence of Cd, Co, Cu, Ni, Pb, Zn, Fe and Mn in different inorganic fertilizers (urea, calcium superphosphate, iron sulphate and copper sulphate) are evaluated and the contribution of these metals to the soils in this way is estimated. The studied zone corresponds to rice farming and is at the north of Albufera Natural Park (Valencia, Spain). The results obtained show that Superphosphate is the fertilizer that contains the highest levels of Cd, Co, Cu and Zn as impurities. Copper sulphate and iron sulphate have the most significant contents of Pb, and are the only fertilizers in which Ni is detected. The most significant additions of impurities that the soil received by agricultural practices were of the heavy metals Mn, Zn, Co and Pb. Two contamination indexes have been applied to provide a basis for comparison for heavy metal potential toxicity. These results denote no potential toxicity for heavy metals in the studied soils.

Introduction

The addition of compounds to the soil has become common practice in present agricultural methods. The yearly consumption of fertilizers in Spain is 2,027,000 tonnes (99.29 kg/ha of the land cultivated surface) with the Valencia Community being the third largest consumer (MOPT, 1991). The main purpose for the use of chemicals is the improvement of the nutrient supply, or crop protection and disease control. These practices may cause chemical degradation of the soils as a result of the accumulation of compounds to undesirable levels. This is because fertilizers are usually not sufficiently purified for economic reasons. They usually contain several impurities of which heavy metals are very common. In addition, the heavy metals contained in pesticides as part of their structural chemical activity are added to the soil.

In this paper, a study of the heavy metal additions to the soil through the use of fertilizers has been made and was carried out in a rice plot in the Albufera Natural Park. The contents of Cd, Co, Cu, Ni, Pb, Zn, Fe and Mn in different inorganic fertilizers (urea, calcium superphosphate, iron sulphate and copper sulphate) were evaluated. The heavy metals added to the soils were assessed according to the amount of these products commonly used in rice farming. Moreover, two contamination indexes, Zn Equivalent and Zn/Cd ratio, were applied to provide a basis for comparison for the heavy metal potential toxicity.

Table 1. Inorganic fertilizers analyzed, riches and proportion added to soil

Fertilizers	Riches	Proportion added
Urea	46% N	300 kg ha^{-1}
Calcium superphosphate	18% P	600 kg ha^{-1}
Iron sulphate 7 hydrate	18.5% Fe	200 kg ha^{-1}
Copper sulphate 5 hydrate	25% Cu	35 kg ha^{-1}

Materials and methods

The study was monitored in a rice plot of 4,000 m^2. The soil was a Calcaric-gleic Fluvisol in saline phase. Fertilizer analyses and the amounts added to the soil are shown in Table 1. A modification of Analytical Methods of the Spanish Ministry of Agriculture, Fishing

Table 2. Heavy metal contents in fertilizers and the soil surface horizon (mg kg^{-1})

Element	Fertilizers (mg kg^{-1})				Soil (mg kg^{-1})
	$CuSO_4$	$FeSO_4$	Urea	Superphosphate	
Cd	0.21	0.03	0.008	2.22	1.83
Co	0.06	1.35	0.05	4.50	5.29
Cu	255[a]	0.30	0.40	12.50	20.35
Ni	0.60	0.50	n.d.[b]	n.d.	22.82
Pb	11	10	n.d.	n.d.	45.96
Zn	21.40	13.30	n.d.	50	55.80
Fe	n.d.	201[a]	n.d.	n.d.	12537
Mn	n.d.	220	n.d.	n.d.	158.10

[a]g kg^{-1}; [b]n.d. No detected.

Table 3. Overall estimate of heavy metal additions from agricultural activities

Element	Fertilizers (mg/ha/year)				Total (g/ha/year)
	$CuSO_4$	$FeSO_4$	Urea	Superphosphate	
Cd	7.14	6.20	2.40	1332	1.35
Co	2.24	270.20	15.30	2700	2.98
Cu	8925[a]	60	120	7500	8932.68
Ni	21	100	–	–	121.00
Pb	385	2000	–	–	2.38
Zn	749	2600	–	30000	33.33
Fe	–	40200[a]	–	–	40200
Mn	–	44000	–	–	44.00

[a]g/ha/year.

and Food (1986) was used for the fertilizer treatments: 10 g of urea, iron sulphate and copper sulphate and 2 g of calcium superphosphate were dissolved in an acid medium (with conc. HCl). The total contents of the heavy metals in the soils were determined by the nitric-perchloric acid digestion method described by Andreu (1993)

The contents of metals were determined by atomic absorption spectrophotometry (AAS) Flame, and for Cd and Co by AAS-Graphite furnace. For Cd the standard addition technique was employed because matrix interferences were observed.

Results and discussion

The metal contents measured for fertilizers and soils are shown in Table 2. For fertilizers, the results show that superphosphate contains the highest levels of Cd, Co, Cu and Zn as impurities. Copper sulphate and iron sulphate have the highest contents of Pb; moreover, they are the only fertilizers in which Ni was detected. No fertilizers showed any amounts of Fe and Mn, except iron sulphate where 220 mg kg^{-1} Mn, and obviously, 209 g kg^{-1} Fe were found.

The results obtained for the phosphate fertilizer compare with the values obtained by different authors (Alloway *et al.*, 1988), are found between these intervals: 0.1–170 mg kg^{-1} Cd, 1–12 mg kg^{-1} Co, 1–300 mg kg^{-1} Cu, 7–38 mg kg^{-1} Ni, 7–225 mg kg^{-1} Pb, 50–1450 mg kg^{-1} Zn and 40–2000 mg kg^{-1} Mn. The same authors established that nitrate fertilizers contain between 0.05–8.5 mg kg^{-1} Cd and 5.4–12 mg kg^{-1} Co, whereas the results obtained for urea are lower (0.008 y 0.051 mg kg^{-1} Cd and Co respectively). Adriano (1985) found that the urea Cu content is 0.6 mg kg^{-1}, a value similar to the 0.4 mg kg^{-1} found by the present authors on the same fertilizer.

The estimate of the heavy metals added to the soil by agricultural practices is shown in Table 3. The most significant additions were Fe and Cu. The highest levels of impurities in the soil were found to be Mn, Zn, Co and Pb, and the lowest were Cd and Ni. Only the amount of Cu added to the soil (9 kg/ha/year) is near to the limit established by European Union legislation (12 kg/ha/year) (CEC, 1986).

To carry out an estimation of the heavy metal added to the soil in comparison to the total heavy metal content of the fertilizer, two contamination indexes have been used: *Zn equivalent* (Chumbley, 1971; Webber, 1972) and *Zn/Cd ratio* (Chaney, 1973). The Agricultural Development and Advisory Service of England and Wales has proposed a general statement for the relative toxicity of Zn, Ni and Cu: Zn Equivalent = $(1 \times Zn) + (2 \times Cu) + (8 \times Ni)$ $\mu g\ g^{-1}$ dry matter. Assuming 1 ha of soil to a depth of 15 cm weighs 2×10^6 kg (Greenland and Hayes, 1981) the Zn equivalent was calculated. The heavy metal addition from agricultural practices according to the Zn equivalent is 9.4 μg/g/year. This represents 2.8% of the total Zn equivalent (332 $\mu g\ g^{-1}$) of the soil surface horizon. It has been recommended that a total of 250 μg Zn equivalent g^{-1} soil is the maximum that can be safely added if there have been no previous additions to the soil and if the pH is maintained at not less than 6.5. This addition may be as a single heavy application or the sum of several small applications spread over a period of 30 years (Greenland and Hayes, 1981). In comparison to this recommendation, the additions found in the soil is low (9.4 $\mu g\ g^{-1}$) and if the time factor is considered the safety limit is surpassed (283 $\mu g\ g^{-1}$).

The Zn/Cd ratio (Chaney, 1973) assesses the antagonism that Zn exerts on Cd assimilation in plants thus decreasing the toxicity of Cd (it must be 200, although Chaney establishes the ideal value close to or over 1000). By considering Chaney's conclusions and applying the same procedure as above, the Zn/Cd ratio obtained is 39.33 for the soil and 24.76 for the addition from fertilizing practice. If we assume that Chaney's first criterion coincides with the one established by legislation in Spain and the European Union (Zn/Cd $\geq$ 200), then it can be said that this addition value is not dangerous. This ratio is very low for the soil and, theoretically, Zn antagonism on Cd can be considered to be effective.

References

Adriano DC (1985) Trace elements in the terrestrial environment. Anonymous (ed) Elsevier, Amsterdam.

Alloway BJ, Thornton I, Smart GA, Sherlock JC and Quinn MJ (1988) Metal availability. Sci Total Environ 75: 41–69.

Andreu V (1993) Contenido y evolución de Cd, Co, Cr, Cu, Ni, Pb y Zn en suelos de las comarcas de L'Horta y La Ribera Baixa (Valencia). Universidad de Valencia (ed) Serv Public Univ Valencia, Valencia.

Chaney RL (1973) Crop and food chain effects of toxic elements in sludges and effluents. In: Recycling municipal sludges and effluents on land. National Associations of State Universities and Land-Grant Colleges, Washington, DC, USA pp 129–141.

Chumbley CR (1971) Permissible levels of toxic metals in sewage used in agricultural land. ADAS Advisory Paper No 10 (ed) MAFF, London.

Council of the European Communities (1986) Directive N 278 of 6/12/1986. Commission of the European Communities, Brussels.

Greenland DJ and Hayes MHD (1981) The chemistry of soil processes. John Wiley and Sons Ltd, London.

Ministerio de Agricultura Pesca y Alimentación (ed) (1986) Métodos oficiales de análisis de suelos, aguas y plantas. Servicio de Publicaciones, Madrid.

MOPT (ed) (1991) Medio Ambiente en España. Centro de Publicaciones Ministerio de Obras Públicas y Transportes, Madrid.

Webber J (1972) Effects of toxic metals in sewage sludge on crops. Water Poll Control 71: 404–413.

C. Rodriguez-Barrueco (ed.), Fertilizers and Environment, 495–498.

Effect of fertilization and irrigation on the nickel content of peas and soybeans

Z. Gyôri, J. Prokisch, B. Kovács & P. Dániel
Central Laboratory, Debrecen Agricultural University, H-4015 Debrecen Pf 36. Hungary

Abstract

The effect of fertilization and irrigation on the nickel content of peas and soybeans was studied in an arable land plot experiment on fertilization effects. It was found that the yearly rainfall has a basic influence on the nickel content. The value of the nickel content of peas and soybeans can be twice in that of a dry year than in a wet year. Irrigation decreases the concentration of nickel, too. Fertilization has an opposite effect. With the acidifying effect of a mineral nitrogen fertilizer, heavy metals become more mobilizable thus leading to significant increases of the nickel content in the seeds. A relationship between the nickel and other microelement content of the seeds was studied. A linear relationship between the nickel and iron contents can be observed. The uptake of other heavy metals such as copper is promoted at the beginning of the increase of nickel concentration, however, as it continues to increase, it is hindered.

Introduction

Nickel, the twenty-fourth element in order of natural abundance in the earth's crust, is widely distributed in the environment. Most nickel compounds are relatively soluble at pH values < 6.5, whereas nickel exists predominantly as insoluble nickel hydroxides at pH values > 6.7. Therefore, soil acidification effects (acidic rain, fertilizers, etc.) have a pronounced tendency to mobilize nickel from soil and to increase the nickel concentration in plants (Wallace *et al.*, 1977; Haele and Ormond, 1982).

Dalton *et al.* (1988) reviewed the evidence for the nickel requirement of plants. In some cases, this micronutrient is needed for urease or hydrogenase activity. For example, several studies have shown that nickel is essential for the formation of active plant ureases (Dixon *et al.*, 1975, 1980; Klucas *et al.*, 1988); the absence of urease activity resulted in accumulation of toxic concentrations of urea, which led to plant necrosis (Dalton *et al.*, 1988). Urease-rich legumes, such as jack beans and soybeans, generally contain high nickel concentration. Soybeans grown on nickel-deficient nutrient solutions accumulate the toxic urea concentrations which result in necrosis of leaflet tips, a characteristic of nickel deficiency (Eskew *et al.*, 1984).

In most plant samples the nickel content is under 0.5 mg/kg, but some plant and food contains more than 1 mg/kg nickel, e.g. wheat bran, bean, soya and soya products, hazelnuts and tea leaves.

Table 1. Fertilizer levels in the experiments

Fertilizer level	N kg/ha	P_2O_5 kg/ha	K_2O kg/ha
1	0	0	0
2	35	30	25
3	70	60	50
4	105	90	75
5	140	120	100

Our aim was to study the effect of fertilization and irrigation on the Ni content of soybeans and peas in an arable land experiment.

Methods

A field plot experiment was carried out at the Látókép Farm 15 km from Debrecen. The soil is Calcareous Chernozem with 2.8–3.0% humus content. The dept of the humus layer is 70–90 cm. The N-content and original P-content of the soil is average, but it is rich in K. Besides macroelements, there is no shortage of

Table 2. Correlation table of the element concentrations in soybean seeds

	Ni	Fe	Cu	Mn	P	N	Zn	K	Na	Mg
Ni	1.0000	.7094**	.5552**	.4276**	.0102	.2166	.5693**	.1182	-.1363	-.1545
Fe	.7094**	1.0000	.3792**	.6625**	-.2692*	.2493*	.3803**	.0623	-.1980	-.0957
Cu	.5552**	.3792**	1.0000	.2799*	.5436**	.0068	.8634**	.4701**	.4191**	-.0274
Mn	.4276**	.6625**	.2799*	1.0000	.106-	-.0584	.3494**	.5100**	.1111	.4200**
P	-.0102	-.2692	.5436**	.1060	1.0000	.0961	-.1729	-,4177**	-.8192**	
N	.2166	.2493*	.0068	-.0584	-.3293**	1.0000	.0961	-.1729	-.4177**	-.8192**
Zn	.5693**	.3803**	.8634**	.3494**	.5330**	.0961	1.0000	.6166**	.2580*	.0358
K	.1182	.0623	.4701**	.5100**	.7401**	-.1729	.6166**	1.0000	.3862**	.4290**
Na	-.1363	-.1980	.4191**	.1111	.5963**	-.4177**	.2580*	.3862**	1.0000	.4985**
Mg	-.1545	-.0957	-.0274	.4200**	.3665**	-.8192**	-.0358	.4290**	.4985**	1.0000

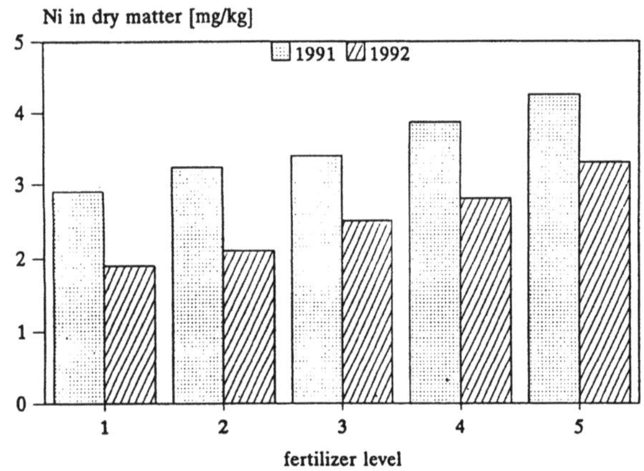

Fig. 1. Effect of fertilization on the nickel content of peas

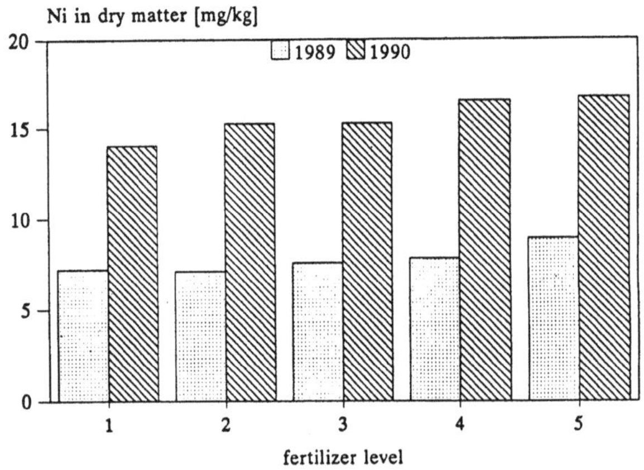

Fig. 2. Effect of fertilization on the nickel content of soybeans

trace elements. The total Ni concentration of this soil is 35 mg/kg and the Lakanen-Erviö soluble Ni content is 9.1 mg/kg (Győri *et al.*, 1993). The Ni concentration of the fertilizer and irrigation water was lower than 200 μg/kg. Our studies were carried out in a long term field experiment. Each treatment consisted of 46 m^2 plots, arranged in a randomized block design with four replications, where the basic treatment was fertilization; the additional treatments were crop rotation, irrigation and cultivation. The applied fertilizer doses are presented in Table 1.

Three levels of irrigation water (0, 60, 100 mm/year) and two types of soil cultivation were used. The element content of the peas and soybeans was determined after digestion with a Labtam 8440M inductively coupled plasma atomic emission spectrometer.

Results

Irrigation, fertilization, cultivation and weather conditions have a significant effect on the Ni content of the seeds. In 1991, the Ni concentration of pea seeds increased from 2.9 ppm to 4.2 ppm, and in 1992 it changed from 2.0 ppm to 3.4 ppm due to fertilization (Fig. 1). Fertilization had a similar effect on the Ni content of soybeans. Thus in 1989 and 1990 it was between 7.7 and 8.9 ppm and 14 and 16.8 ppm, respectively (Fig. 2). Irrigation decreased the nickel concentration in the pease and soybeans (Fig. 3).

The relationship between the Ni and different element concentration in peas and soybeans was also investigated. In the case of peas, the data in the two years studied was very different. This data did not have a normal distribution and therefore we considered only the data for the soybeans. The correlation between the different elements are presented in Table 2. Between

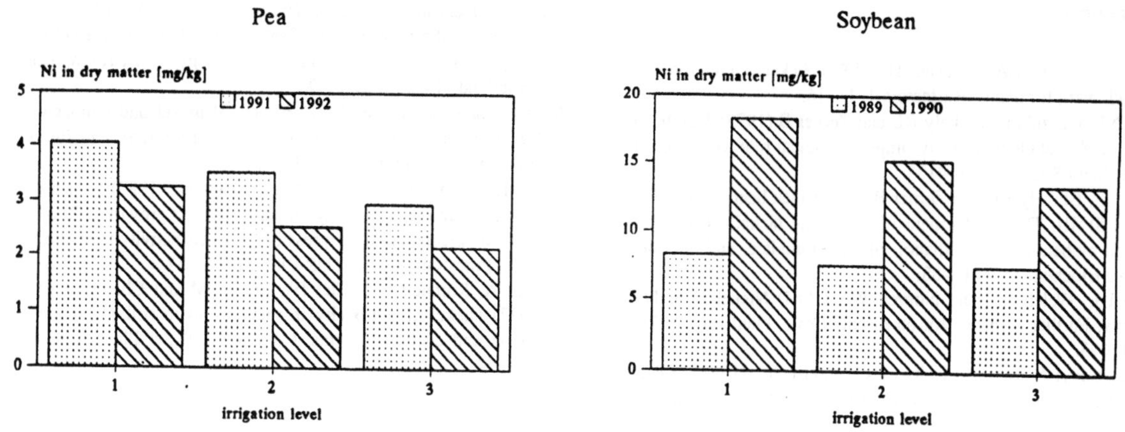

Fig. 3. Effect of irrigation on the nickel content of peas and soybeans

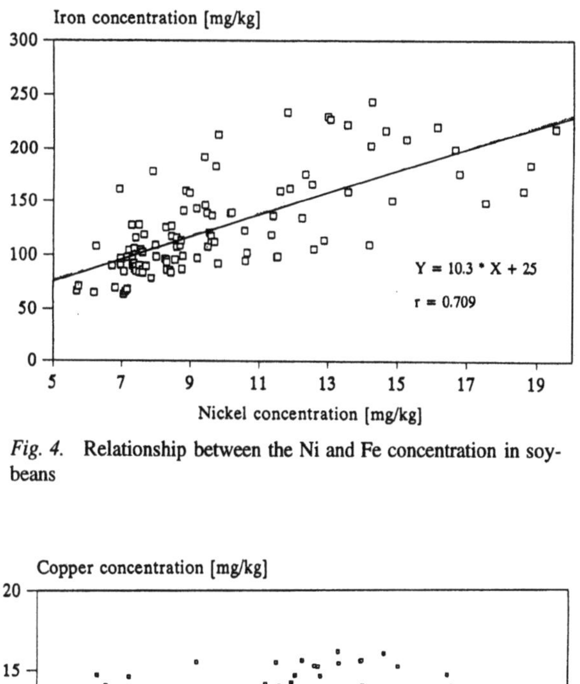

Fig. 4. Relationship between the Ni and Fe concentration in soybeans

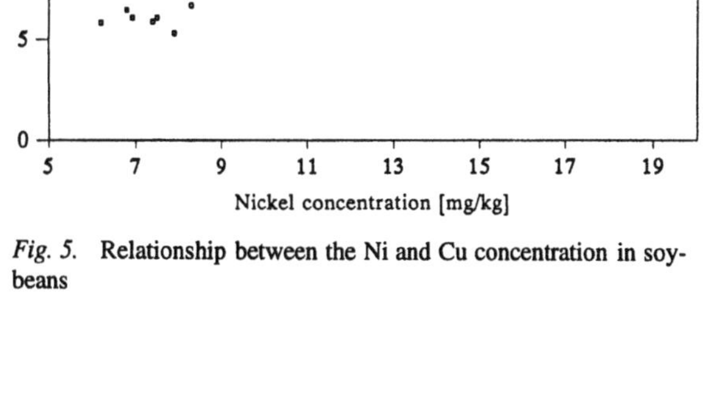

Fig. 5. Relationship between the Ni and Cu concentration in soybeans

the Ni and Fe concentrations an almost linear relationship is shown. (Fig. 4). The relationship between Ni and Cu concentrations can be described as following a maximum curve (Fig. 5)

Discussion

From the data obtained it can be seen that both fertilization and irrigation significantly changes the nickel content in peas and soybeans. In all cases, fertilization increased the content of nickel in the seeds. The presumable cause for this is that the mineral nitrogen fertilizer mobilizes heavy metals through a soil acidification effect, consequently leading to the increase of uptakeable nickel concentrations in the soil solution. There are significant differences between the nickel concentrations measured in seeds from year to year. It is found that the nickel concentrations are much higher in dry years. It can be presumed that the nickel is transported into the plants by the intervention of aerobic *Rhizobium* bacteria. The vital functions of these bacteria are more intensive in more aerated soils and consequently more nickel is transported into the plants. The decreasing effect of irrigation on the nickel content can also be explained in this manner.

The relationship between heavy metals and the nickel content shows that nickel has a promoting effect on the uptake and transport of zinc, copper, iron and manganese at the beginning of the nickel concentration increase. But as the concentration continues to increase, the uptake of the metals is hindered. The relationship between the nickel and copper concentrations was thus described with a maximum curve.

References

Dalton DA, Russel SA and Evans HJ (1988) Nickel as a micronutrient element for plants. Biofactors 1: 11–16.

Dixon NE, Gazzola C, Blakely RL and Zerner B (1975) Jack bean urease. A metalloenzyme: A simple biological role for nickel? J Am Chem Soc 97: 4131–4133.

Dixon NE, Blakely RL and Zerner B (1980) Jack bean urease (EC 3.5.1.5). III. The involvement of active-site nickel ion in inhibition by beta-mercaptoethanol, phosphoramidate, and fluoride. Can J Biochem 58: 481–488.

Eskew DL, Welch RM and Norvell WA (1984) Nickel in higher plants: Further evidence for an essential role. Plant Physiol 76: 691–693.

Gyori Z, Lanyi A, Ruzsanyi L, Kovacs B and Loch J. (1993) Effects of fertilization, irrigation and crop rotation on the transition and toxic element uptake of corn. The Science of the Total Environment Proceedings of the 2nd European Conference on Ecotoxicology – Recent Advances in Ecotoxicology, Amsterdam, The Netherlands 11–15 May 1992.

Heale EL and Ormond P (1982) Effects of nickel and copper on *Acer rubrum*, *Cornus stolonifera*, *Lonicera tatarica*, and *Pinus resinosa*. Can J Bot 60: 2674–2681.

Klucas RV, Hanus FJ, Russell SA and Evans HJ (1983) Nickel: a micronutrient element for hydrogen-dependent growth of Rhizobium japonicum and for expression of urease activity in soybean seeds. Proc Natl Acad Sci USA 80: 2253–2257.

Wallace A, Romney EM, Cha JW, Soufi SM and Chaudhry FM (1977) Nickel phytotoxicity in relationship to soil pH manipulation and chelating agents. Commun Soil Sci Plant Anal 8: 757–764.

C. Rodriguez-Barrueco (ed.), Fertilizers and Environment, 499–502.

The effect of N and P application on the Mn, Cu and Zn content of the winter wheat

Z. Gyôri, L. Ruzsányi, I. Jászberényi, I. Vágó & J. Loch
University of Agriculture, H-4015 Debrecen, P.O. Box 36. Hungary

Introduction

The average yield of winter wheat has increased due to the fertilizer application. Increased growth may lead to a decrease in elemental concentration in the grain due to a dilution effect which results from growth rates which exceed the rate of nutrient uptake (Jarrell and Beverly, 1981). In severe cases this may lead to nutrient deficiency in the crop which limits growth.

It has been demonstrated that copper requirements of a crop increase when nitrogen fertilizer is added (Mulder, 1950). After heavy nitrogen fertilization of a copper deficient soil complete crop failure has been recorded (Fleming and Delaney, 1961).

By contrast, the addition of increasing amounts of nitrogen fertilizer in field experiments resulted in both increased yields and increased concentrations of elements in grain, with the exception of Mn (McGrath, 1985).

In selected treatments of a multifactorial field experiment the effect of N and P application on the Cu, Mn and Zn content of the winter wheat was examined in calcareous chernozem soil under irrigated and non irrigated conditions.

Materials and methods

The multifactorial field experiment was established in 1986 at the Experimental Station of Agronomy Department, Debrecen Agricultural University. In the multifactorial experiment the Box-Wilson design and evaluation methods (Box and Wilson, 1951) were used. The relation between the yield (and other measured values) and the factors examined can be described by a polynomial of the second degree.

The field experiment was projected (Table 2) and statistically evaluated by Tolner and Biczók's computer program (Loch *et al.*, 1987).

Table 1

Soil characteristics	
Humus (%)	2,5
pH-(H_2O)	6,7
pH-(KCl)	5,9
Total N (%)	0,15
Clay (%)	47,8
P_2O_5-AL (mg/kg)	88
K_2O-AL (mg/kg)	150

The plants (winter wheat, cultivar: GK Zombor) were sampled five times during the vegetation period. For Cu, Mn, Zn analysis the aboveground part of plant subsamples and the wholemeal flour of grain were digested with HNO_3-H_2O_2 in a heating block. The elements were measured by ICP-AES (LABTAM 8440M). The correlation between the treatments and the element content was evaluated by linear regression analysis. The trends of the changes are graphically demonstrated.

Results and conclusions

- The Mn, Cu and Zn content in the winter wheat is relatively low due to the clay in and the weak acidity of the chernozem.
- Nitrogen doses (by constant P and K supply) increase the Mn and Zn content in the plant significantly and produce a moderate increase in the Cu content. (Figure 1)
- Phosphorus doses (by constant N and K doses) increase the Mn content of the wheat for period of the first two samplings, but this effect cannot be observed later. Phosphorus reduces the zinc content considerably and produces a moderate decrease in the copper content. (Figure 2)

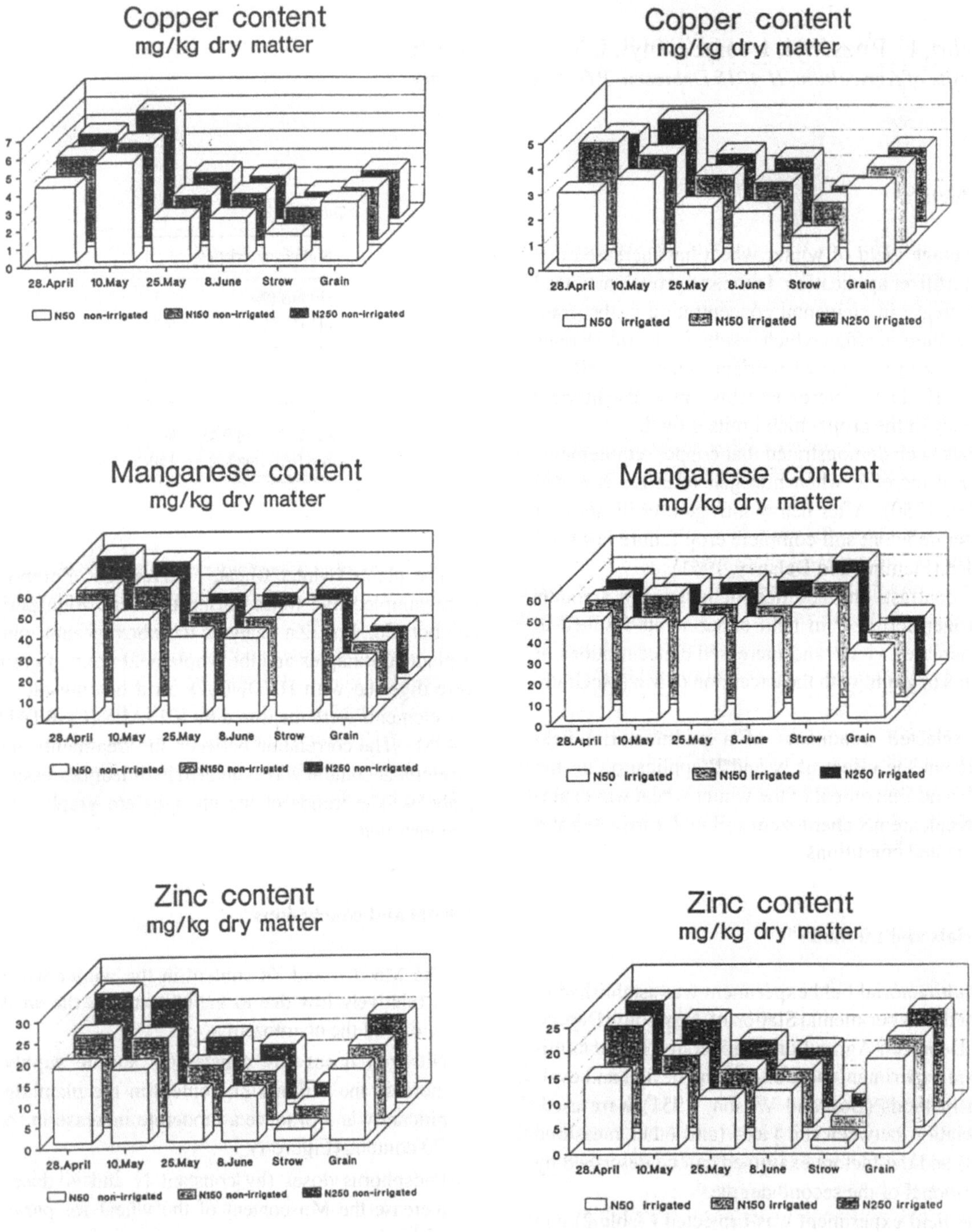
Copper content
mg/kg dry matter
28.April 10.May 25.May 8.June Strow Grain
N50 non-irrigated N150 non-irrigated N250 non-irrigated
Copper content
mg/kg dry matter
28.April 10.May 25.May 8.June Strow Grain
N50 irrigated N150 irrigated N250 irrigated
Manganese content
mg/kg dry matter
28.April 10.May 25.May 8.June Strow Grain
N50 non-irrigated N150 non-irrigated N250 non-irrigated
Manganese content
mg/kg dry matter
28.April 10.May 25.May 8.June Strow Grain
N50 irrigated N150 irrigated N250 irrigated
Zinc content
mg/kg dry matter
28.April 10.May 25.May 8.June Strow Grain
N50 non-irrigated N150 non-irrigated N250 non-irrigated
Zinc content
mg/kg dry matter
28.April 10.May 25.May 8.June Strow Grain
N50 irrigated N150 irrigated N250 irrigated

Fig. 1.

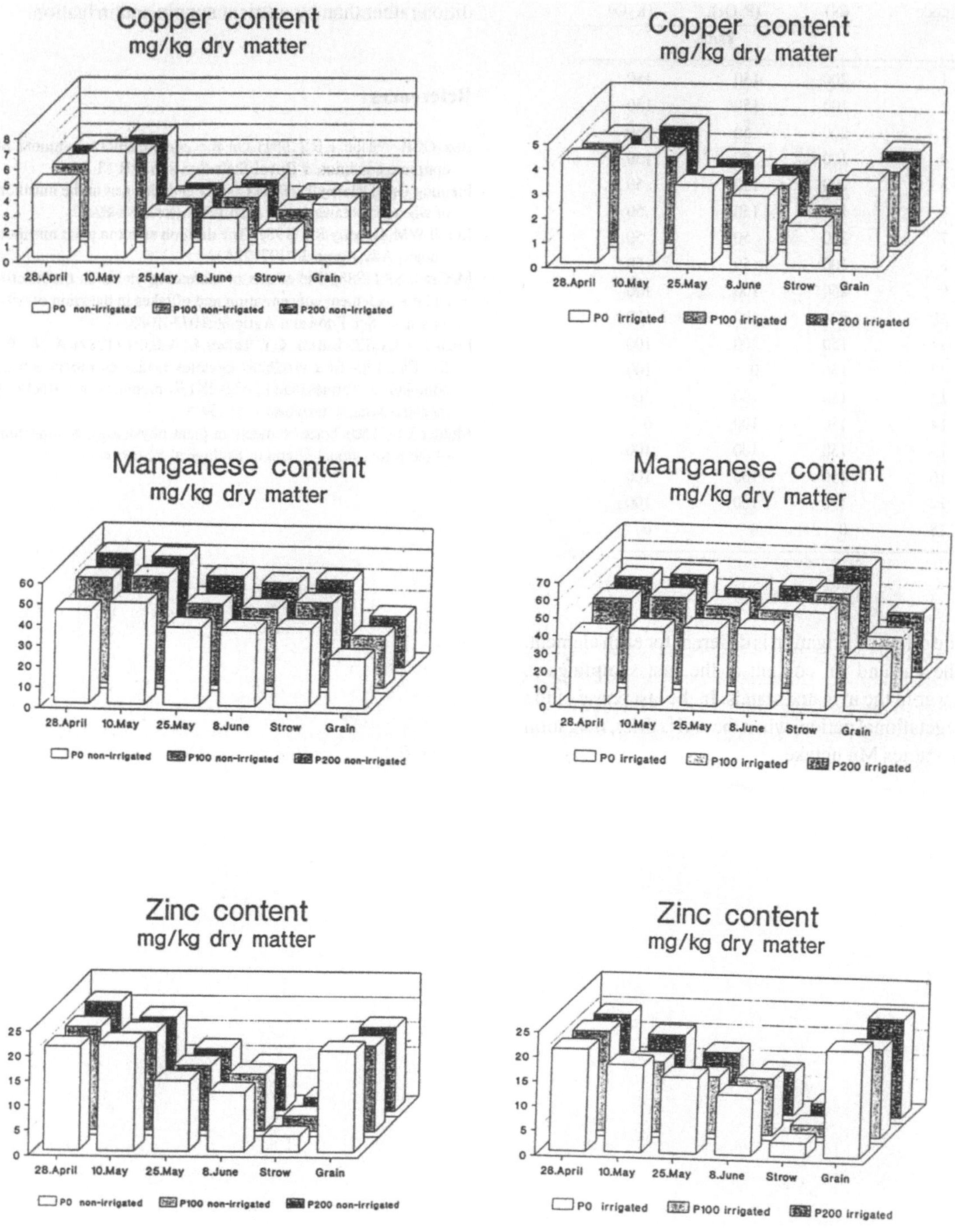
Copper content
mg/kg dry matter
28.April
10.May
25.May
8.June
Strow
Grain
P0 non-irrigated
P100 non-irrigated
P200 non-irrigated
Copper content
mg/kg dry matter
28.April
10.May
25.May
8.June
Strow
Grain
P0 irrigated
P100 irrigated
P200 irrigated
Manganese content
mg/kg dry matter
28.April
10.May
25.May
8.June
Strow
Grain
P0 non-irrigated
P100 non-irrigated
P200 non-irrigated
Manganese content
mg/kg dry matter
28.April
10.May
25.May
8.June
Strow
Grain
P0 irrigated
P100 irrigated
P200 irrigated
Zinc content
mg/kg dry matter
28.April
10.May
25.May
8.June
Strow
Grain
P0 non-irrigated
P100 non-irrigated
P200 non-irrigated
Zinc content
mg/kg dry matter
28.April
10.May
25.May
8.June
Strow
Grain
P0 irrigated
P100 irrigated
P200 irrigated

Fig. 2.

Table 2. Fertiliser treatments

Treatment code	Nitrogen (N)	Phosphorus (P_2O_5)	Potassium (K_2O)
	kg/ha		
1	200	150	150
2	100	150	150
3	200	50	150
4	100	50	150
5	200	150	50
6	100	150	50
7	200	50	50
8	100	50	50
9	250	100	100
10	50	100	100
11	150	200	100
12	150	0	100
13	150	100	200
14	150	100	0
15	150	100	100
16	150	100	100
17	150	100	100
18	0	0	0

– The effect of irrigation is different for each element. The Cu and Zn content at the first samplings is lower in the irrigated stands. In the last phase of the vegetational period, when the soil is drier, irrigation promotes Mn uptake.

It can be concluded that under these experiment conditions the microelement uptake of the winter wheat is determined by soil physical and chemical condition rather than by nutrient supply and irrigation.

References

Box GEP, Wilson KB (1951) On the experimental attainment of optimum condition. J Royal-Stats Soc Series B 13. 1.

Fleming GA, Delaney I (1961) Copper and nitrogen in the nutrition of wheat on cutaway peat. Irish J Agr Res 1:81–82.

Jarrell WM, Beverly RB (1981) The dilution effect in plant nutrition studies. Adv Agron 34:197–224.

McGrath SP (1985) The effects of increasing yields on the macro- and microelement concentration and offtakes in the grain of winter wheat. Sci. Food and Agric 36:1073–1083.

Loch J, Kiss SZ, Biczók GY, Tolner L, Vágó I (1987) A N-, P-, K-, Ca-, Mg- és a vízellátás együttes hatása csernozjomon és homokon. I. Termésadatok. NEVIKI Konferencia, Keszthely. A mezogazdaság kemizálása I. 53–59.

Mulder EJ (1950) Trace elements in plant physiology. Symposium of the International Union of Biological Sciences.

C. Rodriguez-Barrueco (ed.), Fertilizers and Environment, 503–508.

Effect of cadmium on the distribution of micronutrients in *Lactuca* spp., maize and pea plants

L.E. Hernández, I. Ramos, R. Carpena-Ruiz, J.J. Lucena & A. Gárate
Dpto. Química Agrícola, Geología y Geoquímica, Facultad de Ciencias, Universidad Autónoma de Madrid, Canto Blanco E-28049 Madrid, Spain

Key words: Cadmium, *Lactuca*, manganese, metal interaction, micronutrients, *Pisum sativum*, *Zea mays*

Abstract

The effect of different doses of Cd on the uptake and distribution of micronutrients in several species of *Lactuca*, (*Lactuca sativa* cv. Winter yellow; *Lactuca serriola* cv. Hortelano; *Lactuca serriola* cv. Pancalieri), two cultivars of *Zea mays* (cvs. Dakalb XL 72 AA and Dekalb Paolo) and two cultivars of *Pisum sativum* (cvs. Argona and Waverplus) is studied. Plants exposed to Cd concentrations ranging from 0.1 to 15 mg Cd l^{-1} showed more accumulation of Cd in root than in shoot (from 7 to 15 times higher in root). Growth and visual symptoms showed that *Pisum* cultivars were those affected the most. The content of Mn in plant tissues was reduced consistently under Cd treatments. Moreover, the relative concentration of Mn was higher in shoot than in root, which indicates a sort of redistribution among tissues. Fe uptake was also reduced or not by the exposure to Cd in the nutrient solution depending on the species. Moreover, the analysis of other micronutrients (Cu and Zn) revealed that no consistent alteration in their content was achieved.

Introduction

Cadmium is known to be one of the most toxic heavy metals released in the environment and its continuous introduction in the biosphere as a result of industrial activities presents a hazard to mankind and the ecosystem. The use of industrial wastes and sludges as agricultural amendments has surely increased the amount of Cd that is available to plants (van Assche and Clijters, 1990).

The uptake and/or internal distribution of cadmium may differ among plant species or varieties within species exposed to similar external cadmium concentrations (Florijn and van Beusichem, 1993). Most plant species accumulate Cd in root, phenomenon that has been related to a resistance mechanism against Cd toxicity (Leita *et al.*, 1993). Thus, Cd-tolerant populations of *Silene vulgaris* translocated less Cd from root to shoot than a non-tolerant population did (Verkleij and Prast, 1989).

The toxicity of Cd '*per se*' is well established. Among several toxic symptoms described in the literature, the interaction of Cd with other elements in the environment has been observed. It is this interaction which can modulate the resultant metal-toxicity (Singh and Yadava, 1983). Therefore, the absorption of mineral nutrients might be affected by the addition of Cd, as several authors reported (Trivedi *et al.*, 1992; Costa and Morel, 1994).

Some studies proved a reduction in the absorption and transport of Mn in the presence of Cd, as well as a dimininution in the accumulation of Cd when the concentration of Mn was increased (Cataldo *et al.*, 1983; Thys *et al.*, 1991). Moreover, Baszynski *et al.* (1980) found a decrease in the degree of Cd concentration determined in root and leave of tomato plants related to the supply of Mn to the nutrient solution. The content of Fe in plant tissues has also been affected by Cd treatments, as several researchers reported (Haghiri, 1973; Cataldo *et al.*, 1983; Greger and Lindberg, 1987; Burzynski and Buczek, 1989; Thys *et al.*, 1991; Siedlecka and Baszynski, 1993). Less uptake of Zn has also been described (Costa and Morel, 1994).

The effect of Cd upon cultivars of *Lactuca* spp., *Zea mays* and *Pisum sativum* has been investigated.

The incidence of Cd on their micronutrient content has been studied, in order to obtain more information about the effects of Cd on the micronutrient mineral nutrition of higher plants.

Materials and methods

Plant material

Maize (*Zea mays* cvs. Dekalb XL 72 AA and Dekalb Paolo) and pea (*Pisum sativum* cvs. Argona and Waverplus) were germinated in moistened paper for 4 days at 28 °C. Maize (nutrient solution: macronutrients [mmol l^{-1}]: 2.0 KH_2PO_4, 1.5 $MgSO_4$, 0.1 NaCl, 1.0 $Ca(NO_3)_2$, 1.5 KNO_3, and micronutrients [mg l^{-1}]: 2.5 Fe(EDDHA), 1.0 $MnSO_4$, 0.25 $CuSO_4$, 0.5 $ZnSO_4$, 0.5 H_3BO_3, 0.2 $Mo_7O_{24}(NH_4)_6 \cdot 4H_2O$, pH 5.0) and pea (nutrient solution: macronutrients [mmol l^{-1}]: 2.0 $Ca(NO_3)_2$, 1.5 KNO_3, 1.0 $Mg(NO_3)_2$, 1.0 KH_2PO_4, 0.5 $MgSO_4$, 0.1 NaCl, and micronutrients [mg l^{-1}]: 2.5 Fe(EDDHA), 1.0 $MnSO_4$, 0.2 $CuSO_4$, 0.4 $ZnSO_4$, 0.2 $Mo_7O_{24}(NH_4)_6 \cdot 4H_2O$, 0.25 H_3BO_3, pH 6.0) seedlings were cultivated in a controlled environment chamber at 28 °C day/20 °C night, with 16 h of light (10 Sylvania Cool White VHO lamps of 120 W m^{-2} each) and 70–75% of relative humidity. Cd was supplied in 4 concentrations – 0.0 (control), 0.15, 1.5 and 15.0 mg Cd l^{-1} – to maize plants, and in 3 concentrations – 0.0 (control), 1.5 and 6.0 mg Cd l^{-1}) to pea plants. Samples were collected after 15 days of Cd treatment. In the so-called 'shock' treatments, maize plants grown in control nutrient solution were transferred to a nutrient solution with 15 mg Cd l^{-1}, and sampled after 24 and 48 h of treatment.

Lettuce (*Lactuca sativa*, cv. Winter yellow), giant endive (*Lactuca serriola*, cv. Hortelano) and curly endive (*Lactuca serriola*, cv. Pancalieri) seeds were germinated and grown in a mixed substrate of peat and acid-washed sand. Plants were irrigated with a complete 50% diluted nutrient solution which was prepared according to Hewitt and Smith (1975), pH 5.5. Edible-sized plants were cultivated in aerated nutrient solution. Experiments were conducted in a experimental greenhouse with controlled temperature (13–28 °C) and humidity (50–80%). Treatments included: 0.0 (control), 0.1 and 1.0 mg Cd l^{-1}. Samples were collected after 15 days of exposure to Cd.

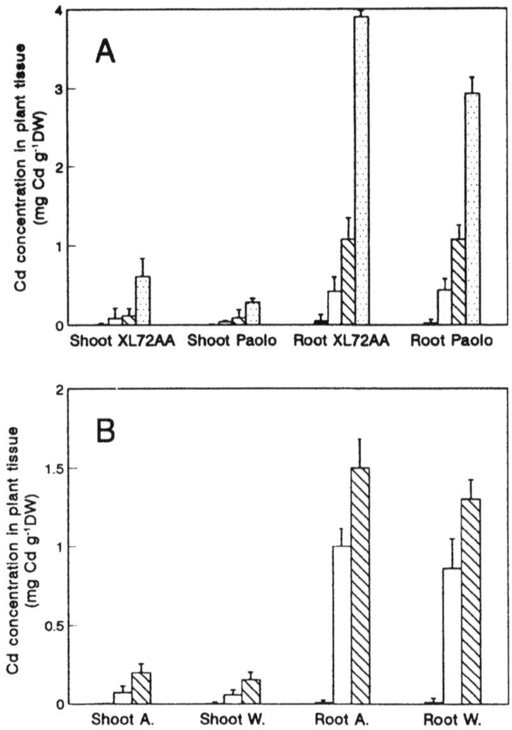

Fig. 1. Concentration of Cd (mg g^{-1}DW) in shoot and root: of *A*, maize (cvs. Dekalb XL72AA and Dekalb Paolo), treated with 0.0 (control, ■), 0.15 (□), 1.5 (▧) and 15.0 mg Cd L^{-1} (■); and B, pea (cvs. Argona -A.- and Waverplus -W.-), treated with 0.0 (control, ■), 1.5 (□) and 6.0 mg Cd L^{-1} (▧).

Cd and micronutrient analysis

Plant tissues were dried at 70 °C until constant weight was achieved. After homogenization, samples were digested with a mixture of acids HNO_3:H_2SO_4:HClO (5:1:2), at < 200 °C. Cd and micronutrients were analysed by spectrophotometric atomic absorption (Perkin-Elmer 4000), as Gárate *et al.* described (1992).

Results and discussion

Cd uptake and distribution

Cd concentrations in root and shoot of maize and pea cultivars are shown in Fig. 1a and 1b, respectively. The increase of Cd in root and shoot are significantly higher when the Cd concentration increased in the nutrient solution. The heavy metal accumulates mainly in the root of maize and pea plants. The Cd concentration in roots found in each variety of maize and pea was

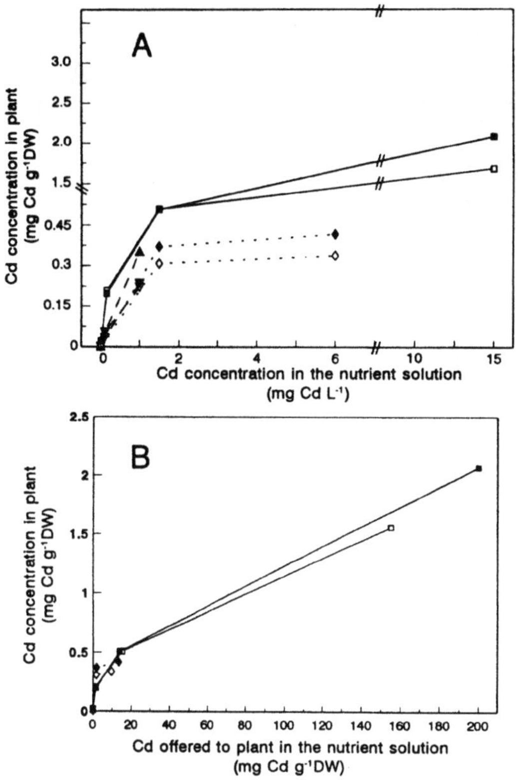

Fig. 2. *A*, concentration of Cd in plant (mg Cd g^{-1}DW) as a function of the concentration of Cd supplied in the nutrient solution (mg Cd L^{-1}) of maize cv. Dekalb XL72AA (■), maize cv. Dekalb Paolo (□), pea cv. Argona (◇), pea cv. Waverplus (◆), lettuce (▲), giant endive (▼) and curly endive (▽). *B*, concentration of Cd (mg Cd g^{-1}DW) of maize and pea plants as a function of the estimated concentration of Cd offered to the plant in the nutrient solution, expressed per dry weight of plant (mg Cd g^{-1}DW).

about five times higher than the one found in shoots. Similar results were observed in *Lactuca*, where the Cd concentration in root was approximately 7 to 10 times higher than in shoot of plants treated with 0.1 and 1.0 mg Cd l^{-1}, respectively (Gárate *et al.*, 1993). Also, similar Cd distribution between root and shoot was observed in maize (Florijn and van Beusichem, 1993) and pea (Hernández *et al.*, 1995), when similar levels of Cd in the nutrient solution were used.

Fig. 2a shows the Cd concentration in *Lactuca*, maize and pea plants for the different Cd levels in the nutrient solution. As it can be observed, a saturation in the accumulation of Cd may occur when the amount of Cd offered to plants increases, although this saturation differs depending on the plant species. Thus, up to 1.5 mg Cd l^{-1} in the nutrient solution, the concentration of Cd in plant was similar for all the plant species studied. However, for Cd concentrations above 1.5 mg Cd l^{-1}, the concentration of Cd in pea plants were lower than in maize plants (approximately 50%). This decline in the relative amount of Cd uptake suggests that a saturation mechanism may be involved in the Cd absorption and it is in agreement with the results of Gárate *et al.* (1993), who observed a relative decrease in uptake of Cd by lettuce and giant endive during the time course of experiment or when the Cd concentration in the nutrient solution was higher than 0.1 mg Cd l^{-1}. Similarly, Fujimoto and Uchida (1979) observed a decrease in Cd uptake by rice plants when the Cd concentration in the nutrient solution was higher than 0.15 mg Cd l^{-1}.

When the concentration of Cd in plant was expressed in terms of Cd offered in the nutrient solution per g dry matter of plant, the relative decrease in the rate of Cd uptake was also observed in maize and pea plants (Fig. 2b). Nevertheless, maize plants tend to accumulate much more Cd than pea plants, although the concentration of Cd offered increased 10 times. Therefore, this suggests that maize plants were more tolerant to the accumulation of Cd.

On the other hand, although similar concentration of Cd was observed in plants of the studied species, for similar levels of Cd supplied to the nutrient solution, maize and *Lactuca* showed less toxic symptoms (e.g. weight, number of secondary roots, darkening of roots) than pea plants (data not shown). Therefore, the sensitivity to Cd might not be only explained by the level of Cd accumulated in plant tissue, but rather as a different distribution within plant tissue compartments (Vazquez *et al.*, 1992).

Effect of Cd on the distribution of micronutrients

The concentration of Mn in plant tissues was the one affected the most by the the presence of Cd in the nutrient solution among the analysed micronutrients. As it can be seen in Fig. 3a, Cd-treated maize plants accumulated less Mn in shoot and root than control plants did. Also, there was a severe decrease in Mn concentration in the root of Cd-treated pea plants (Fig. 3b). On the other hand, pea shoot showed a higher amount of Mn in the 1.5 mg Cd l^{-1} treatment than control plants, although when the supply of Cd increased, Mn-shoot concentration was reduced (Fig. 3b). These results are in accordance with those previously described for several *Lactuca* species (Garate *et al.*, 1992), in which plants treated with 0.1 mg Cd l^{-1} accumulated more Mn in shoot after 15 and 35 days of treatment. This behaviour was also observed in *Lolium perenne* (Jarvis

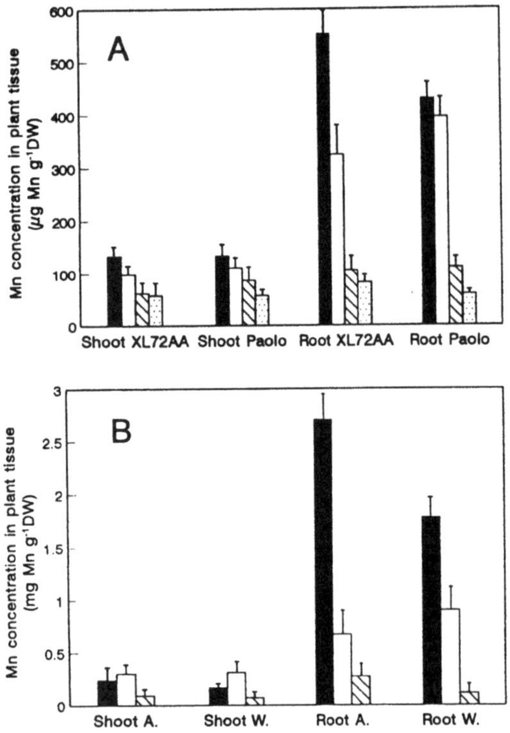

Fig. 3. Manganese concentration in shoot and root: of *A*, maize (μg Mn g^{-1}DW) (cvs. Dekalb XL72AA and Dekalb Paolo), treated with 0.0 (control, ■), 0.15 (□), 1.5 (▨) and 15.0 mg Cd L^{-1} (■); and *B*, pea (mg Mn g^{-1}DW) (cvs. Argona and Waverplus), treated with 0.0 (control, ■), 1.5 (□) and 6.0 mg Cd L^{-1} (▨).

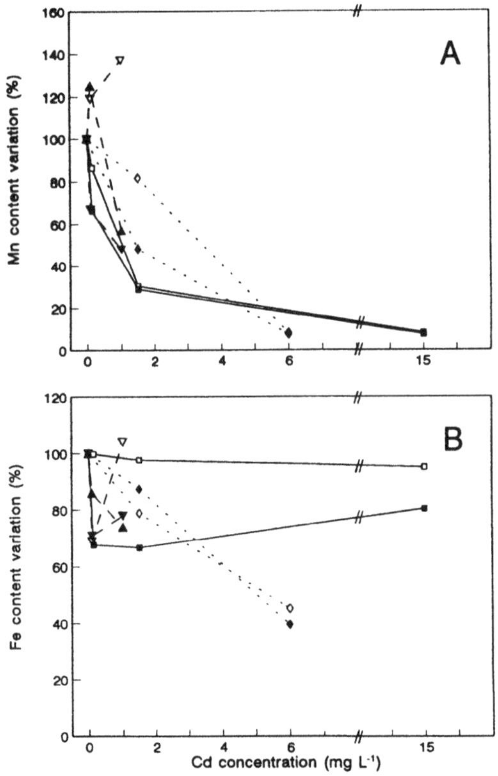

Fig. 4. Relative variation (%) of Mn concentration (*A*) and Fe (*B*) in maize cv. Dekalb XL72AA (■), maize cv. Dekalb Paolo (□), pea cv. Argona (◇), pea cv. Waverplus (◆), lettuce (▲), giant endive (▼) and curly endive (▽) plants in function of the concentration of Cd supplied in the nutrient solution (mg Cd L^{-1}).

et al., 1976), soybean (Cataldo *et al.*, 1983) and N_2-fixing pea (Hernandez *et al.*, 1995). From the data available of *Lactuca*, maize and pea plants, the estimation of the relative variation of the content of Mn was calculated and compared with the supply of increasing levels of Cd in the nutrient solution (Fig. 4a). As it can be observed, that shows a decrease in the three plant species represented. Therefore, these facts support the results previously discussed.

On the contrary, a similar estimation of the relative variation in the content of Fe revealed a less consistent tendency related to the supplied treatments of Cd (Fig. 4b). Moreover, the analysis of Zn and Cu showed a similar behaviour to those found in Fe (data not shown), although the active absorption of Cd decreased in the presence of Zn (Costa and Morel, 1994). A resemblant alteration in the micronutrient content of tomato plants treated with relative high concentrations of Cd was described (Moral *et al.*, 1994). Thys *et al.* (1991) observed a clear antagonism between Cd and Fe in lettuce plants, which caused a higher accumulation of Mn. Pea plants showed a constant reduction in the content of Fe, that was not observed in the other plant species, according to results reported by several authors (Cataldo *et al.*, 1983; Greger and Lindberg, 1987; Sieldlecka and Baszynski, 1993). Nevertheless, the decrease of Fe was not followed by an increment in the concentration of Mn (Fig. 4a), in opposition to the results obtained by Thys *et al.* (1991). These different results might be explained by the levels of Cd applied, the age of the treated plants and the method of application of the heavy metal.

The treatment of maize plants with a shock of Cd (15 mg Cd l^{-1}) caused a significant decrease of Mn concentration in shoot and root, observed even after 24 h of treatment, but no effect on the content of the other studied micronutrients was obtained (data not shown). This tendency was also seen in pea plants that were exposed to a similar treatment (unpublished data).

The relative partitioning of Mn between root and shoot in both maize cultivars indicates that Mn

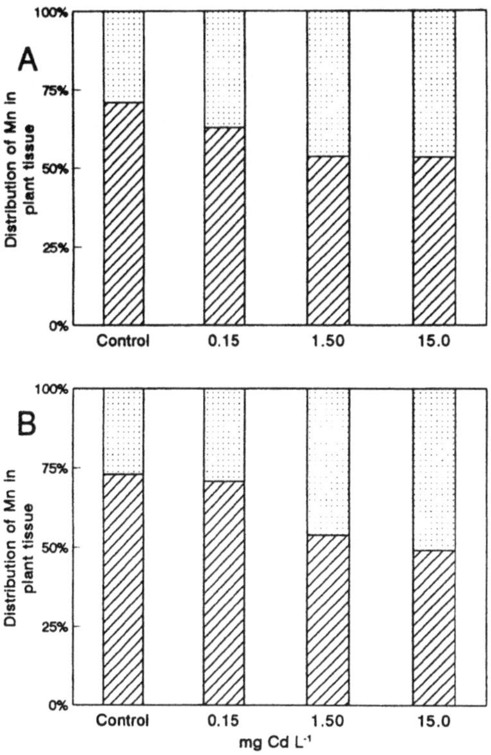

Fig. 5. Relative distribution of Mn (%) among shoot (■) and root (▨) in maize cvs. Dekalb XL72AA (*A*) and Dekalb Paolo (*B*) plants, treated with 0.0 (control), 0.15, 1.5 and 15.0 mg Cd L^{-1}.

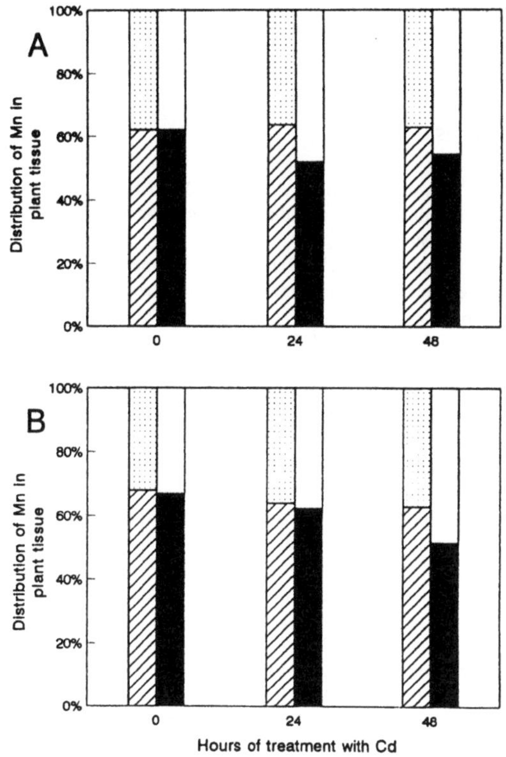

Fig. 6. Relative distribution of Mn (%) among shoot (■) and root (▨) of control plants, and shoot (□) and root (■) of 15 mg Cd L^{-1} treated plants of maize cvs. Dekalb XL72AA (*A*) and Dekalb Paolo (*B*), in the shock exposure to Cd after 0, 24 and 48 h of treatment.

decreased in root and increased in shoot (Fig. 5). Similar results were observed in lettuce (Gárate *et al.*, 1992) and pea (unpublished data). The translocation of Mn to the shoot might be a result of a protection mechanism to avoid the accumulation of Cd in photosynthetic tissues, or of the depletion of Mn in root due to an interaction with Cd for biological positions (e.g. cell wall). The shock treatment showed a fast alteration of Mn distribution in the studied maize cultivars, even after 24 h of treatment, when the relative content in root decreased and increased in shoot (Fig. 6). These data support the hypothesis of an antagonism Cd-Mn, which was not clearly observed with other micronutrients.

Conclusions

Cadmium accumulated more in root than in shoot of all plant species investigated, when different concentrations of Cd were used. This Cd accumulation in roots might be a tolerance mechanism to avoid Cd translocation to aerial – photosynthetic and reproductive – parts. *Pisum sativum* cultivars were the most affected under exposure to Cd. Mn uptake was reduced after the addition of Cd in the plant species studied. In addition, the Mn partitioning between shoot and root revealed that more Mn was accumulated in the aereial parts when Cd concentration in the nutrient solution increased. This might be a resistance response of the plant to diminish the negative effect of Cd on the photosynthetic metabolism (Gárate *et al.*, 1992). Finally, Fe content in *Pisum sativum* decreased similarly to the decline observed for Mn, but this tendency was not proved in *Lactuca* spp. and *Zea mays*.

Acknowledgements

We thank the Spanish CICYT for its finantial support through the projects AGR88–0009, AGR90–0286 and PB92–1097. Dr LE Hernández is also indebted to the Spanish Ministry of Science for its support through a research grant (PN89 3828899). Maize seeds were a gift from Dekalb Ibérica SA

References

Baszynski T, Wajda L, Krol M, Wolinska D and Krupa Z (1980) Photosynthetic activities of cadmium-treated tomato plants. Plant Physiol 48: 365–370.

Burzynsky M and Buczek J (1989) Interaction between cadmium and molybdenum affecting the chorophyll content and accumulation of some heavy metals in the second leaf of *Cucumis sativus* L. Acta Physiol Plant 11: 137–145.

Burzynsky M (1988) The uptake and accumulation of phosphorous and nitrates and the activity of nitrate reductase in cucumber seedlings treated with Pb and Cd. Acta Soci Botan Poloniae 57: 349–359.

Cataldo DA, Garland TR and Wildung RE (1983) Cadmium uptake kinetics in intact soybean plants. Plant Physiol 73: 844–848.

Costa G and Morel JL (1994) Efficiency of H^{+}-ATPase activity on cadmium uptake by four cultivars of lettuce. J Plant Nutri 17: 627–637.

Florijn PJ and van Beusichem ML (1993) Uptake and distribution of cadmium in maize inbred lines. Plant Soil 150: 25–32.

Friberg L, Piscator M, Nordberg G and Kjellstrom T (1974) Cadmium in the environment. CRC Press, Cleveland, USA.

Fujimoto T and Uchida Y (1979) Cadmium absorption by rice plants. I. Mode of the absorption. Soil Sci Plant Nutr 25: 407–415.

Gárate A, Ramos I, Manzanares M and Lucena JJ (1993) Cadmium uptake and distribution in three cultivars of *Lactuca* sp. Bull Environ Contam Toxicol 50: 709–716.

Gárate A, Ramos I and Lucena JJ Efecto del cadmio sobre la absorción y distribución de manganeso en distintas variedades de *Lactuca*. Suelo Planta 2: 581–591.

Greger M and Lindberg S (1987) Effects of Cd and EDTA on young sugar beet (*Beta vulgaris*). II. Net uptake and distribution of Mg, Ca and Fe(II)/Fe(III). Physiol Plant 69: 81–86.

Haghiri F (1973) Cadmium uptake by plants. J Environ Qual 2: 93–96.

Hernández LE, Gárate A and Carpena-Ruiz R (1995) Effect of cadmium on nitrogen fixing pea plants grown in perlite and vermiculite. J Plant Nutri 18: 287–303.

Hewitt EJ and Smith TH (1974) Plant Mineral Nutrition. English Universities Press, London.

Khan DH, Duckett JG, Frankland B and Kirkham JB (1984) An X-ray microanalytical study of the distribution of Cd in roots of *Zea mays* L. J Plant Physiol 115: 19–28.

Jarvis SC, Jones LHP and Hopper MJ (1976) Cadmium uptake from solution by plants and its transport from roots to shoots. Plant Soil 44: 179–191.

Leita L, De Nobili M, Mondini C and Baca-García MT (1993) Response of leguminosae to cadmium exposure. J Plant Nutri 16: 2001–2012.

Leita L, Baca-García MT and Maggioni A (1992) Cadmium uptake by *Pisum sativum*: accumulation and defense mechanism. Agrochimica 36: 253–259.

Moral R, Gomez I, Navarro-Pedreño J and Mataix J (1994) Effects of cadmium on nutrient distribution, yield and growth of tomato grown in soilless culture. J Plant Nutri 17: 953–962.

Siedlecka A and Baszynski T (1993) Inhibition of electron flow around photosystem I in chloroplasts of Cd-treated maize plants is due to Cd-induced iron deficiency. Physiol Plant 87: 199–202.

Singh SP and Yadava V (1983) Cadmium induced inhibition of nitrate uptake in *Anacystis nidulans*: interaction with other divalent cations. J Gen Appl Microbiol 29: 297–304.

Thys C, Vanthomme P, Schrevens E and De Proft M (1991) Interactions of Cd with Zn, Cu, Mn and Fe for lettuce (*Lactuca sativa* L.) in hydroponic culture. Plant Cell Environ 14: 713–717.

Trivedi S and Erdei L (1992) Effects of cadmium and lead on the accumulation of Ca and K, and on the influx and translocation of K in wheat of low and high K status. Physiol Plant 84: 94–100.

van Assche F and Clijsters H (1990) Effects of metals on enzyme activity in plants. Plant Cell Environ 13: 195–206.

Vázquez MD, Poschenrieder Ch and Barceló J (1992) Ultrastructural effects and localization of low cadmium concentrations in bean roots. New Phytol 120: 215–226.

Verkleij JAC and Prast JE (1989) Cadmium tolerance and cotolerance in *Silene vulgaris* (Moench.) Garcke (*S. cucubalus* (L.) Wib.). New Phytol 111: 637–645.

C. Rodriguez-Barrueco (ed.), Fertilizers and Environment, 509–515.

Irrigation with urban wastewater

I. Yield and soil impact

C. Horta-Monteiro[1], F.C. Pinto & J.Q. Santos[2]
[1]*Escola Superior Agrária de Castelo Branco, Quinta da Sr[a] de Mércules, 6000 Castelo Branco, Portugal;* [2] *Instituto Superior de Agronomia, Secção Autónoma de Química Agrícola, Tapada da Ajuda, 1300 Lisboa, Portugal*

Key words: Irrigation, *Lolium multiflorum* Lam, pollution, soil fertility, wastewater

Abstract

With the objective of evaluating the relevance of wastewater use in irrigation, a experiment in pots with *Lolium multiflorum* Lam. was carried out. The soil used (cambisol) was acid and poor in organic matter. The wastewater used in irrigation was treated by an activated sludge process.

It was observed that wastewater used had significant increases in the yields, without really affecting its quality. But its application did not avoid neither fertilization with nitrogen, phosphorus and potassium nor liming.

As far as the soil characteristics analysed after the experiment are concerned, there are several which can be considered more relevant to its fertility. Wastewater increased the electrical conductivity; nitrates, boron, magnesium and sodium levels, whereas the potassium level decreased. Wastewater dilution decreased O.M. and exchangeable Na level in the soil.

Introduction

As there are not many water supplies in Portugal, they should firstly be used to satisfy population' s needs and other demanding quality water purposes.

Agriculture uses large quantities of water. Although it does not mean the use of high quality water, irrigation with treated urban wastewater might be an alternative to be taken into account in our country.

The correct use of irrigation with treated urban wastewater leads to water preservation and nutrients recycling. It avoids surface and ground water pollution and it can also be used as a restoring method for ground water (Chopp *et al.*, 1982; Who, 1989; Asano *et al.*, 1992).

In this experiment we aimed at the use of wastewater in irrigation.

Materials and methods

An experiment in pots using *Lolium multiflorum* Lam. was carried out. The soil used was a cambisol. Table 1 shows the main characteristics of this soil, where it is evident its acidity; poverty in organic matter; low level of available phosphorus and medium level in available potassium; low cation exchange capacity and low degree of bases saturation. Copper, zinc, manganese and boron levels did not seem to lead to phytotoxic risks.

The methods used in the analysis of the soil are: the organic matter was calculated by multiplying the percentage of organic carbon by the factor 1.724. Organic carbon was measured by dry combustion at 1200 °C in a Ströhlein apparatus. Available phosphorus and potassium in the soil, were determined by colorimetric and flame emission photometric methods, respectively, after extraction by Egner-Riehm method. Electrical conductivity was calculated by a conductimeter apparatus in a 1:5 suspension (soil:water), Kjeldahl nitrogen by Kjeldahl method and nitrates by a NO_3 ion-selective

Table 1. Some analytical results

Parameters	
Texture	Loamy sand
E.C. (mS cm^{-1})	0.058
O.M. (%)	1.29
N-Nk (mg kg^{-1})	340
N-NO_3 (mg kg^{-1})	vest.
pH (H_2O)	4.9
Available Phosphorus (P_2O_5, mg kg^{-1})	16
Avaiable Potassium (K_2O, mg kg^{-1})	100
Exchangeable H^+ [cmol(+) kg^{-1}]	7.43
Exchangeable bases [cmol(+) kg^{-1}]	
Ca	1.07
Mg	0.20
K	0.11
Na	0.19
T [cmol(+) kg^{-1}]	9.09
V (%)	18
Micronutrients (mg kg^{-1})	
Cu	0.5
Fe	31.5
Zn	1.9
Mn	5.1
B	0.441

electrode. The exchangeable bases and micronutrients were calculated by atomic absorption spectrophotometry after extraction by Mehlich and Lakanen methods. The extraction of boron was made by boiling water.

The irrigation water used was either the secondary effluent of the wastewater treatment plant in Beirolas-Lisbon, or deionized water according to the experiment. The wastewater used was treated by an activated sludge process.

Several samples of wastewater were taken during the experiment. Table 2 shows its analytical results.

The methods used in the analysis of wastewater were described in Standard Methods (1980) and AOAC Methods (1980).

After air drying, the original soil was sieved in a 5 mm sieve and it was divided into 12.0 kg portions in twenty-four pots. As required in some treatments, the addition of lime and nutrients was manual. All the pots were irrigated until field capacity with deionized water, before sowing.

The experiment was done in the Instituto Superior de Agronomia, Lisbon, and it was planned with the goal of evaluating the relevance of wastewater use in irrigation as far as agronomic interests, the fertilizer effect of wastewater and the advantages of its dilution were concerned. As the original soil was acid we introduced treatments either with or without lime to assess the importance of this practice simultaneously with wastewater.

The following treatments were used in the present experiment:

- Deionized water irrigation and nitrogen (N), phosphorus (P_2O_5), and potassium (K_2O) fertilization – D + F
- Wastewater irrigation and NPK fertilization – R + F
- Wastewater irrigation without fertilization – R
- 50% wastewater dilution without fertilization – 50% R

All these treatments were made either with or without lime.

When fertilization was done, the amount of nitrogen, phosphorus and potassium introduced was:

- 1.5 g Nitrogen (N)
- 1.5 g Phosphorus (P_2O_5)
- 1.0 g Potassium (K_2O)

Liming was done by the addition of 17.0 g/pot of calcium carbonate which led to an increase in the soil pH level to 6.5.

1 soil × 3 replications × 8 treatments.

Fertilization was made before sowing by the addition of an amonium nitrate solution and potassium dihydrogen phosphate.

The sowing was done by the application of 1.5 g seed to each pot and we made four cuttings in the ryegrass. Cuttings were made when ryegrass was approximately 30 cm high. Forage production was evaluated through the weight obtained after drying each cuttings in a stove at a temperature of 50 °C. After sowing, the pots were put in the open air during the day and in a greenhouse during the night

During the experiment the soil moisture was maintained at 80% of field capacity and wastewater or deionized water irrigation was applied. The position of the pots was changed whenever they were watered in order to offer all of them the same conditions.

After the last cuttings of ryegrass the soil of each pot was taken out, homogeneized and a soil sample was taken for laboratory analysis.

The statistical analysis of data from the different treatments below was made.

Table 2. Wastewater analytical results

parameter	20-11-91	11-12-91	27-12-91	15-1-92	18-1-92	$\bar{x}$	s	min.	max.
TS ($g\ l^{-1}$)	0.60	0.75	0.66	0.55	0.75	0.66	0.09	0.55	0.75
TSV ($g\ l^{-1}$)	0.29	0.08	0.17	0.14	0.30	0.19	0.10	0.08	0.30
TSS ($g\ l^{-1}$)	0.08	0.02	0.03	0.04	0.03	0.04	0.23	0.02	0.08
COD ($mg\ l^{-1}$)	205.90	142.16	125.19	157.44	133.12	152.76	32.03	125.19	205.90
BOD_5 ($mg\ l^{-1}$)	39.05	vest.	vest.	vest.	vest.	39.05	8.18	20.30	43.00
EC_w ($mS\ cm^{-1}$)	0.82	1.00	0.93	0.69	1.01	0.89	0.14	0.69	1.01
pH	7.8	7.6	7.6	7.8	7.9	7.7	0.12	7.6	7.9
N-N_K (N, $mg\ l^{-1}$)	34.50	32.30	30.10	20.30	43.00	32.04	8.18	20.30	43.00
N-NH_4^+ (N, $mg\ l^{-1}$)	–	0.27	26.90	19.70	38.00	27.90	6.54	19.70	38.00
N-NO_3^- (N, $mg\ l^{-1}$)	3.50	0.50	2.10	2.10	33.60	8.36	14.15	0.50	33.60
Cl^- ($mg\ l^{-1}$)	149.10	142.00	135.00	92.00	134.90	130.60	22.36	92.00	149.10
HCO_3^- ($meq\ l^{-1}$)	6.24	7.70	7.30	5.40	7.18	6.76	0.93	5.40	7.70
P_2O_5 ($mg\ l^{-1}$)	4.15	7.25	5.15	1.03	16.18	6.75	5.73	1.03	16.18
K ($mg\ l^{-1}$)	24.50	112.50	40.04	37.50	12.50	45.41	39.09	12.50	112.50
Ca ($mg\ l^{-1}$)	49.00	77.50	102.50	80.00	105.00	82.80	22.69	49.00	105.00
Na ($mg\ l^{-1}$)	103.50	13.75	92.50	50.00	180.00	87.95	62.61	13.75	180.00
Mg ($mg\ l^{-1}$)	6.25	10.00	10.00	5.00	10.00	8.25	2.44	5.00	10.00
Fe ($mg\ l^{-1}$)	4.60	2.53	3.50	2.00	4.00	3.81	1.06	2.00	4.60
Cu ($mg\ l^{-1}$)	–	0.45	0.50	0.50	0.50	0.49	0.10	0.45	0.50
Zn ($mg\ l^{-1}$)	2.05	0.8	30.50	vest.	vest.	0.68	0.85	vest.	2.05
Mn ($mg\ l^{-1}$)	0.10	0.43	vest.	vest.	vest.	0.11	0.19	vest.	0.43
B ($mg\ l^{-1}$)	0.37	1.17	1.45	1.23	2.03	1.25	0.60	0.37	2.03

- Treatments with wastewater or deionized water irrigation, with fertilization and with or without liming (D + FC0; D + FC1; R + FC0; R + FC1).
- Treatments irrigated with wastewater with or without fertilization and with or without liming (R + FC0; R + FC1; RC0; RC1).
- Treatments without fertilization and irrigated with wastewater with or without dilution and with or without liming (RC0; RC1; 50% RC0; 50% RC1).

The first approach allowed us to evaluate the relevance of wastewater use in irrigation. The second one showed the fertilizing effect of wastewater and, finally, the third indicated the advantages of wastewater dilution. Simultaneously the effect of wastewater on soil pH value was studied either by applying lime or not applying it.

Statistical data analysis

The experimental design was completely randomized. The data analyses were performed in the statistical package Statgraphics 5.0, by the analysis of variance using the linear fixed additive model, incomplete and bifactorial. The multiple mean comparisions were made by the Tukey test, with a probability of type I error of 50% ($p \leq 0.05$).

Results and discussion

Forage yield

The treatment R + F on ryegrass showed higher *yields* than those where deionized water was applied (D + F) (Fig. 1); the differences among these treatments were significant in the second and third cuttings ($p \leq 0.001$). On the other hand, the forage yields decreased significantly ($p \leq 0.001$) in the second, third and fourth cuttings when wastewater without fertilization (NPK) was used (R) (Fig. 1).

The wastewater irrigation showed some advantages to the forage yields, mainly because it is a rich nutrient source, having the same effects as continuous fertirrigation with a diluted concentration of nutrients, according to what Neilsen *et al.* (1989) reported.

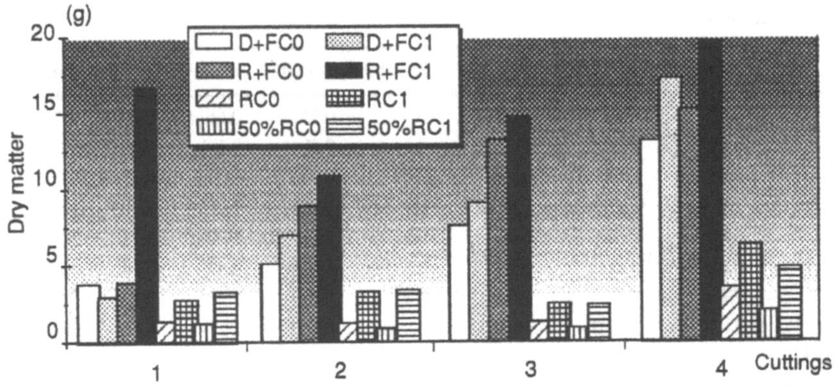

Fig. 1. Dry matter yields of ryegrass in the four cuttings.

Where fertilizers were not applied, forage yields decreased significantly which led us to draw the conclusion that this wastewater did not have enough nutrients to replace the application of fertilizers.

As it was previously mentioned, in this particular soil, poor in phosphorus and in organic matter, and consequently in nitrogen too, the use of this wastewater in irrigation did not replace the use of fertilizers (NPK) because it lacked the required nutrients as it was observed in the decrease of forage yield which underwent this treatment. Therefore, it seems that the main forage yield limiting factors are nitrogen and phosphorus elements in the soil.

Wastewater dilution led to a decrease in the forage yield. This decrease was significant ($p \leq 0.001$) at the fourth cuttings (Fig. 1) and it was due to a low level of nutrients in the wastewater and in the soil.

The plants of these last treatments (irrigated with wastewater or diluted wastewater both of them without NPK fertilizers) had low forage yields and had thin leaves with a yellowish colour which are symptoms of lack of nutrients, mainly nitrogen.

Liming had a significant increase in forage yield and the highest yields were obtained in the treatment where lime, NPK fertilizers in the soil and wastewater in irrigation were applied. Liming determined the best forage yields because it improved the mineralization of organic matter and nitrification, and therefore more nutrients were available and it increased the calcium level in the soil.

As far as the analysed micronutrients are concerned, we can say that although their levels in the soil were not significantly affected, the practice of liming induced a low absorption of micronutrients by plants.

Impact of the use of wastewater in the soil characteristics

The irrigation with wastewater did not significantly affect the organic matter level and pH value in the soil (Fig. 2) as it was observed at the end of the experiment. In spite of adding a small amount of organic matter to the soil through irrigation with wastewater, this was not enough to change its level. On the other hand, mineralization of organic matter in soil probably increased in the presence of micro-organisms and nutrients in wastewater. In those treatments where only wastewater was used in irrigation, the addition of NPK fertilizers to the soil did not significantly interfere with the organic matter level, but it decreased significantly ($p \leq 0.05$) the pH value (Fig. 2), which was due to the acidifying effect of amonium nitrate.

In contrast the dilution of wastewater did not lead to a significant change in pH value although the organic matter level decreased significantly ($p \leq 0.01$) (Fig. 2).

Therefore, we can draw the conclusion that wastewater benefits the organic matter level in the soil, but its effect is not reached because, at the same time, there is an increase in the microorganisms activities.

The irrigation with wastewater does not avoid liming. However when we added it without liming a slight increase in pH value was observed in comparison with the initial pH value in the soil.

The electrical conductivity in soil was significantly risen ($p \leq 0.01$) with the wastewater irrigation, suggesting an ionic enrichment in the soil solution (Fig. 3). The wastewater dilution did not significantly change this parameter, although it tended to decrease. Proba-

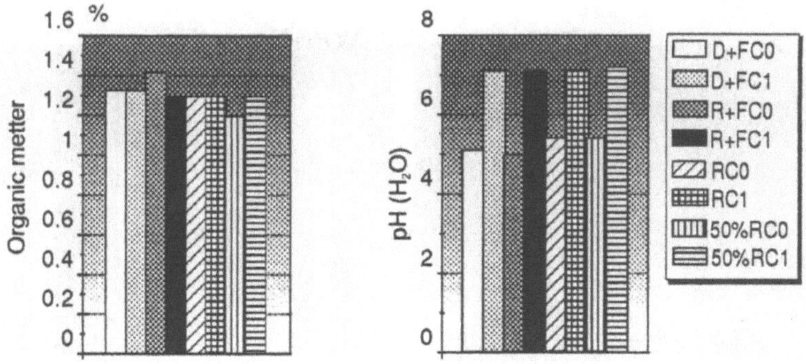

Fig. 2. Organic matter level in the soil (%) an pH (H_2O) values in different treatments.

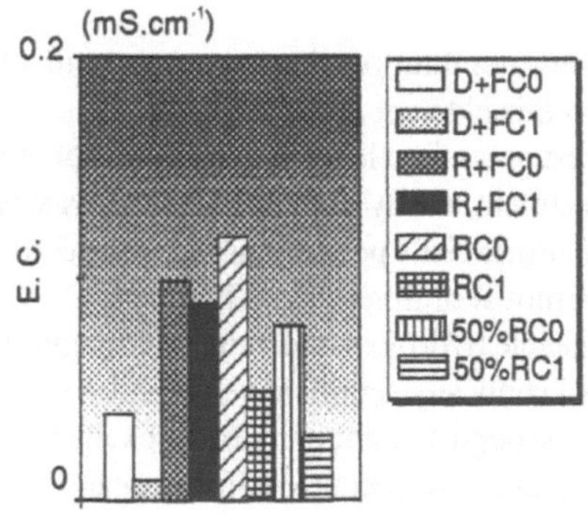

Fig. 3. EC value in the soil.

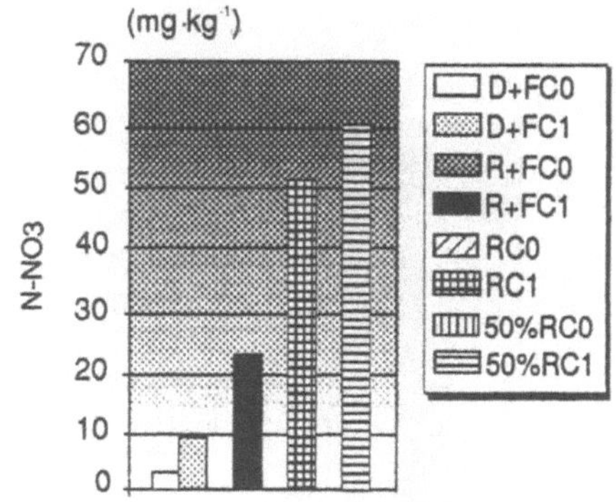

Fig. 5. N-NO_3 level in the soil.

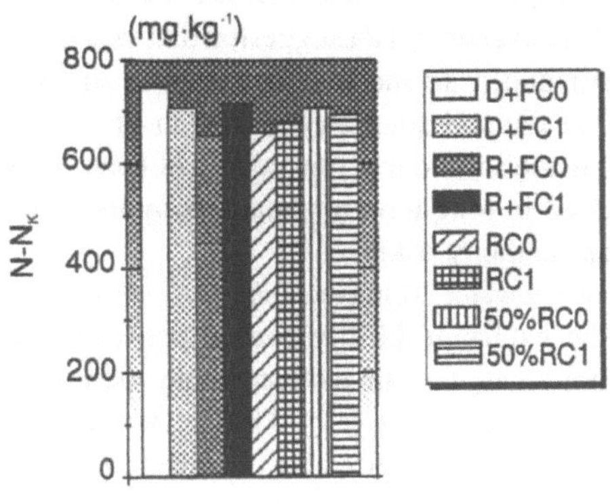

Fig. 4. N-N_k level in the soil.

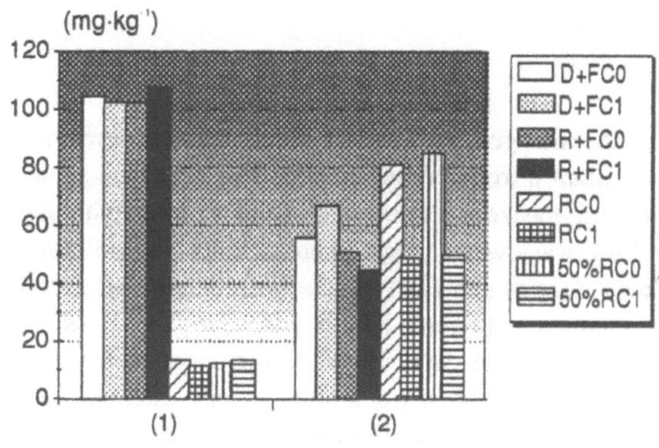

Fig. 6. Available $P^{(1)}$ and $K^{(2)}$ levels in the soil.

bly this short term experiment did not allow observation of the electrical conductivity reduction in the soil with the use of wastewater dilution. The final electrical conductivity value was not high enough to cause short time salinity risks, while in long term applications this parameter should be monitorized.

When wastewater irrigation was used, Kjeldahl nitrogen level was significantly lower than in those treatments irrigated with deionized water. These results pointed to intense microorganism activities which led to a greater nitrogen mineralization/nitrification as well as a greater nitrogen absorption by plants. In contrast, the wastewater dilution did not interfere with the kjeldahl nitrogen in the soil (Fig. 4).

The wastewater irrigation induced an significant increase ($p \leq 0.01$) in the nitrate level in the soil, caused by the nitrogen level in the wastewater and

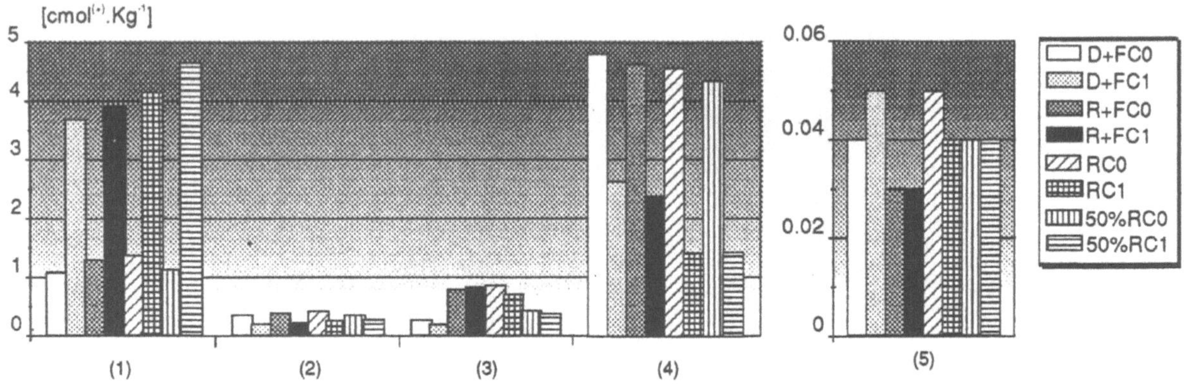

Fig. 7. Exchangeable Ca(1), Mg(2), Na(3), Hydrogen(4) and K(5) level in the soil.

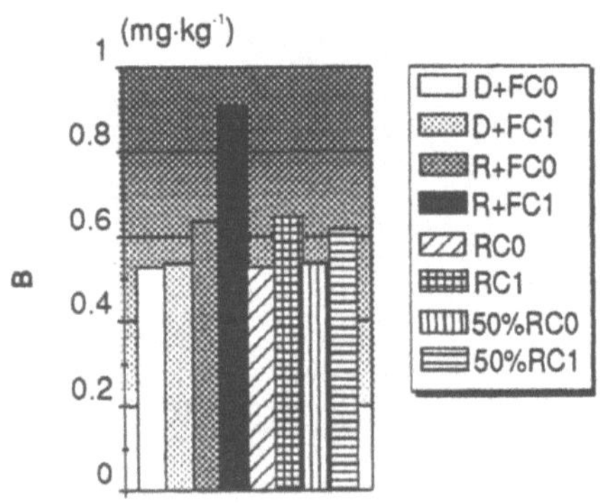

Fig. 8. Boron level in the soil.

by the nitrogen fertilization which were excessive for the plant nitrogen requirements (Fig. 5). Once again, we emphasize the relevance of the accurate planning of the wastewater irrigated areas, since soil pollution with nitrates can easily occur with disastrous effects on human and animal health.

This increase in the level of nitrates in the soil was more evident when we simultaneously used wastewater irrigation and liming (R + FC1). The wastewater dilution did not affect this parameter.

As far as the use of wastewater or deionized water in irrigation are concerned, the phosphorus level in the soil did not show any significative differences. However, the removal of this nutrient was superior in the wastewater irrigated plants. As expected, phosphorus fertilization increased its level in the soil, suggesting that this wastewater did not replace this kind of fertilization. On the other hand, the unfertilized plants showed a low phosphorus level, suggesting that this element could be considered as limiting forage yield.

Finally, wastewater dilution did not cause significative variations on this element in the soil (Fig. 6).

The lowest potassium level in the soil was obtained in R + F treatment, easily explained by the wastewater and soil insufficient potassium supply and to the increasing removal of this element by plants.

Exchangeable hydrogen, calcium, and magnesium levels did not show any significative variation in R + F treatments, although a slight increase in exchangeable calcium and magnesium occurred. In this treatment, we could still observe a significant increase in exchangeable sodium level ($p \leq 0.001$), while the exchangeable potassium (Fig. 7) level had a significative decrease ($p \leq 0.01$). The available data suggested that wastewater irrigation induced an increase not only in the sodium level at the soil solution but also in the sodium adsorbed in the soil. The consequences of these would be expressed in long term applications by the rise of salinity and sodicity risks.

The wastewater dilution did not significantly change the exchangeable hydrogen, potassium and magnesium levels but the exchangeable sodium level decreased significantly ($p \leq 0.001$). Therefore, the wastewater dilution was effective in preventing salinity and sodicity risks.

The levels of the cationics micronutrients in the soil (copper, iron, zinc and manganese) were not significantly affected by the use of wastewater and wastewater dilution. This fact occurred due to the low level of micronutrients in the wastewater and to the short term of the experiments. At long term applications of wastewater scrutinizing the introduction of these elements should be considered in order to prevent pollution risks.

The boron level in the soil was significantly risen ($p \leq 0.001$) when wastewater was applied (Fig. 8). When we used simultaneously wastewater irrigation and liming in the soil, this increase was more evident and benefited deficient boron soils. As the range between phytoxicity and defficiency levels are tight the use of this kind of water should be correctly managed to avoid phytotoxicity risks.

Conclusions

Up to now the available data have suggested that the use of wastewater in irrigation is safe, because it supplies plants both with nutrients and water. However, its biological and chemical characteristics may have a negative effect on crops, on soil, on ground water, on farm workers and on consumers. Therefore, a correct management should be carefully applied in the wastewater irrigated areas.

References

Asano T, Richard D, Crites RW and Tchobanoglous G (1992) Evolution of tertiary treatment requirements in California. Water Environ & Techn 4(2): 36–41.

Chopp KM, Clapp CE and Schimdt EL (1982) Ammonia – oxidizing bacteria populations and activities in soils irrigated with municipal wastewater effluent. J Environ Qual 11: 221–226.

Horta-Monteiro, C. (1994) Utilização de água residual urbana na cultura de azevém (*Lolium multiflorum* Lam.) Trabalho realizado para obtenção do grau de Mestre em Nutrição Vegetal, Fertilidade do Solo e Fertilização. apresentado no Instituto Superior de Agronomia, Lisboa.

Neilsen GH, Stevenson DS, Fitzpatrick JJ and Brownlee CH (1989) Nutrition and yield of young apple trees irrigated with municipal wastewater. J Am Soc Hort Sci 114: 377–383.

WHO (1989) Health Guidelines for the Use of Wastewater in Agriculture and Aquaculture. Report of a WHO Scientific Group, World Techni 778, WHO, Geneve, Switzerland.

C. Rodriguez-Barrueco (ed.), Fertilizers and Environment, 517–521.

City refuse compost as a source of micronutrients for plants

E. Iglesias-Jiménez
Instituto de Recursos Naturales y Agrobiología de Salamanca, CSIC, Cordel de Merinas 40–52, 37008 Salamanca, Spain

Key words: Compost, organic matter, micronutrients, plant nutrition, phytotoxicity, maturity degree

Abstract

The effectiveness of city refuse compost (CRC) with a high maturity degree for supplying Mn, Cu and Zn to a plant test in comparison with inorganic sources of micronutrients (MF) was deduced in a greenhouse pot bioassay. Highly matured CRC was applied at rates equivalent to 10, 20, 30, 40 and 50 t ha^{-1} to a variable-charge soil from Tenerife, Canary Islands. $MnSO_4 \cdot H_2O$, $CuCl_2 \cdot H_2O$ and $ZnSO_4 \cdot 7H_2O$ were applied at rates equivalent to the *aqua regia*-extractable content of Mn, Cu and Zn of the CRC treatments. Perennial ryegrass (*Lolium perenne* L.) was grown in 3 kg pots and the plants were harvested at regular intervals. Evolution of micronutrient concentration and plant uptake were determined. The relative efficiency (RE) was also established: the ratio of the slopes of the regression equations which describe the relation between the quantity of a micronutrient applied as CRC or MF sources and the quantity removed by the plant.

CRC promoted a significant increase in Mn and Zn plant concentrations in relation to MF treatments, but total removal was lower. The RE values were 61% Mn, 13% Cu and 40% Zn (the RE values for N, P, K, Ca, Mg and Fe were 15%, 43%, 23%, 24%, 28% and 27% respectively). These results imply that CRC could be a suitable source of Mn and Zn for agricultural soils. However, this facility to supply Mn and Zn to the plant involves a potential phytotoxicity problem when the CRC is used at high rates and above all in strongly-acid soils.

Introduction

Land application of city refuse compost (CRC), produced by the aerobic-thermophilic digestion of municipal solid wastes, represents an attractive alternative for the disposal of these wastes, currently land-filled or incinerated. The CRC normally exerts a positive influence on crops yield and improves some physical and chemical soil properties such as porosity, aggregates stability, water-holding capacity, pH-buffering capacity, cation-exchange capacity, and releases nutrients gradually (Gallardo-Lara and Nogales, 1987; Piccolo and Mbagwu, 1990; Iglesias-Jiménez *et al.*, 1993).

Richness in plant micronutrients is one of the main characteristics of CRC and this material may therefore be an important source of micronutrients for crops. However, high rates of CRC applied to acid soils or added repeatedly may induce soil pollution and eventually phytotoxicity problems either due to an increase in the amount of certain essential micronutrients to toxic levels, mainly Zn toxicity, or to an increase in the bioavailability of non-essential heavy metals (Petruzzelli, 1989). This problem is especially important when immatured CRC is applied to agricultural soils, i.e. insufficiently stabilized as regards mineralization and 'humification' (Iglesias-Jiménez and Pérez-García, 1989).

The objective of this study was to evaluate, in a moderately acid soil, the effectiveness of a CRC with a high degree of maturity for supplying Mn, Cu and Zn to a plant test, using normal dressing rates.

Materials and methods

The experiment was conducted under a controlled greenhouse environment in pots with 3 kg of air-dried soil (Andeptic Paleudult, sand: 26%, silt: 34%, clay: 40%, pH-H_2O: 5.8, EC: 1.00 dS m^{-1}, organic C: 1.65%, CEC: 21.2 cmol(+)kg^{-1}, C/N ratio: 11.6).

The CRC was obtained in a controlled pile-composting trial from the organic fraction of the

Table 1. Total uptake of Mn, Cu and Zn (ug pot^{-1}) by the aerial part of ryegrass in soil amended with city refuse compost (CRC) and mineral fertilizers (MF).

Control	Mn 687 g	Cu 103 f	Zn 322 f
10 CRC	1119 f	150 f	568 e
20 CRC	1657 e	169 ef	688 de
30 CRC	2595 d	184 ef	860 d
40 CRC	3393 d	192 def	1119 c
50 CRC	4123 c	208 de	1159 c
10 MF	1686 f	241 d	858 d
20 MF	2959 de	427 c	1159 c
30 MF	4042 c	555 b	1729 b
40 MF	5345 b	766 a	2182 a
50 MF	6450 a	788 a	2385 a

Values within the same columns followed by the same letter do not differ significantly at $p < 0.01$ level (Duncan's Multiple Range Test).

Table 2. Correlations for Mn, Cu and Zn uptake (ug pot^{-1}) versus micronutrient applied (mg pot^{-1}) as city refuse compost (CRC) and mineral fertilizers (MF).

Element	Treatment	Equation	r	RE
Mn	MF	Y = 595.5 + 244.05 X	0.999***	61.0%
	CRC	Y = 472.5 + 148.95 X	0.994***	
Cu	MF	Y = 112.6 + 21.28 X	0.986***	13.0%
	CRC	Y = 120.0 + 2.76 X	0.952**	
Zn	MF	Y = 374.9 + 27.25 X	0.993***	40.4%
	CRC	Y = 356.2 + 11.02 X	0.989***	

RE: Relative efficiency, the ratio of the slopes of the regression equations as percent.
Level of significance: $**p < 0.01$, $***p < 0.001$

municipal solid waste of Santa Cruz de Tenerife (Canary Islands) after a bio-oxidative period of 75 days and a 90-day complementary maturity process (C/N ratio -solid phase: 9.9, C/N ratio -water-soluble phase 1:5 w:v-: 5.4, CEC: 80.1 cmol(+)kg^{-1}, HA/FA ratio: 1.9).

The experimental design was complete randomized blocks. Five compost treatments were set up, equivalent to 10, 20, 30, 40 and 50 t ha^{-1} (CRC treatments). At the same time five treatments with mineral sources of bioelements were also established (MF treatments) on the basis of the N, P, K, Ca, Mg, S, Fe, Mn, Cu and Zn content of the CRC treatments. $MnSO_4 \cdot H_2O$, $CuCl_2 \cdot H_2O$ and $ZnSO_4 \cdot 7H_2O$ were used as Mn, Cu and Zn sources. A control was also established. The total number of pots used was 55 (11 treatments × 5 replications). Perennial ryegrass (*Lolium perenne* L.) was used as plant test. All pots were harvested at 30, 60, 90, 120, 150 and 180 days after seedling emergence.

The sample mineralization was conducted by dry-ashing to a 1 g dried sample in a muffle furnace at 480 °C and ash was redissolved with 6 *N* HCl. From this solution P was determined according to the molybdivanadophosphoric acid procedure. Plant Cu, Mn, Zn, Fe, K, Ca, and Mg were determined by AAS. Plant N was determined titrimetrically following a semimicro-Kjeldahl digestion method.

Results and discussion

The CRC did not significantly affect plant Cu (Fig. 1). The concentration varied from 9 to 21 mg kg^{-1}, values which are within the normal range reported for *Lolium perenne*: 2 to 30 mg kg^{-1} (Whitehead, 1987). On the contrary, CRC promoted a significant increase in Mn and Zn concentrations in relation to the MF treatments. This increase was generally proportional to the applied rate of CRC. Besides, the concentrations with the CRC treatments were higher than those considered to be normal for *Lolium*: 40–60 mg Mn kg^{-1}, 14-20 mg Zn kg^{-1} (Benton Jones *et al.*, 1991), although they did not reach phytotoxic levels.

Total uptake of Mn, Cu and Zn was higher with MF treatments (Table 1) because the amount of biomass (dry-matter yield) was higher than with CRC treatments. Compost-N is especially the element that determines a lower growth rate compared to the MF treatments but CRC treatments were not supplemented voluntarily with N-fertilizer in order to obtain information on the suplying capacity of CRC used as the sole source of nutrients (Iglesias-Jiménez and Alvarez, 1993). Moreover, some evidence exists that the combined addition of compost and N-fertilizer may reduce the concentration of micronutrients in the plant (Gallardo-Lara and Nogales, 1987).

The RE (relative efficiency) value (Table 2) for Mn (61%) and Zn (40%) were higher than that of N (15%), K (23%), Ca (24%), Mg (28%), Fe (27%), and Cu (13%). (RE value for P was 43%). These higher percentages imply a relatively high capacity of the CRC for supplying Mn and Zn to plants. The RE value of Cu was lower than that of the other bioelements, and even lower than N. This result suggests that (i) a large

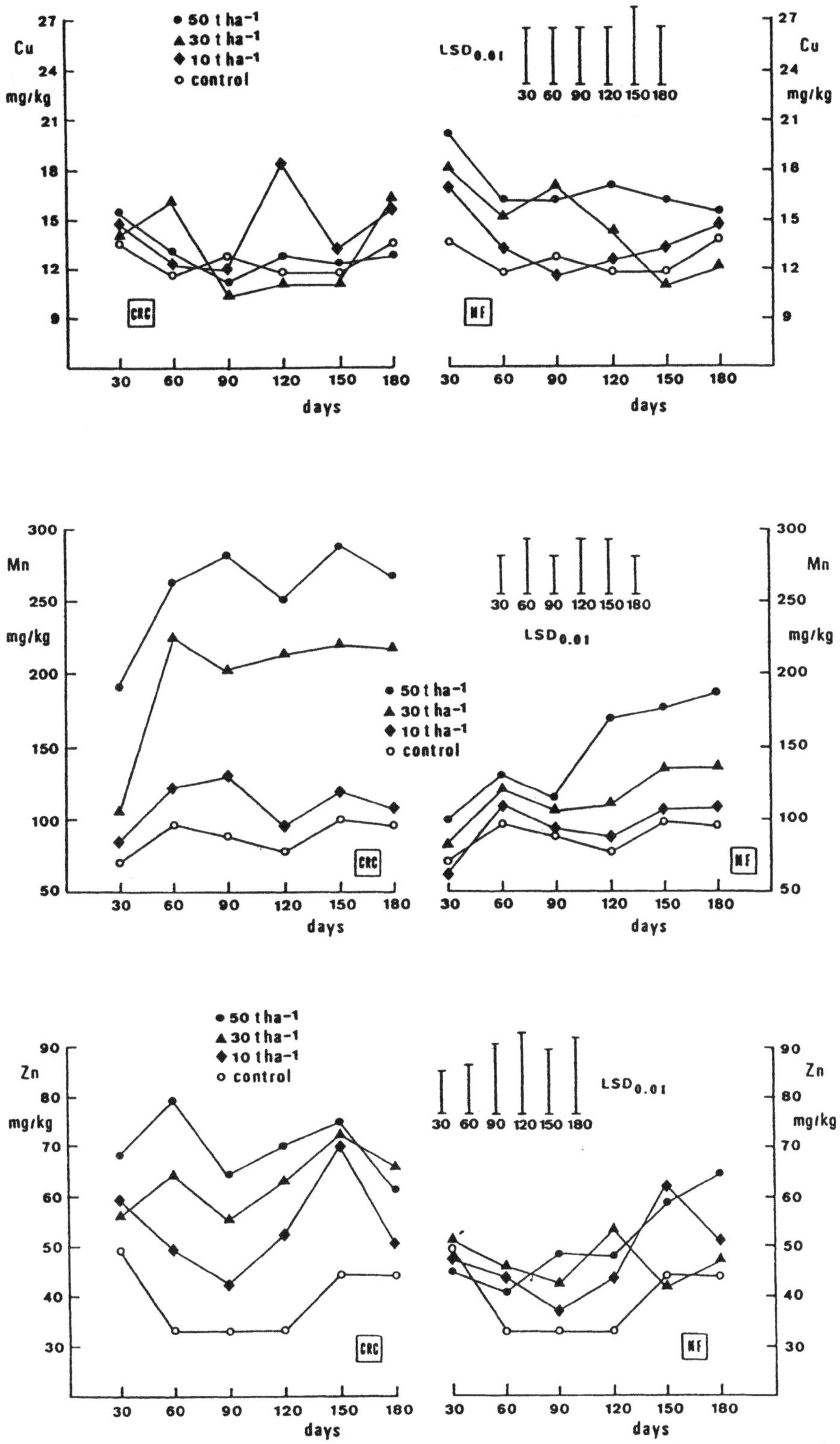

Fig. 1. Changes in Cu, Mn and Zn concentrations with time in the aerial part of ryegrass for city refuse compost (CRC) and mineral sources of bioelements (MF).

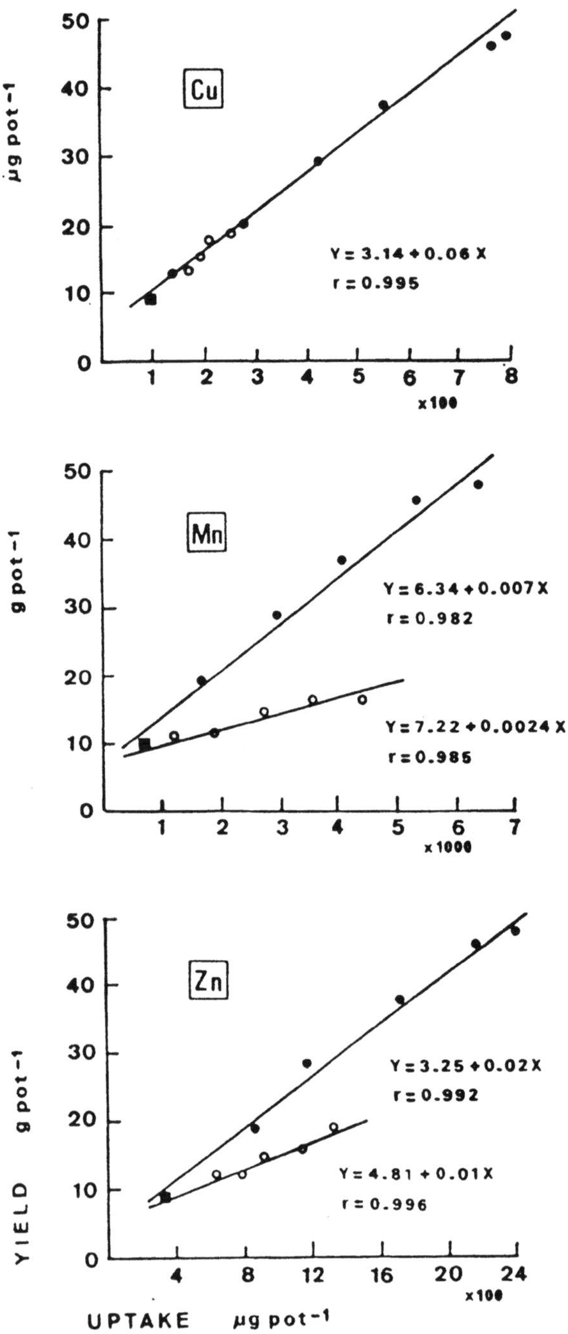

Fig. 2. Relationships between dry-matter yield and micronutrients removal for city refuse compost (o) and mineral sources of bioelements (•).

part of the Cu remains bound in the root with CRC treatments (smaller translocation degree); (ii) Cu is incorporated with the highly matured CRC in the form of primarily insoluble Cu-'humic' complexes and thus Cu is poorly taken up; (iii) available-Cu from CRC may be immobilized partially by the soil humified organic matter, owing to the formation of secondarily insoluble metal-humic complexes (Stevenson, 1982). The high affinity existing between Cu and humic acid functional groups is a well-known phenomenon. However, the behaviour of Cu in compost-amended soils is contradictory because other authors reported significant increases in soil-available Cu and plant uptake (Purves and Mackenzie, 1973; Chu and Wong, 1987). This finding may be attributed to the high available-Cu content of some composts, as a result of an insufficient degree of maturity (Chu and Wong, 1987).

Mn and Zn removal per unit of dry-matter weight was greater with CRC source than with MF source (Fig. 2). For Cu was the same under our experimental conditions, as well as for N, K, Ca, Mg and Fe (for P was slightly lower). The ratio of the slopes of the regression lines was 34% for Mn and 50% for Zn. Thus, ryegrass removed overall three times more Mn and twice more Zn per unit of dry-matter weight from CRC treatments than from inorganic sources, indicating the high capacity of CRC to supply these elements to plants in a short-time period.

Acknowledgements

Special thanks to Pauline Agnew (IPNA-CSIC, Tenerife, Canary Islands) for translating the original manuscript into English.

References

Benton Jones J, Wolf B and Mills HA (1991) Plant Analysis Handbook. Micro-Macro Publishing Inc. Athens, Georgia, USA.

Chu LM and Wong MH (1987) Heavy metal content of vegetable crops treated with refuse compost and sewage sludge. Plant Soil 103: 191–197.

Gallardo-Lara F and Nogales R (1987) Effect of the application of town refuse compost on the soil-plant system: A review. Biol. Wastes 19: 35–62.

Iglesias-Jiménez E and Pérez-García V (1989) Evaluation of city refuse compost maturity. A review. Biol Wastes 27: 115–142.

Iglesias-Jiménez E and Alvarez CE (1993) Apparent availability of nitrogen in composted municipal refuse. Biol Fertil Soils 16: 313–318.

Iglesias-Jiménez E, Pérez-García V, Espino M and Hernández Moreno JM (1993) City refuse compost as a phosphorus source to overcome the P-fixation capacity of sesquioxide-rich soils. Plant Soil 148: 115–127.

Petruzzelli G (1989) Recycling wastes in agriculture: heavy metal bioavailability. Agric Ecosystems Environ 27: 493–503.

Piccolo A and Mbagwu JS (1990) Effects of different organic waste amendments on soil microaggregates stability and molecular sizes of humic substances. Plant Soil 123: 27–37.

Purves D and Mackenzie E (1973) Effects of applications of municipal compost on uptake of copper, zinc and boron by garden vegetables. Plant Soil 39: 361–371.

Stevenson FJ (1982) Humus Chemistry. John Wiley & Sons Inc, New York/Chichester/Brisbane/Toronto/Singapore.

Whitehead DC (1987) Some soil-plant and root-shoot relationships of copper, zinc and manganese in white clover and perennial ryegrass. Plant Soil 97: 47–56.

C. Rodriguez-Barrueco (ed.), Fertilizers and Environment, 523–526.

Agrochemical and technological characteristic features of sewage sludges applications in Russia

V. Kasatikov
All-Russia Research Institute of Organic Fertilizers and Peat, Vladimir region 601242, Russia

Key words: Sewage sludges, heavy metals, fertilizers, rates of application

Abstract

From 4 mln tons (in terms of dry matter) of sewage sludges (SS) produced in Russia annually, only 15 to 20 percent are recovered. Stockpiling SS at sewage processing plants bears the risk of contaminating the biosphere of towns and their suburbs. The suggestion is made to use most of the SS produced as fertilizers for commercial and feed crops. The method of classifying SS in accordance with their applicability for agriculture is suggested. Requirements to agroecological properties of SS have been worked out.

Introduction

One of the by-products of the antropogenic activities are SS. Physical, physicochemical and biological properties of SS produced in Russia in quantities of 4 mln tons per year differ in wide range depending on processing method, type of a production process and the ratio of municipal and industrial waste in municipal sewage systems. Now several methods are used for SS recovery: using as fertilizers, disposal and burning. The first mentioned recovery method is the most economically beneficial. From aforesaid it follows that laboratory and field experiments, with the aim of evaluating agrochemical effect of SS on soil and plants, is the major way for determining their applicability as fertilizers.

Methods

Soddy-podzolic and chernozem soils with different degrees of fertility and geochemical backgrounds as well as sludges from different industries were subject of the present study. For agrochemical evaluation of soils from collective farms prior to SS application, soil samples were taken from plots with the area from 6 to 8 ha, following the routine practice of sampling for making up agrochemical maps. Sheaves of grain and grass crops were used as plant samples. Field experiments were carried on 3 to 6 and 50 to 60 m^2 plots with 4 to 6 tire. Organic and organic-mineral SS based fertilizer systems were evaluated comparatively to mineral based ones.

Results

SS produced in towns with 100 thousand population or more have very complex macro- and microelement composition and as a result of this rather large quantities of organic matter, nitrogen and ashy products are stockpiled and excluded from exchange processes. According to the method proposed by the authors all SS are divided into types and kinds depending on the method they were produced. The major SS types in Russia are thermophylically- and mezophylically fermented and oxibiotically stabilized ones. From the three types mentioned above oxibiotically stabilized prevail ptrevail because of the high energy consumption necessary in case of production of the fist two types. SS of each type are divided into two kinds: floculated and non-floculated. Agrochemical characteristics of SS from several Russian towns are given in Tables 1 and 2. Owing to the presence of organic matter and because of high N, P and other microelements content in SS, they are competitive to mineral fertilizers, provided heavy metal concentration does not exceed standard levels. In the circumstances of the present economic situation in Russia, SS becomes more attractive for agricultural application. The major

Table 1. Agrochemical characteristics of SS from some towns in Russia

Characteristics	Novosibirsk	Moscow	Stcholkovo
Moisture content, %	40–65**	50–60	60–70
	40–55***	40–60	15–30
Ash content, %	43–63	65–75	65–63
	58–77	50–60	50–70
N*, %	1.3–2.7	1.8–2.0	2.0–2.3
	0.8–1.4	0.8–2.0	1.5–2.0
P_2O_5*, %	2.9–6.0	1.4–3.0	3.5–4.1
	1.1–2.8	1.4–2.3	2.7–2.9
K_2O*, %	0.5–1.0	0.5–0.7	0.4–0.6
	0.2–0.7	0.1–0.2	0.1–0.2
pH_{KCl}	6.7–7.6	6.5–6.8	6.9–7.9
	8.0–12.3	11.9–12.5	7.8–8.5

* in terms of dry matter.
** - non-floculated.
*** - floculated.

Table 2. Geochemical characteristics of SS from some towns in Russia

Element	Novosibirsk	Moscow	Stcholkovo
B	30*/13**	73/40	35/14
V	64/48	33/25	70/13
Wi	26/14	60/12	52/17
Cd	31/11	91/13	85/40
Co	15/6	12/3	42/15
Mn	365/213	418/497	850/1100
Cu	1863/427	1600/230	2100/720
Mo	15.7/3.5	18/1.2	6/2
Ni	293//85	365/58	840/130
Sn	293/121	120/12	200/43
Pb	291/142	611/73	127/66
Ag	27/5.5	60/10	200/130
Cr^{3+}	1625/427	2800/320	700/170
Zn	3462/355	4000/1400	2900/680

* – non-floculated.
**–floculated.

limiting factor for the wider use of SS as fertilizers is their possible toxic effect on field agrocenosis.

The critical factors determining wider use of SS shown in Table 2 are decrease of content of such metals as Cd, Sn, Cr, Zn, Cu and Ni and change to the production of floculated sludges. Agrochemical studies of soddy-podzolic soils showed high intensity of metals migration in the sludge-soil-plant system. At the same time, in bleached chernozem soils with endo-humus content from 6 to 8 percent, plants receive less amount of metals as the result of their absorption by the soil. In case of chernozem soils, metal content in grain and feed crops with higher yielding ability increases to a less degree due to the so called "growth dilution" effect. By and large, in both cases (at a specific metal content in SS and natural background) organic and SS-based organic-mineral fertilizer systems, applied at 20 to 30 t/ha in terms of dry matter, and a mineral system made up the following decreasing sequence as for the metal content increase in field agrocenosis: organic-mineral > organic > mineral system. In cases where floculated sludges were used for a single soil dressing, their effect on microelement content in plants was equal to that of mineral fertilizers.

Based on the results of experiments carried out in different regions of Russia the agrochemical and technological concepts of SS usage as fertilizers were worked out. In accordance to them, SS and composts based thereon should be used for the crops giving the highest yield gain at minimal absorption of heavy metals, arsenic, and benz(a)perene. With respect to above mentioned elements content, SS were divided into six main groups (Table 3). The application technology for specific SS types use for seed production, gardening and flower growing and in nurseries was suggested proceeding from their belonging to particular groups. When using SS for dressing soils with $pH_{KCl} < 6.5$, chaptalization with 1.5 to 2 unit doses is necessary prior to that. It leads to decreased availability of heavy

Table 3. Crops, groups and threshold limits of heavy metal content in SS, mg/kg in terms of dry matter Threshold limits

Characteristics	Group 1	Group 2	Group 3	Group 4		Group 5	Group 6
growing, Characteristics	Grain, feed, industrial crop, per year	Grain, feed, industrial crop, per year	Grain, feed, industrial crop, 5–10t/ha once in 3–5 years	Grain, feed, industrial crop, posting	Flower growing, landscape gardening For composting	Reclamation of industrial disposal areas	Other reclamation methods
Sn	< 75	75–100	100–200	200–300	300–400	400–500	> 500
Pb	< 200	200–500	500–800	800–1000	1000–12000	1200–1500	> 1500
Co	< 40	40–60	60–80	80–100	100–150	150–200	> 200
Hg	< 4	4–6	6–8	8–10	10–15	15–20	> 20
Cd	< 5	5–10	10–30	30–60	60–100	100–150	> 150
Ni	< 100	100–150	150–300	300–400	400–500	500–600	> 600
Cr^{3+}	< 400	400–600	600–1200	1200–2000	2000–3000	3000–4000	> 4000
Mn	< 500	500–1000	1000–1500	1500–2000	2000–3000	3000–4000	> 4000
Zn	< 800	800–2000	2000–4000	4000–5000	5000–6000	6000–7000	> 7000
Cu	< 250	250–500	500–1500	1500–3000	3000–4000	4000–5000	> 5000
As	< 10	10–15	15–20	20–30	30–50	50–60	> 60
Benz(a)perene	< 5	2–8	2–8	2–10	10–12	10–15	> 15

metals, except Cr^{3+}, to plants and intensity of their migration in the soil. Chaptalization with 1 unit dose decreased Cd, Cu, Ni and Zn concentration in barley grain correspondingly by 60, 22, 61 and 49 percent compared to SS background concentration of 60 t/ha in terms of dry matter. Zc coefficient values were decreased by 2 times.

SS of group 1 and 2 may be used for soils of types and kinds available in Russia, those belonging to groups 3 to 5 – for medium- and heavy loamy soils formed under conditions similar to those for soddy-podzolic, for gray forest and chernozem soils littered from the depth of 1 m or less with clay and loamy soils. SS from groups 4 and 5 may be used only as composts with SS to organic filler ratio from 1:1 to 1:(1–3), provided there is close control of metal content after composting.

Maximum rate of a single SS application, irrespective of toxic substances concentration, was calculated from their bulk content, and monitoring of the rate of SS application was carried out on the basis of Cd, Cu, Zn, Ni, Cr^{3+} and Co mobile forms content in SS. Calculation of tolerable rates of SS application was based on the proposition that after dressing cumulative concentration of toxic substances in soils, with respect to their scattering in the plow horizon, should not exceed threshold limits (TSL):

$$F + D \leq TSL,$$

where: F – toxic substances content before SS application, D – additional amounts of toxic substances introduced with SS into the plow horizon.

Discussion

Research on the SS-soil-plant agroecological system confirmed universal character of the method proposed for determining rates of SS application for crops. In this case, contrary to other methods (Process., 1983; Elliot, 1986; Maclean and Dekker, 1978), such characteristic as nitrogen content, rate of cation exchange and TSL values for heavy metal content, calculated from their zinc equivalents, are not used for SS dosage monitoring for the reason that, compared to traditional organic fertilizers, SS have lower content of nitrogen available to plants, when mobile element forms (ammonia-acetate extract, pH = 4.8) concentration is used for monitoring rates of SS application. But if anhydrous ammonia or 25 percent water solution of ammonia is used for SS decontamination, maximum – tolerable rate of SS application should be confined to annual tolerable addition of nitrogen, which is calculated with allowance made for N-NO_3 TSL values.

References

Process Design Manual for Land Application of Municipal sludge (1983) US Environmental Protection Agency 200.
Elliot HA (1986) Journal of Soil and Water Conservation 41: 5–10.
Maclean A and Dekker A (1978) Canad J Soil Sci 58: 381–388.

C. Rodriguez-Barrueco (ed.), Fertilizers and Environment, 527–531.

Residual effects of sugar beet vinasse on plant growth

P. Martín-Olmedo, F. Cabrera, R. Lopez & J.M. Murillo
Instituto de Recursos Naturales y Agrobiología de Sevilla (CSIC), Apartado 1052, 4180 Sevilla, Spain

Key words: Organic wastes, vinasse, plant nutrition, micronutrients assimilability, tall fescue

Abstract

Residual effect of a concentrated and depotassified beet vinasse on growth and mineral composition of tall fescue (*Festuca arundinacea*) was studied. Results were compared with those obtained for a mineral fertilizer treatment and a control. Vinasse and inorganic fertilizer had been previously applied to two different soils at two rates for five years. High doses of vinasse produced the highest yield and the highest concentrations of P and K in tissue plant, while decreasing Na, Mn, and Zn assimilability.

Introduction

Beet vinasses are effluents from the alcoholic fermentation of sugar beet molasses and their later ethanol destillation. Vinasse causes considerable pollution problems in rivers and lands near alcohol factories because of its high organic matter and total salt contents (Algur and Kadioglu, 1992; Cabrera *et al.*, 1987).

Land application of wastewater is recognized as an important alternative disposal method, both in recycling nutrients essential for plant growth, and in increasing soil fertility (Ward, 1990).

Agronomical use of vinasses is being studied at present for several authors (Algur and Kadioglu, 1992; López *et al.*, 1990, 1992; Murillo *et al.*, 1993). In all cases it was evaluated the possible consequences, on different aspects of the soil-plant system, after a direct application of vinasse. Information on the residual effect of vinasse over a long period of time is lacking, and we iniciated these studies to obtain this information.

Methods

From 1989 to 1993, a concentrated and depotassified beet vinasse (density 1.3; pH 5; dry matter 54%; total organic matter 40%; N 3.3%; P 0.02%; K 3.5%) was applied to greenhouse containers (ca. 0.42 m^2, 50 cm depth) filled with two topsoils (Table 1): a Xerorthent (loam-clay-sandy soil, S1) and a Xeropsamment (sandy soil, S2).

During this period, five fertilization treatments in a completely randomized design with five replicates per treatment, were assayed: two rates of beet vinasse (V1 and V2); two rates of a mineral fertilizer (MF1 and MF2 supplying the same amounts of N and P as the vinasse treatments); and a control (C), without fertilization. A summary of fertilization during the period 1989–1993 is described elsewhere (López *et al.*, 1990). Ryegrass (*Lolium multiflorum*) was cropped annually and the yield and nutrient contents of tissue plants determined.

In 1994, (study reported here), the residual effect was evaluated and, thus, no organic or mineral fertilizer was applied. Tall fescue (*Festuca arundinacea* cv. Manade) was grown in the same containers.

At 76, 129 and 172 days after sowing, tall fescue was clipped to 3 cm height, weighed and analysed. Plant samples were washed with tap and deionized water, oven dried at 70 °C for 48 h, and ground to pass through a 40 mesh screen. Nitrogen was determined by Kjeldahl digestion. Mineral elements were determined according to Jones *et al.* (1991).

From dried weight and nutritional mineral content in tissue plants, nutrient extraction at each clipping and the total nutrient extraction were determined.

The data were subjected to analysis of variance and the mean separation performed by the Tuckey test. A

Table 1. Soils Characteristics

Soil	Treat.	N	P	K	O.M.	pH	E.C.
		(mg kg^{-1})			(%)		(1:5)
		Initial conditions					
S1	–	536	2.5	170	0.53	7.90	–
S2	–	347	3	47	0.14	8.45	–
		After five years of treatment					
	C	689	3	91	1.29	7.29	0.80
	V1	772	8	91	2.07	7.69	0.48
S1	V2	746	20	125	1.24	7.49	0.59
	MF1	674	9	108	1.52	7.68	0.64
	MF2	746	14	91	1.47	7.78	0.43
	C	646	5	75	0.81	8.04	0.14
	V1	524	8	50	0.62	7.93	0.24
S2	V2	786	33	66	0.93	7.96	0.20
	MF1	495	7	50	0.53	8.07	0.20
	MF2	480	11	42	0.84	7.91	0.23

Table 2. Effects of different treatments and soil type on elemental composition of tall fescue at selected clipping

Clip	Treat.	N	P	K	Na	Mn	Zn
		%				mg kg^{-1}	
		Soil S1					
	C	3.63 a	0.15 a	1.83 ab	1.27 a	97 a	74 a
	V1	3.67 a	0.26 b	1.80 a	1.38 a	69 a	56 a
1	V2	2.75 a	0.27 b	2.30 b	1.11 a	70 a	48 a
	MF1	3.23 a	0.24 b	1.44 a	1.47 a	88 a	64a
	MF2	3.56 a	0.25 b	1.48a	1.40 a	99 a	48 a
		Soil S2					
	C	3.30 b	0.11 a	2.06 a	0.37 a	112 b	61 b
	V1	2.22 a	0.18 b	1.81 a	0.51 ab	67 a	30 a
2	V2	2.14 a	0.25 c	1.98 a	0.43 a	61 a	30a
	MF1	2.35 a	0.16 ab	1.79 a	0.58 b	83 a	39 a
	MF2	2.28 a	0.17 ab	1.73 a	0.48 ab	77 a	30 a

significance level of $p < 0.05$ was considered throughout the study.

Multivariant Discriminant Analysis of total nutrient extraction data was carried out by the Statgraphics computer program (Statgraphics, 1989).

Results and discussion

Yields

Yield of tall fescue was higher in soil S1 (loam-clay-sandy soil) than in soil S2: mean grass weight averaged over all treatments and clippings was 1427 kg ha^{-1} for

Table 3. Standardized discriminant coefficients.

	Soil S1		Soil S2	
Variables	F1	F2	F1	F2
N	1.17674	0.19223	−1.87844	0.74230
P	2.31942	0.28398	2.69114	0.89905
K	−0.71408	−1.59387	1.80433	−0.37005
Na	1.90686	0.53361	0.01545	−0.54487
Ca	−1.03744	0.96591	−3.93211	0.03535
Mg	−0.67652	−0.19034	−0.14766	0.41729
Fe	−0.54362	−1.10567	−0.44199	−0.85139
Cu	−0.88883	−0.01157	0.54742	−0.18079
Mn	0.53283	0.67540	0.08617	0.25144
Zn	−0.96473	0.09336	1.05669	0.38216
Variance (%)	88.48	10.03	48.69	42.91

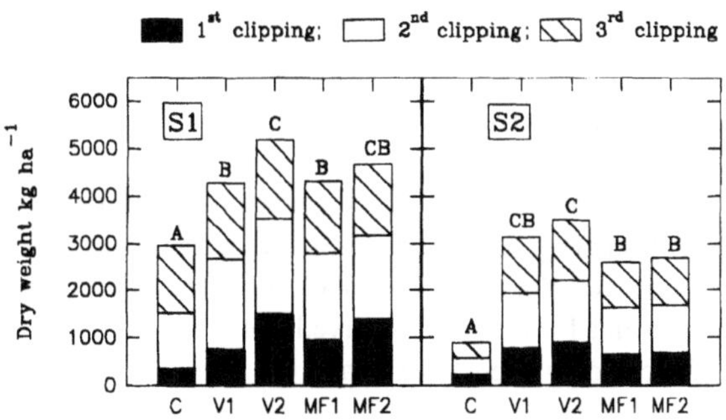

Fig. 1. Dry weight of tall fescue biomass for each treatment in soil S1 and S2.

S1 and 854 kg ha^{-1} for S2. Earlier laboratory incubation studies showed a much higher N release from the mineralization of the residual organic N in soil S1 than in soil S2 after three years of treatment with V, MF and C (Martín-Olmedo *et al.*, 1995). Furthermore, increases in tall fescue growth as the soil clay content increased were also observed by other authors (Devitt *et al.*, 1990; Tester *et al.*, 1982).

The mineralization process, slower in soil S1 than in soil S2 due to a higher protection of the organic matter and soil biomass (Martín-Olmedo *et al.*, 1995), might be responsible for greater differences in soil S2 between organic (V1 and V2) and mineral (MF1 and MF2) fertilizer treatments.

Maximum yields at each harvest clipping (Fig. 1) corresponded, in both soils, to the high vinasse rate (V2), in agreement with previous studies (Martín-Olmedo *et al.*, 1995) in which potentially mineralizable nitrogen, total-N and organic matter were higher in soils treated with vinasse than in those treated with mineral fertilizer or control. V1 and MF1 produced almost equal yields, being both statistically higher than the control.

No phytotoxicity was observed after long-term vinasse application, ratifying anterior experiments (López *et al.*, 1992).

Nutritional stage

The intrinsic soil fertility, greater in S1 than in S2, is responsible for greater levels of macro and micronutrients in plants grown in soil S1 (Table 2).

A decrease of the Fescue Nitrogen Content (FNC) was observed along the growth period. In the first clipping FNC values for all the treatments were within the proposed critical range (2.8–3.4%) (Martin & Matocha, 1973), while in further clippings, FNC values, excepting for the control, were below that range. In general, an inverse relationship between dry matter weight and FNC was observed: lowest FNC was registered for V2 treatment as a dilution effect (Jarrel & Beverly, 1981). Similar results were reported by Eck *et al.* (1981) in fescue and by Lund & Doss (1980) in rye. These low values at the end of the experiment were most likely due to reduced N release from residual N of soils with time, and to the possible presence of dead and senescent tissues in the final harvest (Devitt *et al.*, 1990).

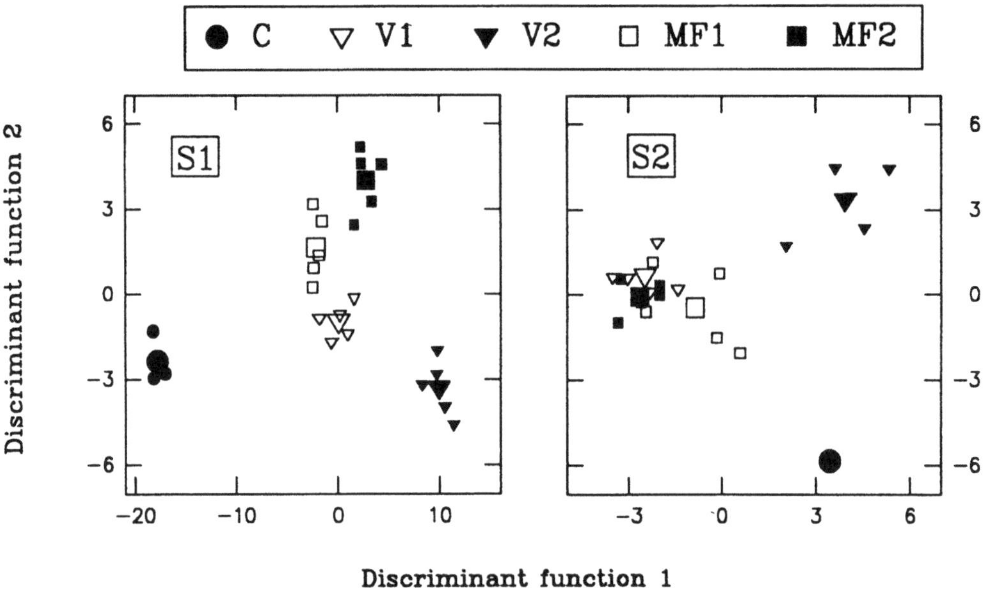

Fig. 2. Discriminant analysis of total nutrient extraction data. (Big symbols correspond to group centroids).

Vinasse treatments showed the highest P and K concentration in fescue. High values of P could be related to the acidic pH of vinasse, while high values of K are related to the high K content of vinasse. Both sets of values were lower than the established critical ranges: 0.26–0.32% P and 2.5–2.8% K (Martin and Matocha, 1973).

Mn and Zn contents were lower for vinasse treatments. This fact could be attributed to the presence of complexing agents in the residual organic matter (Atkinson *et al.*, 1958).

Extraction

Multivariant discriminant analysis shows that function F1 and F2 explain the 98.5% and 91.6% of the variance in soil S1 and soil S2 respectively (Table 3).

Plot F1 vs F2 display a visual separation of the population data of the different treatments (Fig. 2). Analysing the standardized discriminant coefficients for all variables defining function 1 and 2 (Table 3), allows the discernment of the most determinant variables in the separation of the treatments. V2 treatment, which has extracted significantly the greatest amount of N, P, and K, is obviously the best differentiated from the rest of treatments in both soils.

Acknowledgements

The authors acknowledge the Comission Interministerial de Ciencia y Tecnología of Spain for financial support (Project No. AGR91–0600). P. Martín-Olmedo gives thanks for the grant received from the Ministerio de Educación y Ciencia of Spain.

References

Algur OF and Kadioglu A (1992) The effects of vinasse on the growth, biomass, and primary productivity in pea (*Pisum sativum*) and sunflower (*Helianthus annuus*). Agr Ecosyst Environ 39: 139–144.

Atkinson HJ, Giles GR and Desjardins JO (1958) Effect of farmyard manure on the trace elements content of soil and plants grown there-on. Plant Soil 10: 32–36.

Cabrera F, Moreno F, Nacci S and De Arambarri P (1987) Utilization of wastes from olive and sugar beet processing industries in fertirrigation. Proc 4th International CIEC Symposium, Braunschweig, pp 475–483.

Devitt DA, Morris RLI and Bouwman DC (1990) Response of tall fescue to composted sewage sludge used as a soil amended. J Plant Nut 13: 1115–1139.

Eck HV, Martínez T and Wilson GC (1981) Effect of fertilization and harvest frecuency on yield and quality of tall fescue and smooth bromegrass. Commun Soil Sci Plant Anal 12: 537–555.

Jarrel WM and Beverly RB (1981) The dilution effect in plant nutrition studies. Adv Agron 34: 198–224.

Jones Jr JB, Wolf B and Mills HA (1991) Plant Analysis Handbook. G.A. Micro-Macro Publishing, Athens, Georgia, USA.

López R, Cabrera F, Moreno S, Esenarro G and Llorente MM (1992) Efecto de la fertilización con vinaza concentrada sobre la producción y estado nutricional de ryegrass. Proc Cong Nacional Ciencia del Suelo, Pamplona 220–225.

López R, Cabrera F, Murillo JM and Machuca MJ (1990) Utilización como abono de vinaza concentrada. I Intern. Cong de Química de la ANQUE, Tenerife, vol II pp 223–232.

Lund ZF and Doss BD (1980) Residual Effect of Dairy Cattle Manure on Plant Growth and Soil Properties. Agron J 72: 123–130.

Martin WE and Matocha JE (1973) Plant Analysis as an Aid in the Fertilization of Forage Crops. In Soil Testing and Plant Analysis Walsh LM (ed.), Soil Sci Soc Am, Madison, pp 393–425.

Martín-Olmedo P, López R, Cabrera F and Murillo JM (1995) Nitrogen mineralization in soils amended with organic by–products of olive oil and sugar beet processing industries. Fresenius Environ Bull 4: 59–64.

Murillo JM, Cabrera F and López R (1993) Effect of beet vinasse on Germination and seedling performance of ryegrass (*Lolium multiflorum*). J Sci Food Agric 61: 155–160.

Statistical Graphics Corporation (1986) Statgraphics 2.1 User's guide, U.S.A.

Tester CF, Sikora LJ, Taylor JM and Parr JF (1982) Nitrogen utilization by tall fescue from sewage sludge compost amended soils. Agron J 74: 1013–1018.

Ward R (1990) Use of wastes in agriculture. Book of Proceedings 10th World Fertilizer Congress of CIEC, Nicosia 477–489.

C. Rodriguez-Barrueco (ed.), Fertilizers and Environment, 533–537.

Agricultural use of city refuse compost

I. Effect on growth and chemical composition of plants

M.C. Mesquita Dos Santos[1], A.A.G. Réfega[2], R.M.B. Ganho[2]
[1]Escola Superior Agrária de Castelo Branco, Qta da Sra. de Mércules, 6000 Castelo Branco, Portugal; [2] Universidade de Lisboa, Fac. Ciências e Tecnologia, Qta da Torre, 2825 Monte da Caparica, Portugal

Key words: City refuse compost, ryegrass, heavy metals, soil pollution, plant nutrition, phytotoxicity

Abstract

In the present study, an attempt was made to better understand the effect of city refuse compost on the growth and chemical composition of a test crop (*Lolium hybridum.* v. Ariki). A greenhouse pot experiment was conducted on three soils: Cambisols (Cmd), Leptosols (Ps) and Fluvisols (J). Each soil received four rates of city refuse compost (0, 20, 50 and 80 t/ha) and the effects were compared with NPK commercial fertilizer alone and complemented with compost (50 t/ha). Test plants were harvested at regular intervals after seed emergence.

The application of city refuse compost has promoted a positive influence on crop test. Yields were significantly ($p < 0.05$) increased with amount of compost applied, in different soils. Concentration of N, P, K, Ca and Na in plant tissue increased with application of compost. The urban compost did not bring about significant ($p < 0.05$) changes in the concentration of Cr, Fe, Mn, Ni and Cd. Concerning Zn and Cu, the results showed that the city refuse compost tend to increase in the plant concentration in relation to the control. However, levels of every heavy metals analysed were within normal ranges and no toxicity problems were encountered.

However, the application of city refuse compost at high rates and eventually in a strongly acid soils makes possible the potential pollution of soils and the risk of contamination of plants and food chain by heavy metals and for that reason it is necessary to define codes of practice for city waste utilization.

Introduction

Recycling wastes in agriculture has become one of the feasible ways of disposal method and according to waste characteristics can produce valuable soil amendments. However, since heavy metal content in waste materials is often higher than in soil, application rates have to be carefully controlled, to avoid the possible accumulation of toxic metals in agriculture soils and a potential phytotoxicity to crops.

Some research on the effects of the application of urban compost to arable soils has been reported in the literature over the past ten years. However, research reports have produced many conflicting results in part probably due to the complexicity of materials used (compost, soils and plants).

The results obtained from the application of compost to soil are quite variable but, generally, it has been proven that urban compost increases crop production. On the other hand, it has been shown that, in general, the heavy metal accumulation in plant also occurs.

In Portugal, the application of urban wastes in arable land is at the beginning, and so these studies are most suitable to advise the farmers about the rates to be applied in different soils and the Institutions about the necessity to define codes of practice for city waste utilization.

The purpose of this study is to investigate the response of ryegrass to application of urban compost. The objectives of this study were to:

1. Determine the effects of the rate of compost on the yield and on the chemical composition of a test plant (*Lolium hybridum*) and compare these effects with those when a commercial fertilizer, alone or combined with compost are used.

2. Determine the contribution of the heavy metals (Cd, Cr, Cu, Fe, Mn, Ni & Zn) to plant tissue through application of compost to three soils.

Materials and methods

A greenhouse experiment was conducted on three soils to determine the effect of city refuse compost on the growth and chemical composition of a test plant (*Lolium hybridum* cv. Ariki). Surface soil samples (0–20 cm) were collected in fields near Castelo Branco (Portugal). According to the FAO Classification the soils were Cambisols (CMd), Leptosols (LPs) and Fluvisols (J) and according to the Portugal Classification CMd correspond to a Pg soil, LPs to an Ex and J to a Sbl soil. Urban compost was obtained from the Lisbon Municipal Waste Composting Plant. Some analytical characteristics of these soils and the compost are reported in Table 1.

Compost and soils were air-dried and sieved at 2 mm. Pots were filled with about 15 Kg of soil and the compost was added and thoroughly mixed. The treatments included four rates of compost: 0 (*T0*), 20 (*T3*), 50 (*T4*) and 80 (*T5*) metric tons/ha. An N,P,K fertilizer treatment alone (*T1*) and combined with 50 tons/ha (*T2*), provided according to the usual agricultural practices, were also included for comparison.

Ryegrass seed (3 g/pot) was sown in march 1990 and harvested at different times donc at 40, 70 and 100 days after emergence. Tensiometers were installed to monitor the soil moisture regime in pots that were maintained at about 70% of water-holding capacity (WHC).

The experimental design was a randomized complete block with three replications.

After each cutting the plant tissue were washed with distilled water and oven dried at 65 °C and dry weights recorded. Soil samples were collected after final harvest, sieved to 2 mm and retained for analysis.

Organic C was determined after potassium dicromate oxidation; total-N by the Kjeldhal procedure and $N\text{-}NH_4$ and $N\text{-}NO_3$ by a steam destillation method. pH was determined by a potenciometric method. The amount of heavy metals in compost and soils were determined by *aqua regia* extraction with reflux and measured by atomic absorption. The extractable metals were determined on 0.5 M ammonium acetate + EDTA at pH 4.65. Total metals in plant tissue was determined by atomic absorption spectrophotometer after acid digestion (HNO_3-$HClO_4$).

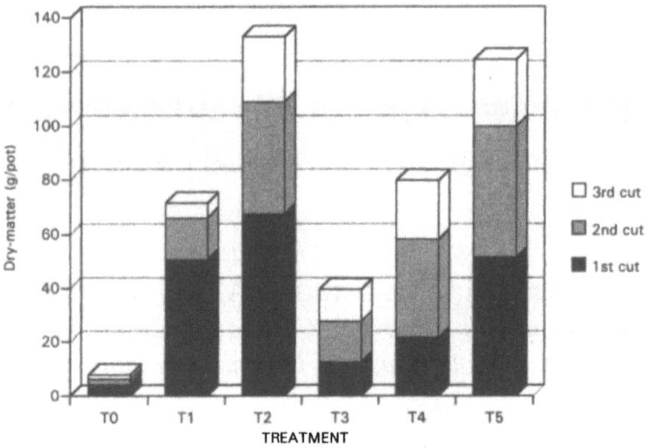

Fig. 1. Effects of urban compost treatment on dry-matter yields of ryegrass in CMd soil.

The data from the experiment were analysed using analysis of variance (ANOVA). Least significant differences (LSD) values were computed for each characteristics found significant at ($p < 0.05$). Some data were analysed using the regression procedures. Homogeneity of variance was evaluated with Bartlett's test.

Results and discussion

Effects of compost on dry-matter yields

The beneficial effects of city refuse compost were evident as indicated by large yield increases in plants grown on compost amended soils as compared with unamended soils (Fig. 1).

In the first cuttings yields was, generally, greater ($p < 0.05$) in the soils which received mineral fertilizer, alone or combined with compost than in soils receiving compost and those unamended. This fact can be attributed almost entirely to a greater amount of N supplied by the fertilizer to plants as reflected by the higher N concentration undern plants grown in this treatment (Fig. 2).

However, after the first cutting the yields in soils of treatment T1 declined dramaticaly and the pratically identical dry-matter yields of T1 and T0 suggest that N supplied by the fertilizer had been exhausted. Low plant N concentration supports this reasoning (Fig. 2). Higher total dry-matter yields with compost treatment could be attributed to a higher level of avaible N, P, K and Ca in those treatments (Mesquita Dos Santos, 1993). On the other hand, the increased yields likely resulted from improvements of soil properties.

Table 1. Analytical characteristics of compost and soils

Parameters	Urban Compost	Soils		
		CMd	LPs	J
Texture		Sandy loam	Silty loam	Sandy loam
Ec (μs/cm) a 25°C	7500.00	80.00	70.00	80.00
Moisture (%)	35.00	–	–	–
pH (H_2O)	7.20	4.90	5.00	5.00
Organic Matter (%)	31.90	1.24	0.98	1.52
Organic Carbonic (%)	18.55	0.72	0.57	0.88
N total (%)	0.86	0.12	0.12	0.08
C/N	21.5	6.2	4.8	10.5
N-NH_4^+ (mg/kg)	97.56	9.62	11.37	8.75
N-NO_3^- (mg/kg)	21.00	8.75	7.00	17.50
P_2O_5 (ppm)	1330.00	102.00	27.00	314.00
K total (%)	1.68	5.04	2.76	4.56
K_2O (ppm)	5000.00	132.00	83.00	157.00
exchang. K (meq/100g)	11.70	0.31	0.22	0.36
Ca total (%)	11.84	0.46	0.08	0.37
exchang. Ca (meq/100g)	22.50	1.23	1.73	3.70
Mg total (%)	1.30	0.43	1.37	0.47
exchang. Na (meq/100g)	4.70	0.13	1.38	0.43
Na total (%)	3.65	1.89	0.76	1.52
exchang. Na (meq/100g)	14.90	0.08	0.10	0.09
S (meq/100g)	–	1.75	3.43	4.58
CTC (meq/100g)	27.50	7.00	11.50	8.50
V (%)	–	25.00	30.00	54.00
Fe_2O_3 (%)	1.82	2.43	6.05	2.75
Cu total (ppm)	231.00	10.00	34.00	19.00
Mn total (ppm)	390.00	215.00	219.00	259.00
Zn total (ppm)	610.00	70.00	88.00	85.00
Cd total (ppm)	2.00	nd	nd	nd
Cr total (ppm)	130.00	19.00	98.00	34.00
Ni total (ppm)	48.00	11.00	41.00	13.00

Higher application rates resulted in higher dry-matter yields and the response was linear.

The results indicated that the application of the compost combined with mineral fertilizer may be more efficient than applications of compost or mineral fertilizers separately.

In general, visual differences in quality of test plant could be observed between compost and control treatments. The plants grown in the compost treatment were greener than those in the check pots and yellowing did not occur on any of compost treatments, perhaps due to additional N being mineralised from compost organic matter.

Effects of urban compost on the chemical composition of plant tissue

In the first cutting, the addition of urban compost produced significantly higher tissue N concentration than that of control soils (Fig. 2) and the N content tend to increase with increasing rates of application. Compared to the control increases of 33, 87 and 66% at CMd soil, 55, 59 and 59% at LPs soil and 52, 68 and

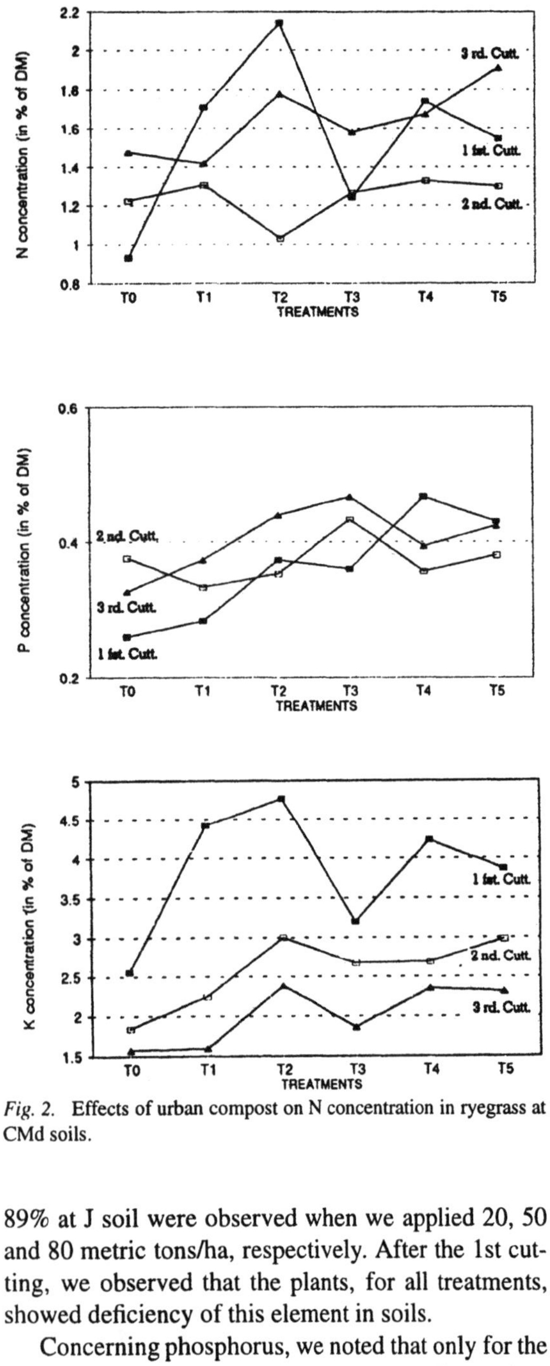

Fig. 2. Effects of urban compost on N concentration in ryegrass at CMd soils.

89% at J soil were observed when we applied 20, 50 and 80 metric tons/ha, respectively. After the 1st cutting, we observed that the plants, for all treatments, showed deficiency of this element in soils.

Concerning phosphorus, we noted that only for the 1st cutting the addition of compost to soils produce significant changes in plant P concentration (Fig. 2). Probably resulting from the improvements in soil properties, especially pH.

With regard to the effects of compost on concentration of K we observed that, for three soils, the levels tend to be greater ($p < 0.05$) in plants grown in soils treated with compost than those grown in controls and these levels tend to increase with rate of application (Fig. 2). These increases can be attributed to the amount of available K supplied by compost and to the increases in exchangeable K caused by addition of compost.

The addition of compost only affected ($p < 0.05$) the Ca levels in plant tissue, at 1st cutting, for LPs soil. At 2nd and 3rd cuttings the treatments had no effect on tissue Ca levels.

The avarage Mg content in tissues tends to decrease ($p < 0.05$) after compost application. We observed that plant Mg concentration decreased with increased compost rate. These tendencies has been attributed to an antagonistic action promoted by the absorption of potassium.

Significant higher amounts of Na were observed in ryegrass tissue from compost treated pots. Compared to the control an increase of 100, 191 and 273% at CMd soil; 170, 320 and 480 at LPs soil and 350, 600 and 917 at J were observed when we applied 20, 50 and 80 tons/ha, respectively. These results can be attributed to the increases in exchangeable Na caused by addition of compost. These results are in agreement with finding of others authors (Juste *et al.*, 1979; François, 1981).

Heavy metal concentration in plant tissue

Data regarding the concentration of Cu by ryegrass reveal that the uptake of this metal was significantly increased in plant tissues under the compost treatments compared with that of plants untreated or treated with mineral fertilizer.

In general, the amounts of Zn in plant tissue did not increase significantly as a result of urban compost. This probably resulted from lower availability of soil Zn associated with higher pH in compost treatment, as the significant ($p < 0.001$) value of the correlation coefficient ($r = -0.606$) between Zn in plant tissue and soil pH show.

It is interesting to note that not all heavy metals increase in plant tissue with compost application. For Cr, Fe and Mn, in all cuttings and soils, levels were lower than those reported from untreated and from mineral fertilizer soils and showed a tendency to reduce with increasing compost rates. This could be due to a negative relationships between pH and these metals.

Neither the application of urban compost nor mineral fertilizer had any influence on the uptake of Ni.

In every instance Cd concentrations were below detection limit (< 0.1).

Conclusions

Land urban may be a feasible disposal method for urban wastes. While containing essential macro and micronutients, the urban compost used in this study did not promote excessively high concentrations of toxic metals in plant tissues. The heavy metals contents found in all plant tissue analyzed in this study are within the normal ranges reported in the literature (Webber *et al.*, 1983).

Speculating, it can be stated that if a high rates of compost are added, the content of these elements in the soil and subsequently in the plants would increase. When conditions exist that promote high solubility and availability of the elements to plants, toxicity may arise.

The uptake of heavy metals from compost treatments by plants may be influenced by pH and texture. In the present study effect of soil texture (silty loam vs. sandy loam) was small.

Once potential environmental hazards are kept under control, compost can be regarded as valuable amendment in soil management practices because of its organic matter content and nutrient supply, particularly in Portugal soils which often contain low quantities of organic matter.

References

François, JM (1981) Utilization agricole des résidus urbaines dans les sols limoneux des plateaux de Brie. Compte rendu des résultats de 1973 à 1981. Convention n° 77–138. Ministère de l'Environnement et du cadre de vie, CETA du sud de l' Aisne.

Juste C, Solda, P and Dureau, P (1979) Étude des possibilités d'utilization des composts d'ordures ménagères comme supports de cultures maraichères. Convention d'étude n°75–135, Ministére de l'Environnement et du cadre de vie, INRA, Bordeaux.

Mesquita dos Santos, MC (1993) Contribuição para o estudo da valorização agrícola do composto urbano: Efeito sobre algumas propriedades de três unidades-solo e sobre o comportamento de uma planta-teste. Dissertação de Mestrado. Fac de Ciências e tecnologia-UNL, Lisboa.

Webber M, Monteith HD and Corneau DGM (1983) Assessment of heavy metals and PCBs at sludge application sites JWPCF: 55: 187–195.

C. Rodriguez-Barrueco (ed.), Fertilizers and Environment, 539–540.

Accumulation and transformation of heavy metals (HM) within the 'soil-plants' system in prolonged agrochemical trials

V.G. Mineev
Moscow State Univ. (MSU), Agrochemical Dept. of the Soil Science Faculty, Moscow, Russia

Key words: Heavy metals, lime, manure, mineral fertilizers, barley, oat

Abstract

Technogenic contamination of biosphere and the increasing application of agrochemical compounds in the modern agriculture may cause high contents in soils of HM in the available forms for plants.

Systematic use of acid forming fertilizers can seriously harm soil properties and due to this fact the contents of HM in mobile forms are increased.

In terms of this, the researches, on developing methodologies to immobilize and reduce uptake of HM in plants, are extremely actual.

The results of these investigations are shown in this article. These investigations were in stationary, agrochemical trials, initiated in 1950, on a clay-texture podzol soil. Four-course rotation included: sunflower, oat with legumes and rye. The forms of used fertilizers have been represented by the chemicals: ammonia nitrate, double superphosphate and muriate potash. Total amount of used fertilizers during 41 years was: N-4080 t, P_2O_5-3220 t, and K_2O-3700 t. Liming of soils took place in 1955, 1961, 1969, 1976 and 1987. Lime value has been taken accordance with value of 1 unit of hydrolytic acidity. Organic fertilizers were applied in 1983 and 1987 in amounts of 90 t. per ha and rotation period. Scheme of applied fertilizers was based on four backgrounds: 1 – acid soil (without liming and applied manure), 2 – soil with applied manure, 3 – soil with applied lime, 4 – soil with applled manure and lime. It was shown that grand total content of different forms of HM depending upon background did not change, and during systematic use of mineral fertilizers there was tendency of HM accumulation in soil. This tendency was stable independently of backgrounds. Maximum accumulation varied where triple mineral fertilizer was used (N, P_2O_5, K_2O). This feature was clearly shown for Cd content in the soil profile.

Mobilization and immobilization of HM as a process also have definite interest, during systematic use of the fertilizers at different agrochemical backgrounds. First of all, it's highlighted that application of double dairy manure has decreased concentration of mobile Cd, but quantities of Pb and Ni have remained without changes. The content of mobile Cd in soil was reduced by 2.5 times at the background with limed soil: in addition to above, in the trial with used manure and lime, the content of mobile Cd was reduced approximately by 4 times. As for Ni, this agrochemical method (lime plus manure) hasn't shown serious influence to any transformation of that element in soil.

Systematic application of acid-forming mineral fertilizers (NK) to acid soils have usually increased HM mobilization of Pb and Ni (2 times more).

At the background with the applied manure, mobilization process of HM has taken place at a much lesser degree, than in the same background without applied manure and lime. Application of triple (NPK) mineral fertilizer caused increase of Pb mobilization. Intensive immobilization of all kinds of HM took place either in limed soils or in soils with applied manure, and it was independent of what was the form of the mineral fertilizer applied.

Transformation of HM in soils under the action of agrochemical methods significantly affected HM accumulation in plants.

For example, looking at oat plus legumes rotation unit at acid background there was visual HM accumulation in the variants with mineral fertilizers (especially for Ni). Ni accumulated in plant tissues 3 times that in control.

In plants grown up at the background with the applied lime and manure there were less amounts of HM, besides of different variants of mineral fertilizers. Meanwhile, in most cases at any backgrounds, there has been a tendency of small HM accumulation in variant with phosphate (as a part of NPK).

Thanks to this, in conditions when biosphere is under the pressure of increasing technogenic contamination, it's necessary to implement complex agrochemical methods in crop farming. It allows a significant increase of plant production. HM cycle and balance in certain agrocenosis is also a most important suborder in agricultural chemistry. We must be aware of all potential sources of contamination of nature including atmosphere by HM, as HM can easely go in to plants from atmosphere to their leaves.

C. Rodriguez-Barrueco (ed.), Fertilizers and Environment, 541–545.

Cadmium in inorganic fertilizers

J. Soler Soler[1] & J. Soler Rovira[2]
[1]*Ministerio de Agricultura, Pesca y Alimentación, P° Infanta Isabel 1, Madrid;* [2]*Escuela Universitaria Ingeniería Técnica Agrícola, Universidad Politécnica Madrid, Spain*

Key words: Heavy metals, cadmium, inorganic fertilizers, phosphatic fertilizers

Abstract

The addition of Cadmium to cultivated land is due to rainfall and aerial depositions, inorganic fertilizers, livestock waste and urban waste. The mean cadmium content in the earth's crust is about 0.11 mg kg^{-1}, and it can be found in raw materials used for phosphatic fertilizers and liming materials production. The Cadmium content in phosphate rocks used as raw materials for phosphatic fertilizer production varies with the geographic origin and the type of rock: 0.15–5 mg Cd kg $rock^{-1}$ in rocks of volcanic origin and 5–300 mg Cd kg $rock^{-1}$ in sedimentary rocks. The amount of Cd can vary considerably in a deposit. The phosphate rocks used in the European Community and Spain are either of volcanic or sedimentary origin. The European Community imports 18.1×10^6 t of phosphate rock; this amount is equivalent to 470 t of cadmium.

In fertilizer industrial processes, some 60–80% of the cadmium in phosphate rocks becomes part of the fertilizer; the amount depends on the type of rock and the P_2O_5 content of the fertilizer. Typical values for the cadmium content in phosphatic fertilizers are between 3 and 110 mg Cd kg $P_2O_5{}^{-1}$. The application of a mean dose of 50 kg P_2O_5 ha^{-1} $year^{-1}$ contributes from 0.15 to 5.5 g of Cd ha^{-1} $year^{-1}$ to the soil. The consumption of phosphatic fertilizers in Spain was 531.7 1×10^3 t P_2O_5 $year^{-1}$ in 1989 and 343.3 1×10^3 t P_2O_5 $year^{-1}$ in 1993.

Introduction

The volume of heavy metals is increasing in the environment as a result of the discharge of metal-rich residues and the application of polluted production factors. Among these heavy metals, cadmium is shown to be very toxic for plants and animals. The issues concerning cadmium in fertilizers are: the soil accumulation of cadmium and the possibility of long term effects on crops and their quality; the plant uptake of cadmium and thus the Cd content in the human diet and the damage to the soil's microbial processes.

The mean cadmium content in the earth's crust is 0.11 mg kg^{-1} (Bowen, 1979). In non-contaminated, non-cultivated soils, the Cd concentration is largely governed by the amount of Cd in the parent material (Lund *et al.*, 1981). Cadmium may be absorbed on, or occluded into soil particles, incorporated into biological structures, or it may be dissolved in soil water. For world soils, Bowen (1979) reported a mean of 0.35 mg kg^{-1} as the background level. Barug *et al.* (1987, 1989) reported higher Cd contents in old arable land than in new broken soils.

Cadmium addition to cultivated land is due to rainfall and aerial depositions, inorganic fertilizers, livestock waste and urban waste. It is lost from the land or made inaccessible by leaching, removal with grazed or harvested crops and bonding to, and in, soil minerals, especially in calcareous soils. Cadmium is fairly immobile in the soil profile. The mobility of cadmium in the soil is controlled by a number of factors among which pH and the concentration are the most important. Andersson (1977) observed higher accumulations of Cd (< 1 mg kg^{-1}) in the upper part of the profiles, paralleling that of humus distribution. Contaminated soils also show the general immobility of Cd in the soil (Jonh *et al.*, 1972).

Table 1. Cadmium content in some important rock phosphates (mg Cd kg^{-1})

Type of rock	Origin	Cadmium content			
		Holland[1]	Sweden[1]	Germany[2]	Other
KhouribgaBPL70/72	Morocco	25	10–20	12.2–15.3	16[3]
KhouribgaBPL75/77		25	10–20		
KhouribgaBPL80/82			50–60		
Youssoufia		30			40[3]
Black Youssoufia			20–60		
Buoraa BPL 75	Spanish Sahara	40	25–30		35[3]
Buoraa BPL 80			25–35		
Taiba	Senegal	90	70–90		
Togo	Togo	100	50–60	59.1–61.9	55[3]
Gafsa phosphate	Tunisia	65	35–60		50[3]
Pebble phosphate BPL 72/74		15			
Pebble BPL 76/77		15			
Louis Gentile		35			
Christmas phosphate	Chrismast Island	15	50–70		31–56[4]
Nauru phosphate	Nauru island		80–90		86–90[4]
Florida phosphate BPL 72/73	USA		3–12	7–9	8[3]
Florida BPL 75/77			12–17		
North Carolina BPL72			30–40		
Idaho BPL 70/72			100–300		
Israel BPL 75	Israel		4–6	18.4– 28.7	20[3]
Israel BPL 70/72			20–30		
Jordan BPL 70/72	Jordan		3–9	5.3–8.2	5[3]
Jordan BPL 73/75			3–5		
Algeria BPL 75	Algeria		10		
Syria BPL 64/69	Syria		35–60	7.5	6[3]
Palfos	South Africa				0.15[3]
Kola phosphate	Russia	5		< 1	0.15[3]

[1] Henkens, 1983; [2]Böhm and Schäfers, 1990; [3]Bockman *et al.*, 1990; [4]David and Williams, 1973.

Table 2. Cadmium content of different phosphatic fertilizers used in different countries (mg Cd kg P_2O_5 $^{-1}$)

Fertilizer	Holland[1]	Belgium[2]	Sweden[3]	Spain[4]	USA	Australia[8]
Superphosphate (18% P_2O_5)	39	22	112 (25% P_2O_5)	71		
Superphosphate (45% P_2O_5)	61–100			59–93	22–44[5] 111–444[5] 191–253[6]	84–106
Diammonium phosphate				55–99	67–111[7] 16–346[6]	
Monoammonium phosphate				55–99		
Phosphoric acid concentrated				55–99		
Compound fertilizers	39–143	3–31	12–122			

[1]Henkens, 1983; [2]Maas, 1980; [3]Strenström and Vahter, 1974; [4]Ministerio de Industrie y Energia, 1991; [5]Mulla *et al.*, 1980; [6]US-EPA, 1974; [7]Mortvedt and Giordano, 1977.

Table 3. Cadmium content in fertilizers and the phosphate rocks used as raw materials (FESA – Fertilizantes Españoles SA., 1991)

Type of rock	Phosphorus content (% P_2O_5)	Cadmium content (mg Cd kg rock^{-1})	Cadmium content (mg Cd kg P_2O_5 $^{-1}$)				
			P.A.C.*	M.A.P.*	D.A.P.*	S.P.45*	S.P.18*
Florida-USA BPL 70/72	32	7	18	18	18	19	23
Kola-URSS BPL 84/85	39	2	3	3	3	3	4
Zin-A-Israel BPL 70/72	32	25	62	62	62	67	80
Djebel-onk-Algeria BPL 75/77	32	10	22	22	22	24	28
Gagsa-Tunisia BPL 65/68	34	30	86	86	86	92	110
Phalfos-S. Africa BPL 79/80	36	2	4	4	4	4	5
Rusaifa Jordan BPL 70/72	32	4	10	10	10	11	13

* P.A.C.: concentrated Phosphoric acid ; M.A.P.: Monoammonium phosphate; D.A.P.: Diammonium phosphate; S.P.45.: Superphosphate 45% P_2O_5; S.P.18: Superphosphate 18% P_2O_5.

Table 4. Importations of phosphate rocks into the European Community (Comite Marche Commun de l'Industrie des Engrais Azotes et Phosphates, 1986)

Origin	Importations (1×10^6 t)	Cadmium content (mg kg^{-1})	Cadmium amount imported (t)
Israel	1.6	23	37
Kola	0.4	2	1
Morocco	6.8	22	150
Senegal	0.8	75	60
Togo	1.6	55	88
Tunisia	0.4	30	12
USA	2.7	10	27
Various	0.5	25	12
EEC-10	14.8	–	387
EEC-12	18.1	–	469

Table 5. Input of Cd to spanish agricultural soils

Type of fertilizer	Consumption 1993 (1×10^3 t P_2O_5)[1]	Cadmium content (mg kg P_2O_5 $^{-1}$)[2]			Cadmium input per year (kg Cd year^{-1})		
		maximum	minimum	mean	maximum	minimum	mean
Superphosphate (18% P_2O_5)	17	110	4	37.6	1870	68	639.2
Superphosphate (45% P_2O_5)	3.4	92	3	31.4	312.8	10.2	106.8
Monoammonium phosphate	62.4	86	3	32.4	5366.4	187.2	2021.8
Diammonium phosphate	220.3	86	3	32.4	18945.8	660.9	7137.7
Total	303.1				26495	926.3	9905.5

[1]Asociación Nacional de Fabricantes de Fertilizantes, 1994; [2]FESA – Fertilizantes Españoles SA, 1991.

Table 6. Production, imports and agricultural comsumption of fertilizers containing phosphorus in Spain in 1993. (Asociación Nacional de Fabricantes de Fertilizantes, 1994)

Type of fertilizer	Production (1×10^3 t)	Importation (1×10^3 t)	Agricultural consumption (1×10^3 t)
Superphosphate ($18\%\ P_2O_5$) (P_2O_5)	21.35	11.11	18.09
Superphosphate ($45\%\ P_2O_5$) (P_2O_5)	–	2.96	3.42
Total superphosphates (P_2O_5)	21.35	14.57	22.00
Monoammonium phosphate	78.53	105.78	62.39
Diammonium phosphate	97.76	124.19	220.29
N-P compounds (without MAP and DAP)	176.29	229.97	282.68
P-K compounds	7.84	1.56	11.32
N-P-K compounds ($\leq 10\%$ N)	184.13	279.72	342.21
N-P-K compounds ($>10\%$ N)	779.17	296.25	1099.61
Total solid compounds (P_2O_5)	219.88	161.58	321.96
Total P_2O_5	241.23	176.15	346.61

Table 7. Evolution of the fertilizer market in Spain (Asociación Nacional de Fabricantes de Fertilizantes, 1994).

Type of fertilizer	Importation consumption				National consumption				Total consumption		93/92 (%)
	1992	%	1993	%	1992	%	1993	%	1992	1993	
Super phosphate ($18\%\ P_2O_5$)	34.62	23.8	61.73	61.4	110.57	76.2	38.78	38.6	145.19	100.52	30.8
Super phosphate ($45\%\ P_2O_5$)	15.53	93.4	6.58	85.5	1.09	6.6	1.12	14.5	16.62	7.70	53.7
Super phosphates	52.90	32.1	71.62	64.2	111.66	67.9	39.90	35.8	164.56	111.52	32.2
Super phosphates (P_2O_5)	13.64	40.1	14.57	66.2	20.39	59.9	7.44	33.8	34.03	22.01	35.3
Monoammonium phoshate	71.37	88.9	57.30	91.8	8.89	11.1	5.09	8.2	80.26	62.39	22.3
Diamonium phosphate	205.10	66.9	119.19	54.1	101.34	33.1	101.10	45.9	306.44	220.29	28.1
Other compounds	576.97	39.1	346.01	29.9	897.58	60.9	813.13	70.1	1474.53	1159.13	21.4
Total P_2O_5 fertilizers	231.96	50.4	148.16	42.7	228.48	49.6	199.20	57.3	460.44	347.36	24.6

The toxicity of cadmium to plants is observed at much lower concentrations than other metals. Phytotoxicity has been observed to be dependent upon plant species as well as the concentration of Cd in the substrate.

Cadmium in phosphatic fertilizers

The level of cadmium in phosphate rocks used as a raw material in the fertilizer industry can vary with the type of rock. Thus phosphate rocks of volcanic origin have a lower cadmium content than sedimentary rock phosphate. The phosphate rocks used in the European Community and Spain are from either volcanic or sedimentary origin. Table 1 shows typical values for the cadmium levels in some important rock phosphates. The rocks from Kola and South Africa are of volcanic origin, and can be appreciated for their low cadmium content.

Table 2 shows the cadmium content of different phosphatic fertilizers used in different countries. As the amount of fertilizers used is a function of its phosphorous content, Table 2 also shows the Cd content as mg Cd kg P_2O_5 $^{-1}$. Thus different phosphatic fertilizers can be compared and the different values found between countries must be due to the different phosphate rock used as raw material. Table 3 shows the content of Cd in spanish fertilizers and the type of rock used. In fertilizer industrial processes, some 60–80% of the cadmium in phosphate rock becomes part of the fertilizer; the amount depends on the type of rock and the P_2O_5 content of the fertilizer.

The amounts of phosphate rock imported by the European Community are described in Table 4. The European Community imports 18.1 1 $\times$ 10^6 t of phosphate rock; this amount is equivalent to 470 t of cadmium.

Phosphatic fertilizers used in agriculture can potentially contribute to environmental pollution of Cd. The input to agricultural soils of Cd in Spain is shown in Table 5. This source of Cd accounts for 10 t Cd year^{-1} (mean) introduced into agricultural soils. The application of a mean dose of 50 kg P_2O_5 ha^{-1} year^{-1} increased the input to the soil from 0.15 to 5.5 g Cd ha^{-1} year^{-1}.

As an increasing amount of the phosphate fertilizer consumption is satisfied by imports; the total Cd input from the use of fertilizer needs to take into account the concentration of these imports as well. Tables 6 and 7 show the imports and production structure of spanish phosphatic fertilizers in 1993 and the evolution of the market between 1992 and 1993.

Tables 6 and 7 show the importance of imported fertilizers and the trend towards the use of compound fertilizers. The Cd inflow through these imports is predominantly due to the compound fertilizers which are 91% of the total. The consumption of phosphatic fertilizers in Spain was 531.7 1$\times$$10^3$ t P_2O_5 year^{-1} in 1989 (Ministerio de Agricultura, Pesca y Alimentaci6n, 1991) and 343.3 10^3 t P_2O_5 year^{-1} in 1993 (Asociaci6n Nacional de Fabricantes de Fertilizantes, 1994).

References

Andersson A (1977) Heavy metals in Swedish soils; on their retention, distribution and amounts. Swedish J Agric Res 7: 7–20.

Asociaci6n Nacional de Fabricantes de Abonos (1994) Evoluci6n del mercado espafiol de fertilizantes. Período enero-noviembre de 1993. Centro de informaci6n y Estudios sobre fertilizantes y afines.

Barug R and Singh BR (1987) The influence of long term use of fertilizers on the content of Cd in soils and plants. Heavy metals in the environment. Internat Conf Vol 1:439–441.

Barug R, Singh BR, Selmer-Olsen AR, Haland A, Myhr K and Steinnes E (1989) Effect of the phosphorous fertilization on Cd content of soils and plants from southern and central parts of Norway. Heavy metals in the environment. Internat Conf Vol 2:32–35.

Berrow ML and Reaves GA (1984) Proc Intl Conf Environ Contamination Edinburgh. pp. 333–340.

Bockman OC, Kaarstad O, Lie OH and Richards I (1990) Agriculture and Fertilizers. Agricultural Group, Norsk Hydro, Oslo, Norway

Böhm E and Schäfers K (1990) Mänahmen zur Minderung des Cadmiumeintrages in die Umwelt. Fraunhofer-Institut für Systemtechnik und Innovationsforschung für den Bund/ Länderausschuss Umweltchemikalien (BLAU), Karlsruhe.

David DJ and Williams CH (1973) The effect of superphosphate on the cadmium-content of soils and plant. Aust J Soil Res 11: 43–56.

FESA – Fertilizantes Españoles SA (1991) El cadmio en los abonos españoles. FESA, Madrid, Spain.

Fleischer M, Sarofim AF, Fasset DW, Hammond P, Shacklette HT, Nisbet IC and Epstein S (1974).

Environ Health Perspec 7: 253–323.

Henkens CH (1983) Beleid ten aanzien van de Cadmiumaanvoer en de akkerbour. III 751/83 Netherland.

John MK, Chuah HH and VanLaerhoven CJ (1972) Cadmium contamination of soil and its uptake by oats. Environ Sci Technol 6: 555–557.

Lund LJ, Betty EE, Page AL and Elliot RA (1981) Occurrence of naturally high cadmium levels in soils and its accumulation by vegetation. J Environ Qual 10: 551–556.

Maas G (1980) Schatting van de hoeveelheid zware metalen, die door de toegediende scheikundige meststoffen op de Belgische landbouwgronden wordt afgezet. Landbouwtijdschrift 33: 326–332.

Ministerio de Industria y Energía (1991) Contenido en cadmio de los abonos fosfatados. Situación en España. Ministerio de Industria y Energía, Madrid, Spain.

Mortvedt JJ and Giordano P M (1977) Biological implications of heavy metals in the environment. CONF-750929. NTIS. Springfield, Virginia.

Mulla DJ, Page AL and Ganje TJ (1980) Cadmium accumulations and bioavailability in soils from long-term phosphorus fertilization. J Environ Qual 9: 408–412.

Stenström T & Vahter M (1974) Cadmium and lead in swedish commercial fertilizers. Ambio 3, 91–92.

US Environmental Protection Agency (US-EPA) (1974) EPA-440/1-74-011-A. Washington, DC.

Williams CH and David DJ (1973) The effect of superphosphate on the cadmium-content of soils and plant. Aust J Soil Res 11:43–56.

C. Rodriguez-Barrueco (ed.), Fertilizers and Environment, 547–550.

Lead uptake and distribution in legume species grown on lead-enriched soils

M.O. Torres, M.M.P.M. Neto, C. Marques Dos Santos & A. De Varennes
Secção Autónoma de Química Agricola, Instituto Superior de Agronomia, Tapada da Ajuda, 1399 Lisboa Codex, Portugal

Key words: Berseem, heavy metal, lead, lupin, plant nutrition

Abstract

Soils can be progressively enriched with lead due to the application of sewage sludge and city refuse or from aerial deposition. Lead incorporated into soils is usually precipitated or tightly bound to colloidal materials and presents low availability to plants. Even so, plants grown on these soils can take up and accumulate increasing amounts of lead. Lead translocation is very restricted, so that roots contain lead at greater concentrations than above-ground tissues. However, translocated lead can be sufficient to be harmful to the animals that eat those plants. In the present work, we studied lead uptake and distribution in lupins (*Lupinus* spp.) and berseem (*Trifolium alexandrinum* L.) grown in nutrient solutions or on soils that were enriched with lead. Lead uptake varied with the external lead concentration, the soil cation-exchange capacity and the plant species. Regression equations were determined for each experiment and compared. The results showed that translocation was restricted in both species, where lead accumulated in roots could represent over 95% of the total lead content. Berseem plants presented higher lead concentrations, both in roots and shoots, than lupins.

Introduction

Lead and other toxic metals can be incorporated into soils by aerial deposition or as a consequence of sewage sludge or city refuse applications. Although soil lead presents low availability to plants, uptake and translocation may be enough to pose serious health risks to animals who feed on those plants. The EEC Directive 86/278/EEC recommends upper limits for total lead content in cultivated soils of 50-300 mg kg^{-1} of soil.

In the present work we studied lead uptake and distribution in two legume species used as animal forage. The plants were grown in nutrient solutions and on soils that were artificially enriched with lead.

Materials and methods

Experiments in nutrient solutions

Berseem (*Trifolium alexandrinum* L.) or lupin (*Lupinus* spp.) seeds were germinated in the dark and grown in nutrient solutions, already described by Santos *et al.* (1993), supplemented with 0 to 350 mg l^{-1} Pb, supplied as lead acetate. Plants were kept in controlled environment with a light regime of 10 h light and 14 h dark with a light intensity of 500 μM foton m^{-2} s^{-1} supplied by day light fluorescent tubes. Temperature was kept at 18 °C with constant 55% humidity. Shoots and roots of 14 days old plants were used for analysis. They were washed with distilled water, oven dried at 105 °C and weighed. The plant material was then burned at 500 °C and the ashes were digested with 3.3% HNO_3 at 100 °C. The clear residue was made up to 100 ml with 3.3% HNO_3 and lead was determined by anodic stripping voltammetry with a hydrodynamic electrode (Neto and Varennes, 1993).

Pot experiments

Pot experiments were carried out in the same controlled environment, with two different soils.

One soil was a clay (eutric vertisol), with pH(H_2O) 6.7 and 3.1% of organic matter and the other was a sandy soil (haplic podzol) with pH(H_2O) 6.3 and 0.94% of organic matter. The soils received a basal dressing of NPK and the sandy soil was also dressed with Mg.

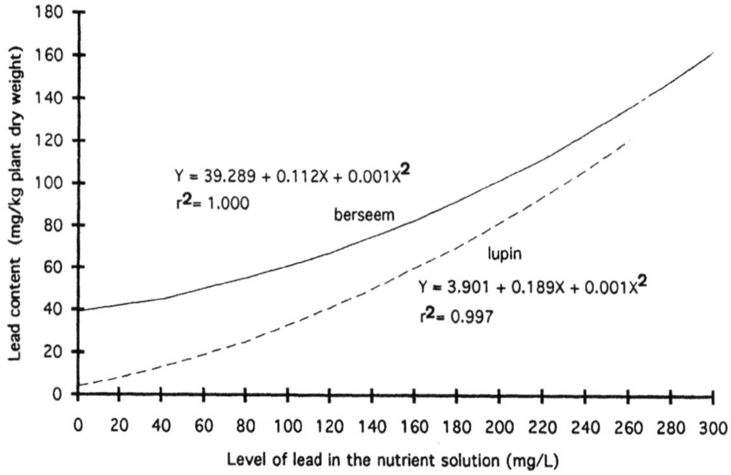

Fig. 1. Accumulation of lead in the shoots of berseem and lupin grown in nutrient solutions.

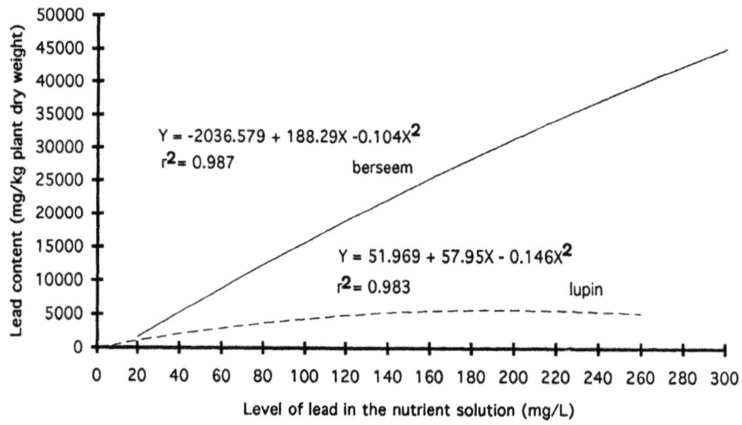

Fig. 2. Accumulation of lead in the roots of berseem and lupin grown in nutrient solutions.

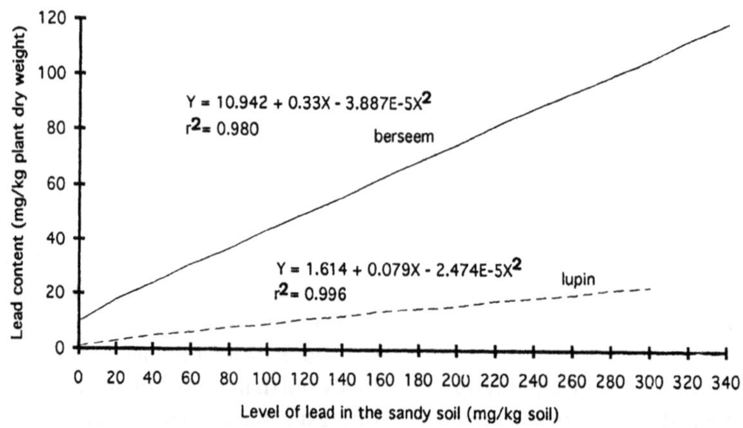

Fig. 3. Accumulation of lead in the shoots of berseem and lupin grown on a sandy soil.

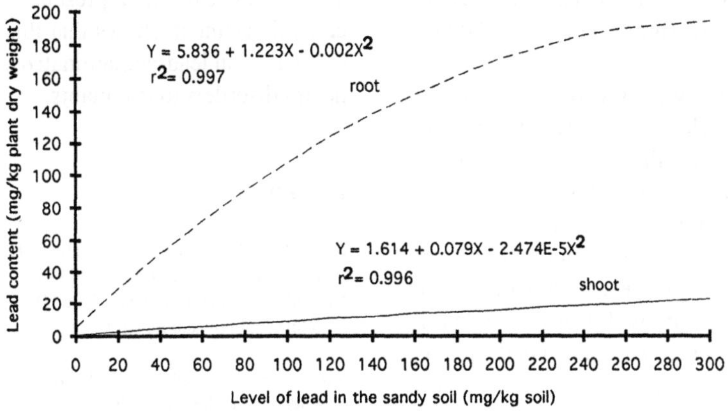

Fig. 4. Accumulation of lead in the roots and shoots of lupin grown on a sandy soil.

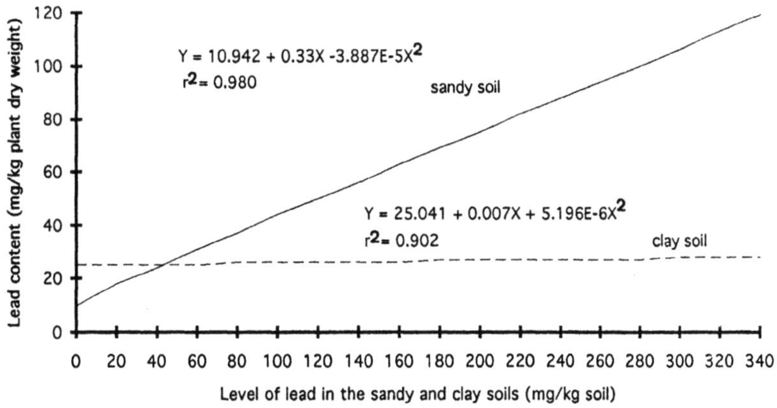

Fig. 5. Accumulation of lead in the shoots of berseem grown on a sandy and clay soils.

Lead was applied to soils, as lead acetate, from 0 to 350 mg kg^{-1}. Twenty five plants of berseem or four plants of lupin were grown per pot. Berseem shoots were collected after one month of growth and lupin plants (roots and shoots) after 100 days of growth. Plant material was processed as already described.

Statistics

At least four replicates were used in each treatment. Regression analysis was performed within each type of experiment in order to estimate plant uptake and distribution response functions to lead applications.

Results and discussion

The regression equations obtained in the experiments carried out in nutrient solutions are represented in Figs 1 and 2. As can be seen, berseem plants accumulated lead to higher levels than lupin plants, both in shoots and roots. Lead accumulated preferentially in the roots, representing over 95% of the total lead content, in both species. Lead accumulation in roots has been previously described for several plants (Chizzola, 1989) and has already been observed in these plants (Neto and Varennes, 1993; Santos *et al.* 1993).

When grown on a sandy soil, the berseem and lupin plants presented lower lead contents than when grown in nutrient solutions (Fig. 3). However, the difference was not remarkable, probably because the soil had low levels of organic matter and thus presented low cation-exchange capacity, and poor capacity to retain lead. Berseem showed again a higher tendency to uptake and translocate lead.

The distribution of lead within lupin plants grown on a sandy soil was similar to the one observed in plants cultivated in nutrient solution. Roots bound high levels of lead preventing translocation. As a consequence, the level of lead present in the shoots remained relatively low (Fig. 4).

When berseem plants were grown on a soil with high binding capacity (a clay soil), the shoots remained relatively free of lead and there was practically no increase in lead contents, when the level of lead in the soil went from 0 to 350 mg kg^{-1} of soil (Fig. 5).

Background levels of lead for lupin and berseem were different and probably represent aerial deposition of the metal, which will vary with leaf area index. A high background content was determined in the plants, probably due to high traffic in a bridge above the trial place.

The values obtained for lead contents in berseem and the high capacity it presents to uptake and translocate this element, shows that this plant should not be cultivated on lead-contaminated soils, as it may cause health disorders to ruminants.

References

Chizzola R (1989) Metallic trace elements in herbs and spices grown in Austria. Acta Hortic 249: 89–96.

Neto M M P M and Varennes A de (1993) Determination of lead in white lupin by anodic stripping voltammetry. Plant and Soil 154: 1–5.

Santos CM dos, Neto MMPM and Varennes A de (1993) Some effects of different levels of lead on berseem. Optimization of Plant Nutrition MAC Fragoso and ML van Beusichem (ed) Kluwer Academic Publishers, Dordrecht. The Netherlands pp 517–521.

C. Rodriguez-Barrueco (ed.), Fertilizers and Environment, 551–553.

Effect of different fertilization systems on soil contamination with heavy metals in long-term trials

M. Ubavić, D. Bogdanović & M. Čuvardić
Faculty of Agriculture, Institute of Field and Vegetable Crops, Novi Sad, Yugoslavia

Key words: Fertilization system, mineral fertilizers, liquid manure, harvest residues, soil fertility, heavy metals

Abstract

Soil samples were taken from long-term trials (lasting from 14 to 28 years) after wheat harvest for the study of the effect of different fertilization systems on the increase of some heavy metals in soils. Compared with unfertilized plots, the application of different fertilization systems increased the content of heavy metals in soil. However, all studied elements have been found to be within normal limits. Considering the increase of heavy metals content, the most unfavourable effects were achieved with mineral fertilizers in combination with liquid manure.

Introduction

Considering our activities and nature, we caused numerous changes which are hazardous for our survival. One of the most endangered natural resources is soil to which, by our different activities, we made more damages than benefits.

From the aspect of natural resources, chernozem of the Pannonian plane represents a real natural resource. According to its chemical, physical and productive properties, this soil is one of the best soils in Europe, even in the world (Živković *et al.*, 1972). However, a number of limiting factors have been denoted recently which can endanger high and stable yields of good quality, one of which is soil contamination by heavy metals (Hadžić *et al.*, 1993). Considering domestic and foreign literature, there are few data on the effect of different fertilization systems including mineral fertilizers, mineral fertilizers and manure, mineral fertilizers and liquid manure, mineral fertilizers and harvest residues incorporated on the changes or with respect to the increase of heavy metals in soil. However, there are many papers on these effects on yield (Broadbent, 1978; Rauhe *et al.*, 1983; Vomel *et al.*, 1980).

The objective of this study was to specify the effect of different systems of fertilization on the increase of some heavy metals in soil at intensive field crop cultivation.

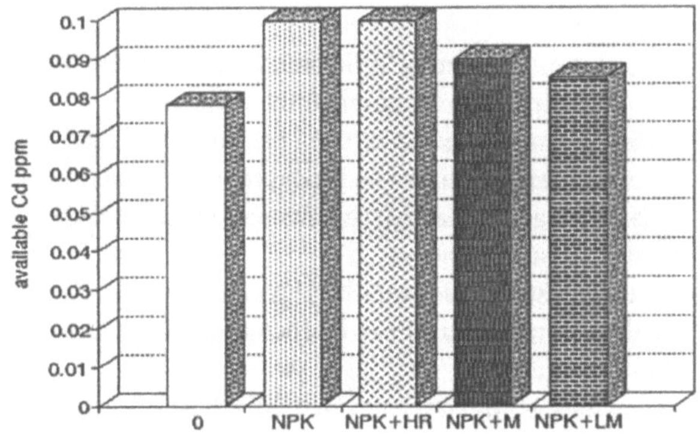

Fig. 1. Changes in Cd content after long-term application of different fertilization systems.

LEGEND
0 - Control
NPK - mineral fertilizer
NPK + HR - mineral fertilizer + harvest rezidues
NPK + M - mineral fertilizer + manure
NPK + LM - mineral fertilizer + liquid manure

Materials and methods

Soil was sampled from long-term field trials after wheat harvest and analyzed for the effect of different systems of fertilization on the increase of some heavy metals.

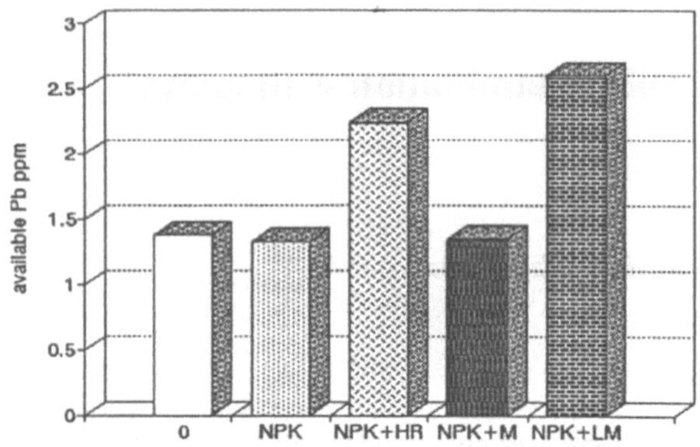

Fig. 2. Changes in Pb content after long-term application of different fertilization systems.

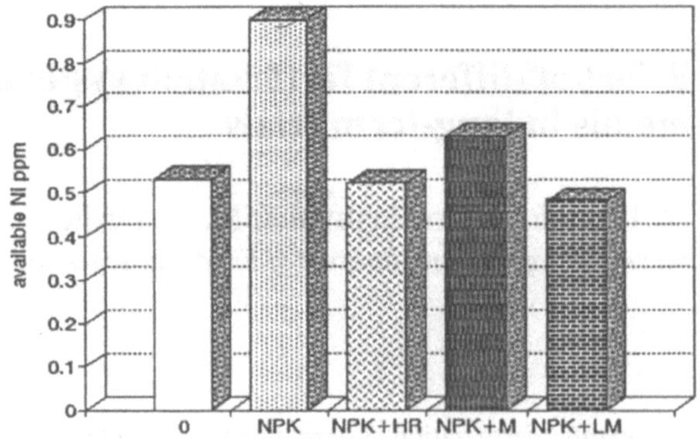

Fig. 3. Changes in Ni content after long-term application of different fertilization systems.

The following fertilization systems have been applied in different plots:

- The plot established on two field system – wheat/maize in 1947/48 has not been fertilized (check);
- The plot established on maize monoculture in 1961/62 has been fertilized with increasing doses of mineral fertilizers and mineral fertilizers and manure;
- The plot established in 1964/65 has been fertilized with increasing doses of mineral fertilizers;
- The plot established in 1965/66 has been fertilized with mineral fertilizers and manure, only mineral fertilizers and mineral fertilizers and incorporated harvest residues and
- The plot established in 1980/81 has been fertilized with mineral fertilizers and mineral fertilizers and liquid manure.

Heavy metals were extracted with 0.005 M DTPA, 0.01 M $CaCl_2$ and 0.1 M TEA (the method of Lindsay and Norvell, 1978).

Results and discussion

The content of heavy metals found in the analyzed soil depended on fertilization systems which have different effect on the content of the studied heavy metals.

The respective highest contents of Cd (Fig. 1); Ni (Fig. 3), Fe (Fig 6) and Mn and Pb (Fig. 2), Zn (Fig. 4) and Cu (Fig. 5) were found in fertilization systems with mineral fertilizer and incorporated harvest residues; mineral fertlizers and mineral fertilizers and liquid manure. The lowest content of Cd, Ni, Fe and Mn was found in the systems with mineral fertilizers and liquid manure and of Pb, Zn and Cu with the application of mineral fertilizers (Table 1).

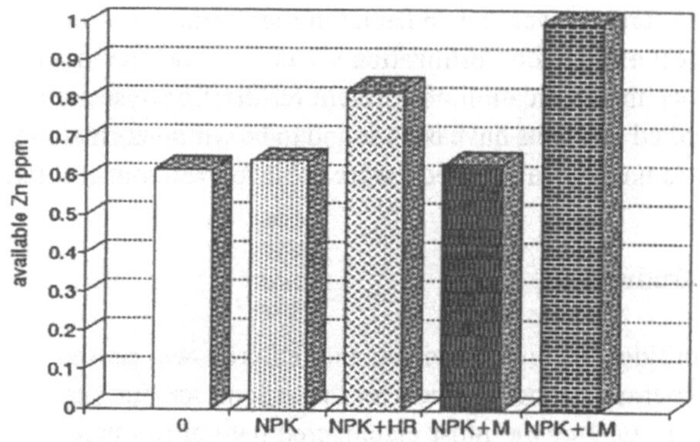

Fig. 4. Changes in Zn content after long-term application of different fertilization systems.

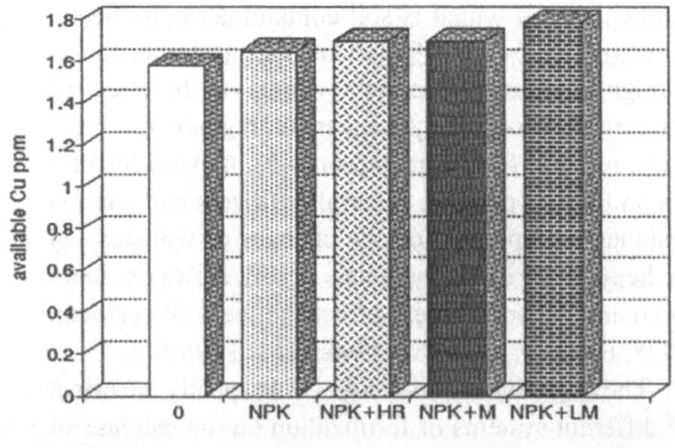

Fig. 5. Changes in Cu content after long-term aplication of different fertilization systems.

Table 1. Heavy metals content in chernozem after long-term application of different fertilization systems

Fertilization systems	Heavy metals content – ppm						
	Cd	Pb	Ni	Zn	Cu	Fe	Mn
Control	0.05–0.09	1.0–1.7	0.27–0.88	0.42–0.9	1.2–1.8	7.9–14.9	12.5–20.0
	0.078	1.38	0.53	0.62	1.58	10.48	15.28
NPK	0.09–1.05	1.0–2.0	0.47–1.5	0.52–0.82	1.55–1.75	9.20–19.70	13.0–20.0
	0.100	1.33	0.90	0.64	1.65	14.30	16.47
Min.fert. + H.R.	0.09– 0.105	2.0–2.5	0.34–0.70	0.65–1.00	1.60–1.80	7.90–13.90	12.60–20.0
	0.10	2.25	0.52	0.825	1.70	10.90	16.30
Min.fert. + manure	0.09–0.09	1.2–1.5	0.43–0.83	0.50–0.77	1.40–2.00	9.70–16.30	12.0–19.20
	0.09	1.35	0.63	0.635	1.70	13.00	15.60
Min.fert. + liquid man.	0.08–0.09	2.5–2.7	0.47–0.50	1.0–1.0	1.72–1.85	6.20–6.20	14.60–15.30
	0.085	2.6	0.48	1.0	1.785	6.20	14.95

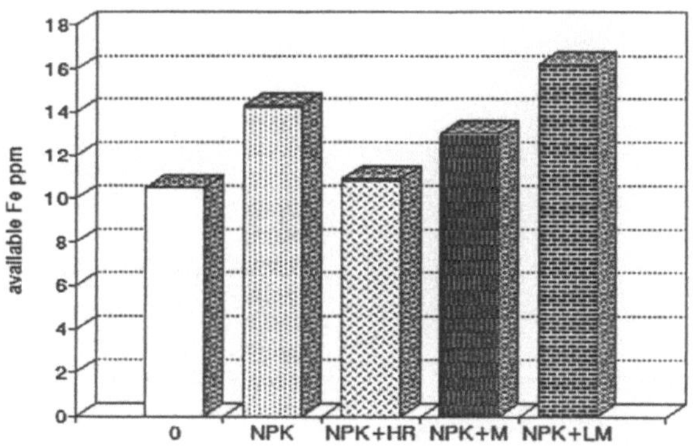

Fig. 6. Changes in Fe content after long-term application of different fertilization systems.

References

Broadbent FE (1975) Nitrogen relase and carbon los during decompozition of adbed plant residues. Soil Sci Soc Am Proc 12: 246–249.

Hadžić V, Ćirović M, Ubavić M, Govedarica M, Dragović S, Verešbaranji I and Kastori R (1993) Control of soil fertility and content of harmful and dangerous matters in soils of the Vojvodina Province. "Zbornik radova", A periodical of scientific research on field and vegetable crops. Institute of Field and Vegetable Crops. Novi Sad 21: 43–48.

Rauhe K, Sieger B and Hoberuck I (1983) Der Wirkungsmech von organischem und mineralischem Dungerstickstoff im System Boden-Pflanze-unter besonderer Berucksichtigung des ^{15}N-Einsatzes in Daueruersuchen auf einem Lo 4 Standart. "Zbornik radova", A periodical of scientific research on field and vegetable crops. Institute of Field and Vegetable Crops. Novi Sad 13: 115–121.

Vomel A and Doll G (1980) Movement of nitrogen fertilizer in combination with organic fertilizers. Soil and fertilizers 43(7): 609.

Živković B, Nejgebauer B, Tanasijević Dj, Miljković N, Stojković C and Drezgić P (1972) Zemljišta Vojvodine. Soils of Vojvodina. Institute of Agricultural Investigation. Novi Sad.

C. Rodriguez-Barrueco (ed.), Fertilizers and Environment, 555–560.

EC policy regarding pollution of agricultural origin: the Nitrates Directive *

Paul Campbell
European Commission Services, DG XI.B.1, rue de la Loi 200, B-1049 Brussels, Belgium

Key words: Nitrate, fertiliser, agriculture, pollution, EC, directive

Abstract

The Common Agricultural Policy has achieved many of its aims, but has created an agriculture which begins to have a heavy environmental impact. One aspect of the EC's policy response to the problem is examined; Council Directive 91/676/EEC concerning the protection of waters against pollution caused by nitrates of agricultural sources (the "Nitrates Directive"). The nature of the nitrate problem is discussed and various policy options considered. The Directive is examined and other agri-environmental measures are briefly discussed.

Introduction

When you look at the European Union today and the range of its common policies and programmes, it is easy to forget that, for a long time, there was only one major area in which policy was truly integrated and which, in many ways, kept the spirit of the European movement alive. The Common Agricultural Policy (CAP) has had many critics over the years, but it has been an unprecedented success on two fronts. Firstly, it has demonstrated that it is possible to have common EC policies and, secondly, it has achieved its policy objectives to an extent which its authors would have found hard to believe. I think it is worth quoting from the Treaty of Rome, Article 39.1.
The objectives of the common agricultural policy shall be :

- to increase agricultural productivity by promoting technical progress and by ensuring the rational development of agricultural production and the optimum utilization of the factors of production, in particular labour;
- thus to ensure a fair standard of living for the agricultural community, in particular by increasing the individual earnings of persons engaged in agriculture;
- to stabilize markets;
- to assure the availability of supplies;
- to ensure that supplies reach consumers at reasonable prices".

We might put a small question mark next to the objective of reasonable prices for consumers and rural wages are still not as high as urban ones, but, otherwise, who could deny that the CAP has been a remarkable success?

The Treaty of Rome was written at a time when food shortages and acute rural poverty were very real issues and this was, quite rightly, reflected in the policy objectives for the CAP.

Where the CAP has been less successful is in relation to two objectives which were *not* included in the Treaty:

- to ensure that the policy should not cost too much, and;
- to ensure that the resulting agricultural practices should not harm the environment.

Chapter 39 of the Treaty of Rome still does not contain anything about ensuring environmentally sustainable agriculture. However, increasingly, agriculture has found itself becoming subject to the same sorts of controls which limit activity in other sectors of the economy.

This is necessary. Let me state categorically that agriculture, as practised in any parts of the EC today, is a polluting activity and is not sustainable.

Excessive or inappropriate use of fertilisers is only one of the problems we must deal with. Intensive live-

* (The views expressed in this article are those of the author and do not necessarily reflect the views or policies of the European Commission)

stock production is the main cause of the fertiliser problem and leads to problems of storage and disposal of manure and slurry. The persistence of pesticide residues in the environment, particularly the aquatic environment is causing increasing concern. Various aspects of modern agriculture, such as overgrazing, land-drainage or intensive use of fertilisers and pesticides can all lead to habitat destruction and a reduction in species diversity.

The problem facing legislators is that agricultural pollution is not as easy to deal with as some other types of pollution. Factory chimneys, car exhausts and sewage outfalls are all very visible point sources which makes legislation, relatively, easy to enforce. However, nitrate pollution, for example, which is not seen in groundwater for, perhaps, decades and which causes eutrophication many miles downstream of the farm is difficult to control at the point where pollution occurs. Moreover, it is almost impossible to demonstrate that any one individual is the polluter. Finally, and this is perhaps the biggest political obstacle, the pollution is very often the result of "normal" agricultural activity; activity which was positively encouraged by a Common Agricultural Policy whose primary objective was "to increase agricultural productivity".

There is absolutely no reason why farmers should not be treated exactly the same as every other sector of industry, but we must recognise that the sort of environmental legislation which has, on the whole, worked for industrial emissions will need to be reconsidered in the context of pollution which tends to be diffuse and difficult to track down to its place of origin.

We must also recognise that many farmers are "innocent polluters" who did not know until recently (and possibly still do not believe) that their activity is damaging the environment and who have been seeking to maximise productivity in line with EC policy.

Despite the difficulties, governments have taken up this challenge and so has the EC. I will mention a number of policies later but, for the present, I will concentrate on one piece of legislation, Council Directive 91/676/EEC of 12 December 1991 concerning the protection of waters against pollution caused by nitrates from agricultural sources, commonly known as the "Nitrates Directive".

The Problem

First of all, let me introduce the problem which the Directive was designed to solve. At its most basic, levels of nitrates in groundwater and surface water were seen to be increasing, posing a potential threat to the drinking water supply and an observed increase in eutrophication of surface waters. The two most obvious sources of this nitrate were urban waste water and agriculture. The Urban Waste Water Treatment Directive (91/271/EEC) was designed to tackle the first source. The Nitrate Directive deals with agricultural sources.

Eutrophication damages the local environment upsetting the balance of flora and fauna and it can destroy the amenity value of rivers, lakes and coasts. It can be a naturally occurring process, but its increasing incidence in recent years is directly attributable to elevated levels of nutrients, including nitrates, in surface water from anthropogenic sources.

The Urban Waste Water Treatment Directive is also aimed at reducing eutrophication and it tackles both nitrogen and phosphorus, but the Nitrates Directive simplifies the issue by looking only at nitrates. The expectation is that measures to tackle nitrate will also have an impact on phosphates, but if this proves to be false, the Commission will consider proposals to cover phosphates as well.

The threat to drinking water supplies brings a human health aspect to the problem. Much of the Community's drinking water supply comes directly from groundwater with only the slightest of treatments before it is supplied to the consumer. The appearance of elevated levels of nitrates in groundwater supplies starts to become a real problem for water companies and authorities when it approaches 50 mg NO_3 as it has done increasingly in recent years. Treatment for nitrate removal is an expensive business and usually involves the blending of several different supplies. This is an expense which, in a perfect world, they should not have to pay. When agriculture starts to pass the costs of pollution to other sectors of society in this way something needs to be done.

I should say a few words at this point about the 50 mg standard. There are many who dispute its value by dismissing the health concerns on which it is based. Admittedly, the threat of stomach cancer has been widely discredited, but the WHO updated their guidelines last year and maintained the 50 mg standard on the basis of the risk to infants of methaemoglobinaemia ("blue baby syndrome"). People regularly point out to me that the incidence of blue baby syndrome in Europe is minimal, but I would point out that the levels of nitrate in groundwater are on a rising trend and that the population has until now been protected from that increase by the water authorities who have a duty under the Drinking Water Directive to blend or treat water to keep within the 50 mg level. Why should water authorities and, inevitably, consumers have to continue to bear that increasing cost?

So, how has agriculture contributed to this problem?

Surface water problems are primarily associated with run off from fields and from farmyards although through flow from groundwater does have a role to play and soil erosion can also introduce nutrients into the river system.

Groundwater problems are more complex and less well understood. The process which transports nitrates from the surface via the soil to groundwater is quite well understood in theory, but in practice our estimates of the impact of particular agricultural activities on future levels of nitrates in particular groundwater reserves are based on informed guesswork. An element of common sense and pragmatism is therefore needed when seeking technical solutions to protect groundwater. As so often in environmental legislation, we find ourselves able to identify a problem before we can identify a definitive solution. Nevertheless, I think that the sort of measures outlined in the Nitrate Directive have received a broad measure of support from those with relevant technical expertise.

Possible solutions

There was nothing inevitable about the Nitrates Directive. Having identified the problem, there were a number of possible options which could have been used to seek a solution.

Society could have taken the view that farmers should carry on as at present and that the clean up costs would be paid by the water consumer and the taxpayer. This is not really an option. Drinking water can be cleaned up (at a cost), but the idea of trying to clean eutrophic rivers and lakes is inconceivable.

Likewise, the idea of charging the individual farmers who are responsible is not a realistic option. Quite apart from the impossible task of identifying the guilty parties the costs of clean up would be out of all proportion to the "crime".

An obvious approach would be to initiate an educational and training campaign designed to raise farmer's awareness of the issues and encourage good agricultural practice. I think it goes without saying that this is an essential element of any policy programme, but it is equally clear that it is not sufficient by itself.

One possible solution would be to look at what happens in a parallel sector. Pesticide residues are causing concerns of a similar nature to those relating to fertilisers. In response, Member States have imposed strict regulations on the licensing composition, labelling, marketing and use of pesticides. I am not convinced that such a strict regime is necessary to solve the nitrate problem though elements of it are already in place and might be built upon if necessary. This option does not tackle the organic elements of the problem.

Another possible solution which looks at the input side and which is a popular suggestion in some quarters is to apply a nitrogen tax to mineral fertilisers. I have seen studies which claim that a tax on fertilisers would perform miracles, but I confess to having my doubts. Like the previous proposal, it does not cover organic fertilisers (though it could be extended to do so, I suppose) and anything that encourages a switch to the use of organic fertilisers could present problems. Also, I think that the rate of tax would have to be very high to have much impact. A fertiliser tax was not included in the EC's policy response, but the idea should not be entirely discarded.

A variation on this is to tax the nutrient loss to the environment. By calculating total nutrient inputs and subtracting nutrient outputs, the surplus can be calculated and taxed. This concept has many supporters, but it is administratively burdensome and its effectiveness depends on the permitted level of surplus and how heavy the taxes are.

Another approach would be to enter into voluntary or compulsory contracts with individual farmers and pay them to restrict their use of fertilisers. I have some sympathy with this concept as a short term, transitional measure, but, in general, I have difficulty with the idea of paying farmers not to pollute.

Finally, there is the possibility of regulating farming activity to varying degrees, the so-called "command and control option". This could range from a limit on the intensity of livestock to complete control over every kilogram of fertiliser added to the soil and every kilogram of product taken off, with intermediate regulations covering requirements for catchcrops and specifying certain types of machinery etc. Bureaucrats are always tempted by strict regulation, but the important thing is to keep a sense of proportion. It would be very easy to invent a regulatory system imposing tight controls over agricultural activity, but, if it is to be effective, the costs in terms of compliance and enforcement can be very high. There is an important place for regulations in any solution to the nitrates problem, but we need to avoid using a sledgehammer to crack a walnut.

Having outlined some of the more important policy options, there is a further point I should make. The scale of the nitrate problem is far from being uniform throughout the Community and the policies adopted must recognise that fact. The sort of rigorous policies required for Flanders are unlikely to be suitable in Scotland.

I said earlier that there was nothing inevitable about the contents of the Nitrates Directive and I have described some of the policy options which were available to the Council. I will now describe those which were actually adopted in 1991. As you will see the Nitrates Directive contains elements of various of the options I have just described.

The Nitrates Directive

There are, essentially, five elements to the Nitrates Directive:

- the monitoring of groundwater and surface water;
- the creation of voluntary codes of good agricultural practice;
- the definition of nitrate vulnerable zones;
- the establishment of compulsory action programmes within the vulnerable zones, and
- the monitoring of results

Member States are obliged to monitor the status of ground and surface waters against criteria laid down in the Directive in order to assess the scale and location of the problem.

Regardless of the results of the monitoring, Member States must produce codes of good agricultural practice in order to inform their farmers of the problem and to suggest ways to minimise it. In principle, farmers are not obliged to follow these codes, but some Member States have chosen to make them compulsory and in others they can be taken into consideration during legal proceedings against farmers accused of breaking environmental regulations. These codes must contain advice on, at least, the following elements:

- periods when the application of fertilisers is inappropriate;
- the recommendation not to apply fertiliser to steeply sloping ground;
- a similar recommendation not to apply fertiliser to water-saturated, flooded, frozen or uncovered ground;
- the maintenance of non-fertilised strips of land alongside water courses;
- the capacity and construction of storage vessels for livestock manure;
- the recommended procedures for the application of fertiliser.

As a rule, farmers do not intentionally damage the environment if they are made aware of the dangers. The codes are intended to be a long term investment to inform farmers about the impact of their activity on the environment and to encourage appropriate changes.

The codes of good agricultural practice are intended to bring the whole of EC agriculture a little closer to sustainability. In many areas the nitrate problem is of minor importance and there is no need to take further action. However, the monitoring exercise will identify particular areas, known as nitrate vulnerable zones, where more concrete action is required to combat the problem. I do not wish today to go into the detail of how the zones are identified, but, essentially, it is based on the catchment area of surface waters or groundwaters which contain, or could contain if no action is taken, higher than 50 mg/l NO_3 and surface waters which are, or could become, eutrophic. Farmers whose land falls within a vulnerable zone may find themselves severely disadvantaged compared to their neighbours just outside the zone and the delineation of the border is therefore a major legal and political problem as well as a technical one. Some Member States have therefore taken the option given in Article 3.5 of the Directive of not identifying vulnerable zones, but of applying the relevant measures over their whole territory. This has the advantage of avoiding discrimination between farmers, but has the potential disadvantage of diluting the enforcement effort and reducing the specificity of the measures.

Having identified vulnerable zones, or taken the option in Article 3.5, Member States have to elaborate action programmes with certain compulsory elements:

- those measures contained in the codes of good agricultural practice;
- periods when the application of fertiliser is prohibited;

- the existence of adequate storage capacity for livestock manure;
- limits on the total amount of fertiliser applied;
- a restriction on the total amount of livestock manure applied, including by the animals themselves, to 170 kg N/ha, and
- any other measures considered necessary to meet, the objectives of the Directive.

The action programme will be the core of the Directive. Much of the detail will be left to Member States to produce a package of measures which are tailored to the situation in each particular vulnerable zone. This will be a very real technical and political challenge for Member States. If the action programmes are to have any effect then they are going to require wholesale changes in some places. It will not simply be a matter of reducing the intensity of agricultural activity; in some parts of the Community existing activities will have to be changed completely or abandoned in favour of less damaging ones. This will not be achieved without genuine political commitment from the Member States.

Finally, the Directive requires the results to be monitored on a regular basis and for a review of the Directive itself by 1998. The monitoring will have to cover not just the state of the environment, but also changes in agricultural activity as a result of the action programmes.

By 1998, if not before, the Commission will review the Directive and I would not like to predict what the result will be. It might propose no changes, it might conclude that stricter measures are required, covering other nutrients and larger vulnerable zones or it might even conclude that the entire Directive has been misguided. At this stage, I suspect that by 1998 we will simply not have enough information upon which to judge the success or failure of the policy.

One reason we will not have this information is that Member States are being conspicuously slow to carry out the actions required of them under the Directive which they unanimously adopted back in 1991. By December last year Member States were supposed to have transposed the Directive into national legislation, carried out the monitoring exercise, published codes of good agricultural practice and designated their vulnerable zones. As of today only one Member State has fulfilled all those obligations, though most of them have fulfilled parts.

The Nitrates Directive is not a perfect piece of legislation, but it does offer a balanced package of measures aimed at a specific problem and targeted on the areas where it is most needed. To be effective it will have to change minds and, in some places, it will have to force changes in agricultural activity. To do that, Member States have to implement the Directive fully and enthusiastically. I hope they will do so.

Other measures

I have talked at length about the Nitrate Directive, because it is my area of responsibility and because it is the aspect of environmental legislation which will probably be of most interest to you. However, I will not be forgiven by my colleagues in DG XI and DG VI if I do not mention in passing various other elements of EC environmental policy which have an impact on agriculture.

The reform of the Common Agricultural Policy was driven primarily by financial considerations, but environmental concerns played a part too. The resulting package has fewer incentives to excessive production and includes limited elements of cross compliance in some of the regimes such as the various cattle subsidies. Accompanying measures have included Council Regulation 2078/92 which allows Member States to establish a number of cofinanced schemes to encourage a more environmentally friendly agriculture. The environmental contribution of set-aside (Council Regulation 1765/92) is more controversial and I will not venture an opinion here.

Overall, environmental policy has had, and will continue to have, an increasing impact on EC agricultural and rural policies. This is as it should be, given that the Treaty on European Union now obliges environmental considerations to be taken into account in all the Community's other policies.

Impact on Fertiliser Industry

The question that will matter most to many of you here today is how all this will affect the fertiliser manufacturing industry.

My answer is that it already is. Fertiliser use in Western Europe is now falling and predicted to continue to do so. Perversely, just as demand is falling, potential supply is increasing as imports from Eastern Europe enter the market. You do not need me to tell you that the industry is facing severe restructuring. However, the future is not entirely bleak. Fertilisers, and in particular efficient, easy to use, mineral fertilisers, will continue to play a major role in a responsible, increasingly sustainable, agriculture. We have a constructive dialogue with the industry and I am sure this will continue. Nevertheless, the trend is clear and farmers and fertiliser manufacturers will have to adapt to new realities and be prepared to work within a society which is looking for and, increasingly, is legislating for a cleaner environment.

C. Rodriguez-Barrueco (ed.), Fertilizers and Environment, 561–566.

Putting the concept of environmentally balanced fertilizer recommendations into practice on the farm

I. Steen
European Fertilizer Manufacturers Association (EFMA), Avenue E. van Nieuwenhuyse 4, 1160 Brussels, Belgium

Key words: balanced fertilizer recommendations, codes of best agricultural practices, fertilizer industry, plant nutrients, nutrient losses

Abstract

The intermediate meeting of the Ministers North Sea Conference was held in Copenhagen during December 1993. The Declaration published as a result of that meeting places particular emphasis on the need to define the exact meaning of balanced fertilizer recommendations and the objective to implement these, preferably by the year 2000, but no later than the year 2002. The progress of this work will be the responsibility of the Oslo and Paris Commissions, OSPARCOM. As a member of OSPARCOM the Commission of the European Union has been specifically asked to play an active role in the work.

In the autumn of 1991, the International Fertilizer Association (IFA) and the European Fertilizer Manufacturers Association (EFMA) organised a conference which was attended by representatives from academic institutions and the fertilizer industry. It resulted in the drawing up of a Code of Best Agricultural Practices to optimize fertilizer use. Central to the code is the efficient use of plant nutrients, which requires fertilizer recommendations to be site specific and to take full account of available nutrient supplies and crop offtakes within the context of prevailing environmental constraints. The fertilizer industry believes that the proper application of Codes of Best Agricultural Practices is a key factor in achieving balanced fertilization.

This paper is a logical progression of the 1991 work. It provides information on the factors to establish a framework for determining fertilizer recommendations which result in the efficient use of plant nutrients. These are:

- the nutrient offtake by the crop;
- the characteristics of each of the available sources of nutrients;
- the biological, chemical and physical processes within the soil which are essential to plant nutrition.

The information provided in the paper demonstrates that to establish recommendations and therefore arrive at balanced fertilization, these need to change in accordance with the varying climatic and soil conditions. Finally, the paper discusses how such a framework can become the basis of the individual farmer's decision-making. Overall, the emphasis is on putting existing research results into practice on the farm.

Introduction

Virtually every human activity has some impact on the environment and agriculture is no exception. It has become increasingly clear over the last 10–15 years that modern agriculture has a significant impact on the environment. Attention has been paid to the impact of excess nutrients on surface, ground, fresh and marine water in particular.

Elevated nutrient concentrations have been observed in the coastal waters of the North Sea, and eutrophication has been reported in coastal areas and estuaries [25]. It has been estimated that nitrogen and phosphorus emanating from agriculture represent, respectively, around 50% and 25% of the total anthropogenic nutrient load to the affected zones of the North Sea [26]. In 1987 the governments of the North Sea countries agreed to "aim to achieve between 1985 and 1995 a substantial reduction (in the order of 50%) in inputs of phosphate and nitrogen to areas where such inputs are likely, directly or indirectly, to cause pollution" [32]. In order to achieve this ambitious target different environmental measures were agreed.

At subsequent International Conferences and Intermediate Ministerial Meetings on the Protection of the North Sea, it was recognised that the required reduction could not be achieved by the year 1995 [25, 26]. It was claimed that this was mainly due to insufficient or insufficiently implemented measures [26]. Therefore, additional measures and means have been suggested. Among the recommendations it was agreed to "aim to achieve balanced fertilization for nitrogen and phosphorus in agricultural production by the year 2000 or the year 2002 at the latest, subject to an agreement on definitions, and having regard to Directive 91/676/EEC" [32]. The Oslo and Paris Commissions, OSPARCOM, has been requested to progress with this work and to adopt all measures necessary to reach the target. The wish to reduce the nutrient load to water has also been declared by the riparian countries of the Baltic Sea, the Rhine and other European waterways. These countries will await the outcome of the OSPARCOM undertakings on balanced fertilization.

Natural processes and unavoidable losses

The inherent character of soil processes, including the nitrogen and phosphorus cycles, are very complex, making nutrient losses inevitable. Losses occur as a result of leaching and run-off into water, emissions to the air and permanent or semipermanent fixation in the soil. The quantity of nutrients lost through these processes depends on the agricultural system practised and, in particular, on climate and soil type [4, 9, 16, 21, 23, 31, 37]. This means that neither 'natural' nor agricultural systems are 100% efficient in their use of nutrients, and this has to be taken into account when defining balanced fertilization. Although there is a need to achieve a tighter cycle of nitrogen and phosphorus in farming systems, it has to be recognised that it is not possible to accomplish a closed system with zero losses.

The nitrogen cycle is one of the foundation stones of life. Indeed, nitrogen is the most important growth promotion factor for plants in every type of eco-system. In the soil, nitrate is continuously formed as a result of the mineralisation of organic nitrogen [21, 22, 23, 31, 39]. This is a natural process, controlled by nature, and dependent on soil moisture, soil temperature, cultivation and the amount of organic matter in the soil. The nitrogen in soil organic matter is converted into ammonium and nitrate nitrogen. Some of the ammonium is taken up by the crop, a very small percentage can be lost to the air as ammonia gas, but most is converted into nitrate nitrogen. The nitrate may be taken up by crops or, in wet conditions, some may be lost either as nitrogen gas or by leaching [17]. Nitrate is water soluble and moves readily through the soil into ground water or with drainage water into rivers, lakes and the sea. The movement of water, and hence nitrate, from the soil zone to ground water is influenced both by the depth of the aquifer and the characteristics of the rocks forming it. There is therefore often a delay of many years before water leaving agricultural land reaches underground water sources. The time lag before changes in land use affect nitrate concentration in ground water supplies can range from 10 to 80 years.

The phosphate cycle is somewhat more complicated than the nitrogen cycle and knowledge about the characteristics of the organic phosphate pool is still limited. Phosphates can enter ground and surface water from several sources [22, 23, 27, 39]. Unlike nitrate, phosphates tend to bind tightly to the soil particles and are therefore much less subject to leaching. However, most phosphate losses are due to particulate movement from soil to water courses by erosion and surface runoff [22].

Nutrient sources to be considered

Various nutrient sources are available within the farming system and there is a considerable variation especially in the availability of organic sources of plant nutrients.

The major organic nutrient source is animal manure, derived indirectly from feed and fodder production on the farm and, very commonly, from purchased feedstuffs [3, 7, 8, 9, 10, 11, 12, 16, 21, 31]. The quantity of nutrients contained in animal manure in the EU and EFTA countries is higher than the amount of nutrients which are currently consumed in the form of mineral fertilizers. Estimations for the year 1988 suggested that the amount of nutrients contained in animal manure, in the EU and EFTA countries was 9.8 million tons of N, 4. 9 million tons of P_2O_5 and 11.1 million tons of K_2O [3, 36]. Recent studies imply that the total volume of animal manure, and thus nutrients, is some 20% than the 1988 estimate [2].

Other important organic sources of nutrients are sewage sludge, green manure crops, crop residues, compost and living organisms in the soil. It has been estimated that sewage 'produced' in the EU and EFTA countries contains some 1.5 million tons of nitrogen and approximately the same amount of phosphorus [2]. However, the most important nutrient source is the organic content of the soil itself, which corresponds to approximately 3–8 tons of nitrogen and 2–5 tons of phosphorus per hectare in the top layer of the soil. Soils rich in humus can contain more than 20 tons of nitrogen per hectare when the organic content is converted to nitrogen. However, only a very small percentage of the nutrients in organic matter in the soil are available to the plants. Properly-managed organic manure is a suitable source of plant nutrients which contributes to soil organic matter, soil structure and soil fertility. Additionally, organic manure offers the opportunity to recycle nutrients. However, since these nutrients, in particular nitrogen, are not released in a controlled manner, the result is often inefficient use by the growing crop and potential damage to the environment.

An inorganic nitrogen source which should not be overlooked is atmospheric wet and dry deposition of NO_X and NH_Y. The amount of nitrogen deposited from the air onto agricultural land varies according to region and climate; however, it is estimated that it should be in the order of 10–60 kg ha^{-1} yr^{-1}.

Efficiency of plant nutrients

In short term fertilizer field trials carried out on a variety of crops, results show that when nitrogen is applied at the optimum rate, the offtake in the harvested crop broadly speaking matches the input [3, 25]. In cereal production up to 85–95% of nutrients applied to crops in the form of mineral fertilizers can be recovered in the grain, but the average recovery rate is in the range of 50–70%. Further, when optimum fertilizer levels are compared to zero nitrogen application, the results from many field trials in several European countries show that only a minute proportion of the fertilizer nitrogen applied to the crop remains in the soil after harvest. Moreover, it is known that the efficiency of nitrogen use can be further enhanced by applying the latest research findings on the farm. These include efficient technology, improved fertilizer timing and catch crops [3, 6, 18, 19, 29, 30]. The use of these and other Best Agricultural Practices will minimise nutrient losses and the environmental impact of fertilizers [4, 7, 8, 9, 10, 11, 12, 16, 29, 35].

Calculations to estimate the recovery level of nutrients contained in animal manure show that 60% of the nitrogen in animal manure is deposited directly in the field during grazing. This would suggest that 40% can be collected and remains available for use. The largest losses of nutrients in animal manure to be considered are through volatilisation and leaching. A recent study estimates that more than 25% of the nitrogen contained in animal manure is lost from stables, storage, spreading and grazing in the form of ammonia [2]. Consequently, less than 10% of all nitrogen, and some 11–25% of all phosphate excreted by animals is actually utilised by the crops.

There are various factors which make it difficult to utilise the nutrient content of animal manures. The nutrient analysis of manure shows considerable variation over time, between livestock species and according to the type and quality of fodder supplied to the animals. Furthermore, the ratio of nutrients contained in animal manure rarely matches that required by the crops. The nitrogen efficiency of animal manure has been estimated and compared with that of mineral nitrogen, the results showing that some 30–50% of the nutrients in animal manure are recovered by the crop to which it is applied [21]. These figures should, however, be interpreted with care; a low recovery figure may indicate loss to water or air, but it may also be accounted for by incorporation of nutrients into the soil's nutrient pool, where they remain available for utilisation by subsequent crops [3, 21, 40].

The quantity of nutrients lost from a farming system therefore depends very much on the balance between the nutrient input and the removal in crops and animal produce. To minimise the risk to the environment, nutrients from all sources require uniform application at a rate which meets the needs of the crop and in a manner which allows the most efficient utilisation [1, 3, 5, 7, 8, 9, 10, 11, 12, 16, 21, 28, 31, 34].

Balanced fertilization; an acceptable definition

OSPARCOM is currently making progress with the work on balanced fertilization. The aim is to minimise nutrient losses from agricultural sources. Definitions put forward vary from zero losses to environmentally acceptable surplus or loss limits. Due to the characteristics of the soil, its biology and chemistry, and the overall influence of the climate, it will prove difficult to specify a uniform European-wide operational definition of balanced fertilization. An acceptable definition would be based on research-based and site-specific recommendations which include permissible losses due to biological processes. Furthermore, any definition must include the introduction of Best Farming Practices, which are the key to achieving sustainable farming systems which are environmentally friendly and economically viable.

A fertilizer industry proposal for the definition of balanced fertilization would be as follows:

- Balanced fertilization implies the minimum supply of nutrients from any source which is, at the same time, sufficient to meet the requirement of the crop and maintain soil fertility.

- Balanced fertilization implies the efficient use of plant nutrients, which is achieved by following research-based sitc-specific recommendations and adopting Best Agricultural Practices.

- Balanced fertilization recognises the existence of natural soil processes which make certain losses unavoidable.

The Commission of the European Union plays an important role in the work on balanced fertilization. Until now some legislative measures to achieve environmental goals have been adopted at EU level. Although these have not been linked to the North Sea in particular, several of them are of significant importance for the ongoing work on the protection of the North Sea. However, there is a requirement that the EC legislation and the OSPARCOM recommendations are in line [13, 14, 15, 38]. The most important is the Nitrate Directive, 91/676/EEC, which outlines measures for improved use of animal manure and which will formally designate vulnerable zones in which action plans must be fully implemented by the year 2000. It also includes national Codes of Good Agricultural Practice to be implemented by 1994. The Agri-Environmental Programme within the 1992 reform of the Common Agricultural Policy encourages farming practices which are compatible with the requirements for the protection of the environment. Additionally, in the Community's Fifth Environmental Action Programme a new strategy is set out for agriculture based on balanced nutrient input/output on the farm level, with specific attention paid to phosphate and ground water.

Progressing a realistic definition of balanced fertilization

The ambition to achieve balanced fertilization is not an unrealistic target provided it is based on a realistic definition. In arable farming systems, where mineral fertilizer is the only nutrient applied, the greater part of it is used according to recommendations [20, 24]. Practical advisory aids, independent agricultural advisory services and crop consultants provide the farmer with recommendations on the appropriate rate and timing for the application of fertilizer to specific crops, taking into account records of crop performance on the farm, soil fertility, yield potential and the availability of organic nutrient sources. Consequently, many farmers already recognise the benefits of a crop management which makes efficient use of applied nutrients and improves farm economics.

Although much has been achieved in the development of recommendations based on long years of research and trials, new knowledge indicates where further improvements can be made. In respect of nutrient management on the farm certain areas in particular need to be improved. These include:

- the monitoring of the nutrient situation of a specific field, in a specific climate and with specific soil conditions;
- the advice given to the farmer on how to determine his specific farming conditions;
- the assessment of yields and the targeting of, in particular nitrogen dressings;
- achieving greater simplicity in recommendations;
- ensuring the cost efficiency of recommended practices.

Implementation

Over the years farmers have developed their farming practices and adopted new technologies according to research findings. The efficiency of inputs used in agricultural production has improved significantly and so has overall farm productivity. For example, in today's cereal production 50% more grain is produced per kg nitrogen compared with cereal production during the fifties. Farmers' experience and knowledge about their own farm, field conditions, suitable crops, yield potentials etc are the most important factors for a successful site-specific implementation of improved recommendations, such as Codes of Good Agricultural Practices.

Given sufficient time and encouragement, farmers will continue to adopt new technologies and change farming practice in accordance with recommendations, provided production remains economically viable. The preferred approach at present is to encourage farmers, by means of awareness campaigns, education and training programmes, to improve the overall efficiency of nutrient use on their farms. Improved technology and adherence to best farming practices will ensure optimal use of plant nutrients and minimise losses to the environment, but will prove successful only if sufficiently promoted.

Industry undertakings

A recent review of practice among the major fertilizer manufacturers within western Europe revealed that the fertilizer industry is very concerned about improving the efficiency of fertilizer use and reducing the environmental impact of its products.

The fertilizer industry has taken initiatives to improve the overall efficiency of nutrient utilisation within agriculture. This work has been carried out in close association with universities, governments and national agricultural advisory services. National industry initiatives include fertilizer recommendations, soil sampling and analysis services, computer modelling and recommendations, highly developed extension programmes and the development of national Codes of Good Agricultural Practices, all aiming at the best farm nutrient management and site-specific use of nutrients, regardless of the source [33].

Given its agronomic research and product-based knowledge, the fertilizer industry wishes to play an active role in developing farming systems which meet new requirements and which make sensible use of the benefits offered by its products. The industry's objective is to help determine:

- the most appropriate areas for the different crops to be grown;
- the right level of fertilizer use;
- the most appropriate fertilizer types for use, and
- the best farm production techniques.

References

1. A Code of Best Agricultural Practices to Optimize Fertilizer Use (1990) The International Fertilizer Industry Association (IFA) / The European Fertilizer Manufacturers Association (EFMA), Paris and Brussels
2. Ammonia Emissions to Air in Western Europe, Technical Report No. 62. (1994) European Centre for Ecotoxicology and Toxicology of Chemicals, (ECETOC), Brussels. ISSN-0773-8072-62
3. Animal manure in Europe, study paper (1991) EFMA, Brussels
4. Bertilsson G (1989) Nitrogen Transformations and Nitrogen Balances in Scandinavian Soils. The Fertiliser Society, Proc. 287
5. Bertilsson G (1992) Environmental consequences of different farming systems using good agricultural practices. The Fertiliser Society, Proc. 332
6. Claesson S, Granstedt A, Gustafson A and Steineck S (1991) Kväveflöden i jordbruket (Nitrogen flows in agriculture). Aktuellt fråan lantbruksuniversitetet 399, Uppsala
7. Code des bonnes pratiques agricoles (1993) Ministèrie de l'Environnement. Journal Officiel de la Rèpublique française, 05/01/1994
8. Code Goede Landbouwpraktijken (1993) Ministerie van Landbouw, Natuurbeheer en Visserij, Ministerie van Volkshuisvesting, Ruimtelijke Ordening en Milieubeheer, en Ministerie van Verkeer en Waterstraat, Den Haag
9. Code of Good Agricultural Practice for the Protection of Water 1991 Ministry of Agriculture, Fisheries and Food (MAFF), and Welsh Office, Agricultural Department, MAFF Publications, London
10. Code of Good Agricultural Practice for the Water Protection from the Nitrogen Pollution of Agricultural Origin (1993) Ministry of Agriculture, Hellenic Republic, Athens
11. Code of Good Agricultural Practice, Prevention of Environmental Pollution from Agricultural Activity (1993) The Scottish Office Agriculture and Fisheries Department (SOADF), Edinburgh
12. Codice di Buona Pratica Agricola per al Protezione delle Acque dai Nitrati (1993) Ministero delle Risorse Agricole, Alimentarie Forestali, Rome
13. EEC Council Directive (1980) The Drinking Water Directive, 80/778/EEC
14. EEC Council Directive (1991) The Nitrate Directive, 91/676/EEC
15. EEC Council Regulation (1992) The Council Regulation on Agricultural Production Methods compatible with the Requirements of the Protection of the Environment and the Maintenance of the Countryside, 2078/92/
16. Fertilizers and Animal Manures, Code of Good Practice for the Environment (1992) The Fertilizer Association of Ireland, Dublin
17. Granli T and Bøockman OC (1994) Nitrous oxide from agriculture. Norw J Agric Sci Suppl. No. 12
18. Granstedt A (1987) Fallstudier av kväveförsörjning i alternativ odling (Case study of nitrogen supply to alternative agricultural production). Dissertation, Sveriges Lantbruksuniversitet, Uppsala, Sweden
19. Greenpeace (1992) Green Fields Grey Future, EC Agriculture Policy at the Crossroads. Greenpeace International, Amsterdam, the Netherlands
20. Hasund KP (1990) Sätt värde påa miljön, Miljöavgifter och andra ekonomiska styrmedel (Put value on the environment, environmental levies and other economical steering methods). SOU 1990:59, Miljödepartementet, Sweden
21. Intensive farming and the impact on the environment and the rural economy of restrictions on the use of chemical and animal

fertilizers (1989) ECSC-EEC-EAEC, Study prepared for the Directorate General for Agriculture, Brussels-Luxembourg
22. Isermann K (1990) Share of Agriculture in Nitrogen and Phosphorous Emissions into the Surface Waters of Western Europe against the Background of their Eutrophication. Fert Res 26: 253–269
23. Isermann K (1991) Territorial, Continental and Global Aspects of C,N,P and S Emissions from Agricultural Ecosystems. Proceedings of NATO Advanced Research Workshop (ARW) on Interactions of C, N, P and S Biochemical Cycles, March 4–8, 1991, Melreux, Belgium
24. Lee J (1987) Land Resources and their use in the European Community. In: Barth M and Hermiler PL (eds) Scientific basis for soil protection in the European Community.
25. Ministerial Meeting of the Oslo and Paris Commissions, Paris, 21–22 September (1992) Publ. Oslo and Paris Commissions, London.
26. Nutrients in the Convention Area (1993) Oslo and Paris Commissions, London
27. Phosphate fertilizers and the environment (1992) International Fertilizer Development Centre (IFDC), Proceedings of an International Workshop, March 23–27, 1992, Tampa, Florida, USA
28. Powlson D (1988) Long-term nitrogen studies at Rothamsted. Arable Farming, July, 1988
29. Rude S and Dubgaard A (1989) Regulering af naereingsstofbelastningen fra landbruget med saerligt henblik påa kvaelstof (Regulation of nutrient load from agriculture especially with illumination on nitrogen). Statens Jordbrugsøkonomiske Institut, Copenhagen
30. Ryding SO (1988) Jordbrukets inverkan påa luft-och vattenmiljö (The influence of agriculture on air and water). LRF, Sweden
31. Solving the nitrate problem, progress in research and development (1993). Ministry of Agriculture, Fisheries and Food (MAFF), MAFF publications, London
32. Statement of Conclusions from the Intermediate Ministerial Meeting, 7–8 December 1993 in Copenhagen. Publ. Secretariat of the Fourth International Conference on the Protection of the North Sea (1995) Copenhagen
33. Steén I and Williams AJ (1993) Reducing the Impact of Crop Nutrients on the Environment. Proceedings of International Conference on Environmental Pollution 2. Publ., the European Centre for Pollution Research, London
34. Sustainable Agricultural Systems for the Twenty-first Century, the Role of Mineral Fertilizers (1992) International Fertilizer Industry Association (IFA), Paris
35. The environment and the management of agricultural resources (1992) Proc. of the 24th seminar of the EAAE, January 1991. Margaret Loseby (ed)
36. The Fertiliser Industry of the European Community; The issues of today, the outlook for tomorrow, (under revision) (1991) CEC (Commission of the European Communities) and EFMA (the European Fertilizer Manufacturers Association), Brussels
37. The role of fertilizers in sustainable agriculture and food production, Position paper (1990) The Canadian Fertilizer Industry (CFI), Ontario, Canada
38. Towards Sustainability; A European Community Programme of Policy and Action in relation to the Environment and Sustainable development (1992) Commission of the European Communities, COM(92) 23/II final, ISBN 92-77-428 19-8, Luxembourg
39. Water Quality; Agriculture's Role (1992) Council for Agricultural Science and Technology (CAST), Task Force Report, lowa, USA
40. Williams AJ (1992) Fertilisers and the European environment - the way ahead. The Fertiliser Society, Proc. 333

C. Rodriguez-Barrueco (ed.), Fertilizers and Environment, 567.

Code of best agricultural practices

Towards a modern agriculture which respects the environment

J.C. Ignazi
COMIFER (French Committee for the Developntent of Rational Fertilization), Paris, France

Key words: agricultural practices, codes

Abstract

In most European countries, agriculture's responsibility as regards environmental quality, in particular that of water, is widely accepted, as also is its determining role in maintaining the countryside. The conservation of water quality seems to be, for each country, a national issue, but it is also international. All the categories of water are concerned; drinking water (surface or underground), with problems as regards nitrate content and pesticide residues, but also bathing water, with specific microbiological problems, and surface waters with the risk of eutrophication.

The discharge into coastal waters of rivers and streams containing undesirable elements extends to the marine environment the concern to limit risks related to agriculture.

In France, a Committee comprising members from diffennt sectors for the control of water pollution resulting from agricultural activities (CORPEN) has been working on the nitrate problem since the beginning of the 1980s. More recently it has been concerned also with pesticides. The avoidance of point pollution (from agro-industry, industrial farms) evidently requires a regulatory approach, whereas diffuse pollution requires a careful study of 'agricultural practices', in order to recommend those which seem to present the least risk in terms of nitrate emissions.

The recommendations cover the following points;

- fertilization must be in accordance with the crop needs in order to avoid excessive applications, taking account of supplies from organic sources (manure, slurry),
- soil must not be lett bare during rainy periods,
- the presence of nitrates in the soil during periods between crops must be limited (through the management of crop residues, planting of 'trap' crops),
- the countryside must be maintained.

These points are common to all the recommendations, and especially to the codes of good agricultural practice drafted in application of the European Directive.

In order to be effective, it is essential that the recommendations should be widely known and accepted. This requires each farmer to receive a clear and consistent message

- adapted to local conditions (climate, soil, crops)
- with an identical content, whatever the origin of the recommendation.

It is along these lines that a series of advisory operations is at present being developed in France in well defined zones, where appropriate advice is given (particularly as regards fertilization) and adopted by all the advisors in the region. This is the 'Fertimieux' operation, which calls upon the voluntary co-operation of farmers. In addition, in applying the nitrate directive, action programmes will be developed for each of the vulnerable zones. These good practice recommendations will then be obligatory. In taking account of the pesticide risks, attention must be paid to the consistency of the 'nitrate' and 'pesticide' recommendations.

Adaptation to the practices of farmers requires a substantial effort in the definition of recommendations and in making them known. All the partners (public authorities, farmers, industry and distributors) must coordinate their efforts. But this is the price for reconciling agricultural activity with maintenance of our way of life, i.e. for ensuring a sustainable agriculture.

C. Rodriguez-Barrueco (ed.), Fertilizers and Environment, 569–572.

Code of Good Agricultural Practice and water pollution

J. Soler Rovira[1], J. Soler Soler[2], P. Soler Rovira[3] & A. Polo Sánchez[3]
[1]*Escuela Universitaria Ingeniería Técnica Agrícola, Universidad Politécnica Madrid;* [2]*Ministerio de Agricultura, Pesca y Alimentación, Po Infanta Isabel 1, Madrid;* [3]*Centro de Ciencias Medioambientales, CSIC, Serrano 115 dpdo, Madrid, Spain*

Key words: Nitrates, nitrogen fertilizers, code of good agricultural practice, legislation

Abstract

Areas having groundwater with a nitrate content higher than 50 mg l^{-1} occur in practically the whole of the twelve large Spanish Hydrographic Basins. Except for the odd region, it has been seen that although the main cause of pollution does not derive from agricultural activities, they always play a more or less relevant role. This is why a series of correct agricultural practices must be promoted to protect water from future pollution. These considerations are included in Council Directive 91/676/1991 of 12 December relating to water protection against pollution caused by nitrates used in agriculture and a series of measures are adopted, like drawing up and implementing a Code of Good Agricultural Practice. This contains provisions relating to measures to be adopted and sets down the bases of the Code and recommendations or actions to be followed and refer to the periods when it is not advisable to apply fertilizers, to apply fertilizers on steep, sloping, flooded, snow covered or frozen fields, near to water courses, the capacity and design of dung tanks, procedures for fertilizer application, soil management and crop alternatives and irrigation system management.

Introduction

Water pollution from nitrates occurs through indiscriminate discharge into soil and, consequently, into water, until reaching the 50 mg l^{-1} maximum admissible concentration, with the results of placing human health in danger, damaging water resources and ecosystems and damaging recreational areas or others for the legitimate enjoyment of water. Eutrophication of surface water occurs through an increase in nutrient concentration (particularly nitrates and phosphates), causing an accelerated growth of algae and higher plants and causing negative upsets in the balance of organisms present in water and in its very quality.

Areas having groundwater with a nitrate content higher than 50 mg l^{-1} occur in practically the whole of the twelve large Spanish Hydrographic Basins. Except for the odd region, it has been seen that although the main cause of pollution does not derive from agricultural activities, they always play a more or less relevant role. This is why a series of correct agricultural practices must be promoted to protect water from future pollution. These practices have been specifically dealt with by limiting the application of nitrogen fertilizers and livestock waste.

Council Directive 91/676/1991 of 12 December relating to water protection against pollution caused by nitrates used in agriculture intends to reduce this pollution and acts in a preventive fashion against future pollution. According to criteria laid down in its Appendix I, water affected and that which could be affected will be determined. Areas whose run-off flows towards this water will be called vulnerable areas. In the first place, a Code of Good Agricultural Practice will be drawn up which farmers will voluntarily implement and which will include a series of measures shown in Appendix II of the said Directive. Action programmes will then be established in the aforementioned vulnerable areas, which will include the compulsory measures of this Directive's Appendix III, apart from the Code of Good Agricultural Practice. Programmes for monitoring the effectiveness of action programmes shall also be drawn up.

Code of Good Agricultural Practice

A Code of Good Agricultural Practice is a regulation whereby farmers and livestock people are shown the general rules they must fulfil in organic and mineral fertilizing of their soils so no nitrate pollution of water occurs (Journal Officiel de la République Française, 1993, 1994; Ministry of Agriculture, Fisheries and Food, 1991). It must define the types of nitrogen fertilizers, the forms of nitrogen present in the fertilizers and their behaviour in soil and plant nutrition: nitrate fertilizers, ammonium fertilizers, fertilizers containing ammonium and nitrate, urea fertilizers, fertilizers with N in organic form, in organic and mineral form, fertilizers with slow release N, inhibitors of enzyme activity, zootechnical effluent, dung and cattle slurry, pig slurry, sheep dung, poultry dung, compost and sewage sludge. It also defines the nitrogen cycle in agricultural soil: nitrogen fixation, conversion of nitrogen in soil, soluble nitrogen uptake by plants and micro-organisms and agronomic circulation and removal of nitrogen. The Code contains provisions relating to the measures to be adopted, establishing the Code bases and recommendations or actions to be taken.

The provisions laid down for the Code of Good Agricultural Practice pertain to the following headings:

Periods when fertilizer application to soils is not advisable

Mineral fertilizing of farm crops is a general practice. With the purpose of increasing their effectiveness, fertilizers should be applied as near as possible timewise to the crop absorbing them to thus prevent loss of nitrogen through leaching.

Although they absorb nitrogen during their whole life cycle, different crops display greater demands in certain phenological states, coinciding with highest growth speed periods and when development phases occur.

Nitrogen will act in soil differently depending on its chemical form in fertilizers: ammonium is retained by the soil's adsorbent complex and loss risks are lesser and, although it may be directly absorbed, it mostly undergoes a nitrification process prior to its uptake. Nitrate is soluble in water and only weakly bonded by soil particles, it is thus a very mobile form of nitrogen in the soil, moving with the flow of water and by diffusion. It acts immediately as it is directly absorbed by plants. Fertilizers containing ammonium and nitrate display intermediate characteristics.

In general, ammonium fertilizers will be used in pre-sowing fertilizings. Urea fertilizers in pre-sowing and first coverage (applying during plant growth period), fertilizers containing ammonium and nitrate in pre-sowing of some crops, like tobacco, beetroot, maize, sorghum, vegetables, etc. and in first coverage. Nitrate fertilizers will always be used in coverage.

Organic fertilizers will be applied to beetroot when it is a first rotation crop well in advance of sowing, to potato in winter, to tobacco in autumn, to vegetables in advance of seedbed preparation, in meadows of seasonal gramineous plants at the end of autumn every two or three years, at the beginning of autumn in fruit trees and repeatedly in citrus trees.

It is not recommendable to fertilize on uncultivated soils with any type of fertilizer during the whole year. Set aside arable lands in application of Community regulations are included in this category.

Application of fertilizers on steep, sloping land

A steep slope is a factor aggravating the risk of nitrogen transport in soil. According to Ministerio de Agricultura, Pesca y Alimentación (1974), seven classes of soil are considered as a function of their slopes and agricultural systems are recommended for controlling erosion: Grade 3 sets the tilling limit with slopes of over 20%.

Fertilizer spreading initially has no definite gradient limit since run-off depends on a series of factors to be taken into account. In general, run-off is limited by the soil's plant cover, coarse texture, good structure (culture practices improving it are recommended: tilling, amendments, etc.), cultural work carried out in a suitable direction so as to favour water retention, and becomes aggravated when rock underlying the soil is not permeable or when the soil in itself is (infiltration rate low).

Liquid fertilizers are more liable to be washed away than solid, and it is recommended that fertilizers be buried and that applications in high rainfall seasons be avoided.

These factors should be studied for each specific case and recommendations should be given as well as keeping drains, hedges, slopes and hill bottoms grassed.

Application of fertilizers on hydromorphic soils and flooded, frozen or snow covered fields

The purpose is to prevent fertilizer being washed away, by taking into account the types of fertilizer and weather conditions. On surface frozen fields, with icing up and melting alternating throughout the day, it will be possible to apply every kind of nitrogen fertilizer (mineral and organic). On completely frozen fields, nitrogen mineral fertilizers, dung, compost and sludge will only be spread in limit cases; applications of slurry or liquid manure is not at all advisable. It is not advisable to apply mineral fertilizers on snow covered fields and only in limit cases are dung, compost and sludge advisable whilst slurry and liquid manure is totally unadvisable. It is not advisable either to apply any kind of nitrogen fertilizer (mineral and organic) on flooded or waterlogged fields (except crops in a water medium like rice).

Conditions for applying fertilizers to land near watercourses

The aim is to prevent the danger of surface water being reached by drift or run-off, when applying fertilizers. Action recommended consists in leaving a strip between 2 and 10 metres wide unfertilized (even without drip or sprinkler) next to all water courses. Likewise, effluent and organic waste should not be applied less than 30–50 m from a fountain, well or perforation supplying water for human consumption or to be used in milking rooms, even though in some cases, a greater distance may be needed. It is recommended to keep banks or sides of water courses with a herbaceous cover.

Capacity and design of dung storage tanks

It is necessary to take measures to prevent water pollution from run-off and filtration in surface or ground water from liquids containing dung and waste from plant products stored as siloed fodder. The aim is to prevent direct removal of liquids with animal stools or effluent of a plant origin in livestock premises so that water pollution from run-off, infiltration or washing is prevented. Volumes to be stored, collection systems and storage systems thus have to be considered. Storage volumes must be such that livestock effluent produced can be contained during the period when it is not advisable to spread it. This volume will depend on the species of animal, feeding, management, etc. and, in addition, a certain margin of safety will be applied. Likewise, endeavour will be made to produce the minimum volume of waste water which will be directed to suitable treatment facilities or farm tanks whilst preventing direct discharge. Leaks and dilution with rain or washing water will be prevented in collection and storage systems.

Application of chemical fertilizers and dung to soils

It is necessary to set procedures for applying chemical fertilizers and dung to land so as to control nutrient losses into water. Determining doses to be applied is important, taking into account the crop's requirements and endeavouring to prevent fertilizer excesses. To do so, a balance must be sought between these requirements and what is supplied by the soil and fertilization, so it is necessary to take into account crop requirements as a function of the yields expected according to the agrological potentiality of plots and cultivation method, nitrogen gains in the ground as a function of the amounts present in the soil when the crop heavily absorbs them, soil nitrogen mineralization during the growing cycle and inputs in organic amendments and via mineral fertilizers. Once the dose is set, it is recommended to fraction the input in order to respond to the needs of the crop according to its different development phases and demands and in order to review the dose downwards if the yield aim set cannot be achieved. Application must be uniform to avoid over-fertilization, keeping to a fertilizer homogeneity and adjustment of application equipment by having it first overhauled at accredited service places.

Soil use management

This makes reference to the crop rotation system and proportion of soil area earmarked to permanent crops in relation to annual crops and to keeping a minimum plant cover during rainy periods, absorbing the soil's N which otherwise could cause water pollution by nitrates.

In annual crops, it is recommended to improve the order of their succeeding each other so that bare soil area is reduced during periods presenting leaching risks. The soil should be conserved with the maximum possible plant cover, by early sowing or intercrops. Crop residues should be buried, which improves the soil, and fallow land be protected by seeding it with rainfed or irrigated leguminous species.

After ploughing up a field, a nitrogen demanding species should be grown, as well as after a legume crop. When sowing is late, endeavour will be made to hold back crop residue mineralization. The bottoms of thalwegs and banks of water courses must be kept grassed, trees and hedges on river banks must be conserved whilst methods of combating soil erosion must be implemented.

Drawing up fertilization plans in keeping with the particular situation of each farm and recording the use of fertilizers in registers

It is recommended that fertilizing plans be established for each plot in farms and the use and application of fertilizers should be recorded in a register according to crops, date of use, amounts and volumes, yields obtained, etc.

Preventing water pollution due to run-off and leaching in irrigation systems

Irrigation may facilitate pollution by the movement of irrigation water both in depth (deep percolation) and by surface run-off. Pollution risks on irrigation lands depend on soil features (permeability, field capacity, depth, slope, water table, organic matter content, etc.), agricultural practices (type of fertilization, crop rotation, tillage, etc.), irrigation method and its use. High risk areas display at least one of these features in an unfavourable fashion. Low risk ones display other parameters with suitable values and moderate risk areas display intermediate features.

Good irrigation practice must endeavour to prevent percolation and surface run-off and achieve high values of efficiency in water distribution where the irrigation method is a determining factor. Irrigation by flooding is not advisable in high and moderate risk areas. In furrow irrigation, care must be taken with sandy soils, shallow soils and shallow root system crops. In sprinkler irrigation, attention will have to be paid to sprinkler distribution, rainfall, wind and vegetation. In fertigation, the fertilizer will be applied after having supplied 20–25% of water and will be completed after having applied 80–90% of the water volume. The aim is to prevent salt concentration in the soil's wet area surface.

References

Council of the European Communities (1991) Council directive relating to water protection against pollution caused by nitrates used in agriculture. Official Journal of the European Communities L375: 1–8.

Journal Officiel de la République Française (1993) Décret No 93-1038 du 27 août 1993 relatif à la protection des eaux contre la pollution par les nitrates d'origine agricole 12417.

Journal Officiel de la République Française (1994) Arrêté du 22 novembre 1993 relatif au code des bonnes pratiques agricoles 287–292.

Ministerio de Agricultura, Pesca y Alimentación (1974) Caracterización Agrológica de los Suelos de España. Dirección General de la Producción Agraria, Madrid, España.

Ministry of Agriculture, Fisheries and Food (1991) Code of Good Agricultural Practice for the Protection of Water. Welsh Office Agriculture Department, London, England.

Subject index

Author index

Developments in Plant and Soil Sciences

1. J. Monteith and C. Webb (eds.): *Soil Water and Nitrogen in Mediterranean-type Environments.* 1981 ISBN 90-247-2406-6
2. J. C. Brogan (ed.): *Nitrogen Losses and Surface Run-off from Landspreading of Manures.* 1981 ISBN 90-247-2471-6
3. J. D. Bewley (ed.): *Nitrogen and Carbon Metabolism.* 1981 ISBN 90-247-2472-4
4. R. Brouwer, I. Gašparíková, J. Kolek and B. C. Loughman (eds.): *Structure and Function of Plant Roots.* 1981 ISBN 90-247-2510-0
5. Y. R. Dommergues and H. G. Diem (eds.): *Microbiology of Tropical Soils and Plant Productivity.* 1982 ISBN 90-247-2624-7
6. G. P. Robertson, R. Herrara and T. Rosswall (eds.): *Nitrogen Cycling in Ecosystems of Latin America and the Caribbean.* 1982 ISBN 90-247-2719-7
7. D. Atkinson, K. K. S. Bhat, M. P. Coutts, P. A. Mason and D. J. Read (eds.): *Tree Root Systems and Their Mycorrhizas.* 1983 ISBN 90-247-2821-5
8. M. R. Sarić and B. C. Loughman (eds.): *Genetic Aspects of Plant Nutrition.* 1983 ISBN 90-247-2822-3
9. J. R. Freney and J. R. Simpson (eds.): *Gaseous Loss of Nitrogen from Plant-Soil Systems.* 1983 ISBN 90-247-2820-7
10. United Nations Economic Commission for Europe (ed.): *Efficient Use of Fertilizers in Agriculture.* 1983 ISBN 90-247-2866-5
11. J. Tinsley and J. F. Darbyshire (eds.): *Biological Processes and Soil Fertility.* 1984 ISBN 90-247-2902-5
12. A. D. L. Akkermans, D. Baker, K. Huss-Danell and J. D. Tjepkema (eds.): Frankia *Symbioses.* 1984 ISBN 90-247-2967-X
13. W. S. Silver and E. C. Schröder (eds.): *Practical Application of* Azolla *for Rice Production.* 1984 ISBN 90-247-3068-6
14. P. G. L. Vlek (ed.): *Micronutrients in Tropical Food Crop Production.* 1985 ISBN 90-247-3085-6
15. T. P. Hignett (ed.): *Fertilizer Manual.* 1985 ISBN 90-247-3122-4
16. D. Vaughan and R. E. Malcolm (eds.): *Soil Organic Matter and Biological Activity.* 1985 ISBN 90-247-3154-2
17. D. Pasternak and A. San Pietro (eds.): *Biosalinity in Action.* Bioproduction with Saline Water. 1985 ISBN 90-247-3159-3
18. M. Lalonde, C. Camiré and J. O. Dawson (eds.): Frankia *and Actinorhizal Plants.* 1985 ISBN 90-247-3214-X
19. H. Lambers, J. J. Neeteson and I. Stulen (eds.): *Fundamental, Ecological and Agricultural Aspects of Nitrogen Metabolism in Higher Plants.* 1986 ISBN 90-247-3258-1
20. M. B. Jackson (ed.): *New Root Formation in Plants and Cuttings.* 1986 ISBN 90-247-3260-3
21. F. A. Skinner and P. Uomala (eds.): *Nitrogen Fixation with Non-Legumes* (Proceedings of the 3rd Symposium, Helsinki, 1984). 1986 ISBN 90-247-3283-2
22. A. Alexander (ed.): *Foliar Fertilization.* 1986 ISBN 90-247-3288-3
23. H. G. v.d. Meer, J. C. Ryden and G. C. Ennik (eds.): *Nitrogen Fluxes in Intensive Grassland Systems.* 1986 ISBN 90-247-3309-X
24. A. U. Mokwunye and P. L. G. Vlek (eds.): *Management of Nitrogen and Phosphorus Fertilizers in Sub-Saharan Africa.* 1986 ISBN 90-247-3312-X
25. Y. Chen and Y. Avnimelech (eds.): *The Role of Organic Matter in Modern Agriculture.* 1986 ISBN 90-247-3360-X
26. S. K. De Datta and W. H. Patrick Jr. (eds.): *Nitrogen Economy of Flooded Rice Soils.* 1986 ISBN 90-247-3361-8
27. W. H. Gabelman and B. C. Loughman (eds.): *Genetic Aspects of Plant Mineral Nutrition.* 1987 ISBN 90-247-3494-0
28. A. van Diest (ed.): *Plant and Soil: Interfaces and Interactions.* 1987 ISBN 90-247-3535-1

Developments in Plant and Soil Sciences

29. United Nations Economic Commission for Europe and FAO (eds.): *The Utilization of Secondary and Trace Elements in Agriculture.* 1987 ISBN 90-247-3546-7
30. H. G. v.d. Meer, R. J. Unwin, T. A. van Dijk and G. C. Ennik (eds.): *Animal Manure on Grassland and Fodder Crops.* Fertilizer or Waste? 1987 ISBN 90-247-3568-8
31. N. J. Barrow: *Reactions with Variable-Charge Soils.* 1987 ISBN 90-247-3589-0
32. D. P. Beck and L. A. Materon (eds.): *Nitrogen Fixation by Legumes in Mediterranean Agriculture.* 1988 ISBN 90-247-3624-2
33. R. D. Graham, R. J. Hannam and N. C. Uren (eds.): *Manganese in Soils and Plants.* 1988 ISBN 90-247-3758-3
34. J. G. Torrey and J. L. Winship (eds.): *Applications of Continuous and Steady-State Methods to Root Biology.* 1989 ISBN 0-7923-0024-6
35. F. A. Skinner, R. M. Boddey and I. Fendrik (eds.): *Nitrogen Fixation with Non-Legumes* (Proceedings of the 4th Symposium, Rio de Janeiro, 1987). 1989 ISBN 0-7923-0059-9
36. B. C. Loughman, O. Gašparíková and J. Kolek (eds.): *Structural and Functional Aspects of Transport in Roots.* 1989 ISBN 0-7923-0060-2; Pb 0-7923-0061-0
37. P. Plancquaert and R. Haggar (eds.): *Legumes in Farming Systems.* 1990 ISBN 0-7923-0134-X
38. A. E. Osman, M. M. Ibrahim and M. A. Jones (eds.): *The Role of Legumes in the Farming Systems of the Mediterranean Areas.* 1990 ISBN 0-7923-0419-5
39. M. Clarholm and L. Bergström (eds.): *Ecology of Arable Land – Perspectives and Challenges.* 1989 ISBN 0-7923-0424-1
40. J. Vos, C. D. van Loon and G. J. Bollen (eds.): *Effects of Crop Rotation on Potato Production in the Temperate Zones.* 1989 ISBN 0-7923-0495-0
41. M. L. van Beusichem (ed.): *Plant Nutrition – Physiology and Applications.* 1990 ISBN 0-7923-0740-2
42. N. El Bassam, M. Dambroth and B.C. Loughman (eds.): *Genetic Aspects of Plant Mineral Nutrition.* 1990 ISBN 0-7923-0785-2
43. Y. Chen and Y. Hadar (eds.): *Iron Nutrition and Interactions in Plants.* 1991 ISBN 0-7923-1095-0
44. J. J. R. Groot, P. de Willigen and E. L. J. Verberne (eds.): *Nitrogen Turnover in the Soil-Crop System.* 1991 ISBN 0-7923-1107-8
45. R. J. Wright, V.C. Baligar and R. P. Murrmann (eds.): *Plant-Soil Interactions at Low pH.* 1991 ISBN 0-7923-1105-1
46. J. Kolek and V. Kozinka (eds.): *Physiology of the Plant Root System.* 1992 ISBN 0-7923-1205-8
47. A. U. Mokwunye (ed.): *Alleviating Soil Fertility Constraints to Increased Crop Production in West Africa.* 1991 ISBN 0-7923-1221-X; Pb 0-7923-1222-8
48. M. Polsinelli, R. Materassi and M. Vincenzini (eds.): *Nitrogen Fixation* (Proceedings of the 5th Symposium, Florence, 1990). 1991 ISBN 0-7923-1410-7
49. J.K. Ladha, T. George and B.B. Bohlool (eds.): *Biological Nitrogen Fixation for Sustainable Agriculture.* 1992 ISBN 0-7923-1774-2
50. P.J. Randall, E. Delhaze, R.A. Richards and R. Munns (eds.): *Genetic Aspects of Plant Mineral Nutrition.* 1993 ISBN 0-7923-2118-9
51. K.S. Kumarasinghe and D.L. Eskew (eds.): *Isotopic Studies of Azolla and Nitrogen Fertilization of Rice.* 1993 ISBN 0-7923-2274-6
52. F.A. Bliss and G. Hardarson (eds.): *Enhancement of Biological Nitrogen Fixation of Common Baen in Latin America.* 1993 ISBN 0-7923-2451-X
53. M.A.C. Fragoso and M.L. van Beusichem (eds.): *Optimization of Plant Nutrition.* 1993 ISBN 0-7923-2519-2
54. N.J. Barrow (ed.) *Plant Nutrition - From Genetic Engineering to Field Practice.* 1993 ISBN 0-7923-2540-0
55. A.D. Robson (ed.): *Zinc in Soils and Plants.* 1993 ISBN 0-7923-2631-8
56. A.D. Robson, L.K. Abbott and N. Malajczuk (eds.): *Management of Mycorrhizas in Agriculture, Horticulture and Forestry.* 1994 ISBN 0-7923-2700-4

Developments in Plant and Soil Sciences

57. P.H. Graham, M.J. Sadowsky and C.P. Vance (eds.): *Symbiotic Nitrogen Fixation.* 1994 ISBN 0-7923-2781-0
58. F. Baluška, M. Čiamporová, O. Gašparíkova and P.W. Barlow (eds.): *Structure and Function of Roots.* 1995 ISBN 0-7923-2832-9
59. J. Abadía (ed.): *Iron Nutrition in Soils and Plants.* 1995 ISBN 0-7923-2900-7
60. P.S. Curtis, E.G. O'Neill, J.A. Teeri, D.R. Zak and K.S. Pregitzer (eds.): *Belowground Responses to Rising Atmospheric* CO_2. Implications for Plants, Soil Biota, and Ecosystem Processes. 1994 ISBN 0-7923-2901-5
61. P.C. Struik, W.J. Vredenberg, J.A. Renkema and J.E. Parlevliet (eds.): *Plant Production on the Threshold of a New Century.* 1994 ISBN 0-7923-2903-1
62. L.O. Nilsson, R.F. Hüttl and U.T. Johansson (eds.): *Nutrient Uptake and Cycling in Forest Ecosystems* (CEC/IUFRO Symposium, Halmstad, Sweden, 1993). 1995 ISBN 0-7923-3030-7
63. H.P. Collins, G.P. Robertson and M.J. Klug (eds.): *The Significance and Regulation of Soil Biodiversity* (International Symposium, East Lansing, 1993). 1995 ISBN 0-7923-3138-9
64. R.A. Date, N.J. Grundon, G.E. Rayment and M.E. Probert (eds.): *Plant-Soil Interactions at Low pH: Principles and Management* (3rd International Symposium, Brisbane, 1993). 1995 ISBN 0-7923-3198-2
65. J.K. Ladha and M.B. Peoples (eds.): *Management of Biological Nitrogen Fixation for the Development of More Productive and Sustainable Agricultural Systems.* (15th Congress of Soil Science, Acapulco, Mexico, 1994) 1995 ISBN 0-7923-3413-2; Pb 0-7923-3414-0
66. C. Rodriguez-Barrueco (ed.): *Fertilizers and Environment.* 1996 ISBN 0-7923-3729-8
67. B. Buerkert, B.E. Allison and M. von Oppen (eds.): *Wind Erosion in Niger.* Implications and Control Measures in a Millet-based Farming System. 1996 ISBN 0-7923-3885-5
68. O. van Cleemput, G. Hofman and A. Vermoesen (eds.): *Progress in Nitrogen Cycling.* 1996 ISBN 0-7923-3962-2
69. N. Ahmad (ed.): *Nitrogen Economy in Tropical Soils.* 1996 ISBN 0-7923-4094-9

Kluwer Academic Publishers – Dordrecht / Boston / London

MIX
Papier aus verantwortungsvollen Quellen
Paper from responsible sources
FSC® C105338

If you have any concerns about our products,
you can contact us on
ProductSafety@springernature.com

In case Publisher is established outside the EU,
the EU authorized representative is:
Springer Nature Customer Service Center GmbH
Europaplatz 3, 69115 Heidelberg, Germany

Printed by Libri Plureos GmbH
in Hamburg, Germany